Sectional Anatomy
PET/CT and SPECT/CT

Sectional Anatomy

PET/CT and SPECT/CT

Edited by

E. Edmund Kim, MD, MS
Professor of Radiology and Medicine, Department of Nuclear Medicine, University of Texas M.D. Anderson Cancer Center and Medical School, Houston, Texas

Martha V. Mar, BBA, CNMT, PET
Nuclear Medicine/PET and CT Technologist, Department of Nuclear Medicine, University of Texas M. D. Anderson Cancer Center, Houston, Texas

Tomio Inoue, MD, PhD
Professor and Chairman, Department of Radiology, Yokohama City University, Graduate School of Medicine, Yokohama University Medical Center, Yokohama, Japan

June-Key Chung, MD, PhD
Professor, Department of Nuclear Medicine, Seoul National University and Hospital, Seoul, South Korea

 Springer

E. Edmund Kim, MD, MS
Professor of Radiology and Medicine
Department of Nuclear Medicine
University of Texas M.D. Anderson
 Cancer Center and Medical School
Houston, TX 77030, USA

Tomio Inoue, MD, PhD
Professor and Chairman
Department of Radiology
Yokohama City University
Graduate School of Medicine
Yokohama University Medical Center
Yokohama 236-0004, Japan

Martha V. Mar, BBA, CNMT, PET
Nuclear Medicine/PET and CT Technologist
Department of Nuclear Medicine
University of Texas M. D. Anderson
 Cancer Center
Houston, TX 77030, USA

June-Key Chung, MD, PhD
Professor
Department of Nuclear Medicine
Seoul National University and Hospital
Seoul, South Korea

Library of Congress Control Number: 2006933868

ISBN: 978-0-387-38296-8   e-ISBN: 978-0-387-38297-5

Printed on acid-free paper.

9 8 7 6 5 4 3 2 1

springer.com

Preface

Combined positron emission tomography (PET) or single photon emission computed tomography (SPECT) with computed tomography (CT) has developed into the fastest-growing imaging modality largely because combined PET or SPECT with CT data acquisition is highly synergistic in diagnosis and therapeutic evaluation. All currently available data indicate that integrated PET or SPECT/CT is more sensitive and specific than either of its constituent imaging methods alone and probably more so than images obtained from separate systems viewed side by side. The PET or SPECT/CT provides precise localization of the lesions, thereby increasing diagnostic specificity particularly by reducing false-positive findings. The other advantage of adding CT is that the transmission data obtained with the CT component is useful for attenuation correction of the emission data. This makes PET/CT 25%–30% faster than PET alone with standard attenuation–correction method, leading to higher patient throughput and a more comfortable examination, which typically lasts 30 minutes or less.

With hybrid imaging, unique physiologic information benefits from a precise topographic localization. Simultaneous evaluation of metabolic and anatomic information about normal and disease processes is needed to answer complex clinical questions and also raise the level of confidence of scan interpretation.

It is always difficult to consider three dimensions in the mind's eye and view the relationship of the viscera and fascial planes in transverse and vertical sections of the body's structure. The introduction of modern imaging techniques has enormously expanded the already considerable importance of sectional anatomy. This text was written to address the needs of today's practicing PET and SPECT, with Part I demonstrating normal anatomy and physiologic distribution of various radiopharmaceutical and Part II illustrating physiologic variation artifact and certain pathology.

The nuclear physicians, diagnostic radiologists, oncologists, neurologists, internists, and surgeons have had to reeducate themselves for better understanding of functional processes at precise location. Our goal has been to create a text in an easy-to-use, yet comprehensive format to interpret PET or SPECT/CT.

E. Edmund Kim, MD, MS
Martha V. Mar, BBA, CNMT, PET
Tomio Inoue, MD, PhD
June-Key Chung, MD, PhD

Acknowledgments

We express our gratitude to Ms. Sharon Davis for typing and editing the manuscripts. We are also deeply grateful to our colleagues for their support of our work. Finally, we thank Mr. Robert Albano and his assistants at Springer Science+Business Media in New York who supported in the creation and editing of this book.

E. Edmund Kim, MD, MS
Martha V. Mar, BBA, CNMT, PET
Tomio Inoue, MD, PhD
June-Key Chung, MD, PhD

Contents

PART I   NORMAL ANATOMY OF PET/CT AND SPECT/CT

**PART II   ANATOMIC VARIATIONS AND ARTIFACTS
          OF PET/CT AND SPECT/CT**

Contributors

Gi-Jeong Cheon, MD, PhD
Chairman, Department of Nuclear Medicine, Korea Institute of Radiological and Medical Sciences, Seoul, South Korea

Won Joon Kang, MD
Assistant Professor, Department of Nuclear Medicine, Seoul National University Hospital, Seoul, South Korea

Keon Wook Kang, MD, PhD
Chairman, Department of Nuclear Medicine, National Cancer Center, Goyang, Gyeonggi, Korea

Ryogo Minamimoto, MD
Department of Radiology, Yokohama City University, Graduate School of Medicine, Yokohama, Japan

Tetsu Niwa, MD
Department of Radiology, Yokohama City University, Graduate School of Medicine, Yokohama, Japan

Part I
Normal Anatomy of PET/CT and SPECT/CT

1 FDG PET/CT

Combined positron emission tomography (PET)/computed tomography (CT) allows for acquiring molecular and anatomic information in a single examination without moving the patient in between, thus minimizing any intrinsic spatial misalignment. Fluorodeoxyglucose (FDG) distributes throughout the body in proportion to glucose metabolism of tissues. Normal tissues utilizing large amounts of glucose include brain, working muscles, mucous membranes, and liver. The renal excretion of FDG implies that considerable tracer activity accumulates in the renal collecting system and in the urinary bladder. There are obvious clinical advantages of PET/CT over PET alone. PET/CT allows the localization of molecular alteration of cancer tissue, and the exact lesion's localization can reduce the number of false-positive and false-negative PET findings.[1] It is evident that patients for whom surgical or radiation treatment is contemplated need accurate localization of lesions. The level of mediastinal lymph node involvement in lung cancer patients cannot be determined reliably with PET alone. The appropriate localization of hypermetabolic foci to chest wall versus lung, lung base versus liver, neck versus superior mediastinum, and others might have some impact on patient management. Clinical data are emerging that demonstrate an incremental value of PET/CT over PET alone in a variety of cancers.[2]

Section 1: Coronals

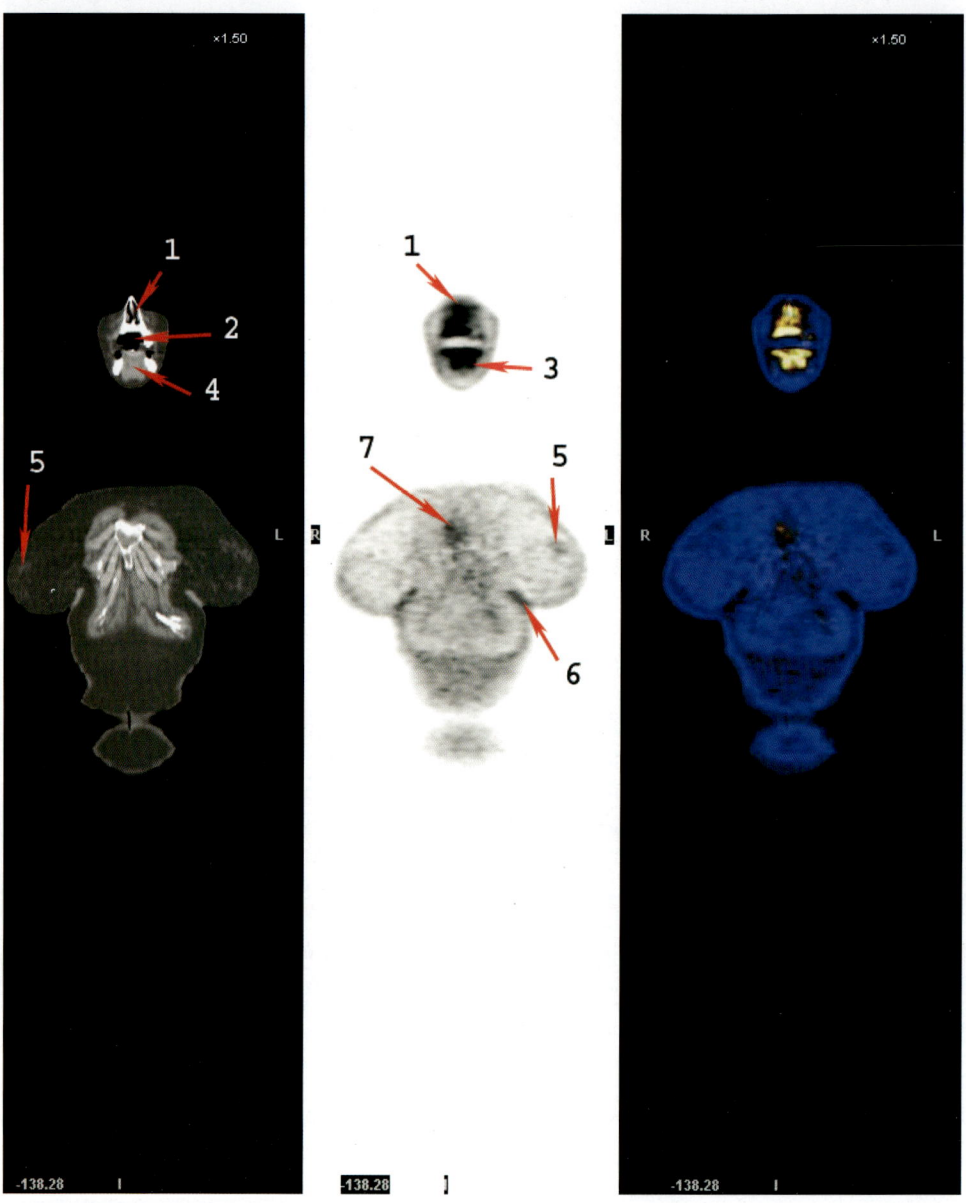

FIGURE 1-1: Coronal 1

1. Nasopharynx
2. Oral cavity
3. Lingual tonsil
4. Tongue
5. Breast nipple
6. Skin fold
7. Sternum

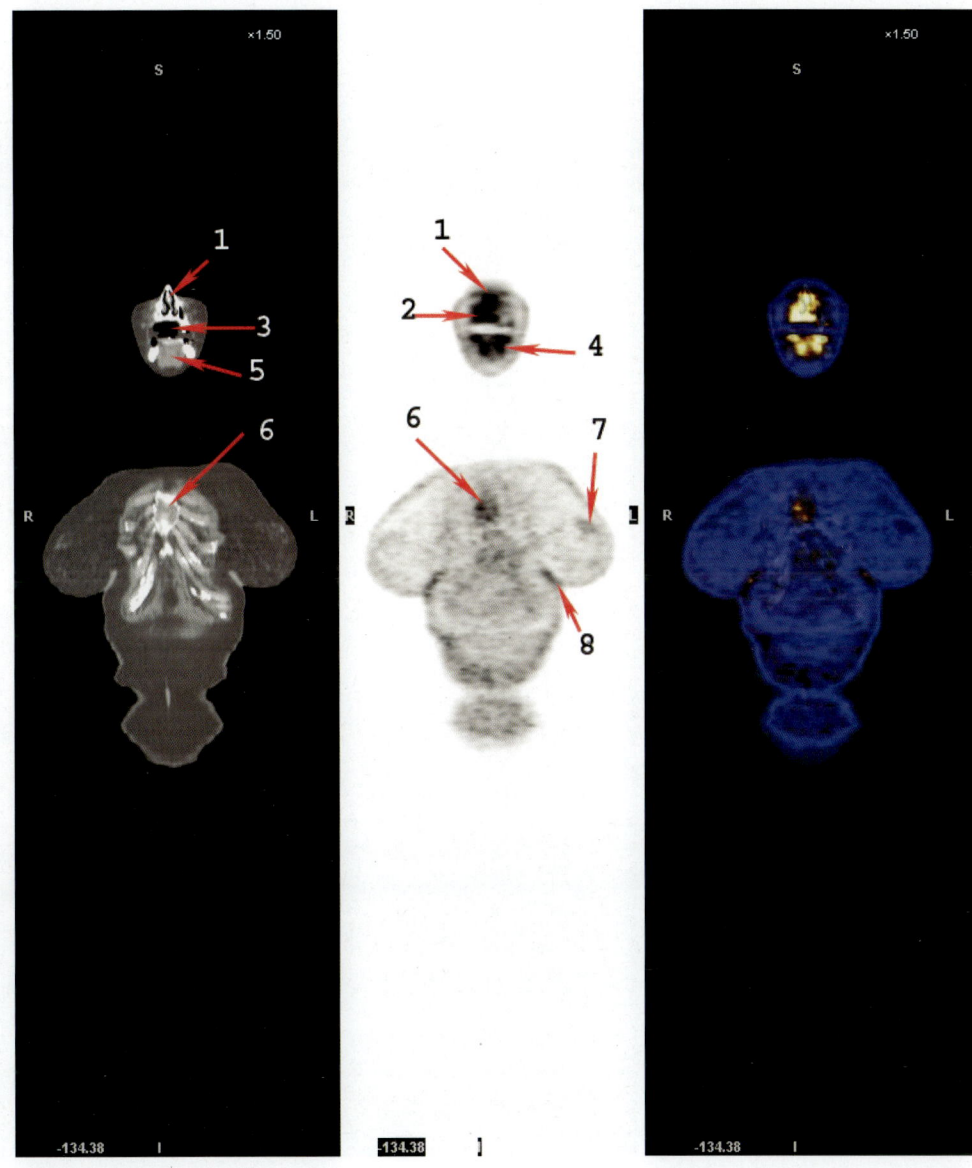

FIGURE 1-2: Coronal 2

1. Nasopharynx
2. Right maxillary sinusitis
3. Oral cavity
4. Lingual tonsil
5. Tongue
6. Sternum
7. Breast nipple
8. Skin fold

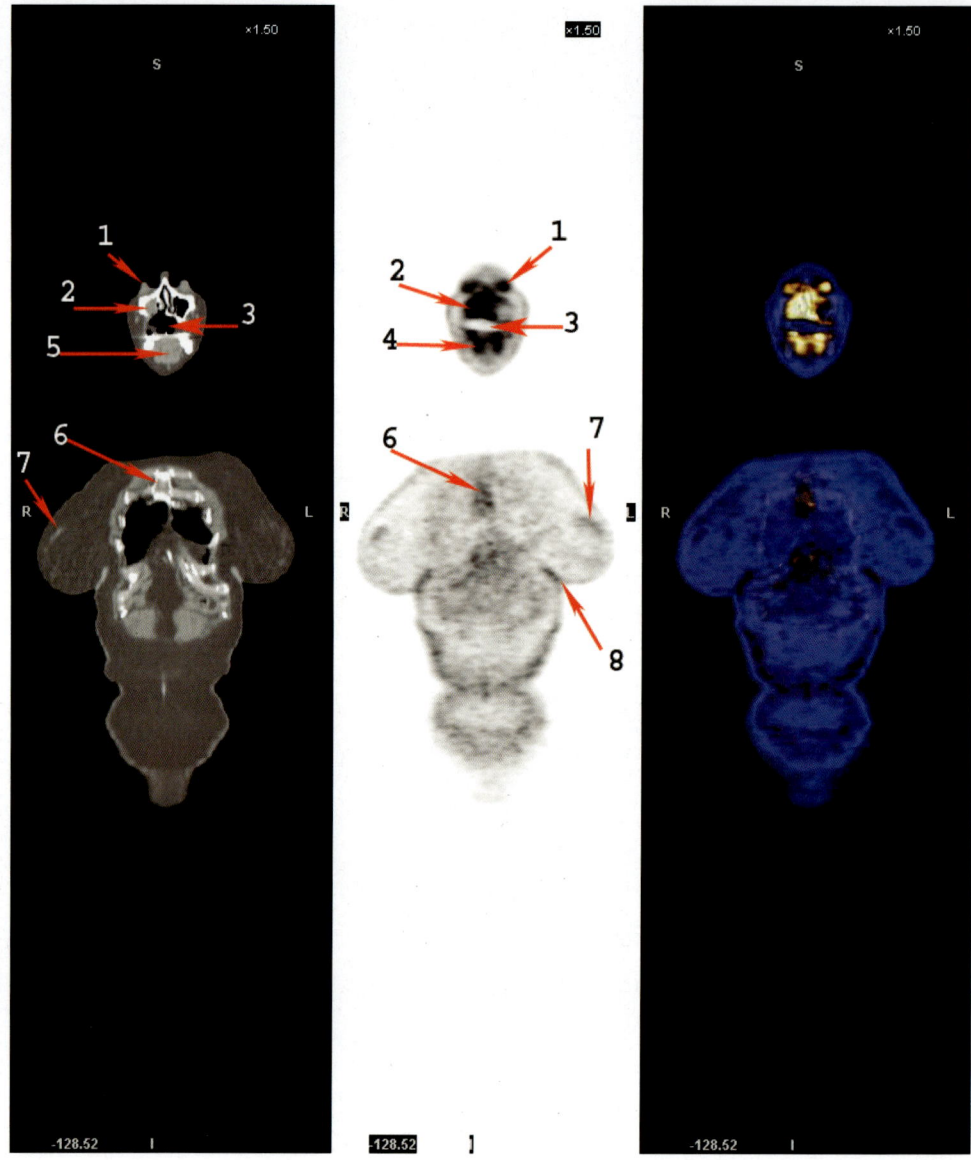

FIGURE 1-3: Coronal 3

1. Orbital fat or muscle
2. Right maxillary sinusitis
3. Oral cavity
4. Lingual tonsil
5. Tongue
6. Sternum
7. Breast nipple
8. Skin fold

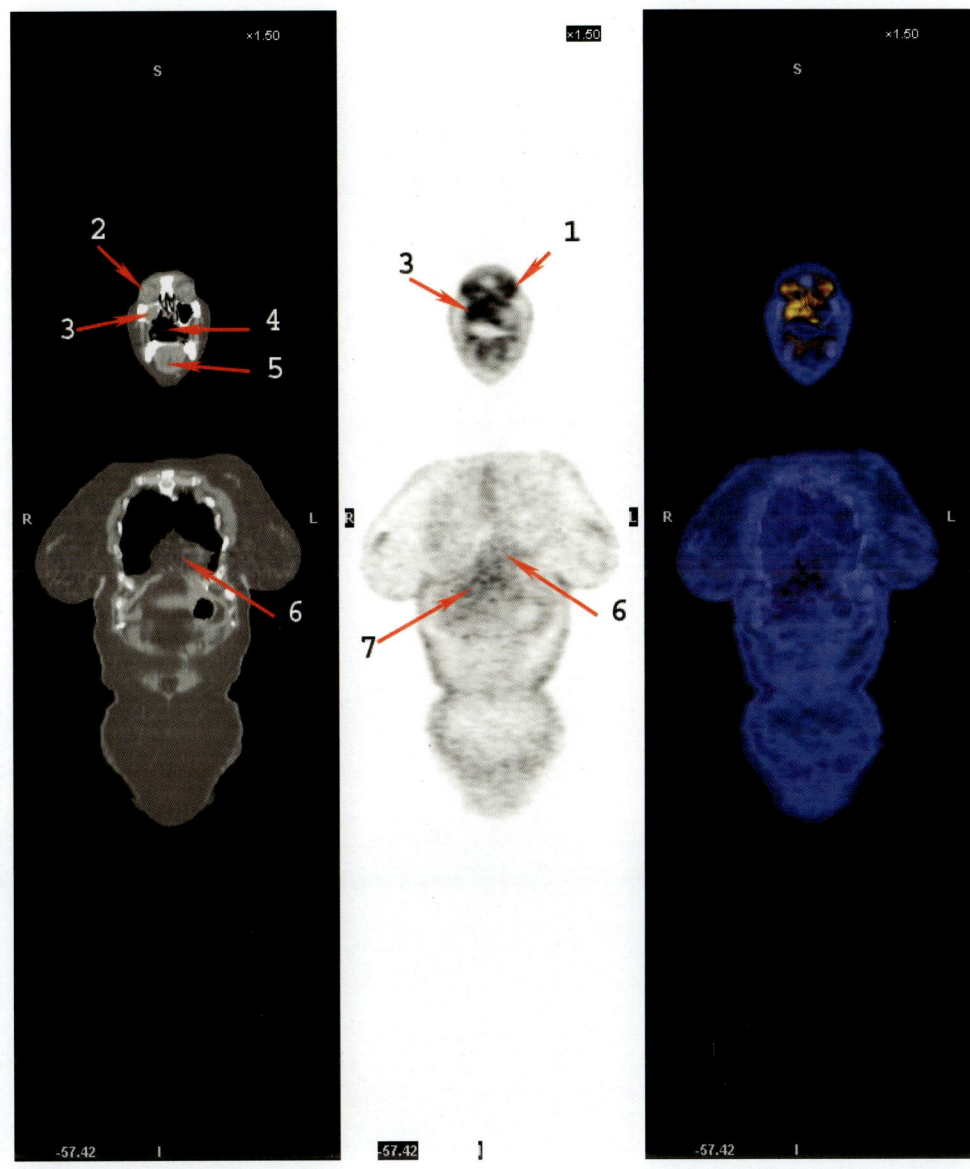

FIGURE 1-4: Coronal 4

1. Periorbital fat or muscle
2. Orbital globe
3. Right maxillary sinusitis
4. Oral cavity
5. Tongue
6. Heart
7. Liver

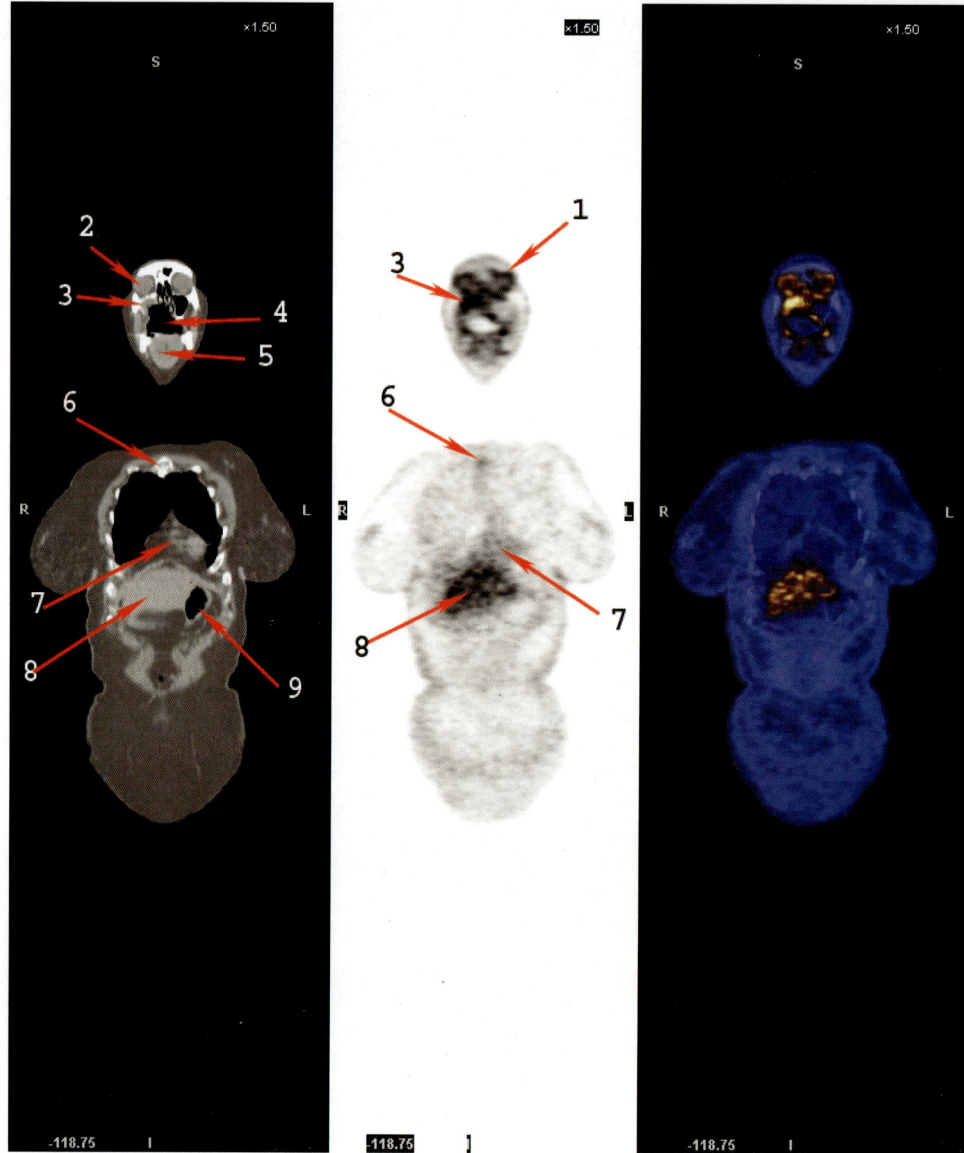

FIGURE 1-5: Coronal 5

1. Periorbital fat or muscle	4. Oral cavity	7. Heart
2. Orbital globe	5. Tongue	8. Liver
3. Right maxillary sinusitis	6. Manubrium	9. Stomach

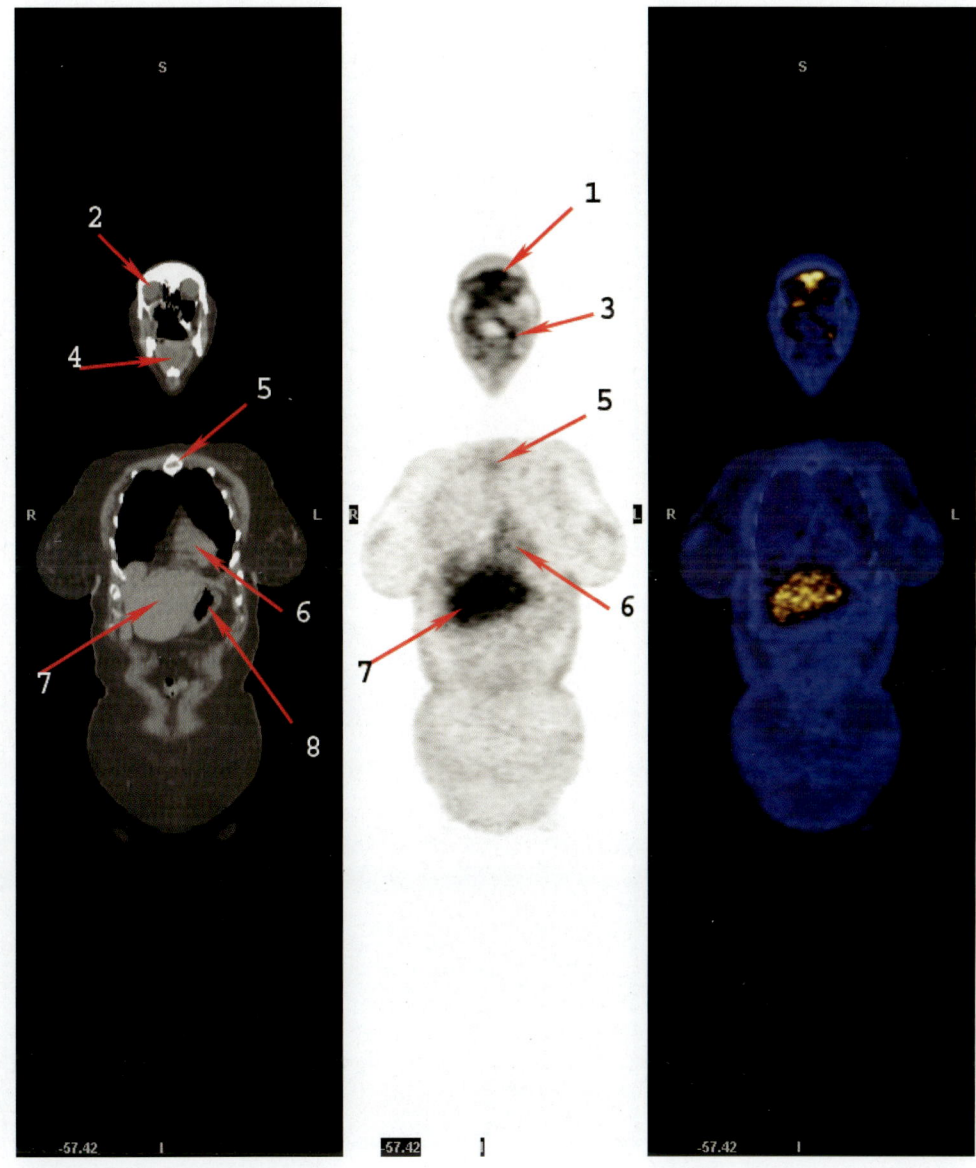

FIGURE 1-6: Coronal 6

1. Frontal sinusitis or brain
2. Orbital globe
3. Palatine tonsil
4. Tongue
5. Manubrium
6. Heart
7. Liver
8. Stomach

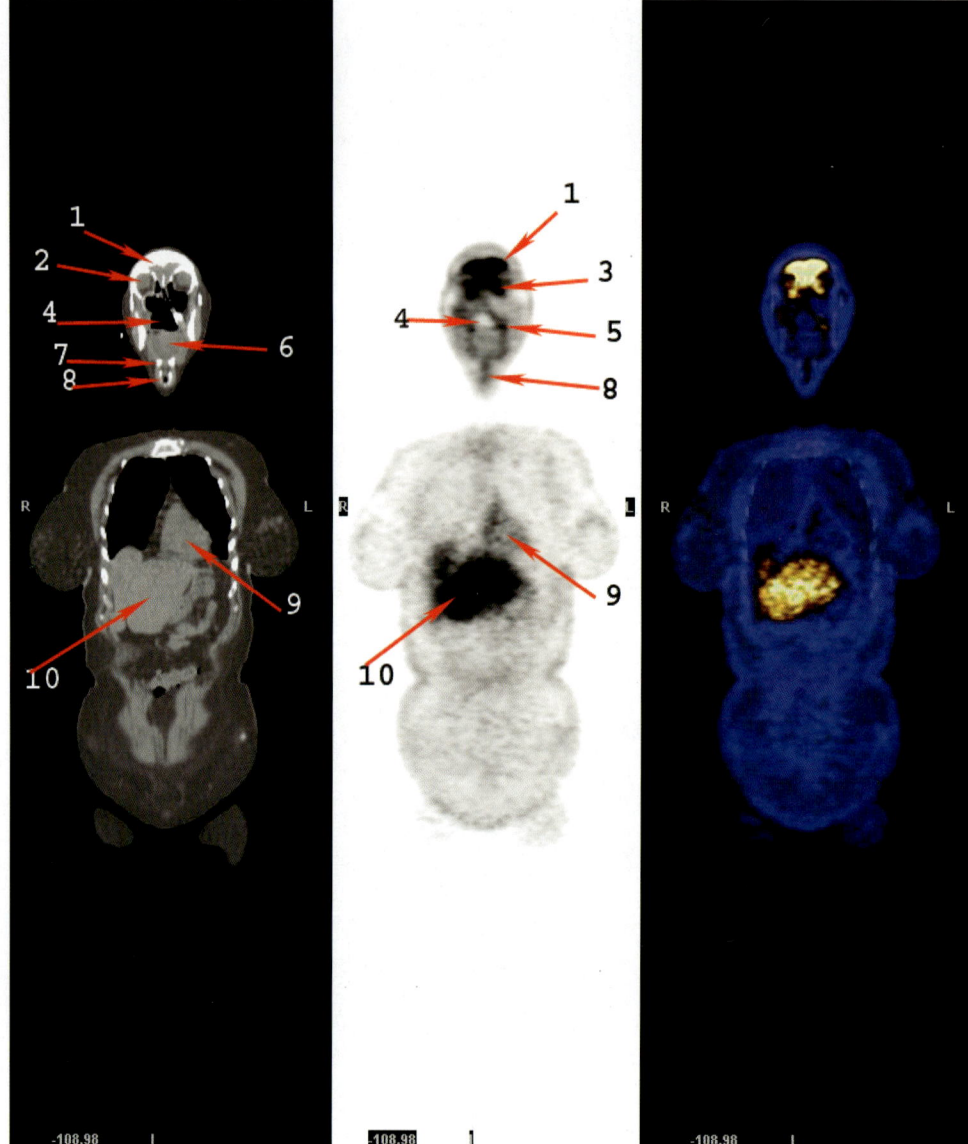

FIGURE 1-7: Coronal 7

1. Frontal lobe (gyrus) of brain	4. Oropharynx	8. Vocal muscle
2. Orbital globe	5. Palatine tonsil	9. Heart
3. Maxillary sinus	6. Tongue	10. Liver
	7. Hyoid	

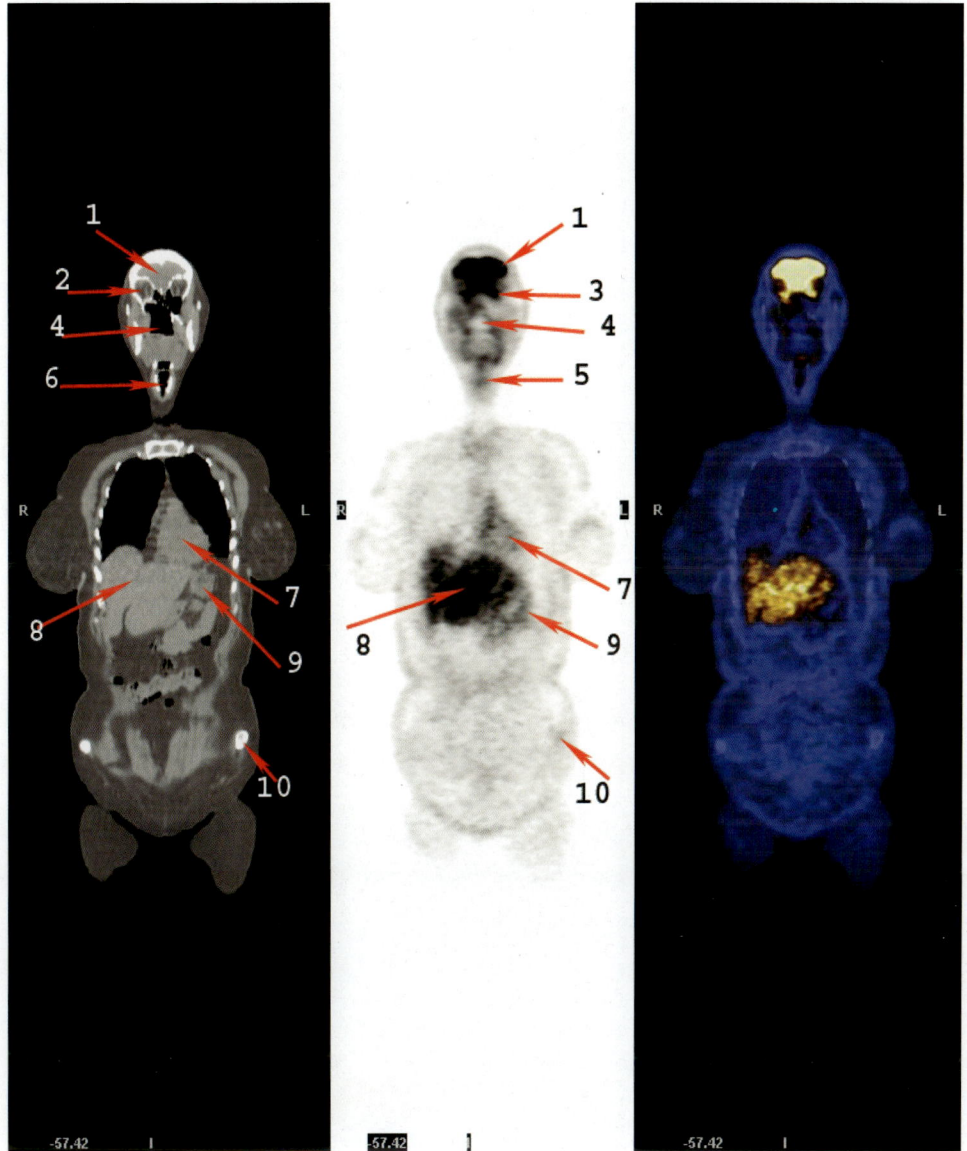

FIGURE 1-8: Coronal 8

1. Frontal lobe (gyrus) of brain
2. Orbital globe
3. Maxillary sinus
4. Oropharynx
5. Vocal muscle
6. Larynx
7. Heart
8. Liver
9. Stomach
10. Iliac crest

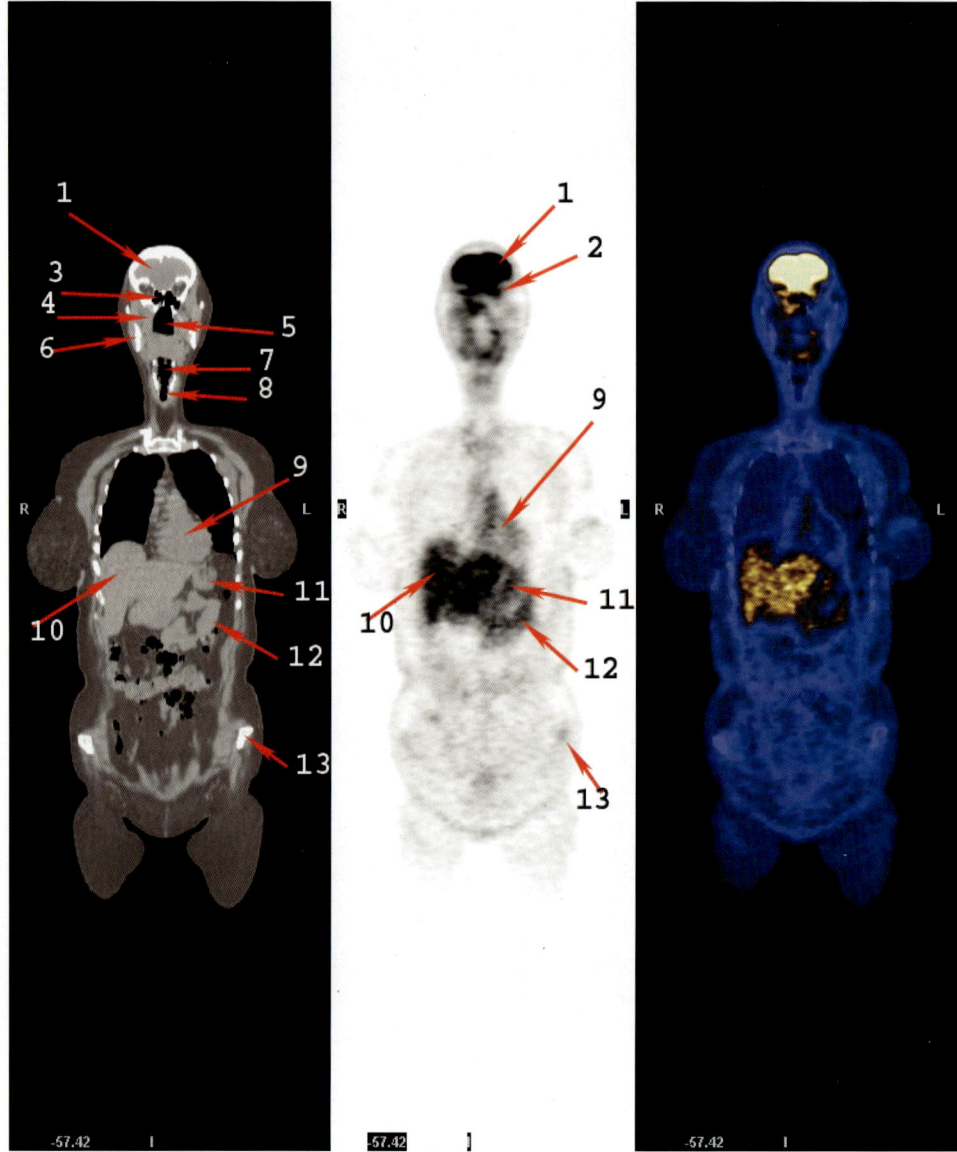

FIGURE 1-9: Coronal 9

1. Frontal lobe (gyrus) of brain
2. Periorbital fat or muscle
3. Sphenoid sinus
4. Pterygoid muscle
5. Oropharynx
6. Mandible
7. Hypopharynx
8. Larynx
9. Heart
10. Liver (segment VIII)
11. Stomach
12. Jejunum
13. Iliac crest

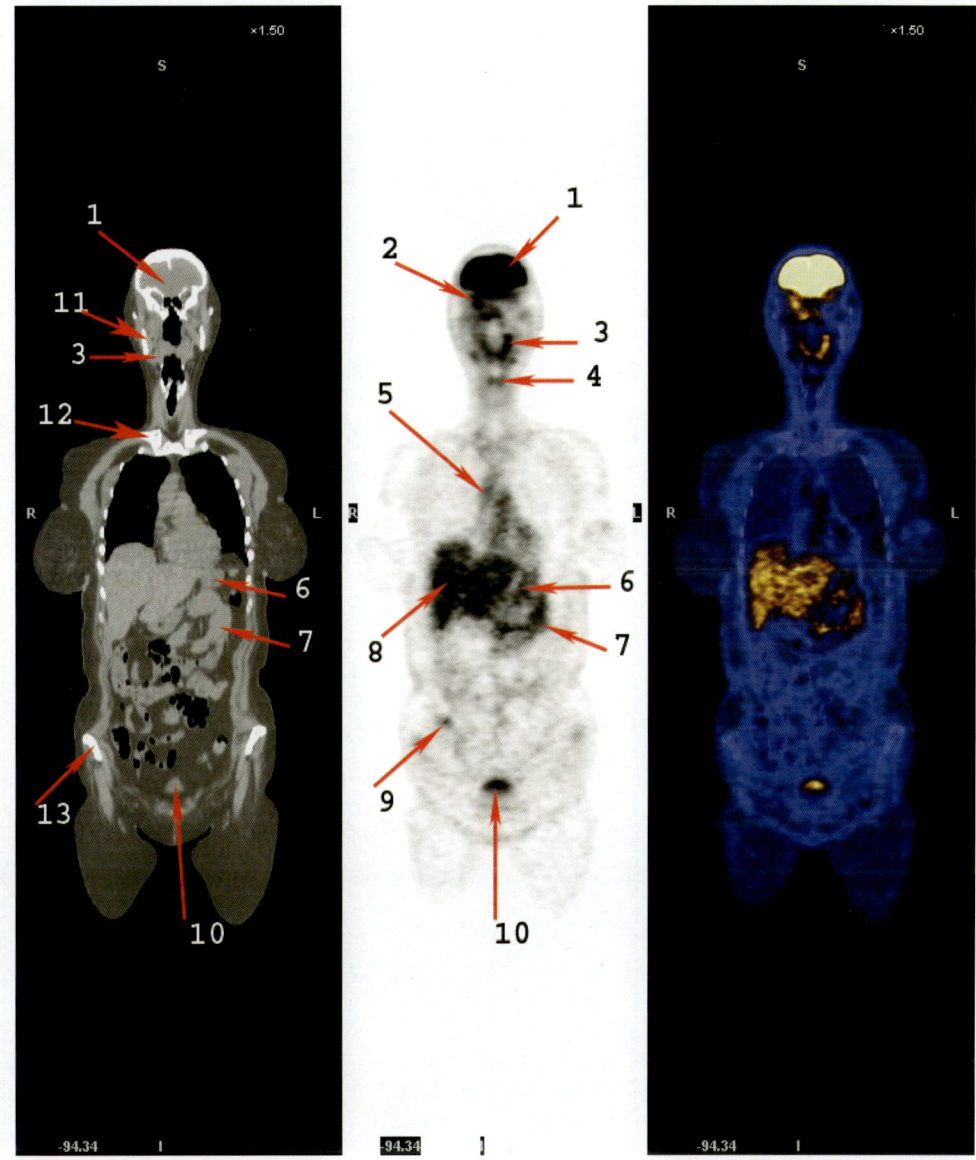

FIGURE 1-10: Coronal 10

1. Frontal cerebral gyrus
2. Temporal cerebral gyrus
3. Mylohyoid muscle
4. Vocal cord muscle
5. Ascending aorta
6. Gastric fundus
7. Jejunum
8. Liver
9. Cecum
10. Bladder
11. Mandible
12. Clavicle
13. Iliac crest

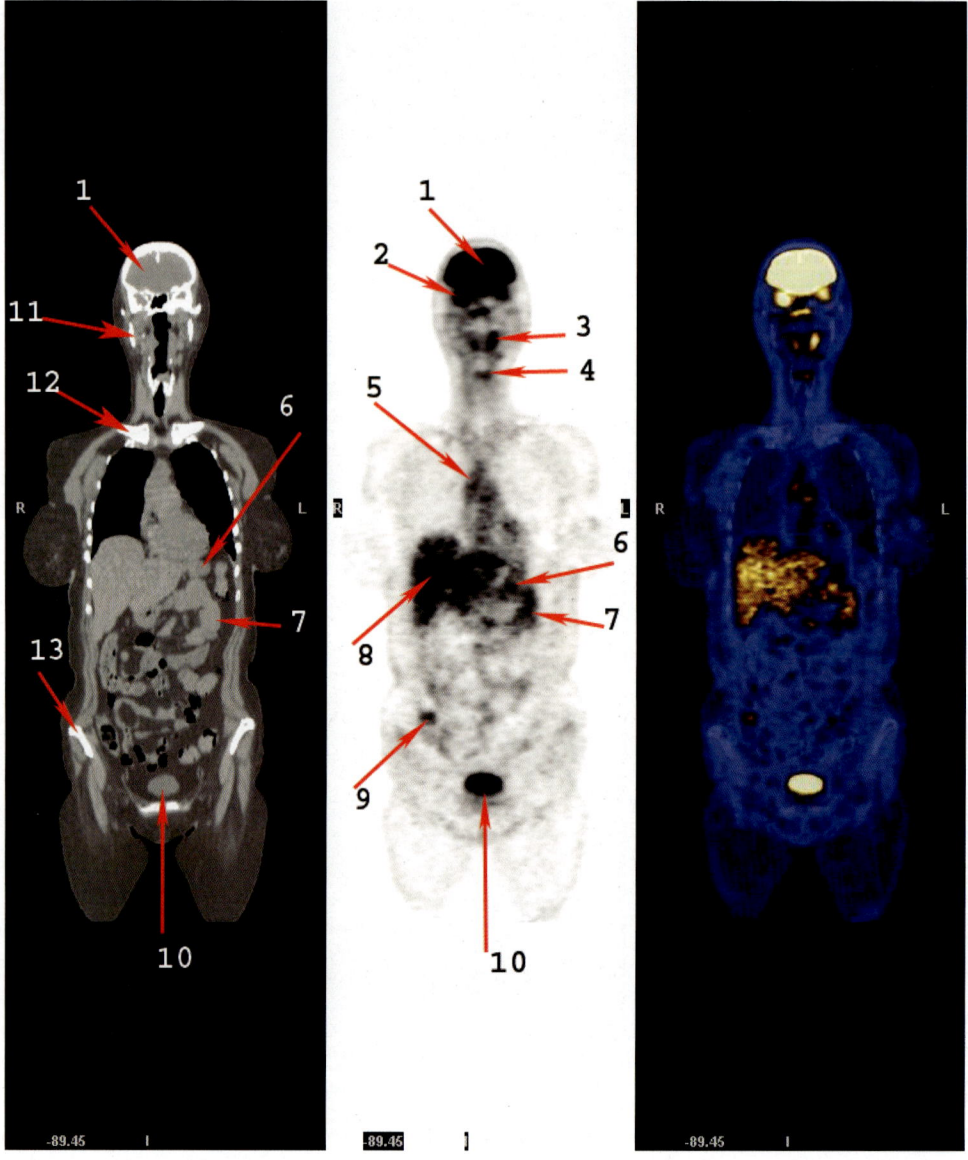

FIGURE 1-11: Coronal 11

1. Frontal cerebral gyrus	6. Gastric fundus	11. Mandible
2. Temporal cerebral gyrus	7. Jejunum	12. Clavicle
3. Genioglossus muscle	8. Liver	13. Iliac crest
4. Vocal cord muscle	9. Cecum	
5. Ascending aorta	10. Bladder	

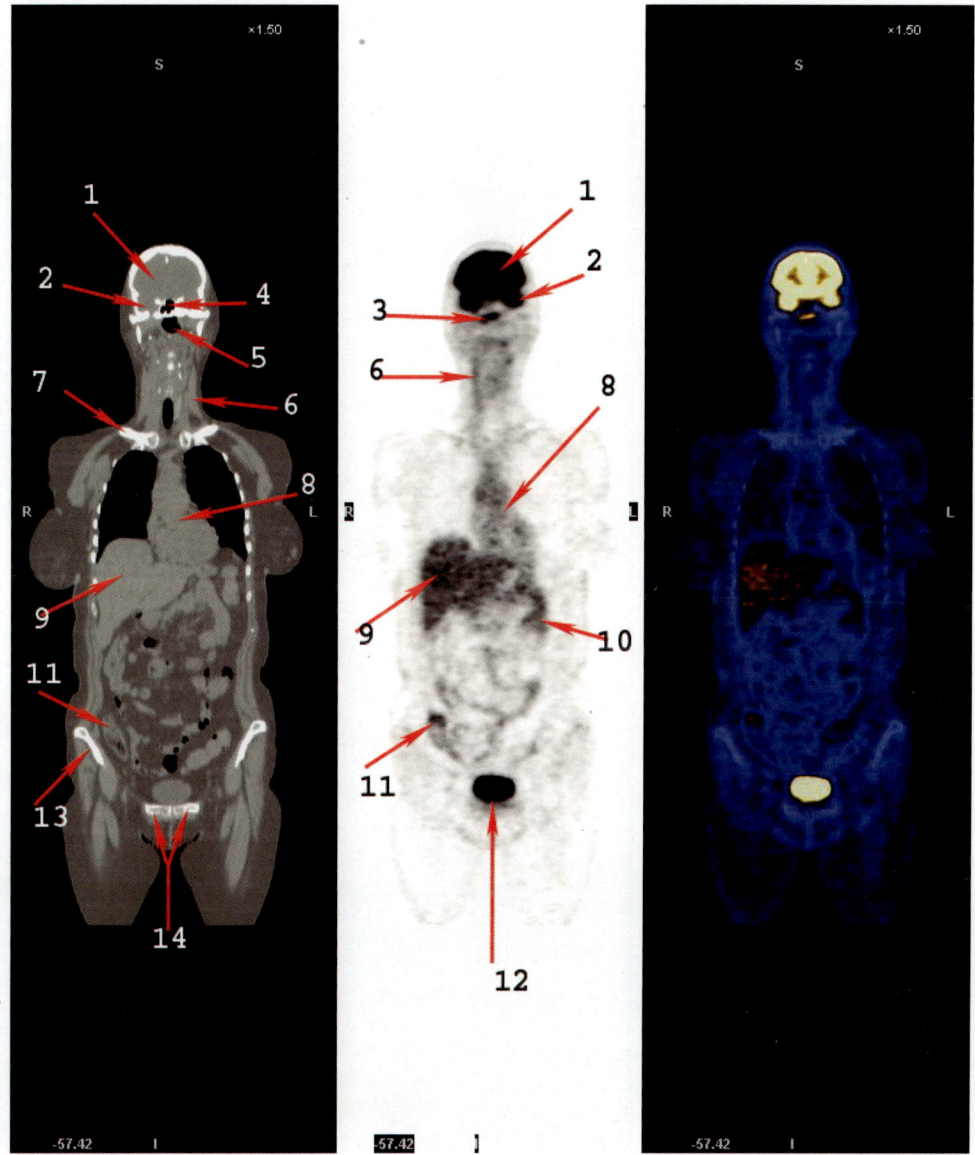

FIGURE 1-12: Coronal 12

1. Frontal cerebral gyrus	6. Sternocleidomastoid muscle	10. Jejunum
2. Temporal cerebral gyrus		11. Cecum
3. Pharyngeal muscle or fat	7. Clavicle	12. Bladder
4. Sphenoid sinus	8. Heart	13. Iliac crest
5. Oropharynx	9. Liver (segment VIII)	14. Pubis

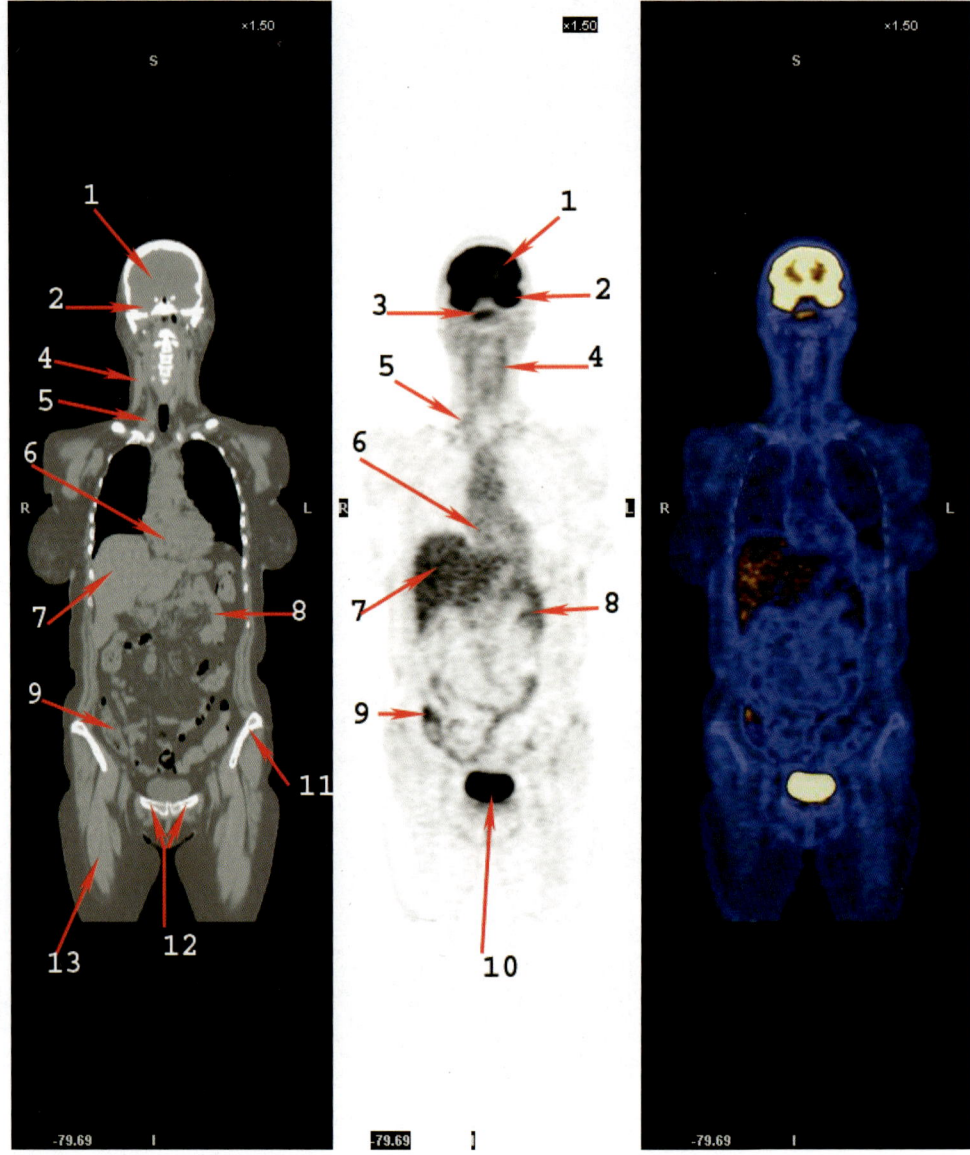

FIGURE 1-13: Coronal 13

1. Parietal cerebral gyrus
2. Temporal cerebral gyrus
3. Pharyngeal muscle
4. Sternocleidomastoid
   muscle
5. Anterior scalene muscle
6. Heart (right ventricle)
7. Liver (segment VIII)
8. Jejunum
9. Cecum
10. Bladder
11. Iliac crest
12. Pubis
13. Rectus femoris muscle

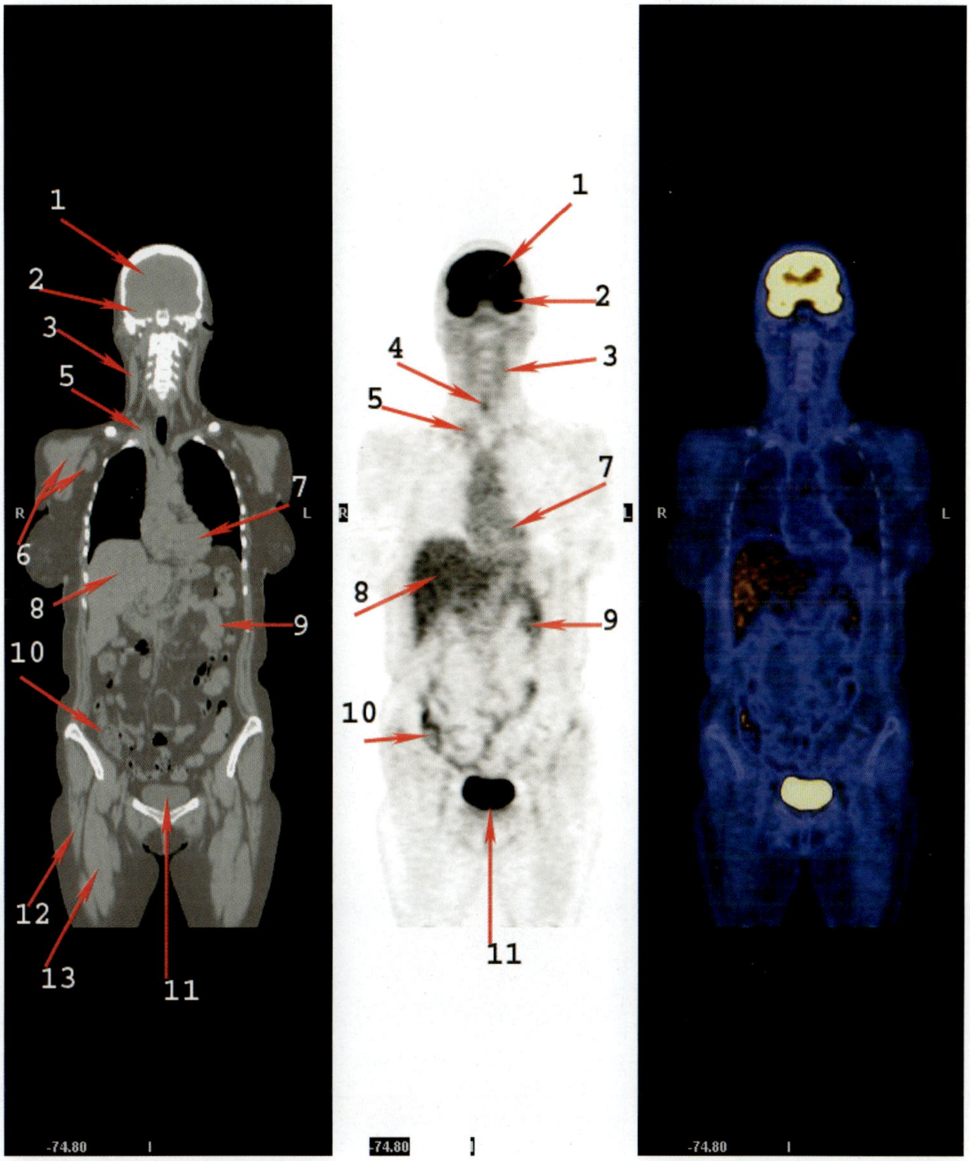

FIGURE 1-14: Coronal 14

1. Parietal cerebral gyrus	6. Shoulder cuff muscles	11. Bladder
2. Temporal cerebral gyrus	7. Heart (left ventricle)	12. Tensor fascia lata
3. Longus colli muscle	8. Liver (segment VIII)	muscle
4. Vocal cord muscle	9. Jejunum	13. Rectus femoris muscle
5. Scalene muscle	10. Cecum	

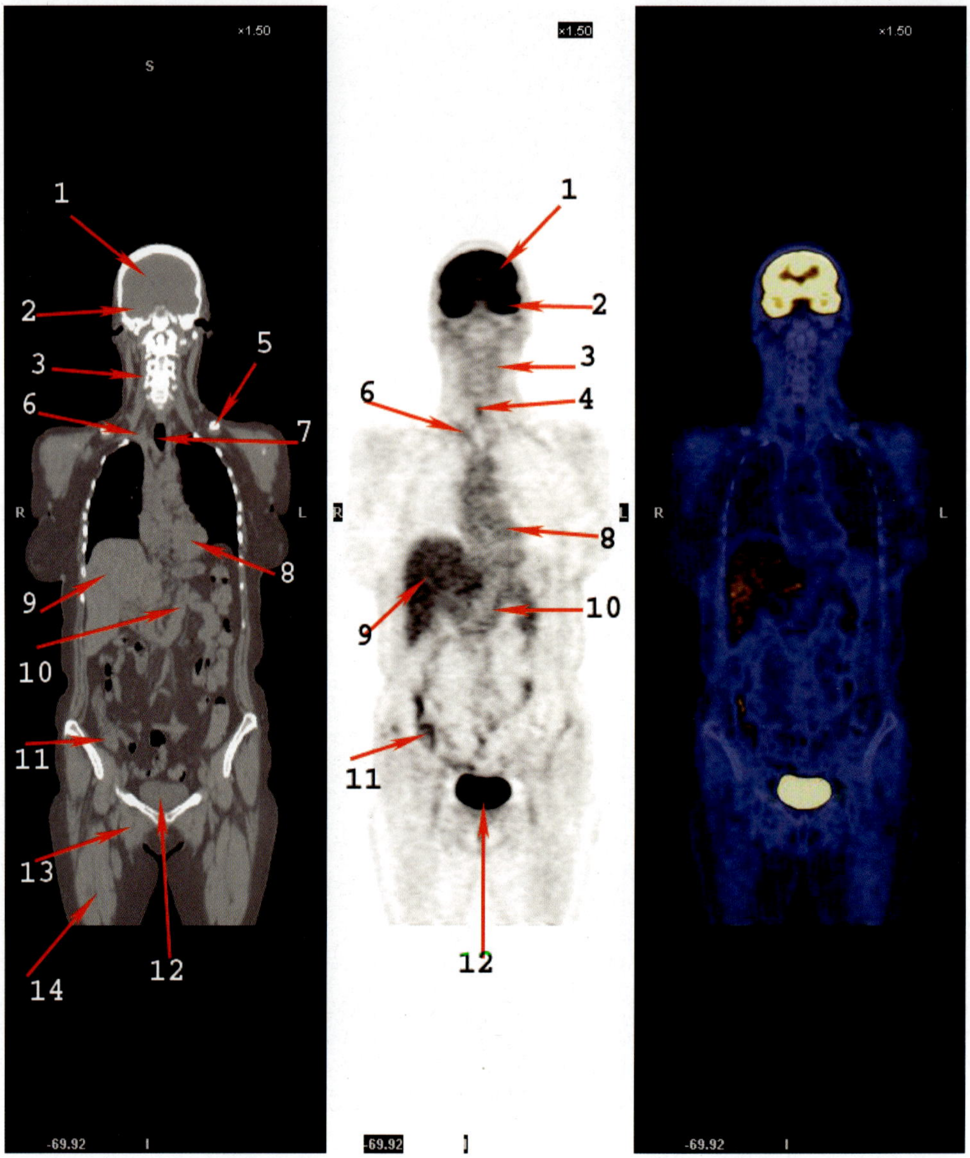

FIGURE 1-15: Coronal 15

1. Parietal cerebral gyrus	6. Scalene muscle	11. Cecum
2. Temporal cerebral gyrus	7. Larynx	12. Bladder
3. Longus colli muscle	8. Heart (left ventricle)	13. Adductor muscles
4. Vocal cord muscle	9. Liver (segment VIII)	14. Rectus femoris muscle
5. Scapula	10. Transverse colon	

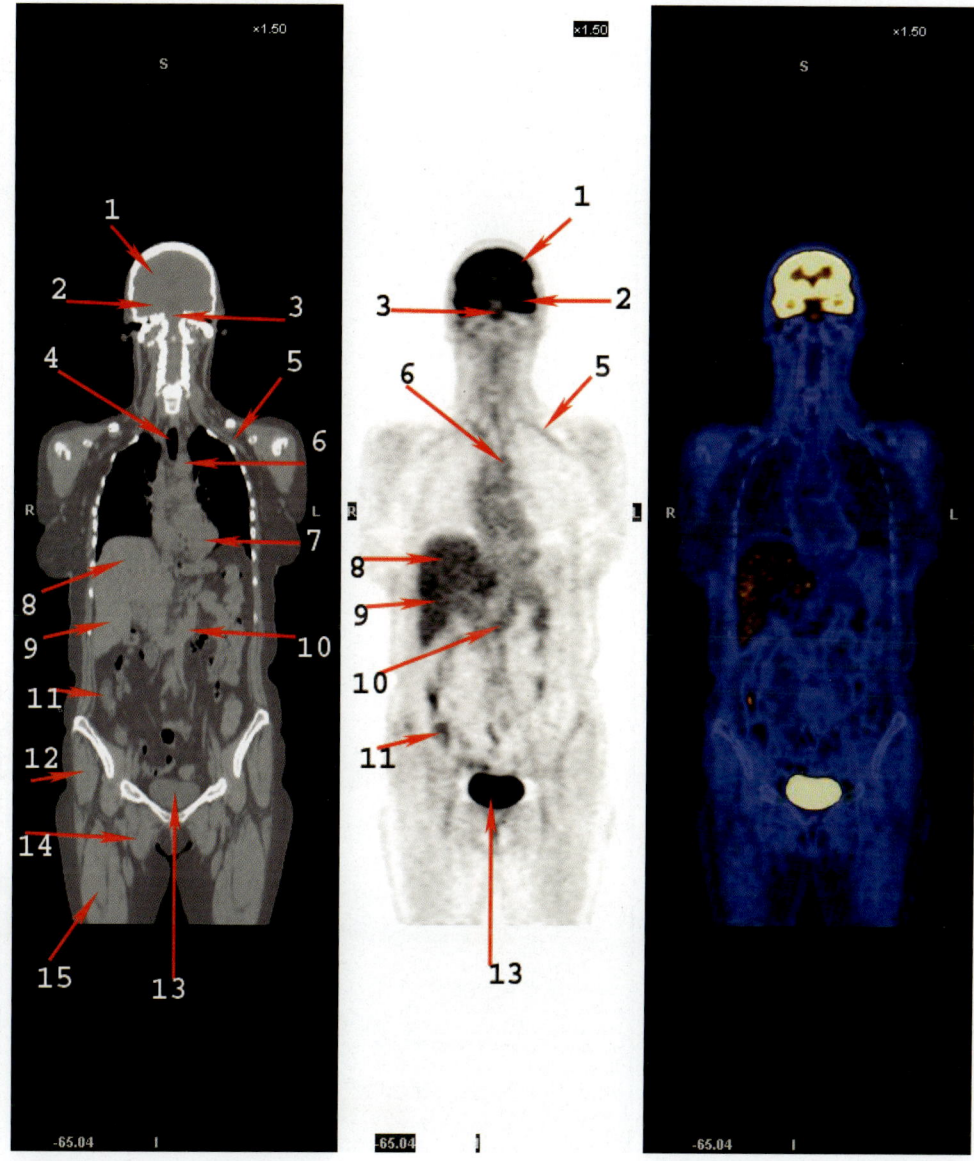

FIGURE 1-16: Coronal 16

1. Parietal cerebral gyrus
2. Temporal cerebral gyrus
3. Pons
4. Trachea
5. Serratus anterior muscle
6. Aortic arch
7. Left ventricle
8. Liver (segment VII)
9. Liver (segment VI)
10. Transverse colon
11. Cecum
12. Iliopsoas muscle
13. Bladder
14. Adductor muscles
15. Rectus femoris muscle

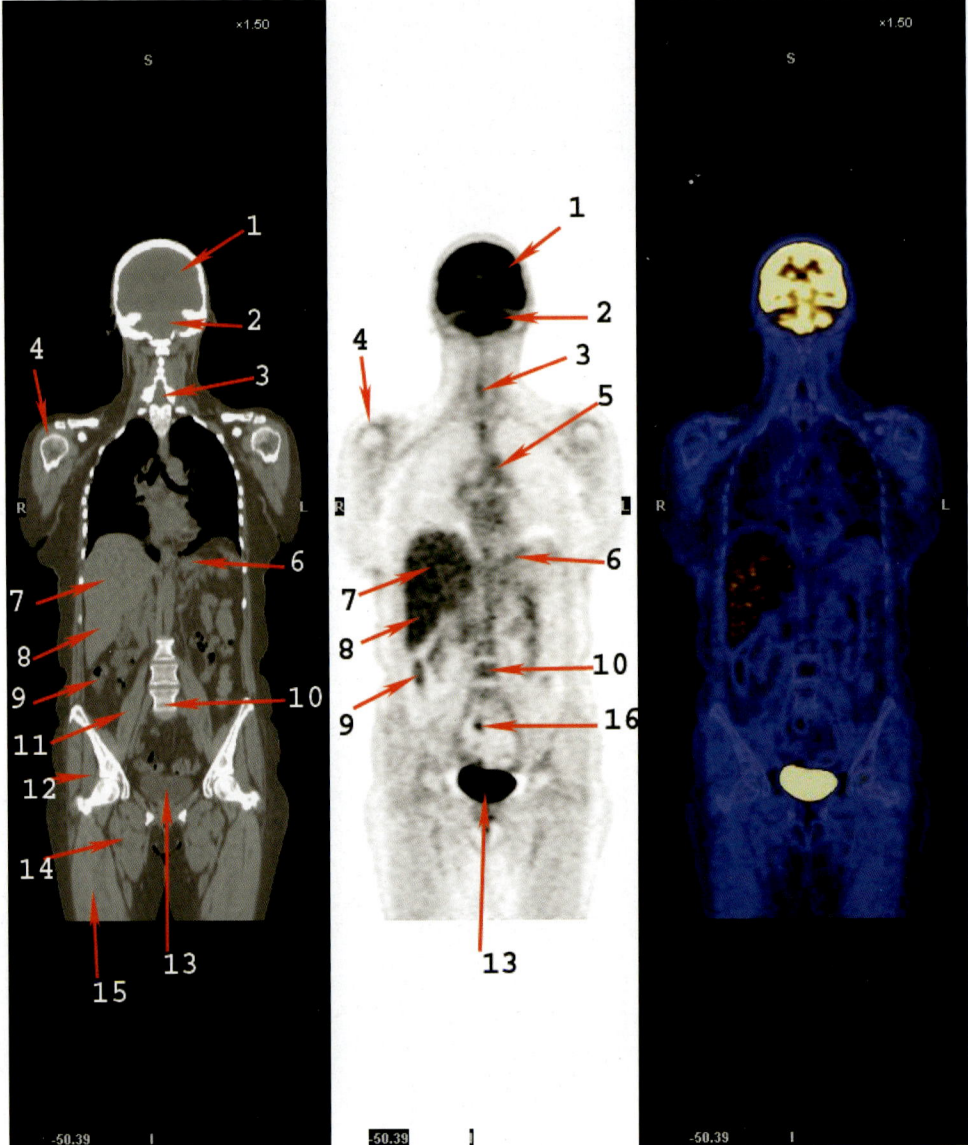

FIGURE 1-17: Coronal 17

1. Parietal cerebral gyrus
2. Cerebellum
3. Spinal cord
4. Synovial membrane of shoulder joint
5. Aortic arch
6. Left hepatic lobe (segment II)
7. Right hepatic lobe (segment VII)
8. Right hepatic lobe (segment VI)
9. Ascending colon
10. L5 marrow
11. Psoas muscle
12. Iliopsoas muscle
13. Bladder
14. Adductor muscles
15. Rectus femoris muscle
16. Sigmoid colon

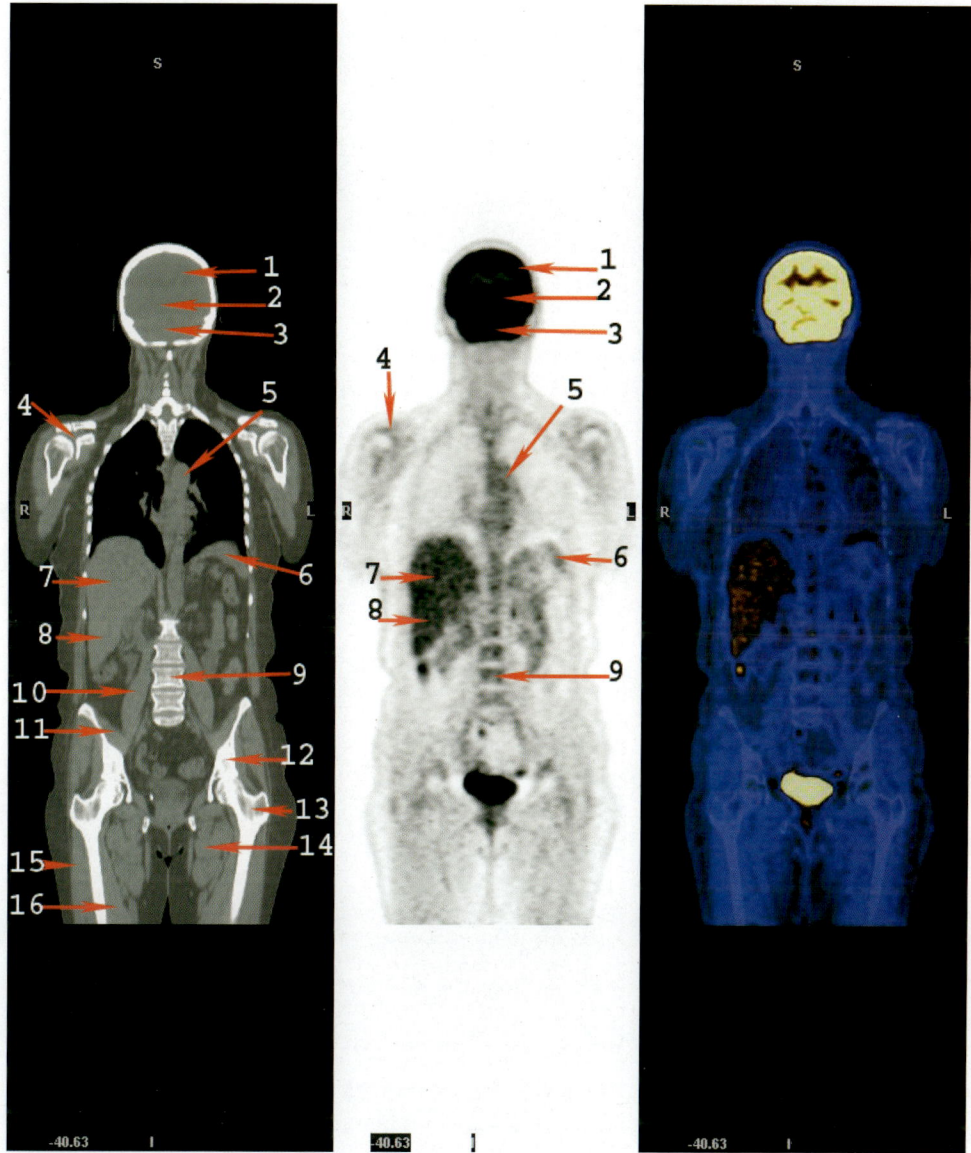

FIGURE 1-18: Coronal 18

1. Parietal cerebral gyrus
2. Occipital cerebral gyrus
3. Cerebellar gyrus
4. Synovial membrane of shoulder joint
5. Aortic arch
6. Spleen
7. Right liver (segment VII)
8. Right liver (segment VI)
9. L4 marrow
10. Psoas muscle
11. Iliacus muscle
12. Acetabulum
13. Femur
14. Adductor muscles
15. Vastus lateralis muscle
16. Vastus medialis muscle

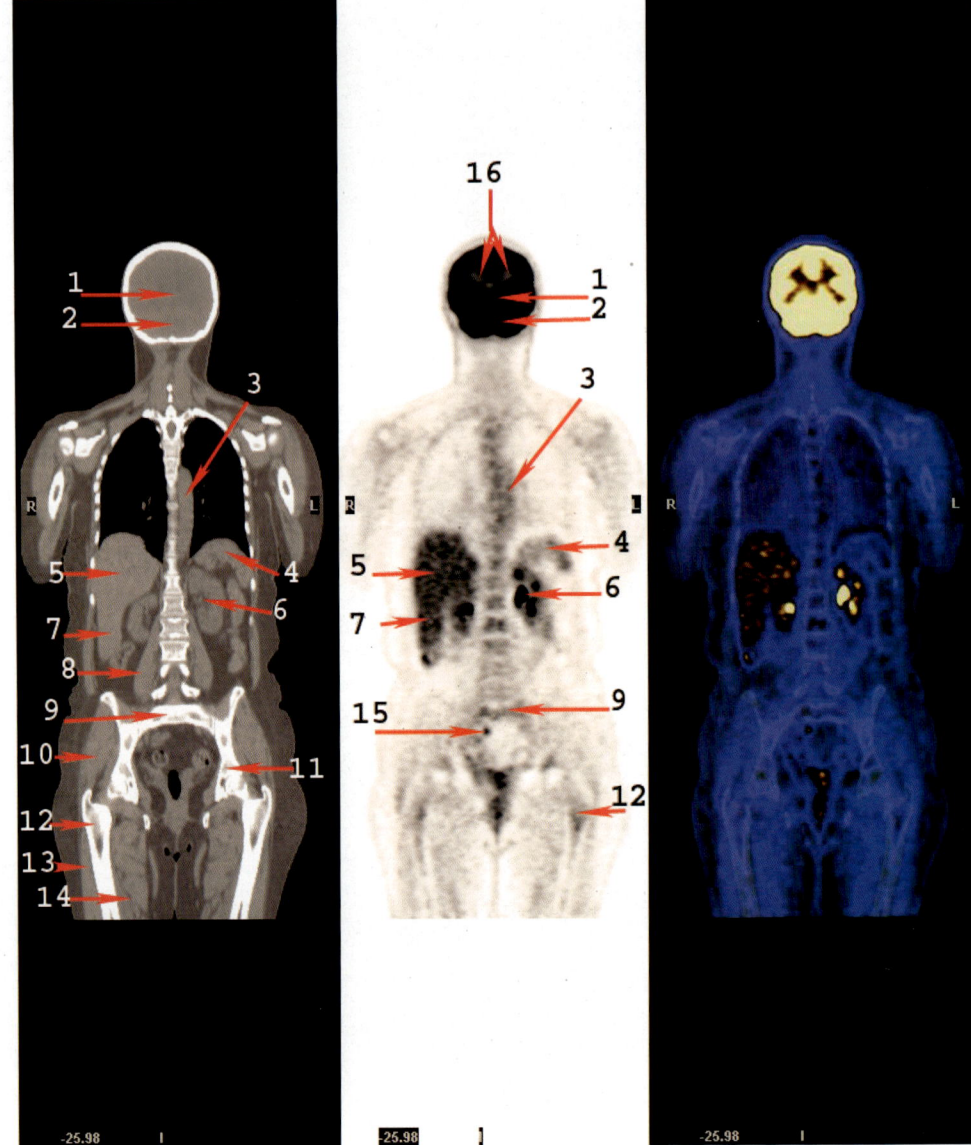

FIGURE 1-19: Coronal 19

1. Occipital cerebral gyrus
2. Cerebellar gyrus
3. Descending aorta
4. Spleen
5. Right liver (segment VII)
6. Renal pelvis
7. Right liver (segment VI)
8. Psoas muscle
9. Sacrum
10. Gluteus muscle
11. Acetabulum
12. Femur
13. Vastus lateralis muscle
14. Vastus medialis muscle
15. Sigmoid colon
16. Lateral ventricle

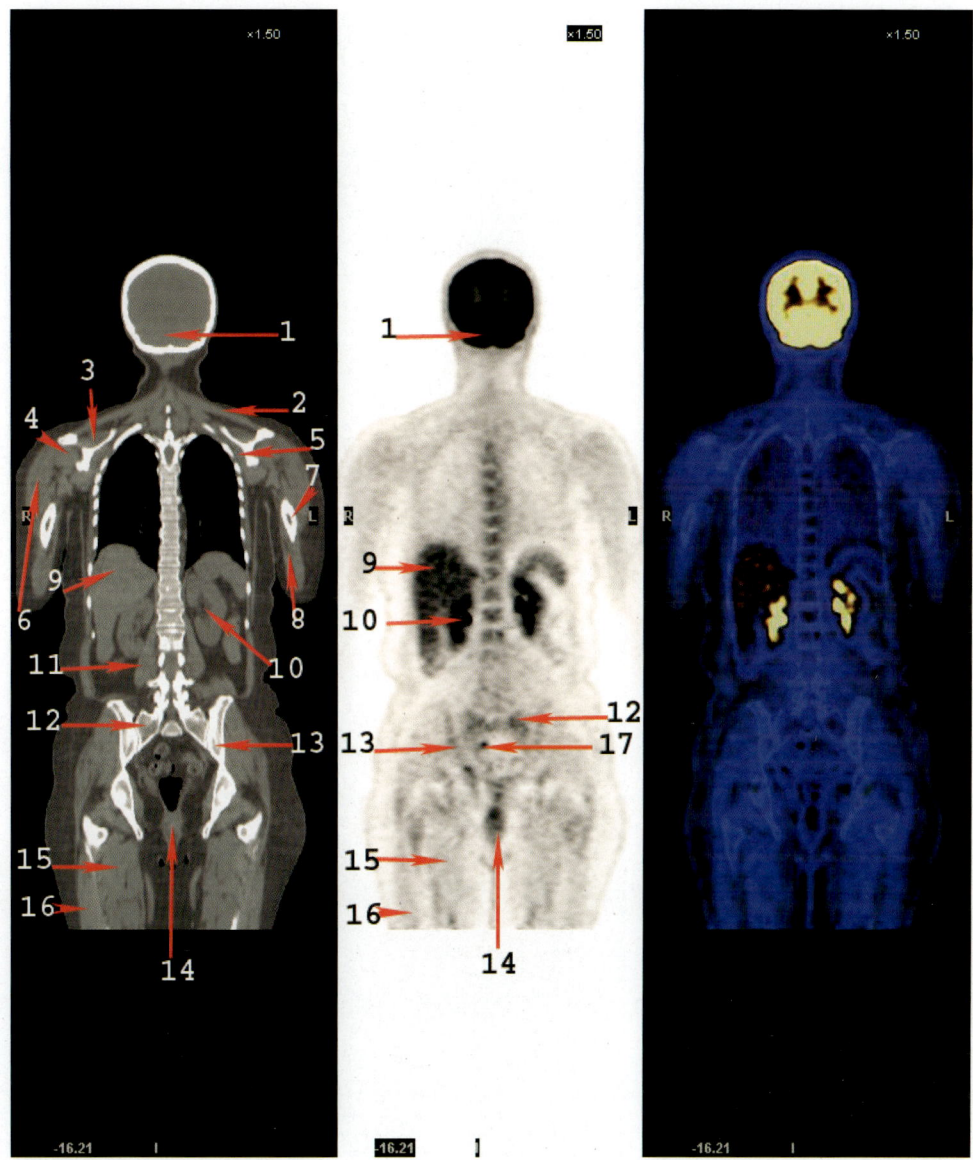

FIGURE 1-20: Coronal 20

1. Cerebellar gyrus
2. Trapezius muscle
3. Supraspinatus muscle
4. Infraspinatus muscle
5. Subscapularis muscle
6. Deltoid muscle
7. Humerus
8. Triceps brachii muscle
9. Right liver (segment VII)
10. Renal pelvis
11. Psoas muscle
12. Sacral ala
13. Iliac tuberosity
14. Rectum
15. Vastus medialis muscle
16. Vastus lateralis muscle
17. Sigmoid colon

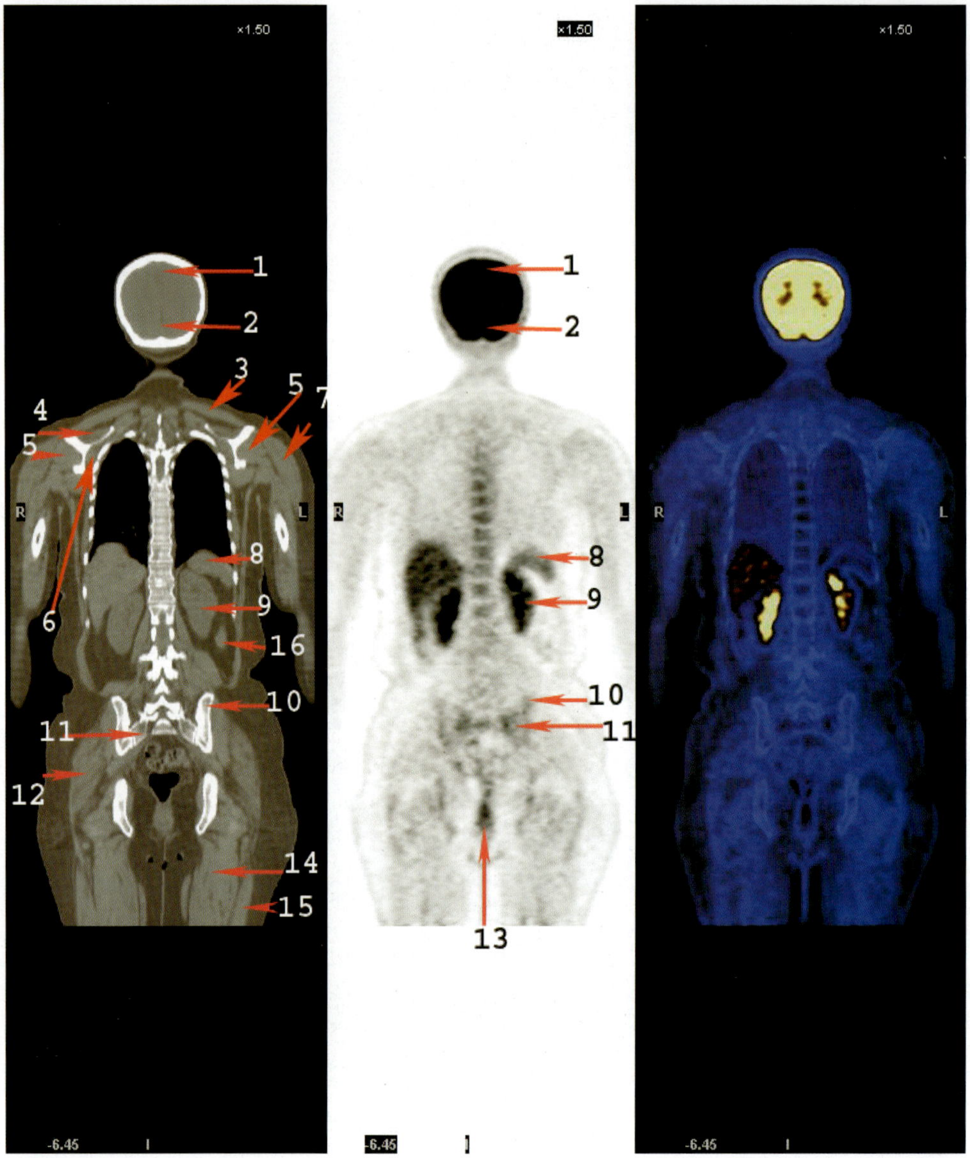

FIGURE 1-21: Coronal 21

1. Parietal cerebral gyrus
2. Cerebellar gyrus
3. Trapezius muscle
4. Supraspinatus muscle
5. Infraspinatus muscle
6. Subscapularis muscle
7. Deltoid muscle
8. Spleen
9. Kidney
10. Iliac tuberosity
11. Sacral ala
12. Gluteus muscle
13. Rectum
14. Semimembranosus muscle
15. Biceps femoris muscle
16. Descending colon

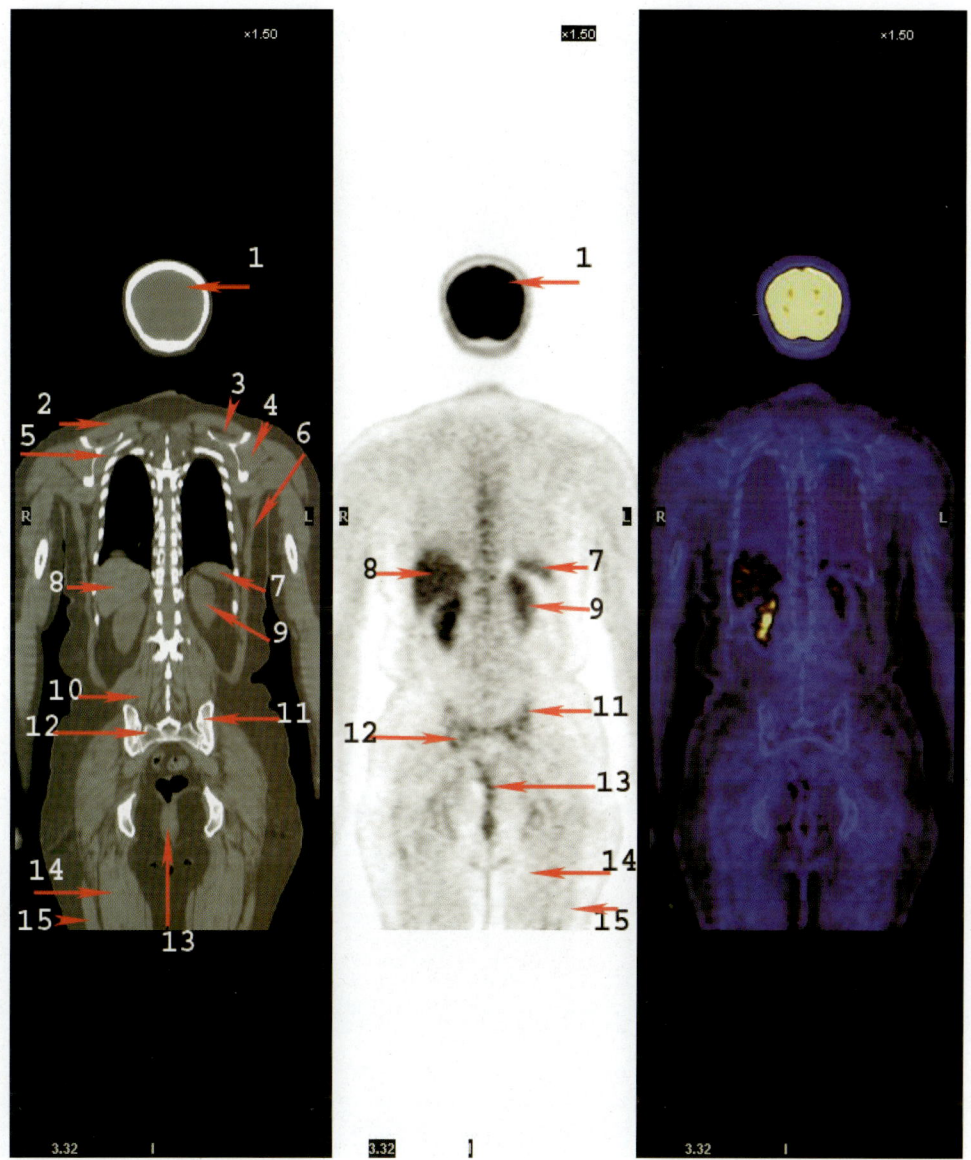

FIGURE 1-22: Coronal 22

1. Cerebral gray matter
2. Trapezius muscle
3. Supraspinatus muscle
4. Infraspinatus muscle
5. Subscapularis muscle
6. Latissimus dorsi muscle
7. Spleen
8. Right liver (segment VII)
9. Kidney, left
10. Psoas muscle
11. Iliac tuberosity
12. Sacral ala
13. Rectum
14. Semimembranosus muscle
15. Biceps femoris muscle

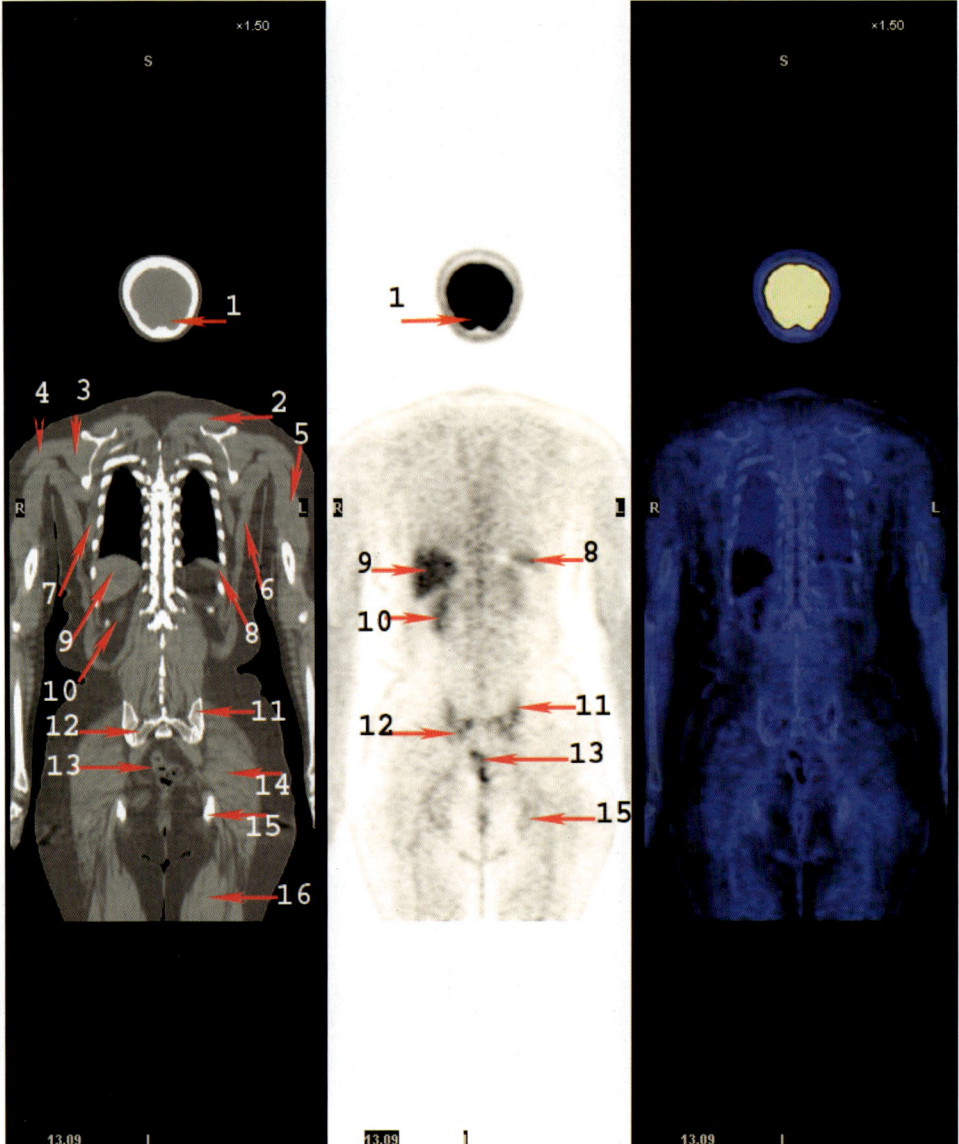

FIGURE 1-23: Coronal 23

1. Cerebellar gray matter
2. Supraspinatus muscle
3. Infraspinatus muscle
4. Deltoid muscle
5. Biceps brachii muscle
6. Latissimus dorsi muscle
7. Serratus anterior muscle
8. Spleen
9. Right liver (segment VII)
10. Right kidney
11. Iliac tuberosity
12. Sacral ala
13. Sigmoid
14. Gluteus muscles
15. Ischium
16. Semimembranosus muscle

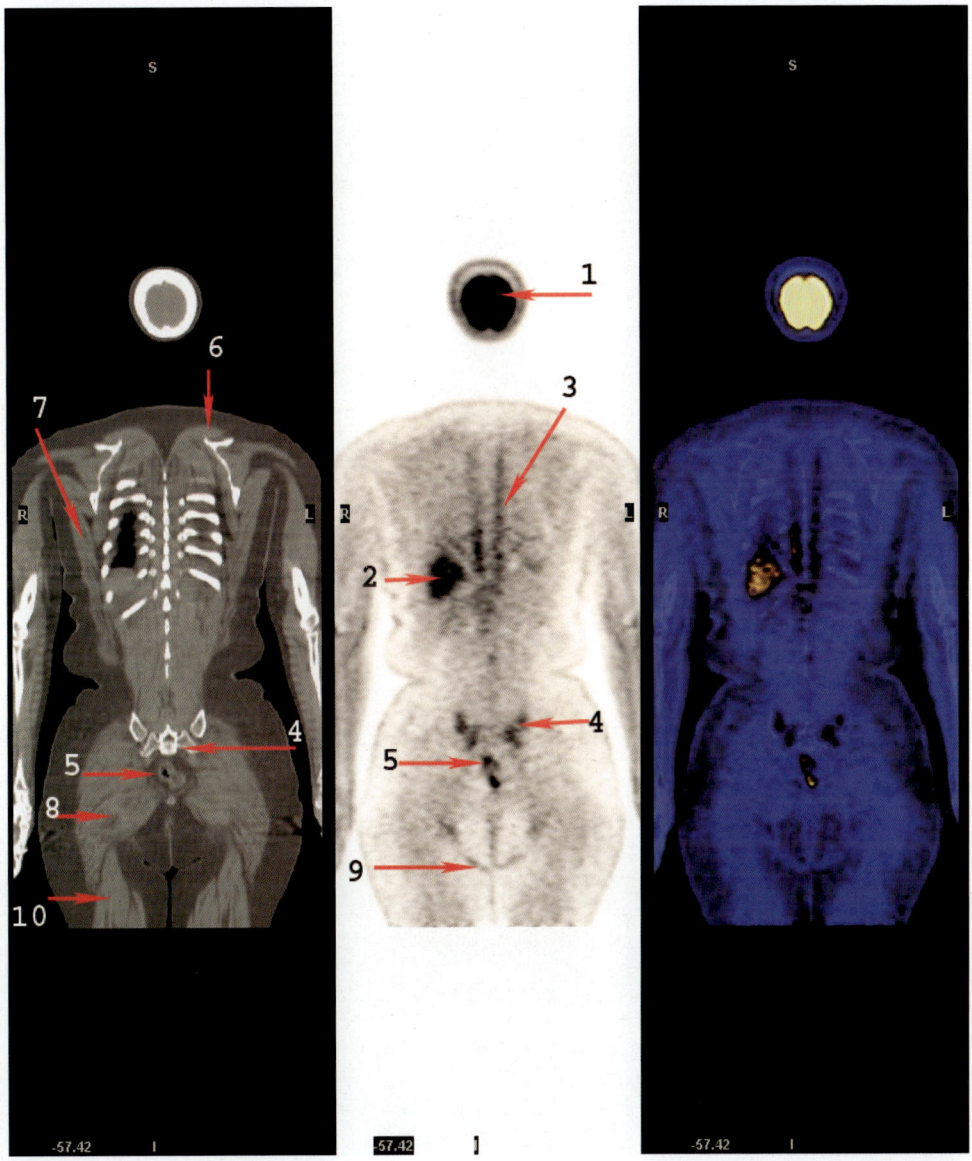

FIGURE 1-24: Coronal 24

1. Cerebral gray matter
2. Right hepatic lobe, superior segment (#7)
3. Erector spinae muscle/ brown fat
4. Sacral marrow
5. Rectosigmoid colon
6. Trapezius muscle
7. Latissimus dorsi muscle
8. Gluteus muscles
9. Skin fold
10. Semimembranosus muscle

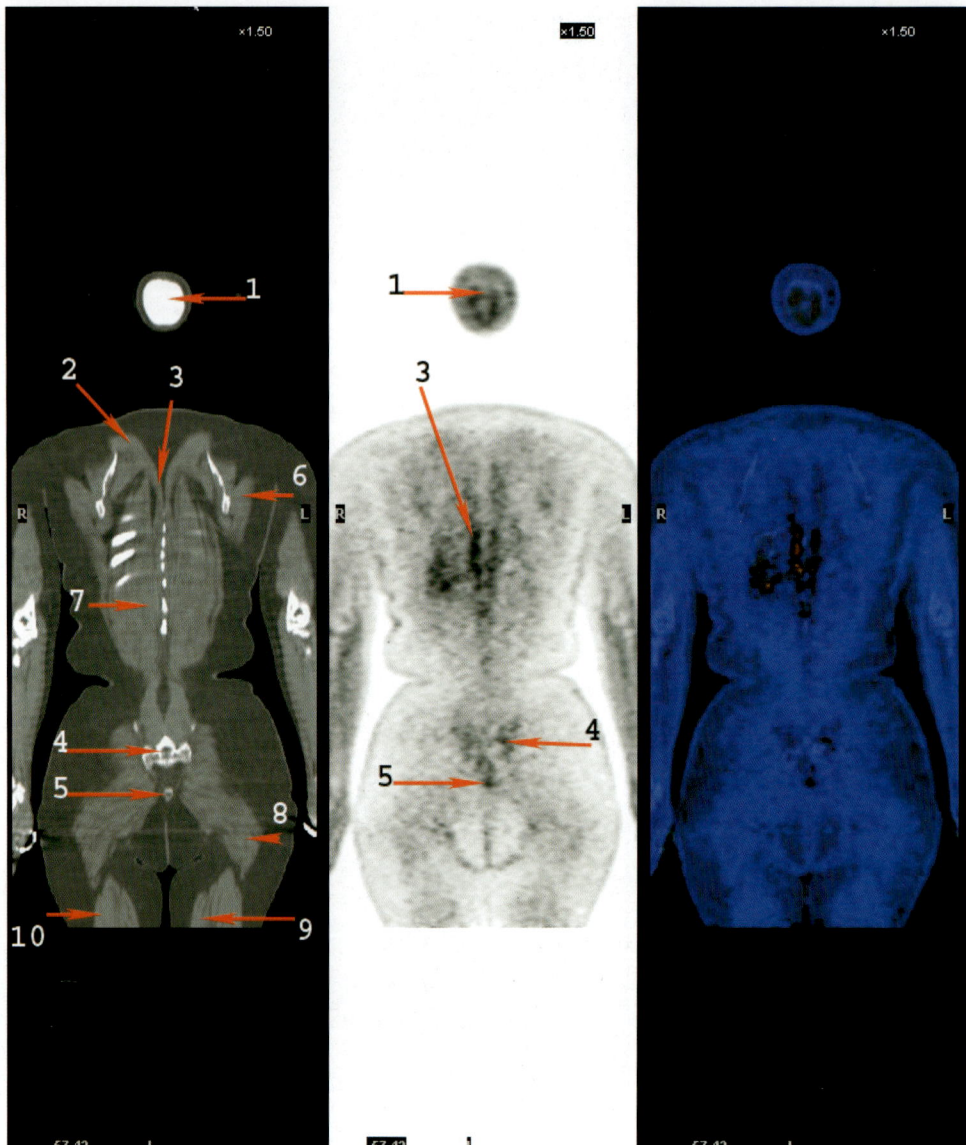

FIGURE 1-25: Coronal 25

1. Cerebral gray matter
2. Trapezius muscle
3. Erector spinae muscle/
   brown fat
4. Sacral marrow
5. Rectosigmoid colon
6. Latissimus dorsi muscle
7. Longissimus dorsi muscle
8. Gluteus muscles
9. Semimembranosus
   muscle
10. Semitendinosus muscle

Section 2: Sagittals

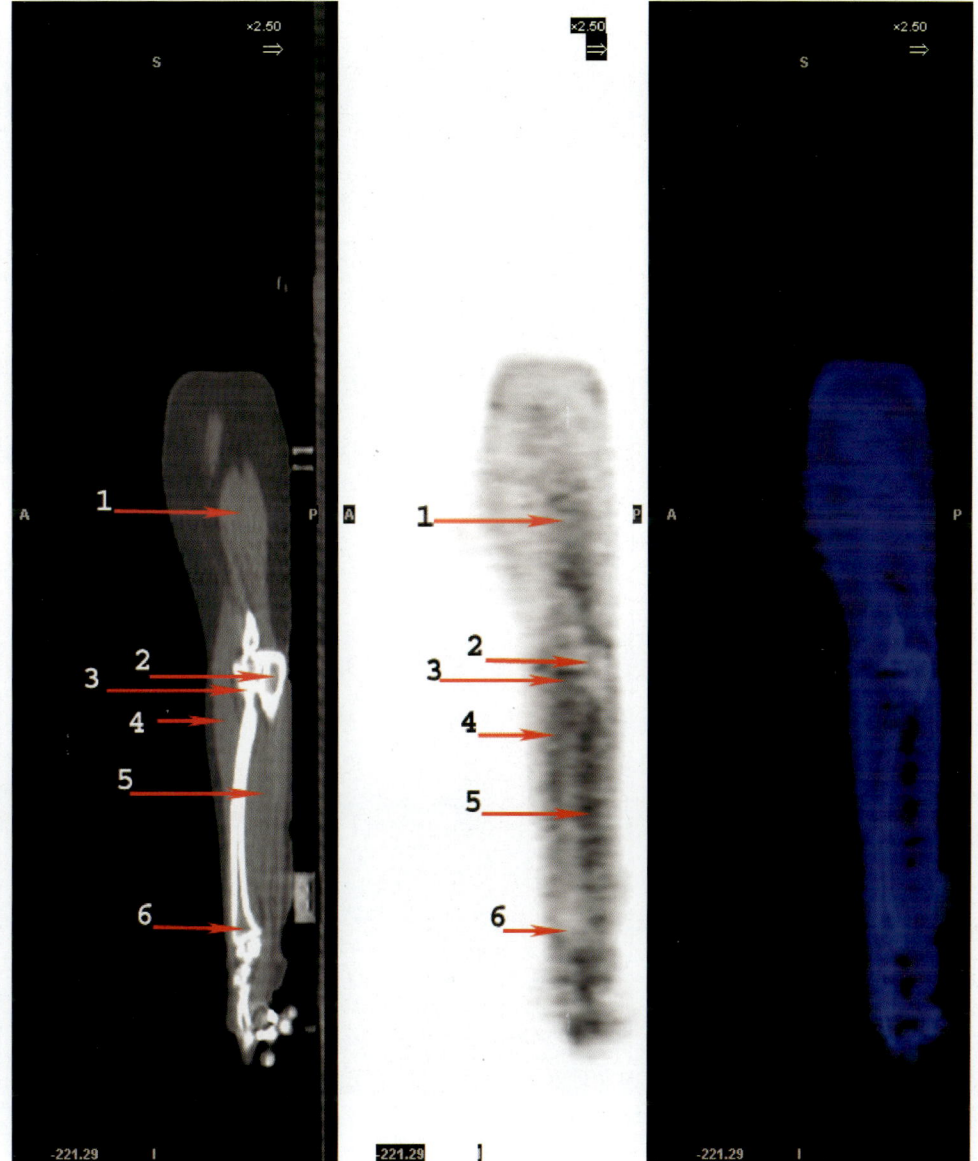

FIGURE 1-26: Sagittal.1

1. Triceps brachii muscle
2. Olecranon process of ulna
3. Radial head
4. Brachioradialis muscle
5. Flexor digitonin profundus muscle
6. Distal radius

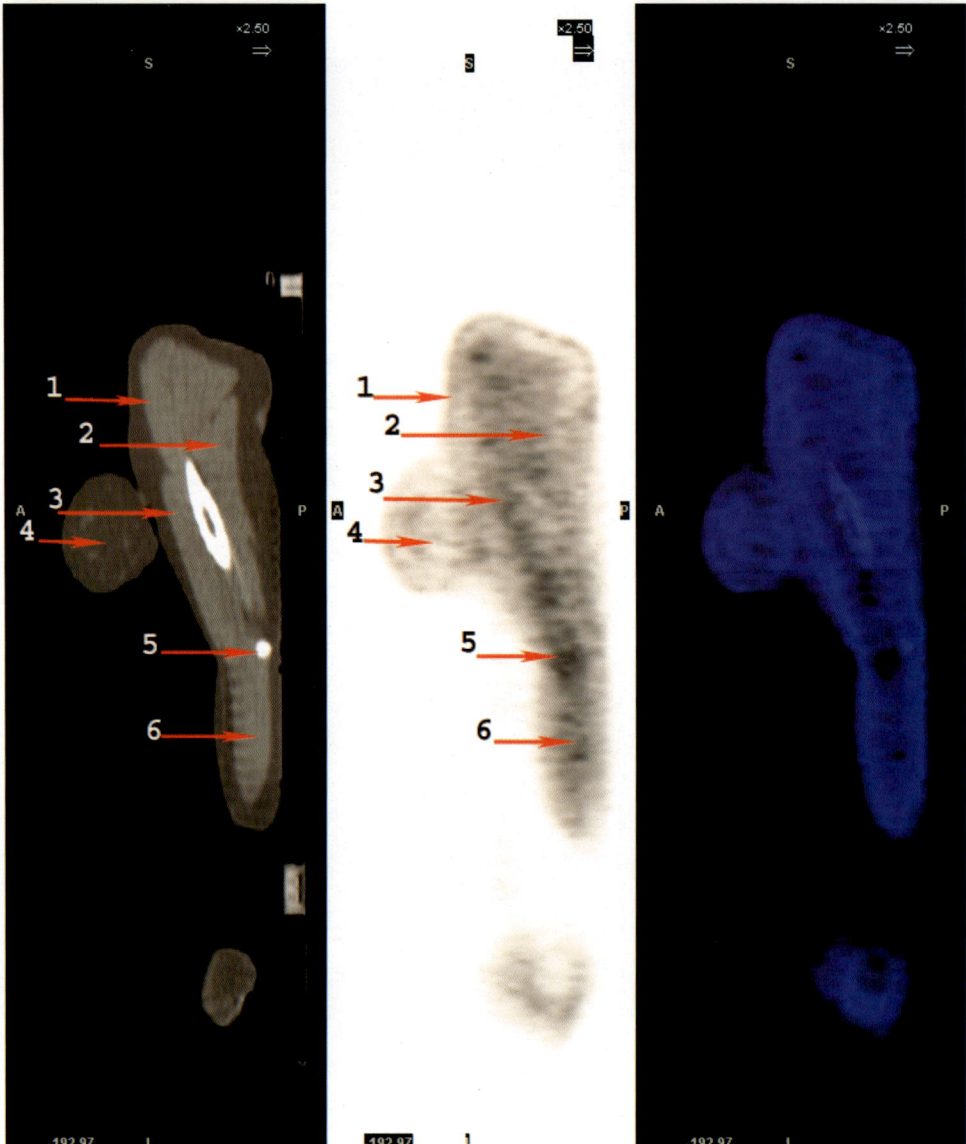

FIGURE 1-27: Sagittal 2

1. Deltoid muscle
2. Triceps brachii muscle
3. Biceps brachii and brachialis muscles
4. Right breast
5. Ulnar olecranon
6. Flexor digitorum superficialis muscle

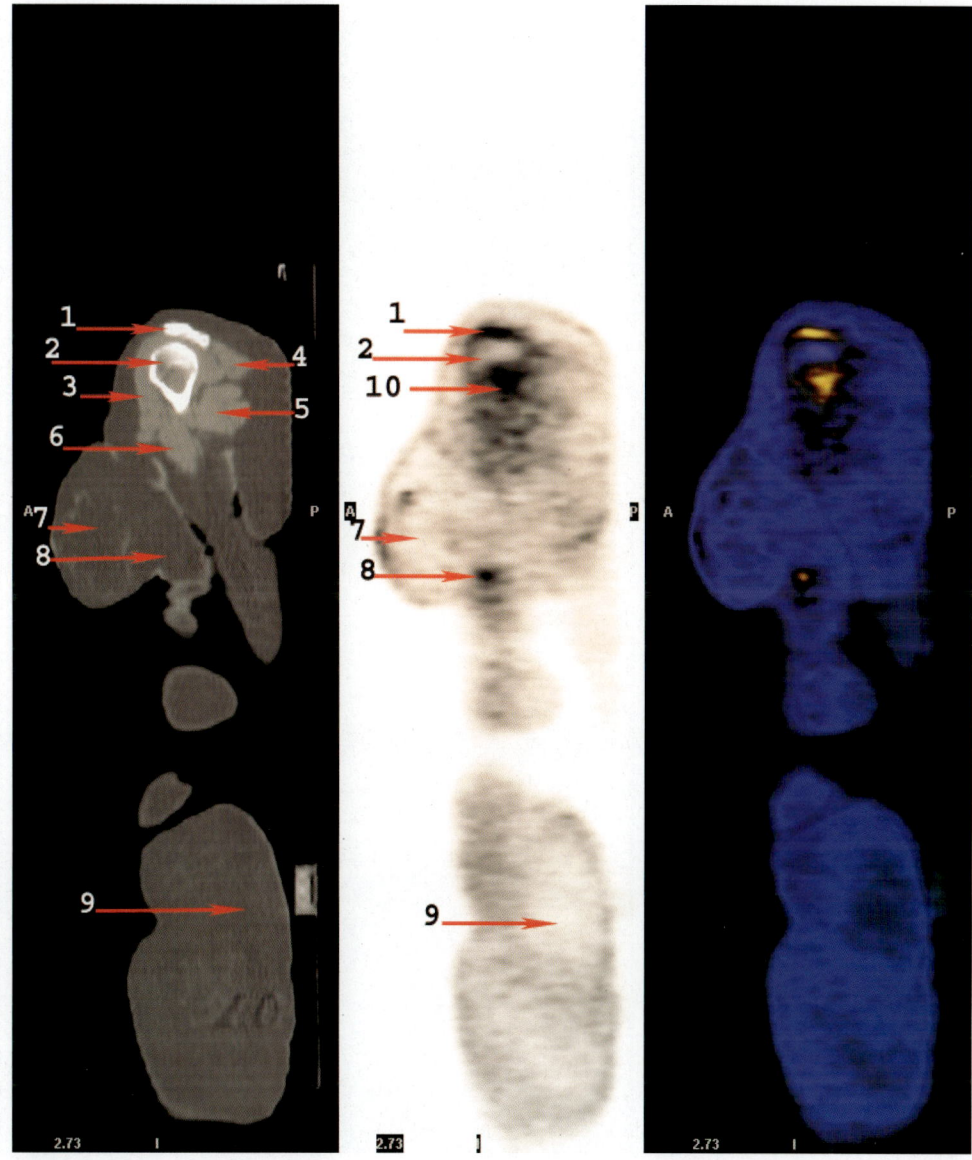

FIGURE 1-28: Sagittal 3

1. Right distal clavicle/joint capsule
2. Humeral head
3. Deltoid muscle
4. Biceps brachii and coracobrachialis muscles
5. Teres muscles
6. Biceps brachii muscle
7. Breast fat
8. Axillary vein
9. Subcutaneous fat in right buttock
10. Joint capsule activity

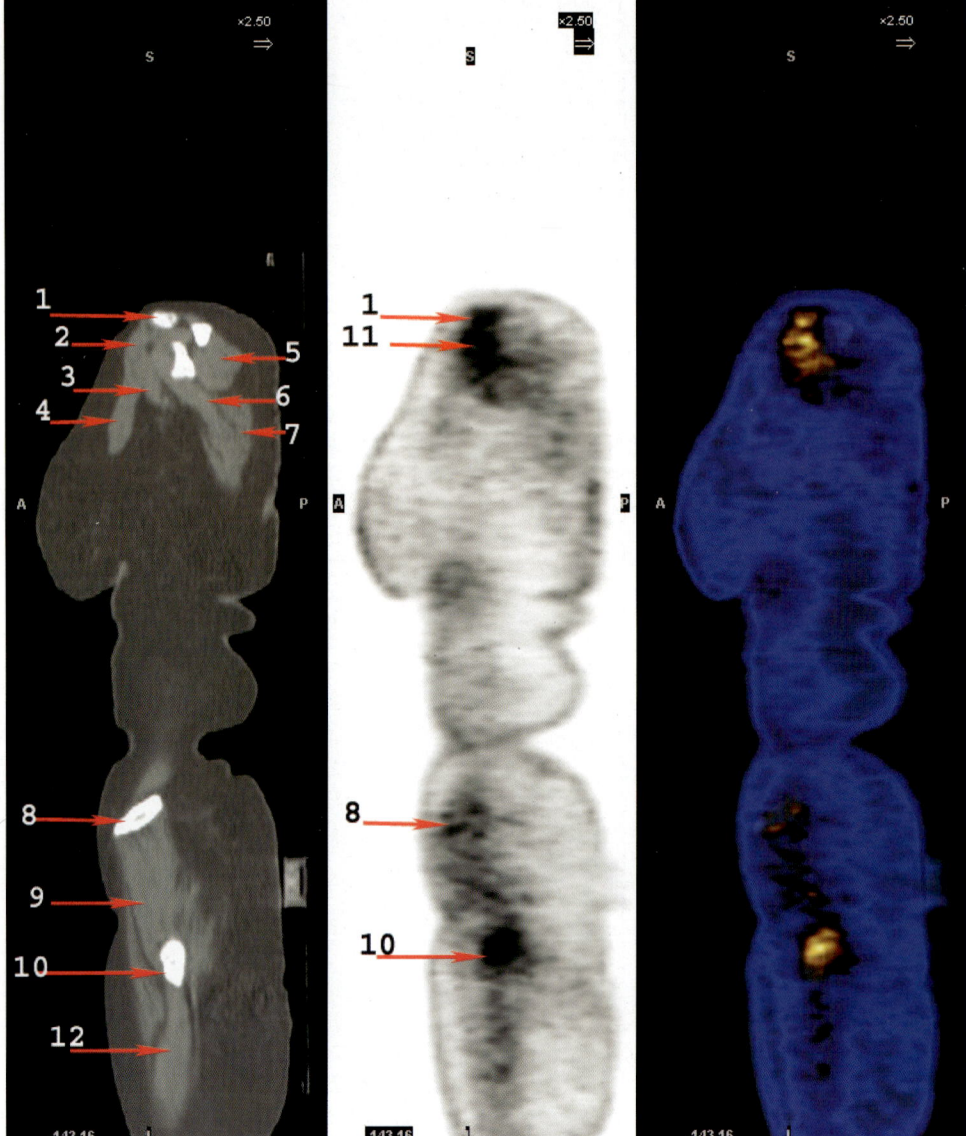

FIGURE 1-29: Sagittal 4

1. Acromion
2. Deltoid muscle
3. Pectoralis minor muscle
4. Pectoralis major muscle
5. Teres muscles
6. Subscapularis muscle
7. Infraspinatus muscle
8. Iliac crest
9. Gluteus minimus muscle
10. Proximal femur
11. Supraspinatus muscle/
    joint capsule
12. Rectus femoris muscle

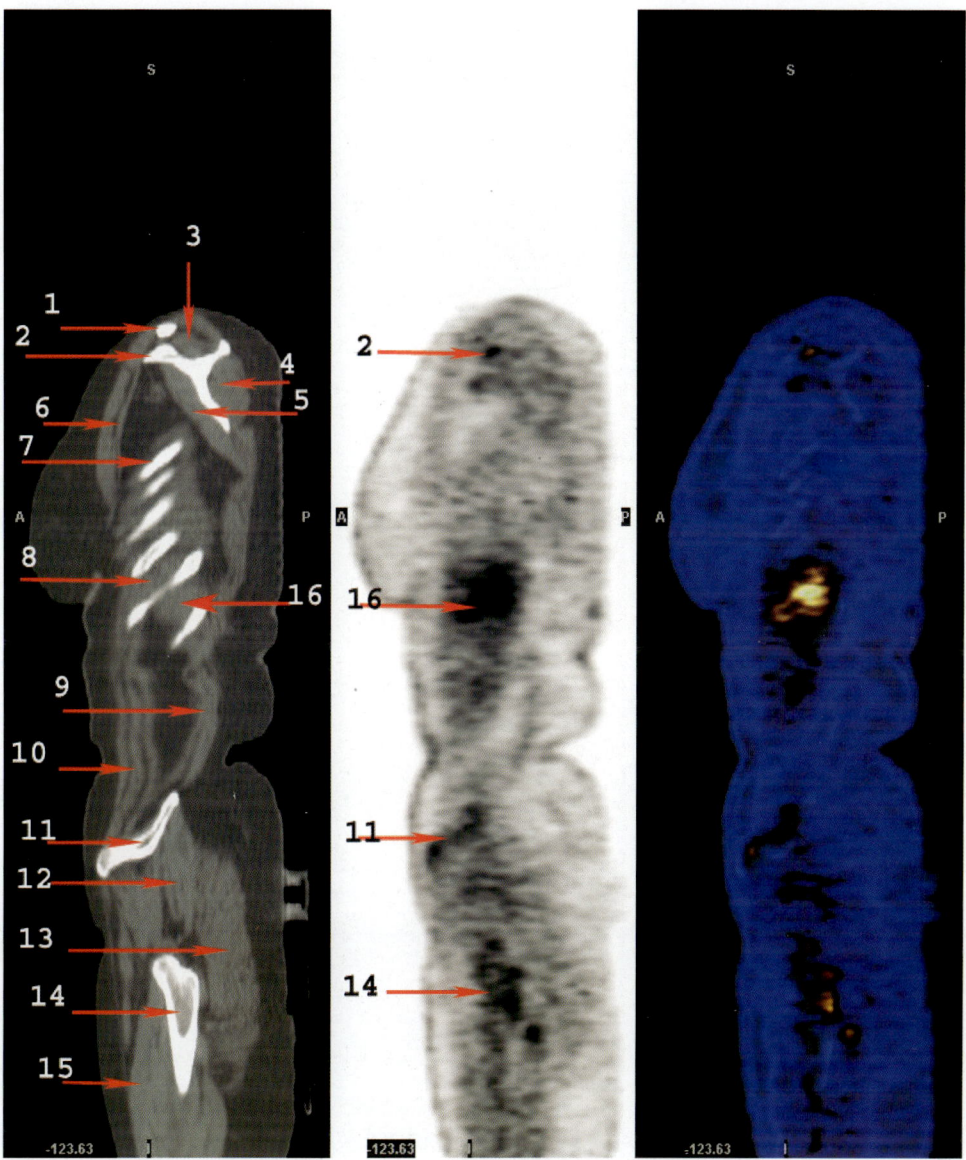

FIGURE 1-30: Sagittal 5

1. Acromion
2. Coracoid process
3. Supraspinatus muscle
4. Infraspinatus muscle
5. Subscapularis muscle
6. Pectoralis muscles
7. Rib
8. Intercostal muscle
9. Latissimus dorsi muscle
10. External oblique muscle
11. Ilium
12. Gluteus medius muscle
13. Gluteus maximus muscle
14. Proximal femur
15. Rectus femoris muscle
16. Liver

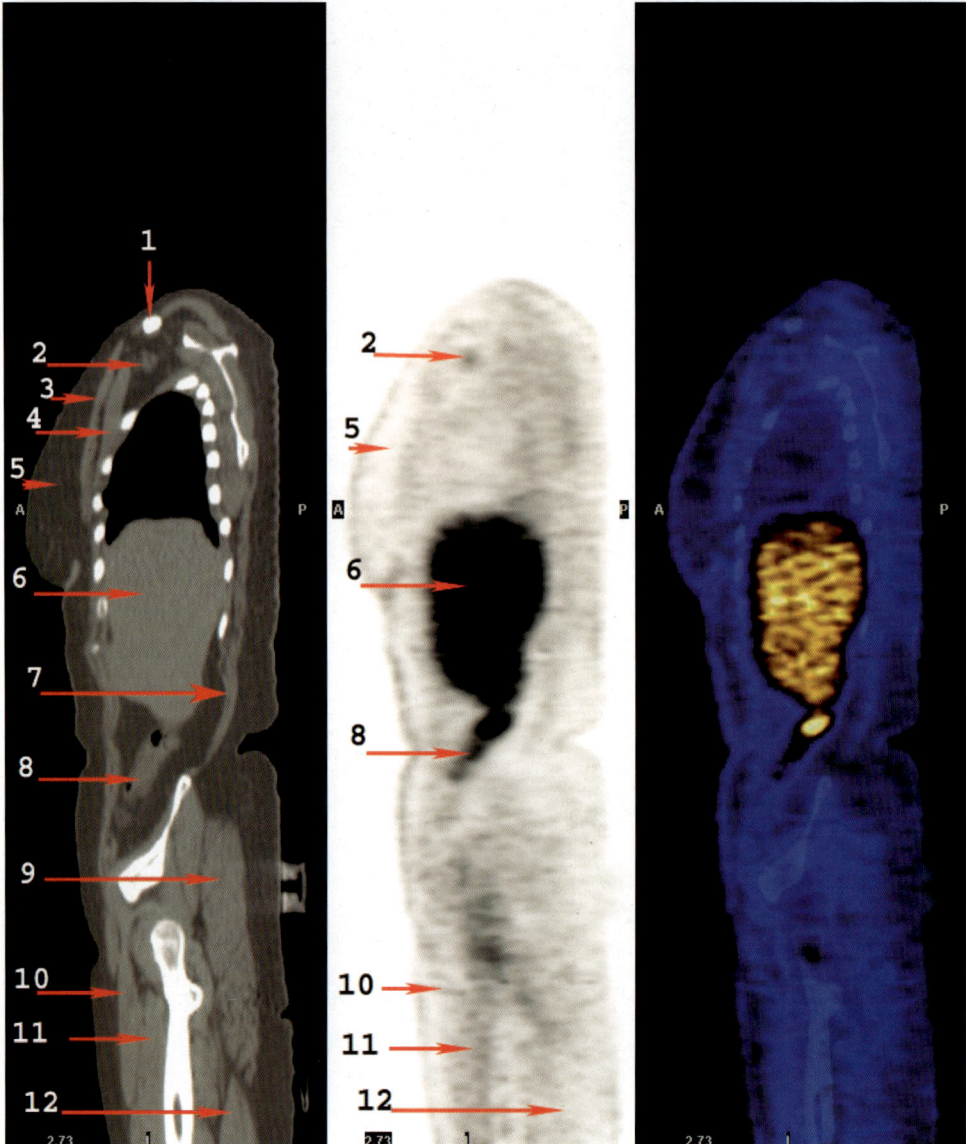

FIGURE 1-31: Sagittal 6

1. Right clavicle
2. Right axillary vein
3. Pectoralis major muscle
4. Pectoralis minor muscle
5. Right breast fat
6. Right hepatic lobe
7. Latissimus dorsi muscle
8. Ascending colon
9. Gluteus maximus muscle
10. Rectus femoris muscle
11. Vastus intermedius muscle
12. Biceps femoris muscle

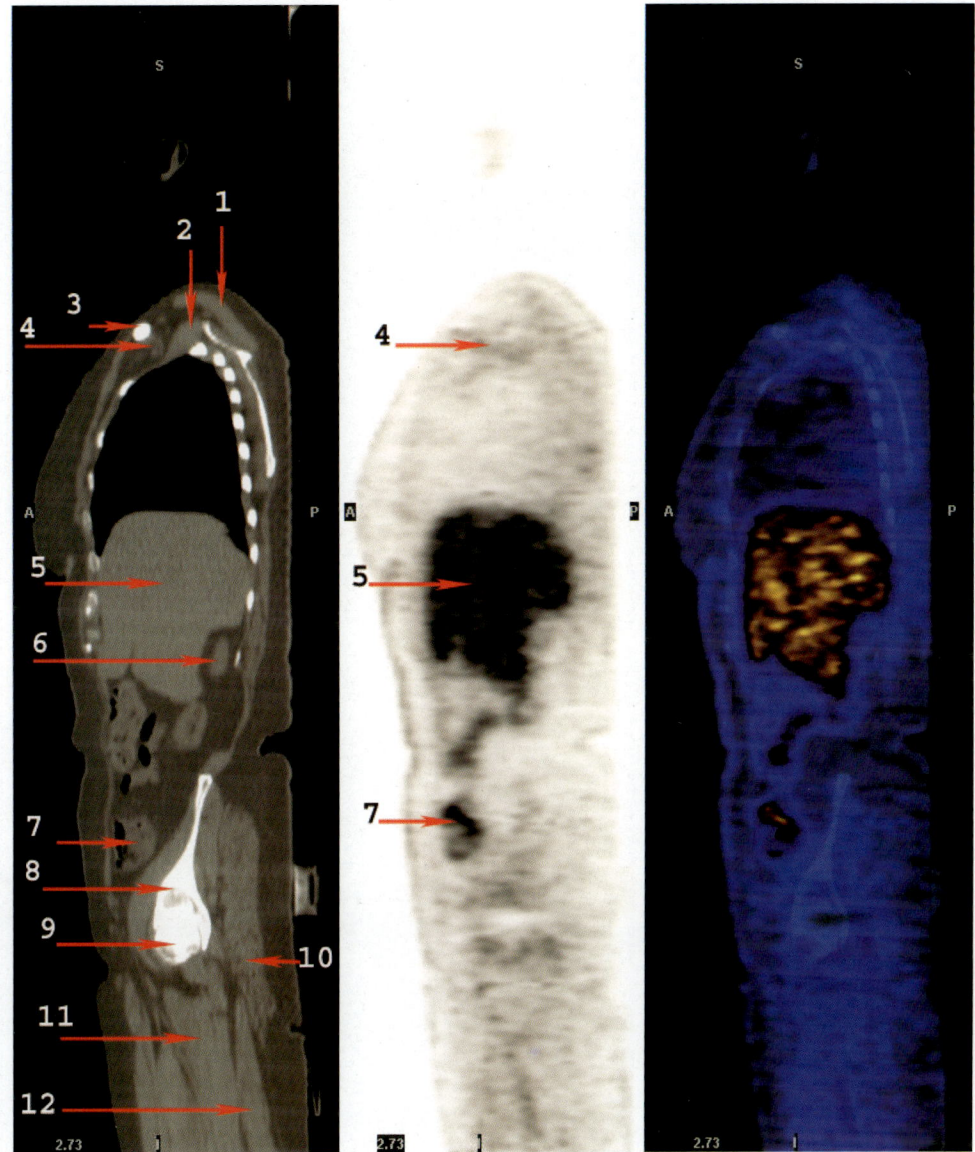

FIGURE 1-32: Sagittal 7

1. Trapezius muscle
2. Supraspinatus muscle
3. Right clavicle
4. Right subclavian vein
5. Right liver
6. Right kidney
7. Ascending colon
8. Right acetabulum
9. Right femoral head
10. Gluteus maximus muscle
11. Adductor magnus muscle
12. Biceps femoris muscle

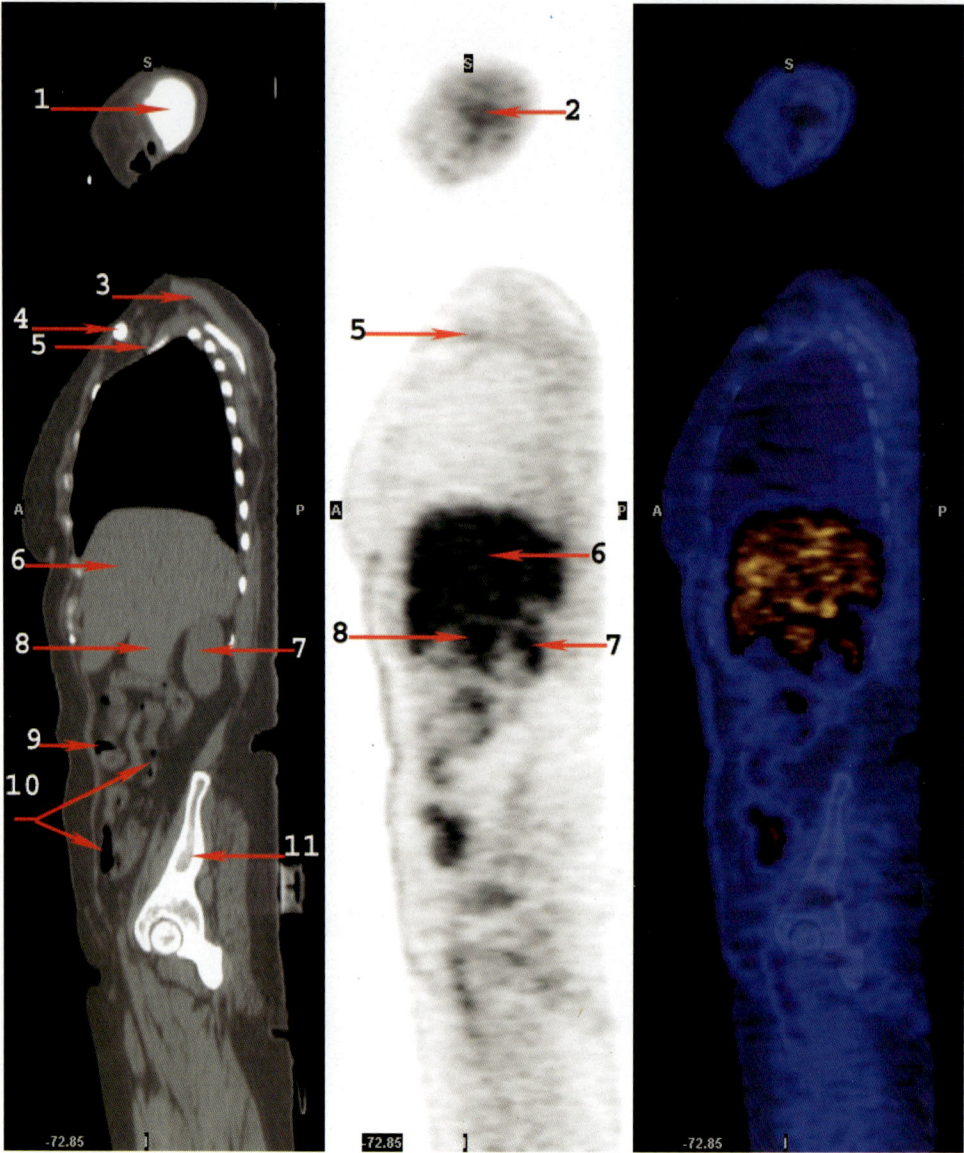

FIGURE 1-33: Sagittal 8

1. Temporal skull
2. Temporalis muscle
3. Trapezius muscle
4. Right clavicle

5. Right subclavian vein
6. Right liver
7. Right kidney
8. Duodenum

9. Jejunum
10. Ileum
11. Ilium

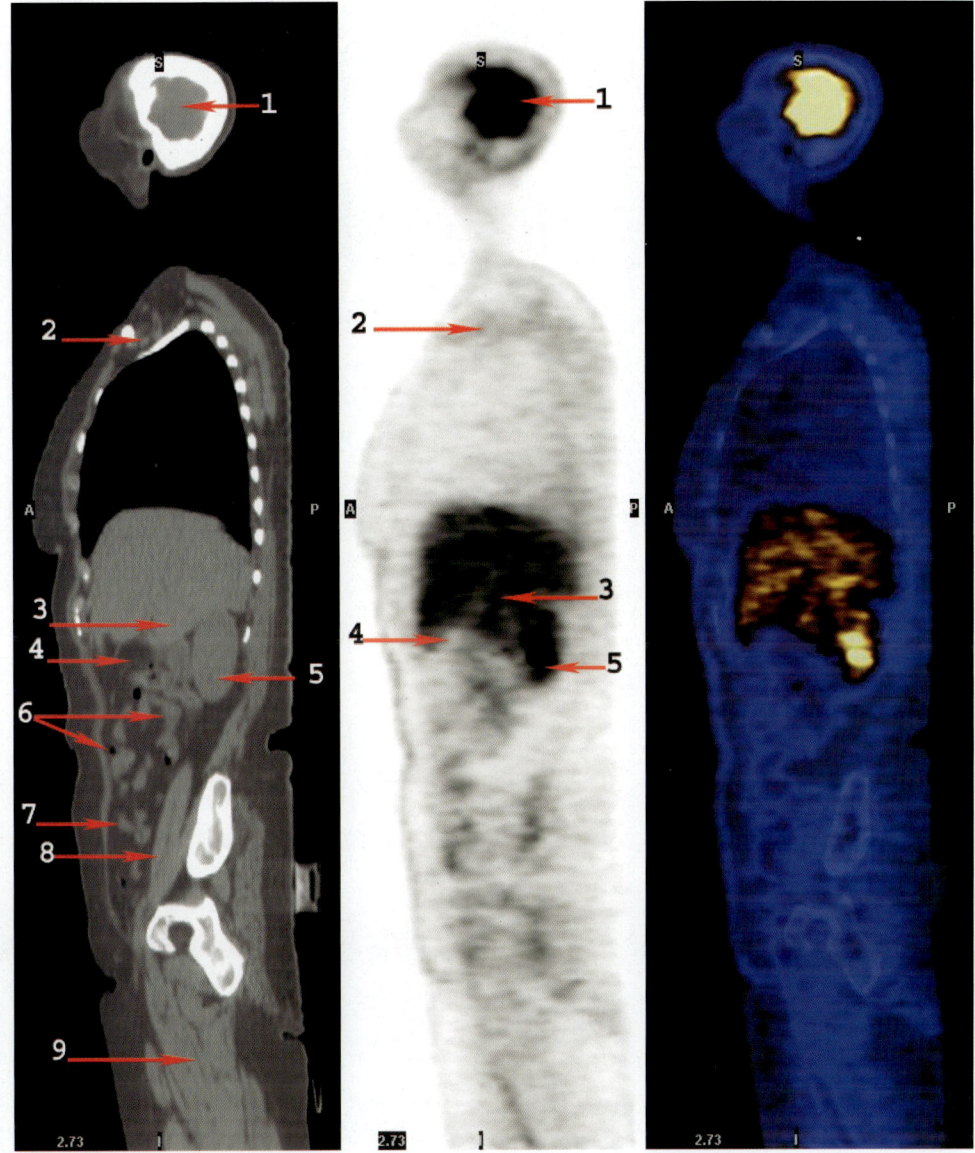

FIGURE 1-34: Sagittal 9

1. Temporal gray matter	4. Gallbladder fossa	7. Ileum
2. Right subclavian vein	5. Right kidney	8. Iliacus muscle
3. Duodenum	6. Jejunum	9. Adductor muscle

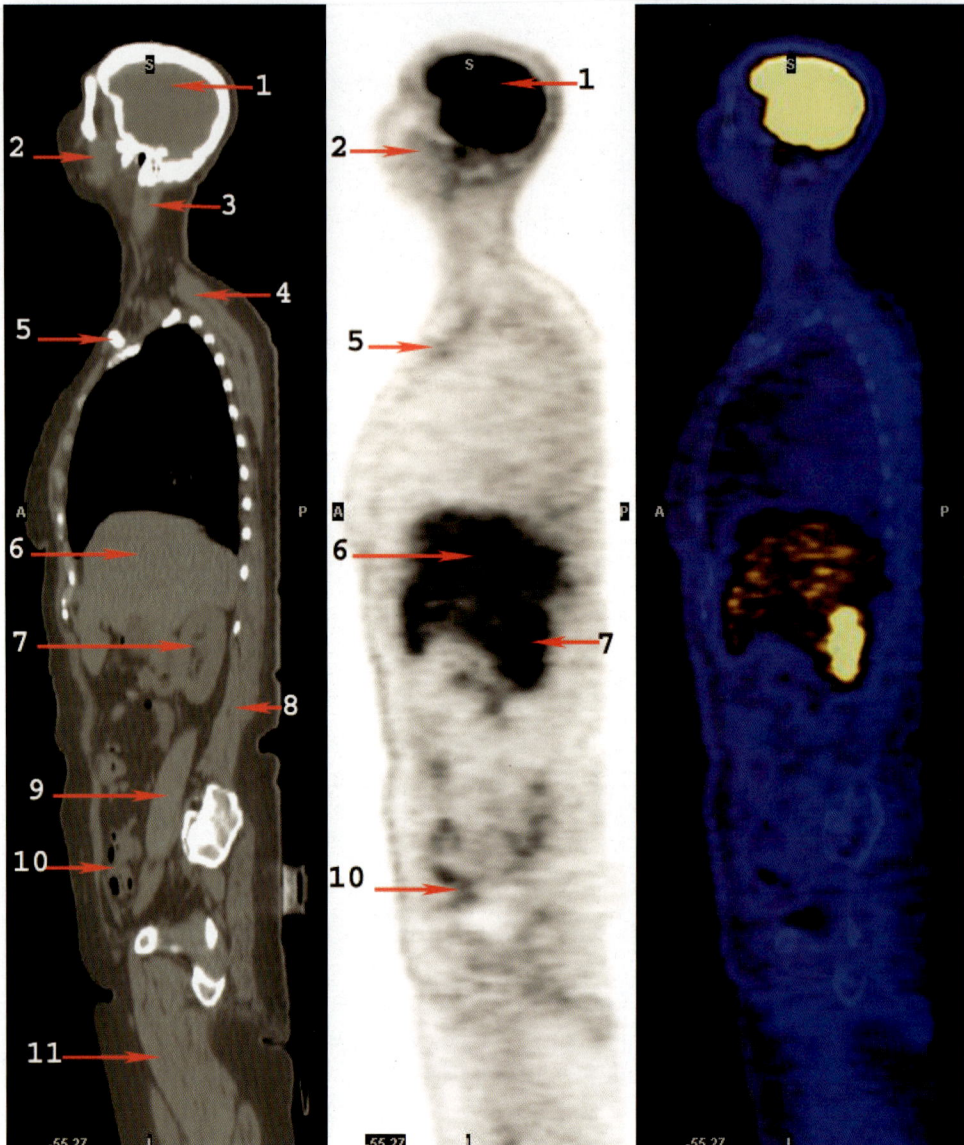

FIGURE 1-35: Sagittal 10

1. Parietal gray matter
2. Masseter muscle
3. Splenius capitis muscle
4. Trapezius muscle
5. Right clavicle
6. Right hepatic lobe (segment VII)
7. Right kidney
8. Erector spinal muscle
9. Psoas muscle
10. Ileum
11. Adductor muscles

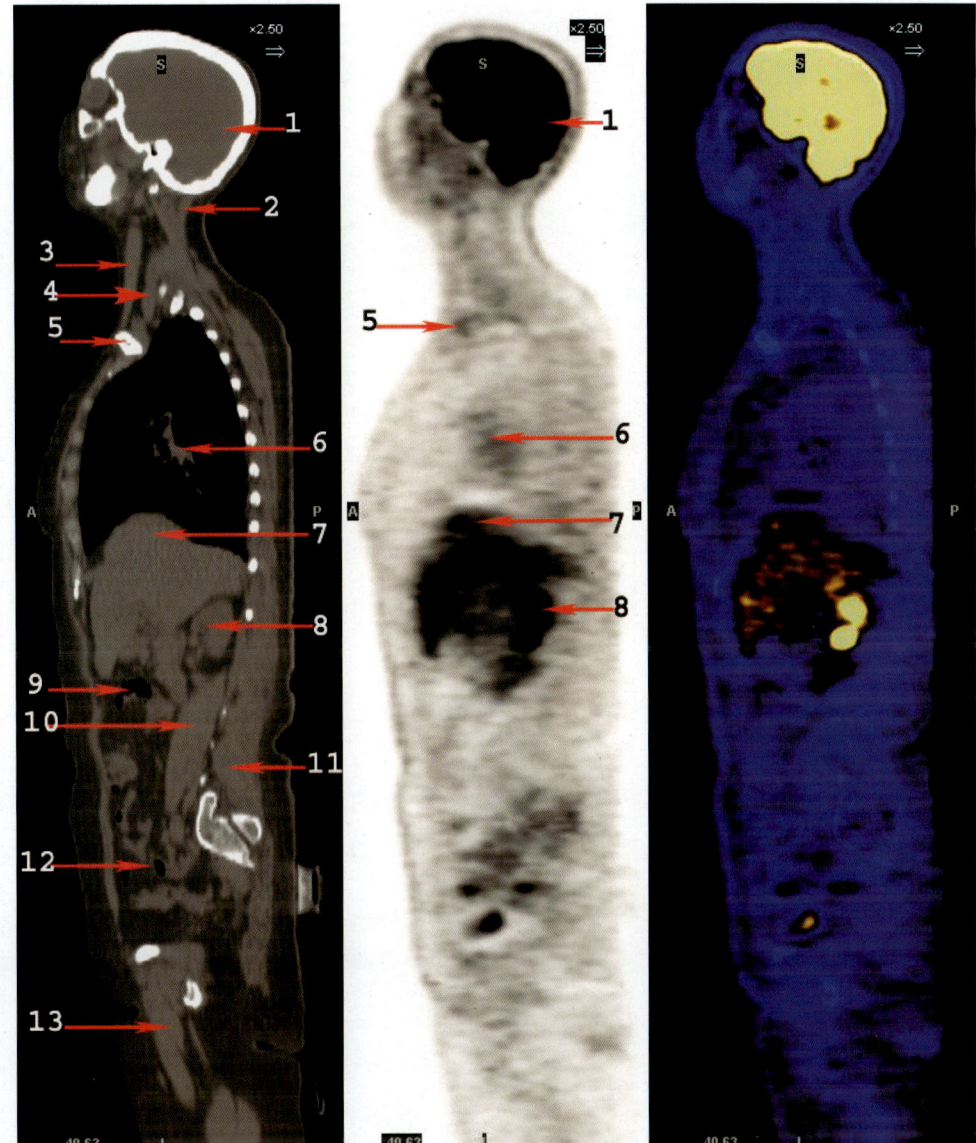

FIGURE 1-36: Sagittal 11

1. Occipital brain
2. Splenius capitis muscle
3. Sternocleidomastoid muscle
4. Scalene muscles
5. Right manubrium
6. Pulmonary vessels in right hilum
7. Right liver (segment VIII)
8. Right kidney
9. Transverse colon
10. Psoas muscle
11. Erector spinae muscle
12. Ileum
13. Adductor muscles

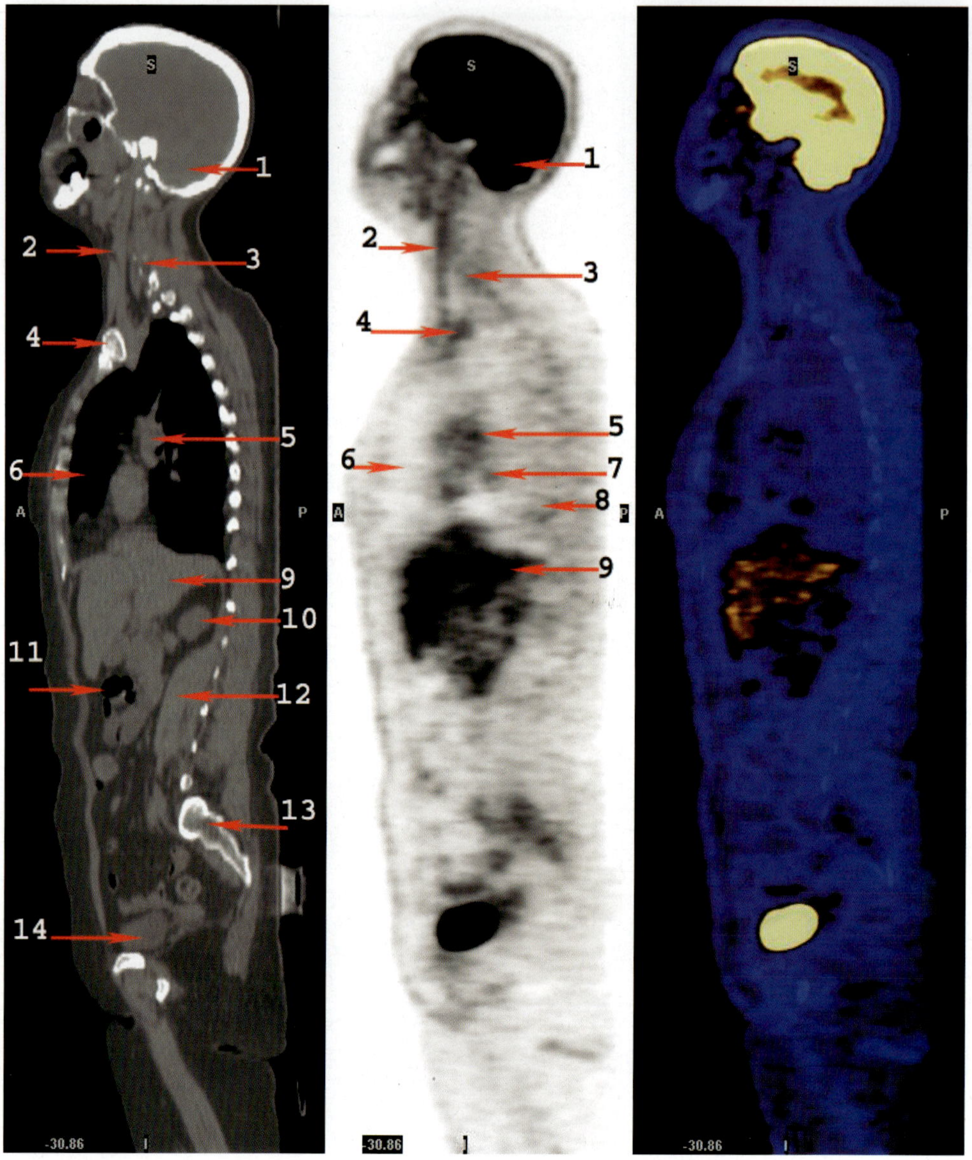

FIGURE 1-37: Sagittal 12

1. Right cerebellum
2. Sternocleidomastoid muscle
3. Scalene muscles
4. Right manubrium
5. Right pulmonary artery
6. Right middle lobe of lung
7. Right atrium
8. Right lower lobe of lung
9. Lateral segment (II) of left hepatic lobe
10. Right kidney
11. Transverse colon
12. Psoas muscle
13. Right sacral ala
14. Bladder

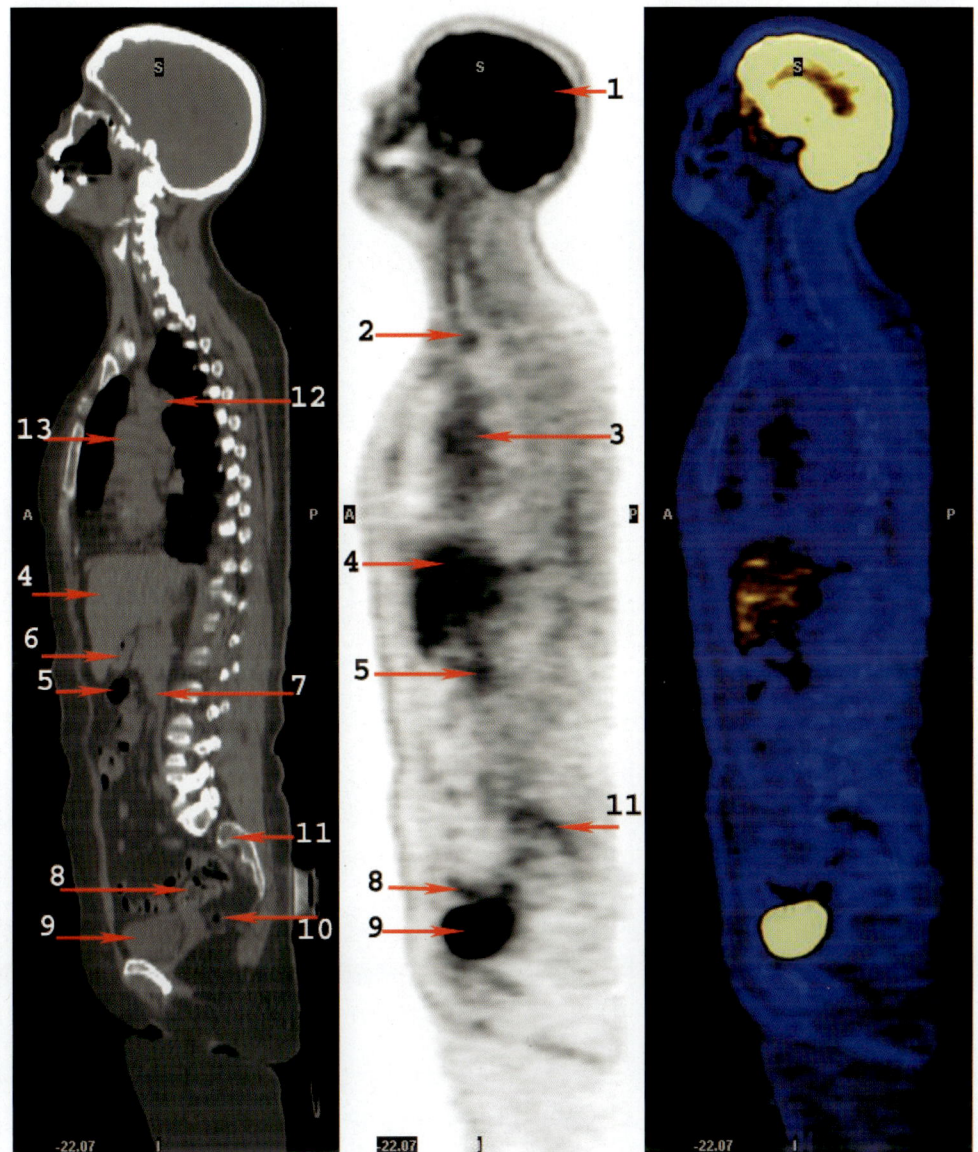

FIGURE 1-38: Sagittal 13

1. Parietal gray matter
2. Manubrioclavicular joint
3. Main pulmonary vessels
4. Liver
5. Transverse colon
6. Stomach
7. Pancreatic head
8. Ileum
9. Bladder
10. Sigmoid
11. Sacral marrow
12. Arch of azygos vein
13. Superior vena cava

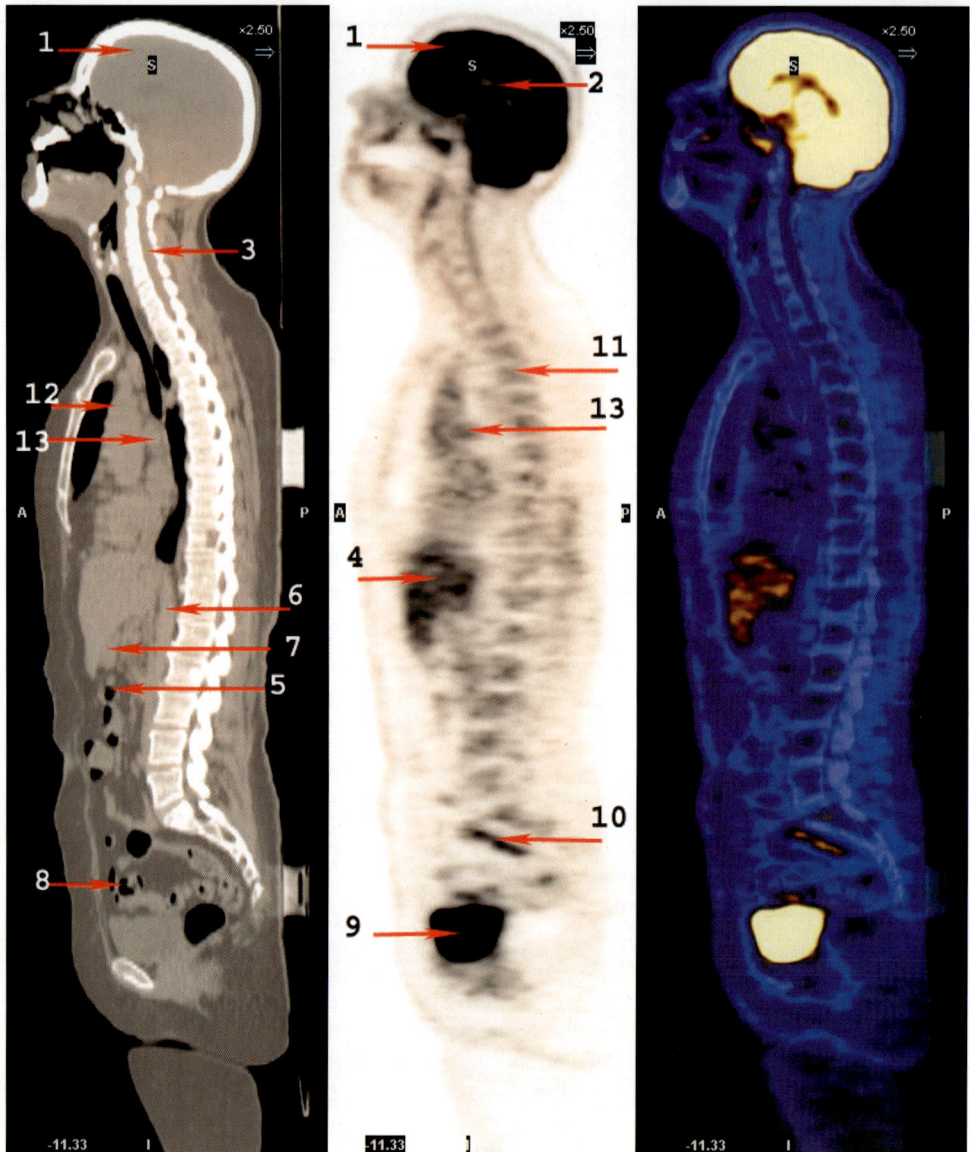

FIGURE 1-39: Sagittal 14

1. Cerebral gray matter
2. Lateral ventricle
3. Spinal cord
4. Liver
5. Transverse colon
6. Inferior vena cava
7. Pancreatic head
8. Ileum
9. Bladder
10. Sigmoid
11. Vertebral marrow (T2)
12. Superior vena cava
13. Right pulmonary artery

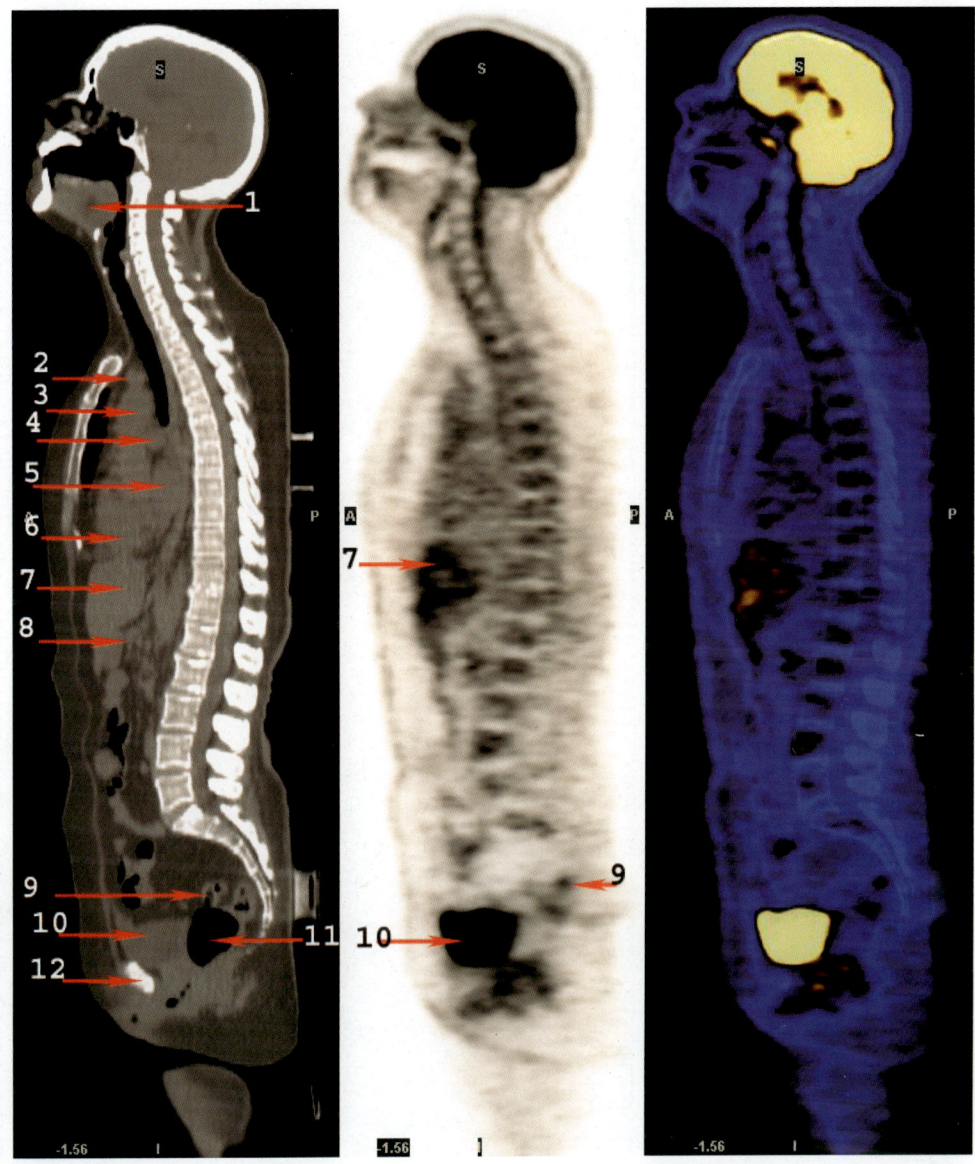

FIGURE 1-40: Sagittal 15

1. Tongue
2. Left innominate vein
3. Aortic arch
4. Right pulmonary artery
5. Left atrium
6. Right ventricle
7. Left hepatic lobe
   (segment III)
8. Pancreas
9. Sigmoid colon
10. Bladder
11. Rectum
12. Symphysis pubis

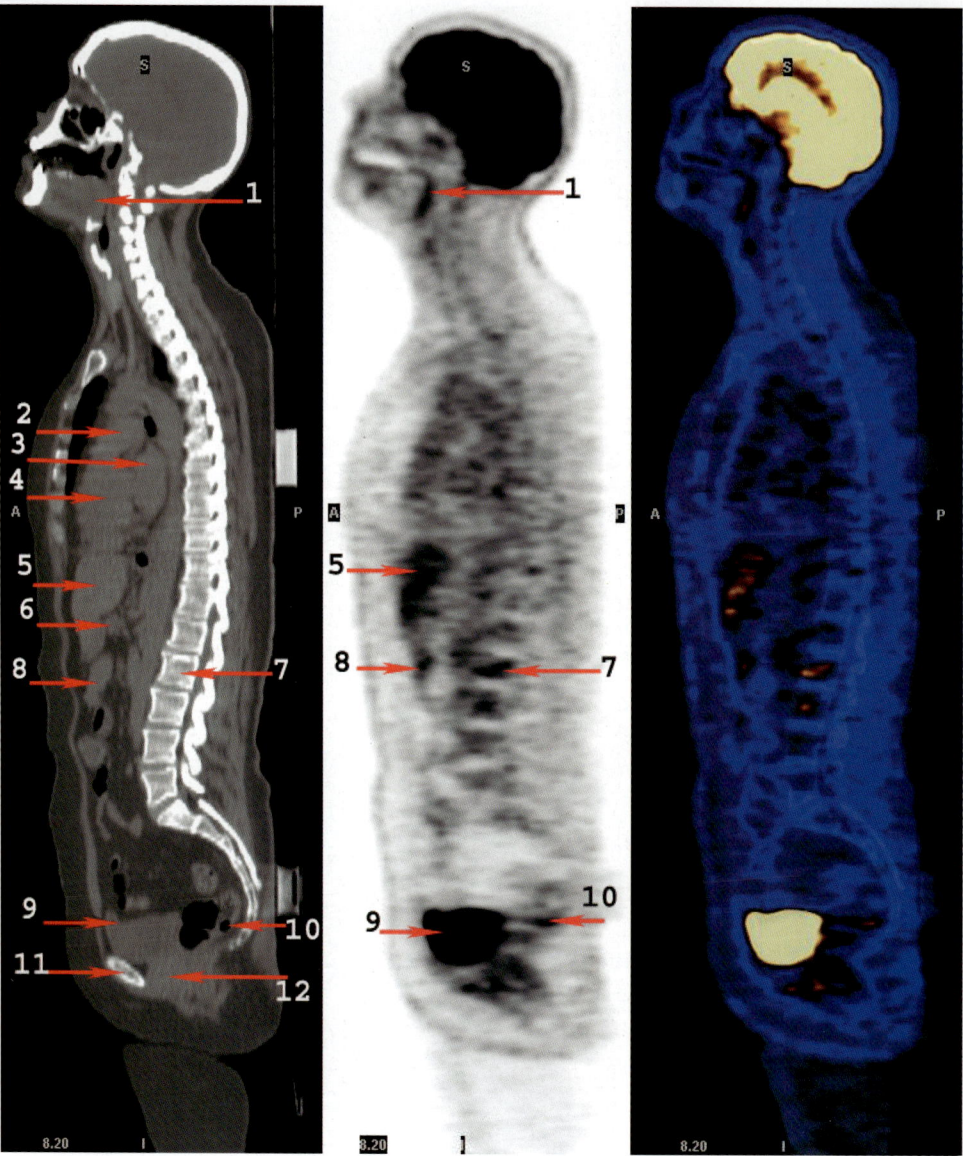

FIGURE 1-41: Sagittal 16

1. Tongue base	5. Left liver (segment III)	9. Bladder
2. Aortic arch	6. Pancreas	10. Rectosigmoid colon
3. Left atrium	7. L2 marrow	11. Symphysis pubis
4. Right ventricle	8. Jejunum	12. Prostate

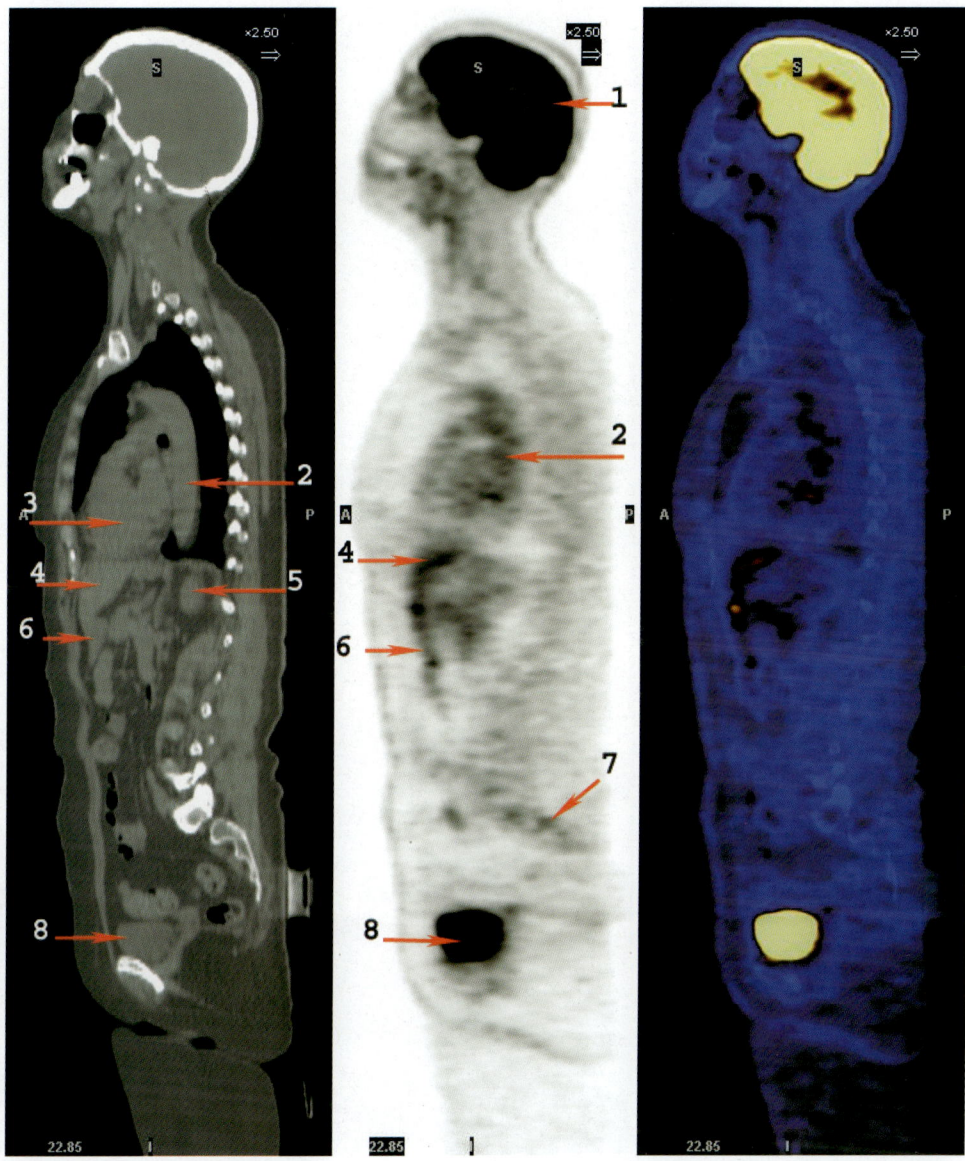

FIGURE 1-42: Sagittal 17

1. Cerebral gray matter	4. Left hepatic lobe	7. Sacral marrow
2. Descending aorta	5. Spleen	8. Bladder
3. Left ventricle	6. Transverse colon	

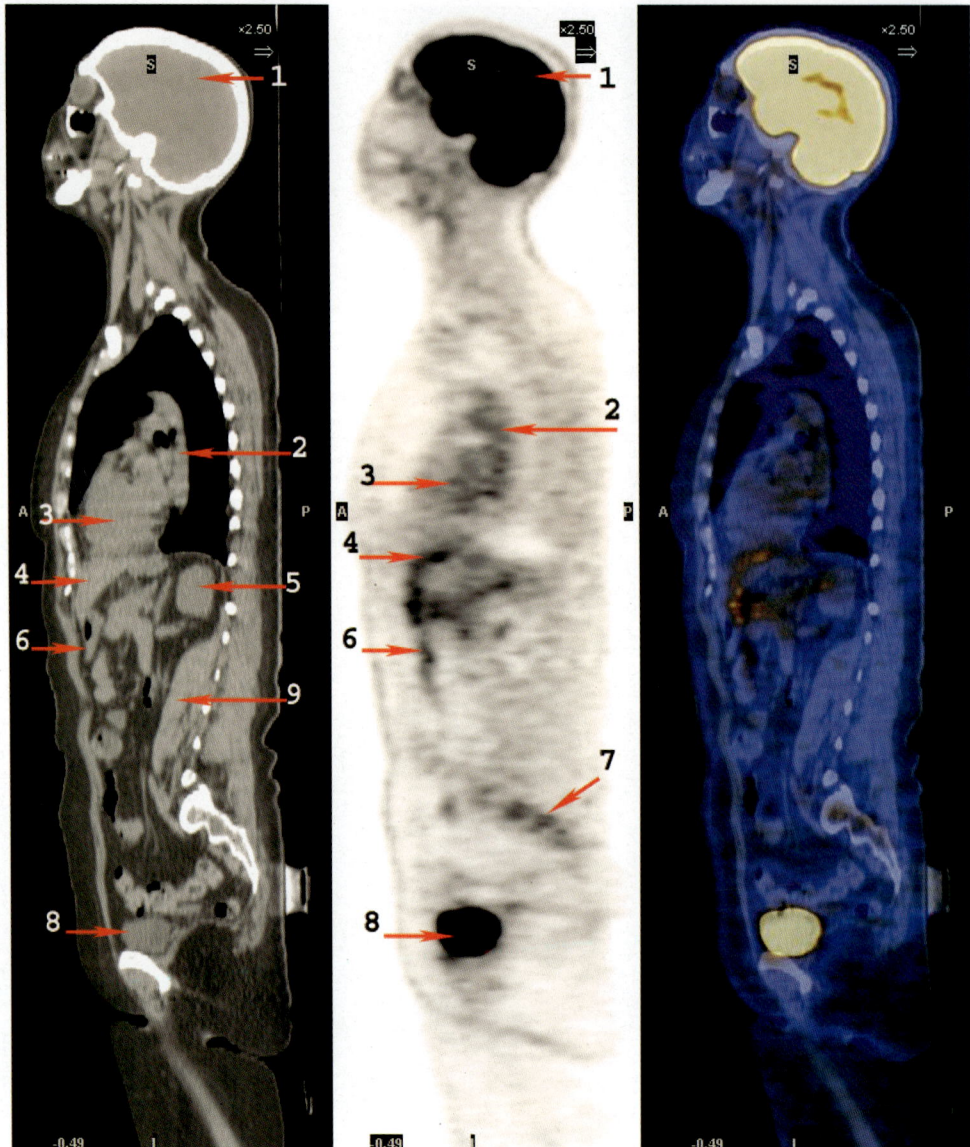

FIGURE 1-43: Sagittal 18

1. Cerebral gray matter
2. Descending aorta
3. Left ventricle
4. Left hepatic lobe
5. Spleen
6. Transverse colon
7. Sacral marrow
8. Bladder
9. Psoas muscle

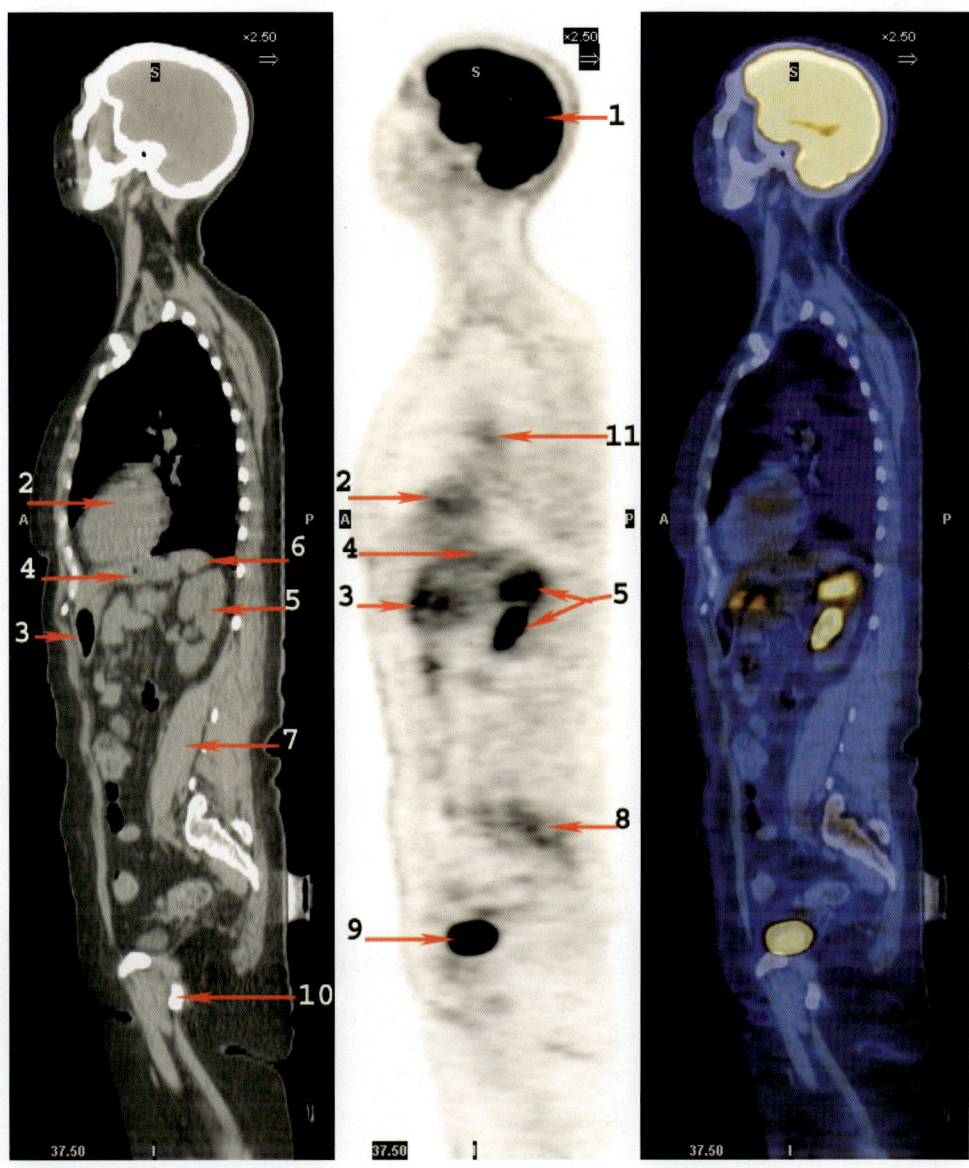

FIGURE 1-44: Sagittal 19

1. Cerebral gray matter	5. Left kidney	9. Bladder
2. Left ventricular wall	6. Spleen	10. Ischium
3. Stomach	7. Psoas muscle	11. Left pulmonary vein
4. Left hepatic lobe	8. Sacral marrow	

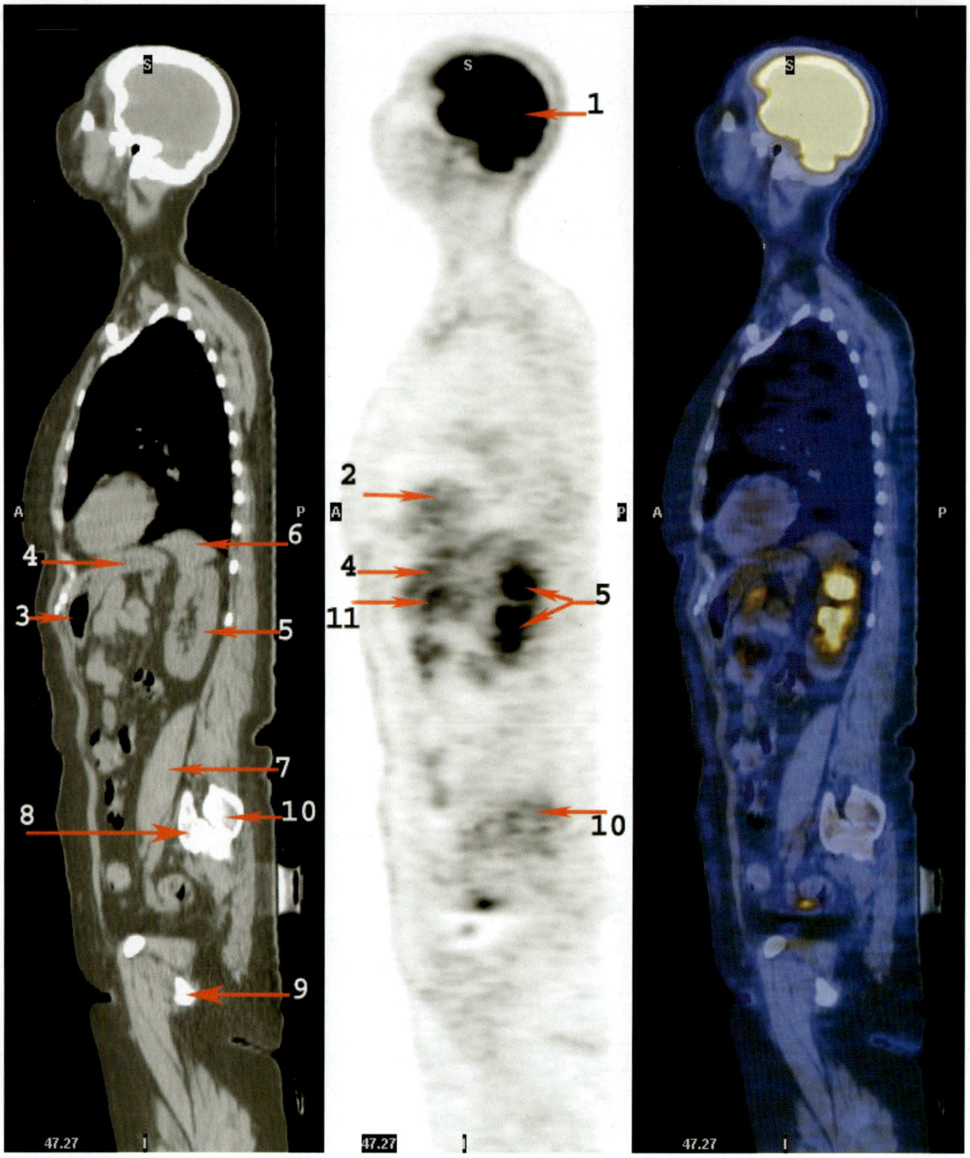

FIGURE 1-45: Sagittal 20

1. Cerebral gray matter
2. Left ventricular wall
3. Stomach
4. Left hepatic lobe
5. Left kidney
6. Spleen
7. Psoas muscle
8. Sacral marrow
9. Ischial tuberosity
10. Iliac tuberosity
11. Duodenum

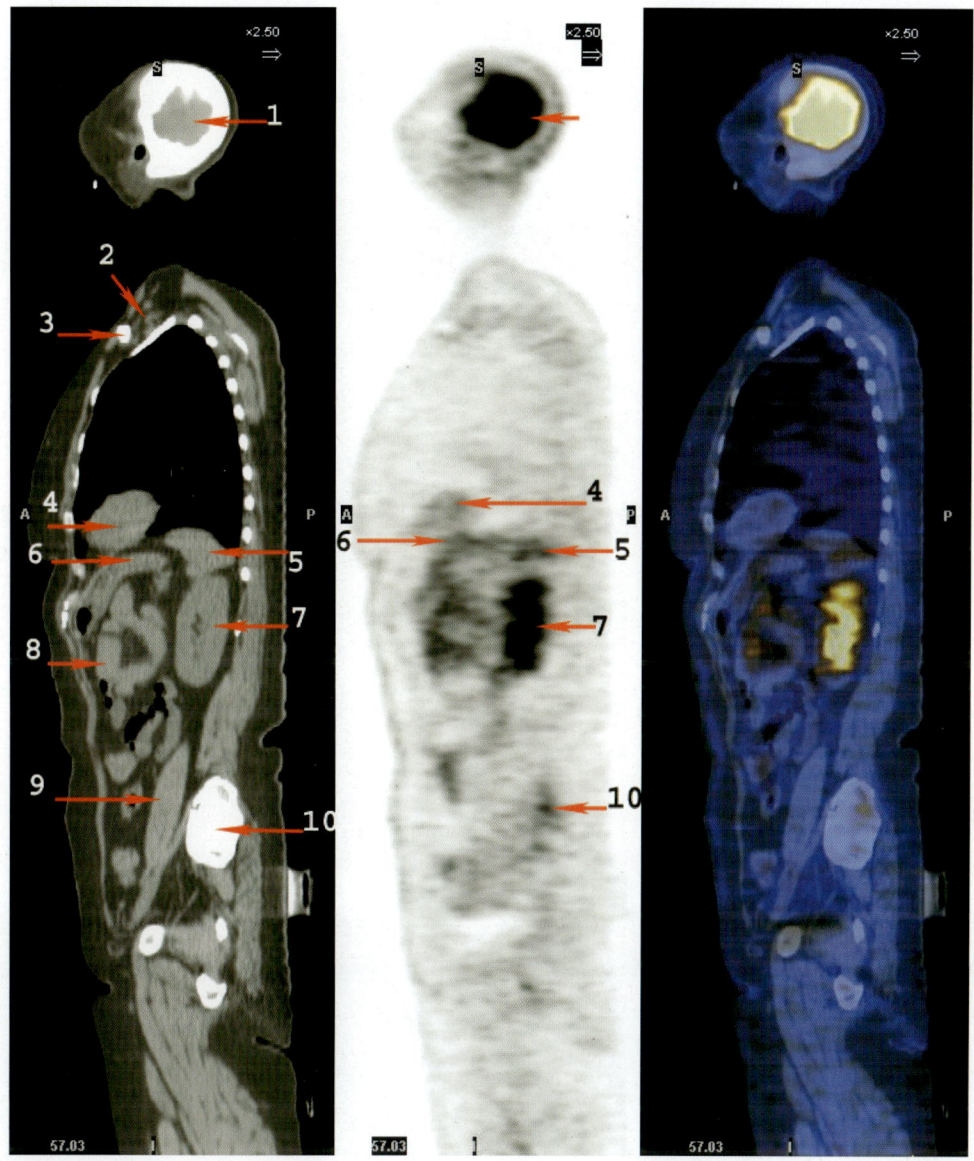

FIGURE 1-46: Sagittal 21

1. Left parietal gray matter
2. Left brachial plexus
3. Left clavicle
4. Left ventricle
5. Spleen
6. Left liver (segment II)
7. Left kidney
8. Jejunum
9. Psoas muscle
10. Iliac tuberosity

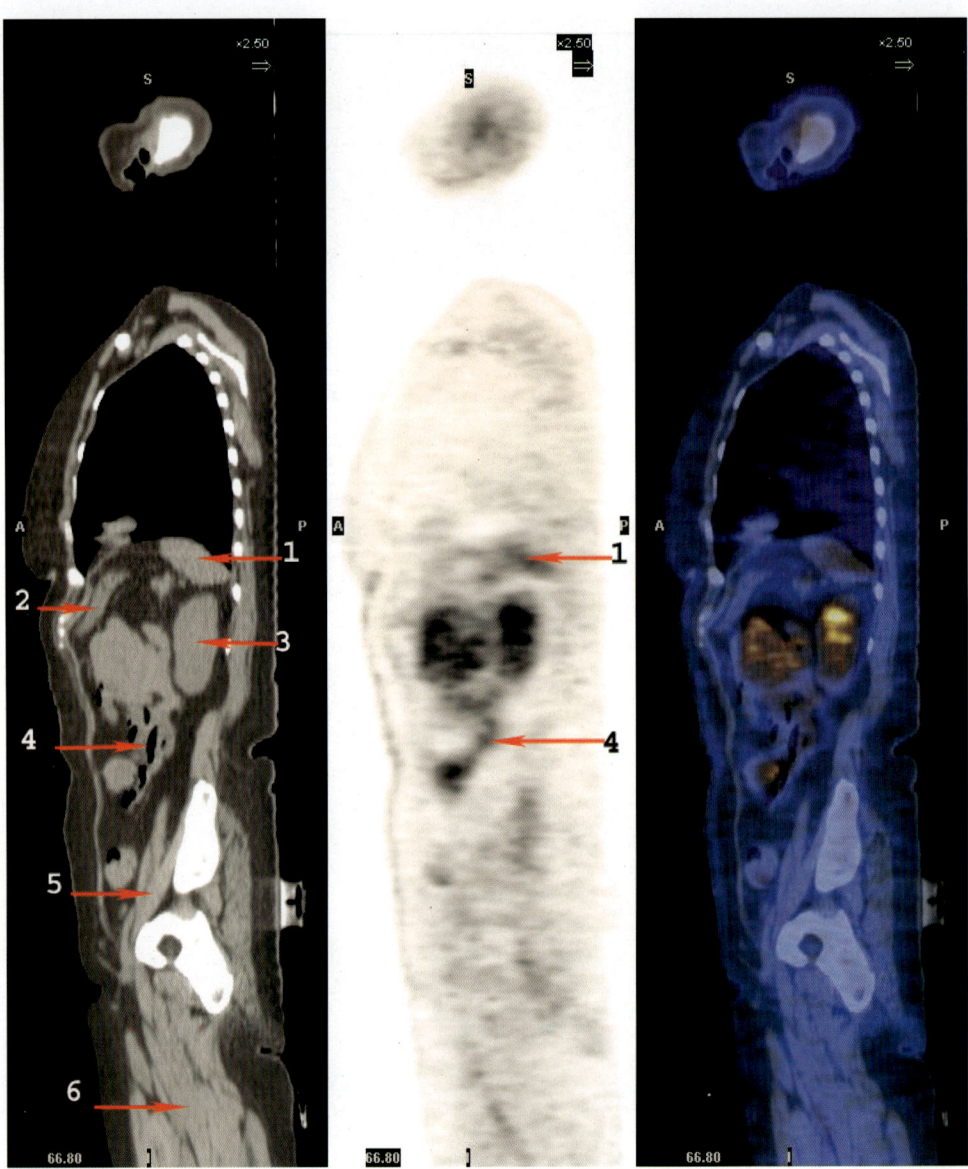

FIGURE 1-47: Sagittal 22

1. Spleen	3. Left kidney	5. Iliacus muscle
2. Stomach	4. Jejunum	6. Adductor muscles

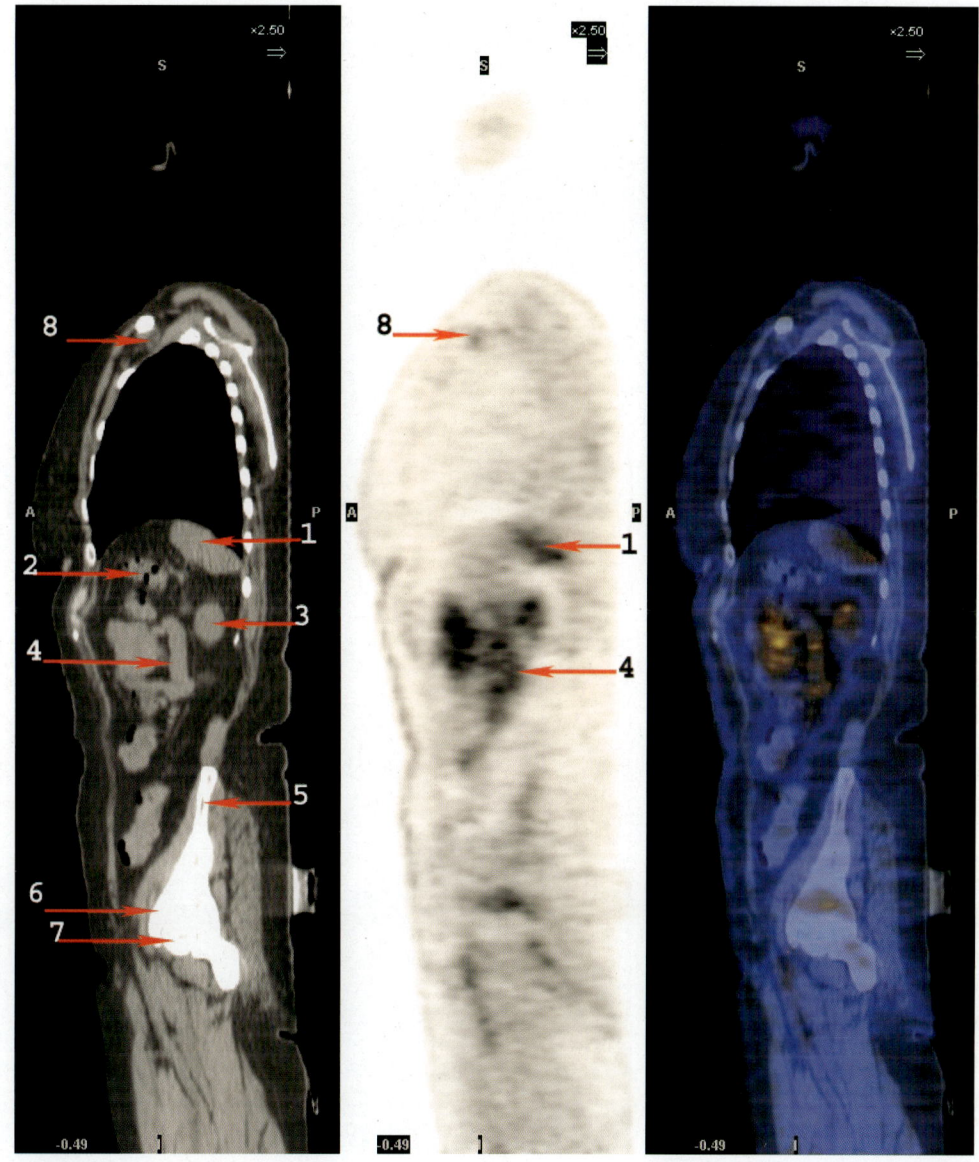

FIGURE 1-48: Sagittal 23

1. Spleen
2. Stomach
3. Left kidney
4. Jejunum
5. Ilium
6. Left acetabulum
7. Left femoral head
8. Left subclavian vein

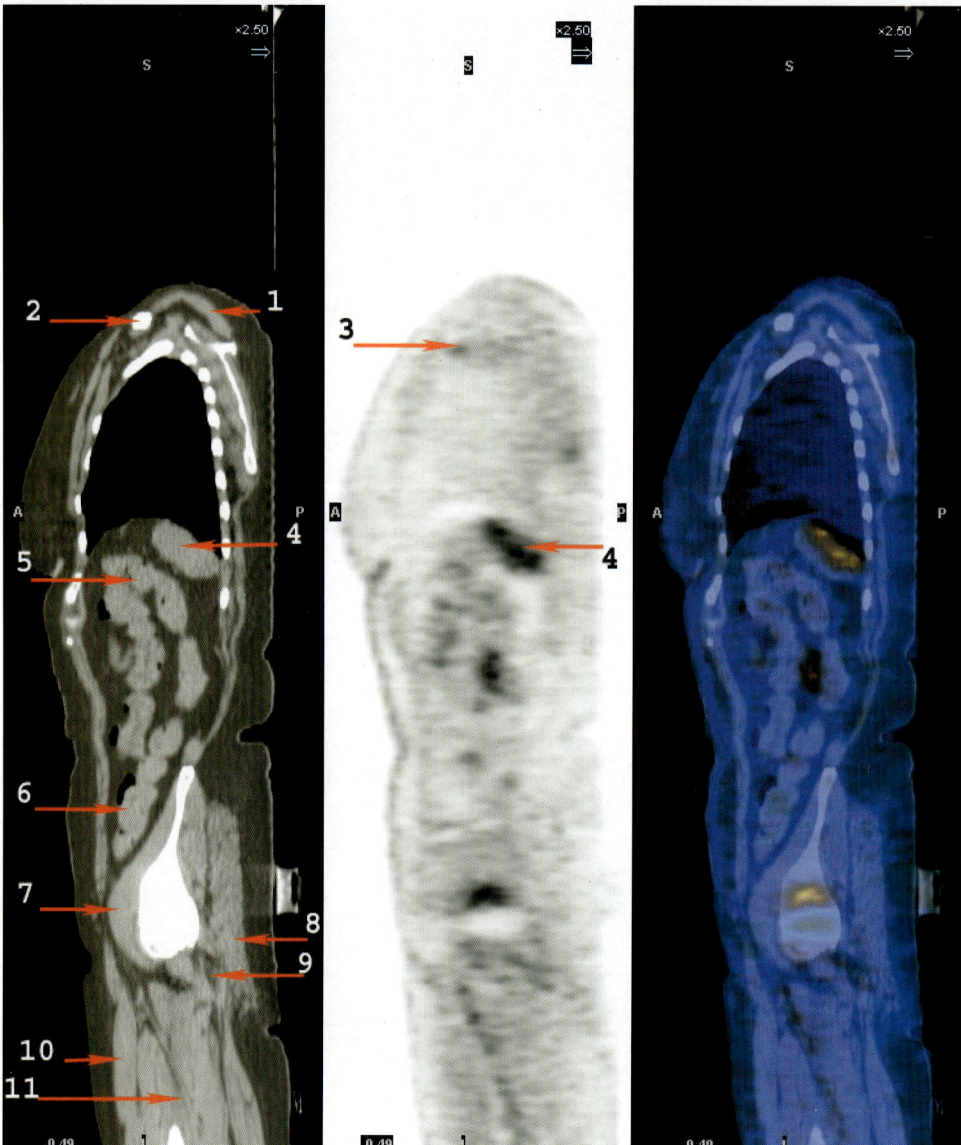

FIGURE 1-49: Sagittal 24

1. Trapezius muscle
2. Left clavicle
3. Left subclavian vein
4. Spleen
5. Splenic flexure of colon
6. Descending colon
7. Iliacus muscle
8. Gluteus maximus muscle
9. Sciatic nerve and interior gluteal vein
10. Rectus femoris muscle
11. Adductor muscles

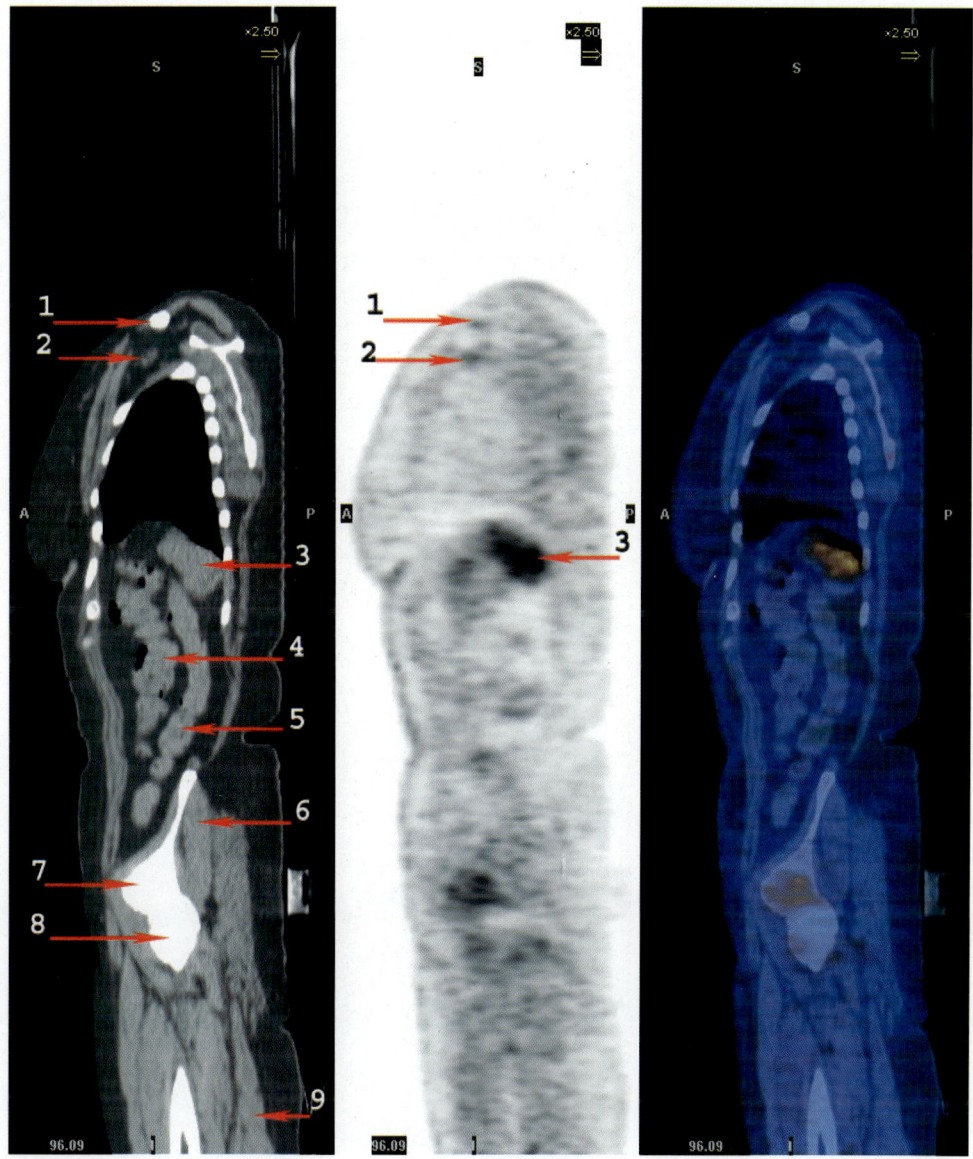

FIGURE 1-50: Sagittal 25

1. Left clavicle
2. Left subclavian vein
3. Spleen
4. Transverse colon
5. Descending colon
6. Gluteus medius muscle
7. Left acetabulum
8. Left femoral head
9. Semimembranosus and semitendinosus muscles

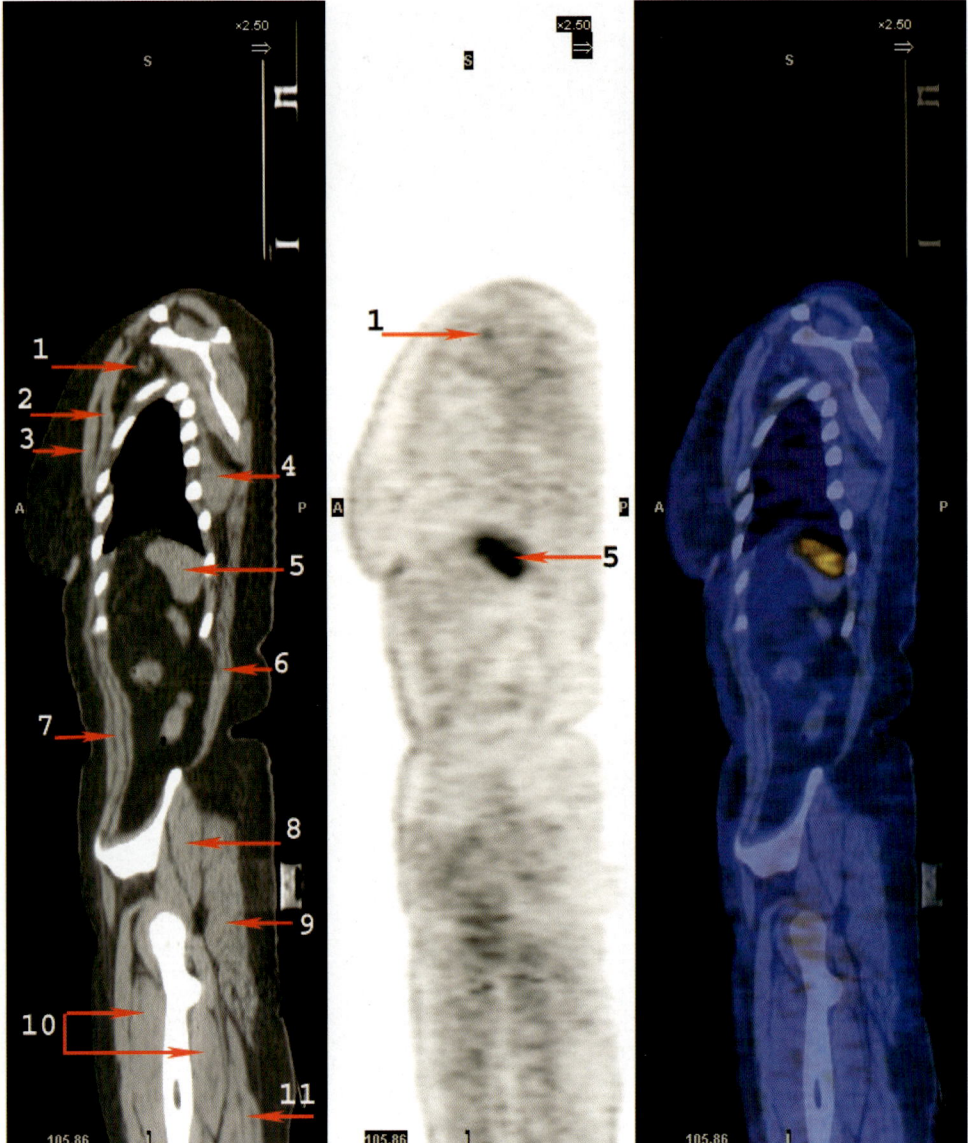

FIGURE 1-51: Sagittal 26

1. Left subclavian vein
2. Pectoralis minor muscle
3. Pectoralis major muscle
4. Serratus anterior muscle
5. Spleen
6. Latissimus dorsi muscle
7. Oblique muscles
8. Gluteus medius muscle
9. Gluteus maximus muscle
10. Vastus intermedius muscle
11. Biceps femoris muscle

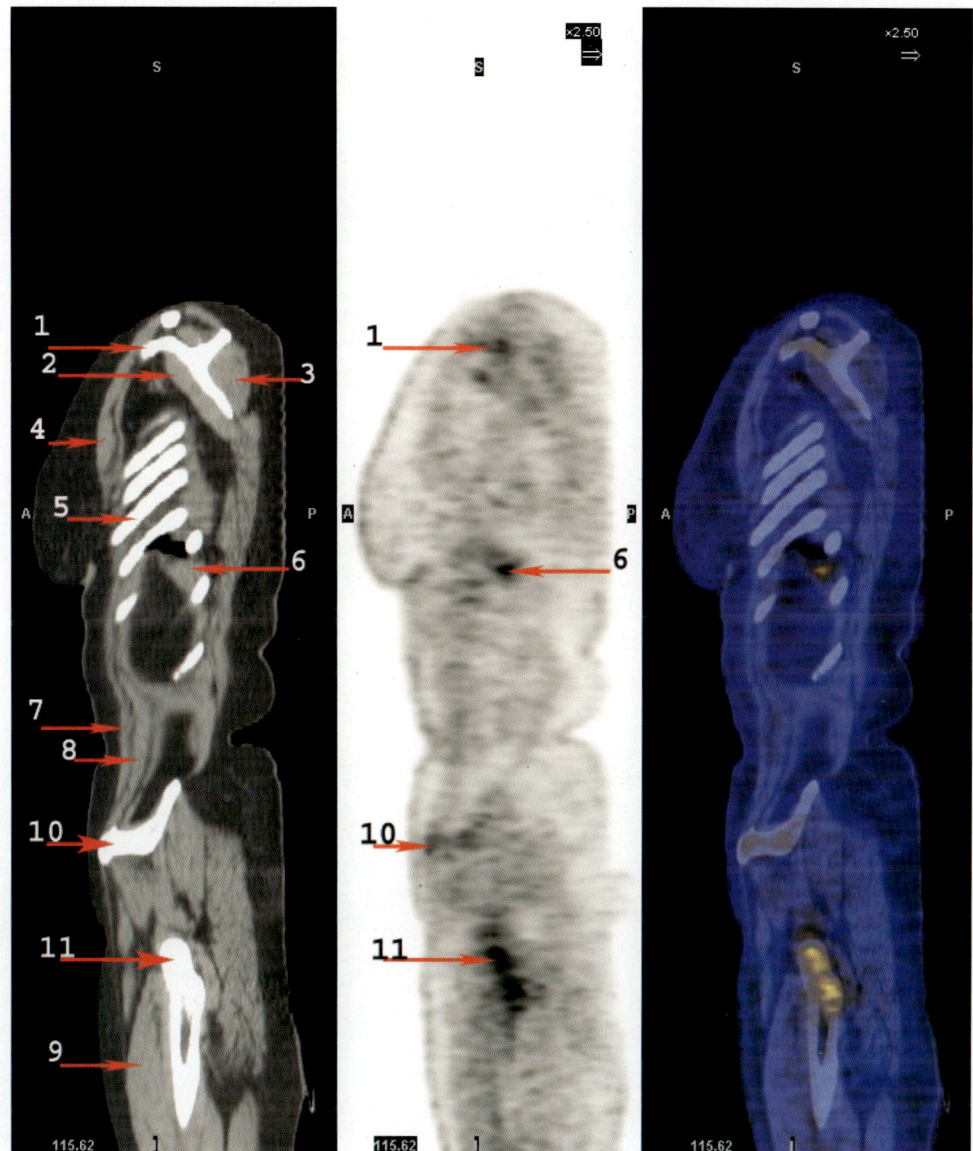

FIGURE 1-52: Sagittal 27

1. Coracoid process
2. Subscapularis muscle
3. Infraspinatus muscle
4. Pectoralis major muscle
5. Rib
6. Spleen
7. External oblique muscle
8. Internal oblique muscle
9. Rectus femoris muscle
10. Ilium
11. Femur

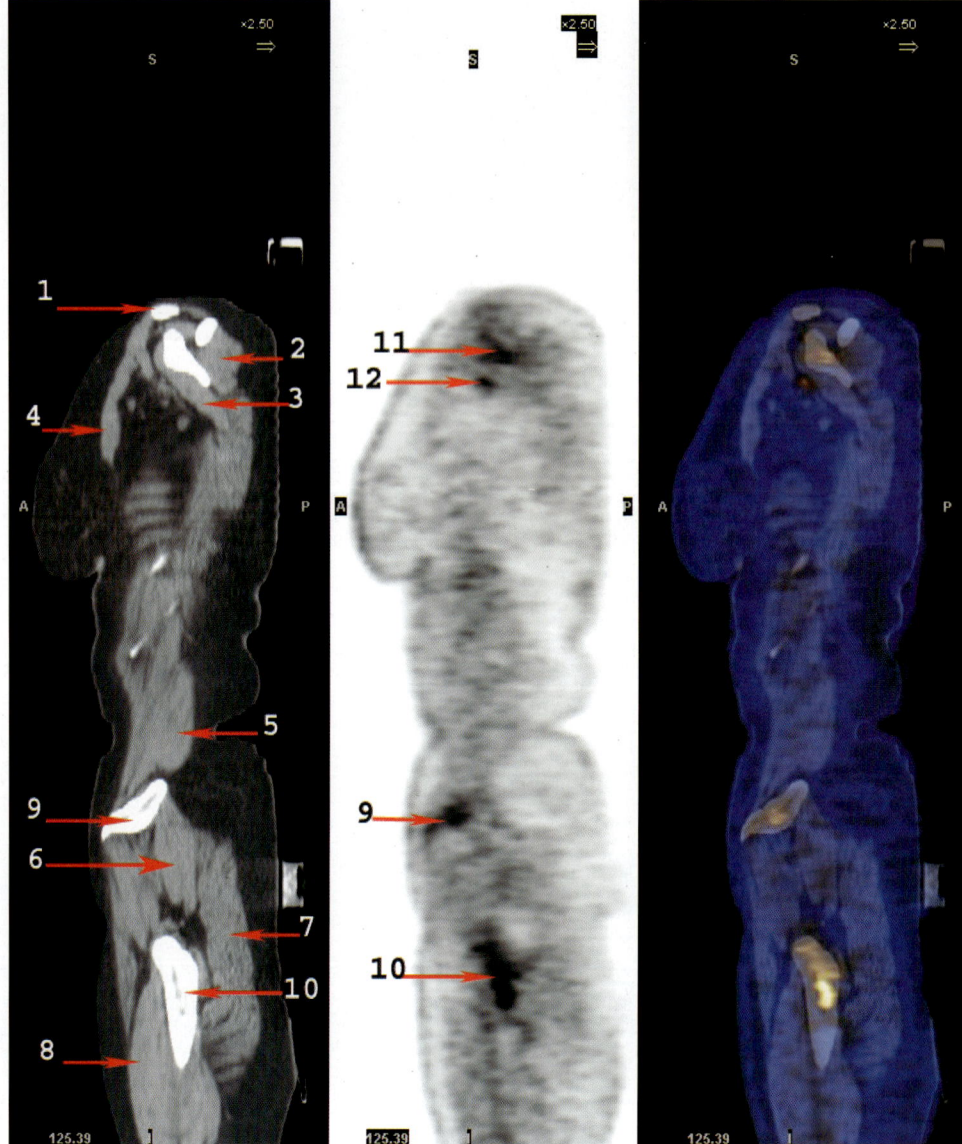

FIGURE 1-53: Sagittal 28

1. Acromion	5. Psoas muscle	9. Ilium
2. Infraspinatus muscle	6. Gluteus medius muscle	10. Femur
3. Subscapularis muscle	7. Gluteus maximus muscle	11. Scapular marrow
4. Pectoralis major muscle	8. Rectus femoris muscle	12. Axillary vein

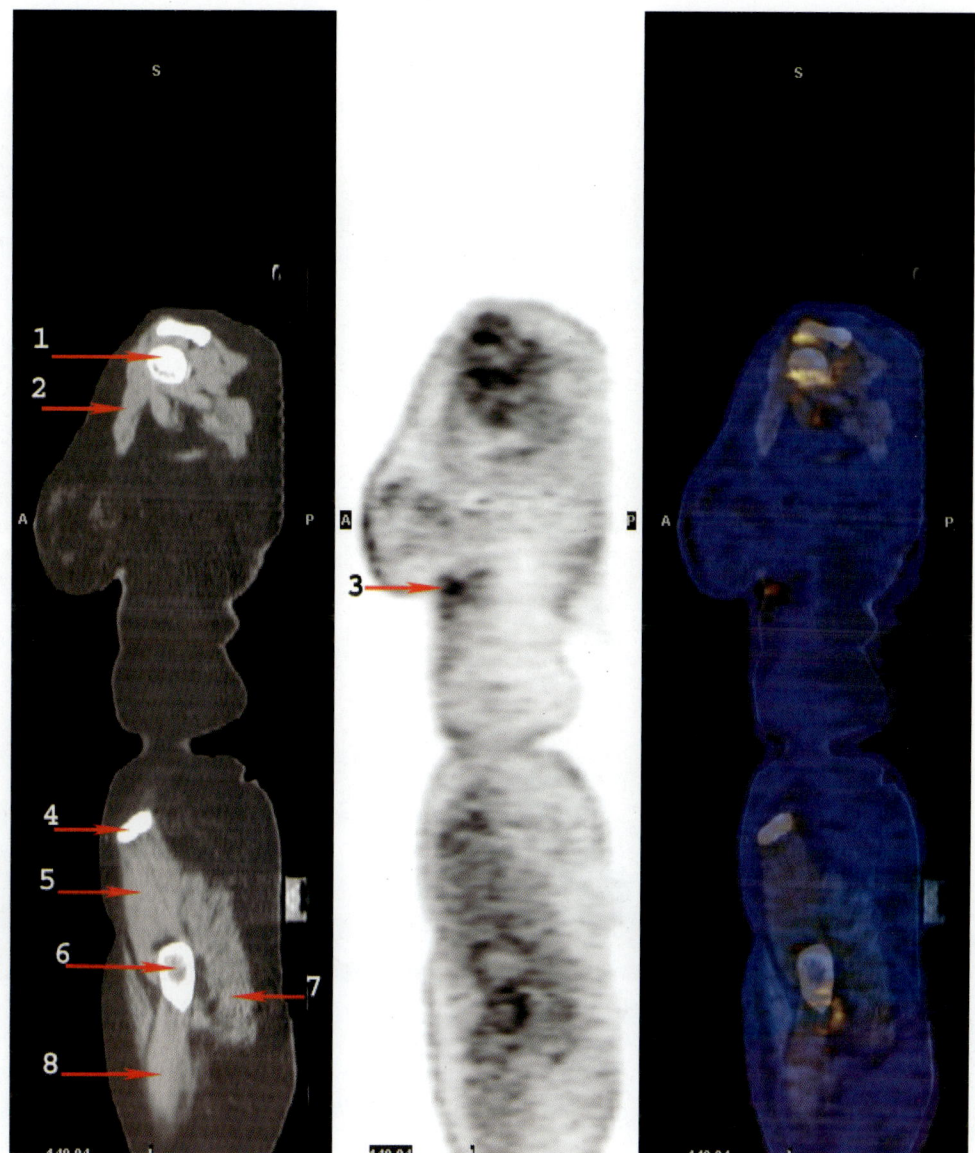

FIGURE 1-54: Sagittal 29

1. Left humeral head
2. Pectoralis major muscle
3. Skin fold
4. Ilium
5. Gluteus minimus muscle
6. Left femur
7. Gluteus maximus muscle
8. Rectus femoris muscle

Section 3: Transaxials

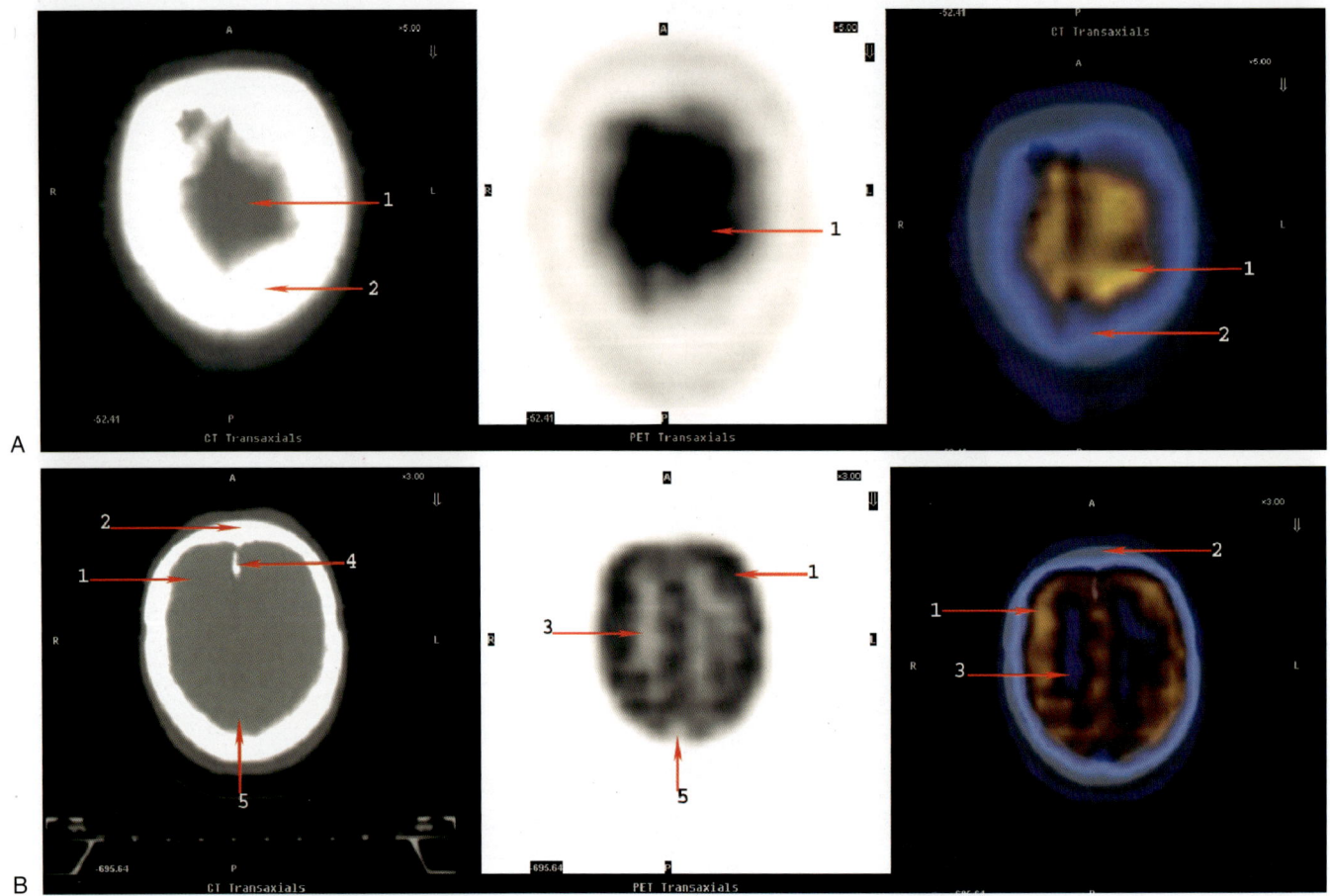

FIGURE 1-55: (A) Transaxial 1 (B) Transaxial 2

1. Cortical gray matter	3. Cerebral white matter	5. Superior sagittal sinus
2. Calvarium	4. Falx cerebri calcification	

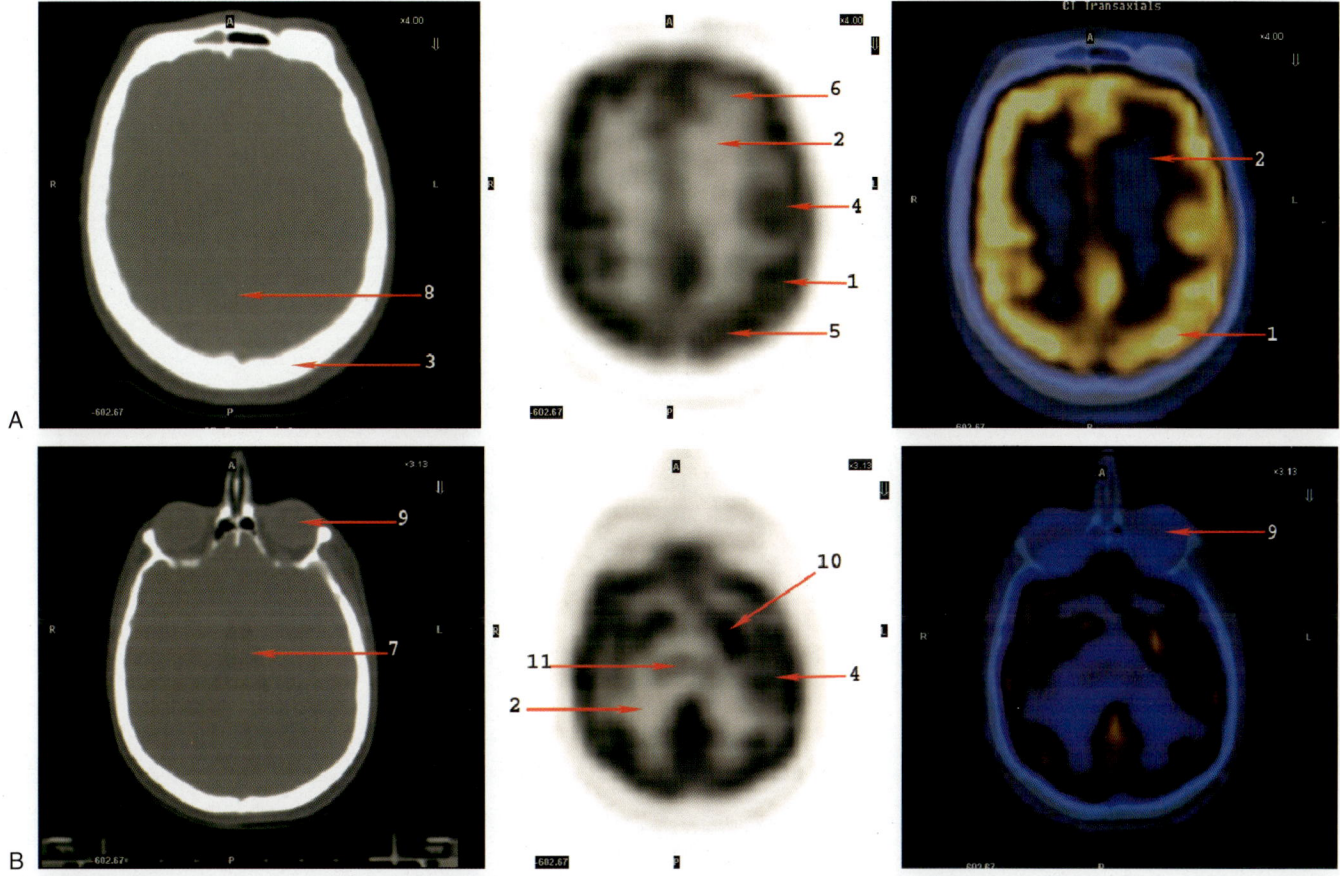

FIGURE 1-56: (A) Transaxial 3 (B) Transaxial 4

1. Cerebral gray matter
2. Cerebral white matter
3. Calvarium
4. Parietal lobe
5. Occipital lobe
6. Frontal lobe
7. Lateral ventricle
8. Falx cerebri in midline
9. Optic globe
10. Caudate nucleus
11. Thalamus

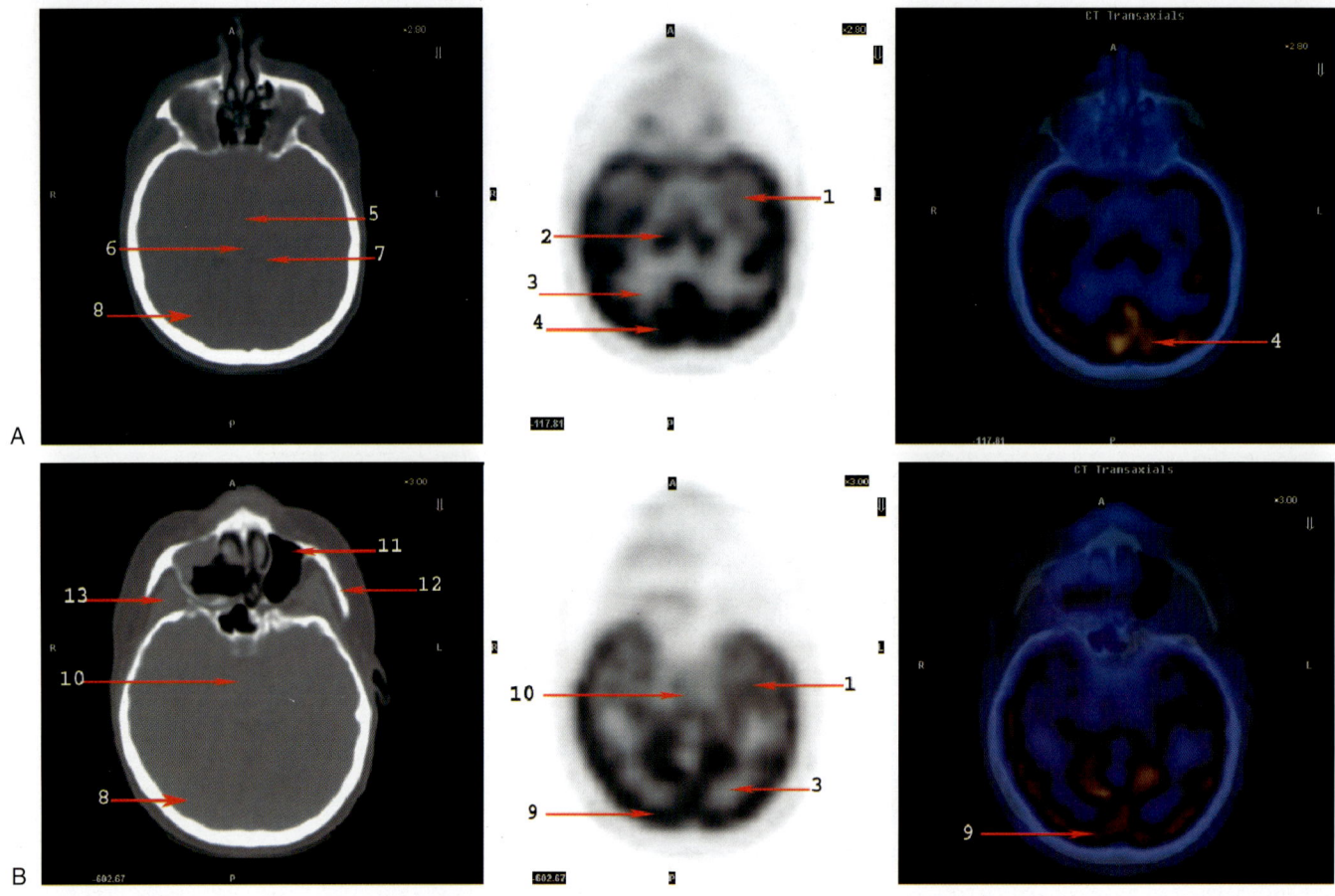

FIGURE 1-57: (A) Transaxial 5 (B) Transaxial 6

1. Temporal lobe
2. Thalamus
3. Occipital white matter
4. Occipital gray matter
5. Third ventricle

6. Pineal gland
7. Choroid plexus
8. Occipital lobe
9. Occipital gyrus
10. Brain stem

11. Maxillary sinus
12. Zygomatic arch
13. Temporalis muscle

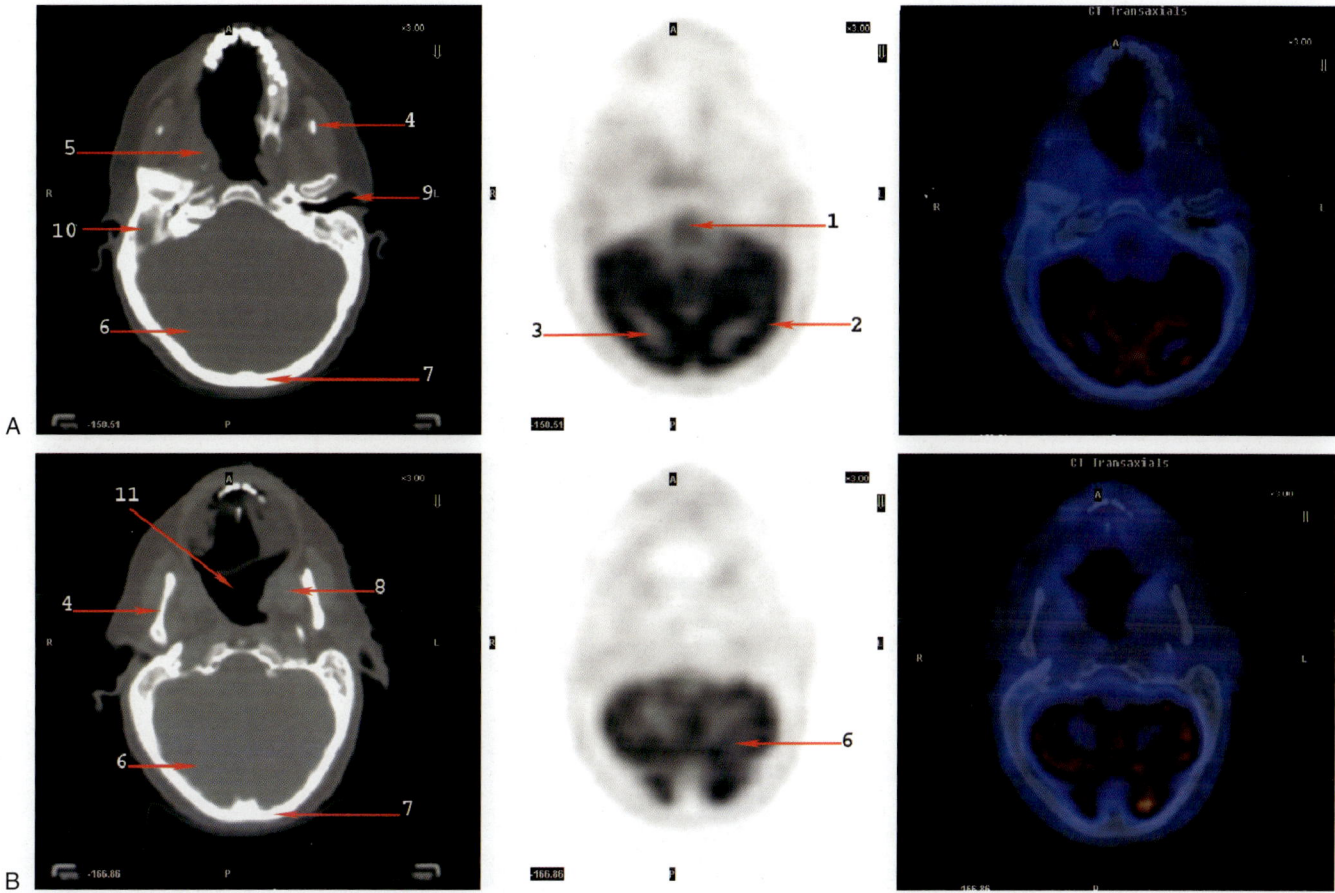

FIGURE 1-58: (A) Transaxial 7 (B) Transaxial 8

1. Pons
2. Cerebellar gray matter
3. Cerebellar white matter
4. Mandible
5. Longus capitis muscle
6. Cerebellum
7. Occipital skull
8. Pterygoid muscles
9. External auditory canal
10. Mastoid air cells
11. Pharynx

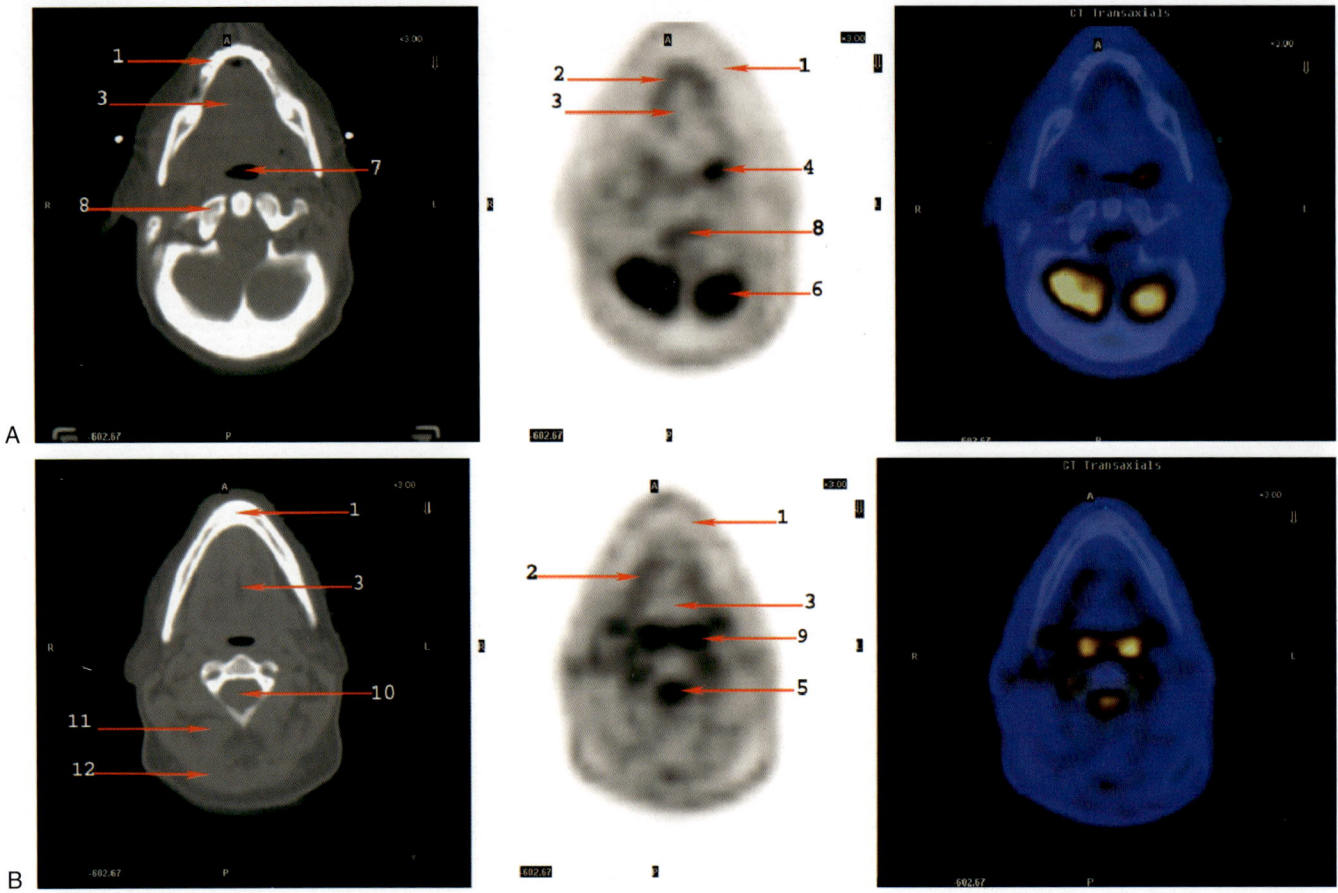

FIGURE 1-59: (A) Transaxial 9 (B) Transaxial 10

1. Mandible
2. Mylohyoid muscle
3. Genioglossus muscle
4. Palatine tonsil
5. Post tubercle of spinous process
6. Cerebellum
7. Oropharynx
8. Atlas
9. Lingual tonsil
10. Spinal cord
11. Rectus capitis muscle
12. Splenius capitis muscle

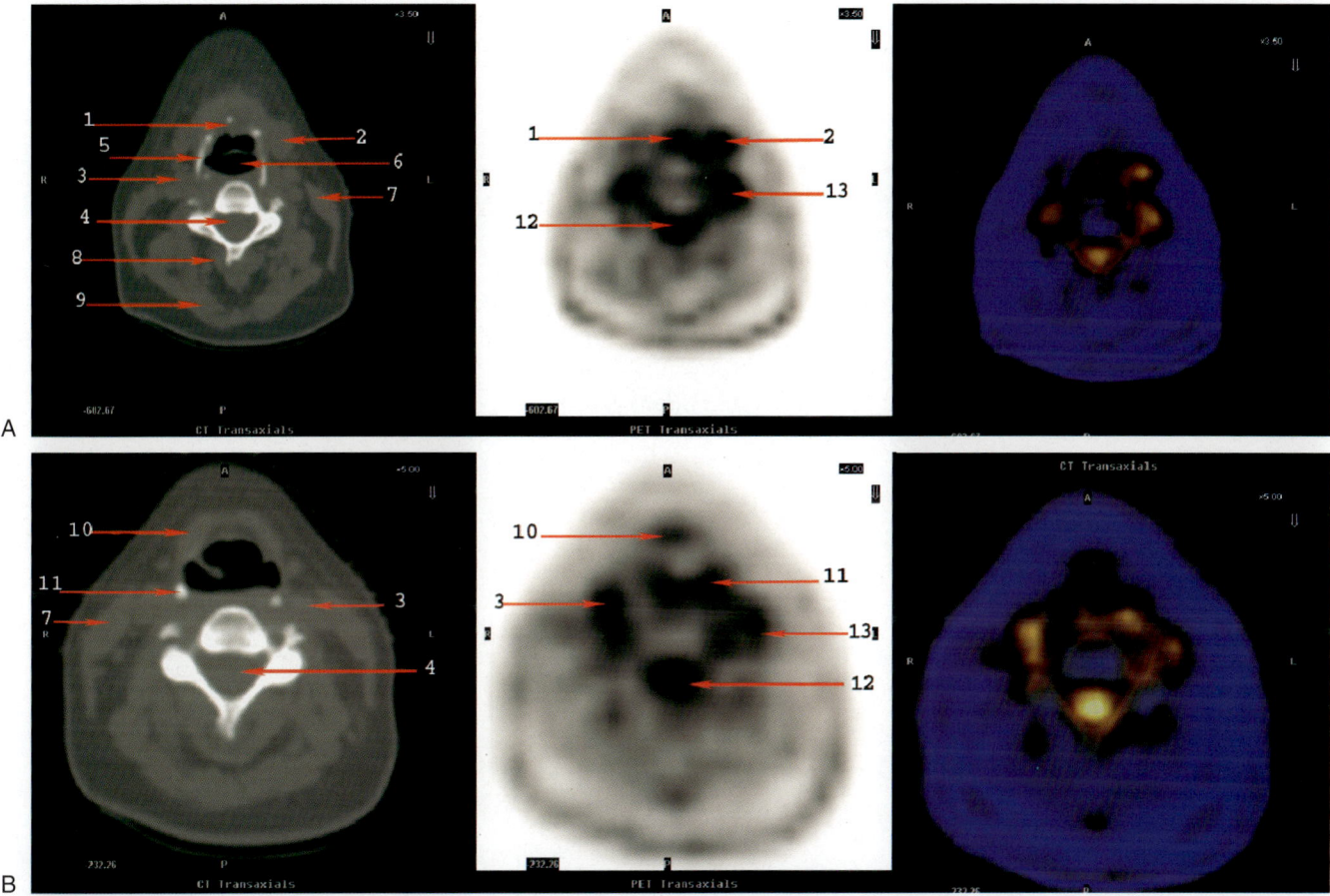

FIGURE 1-60: (A) Transaxial 11 (B) Transaxial 12

1. Tongue base
2. Submandibular gland
3. Internal jugular vein
4. Spinal cord
5. Hyoid bone

6. Oropharynx
7. Sternocleidomastoid muscle
8. Semispinalis cervicis muscle
9. Semispinalis capitis muscle
10. Infrahyoid muscle

11. Prevertebral lymphoid tissue
12. Post tubercle of spinous process
13. Facet joint in articular processes

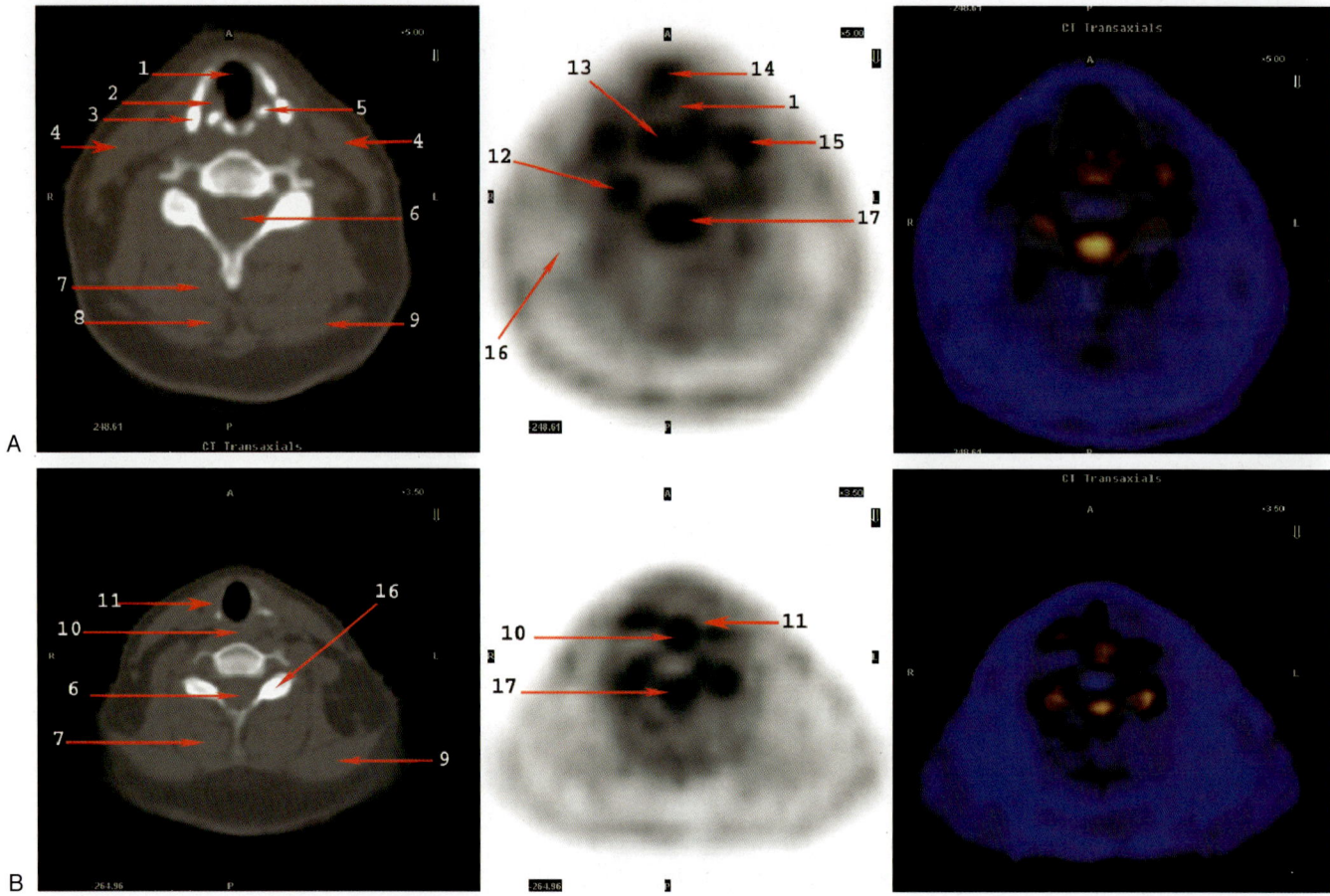

FIGURE 1-61: (A) Transaxial 13 (B) Transaxial 14

1. Fissure of glottis
2. Vocal cord fold
3. Thyroid cartilage
4. Sternocleidomastoid muscle
5. Arytenoid cartilage
6. Spinal cord
7. Semispinalis cervicis
8. Semispinalis capitis
9. Trapezius muscle
10. Esophagus and vertebral body
11. Infrahyoid muscle
12. Vertebral artery
13. Prevertebral lymphoid tissue and vertebral body
14. Vocal cord
15. Jugular vein
16. Facet joint of articular processes
17. Post tubercle of spinous process

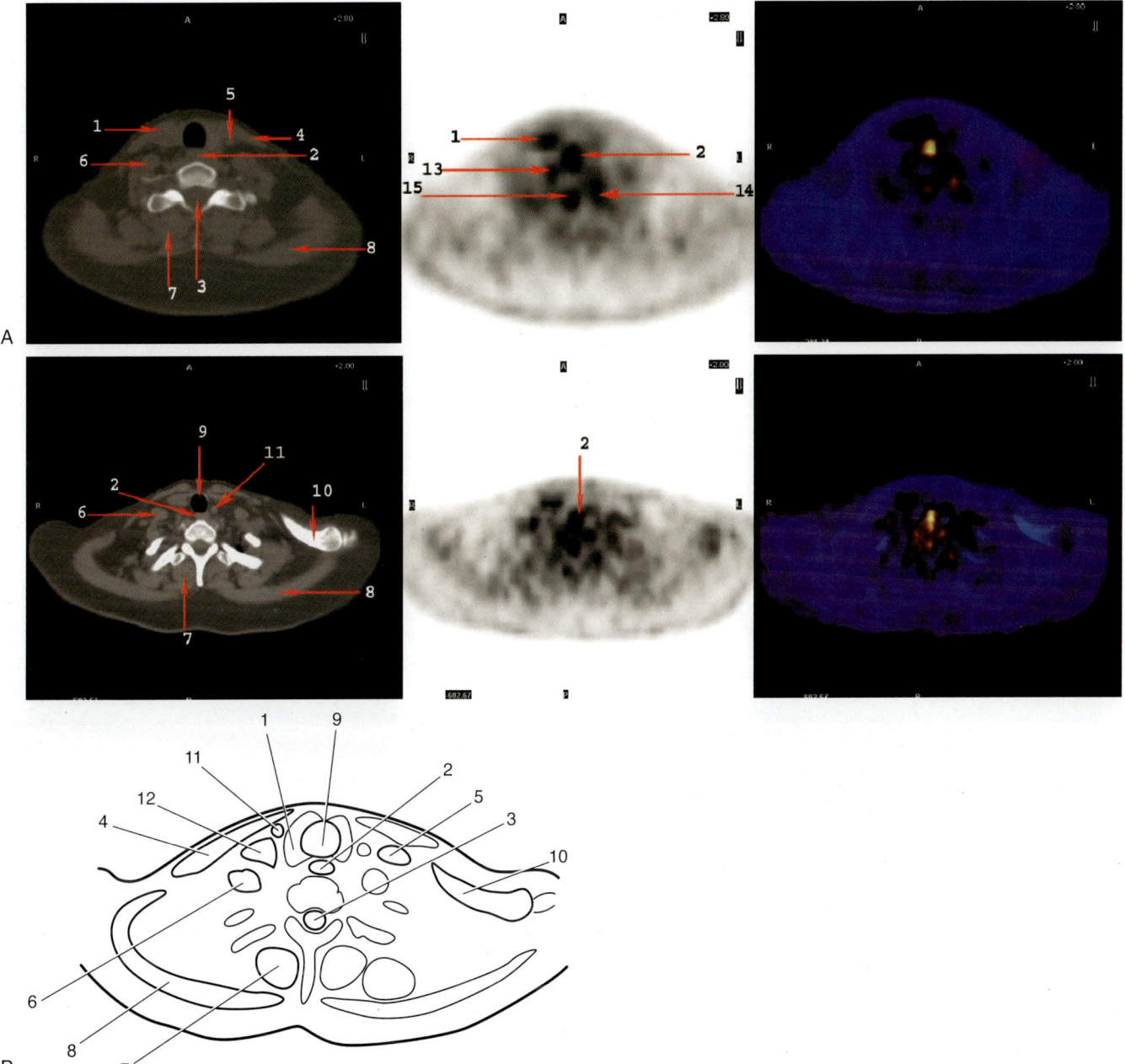

FIGURE 1-62: (A) Transaxial 15 (B) Transaxial 16 (Part A: Copyright 2006 The University of Texas M. D. Anderson Cancer Center. Used with permission.)

1. Right thyroid lobe
2. Esophagus and vertebral body
3. Spinal cord
4. Sternocleidomastoid muscle
5. Left internal jugular vein
6. Right anterior scalene muscle
7. Semispinalis cervicis muscle
8. Trapezius muscle
9. Trachea
10. Clavicle
11. Common carotid artery
12. Internal jugular vein
13. Right scalene muscle
14. Facet joint in articular processes
15. Post tubercle of spinous processes

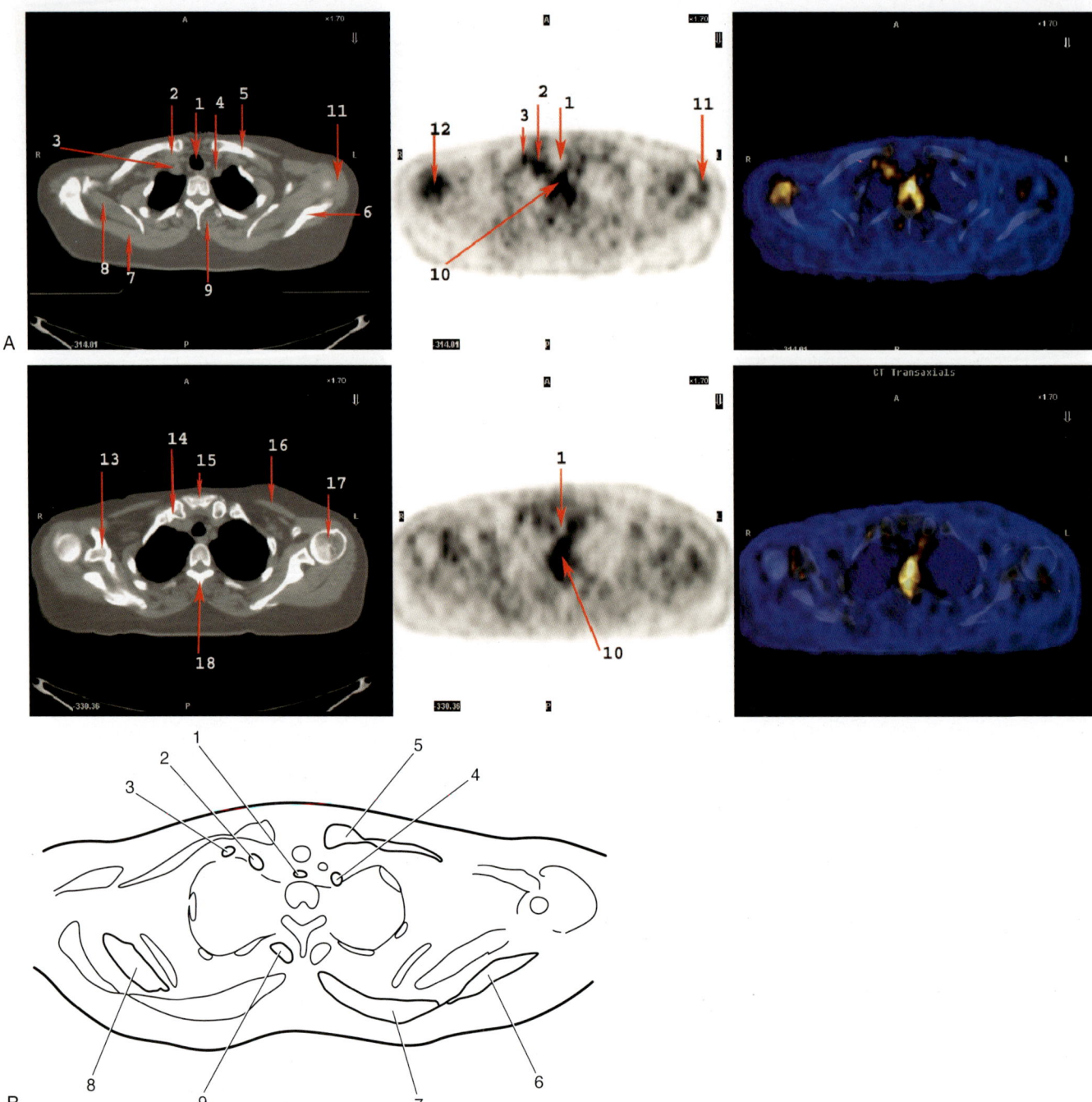

FIGURE 1-63: (A) Transaxial 17 (B) Transaxial 18 (Part A: Copyright 2006 The University of Texas M. D. Anderson Cancer Center. Used with Permission.)

1. Trachea
2. Right brachiocephalic vein
3. Right subclavian vein
4. Left common carotid artery
5. Clavicle
6. Scapula
7. Trapezius muscle
8. Infraspinatus muscle
9. Semispinalis muscle
10. T2 vertebral body
11. Deltoid muscle
12. Scapular marrow
13. Right scapula (coracoid)
14. Right clavicle
15. Manubrium
16. Pectoralis major muscle
17. Left humeral head
18. Spinous process

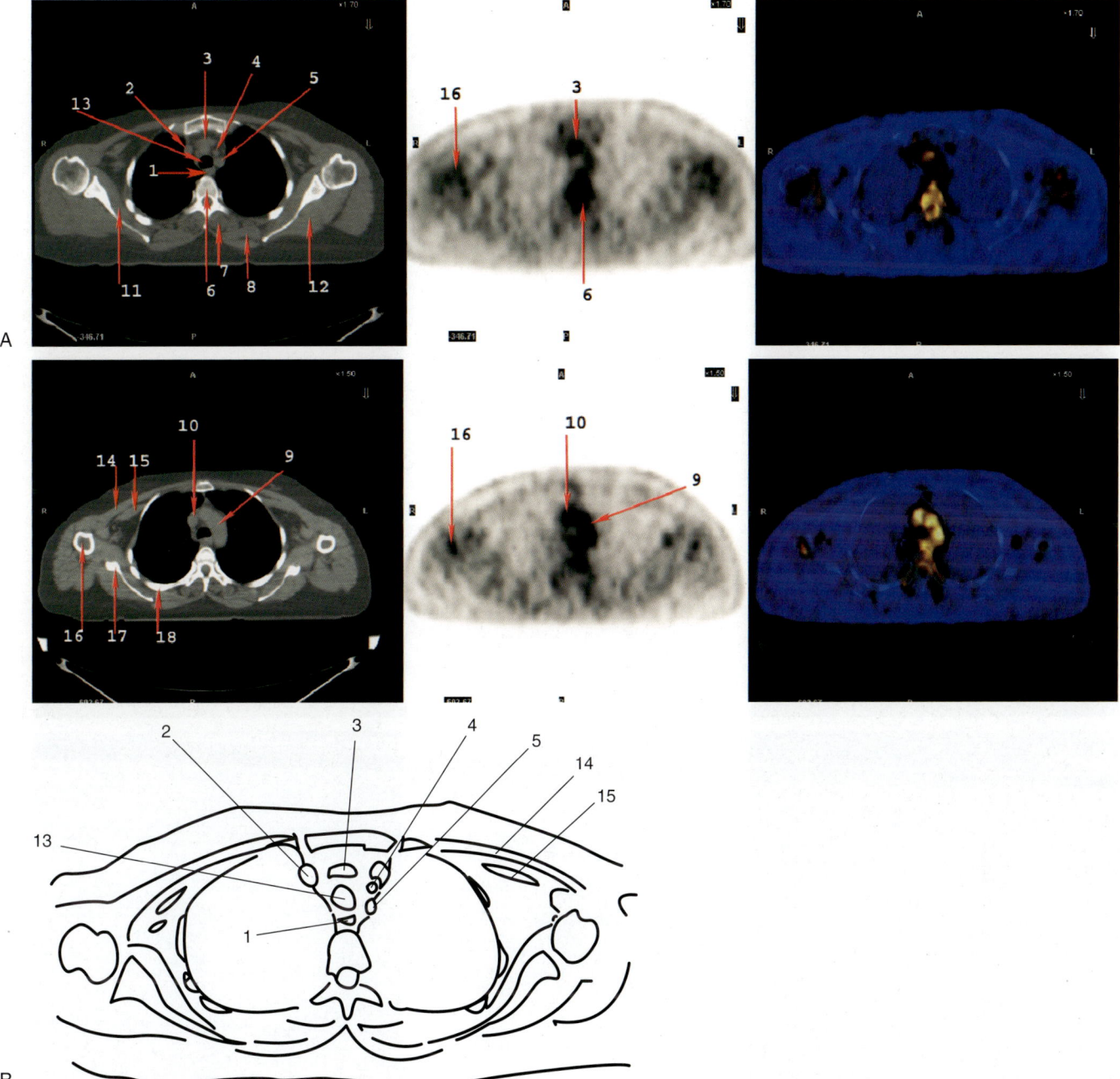

FIGURE 1-64: (A) Transaxial 19 (B) Transaxial 20 (Part A: Copyright 2006 The University of Texas M. D. Anderson Cancer Center. Used with Permission.)

1. Esophagus
2. Right brachiocephalic vein
3. Brachiocephalic artery
4. Left common carotid artery
5. Left subclavian artery
6. Vertebral marrow

7. Semispinalis muscle
8. Trapezius muscle
9. Aortic arch
10. Superior vena cava
11. Subscapularis muscle
12. Infraspinatus muscle

13. Trachea
14. Pectoralis major muscle
15. Pectoralis minor muscle
16. Right humerus
17. Right scapula
18. Right posterior 4th rib

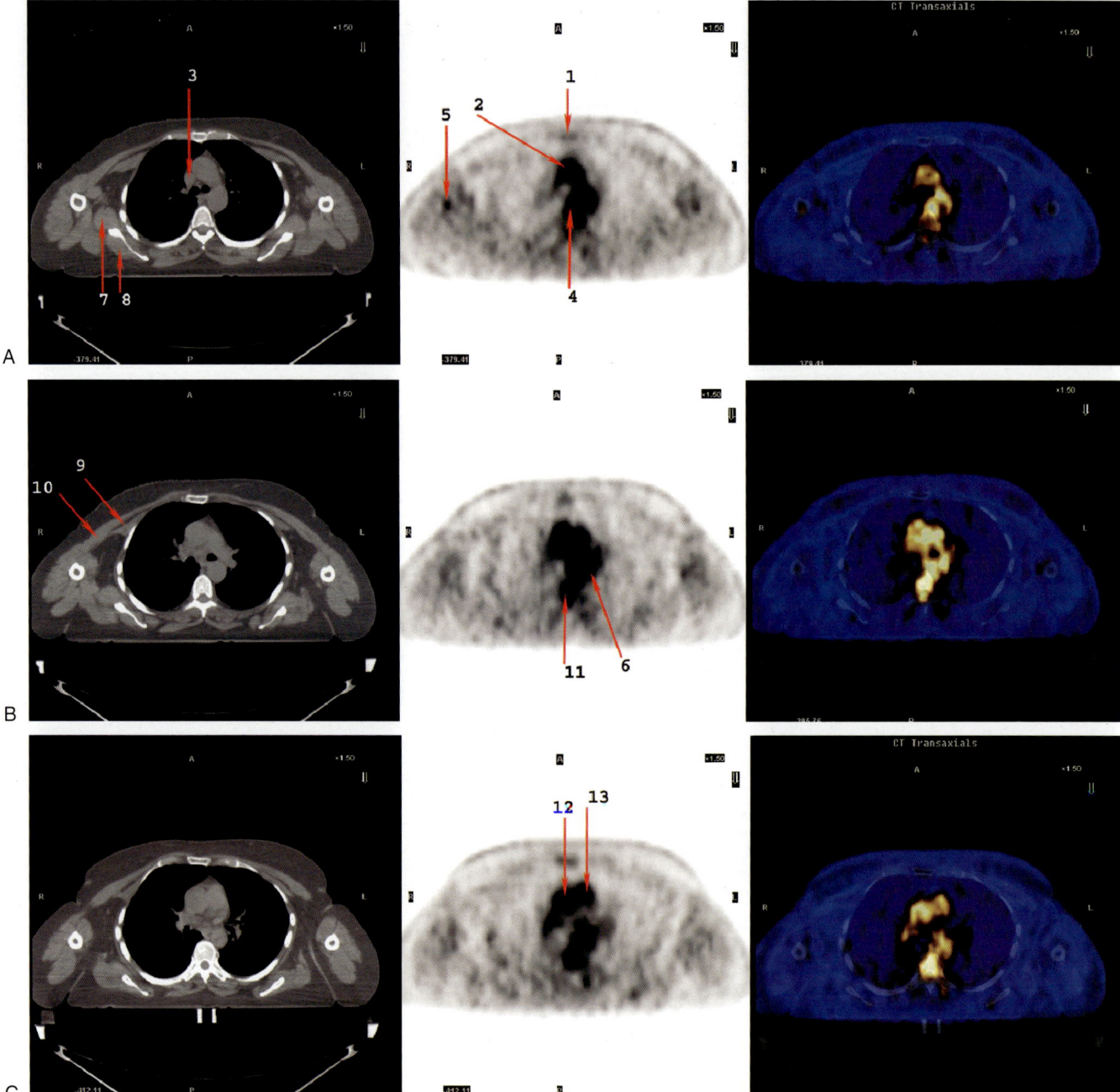

FIGURE 1-65: (A) Transaxial 21 (B) Transaxial 22 (C) Transaxial 23

1. Sternum
2. Aortic arch
3. Superior vena cava
4. Esophagus
5. Humeral marrow
6. Descending aorta
7. Subscapularis muscle
8. Infraspinatus muscle
9. Pectoralis minor muscle
10. Pectoralis major muscle
11. T5 vertebral body
12. Ascending aorta
13. Main pulmonary artery

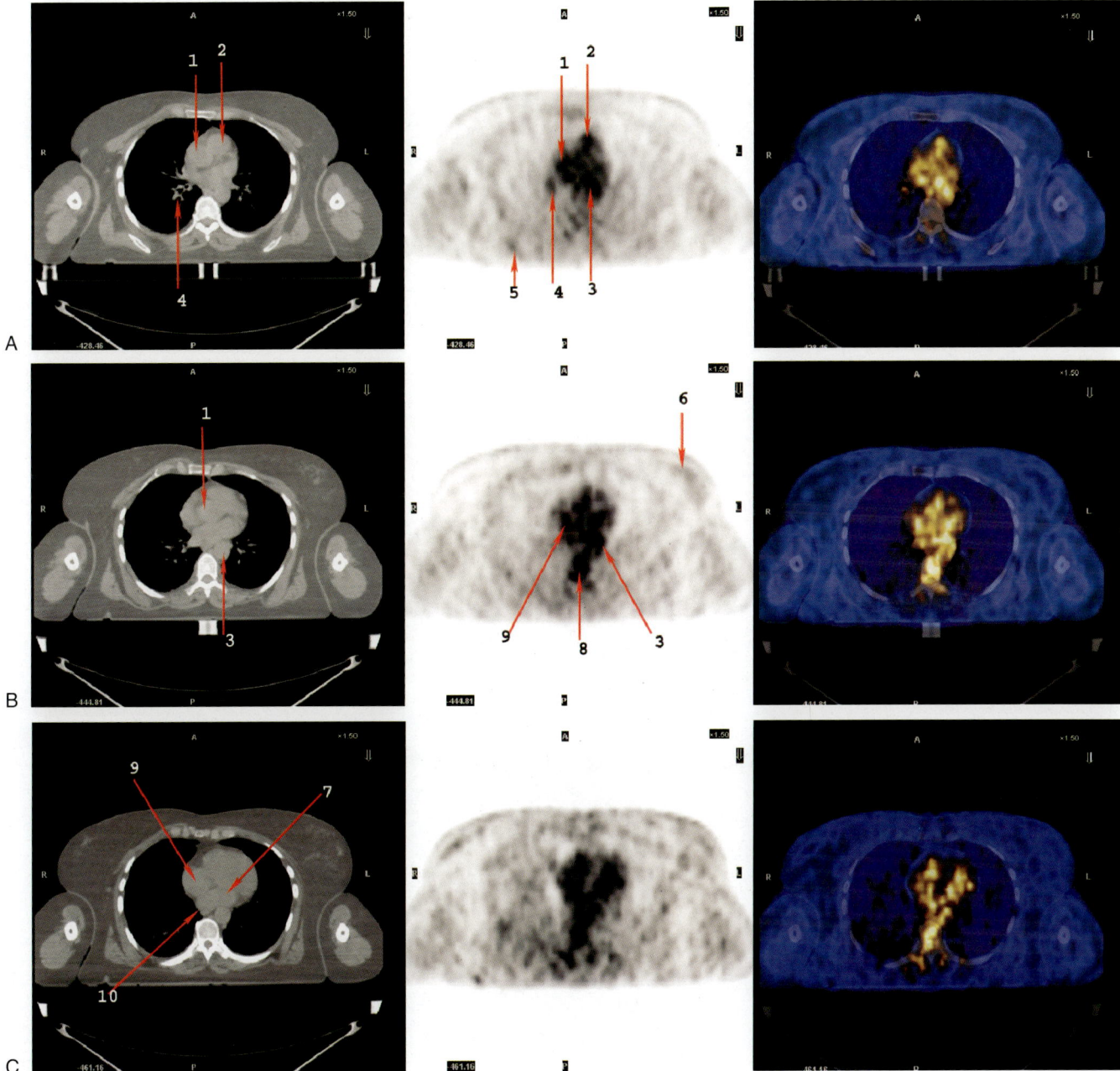

FIGURE 1-66: (A) Transaxial 24 (B) Transaxial 25 (C) Transaxial 26

1. Ascending aorta
2. Main pulmonary artery
3. Descending aorta
4. Right pulmonary vein
5. Scapula
6. Left breast
7. Left atrium
8. T7 marrow
9. Superior vena cava
10. Azygos vein

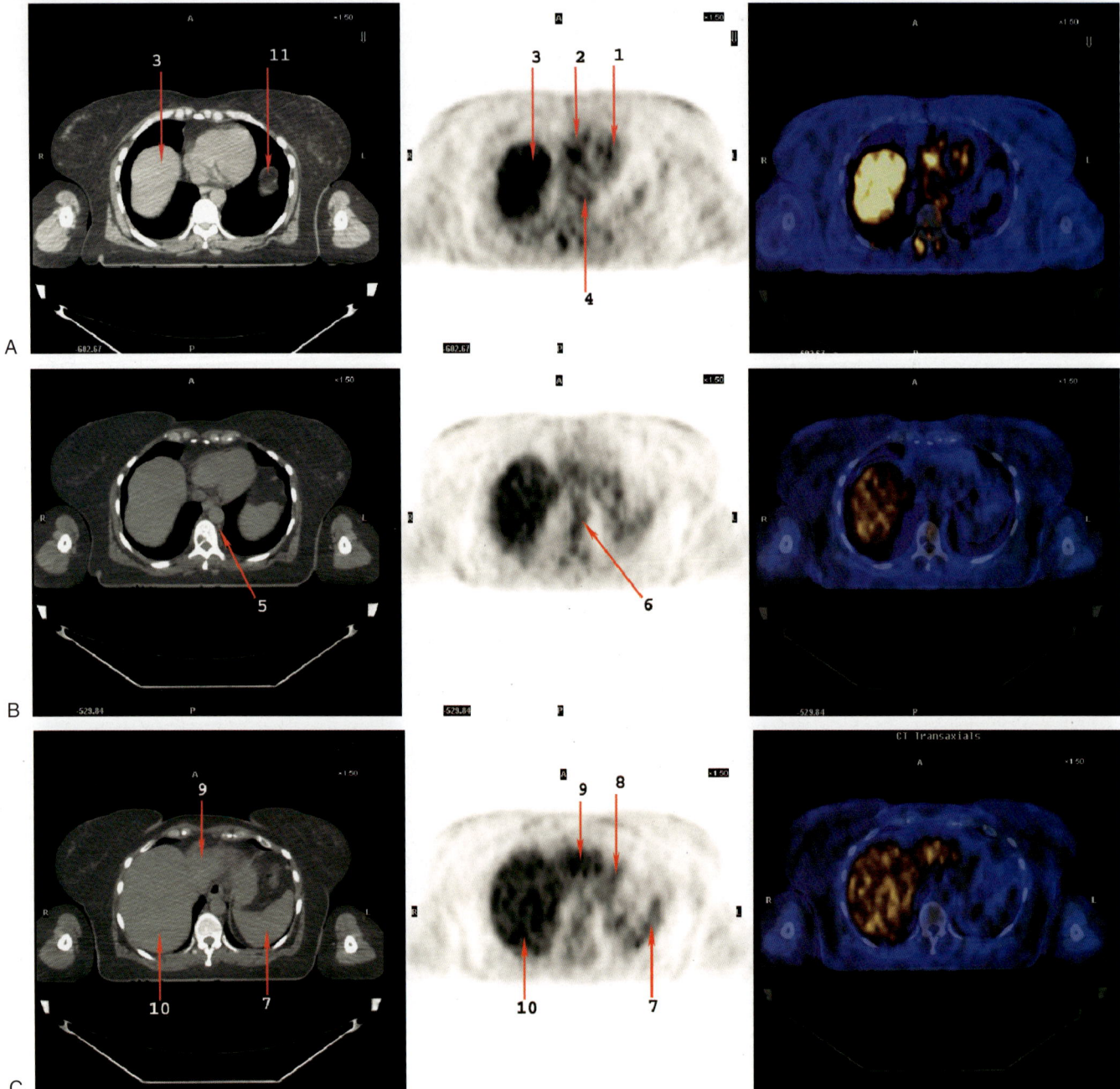

FIGURE 1-67: (A) Transaxial 27 (B) Transaxial 28 (C) Transaxial 29

1. Left ventricle
2. Right ventricle
3. Right hepatic lobe (segment VIII)
4. Descending aorta
5. Hemiazygos vein
6. Esophagus
7. Spleen
8. Stomach
9. Left hepatic lobe (segment III)
10. Right hepatic lobe (segment VII)
11. Gastric fundus

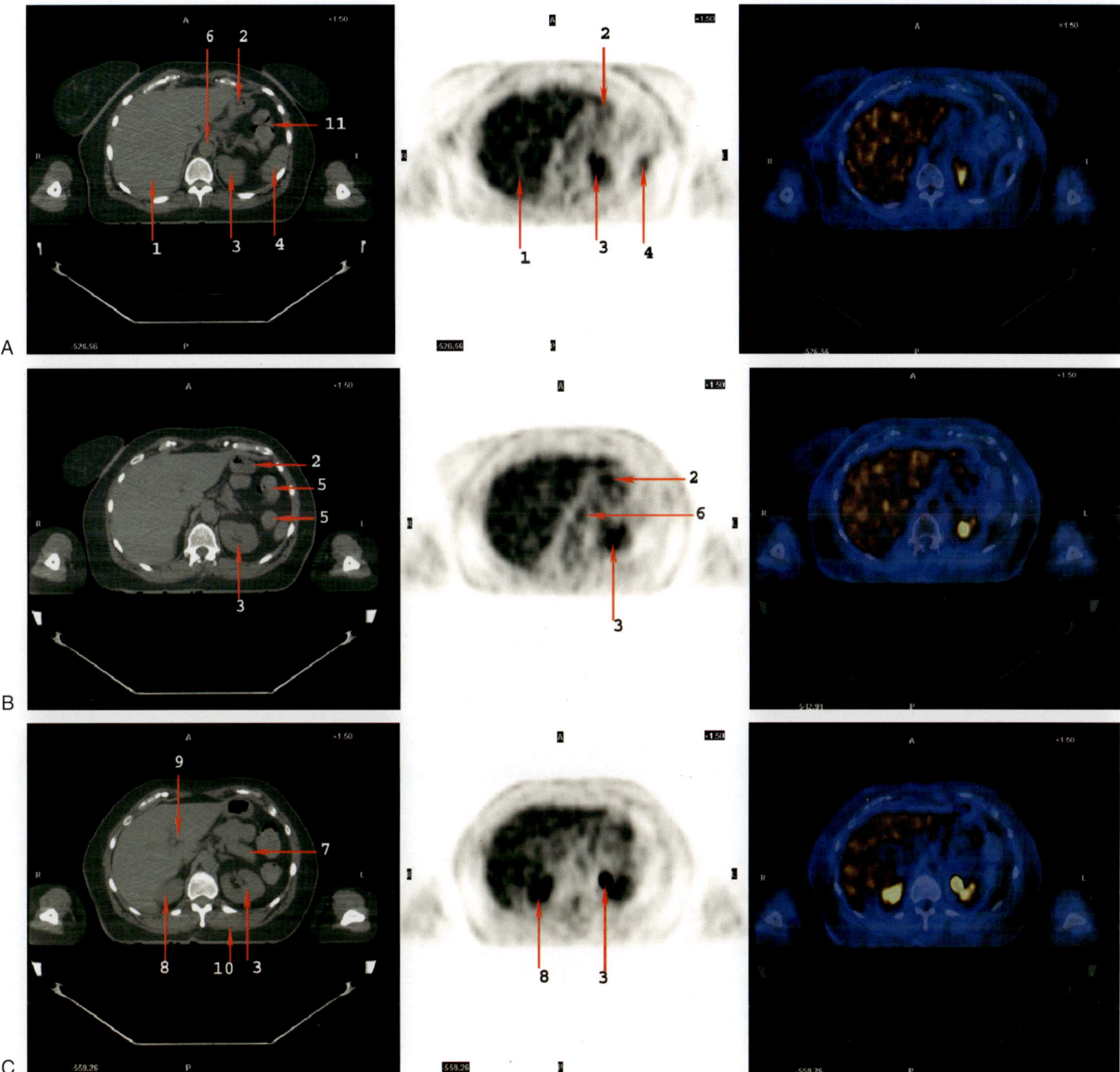

FIGURE 1-68: (A) Transaxial 30 (B) Transaxial 31 (C) Transaxial 32

1. Right hepatic lobe (segment VII)
2. Stomach
3. Left kidney
4. Spleen
5. Colon
6. Abdominal aorta
7. Pancreas
8. Right kidney
9. Portal vein
10. Erector spinae muscle
11. Splenic flexure of colon

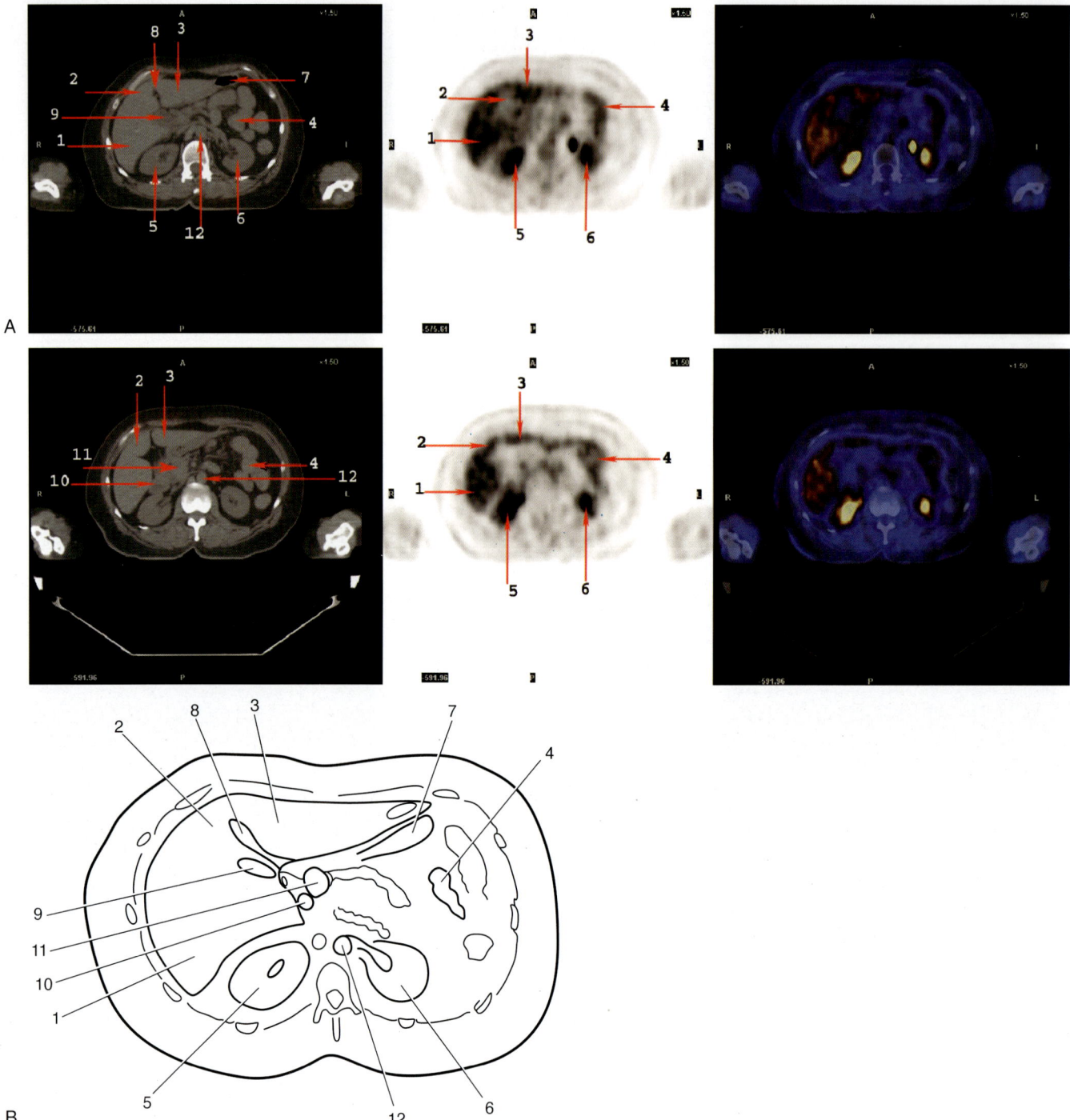

FIGURE 1-69: (A) Transaxial 33 (B) Transaxial 34 (Part A: Copyright 2006 The University of Texas M. D. Anderson Cancer Center. Used with permission.)

1. Right hepatic lobe, inferior segment (#6)
2. Right hepatic lobe, medial segment (#4)
3. Left hepatic lobe, lateral segment (#3)
4. Colon
5. Right kidney
6. Left kidney
7. Stomach
8. Ligamentum teres
9. Gallbladder
10. Duodenum, second portion
11. Pancreatic head
12. Aorta

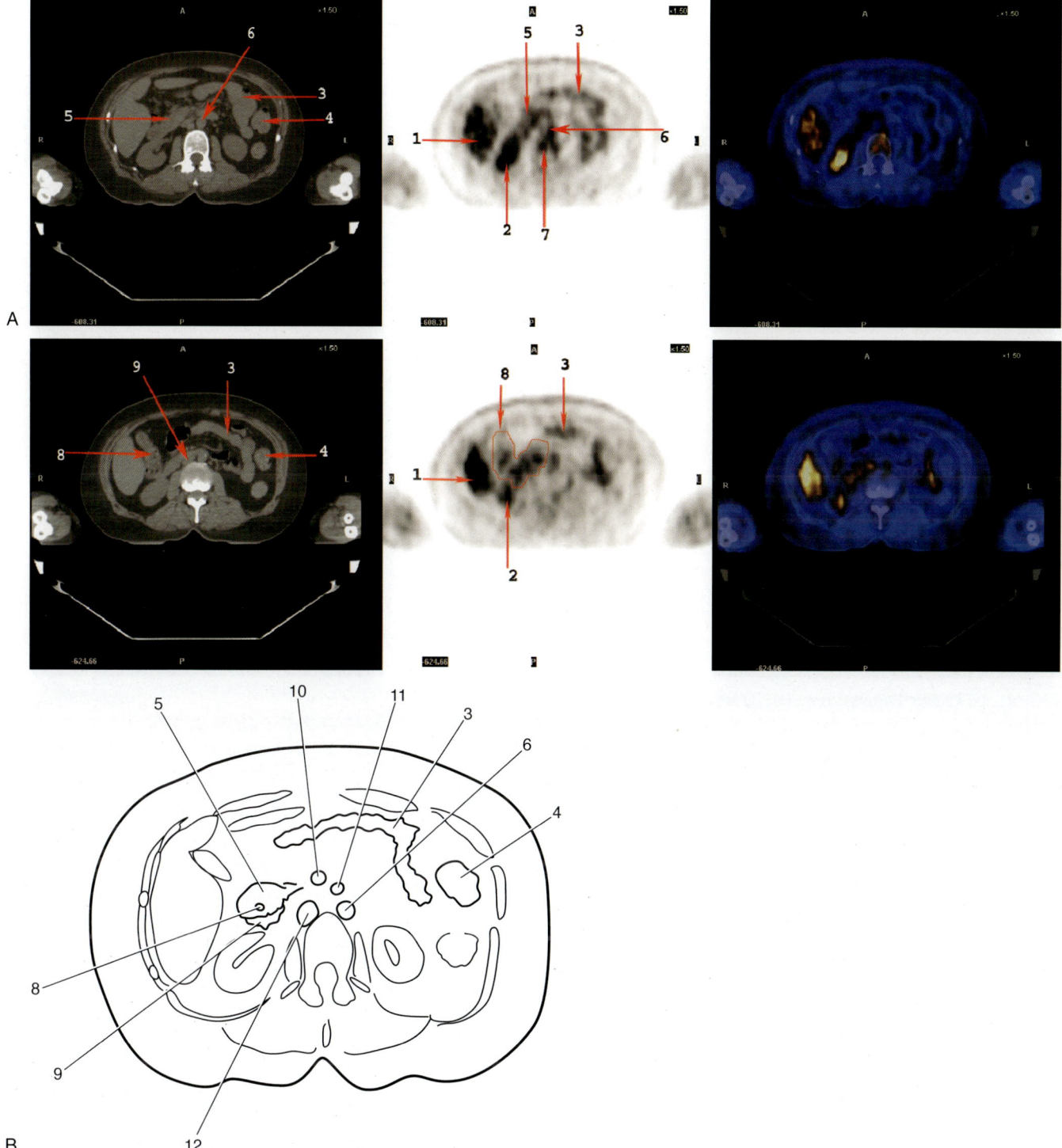

FIGURE 1-70: (A) Transaxial 35 (B) Transaxial 36 (Part A: Copyright 2006 The University of Texas M. D. Anderson Cancer Center. Used with permission.)

1. Right hepatic lobe (segment VI)
2. Right kidney
3. Jejunum
4. Colon
5. Pancreatic head
6. Aorta
7. L2 marrow
8. Duodenum
9. Inferior vena cava

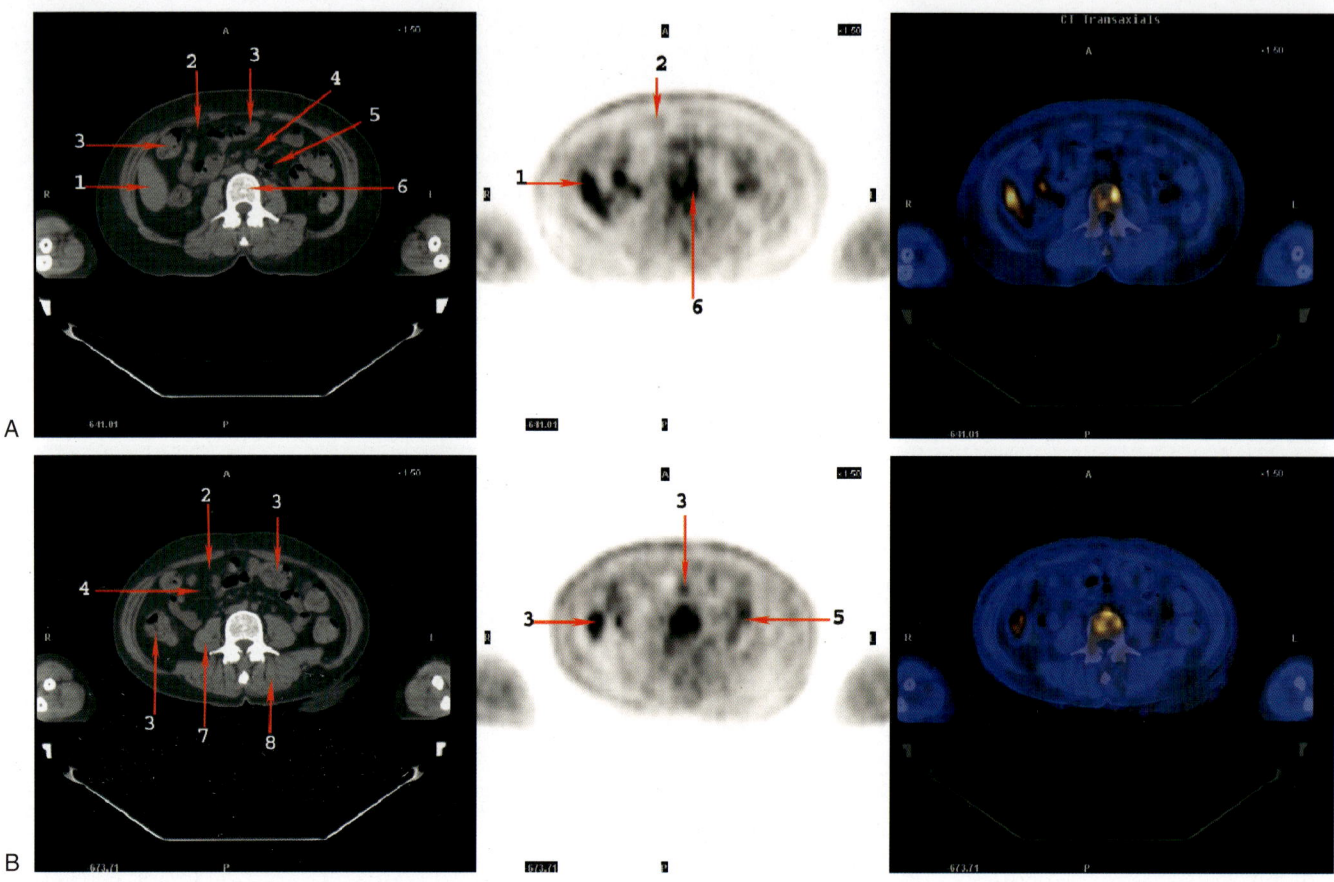

FIGURE 1-71: (A) Transaxial 37 (B) Transaxial 38

1. Tip of right liver
2. Omentum
3. Colon
4. Mesenteric fat
5. Jejunum
6. L3 body
7. Psoas muscle
8. Multifidus muscle

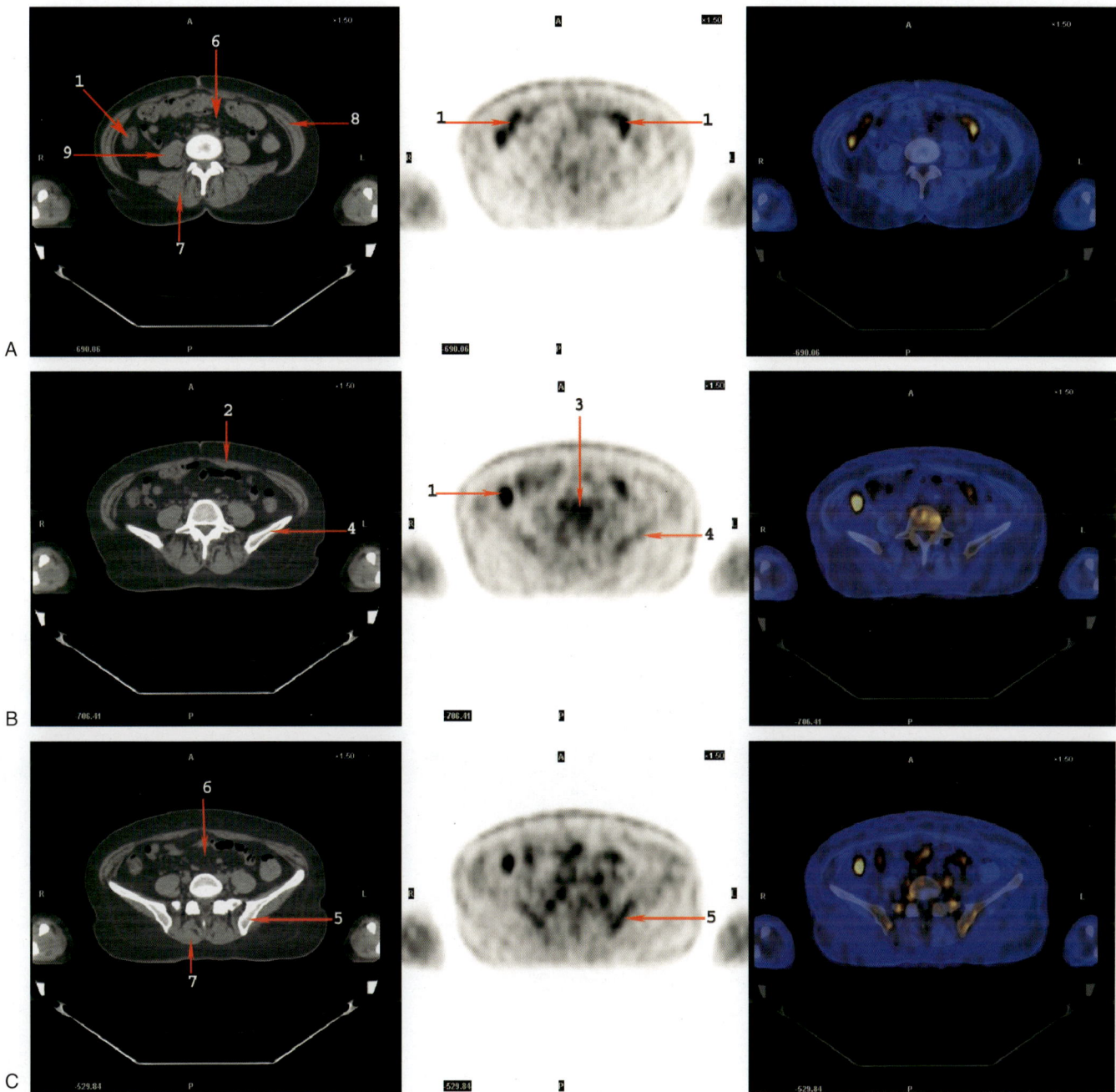

FIGURE 1-72: (A) Transaxial 39 (B) Transaxial 40 (C) Transaxial 41

1. Ascending colon
2. Rectus abdominis muscle
3. L$_5$ marrow
4. Iliac crest
5. Iliac tuberosity
6. Mesentery
7. Multifidus muscle
8. Oblique muscle
9. Psoas muscle

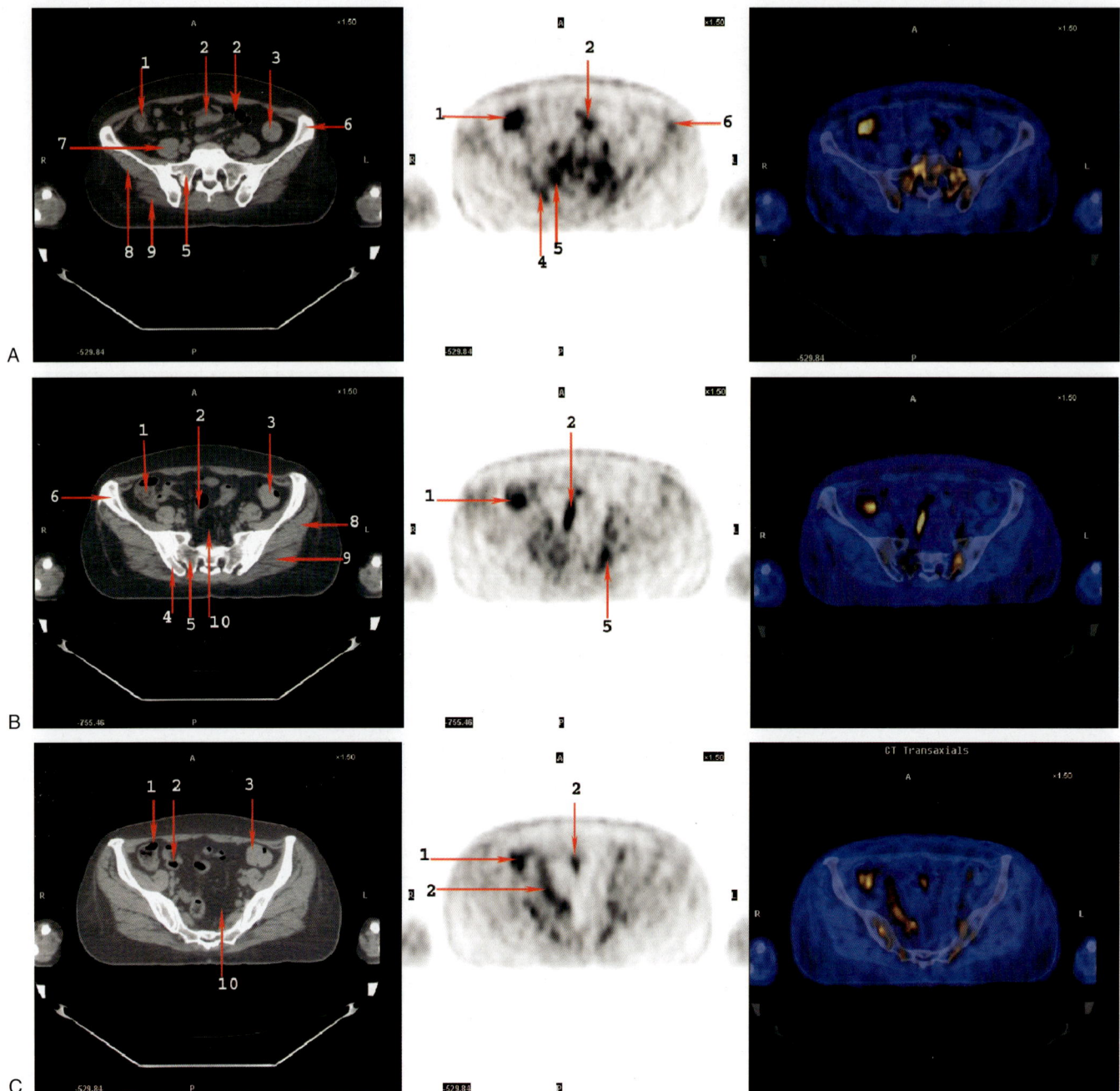

FIGURE 1-73: (A) Transaxial 42 (B) Transaxial 43 (C) Transaxial 44

1. Cecum
2. Small bowel
3. Descending colon
4. Iliac tuberosity
5. Sacral wing
6. Anterior iliac crest
7. Psoas muscle
8. Gluteus medius muscle
9. Gluteus maximus muscle
10. Mesenteric fat in presacral space

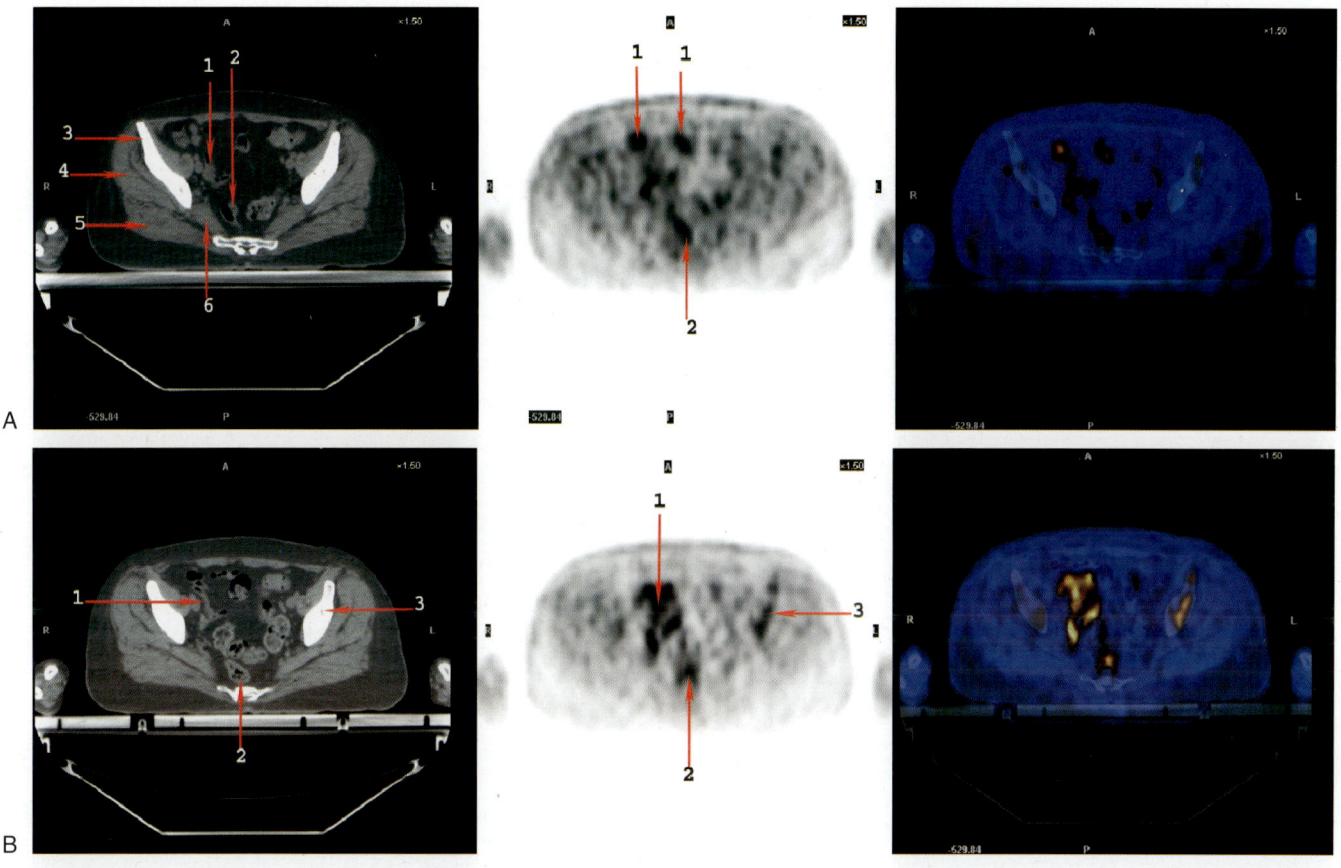

FIGURE 1-74: (A) Transaxial 45 (B) Transaxial 46

1. Ileum	3. Iliac crest	5. Gluteus maximus muscle
2. Sigmoid	4. Gluteus medius muscle	6. Piriformis muscle

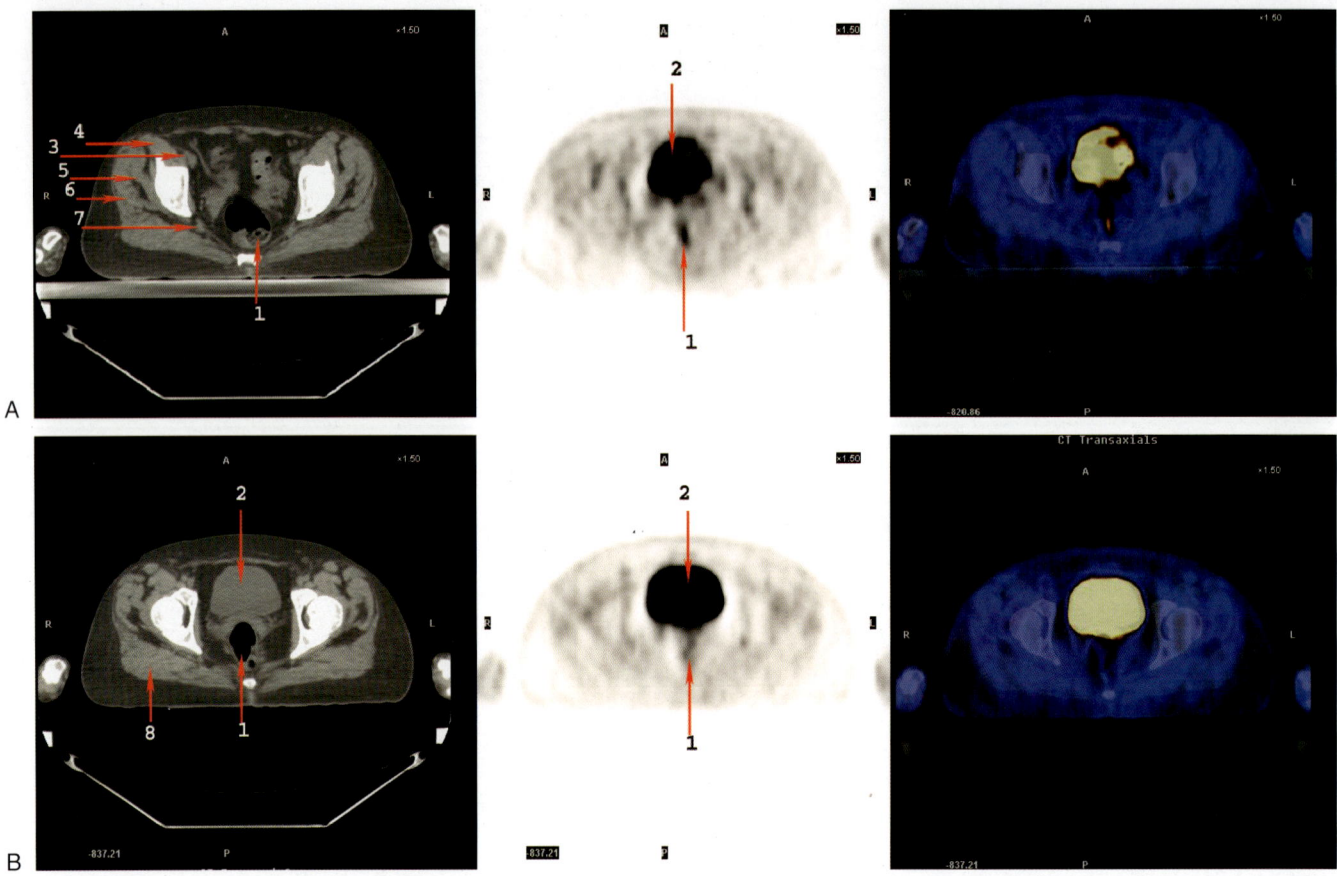

FIGURE 1-75: (A) Transaxial 47 (B) Transaxial 48

1. Rectum
2. Bladder
3. External iliac vein
4. Iliopsoas muscle
5. Gluteus minimus muscle
6. Gluteus medius muscle
7. Inferior gluteal vein
8. Gluteus maximis muscle

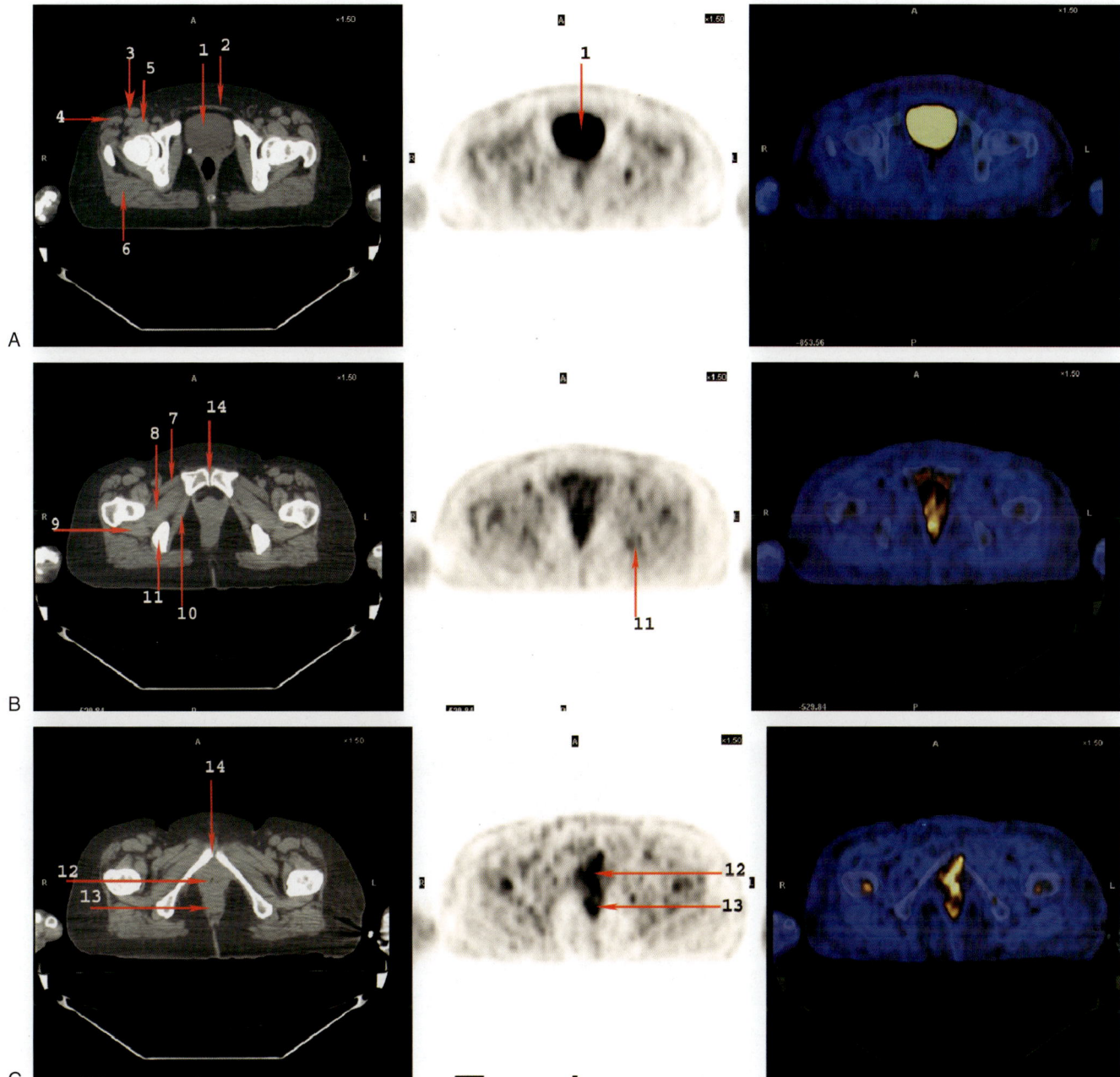

FIGURE 1-76: (A) Transaxial 49 (B) Transaxial 50 (C) Transaxial 51

1. Bladder
2. Rectus sheath
3. Sartorius muscle
4. Tensor fascia lata muscle
5. Iliopsoas muscle
6. Gluteus maximus muscle
7. Pectineus muscle
8. Obturator externus muscle
9. Quadratus femoris muscle
10. Obturator internus muscle
11. Ischial tuberosity
12. Urethra
13. Anal canal
14. Symphysis pubis

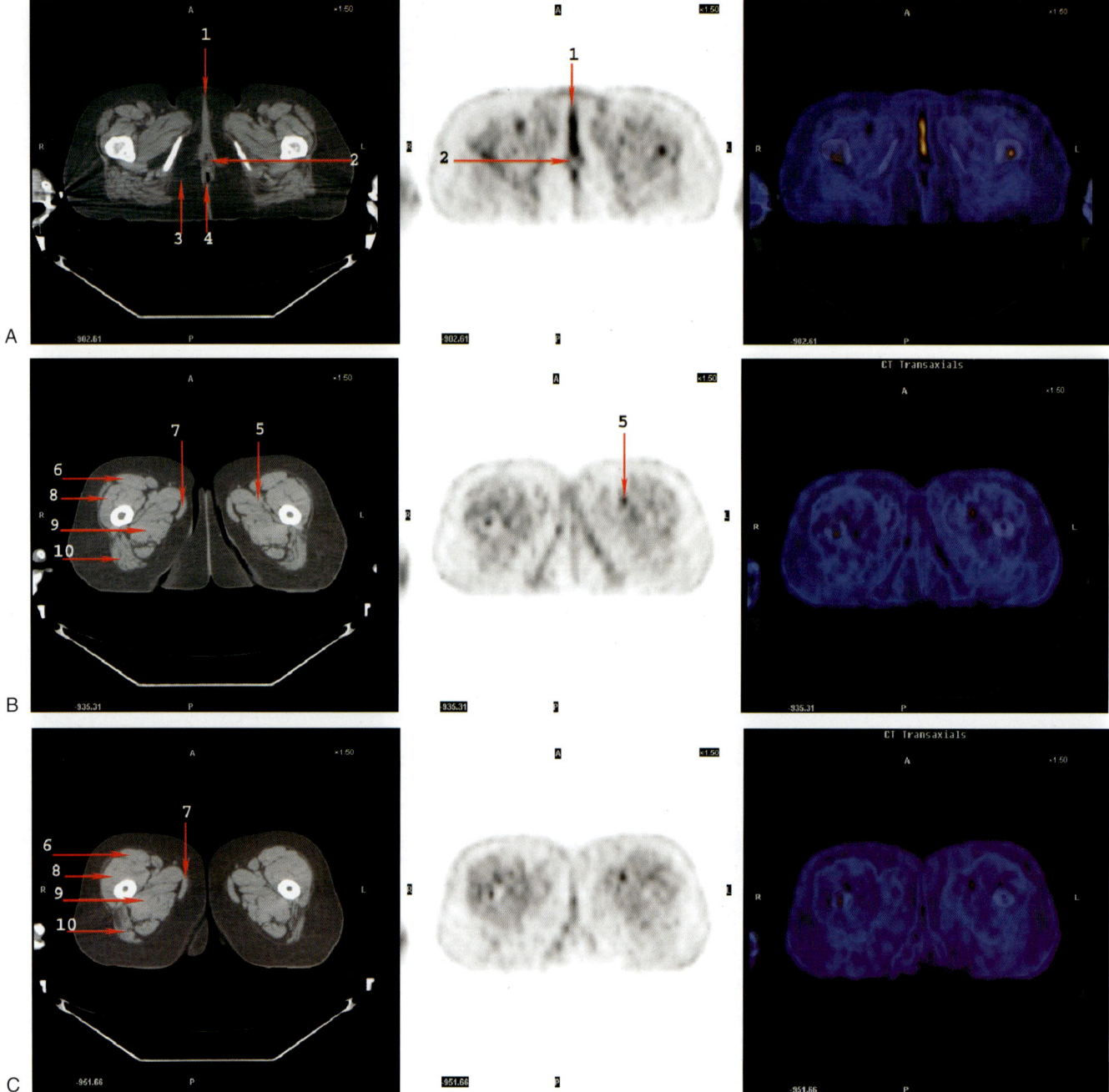

FIGURE 1-77: (A) Transaxial 52 (B) Transaxial 53 (C) Transaxial 54

1. Urethra
2. Vagina
3. Ischiorectal fat
4. Anus

5. Superficial femoral vein
6. Rectus femoris muscle
7. Gracilis muscle
8. Vastus intermedius muscle

9. Adductor magnus muscle
10. Semimembranous muscle

Section 4: Male Pelvis

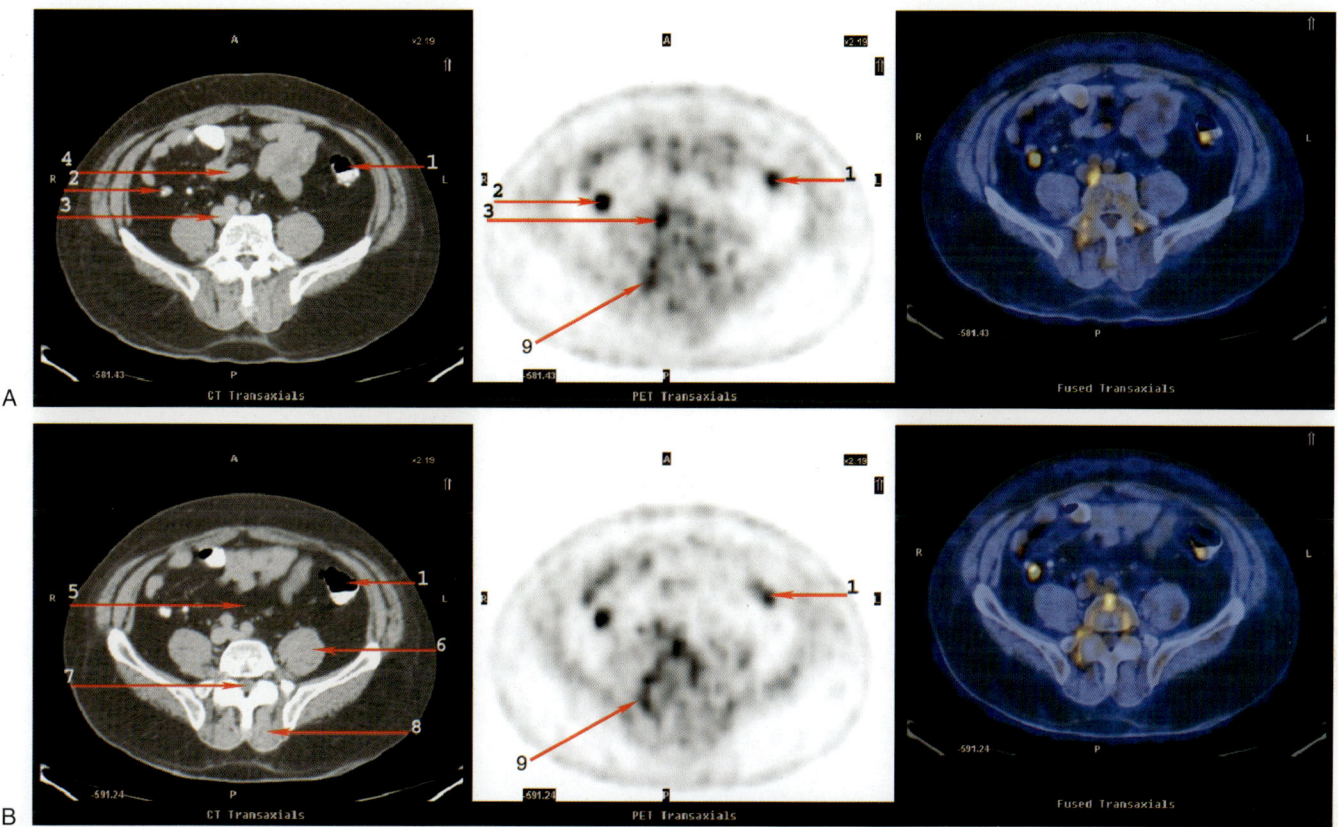

FIGURE 1-78: (A) Male Pelvis Transaxial 1 (B) Male Pelvis Transaxial 2

1. Descending colon
2. Ascending colon
3. Common iliac artery and vein
4. Ileum
5. Mesenteric fat
6. Psoas muscle
7. Cauda equina
8. Multifidus muscle
9. Facet joint in articular processes

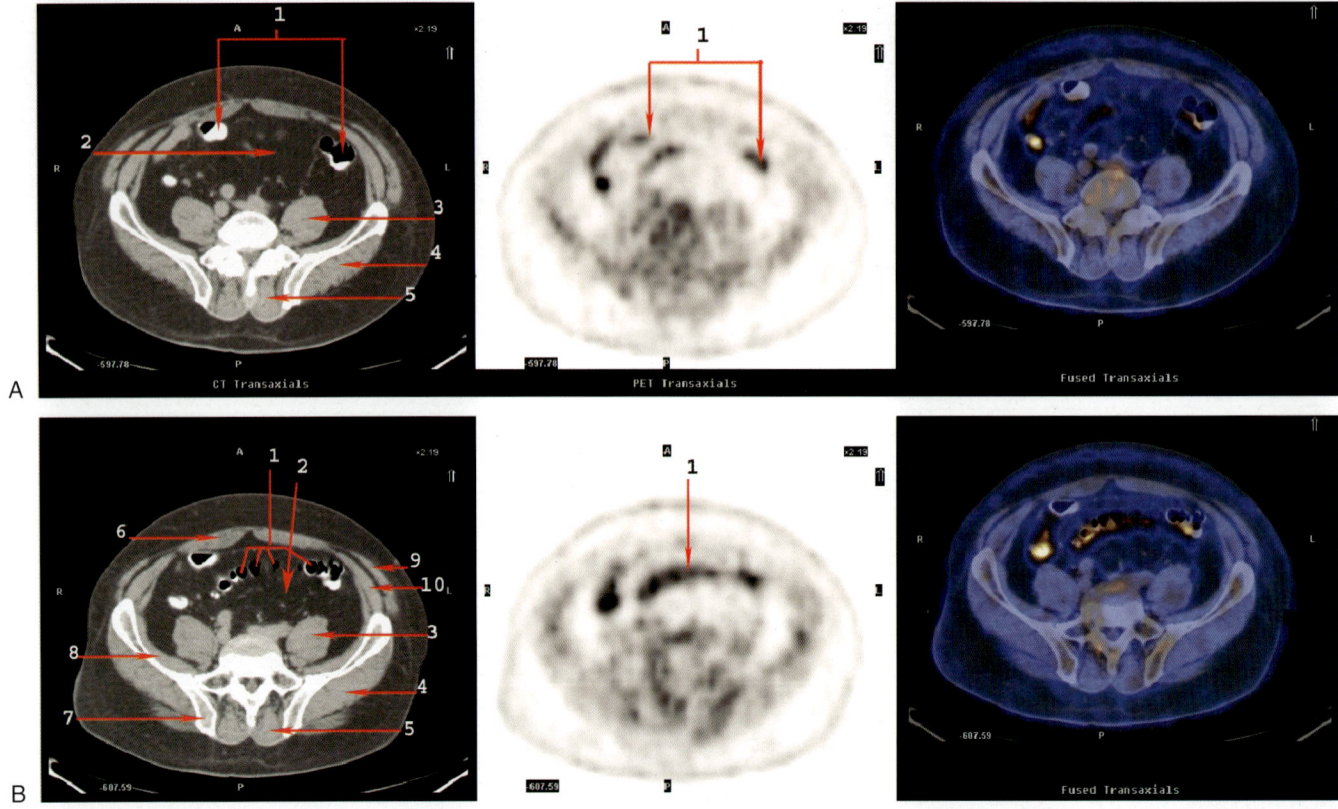

FIGURE 1-79: (A) Male Pelvis Transaxial 3 (B) Male Pelvis Transaxial 4

1. Transverse colon
2. Mesenteric fat
3. Psoas muscle
4. Gluteus medius muscle

5. Multifidus muscle
6. Rectus abdominis muscle
7. Iliac tuberosity
8. Iliacus muscle

9. External oblique muscle
10. Internal oblique muscle

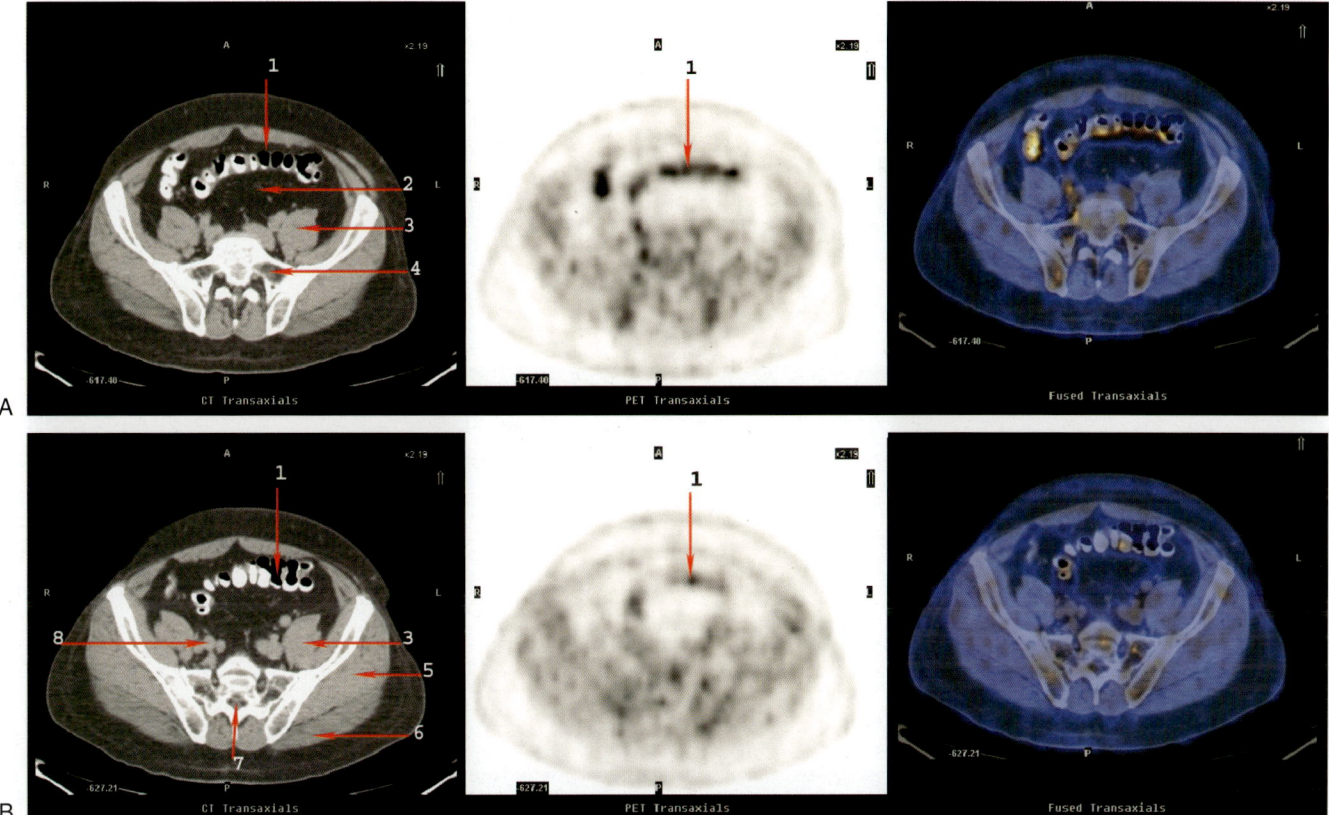

FIGURE 1-80: (A) Male Pelvis Transaxial 5 (B) Male Pelvis Transaxial 6

1. Transverse colon
2. Mesenteric fat
3. Psoas muscle
4. Sacral ala
5. Gluteus medius muscle
6. Gluteus maximus muscle
7. Sacral canal
8. Common iliac vein

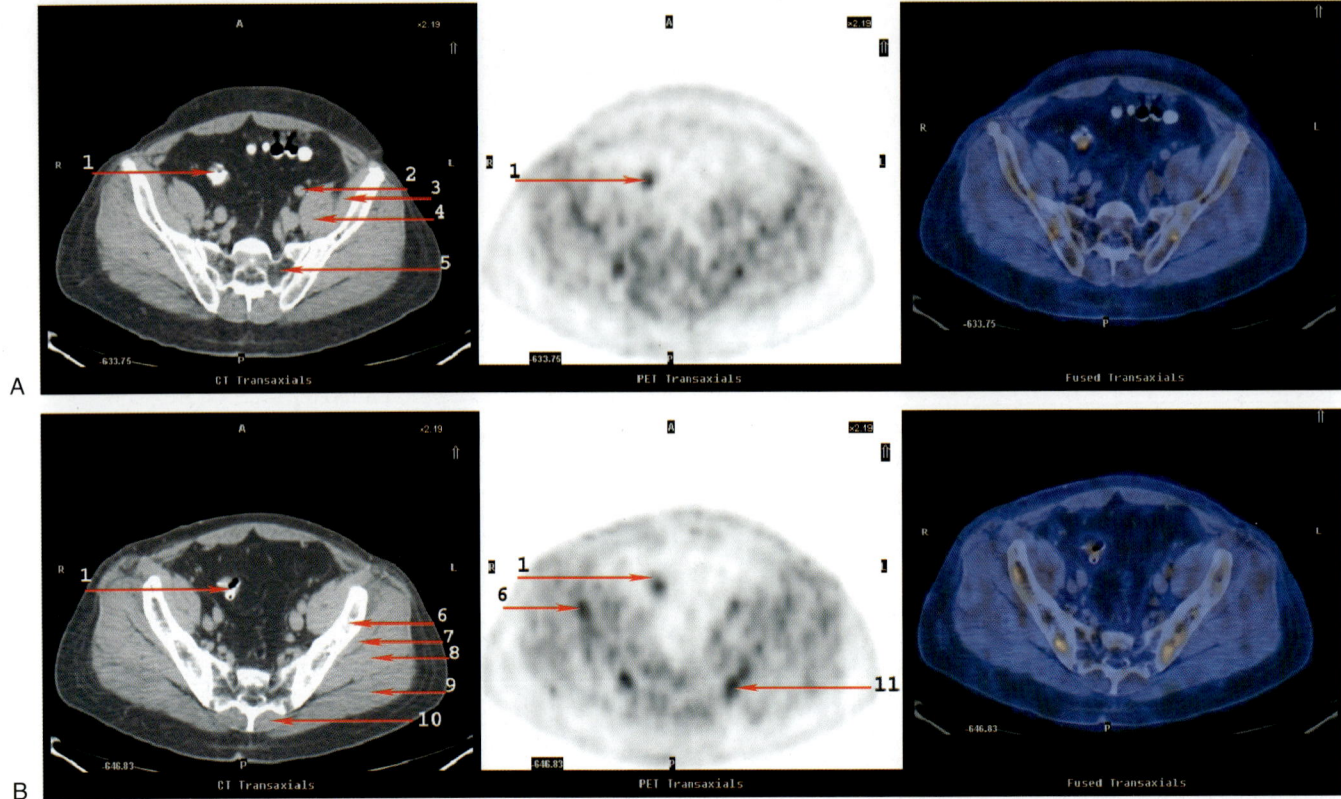

FIGURE 1-81: (A) Male Pelvis Transaxial 7 (B) Male Pelvis Transaxial 8

1. Colon
2. External iliac artery
3. Iliacus muscle
4. Psoas muscle

5. Sacral ala
6. Ilium
7. Gluteus minimus muscle
8. Gluteus medius muscle

9. Gluteus maximus muscle
10. Multifidus muscle
11. Iliac tuberosity

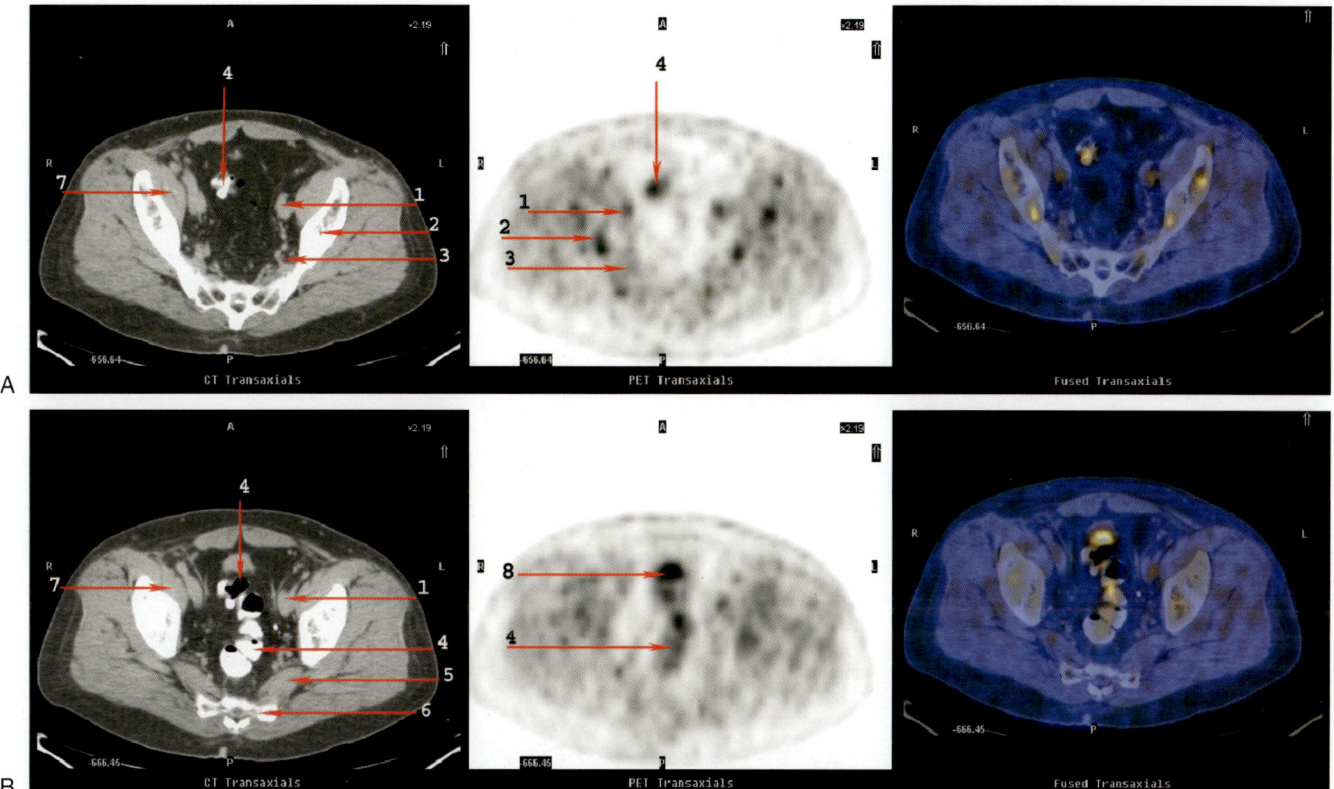

FIGURE 1-82: (A) Male Pelvis Transaxial 9 (B) Male Pelvis Transaxial 10

1. External iliac vein
2. Ilium
3. Internal iliac vein
4. Sigmoid colon
5. Piriformis muscle
6. Sacrum
7. Iliopsoas muscle
8. Bladder

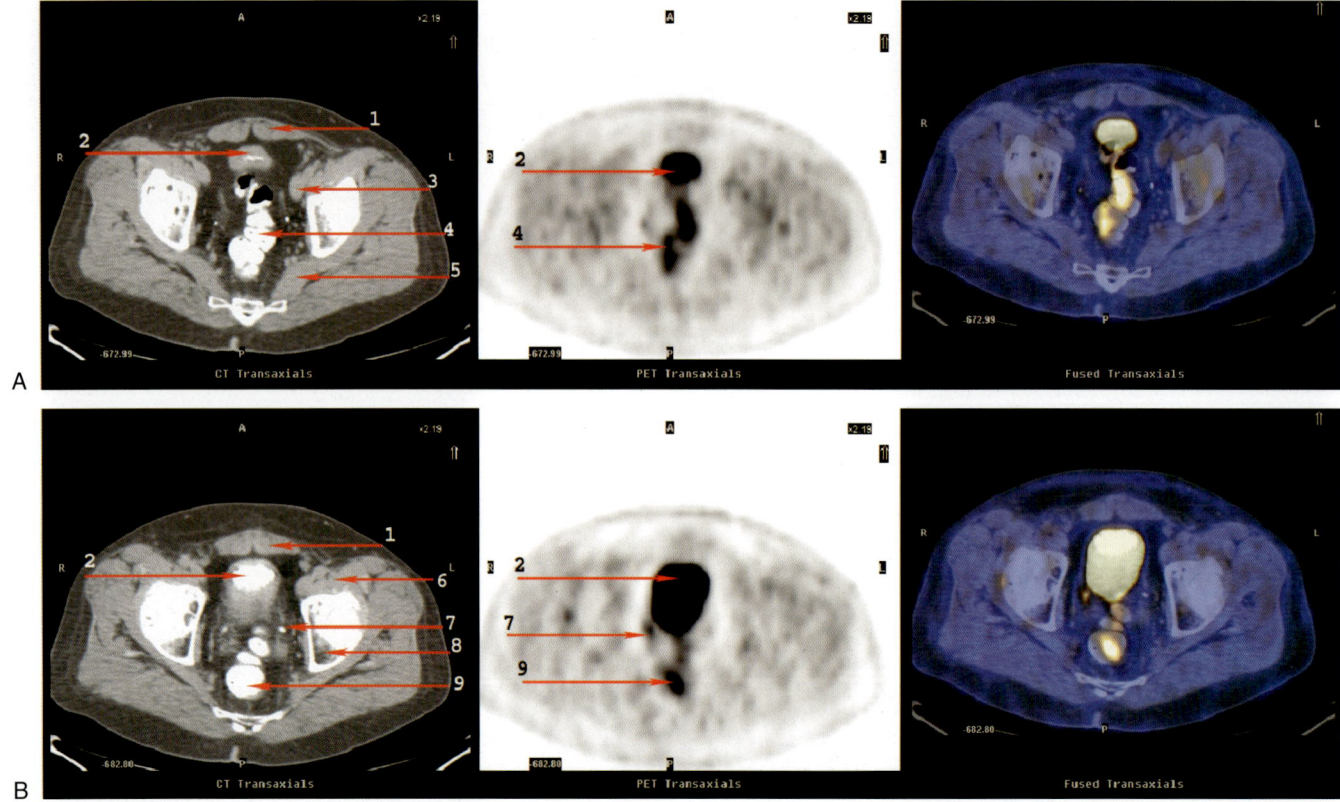

FIGURE 1-83: (A) Male Pelvis Transaxial 11 (B) Male Pelvis Transaxial 12

1. Rectus abdominis muscle
2. Bladder
3. External iliac vein
4. Sigmoid colon
5. Piriformis muscle
6. Iliopsoas muscle
7. Ureter
8. Ilium (posterior lip of acetabulum)
9. Rectum

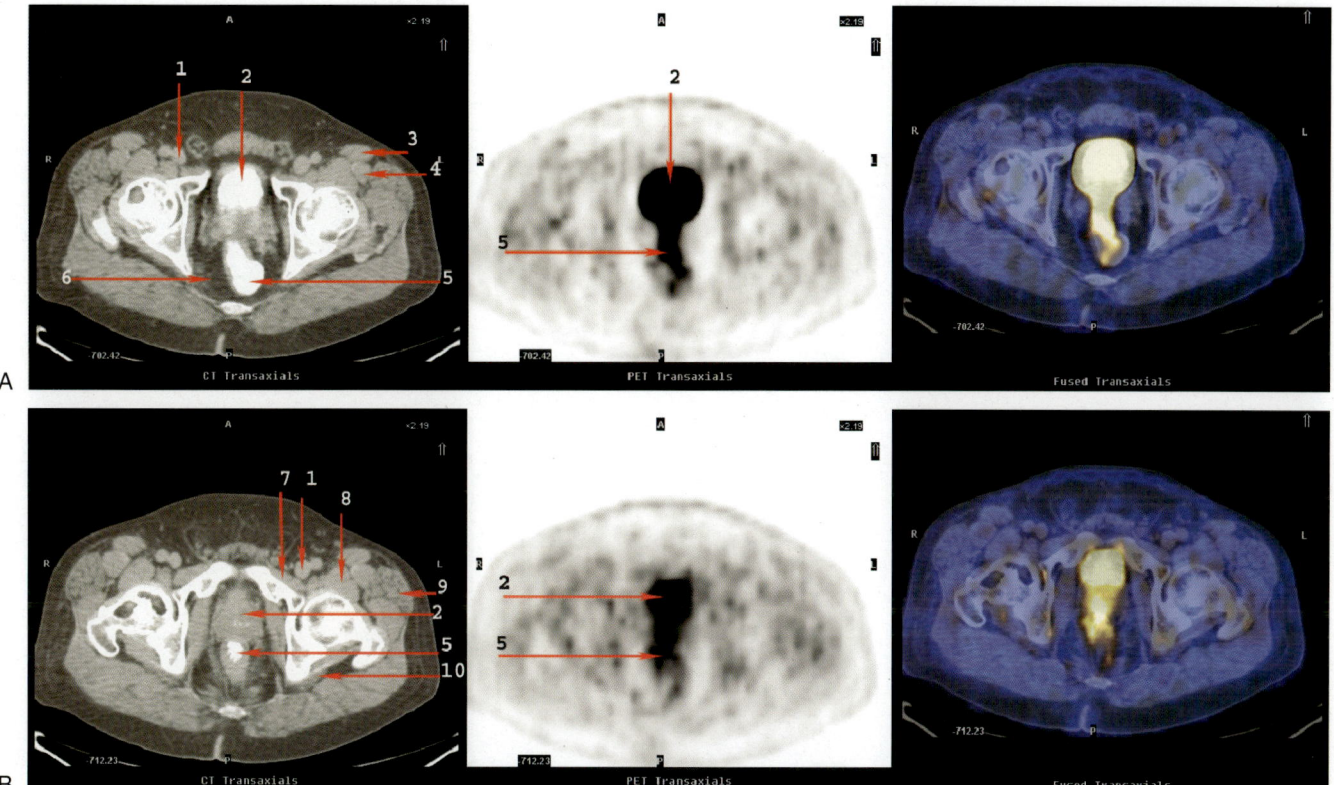

FIGURE 1-84: (A) Male Pelvis Transaxial 13 (B) Male Pelvis Transaxial 14

1. Common femoral vein
2. Bladder
3. Sartorius muscle
4. Rectus femoris muscle
5. Rectum
6. Ischiorectal fossa (fat)
7. Pectineus muscle
8. Iliopsoas muscle
9. Tensor fasciae latae muscle
10. Sciatic nerve

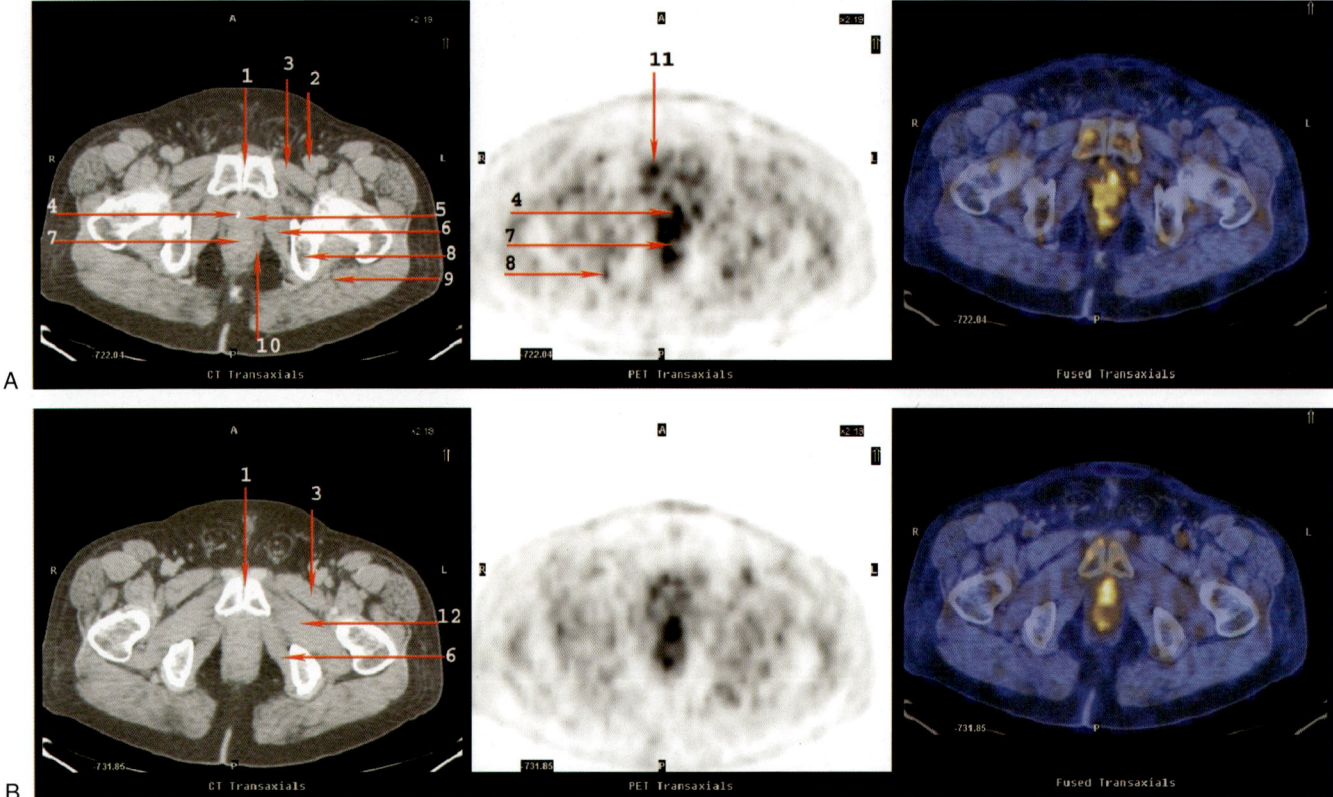

FIGURE 1-85: (A) Male Pelvis Transaxial 15 (B) Male Pelvis Transaxial 16

1. Symphysis pubis
2. Common femoral vein
3. Pectineus muscle
4. Prostatic urethra
5. Prostate
6. Obturator internus muscle
7. Rectum
8. Ischial tuberosity
9. Sciatic nerve
10. Levator ani muscle
11. Right pubis
12. Obturator externus muscle

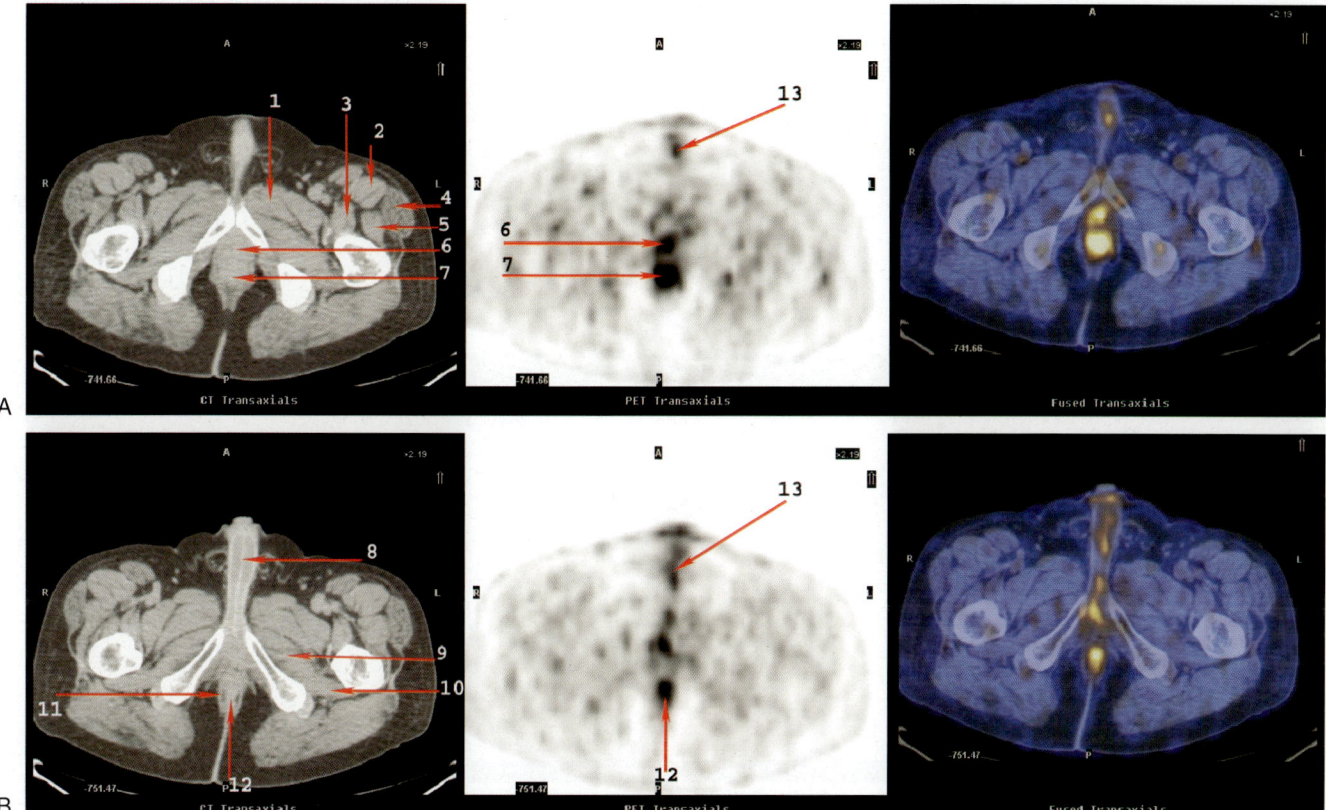

FIGURE 1-86: (A) Male Pelvis Transaxial 17 (B) Male Pelvis Transaxial 18

1. Pectineus muscle
2. Sartorius muscle
3. Iliacus muscle
4. Tensor fasciae latae muscle
5. Rectus femoris muscle
6. Prostatic urethra
7. Rectum
8. Corpora cavernosum and spongiosum
9. Obturator externus muscle
10. Quadratus femoris muscle
11. Levator ani muscle
12. Anus
13. Penus

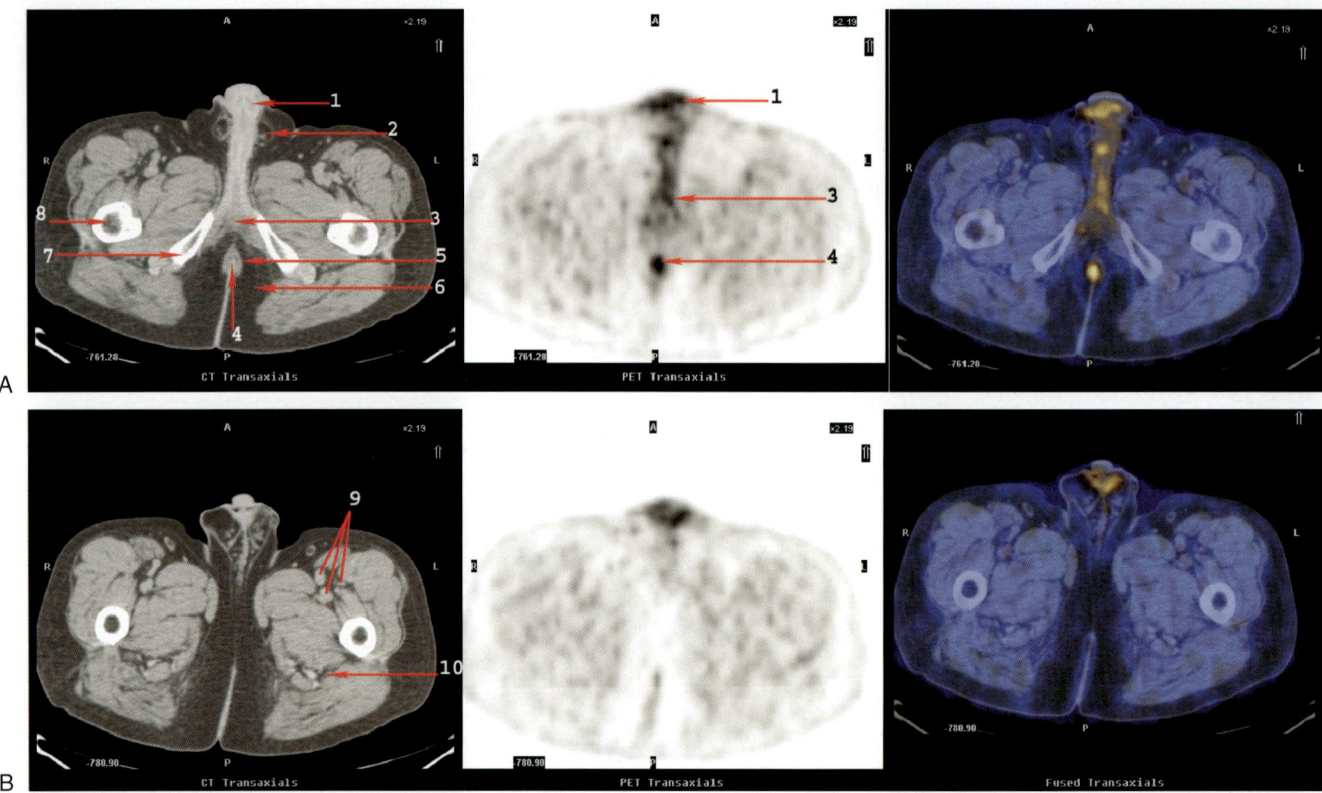

FIGURE 1-87: (A) Male Pelvis Transaxial 19 (B) Male Pelvis Transaxial 20

1. Corpora cavernosum and
   spongiosum
2. Spermatic cord
3. Cavernous urethra

4. Anus
5. Levator ani muscle
6. Ischiorectal fat
7. Ischium

8. Femur
9. Superficial femoral artery, vein and
   nerve
10. Sciatic nerve

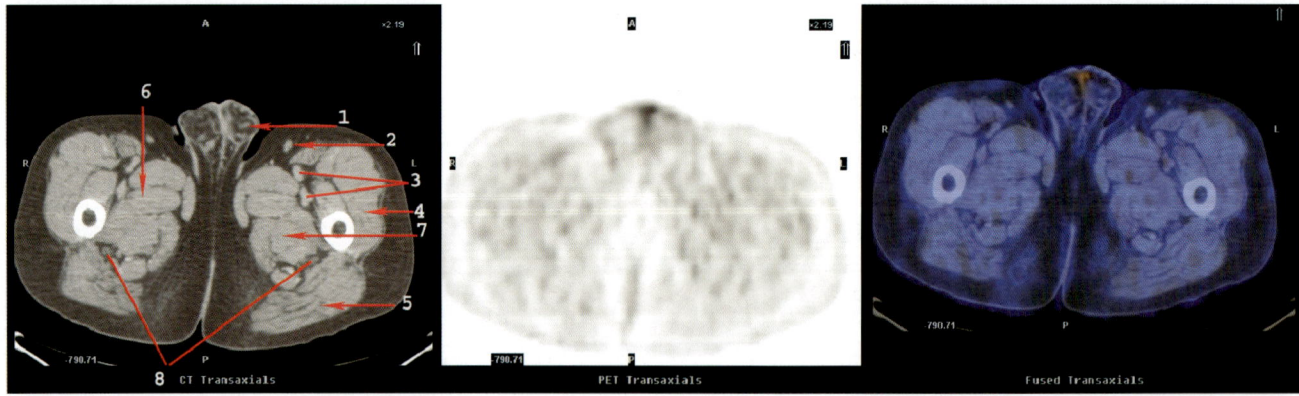

FIGURE 1-88: Male Pelvis Transaxial 21

1. Ductus deferens and spermatic cord
2. Greater saphenous vein
3. Superficial femoral artery and vein

4. Vastus lateralis muscle
5. Gluteus maximus muscle
6. Adductor longus muscle

7. Adductor brevis and minimus
   muscle
8. Sciatic nerve

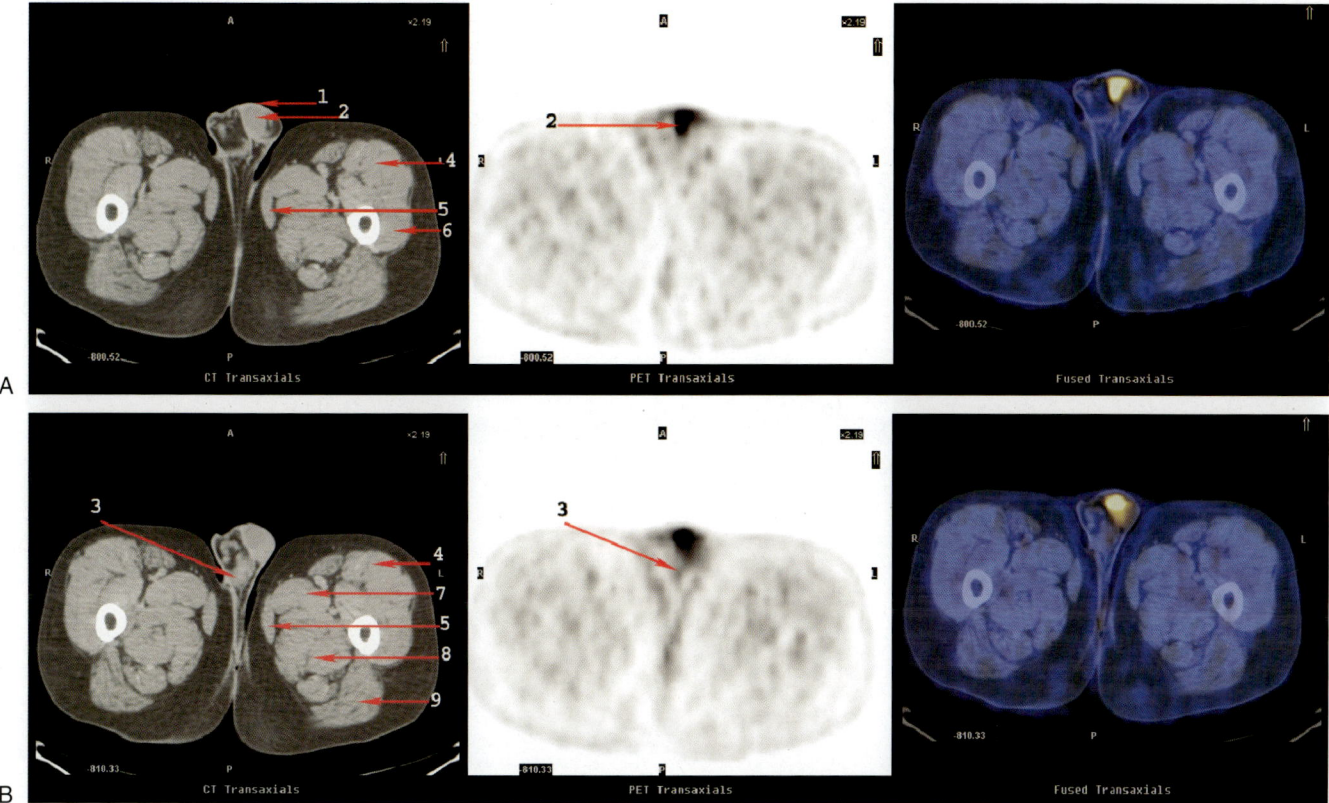

FIGURE 1-89: (A) Male Pelvis Transaxial 22 (B) Male Pelvis Transaxial 23

1. Scrotum
2. Testis
3. Penile bulb

4. Rectus femoris muscle
5. Gracilis muscle
6. Vastus lateralis muscle

7. Adductor longus muscle
8. Adductor brevis muscle
9. Gluteus maximus muscle

Section 5: Female Pelvis

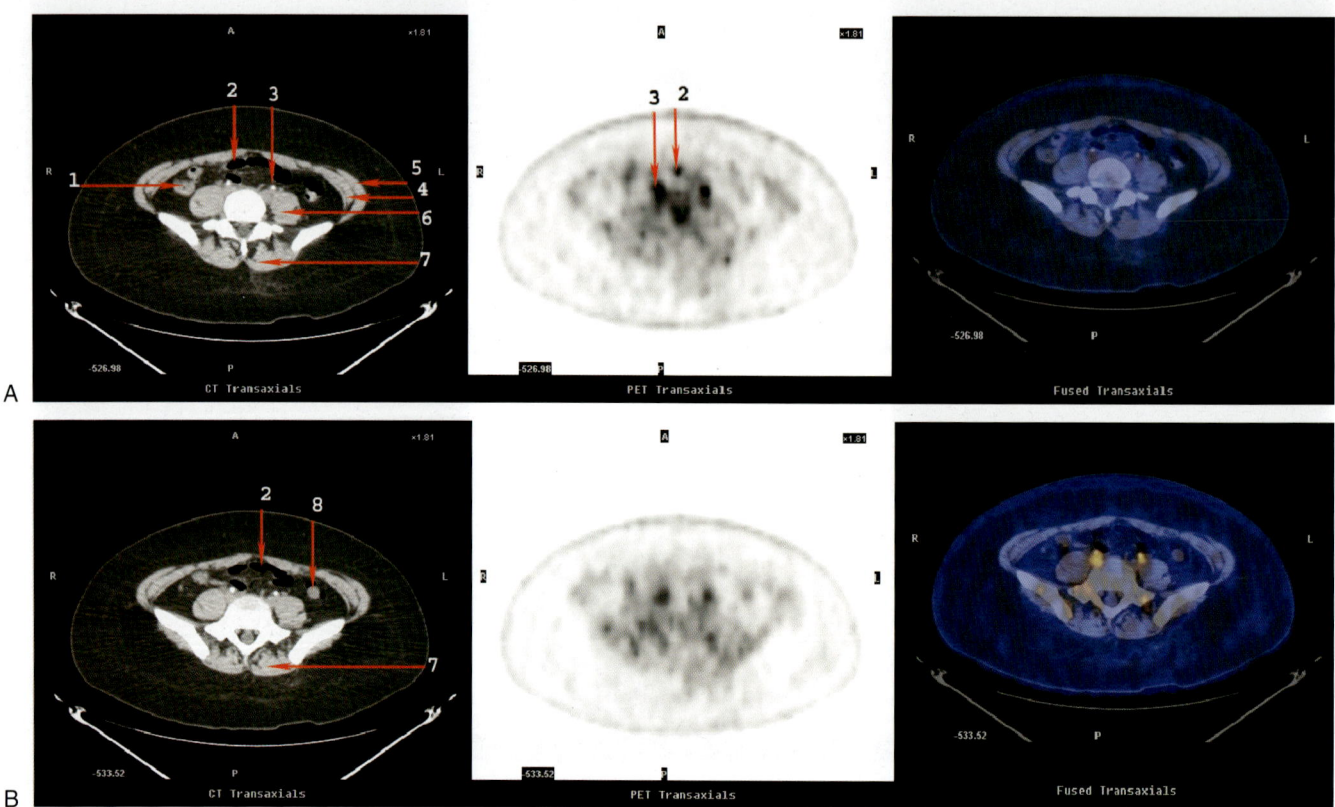

FIGURE 1-90: (A) Female Pelvis Transaxial 1 (B) Female Pelvis Transaxial 2

1. Ascending colon
2. Ileum
3. Ureter
4. Internal oblique muscle
5. External oblique muscle
6. Psoas muscle
7. Multifidus muscles
8. Descending colon

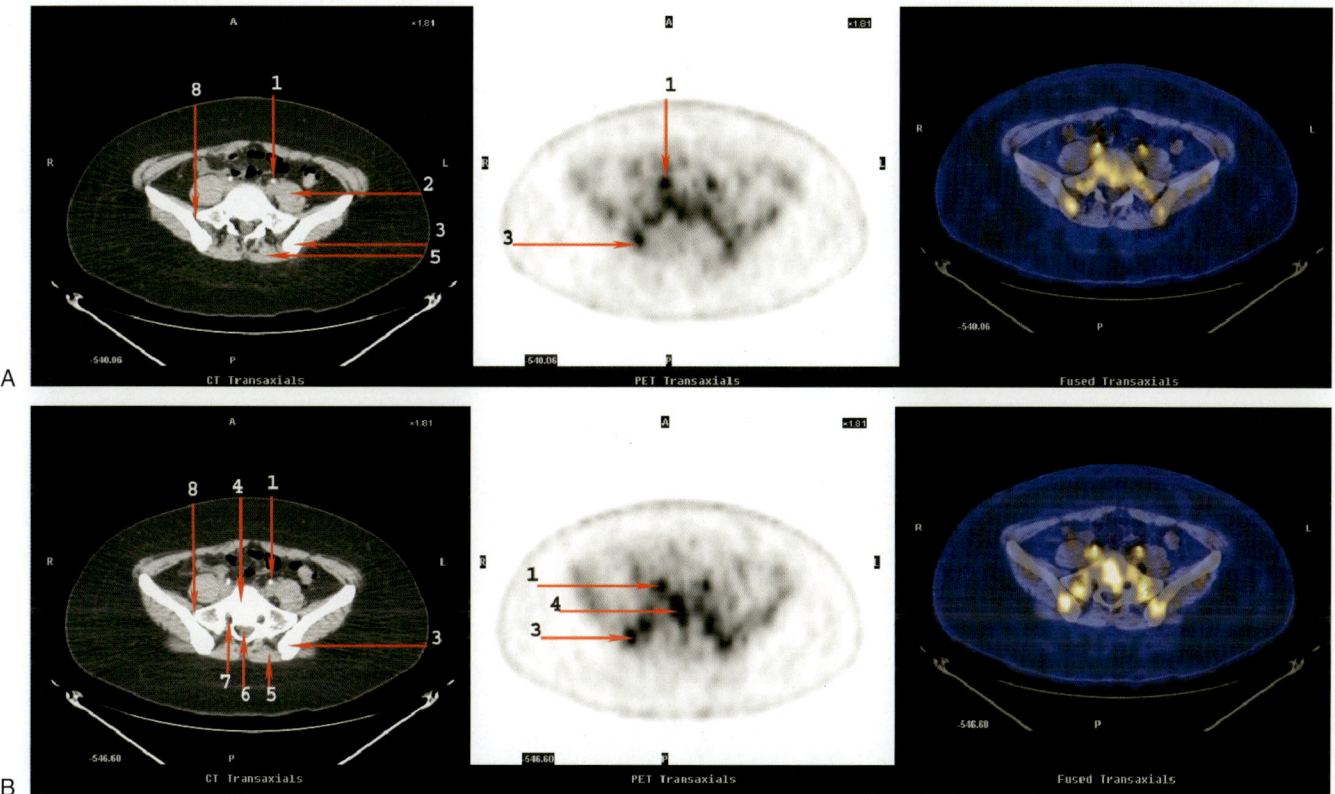

FIGURE 1-91: (A) Female Pelvis Transaxial 3 (B) Female Pelvis Transaxial 4

1. Ureter
2. Psoas muscle
3. Iliac tuberosity

4. Sacrum
5. Multifidus muscle
6. Sacral canal

7. Sacral foramen
8. Sacroiliac joint

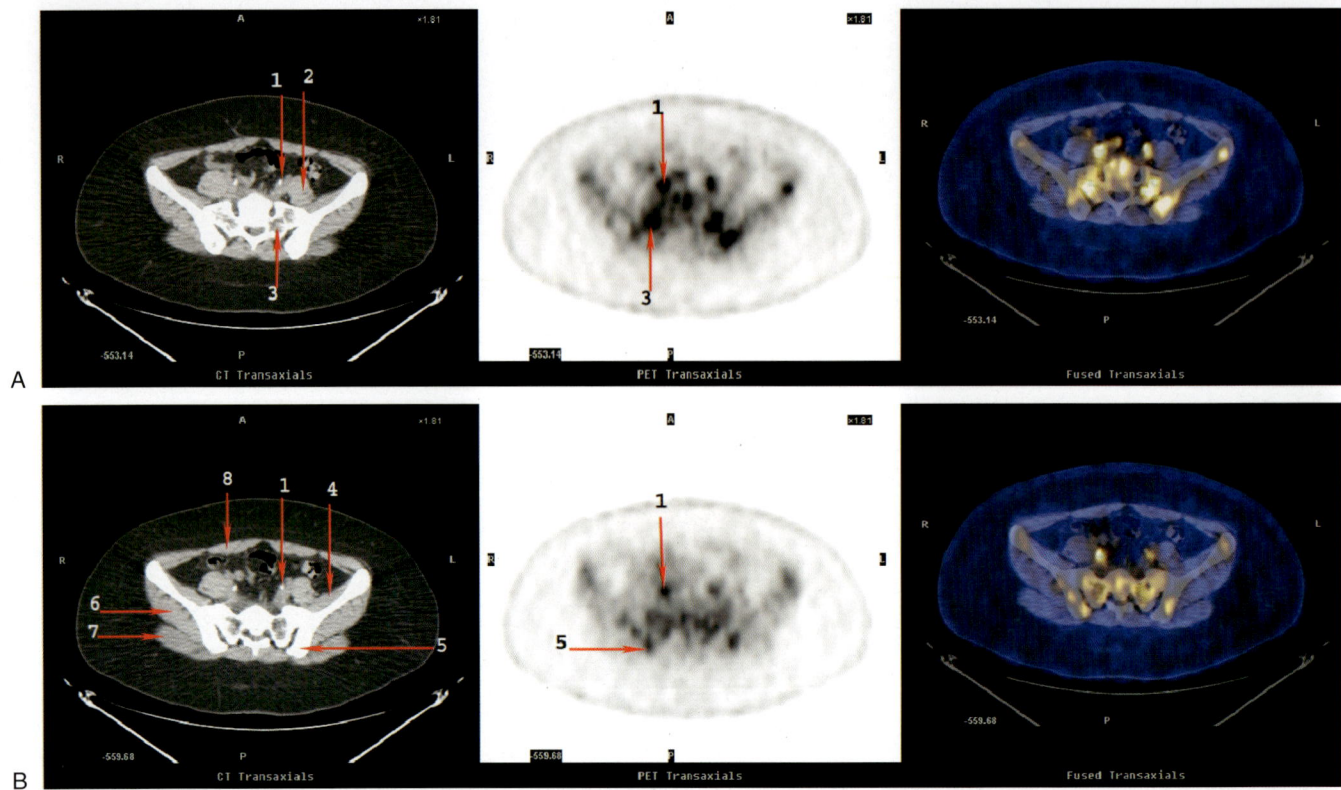

FIGURE 1-92: (A) Female Pelvis Transaxial 5 (B) Female Pelvis Transaxial 6

1. Ureter
2. Psoas muscle
3. Sacral ala
4. Iliacus muscle
5. Iliac tuberosity
6. Gluteus medius muscle
7. Gluteus maximus muscle
8. Rectus abdominis muscle

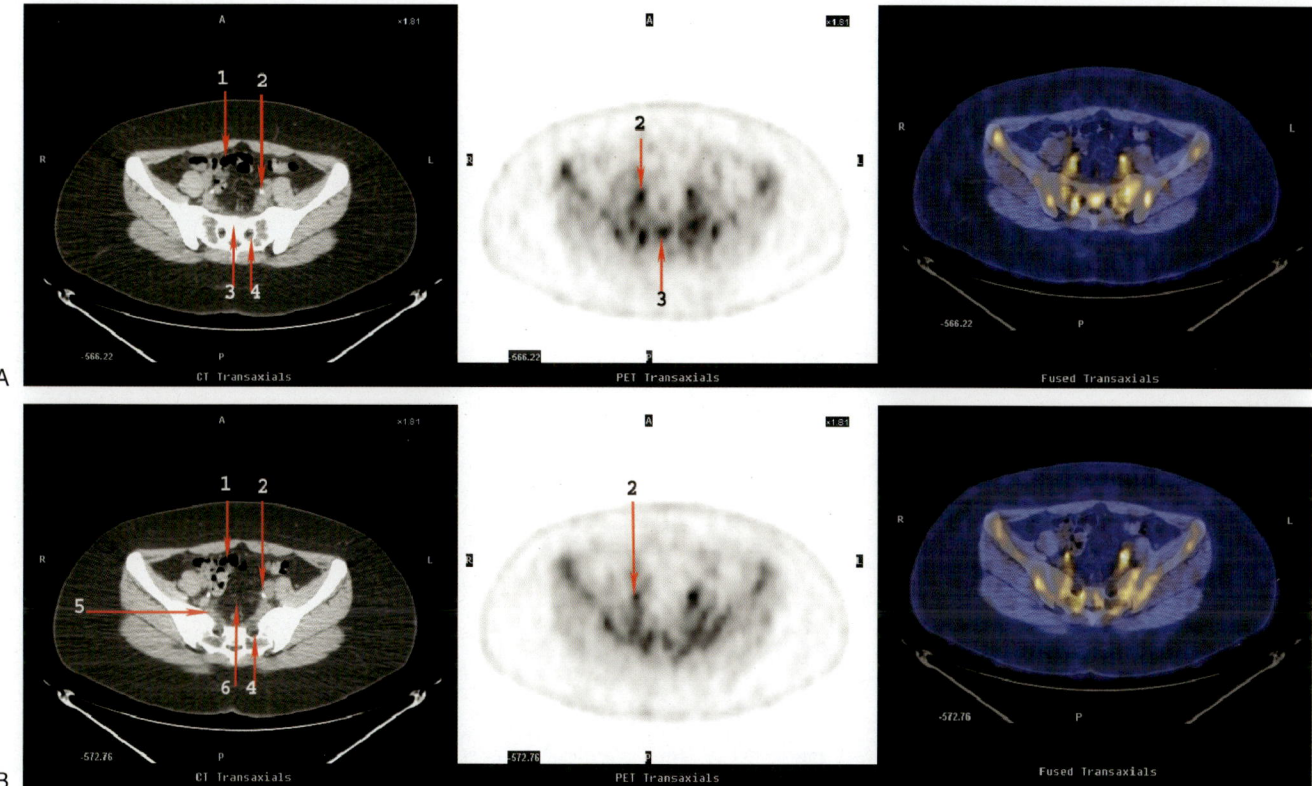

FIGURE 1-93: (A) Female Pelvis Transaxial 7 (B) Female Pelvis Transaxial 8

1. Ileum
2. Ureter
3. Sacrum
4. Sacral foramen
5. Common iliac artery and vein
6. Mesenteric fat

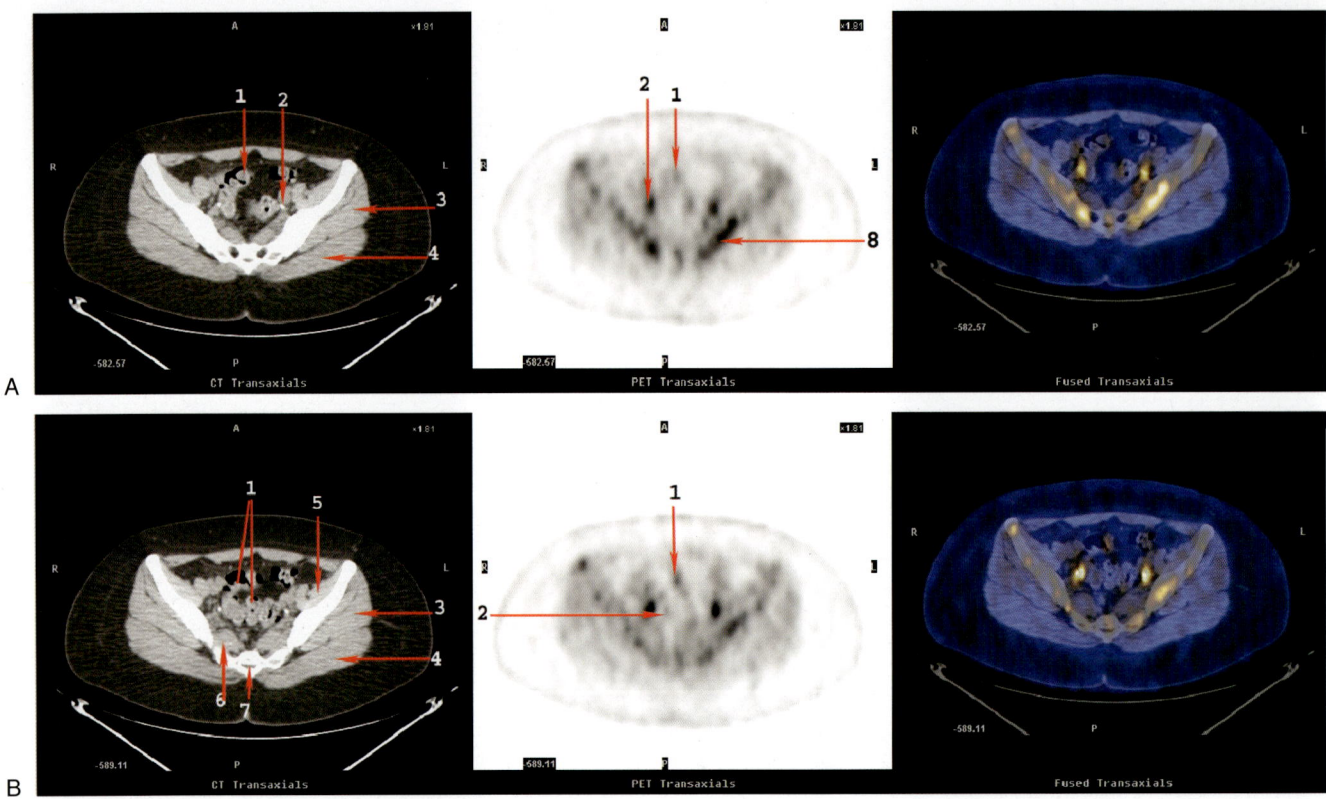

FIGURE 1-94: (A) Female Pelvis Transaxial 9 (B) Female Pelvis Transaxial 10

1. Ileum
2. Ureter
3. Gluteus medius muscle

4. Gluteus maximus muscle
5. Iliacus muscle
6. Psoas muscle

7. Sacral canal
8. Iliac crest

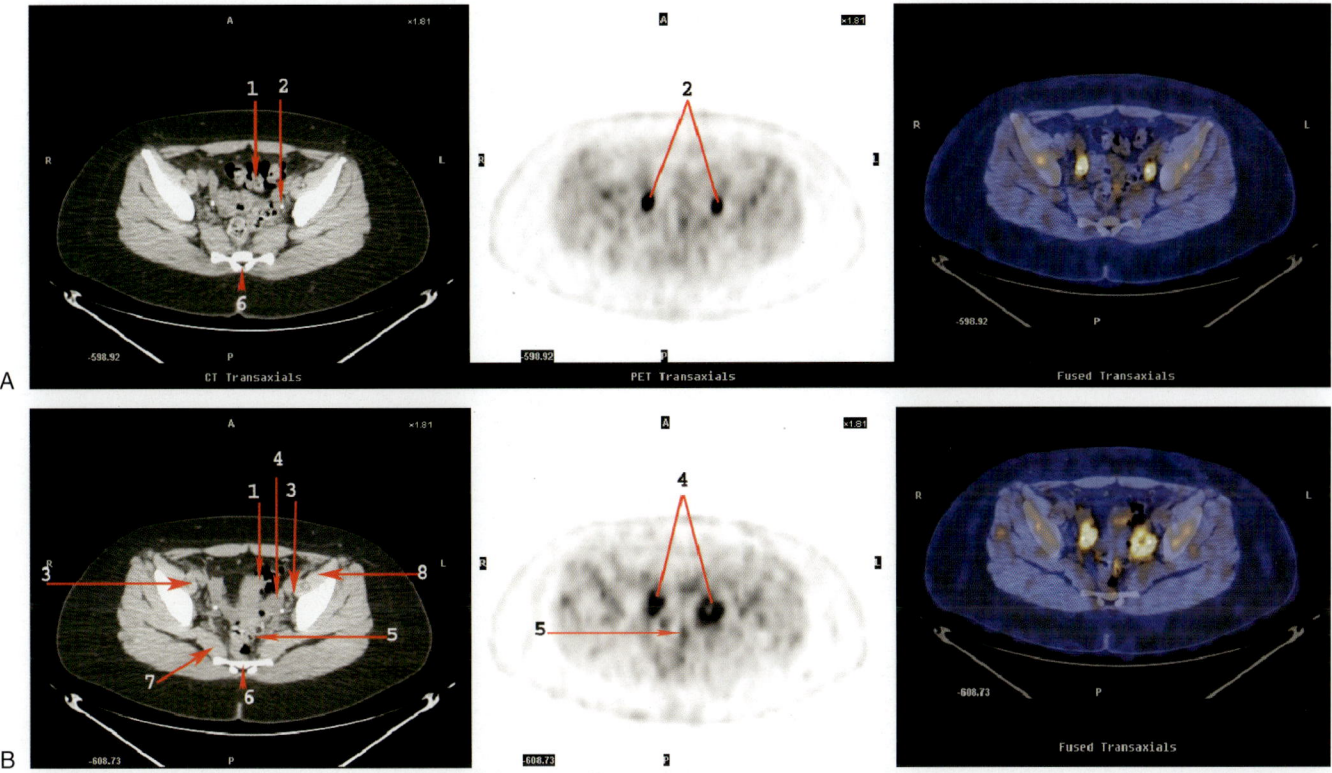

FIGURE 1-95: (A) Female Pelvis Transaxial 11 (B) Female Pelvis Transaxial 12

1. Ileum
2. Ureter
3. External iliac artery and vein
4. Ovary
5. Sigmoid
6. Sacrum
7. Piriformis muscle

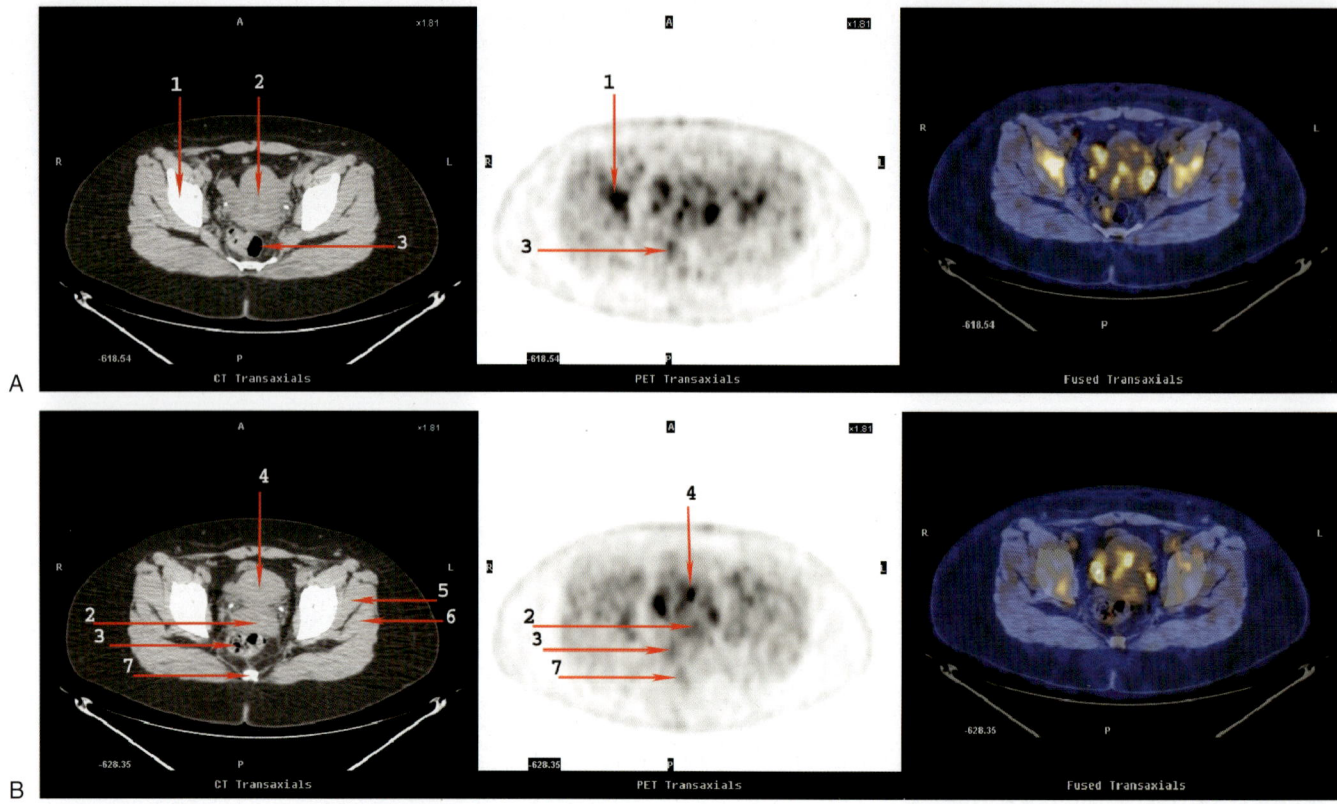

FIGURE 1-96: (A) Female Pelvis Transaxial 13 (B) Female Pelvis Transaxial 14

1. Lower ilium
2. Endometrial cavity of uterus
3. Rectum
4. Bladder
5. Gluteus minimus muscle
6. Gluteus medius muscle
7. Coccyx

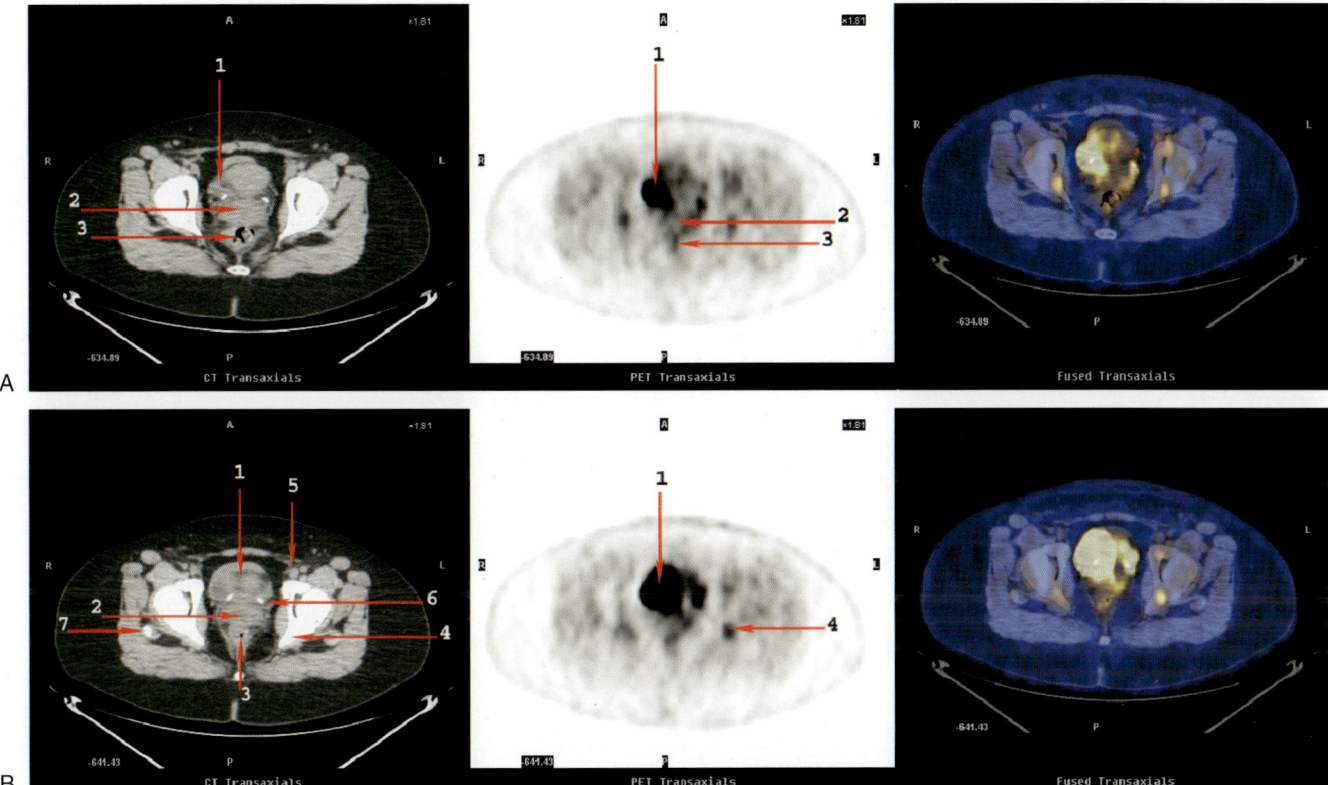

FIGURE 1-97: (A) Female Pelvis Transaxial 15 (B) Female Pelvis Transaxial 16

1. Bladder
2. Uterine cervix
3. Rectum
4. Ischium
5. Common femoral vein
6. Ureter
7. Greater trochanter of femur

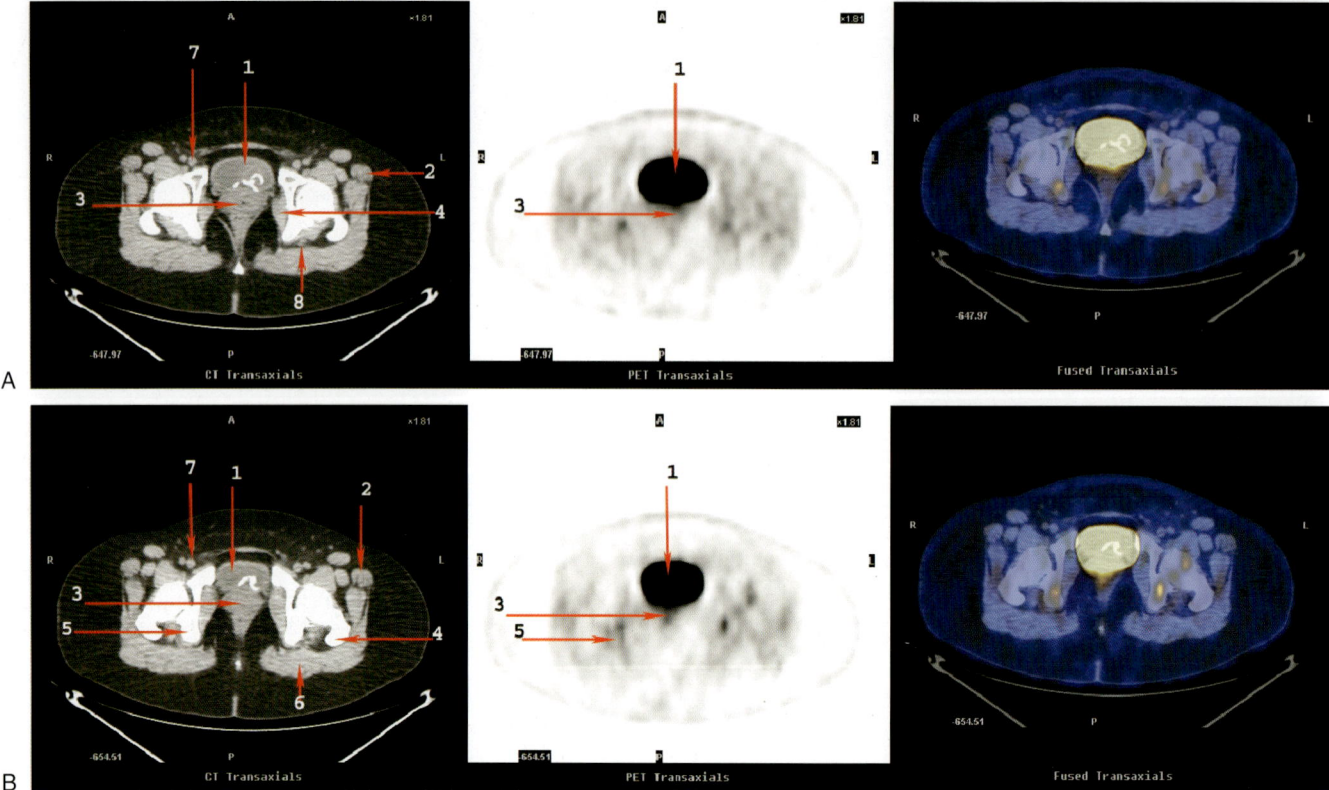

FIGURE 1-98: (A) Female Pelvis Transaxial 17 (B) Female Pelvis Transaxial 18

1. Bladder
2. Tensor fasciae latae muscle
3. Vagina
4. Obturator internus muscle
5. Ischium (posterior lip of acetabulum)
6. Gluteus maximus muscle
7. Femoral artery
8. Sciatic nerve

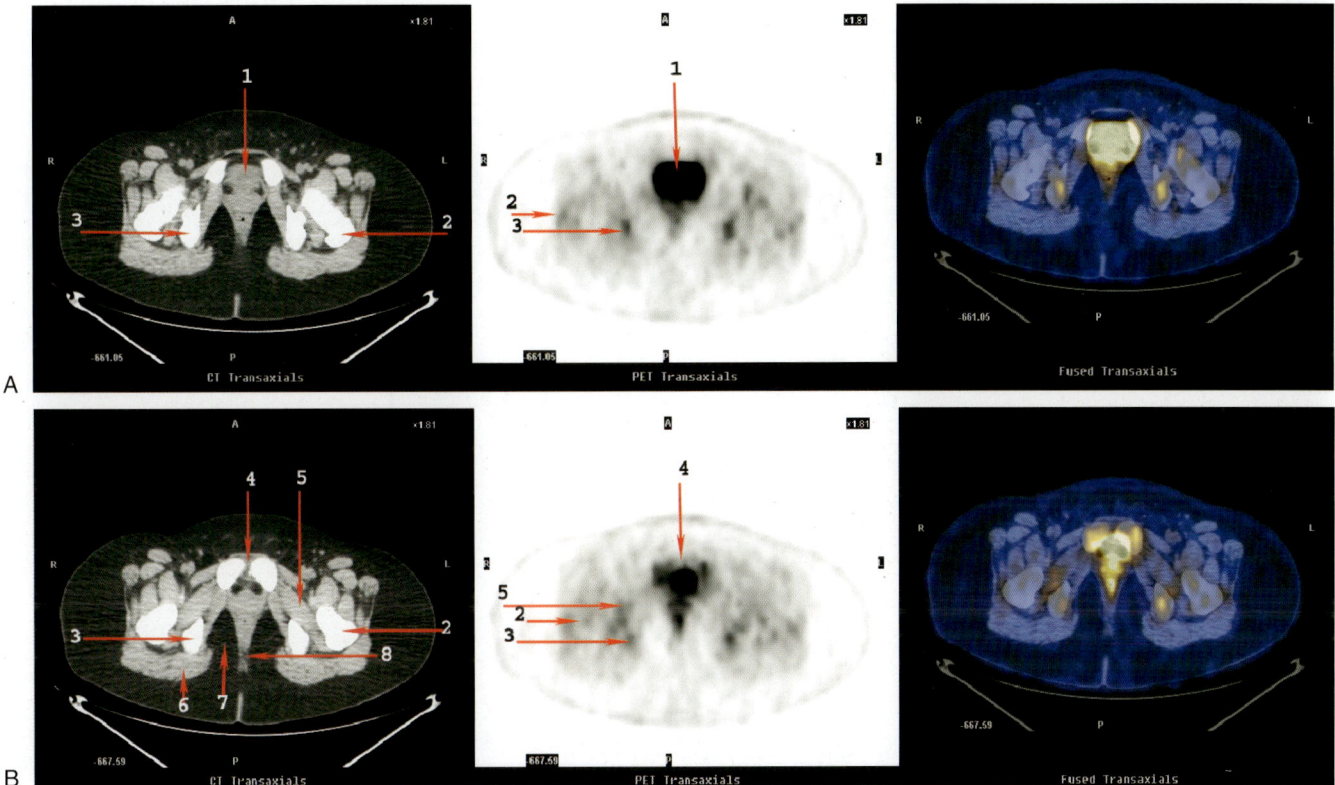

FIGURE 1-99: (A) Female Pelvis Transaxial 19 (B) Female Pelvis Transaxial 20

1. Bladder
2. Greater trochanter
3. Ischial tuberosity
4. Symphysis pubis
5. Obturator externus muscle
6. Gluteus maximus muscle
7. Ischiorectal fat
8. Rectum

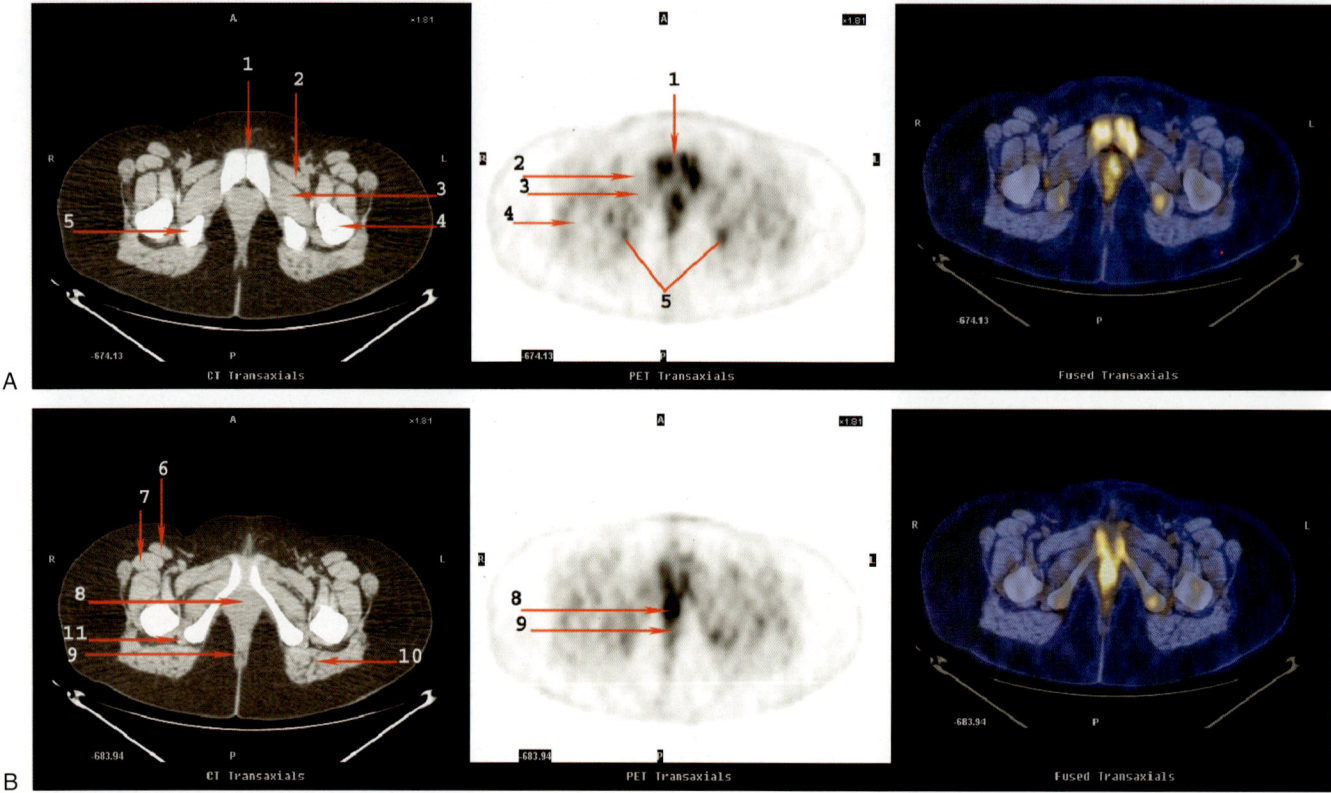

FIGURE 1-100: (A) Female Pelvis Transaxial 21 (B) Female Pelvis Transaxial 22

1. Symphysis pubis
2. Pectineus muscle
3. Obturator externus muscle
4. Femur
5. Ischial tuberosity
6. Sartorius muscle
7. Rectus femoris muscle
8. Bladder neck
9. Rectum
10. Gluteus maximus muscle
11. Sciatic nerve

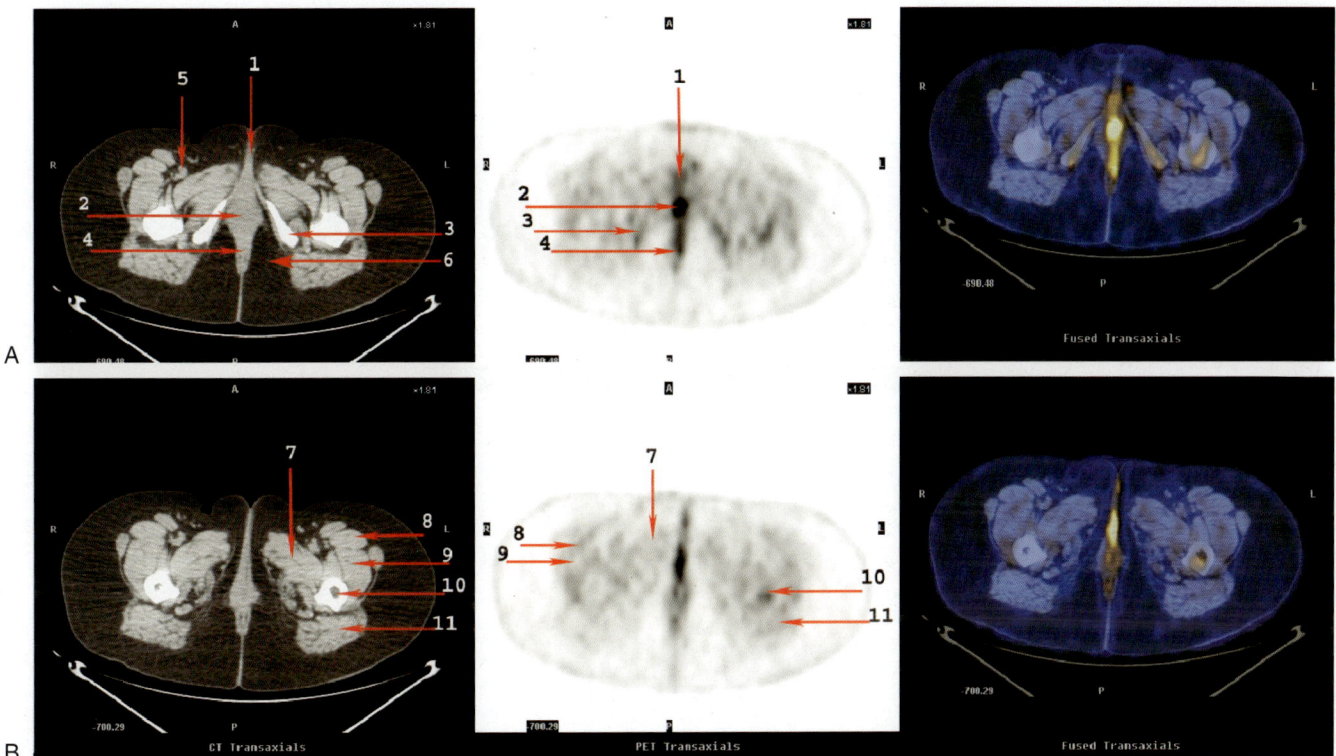

FIGURE 1-101: (A) Female Pelvis Transaxial 23 (B) Female Pelvis Transaxial 24

1. Urethra
2. Vagina
3. Ischium
4. Anal canal
5. Superficial femoral vein
6. Ischiorectal fossa
7. Adductor muscle
8. Rectus femoris muscle
9. Vastus lateralis muscle
10. Femur
11. Gluteus maximus muscle

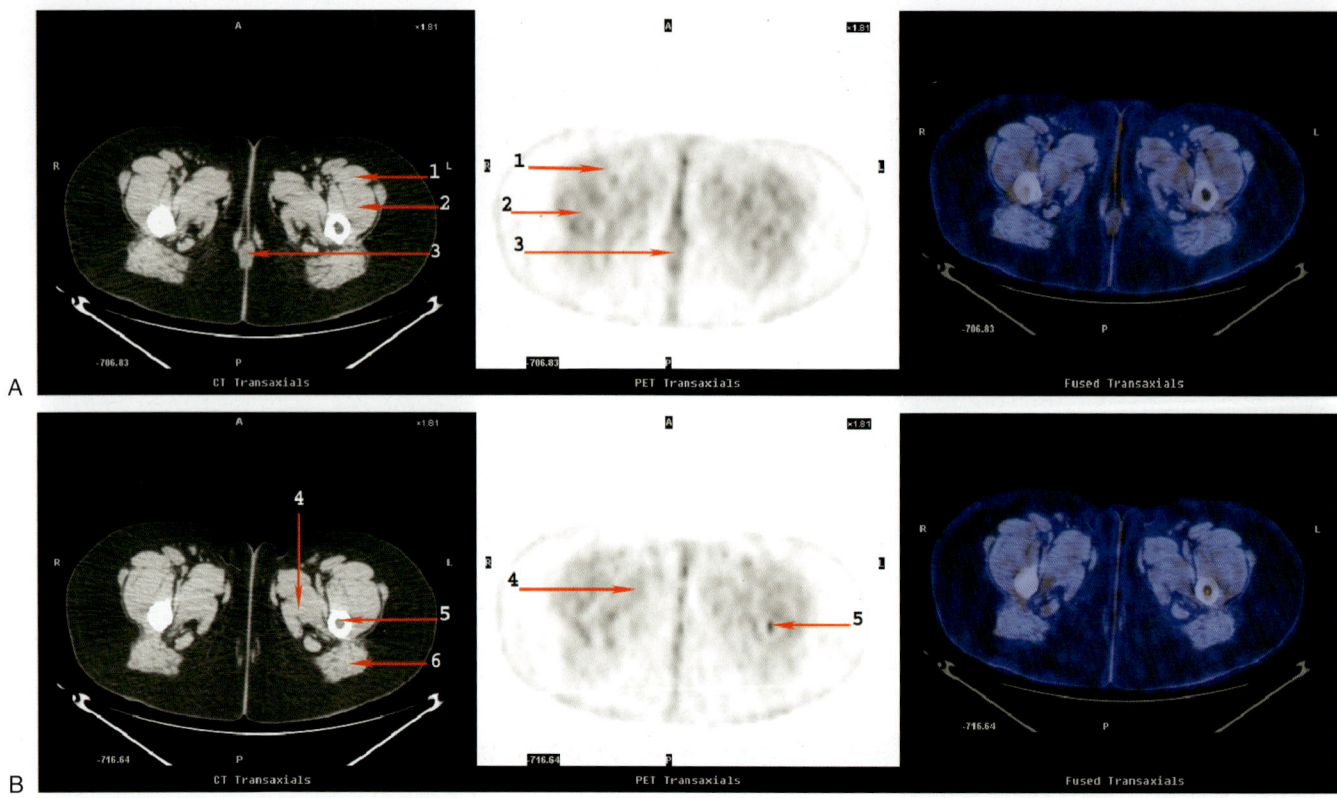

FIGURE 1-102: (A) Female Pelvis Transaxial 25 (B) Female Pelvis Transaxial 26

1. Rectus femoris muscle
2. Vastus lateralis muscle
3. Anal canal
4. Adductor magnus muscle
5. Femur
6. Gluteus maximus muscle

2 Non-FDG PET/CT

Increased transport and utilization of amino acids are common in cancers. The use of L-methionine in cancer imaging is based on this conversant and on the increased activity of the transmethylation in some cancers. There is normally substantial uptake of C-11 methionine in the pancreas, salivary glands, liver, and kidneys. As a natural amino acid, there is some metabolism of L-methionine in the bloodstream. This tracer mostly has been used in imaging of brain tumors, head and neck cancers, lymphoma, and lung cancers.[1] C-11 methionine typically shows fluorodeoxyglucose (FDG). The diagnostic accuracy of positron emission tomography (PET)/computed tomography (CT) seems to be superior to that of PET alone.

Cellular proliferation is one of the key factors determining tumor biology and response to therapy. Thymidine labeled with C-11 has been tested as a PET tracer of thymidine incorporation into DNA.[2]

C-11 acetate has been proposed as an alternative for the assessment for residual oxidative anabolism in patients with severe coronary artery disease. It also may be useful in the assessment of tissue viability in chronic stable coronary artery disease.[3]

The role of PET in the evaluation and management of skeletal disorders is increasing. F-18 fluoride can be used to evaluate bone metastasis both qualitatively and, for a number of focal and systemic skeletal disorders, quantitatively.[4] PET has the advantage of superior quantitative accuracy over planar or single photon emission computed tomography bone scintigraphy because of its ability to measure regional skeletal kinetic parameters in absolute numbers.

Section 1: PET with C-11 Methionine

Coronals

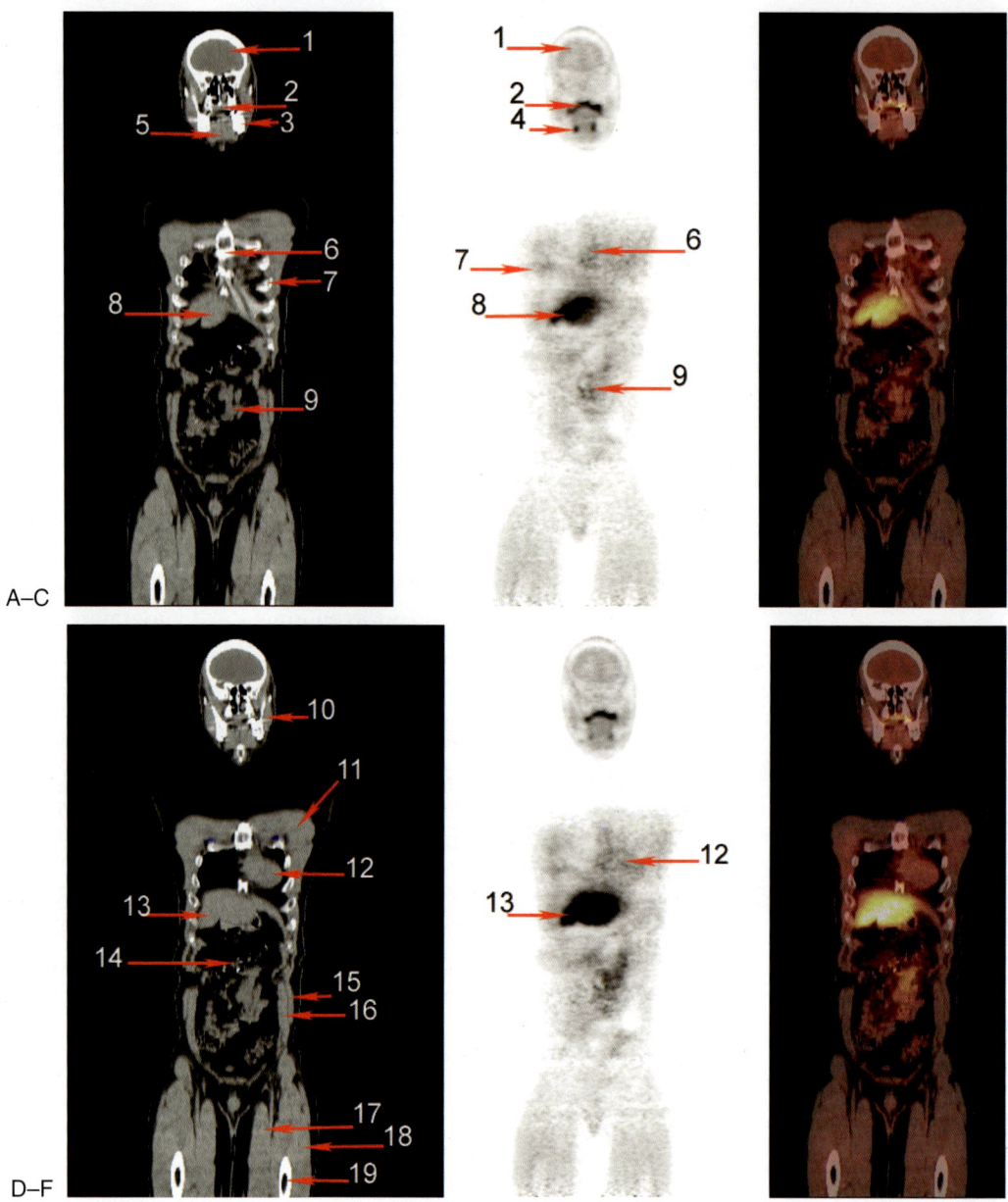

A–C

D–F

FIGURE 2-1:

1. Frontal gyrus
2. Soft palate
3. Mandible
4. Lingual tonsil
5. Tongue
6. Manubrium
7. Rib
8. Left hepatic lobe
9. Jejunum
10. Masseter muscle
11. Pectoralis major muscle
12. Right ventricle
13. Right hepatic lobe
14. Transverse colon
15. External oblique muscle
16. Internal oblique muscle
17. Adductor muscle
18. Vastus lateral muscle
19. Femur

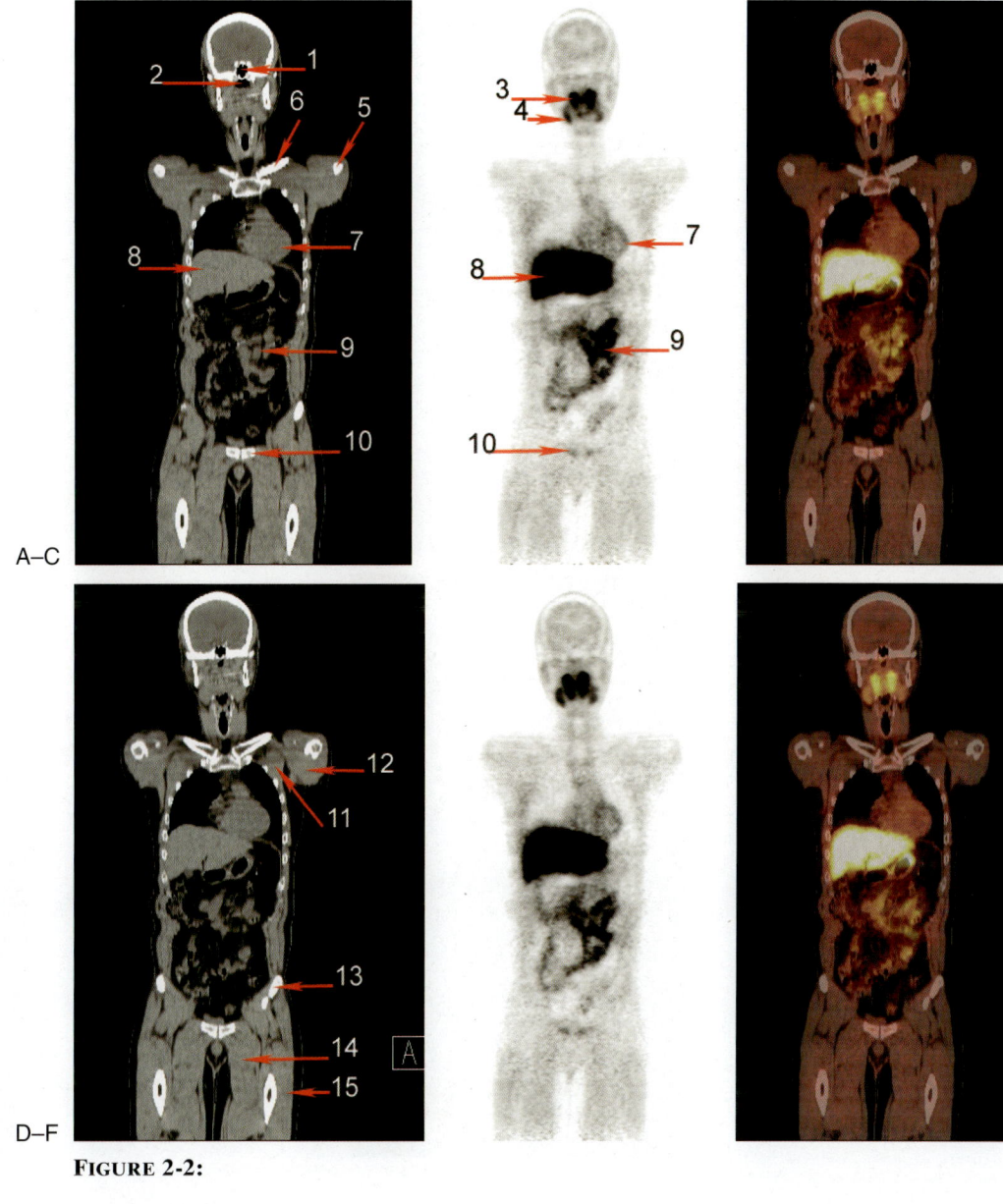

A–C

D–F

FIGURE 2-2:

1. Sphenoidal sinus
2. Nasal cavity
3. Lingual tonsil
4. Submandibular gland
5. Humeral head
6. Clavicle
7. Left ventricle
8. Right hepatic lobe
9. Jejunum
10. Superior ramus of pubis
11. Pectoralis muscle
12. Triceps brachii muscle
13. Iliac crest
14. Adductor muscles
15. Vastus lateralis muscle

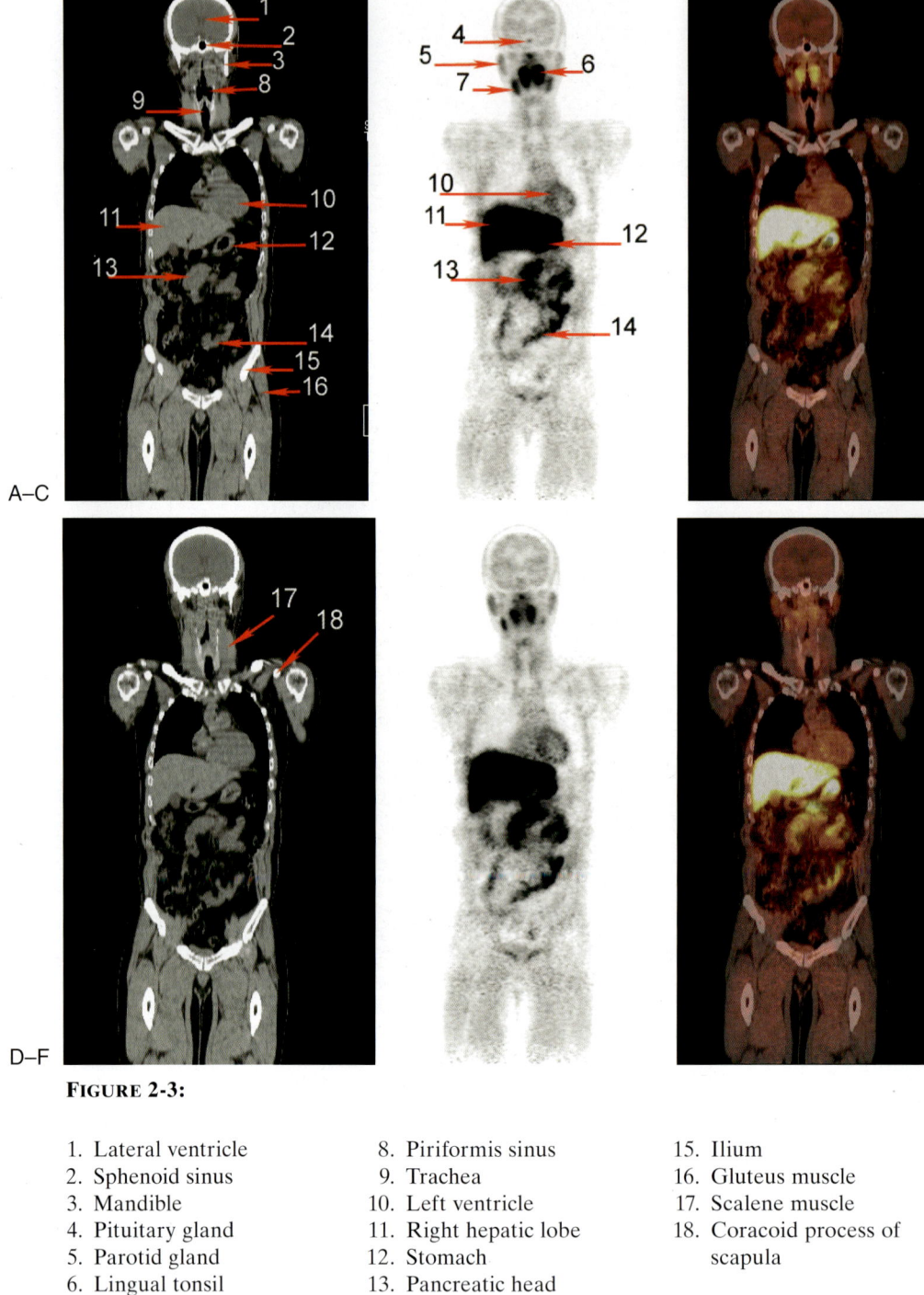

A–C

D–F

FIGURE 2-3:

1. Lateral ventricle
2. Sphenoid sinus
3. Mandible
4. Pituitary gland
5. Parotid gland
6. Lingual tonsil
7. Submandibular gland
8. Piriformis sinus
9. Trachea
10. Left ventricle
11. Right hepatic lobe
12. Stomach
13. Pancreatic head
14. Jejunum
15. Ilium
16. Gluteus muscle
17. Scalene muscle
18. Coracoid process of scapula

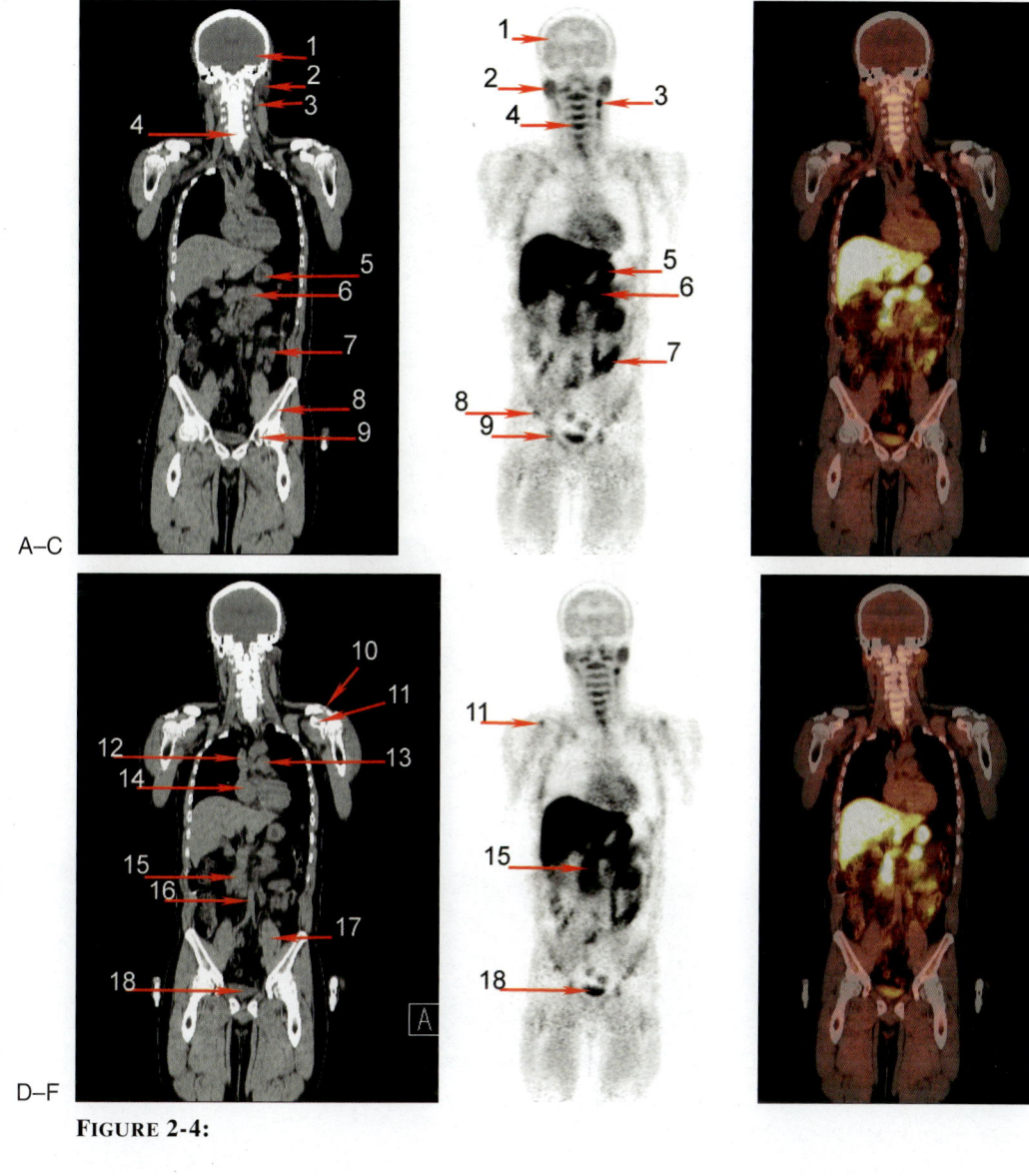

A–C

D–F

FIGURE 2-4:

1. Temporal gyrus
2. Parotid gland
3. Upper jugular node (level II)
4. Vertebral body
5. Stomach
6. Pancreatic body
7. Jejunum
8. Iliac wing
9. Acetabulum
10. Acromioclavicular joint
11. Coracoid process of scapula
12. Superior vena cava
13. Pulmonary artery
14. Right atrium
15. Pancreatic head
16. Abdominal aorta
17. Psoas muscle
18. Bladder

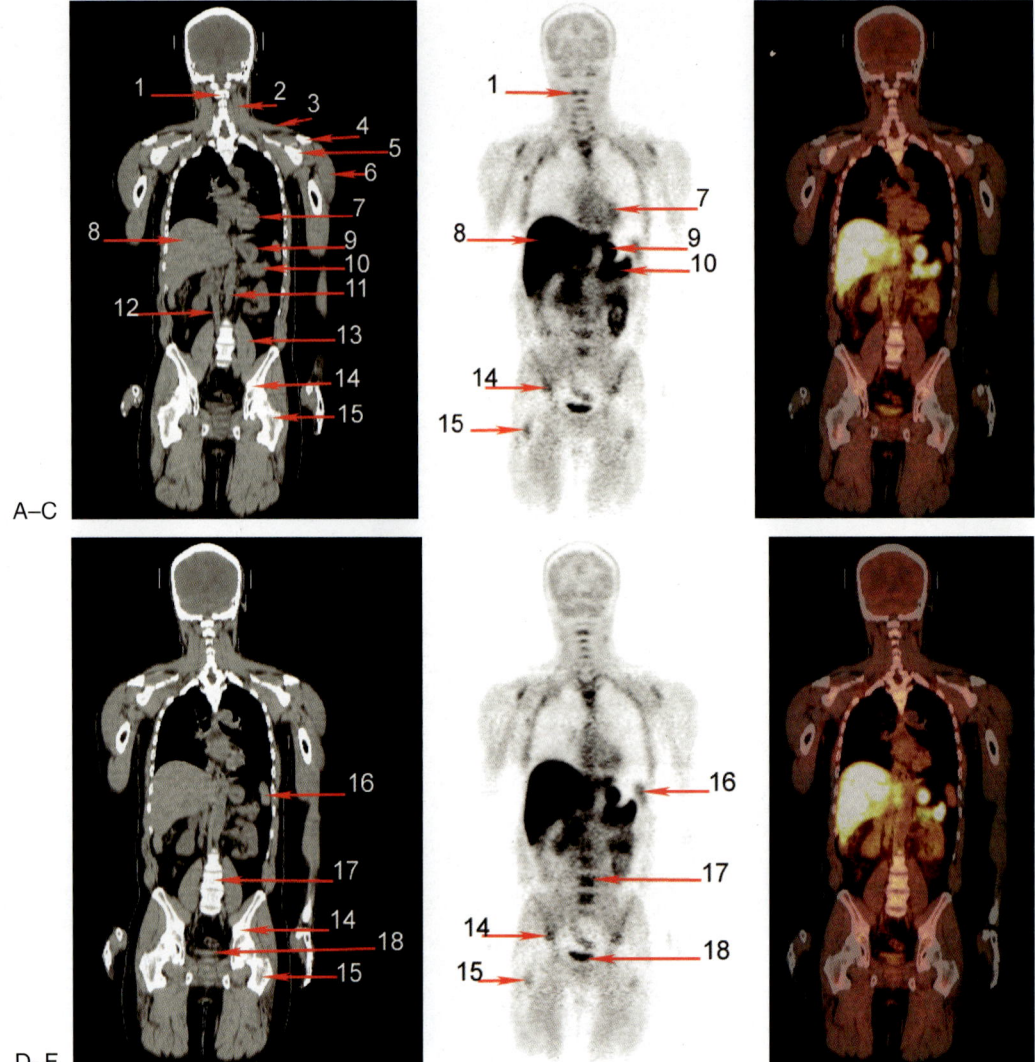

FIGURE 2-5:

1. Vertebral pedicle
2. Sternocleidomastoid
   muscle
3. Trapezius muscle
4. Acromion
5. Coracoid process
6. Deltoid muscle

7. Left ventricle
8. Right hepatic lobe
9. Stomach
10. Pancreatic tail
11. Abdominal aorta
12. Inferior vena cava
13. Psoas muscle

14. Acetabulum
15. Greater trochanter
16. Spleen
17. Vertebral body
18. Bladder

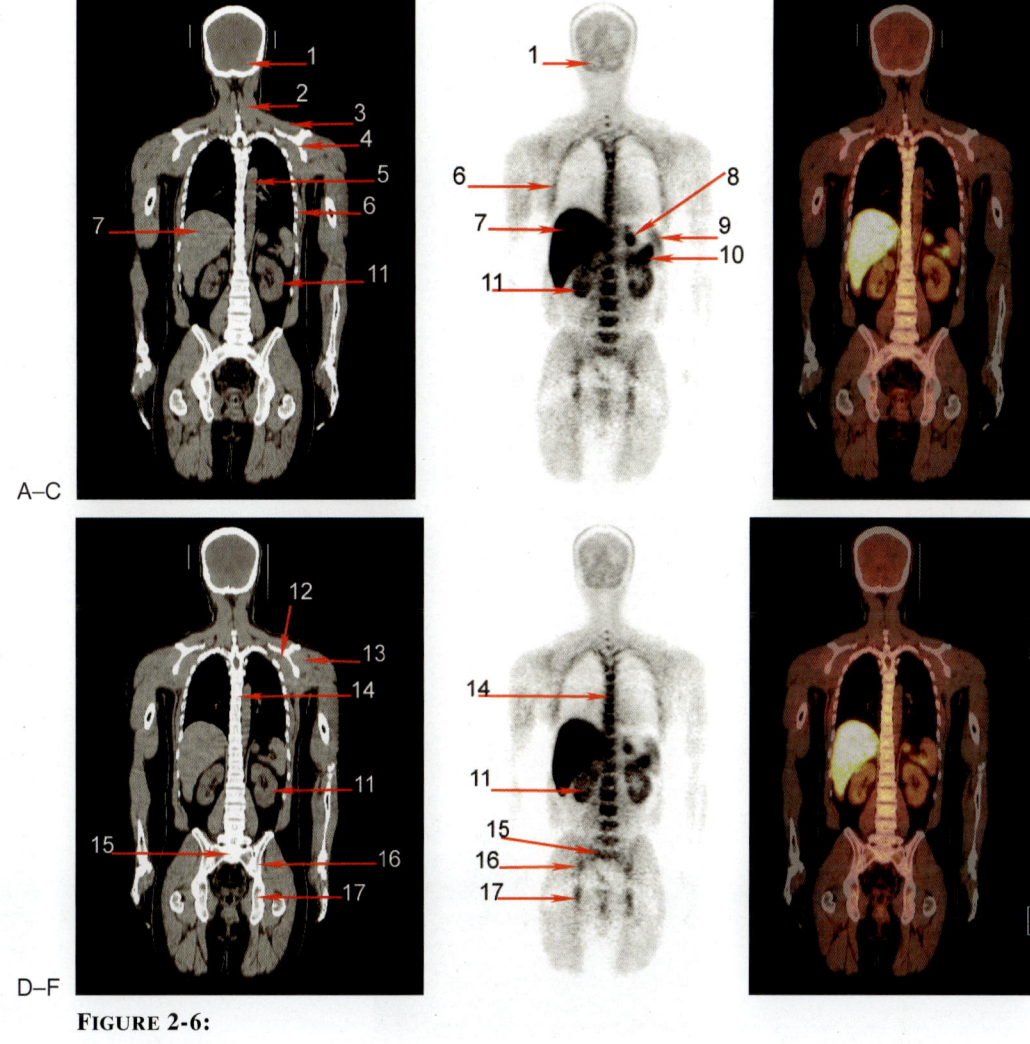

A–C

D–F

FIGURE 2-6:

1. Cerebellum
2. Erector spinae muscle
3. Supraspinatus muscle
4. Scapula
5. Descending aorta
6. Rib
7. Right hepatic lobe
8. Stomach
9. Spleen
10. Pancreatic tail
11. Kidney
12. Subscapularis muscle
13. Infraspinatus muscle
14. Vertebral body
15. Sacrum
16. Iliac tuberosity
17. Ischium

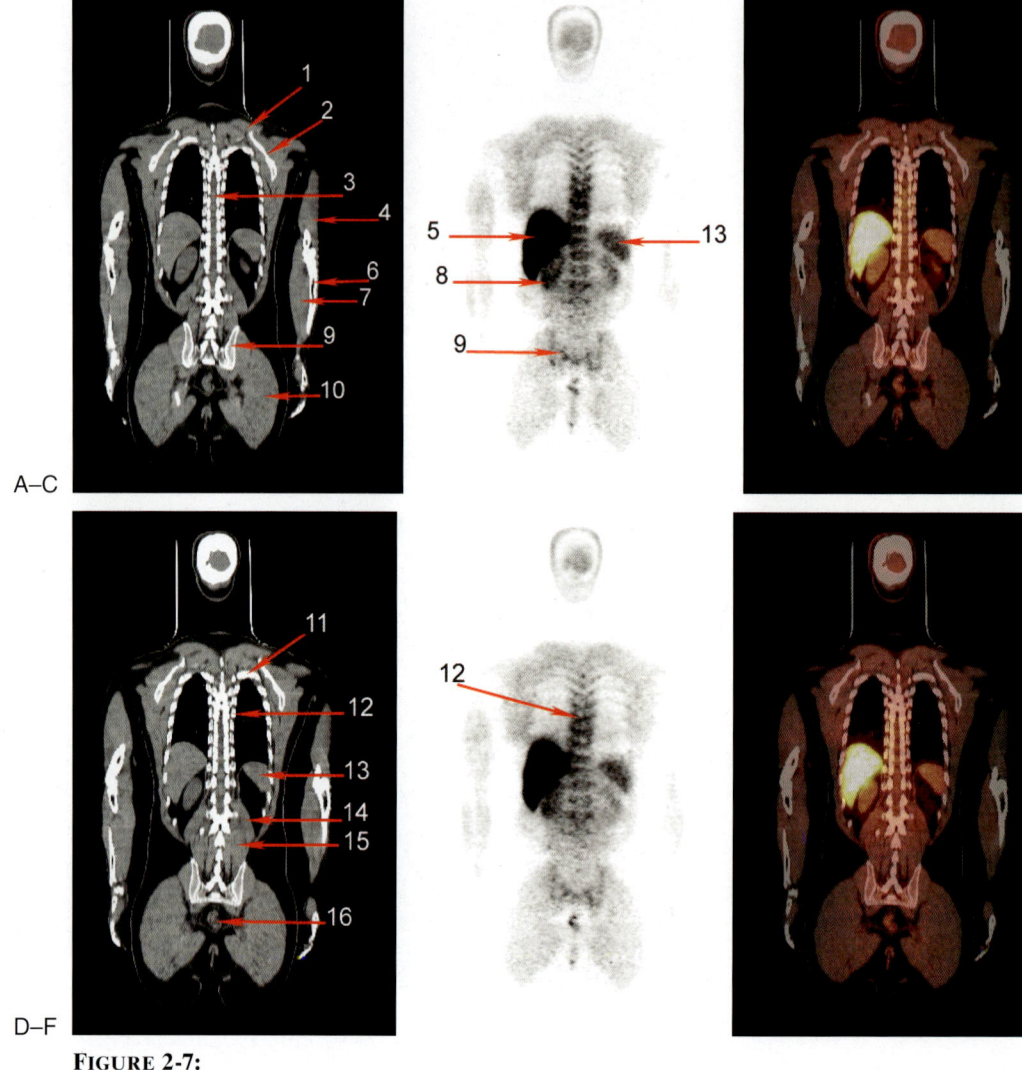

A–C

D–F

FIGURE 2-7:

1. Trapezius muscle
2. Scapula
3. Spinal canal
4. Triceps brachii muscle
5. Liver
6. Radius
7. Flexor digitorum profundus muscle
8. Kidney
9. Iliac tuberosity
10. Gluteus muscle
11. Rib
12. Vertebral pedicle
13. Spleen
14. Psoas muscle
15. Quadratus lumborum muscle
16. Sigmoid

Transaxials

Head

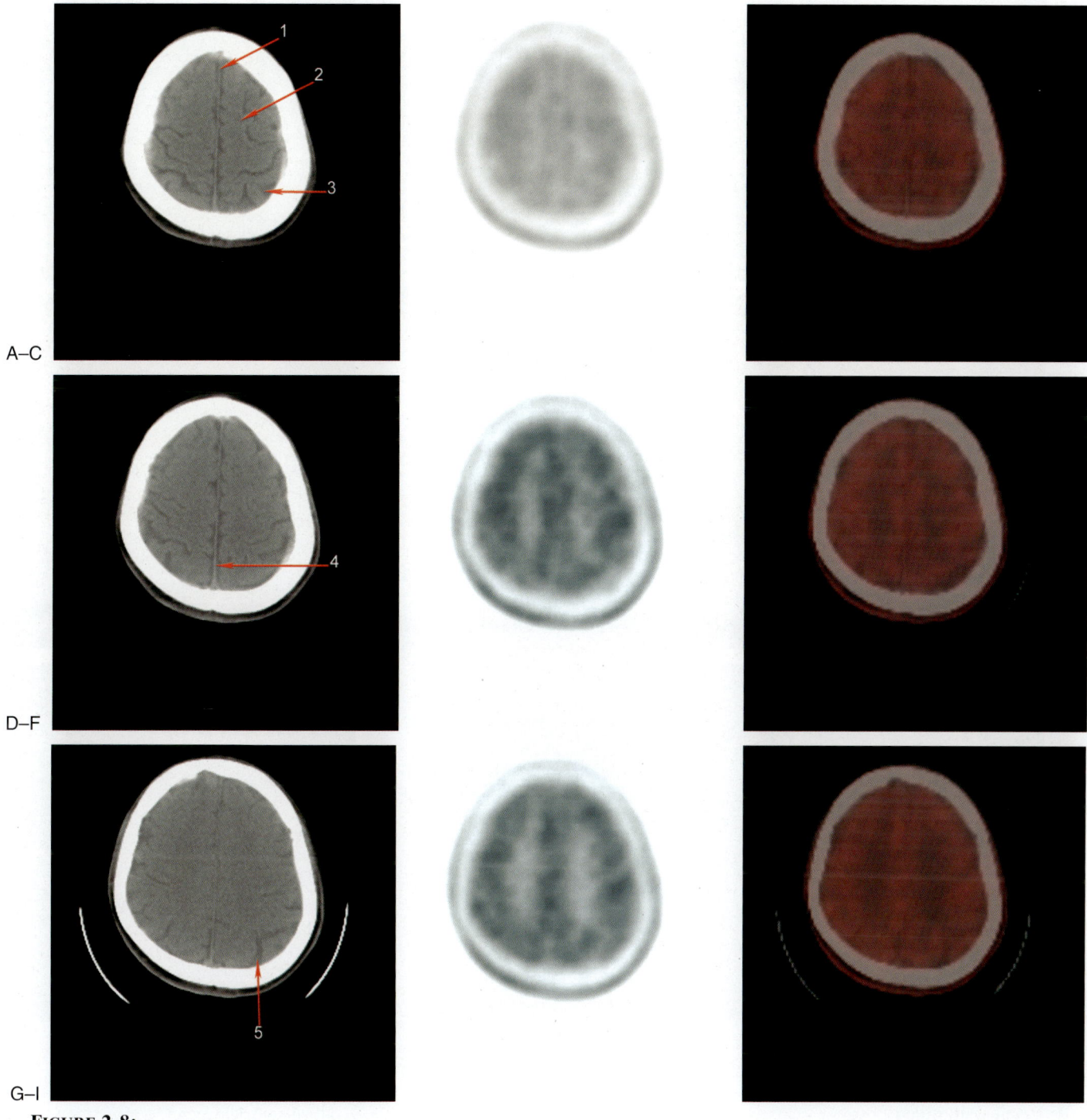

A–C

D–F

G–I

FIGURE 2-8:

1. Superior sagittal sinus
2. Cortical white matter
3. Cortical gray matter
4. Falx cerebri
5. Subarachnoid space

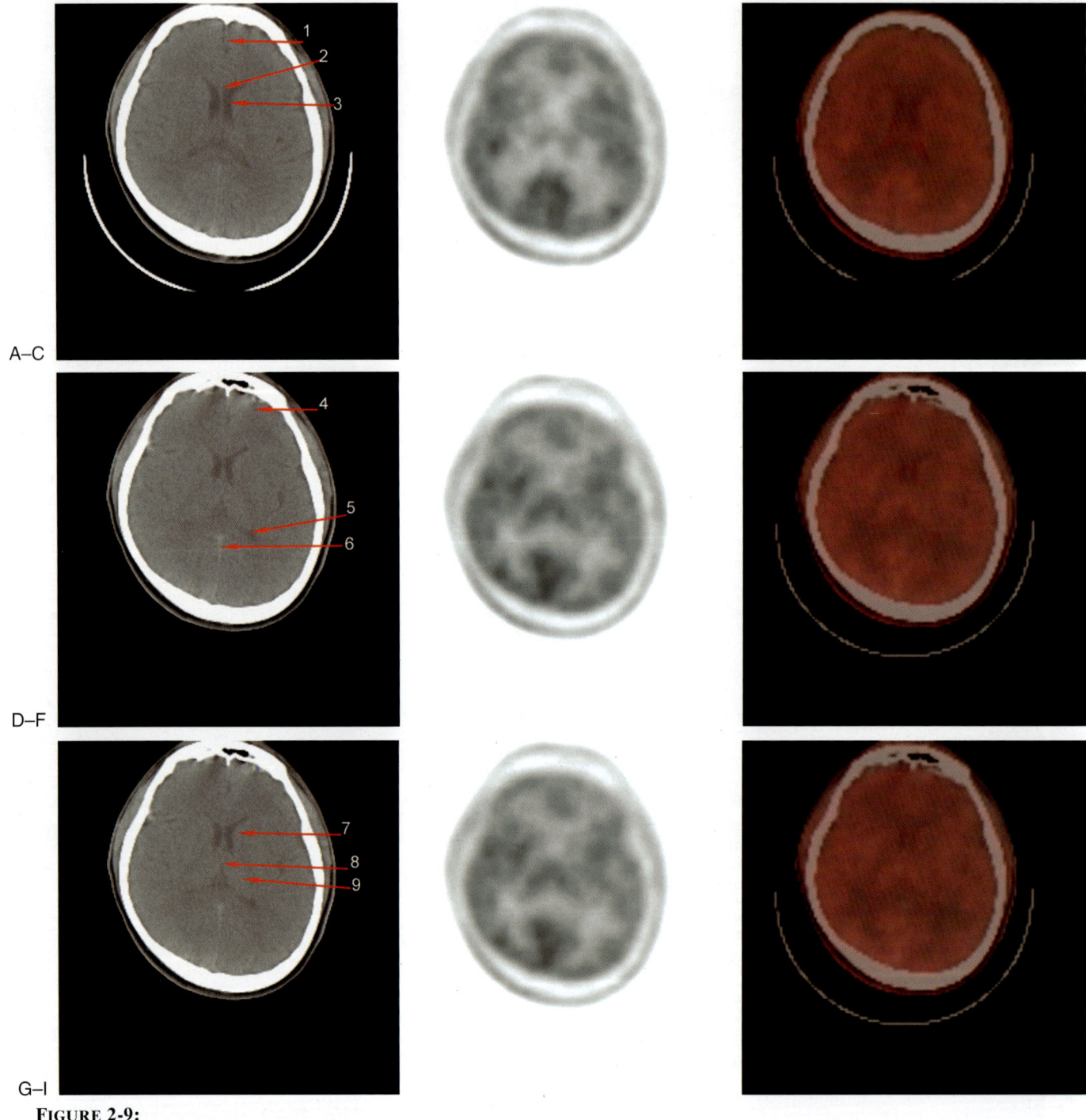

FIGURE 2-9:

1. Interhemispheric fissure
2. Corpus callosum
3. Lateral ventricle

4. Frontal lobe of brain
5. Choroid plexus
6. Straight sinus

7. Caudate nucleus
8. Third ventricle
9. Thalamus

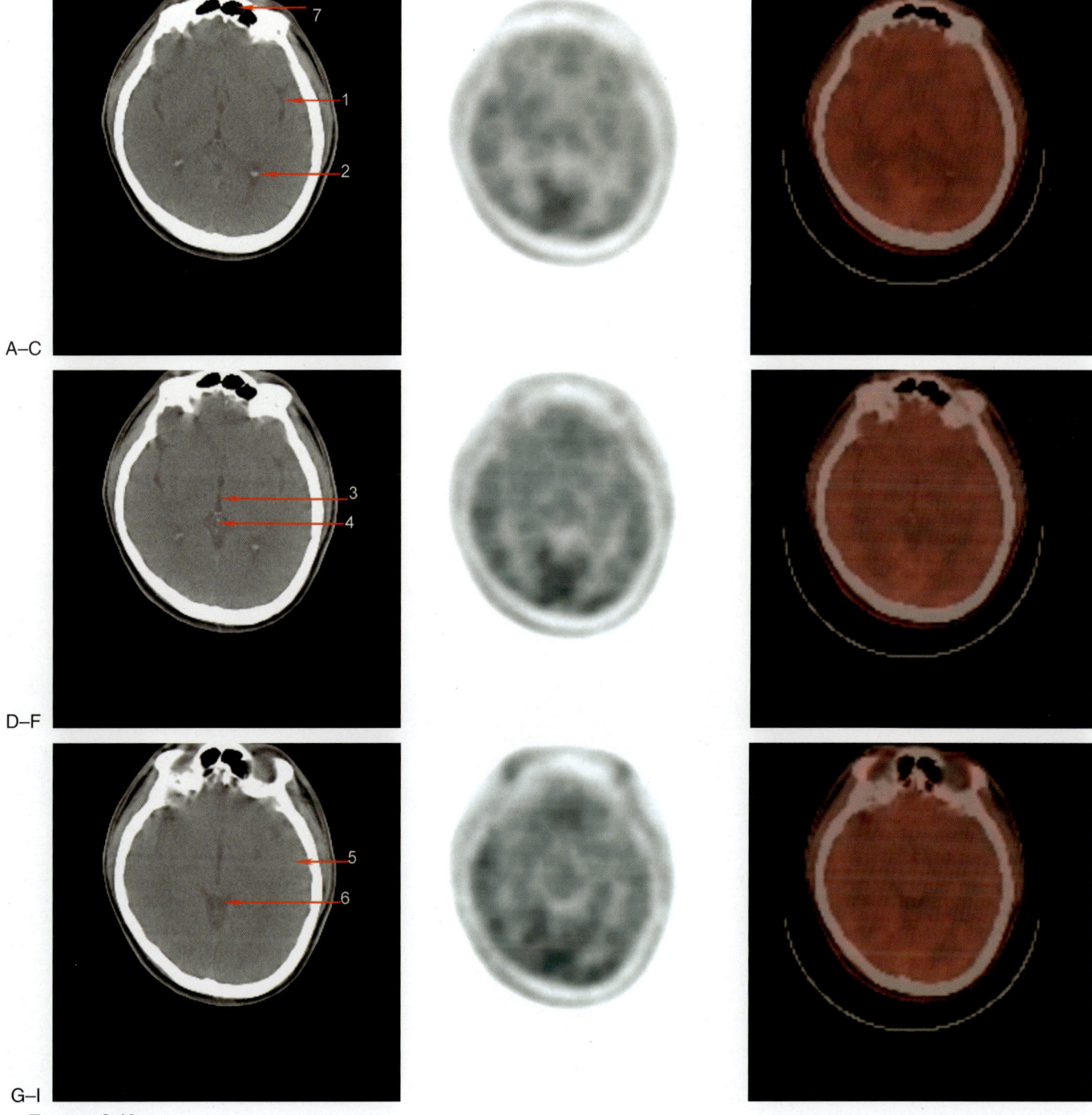

FIGURE 2-10:

1. Sylvian fissure
2. Choroid plexus
3. Third ventricle
4. Pineal gland
5. Temporal lobe of brain
6. Superior cerebellar cistern
7. Frontal sinus

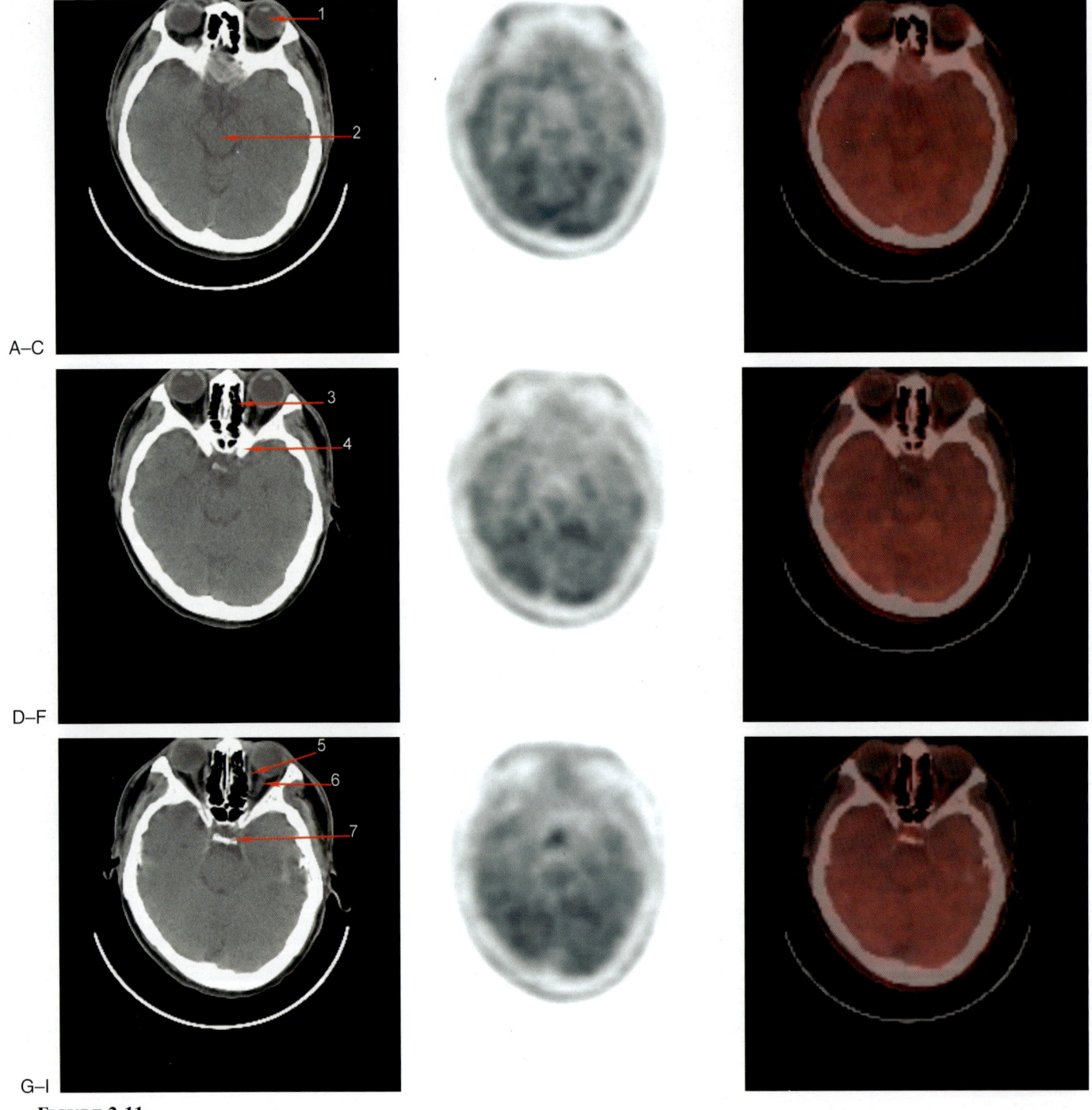

FIGURE 2-11:

1. Orbital globe
2. Brain stem
3. Ethmoidal sinus
4. Anterior clinoid process
5. Medial rectus muscle
6. Optical nerve
7. Dorsum sellae

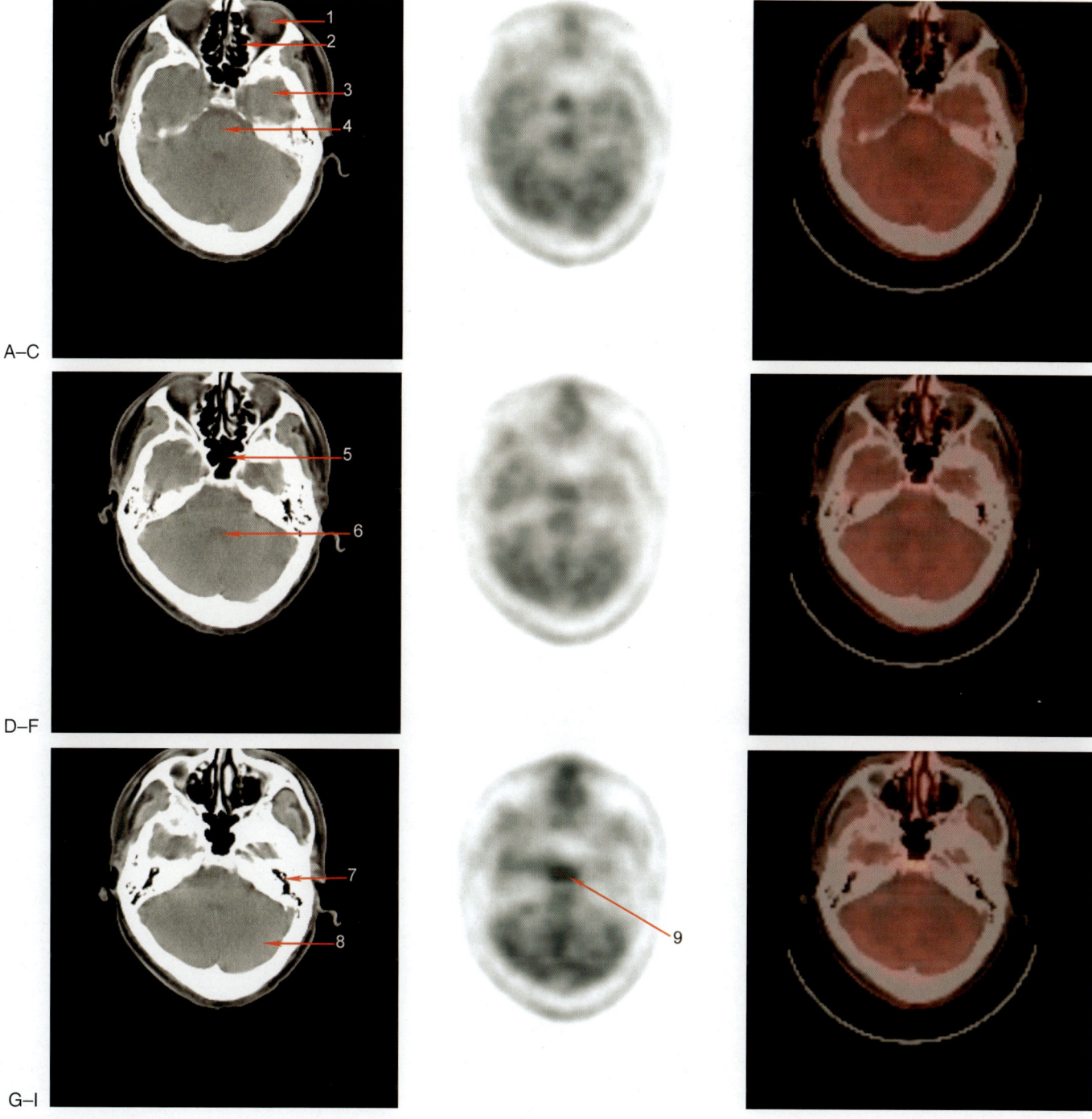

A–C

D–F

G–I

FIGURE 2-12:

1. Orbital globe
2. Ethmoidal sinus
3. Temporal lobe
4. Midbrain
5. Pharynx
6. Fourth ventricle
7. Mastoid air cells
8. Cerebellum
9. Clivus of occipital bone

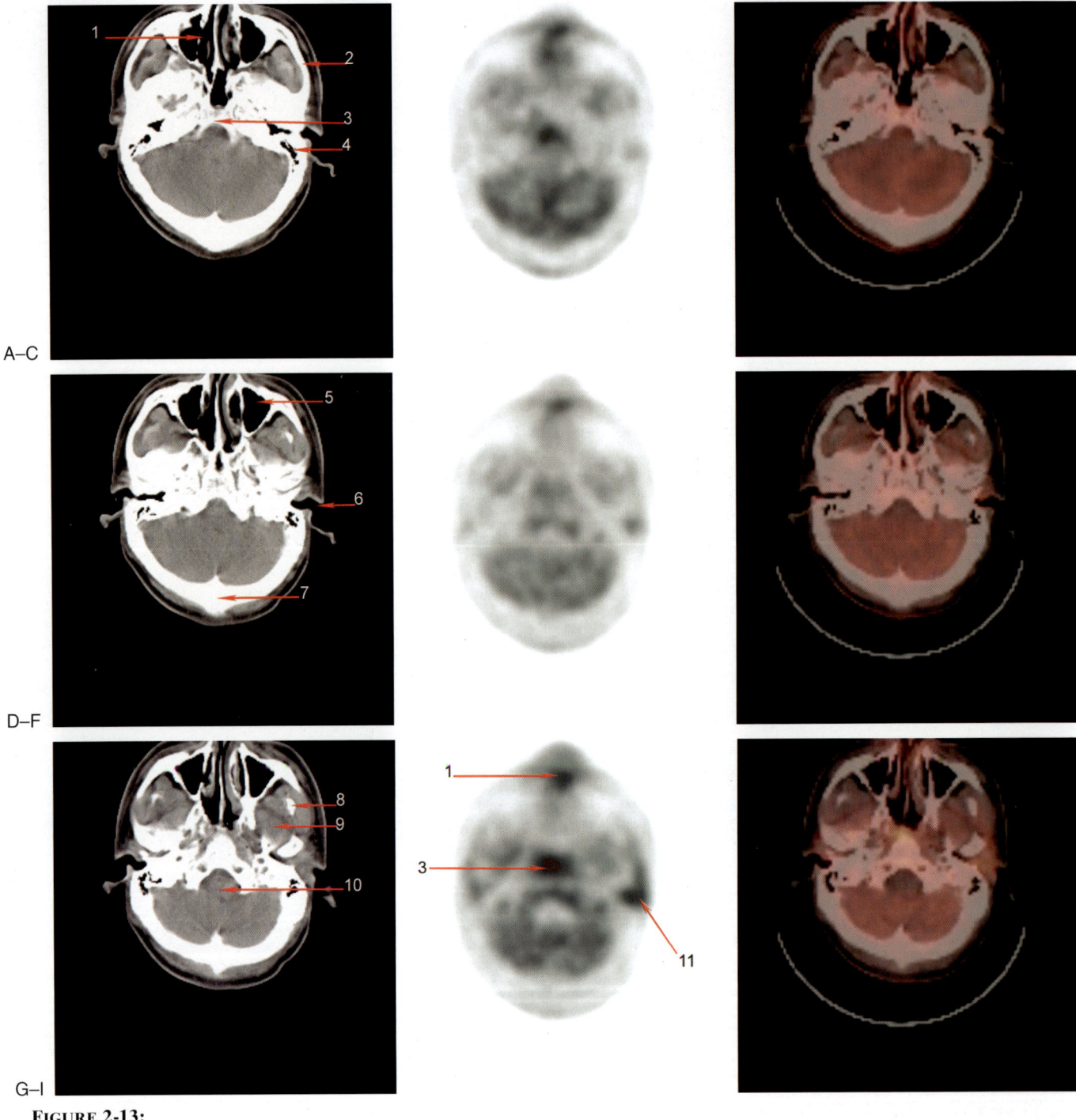

FIGURE 2-13:

1. Nasopharynx	5. Maxillary sinus	9. External pterygoid muscle
2. Zygomatic arch	6. External auditory canal	10. Medulla oblongata
3. Clivus of occipital bone	7. Occipital skull	11. Parotid gland
4. Mastoid air cells	8. Mandible	

Neck

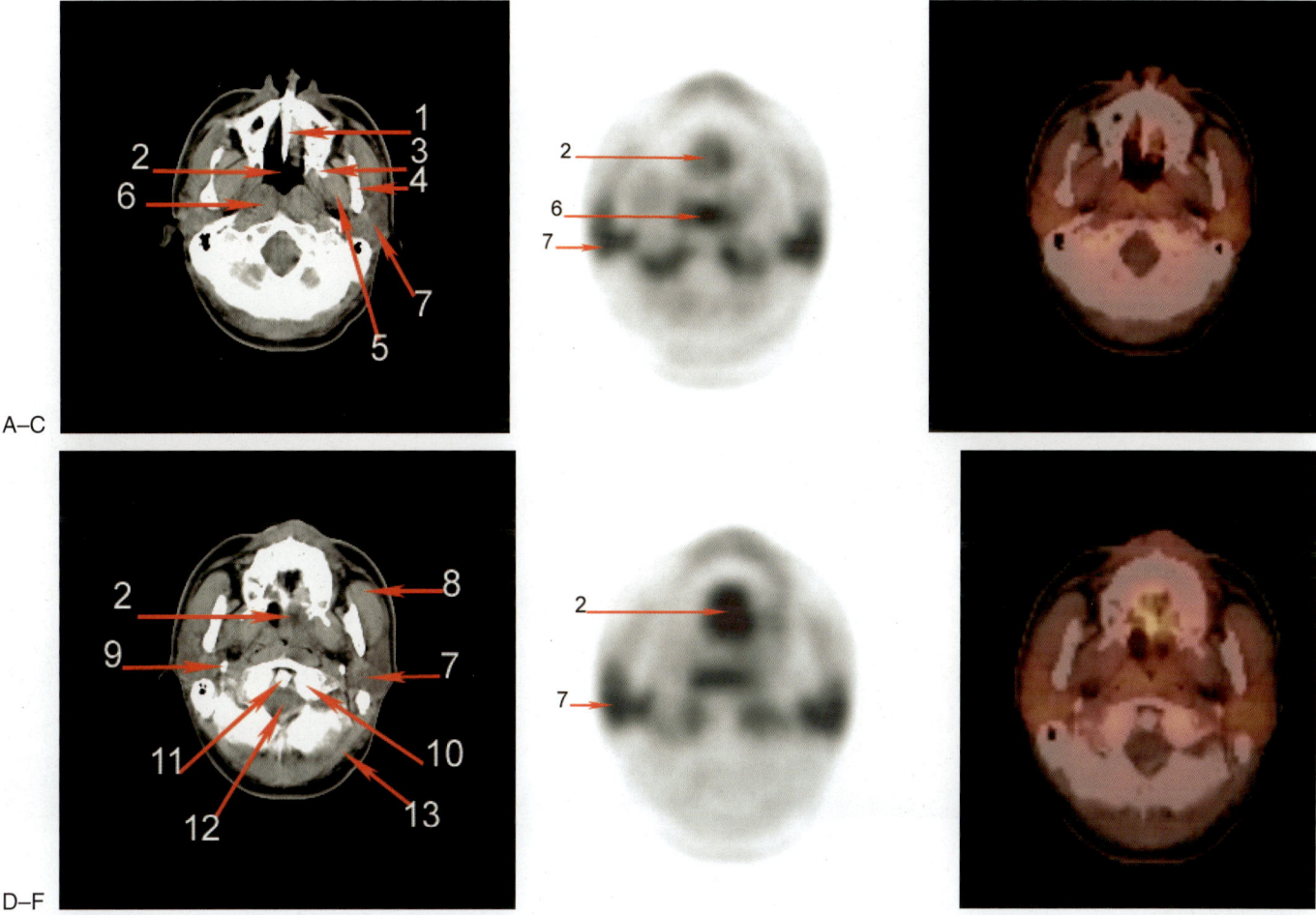

A–C

D–F

FIGURE 2-14:

1. Maxilla
2. Nasopharynx
3. Pterygoid plate
4. Mandible
5. Pterygoid muscle
6. Longus capitis muscle
7. Parotid gland
8. Masseter muscle
9. Styloid process
10. Atlas (C1)
11. Odontoid process (C2)
12. Spinal cord
13. Trapezius muscle

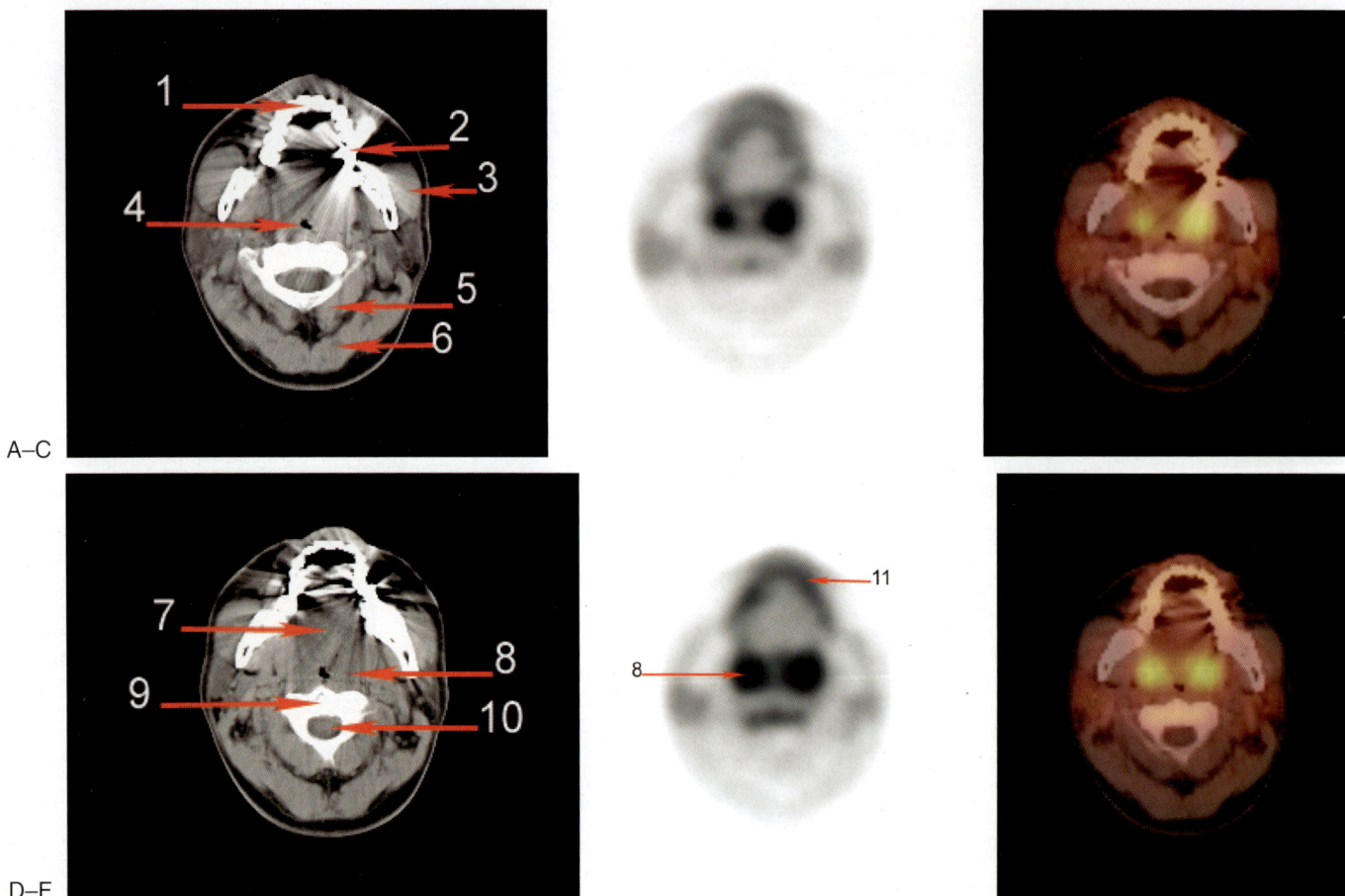

A–C

D–F

FIGURE 2-15:

1. Maxilla
2. Metallic artifacts by gold filling
3. Masseter muscle
4. Oropharynx
5. Semispinalis cervis muscle
6. Semispinalis capitis muscle
7. Tongue (genioglossus muscle)
8. Lingual tonsil
9. Vertebral body
10. Spinal cord
11. Periodontal tissue

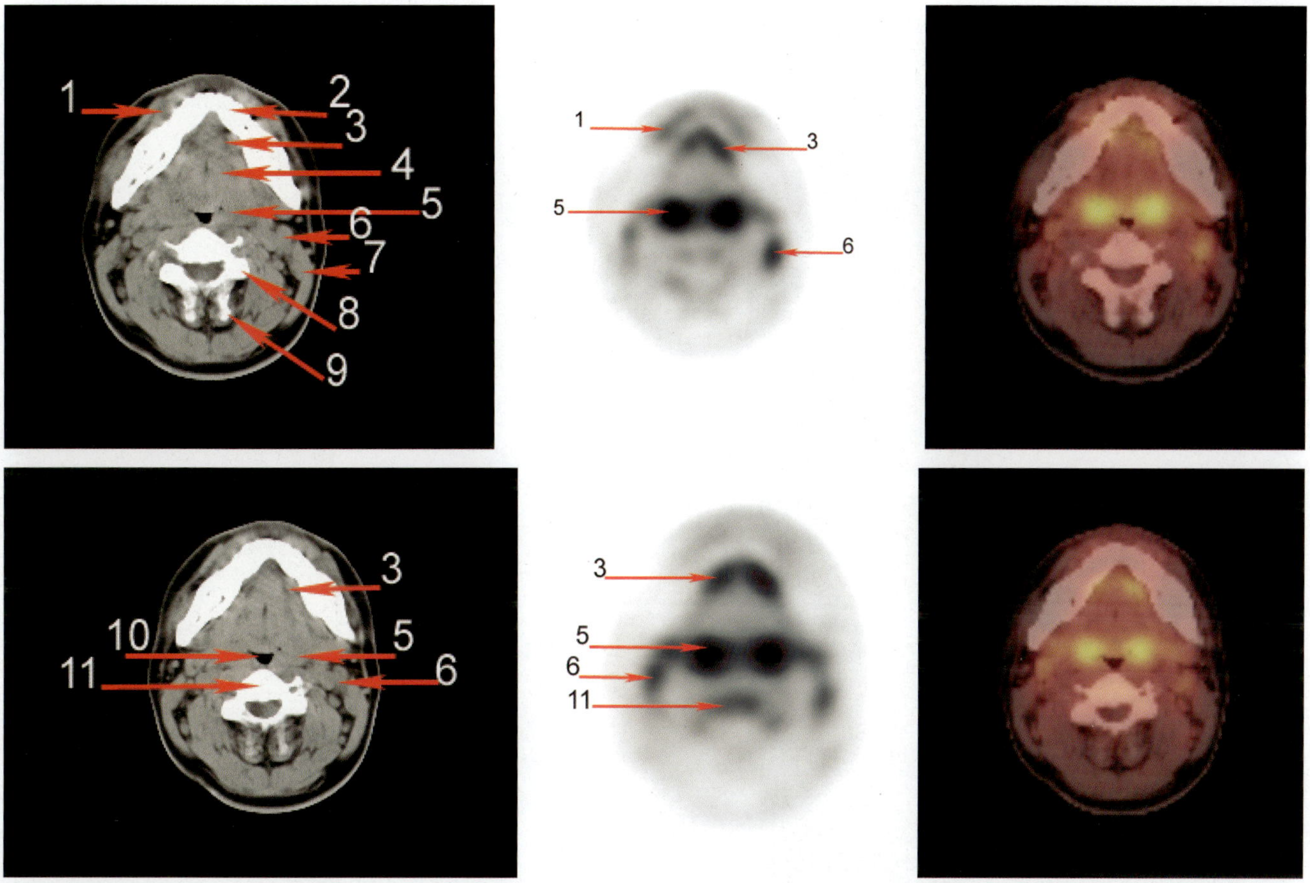

A–C

D–F

FIGURE 2-16:

1. Perioral muscles
2. Mandible
3. Mylohyoid muscle of tongue
4. Genioglossus muscle of tongue
5. Lingual tonsil
6. Parotid gland
7. Sternocleidomastoid muscle
8. Articular pillar (process)
9. Spinous process
10. Oropharynx
11. Vertebral body

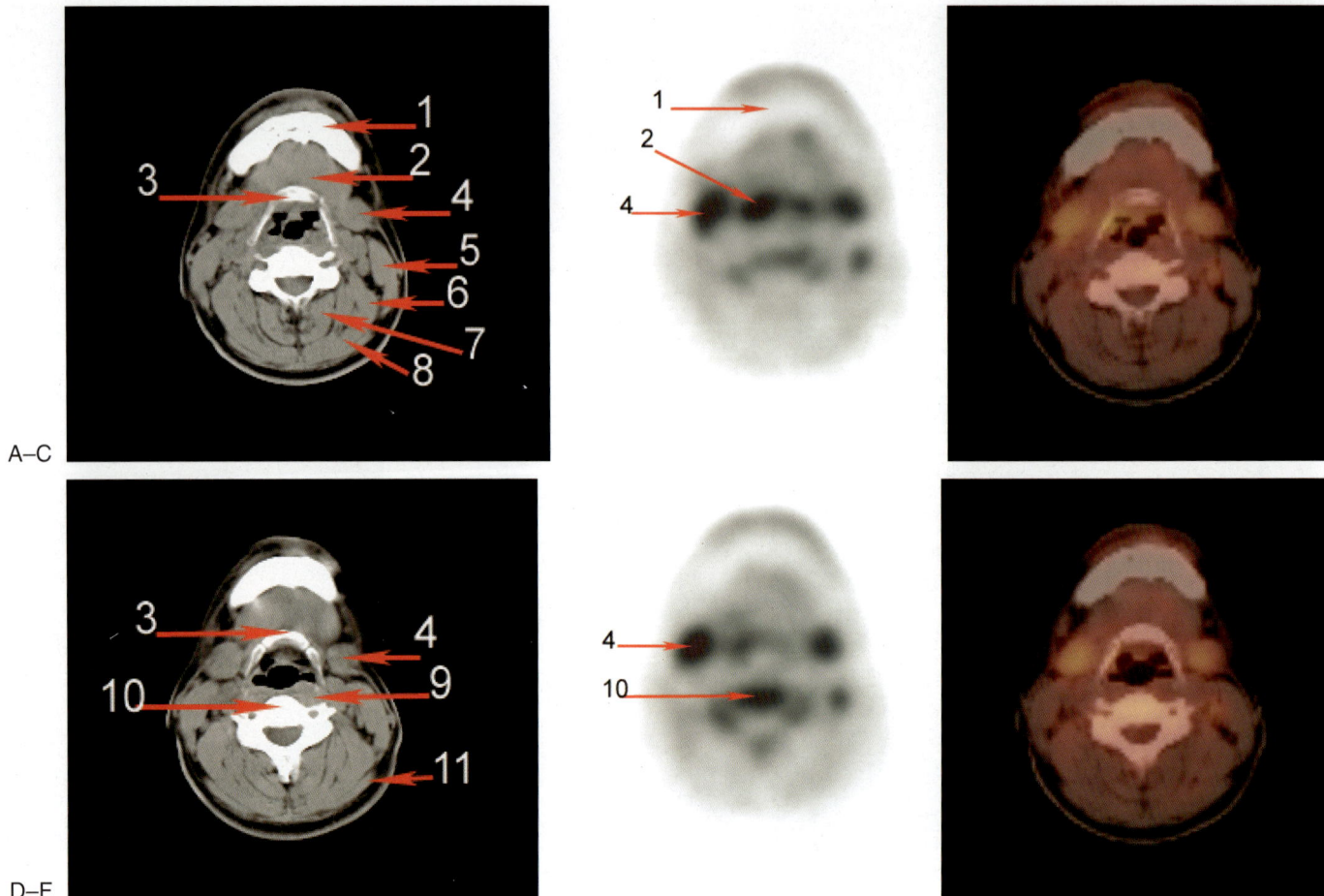

FIGURE 2-17:

1. Mandible
2. Lingual tonsil (tongue base)
3. Hyoid bone
4. Submandibular gland
5. Sternocleidomastoid muscle
6. Levator scapulae muscle
7. Semispinalis cervicis muscle
8. Semispinalis capitis muscle
9. Longus capitis muscle
10. Vertebral body
11. Trapezius muscle

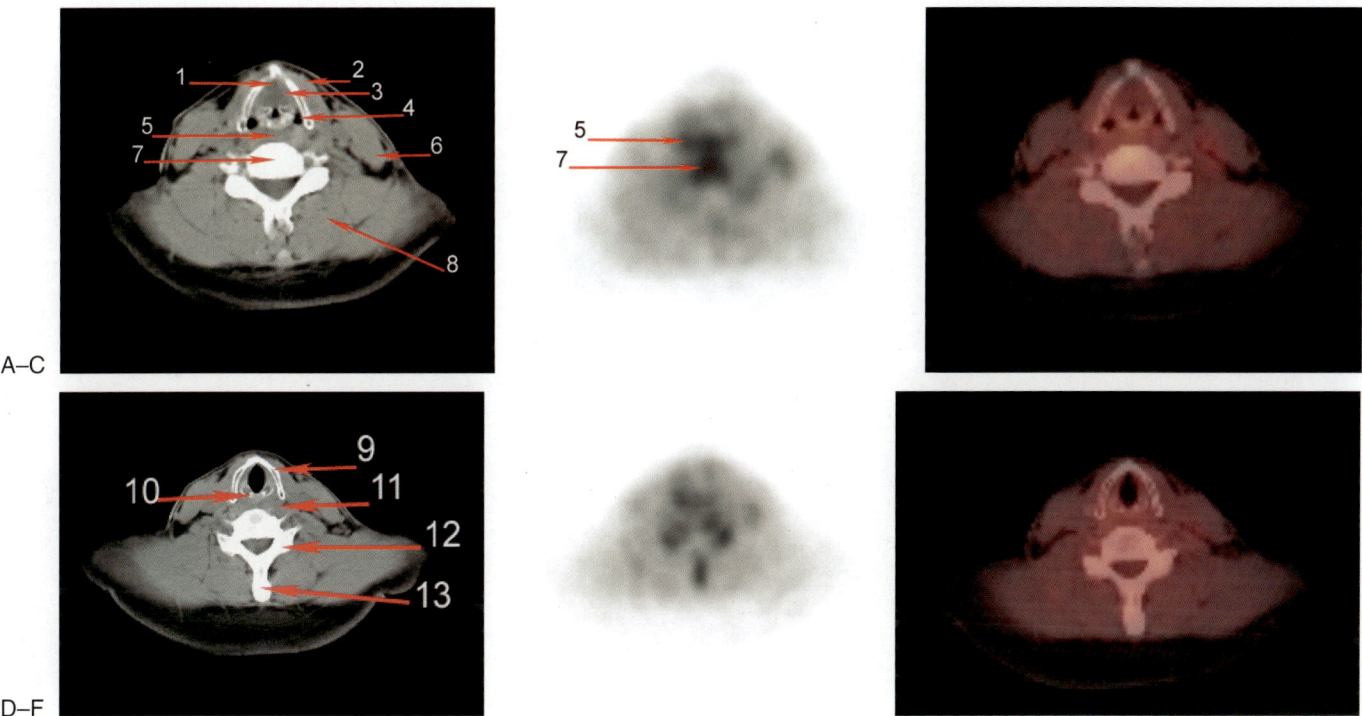

FIGURE 2-18:

1. Anterior laryngeal commissure
2. Infrahyoid muscles
3. Vocalis muscle
4. Piriform sinus
5. Esophagus

6. Sternocleidomastoid muscle
7. C5 vertebral body
8. Erector spinae muscle
9. Thyroid cartilage
10. Cricoid cartilage

11. Anterior scalene muscle
12. Articular pillar (process)
13. Spinous process

Chest

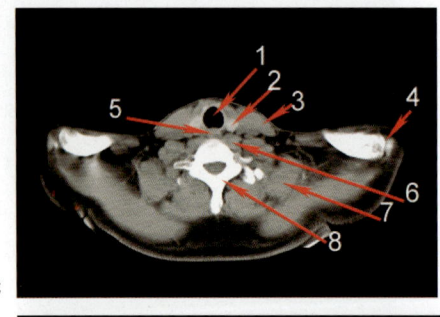
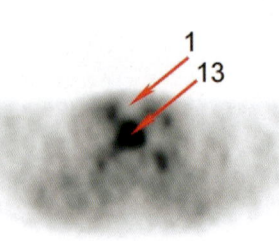
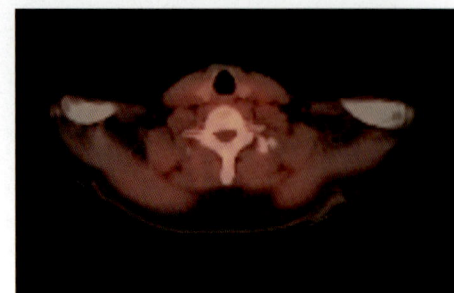

A–C

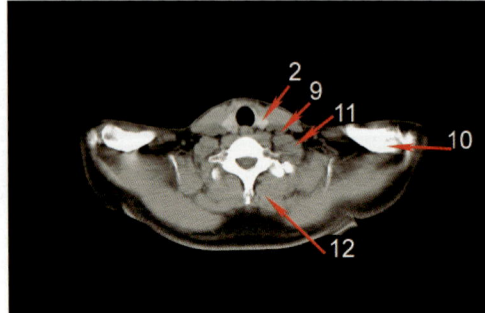
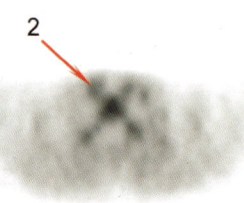

D–F

FIGURE 2-19:

1. Trachea
2. Thyroid
3. Sternocleidomastoid muscle
4. Acromioclavicular joint
5. Esophagus

6. Longus colli muscle
7. Levator scapulae muscle
8. Lamina of C6
9. Internal jugular vein

10. Distal clavicle
11. Anterior scalene muscle
12. Erector spinae muscle
13. Vertebral body

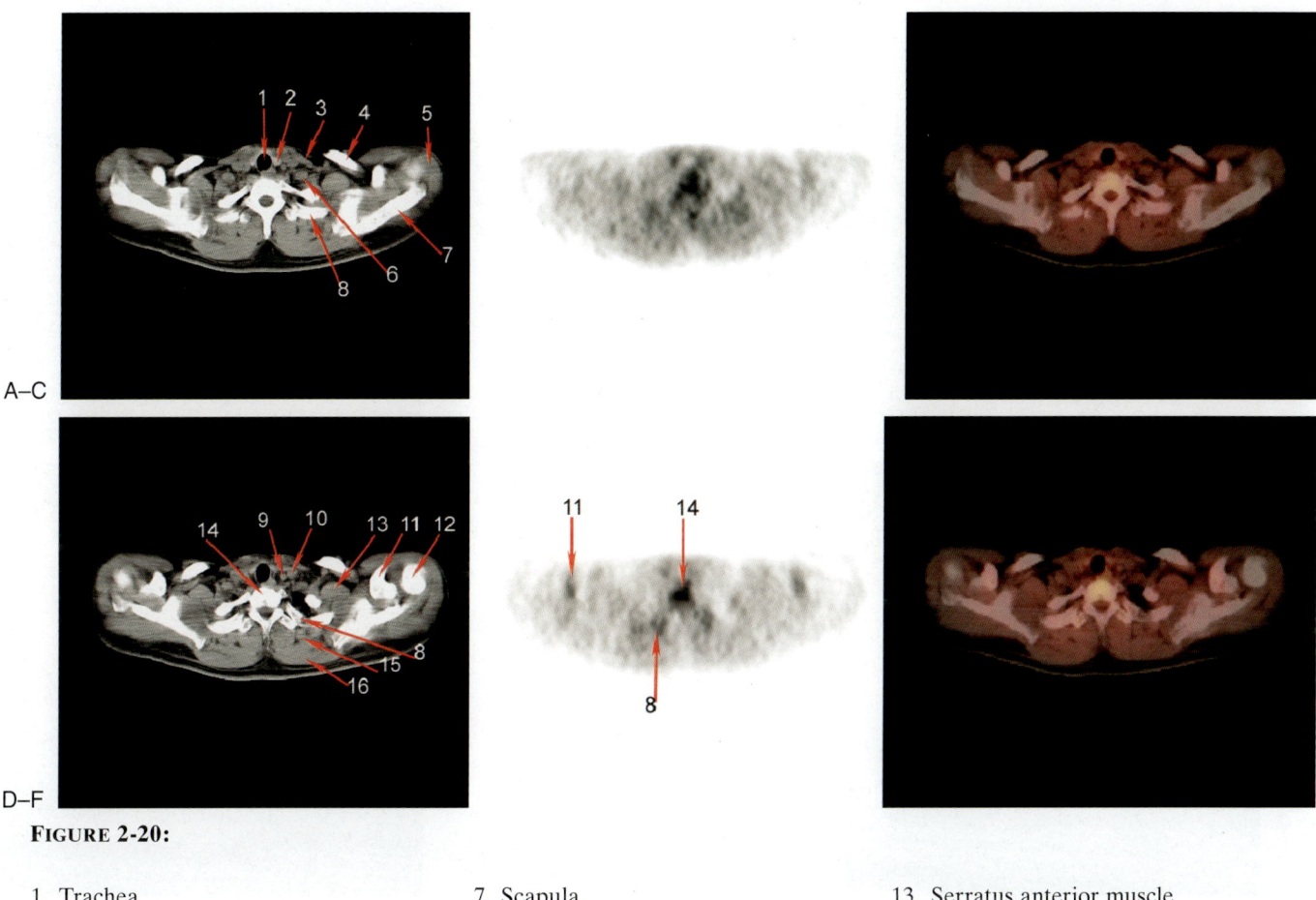

FIGURE 2-20:

1. Trachea
2. Thyroid
3. Sternocleidomastoid muscle
4. Clavicle
5. Deltoid muscle
6. Anterior scalene muscle

7. Scapula
8. Rib
9. Carotid artery
10. Internal jugular vein
11. Coracoid process of scapula
12. Humeral head

13. Serratus anterior muscle
14. Vertebral body
15. Rhomboideus muscle
16. Trapezius muscle

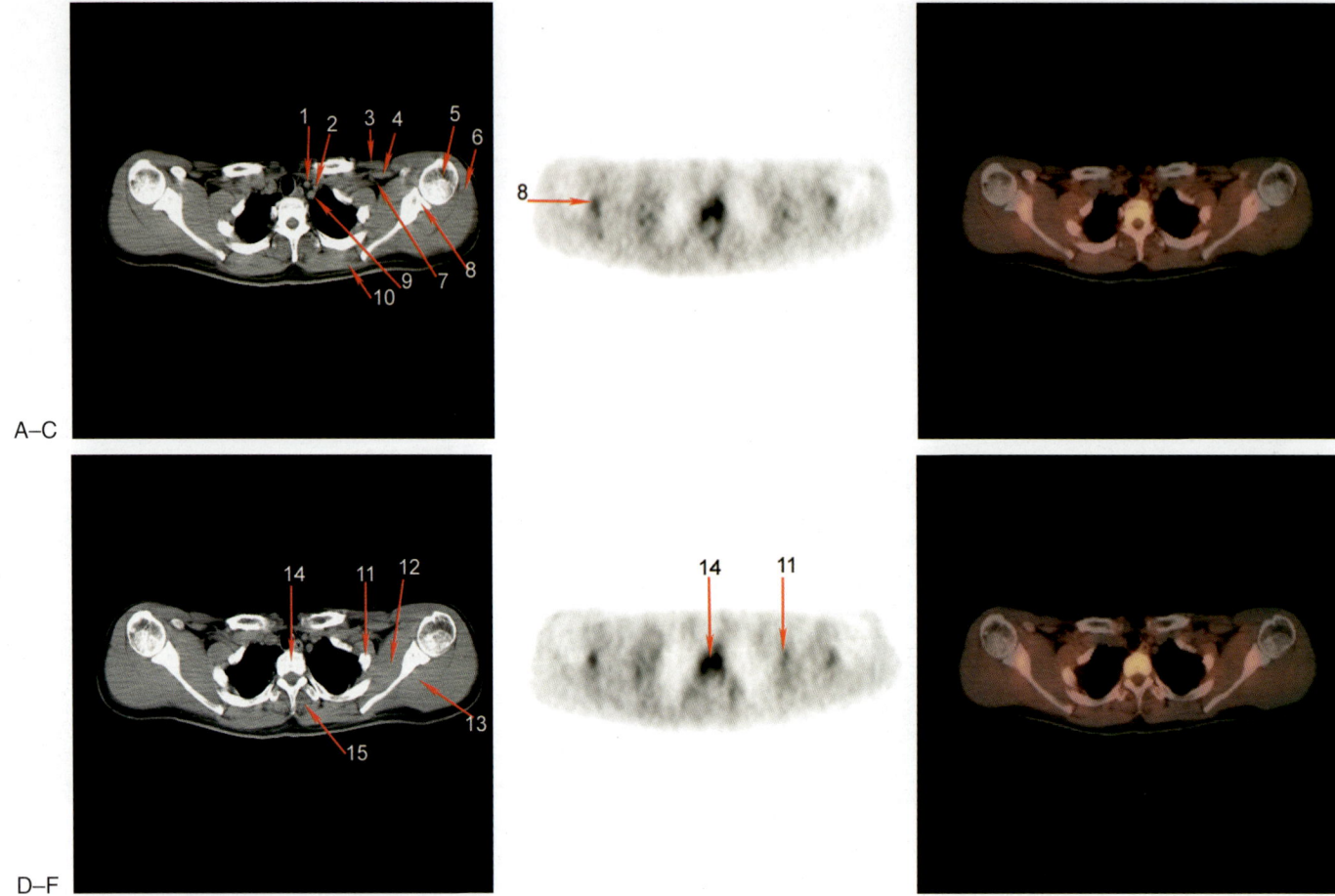

A–C

D–F

FIGURE 2-21:

1. Left carotid artery
2. Left internal jugular vein
3. Pectoralis major muscle
4. Pectoralis minor muscle
5. Humeral head
6. Deltoid muscle
7. Left axillary vessels
8. Glenoid of scapula
9. Left subclavian artery
10. Trapezius
11. Rib
12. Subscapularis muscle
13. Infraspinatus muscle
14. Vertebral body
15. Erector spinae muscle

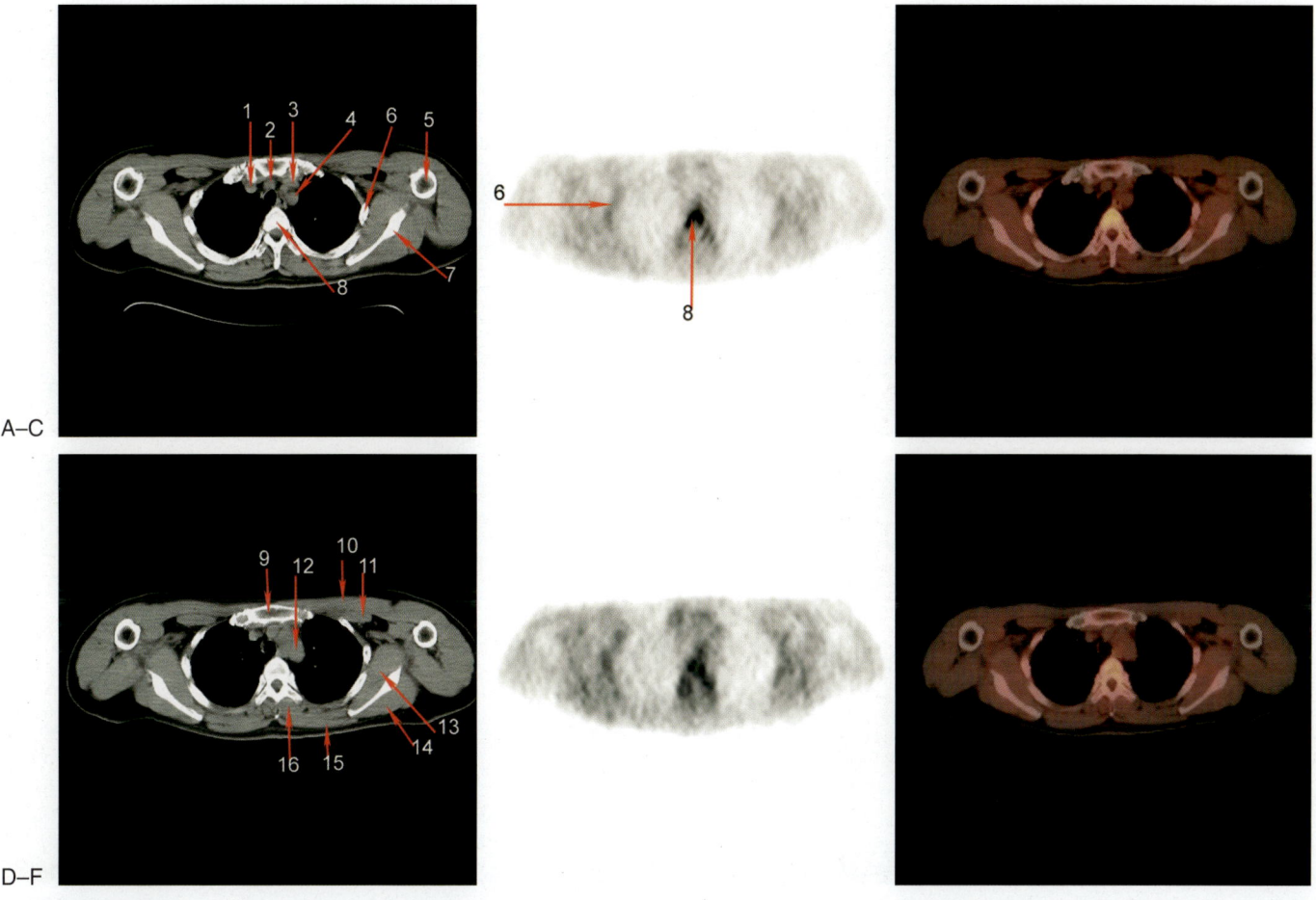

FIGURE 2-22:

1. Right brachiocephalic vein
2. Brachiocephalic artery
3. Left brachiocephalic vein
4. Left subclavian artery
5. Humerus
6. Rib
7. Scapula
8. Vertebral body
9. Manubrium
10. Pectoralis major muscle
11. Pectoralis minor muscle
12. Aortic arch
13. Subscapularis muscle
14. Infraspinatus muscle
15. Trapezius muscle
16. Erector spinae muscle

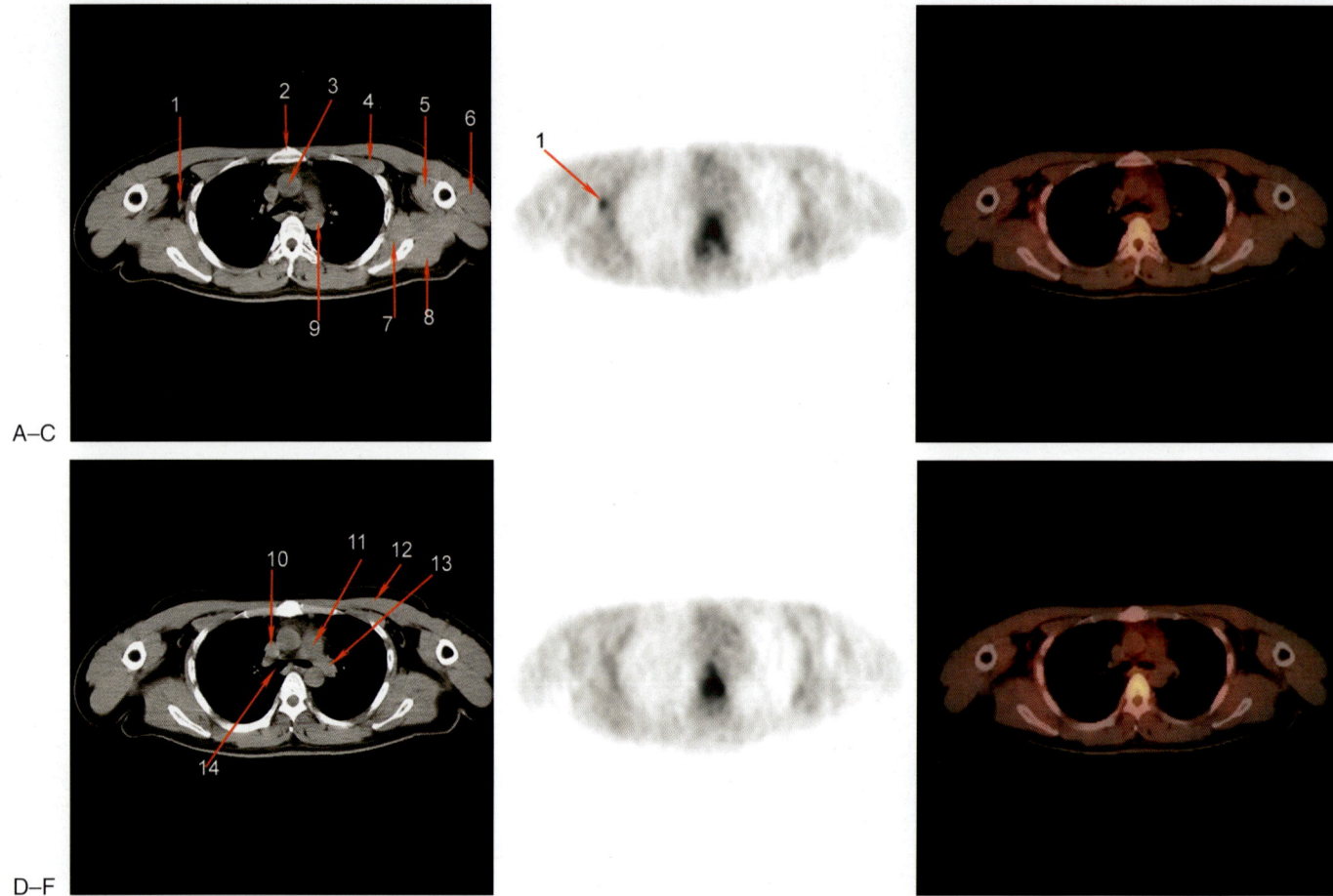

FIGURE 2-23:

1. Axillary vein
2. Sternum
3. Ascending aorta
4. Pectoralis minor muscle
5. Coracobrachialis and biceps brachii muscles
6. Deltoid muscle
7. Subscapularis muscle
8. Infraspinatus muscle
9. Descending aorta
10. Superior vena cava
11. Main pulmonary artery
12. Pectoralis major muscle
13. Left pulmonary artery
14. Right main bronchus

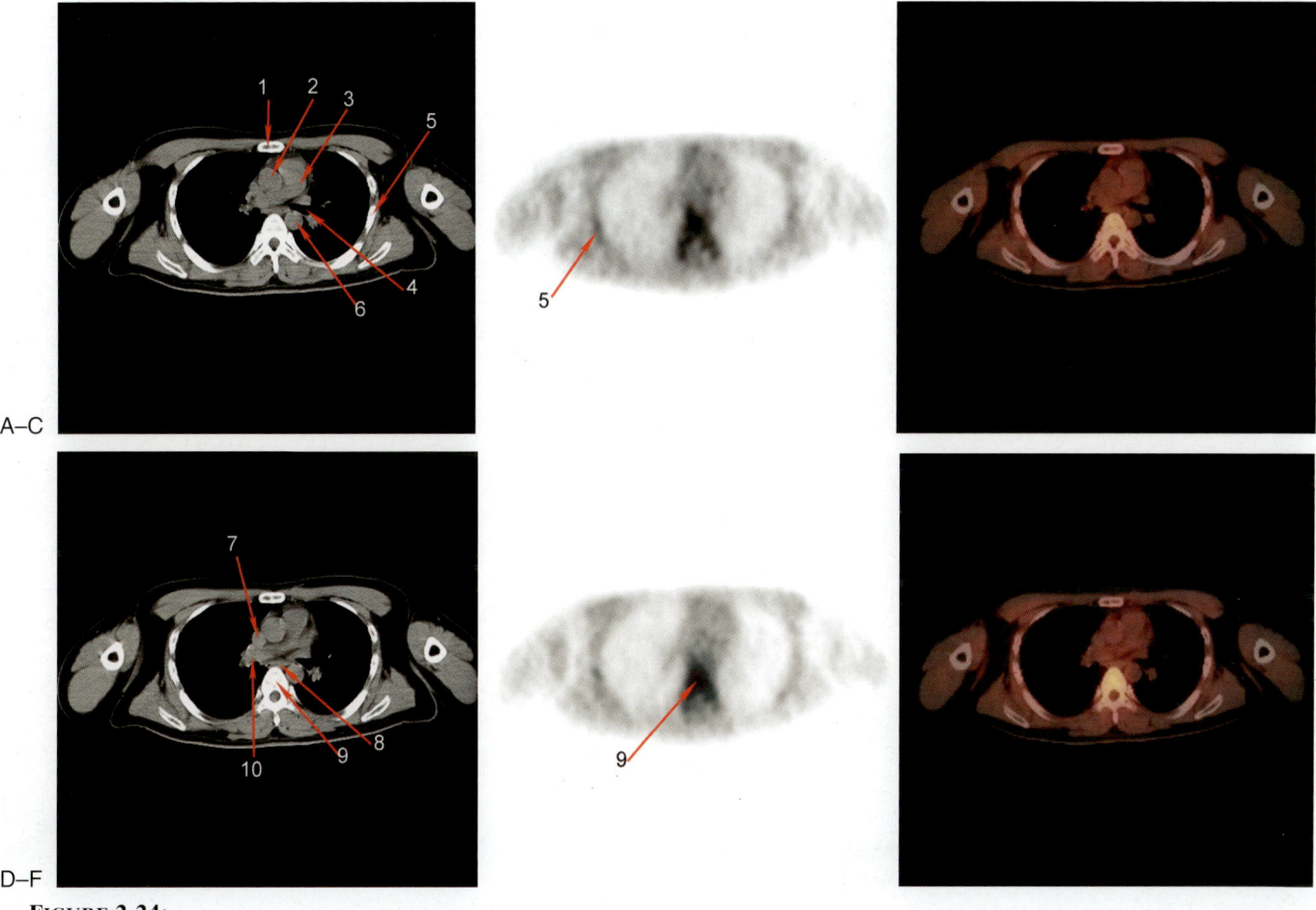

FIGURE 2-24:

1. Sternum
2. Ascending aorta
3. Main pulmonary artery
4. Left main bronchus
5. Rib
6. Descending aorta
7. Superior vena cava
8. Esophagus
9. Vertebral body
10. Right pulmonary artery

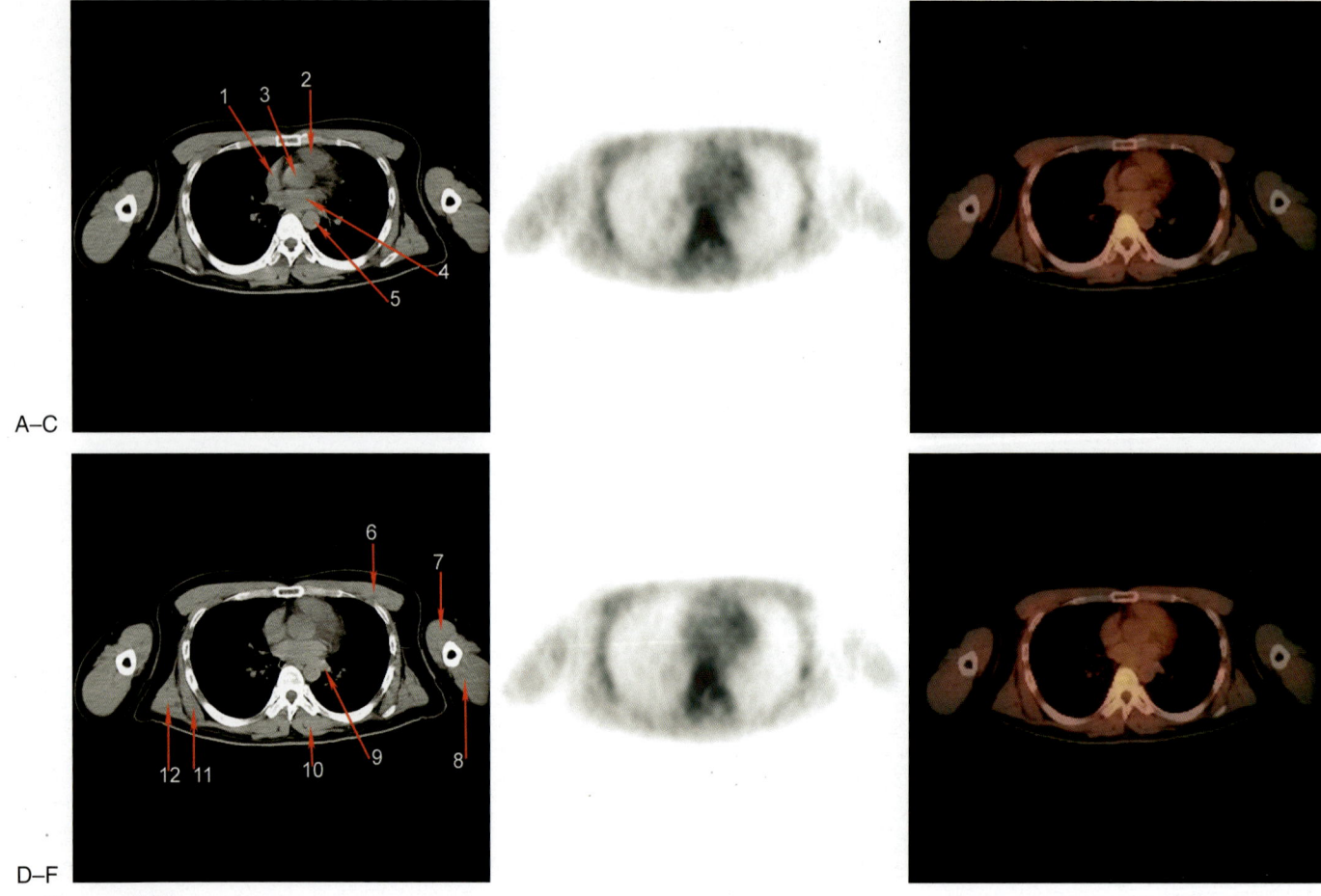

A–C

D–F

FIGURE 2-25:

1. Right atrium
2. Right ventricle
3. Ascending aorta
4. Left atrium
5. Descending aorta
6. Pectoralis muscle
7. Biceps brachii muscle
8. Triceps brachii muscle
9. Segmental pulmonary vein
10. Erector spinae muscle
11. Serratus anterior muscle
12. Latissimus dorsi muscle

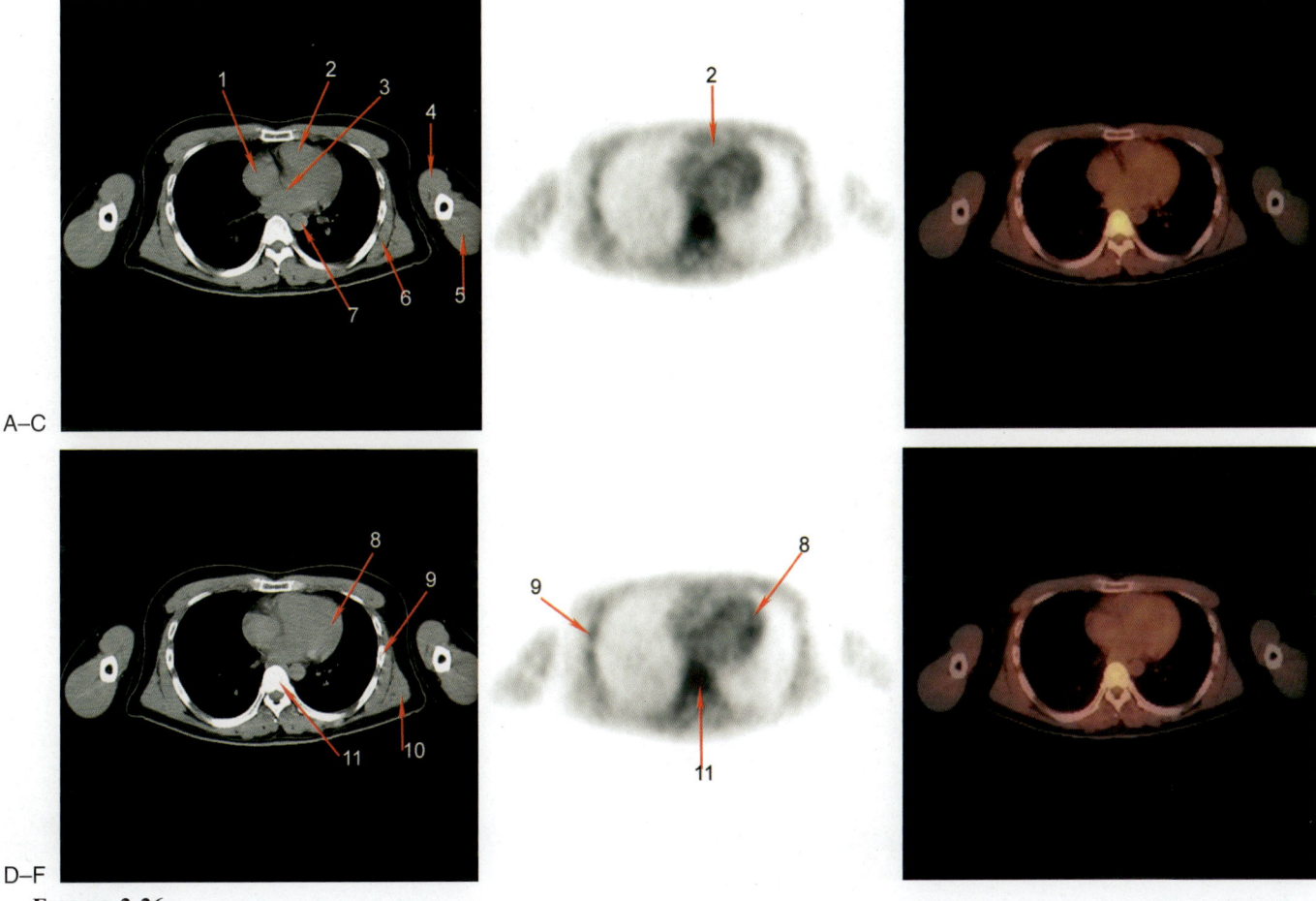

FIGURE 2-26:

1. Right atrium
2. Right ventricle
3. Left atrium
4. Biceps brachii muscle
5. Triceps brachii muscle
6. Serratus anterior muscle
7. Descending aorta
8. Left ventricle
9. Rib
10. Latissimus dorsi muscle
11. Vertebral body

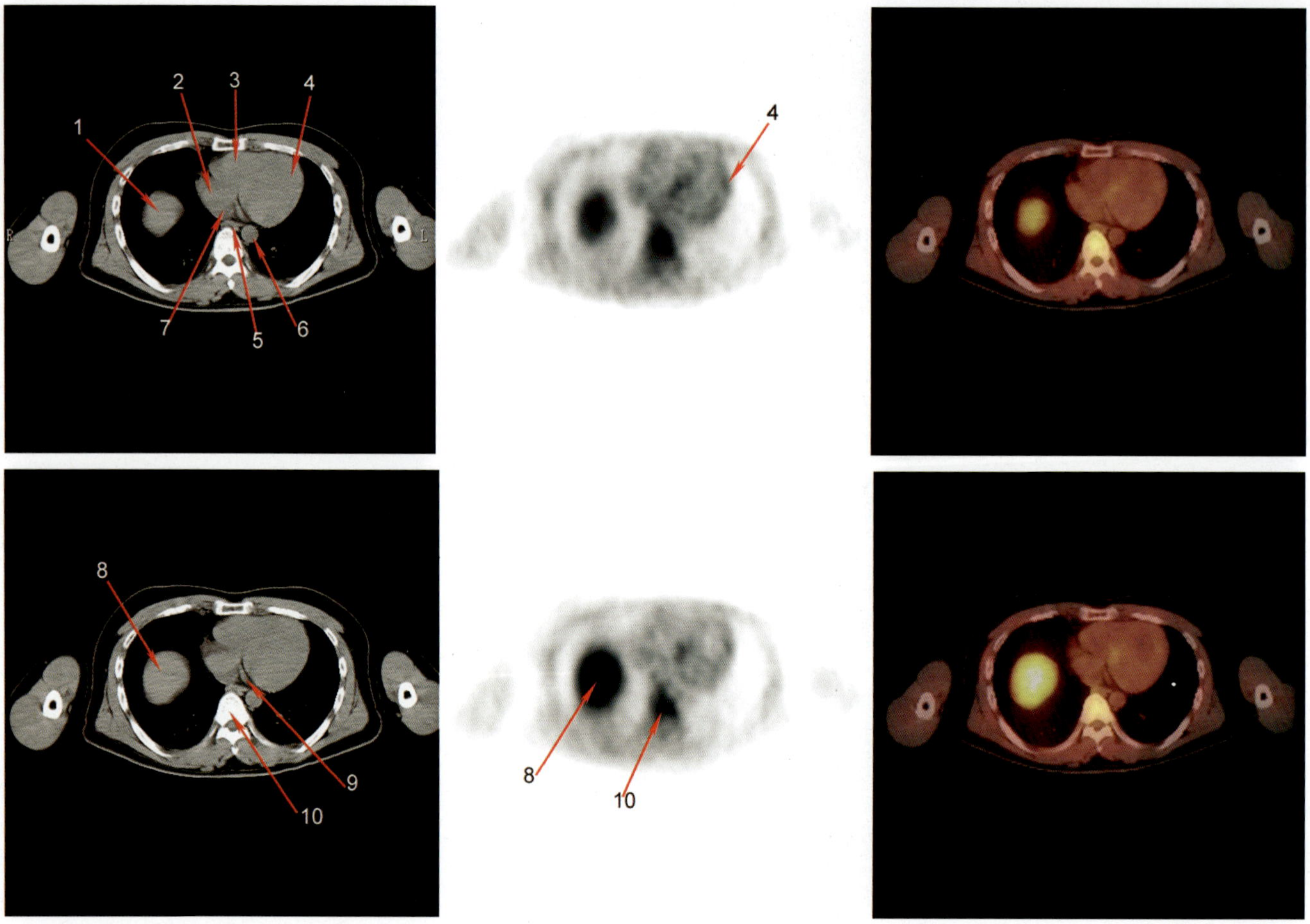

A–C

D–F

FIGURE 2-27:

1. Hepatic dome (cupula of diaphragm)
2. Right atrium
3. Right ventricle
4. Left ventricle

5. Esophagus
6. Descending aorta
7. Inferior vena cava

8. Upper segment of right hepatic lobe (#VIII)
9. Coronary sinus
10. Vertebral body

Abdomen

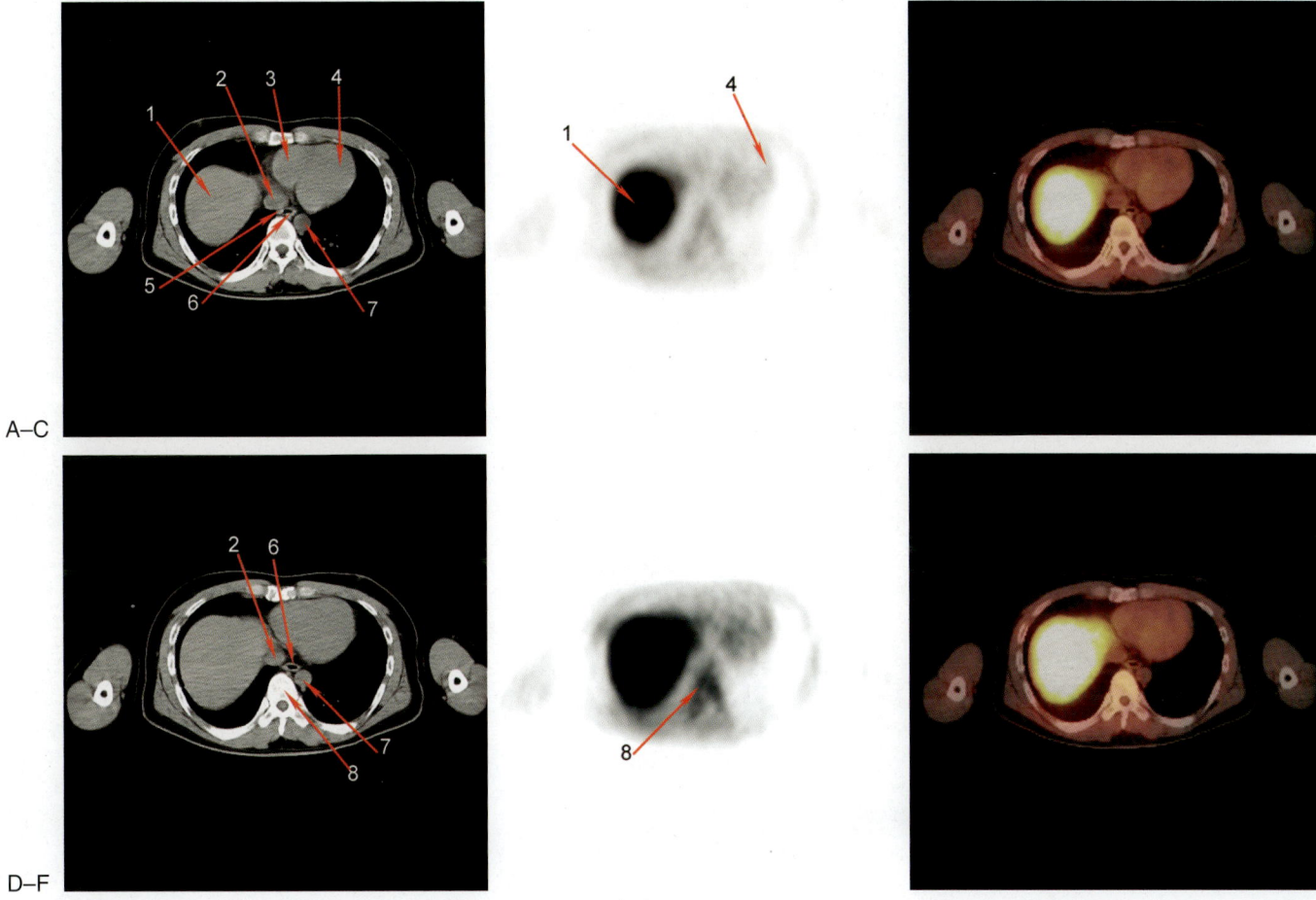

A–C

D–F

FIGURE 2-28:

1. Right hepatic lobe (#VIII)
2. Inferior vena cava
3. Right ventricle
4. Left ventricle
5. Azygos vein
6. Esophagus
7. Descending aorta
8. Vertebral body

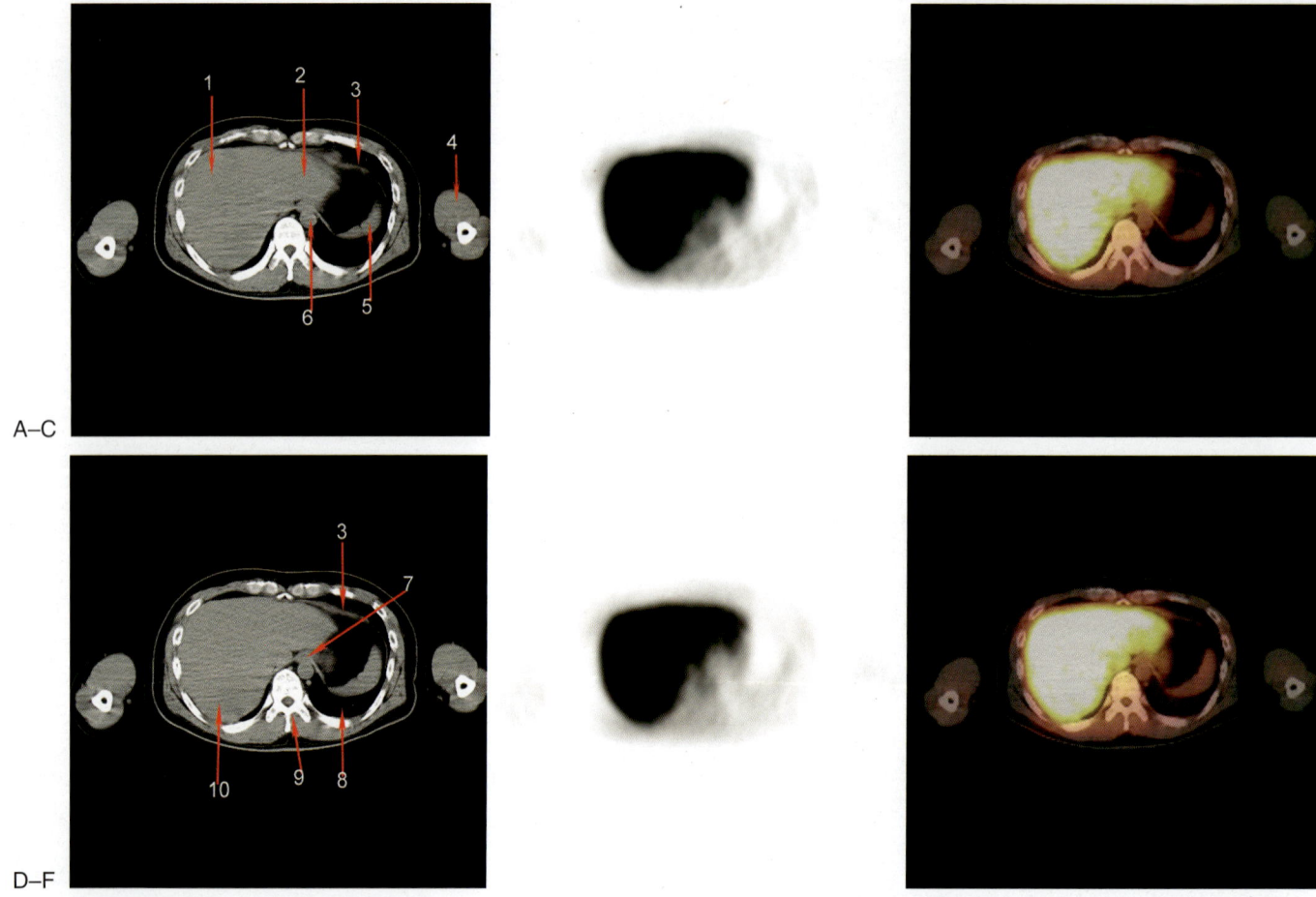

A–C

D–F

FIGURE 2-29:

1. Superior segment of right hepatic
   lobe (#VIII)
2. Left hepatic lobe (#III)
3. Left diaphragm

4. Biceps brachii muscle
5. Spleen
6. Abdominal aorta
7. Esophagogastric junction

8. Left lung base
9. Vertebral lamina
10. Right hepatic lobe (#VII)

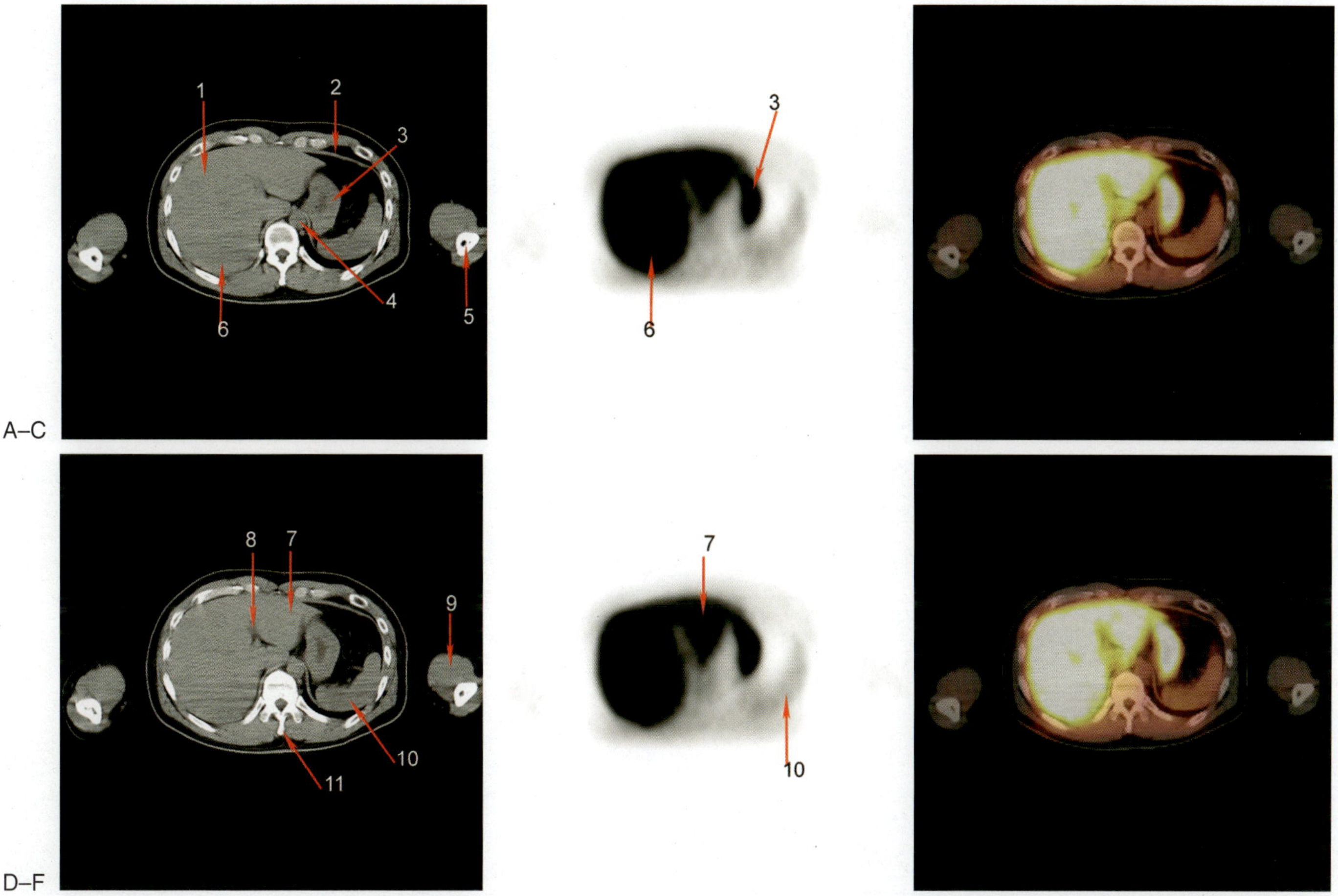

A–C

D–F

FIGURE 2-30:

1. Right hepatic lobe (#VIII)	5. Humerus	9. Biceps brachii muscle
2. Left diaphragm	6. Right hepatic lobe (#VII)	10. Spleen
3. Stomach	7. Left hepatic lobe (#III)	11. Spinous process
4. Abdominal aorta	8. Falciform ligament	

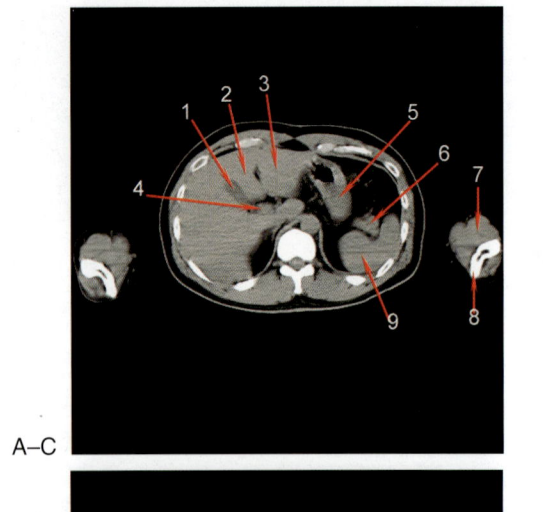
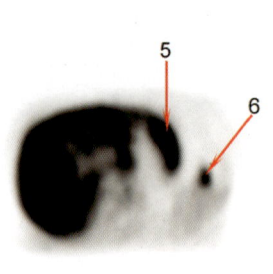
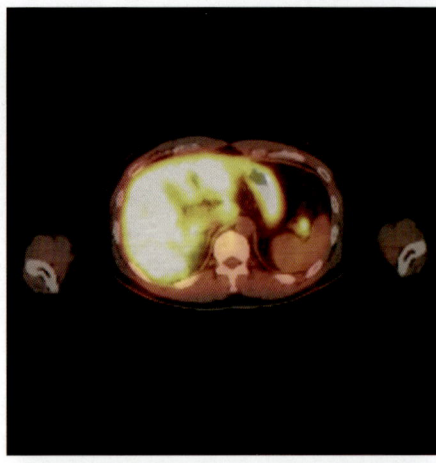

A–C

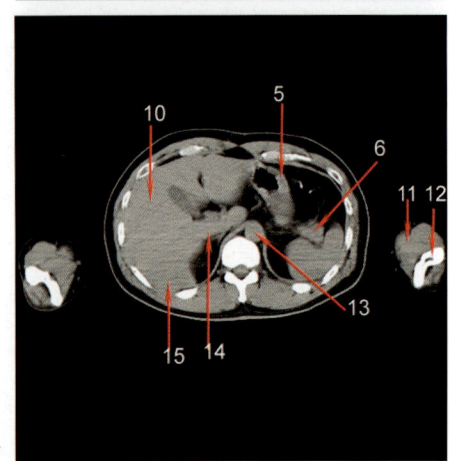
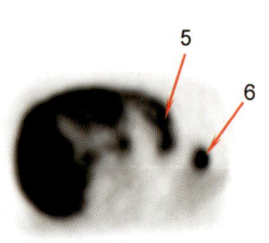
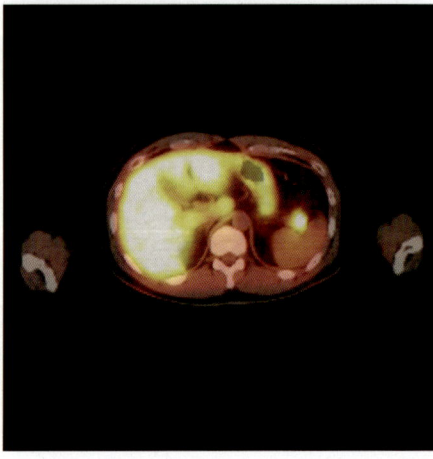

D–F

FIGURE 2-31:

1. Gallbladder
2. Medial segment of left hepatic lobe (#IV)
3. Lateral segment of left hepatic lobe (#III)
4. Portal vein in porta hepatis
5. Stomach
6. Pancreatic tail
7. Brachioradialis muscle
8. Medial epicondyle of humerus
9. Spleen
10. Inferior segment of right hepatic lobe (#V)
11. Brachialis muscle
12. Lateral epicondyle of humerus
13. Abdominal aorta
14. Caudate hepatic lobe (#I)
15. Inferior segment of right hepatic lobe (#VI)

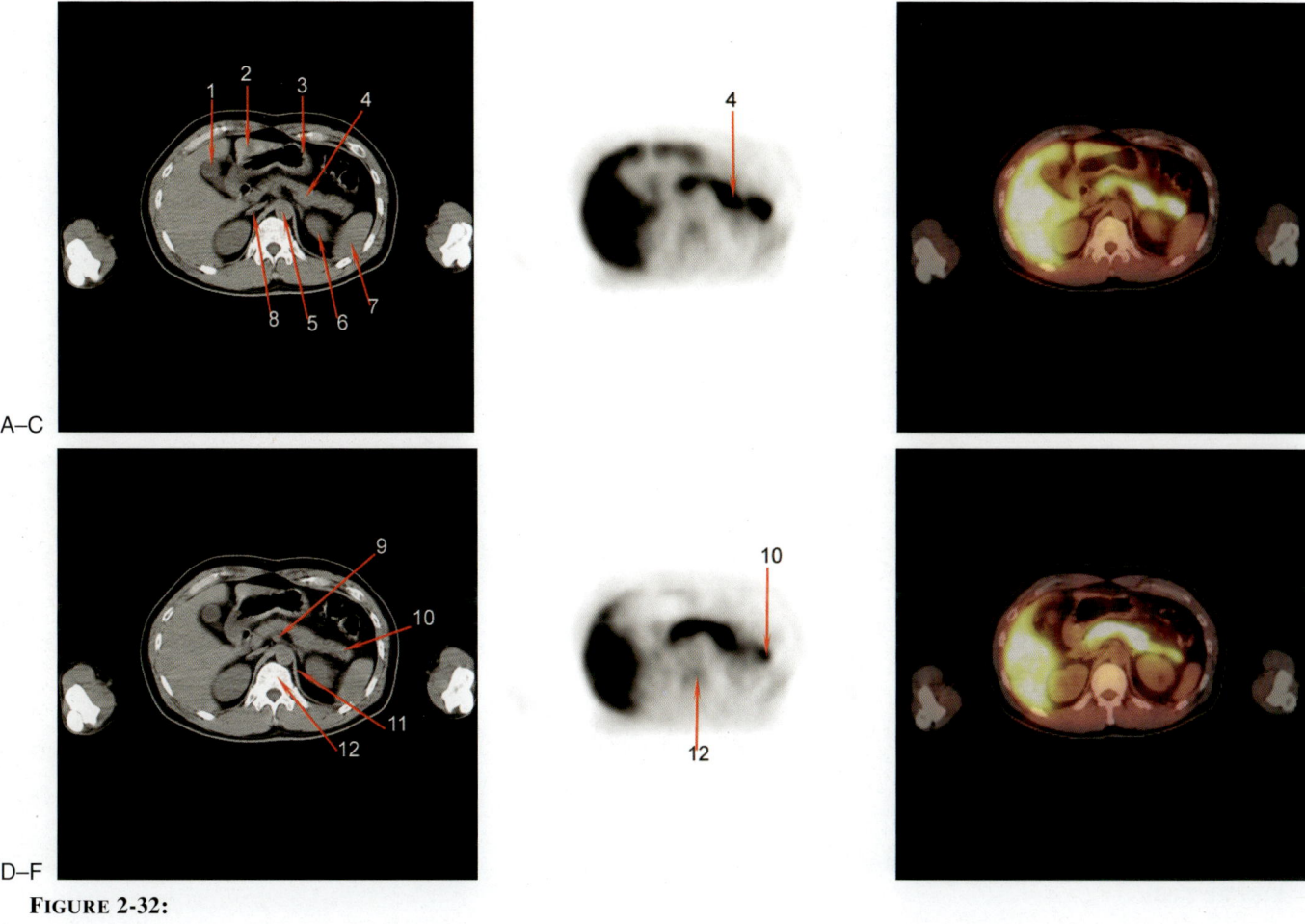

A–C

D–F

FIGURE 2-32:

1. Gallbladder
2. Lateral segment of left hepatic lobe (#III)
3. Stomach
4. Pancreatic body
5. Abdominal aorta
6. Kidney
7. Spleen
8. Inferior vena cava
9. Superior mesenteric artery
10. Pancreatic tail
11. Diaphragmatic crus
12. Vertebral body

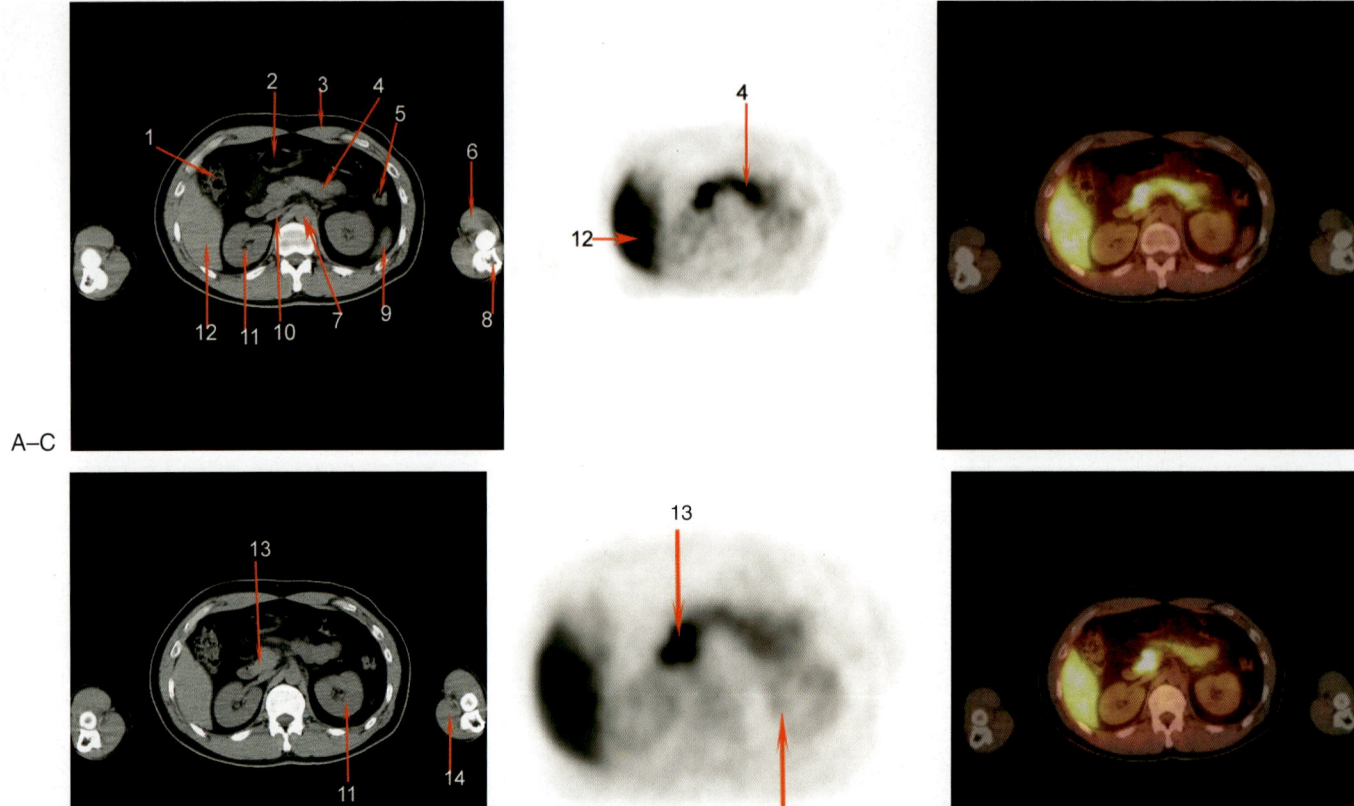

FIGURE 2-33:

1. Colon (hepatic flexure)
2. Stomach
3. Rectus abdominis muscle
4. Pancreatic body
5. Colon (splenic flexure)
6. Brachioradialis muscle
7. Abdominal aorta
8. Olecranon process of ulna
9. Spleen
10. Inferior vena cava
11. Kidney
12. Inferior segment of right hepatic lobe
13. Pancreatic head
14. Brachialis muscle

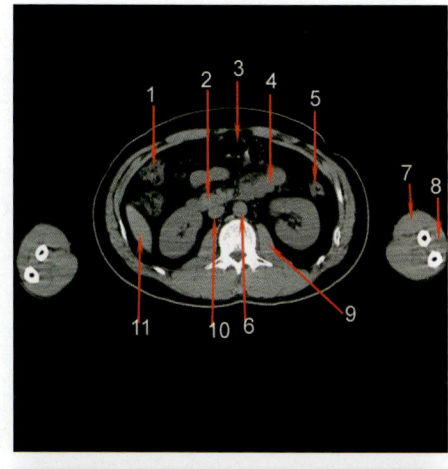
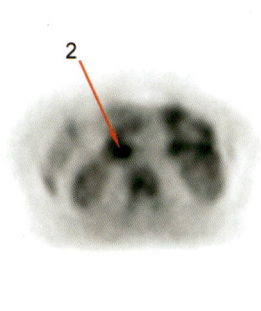
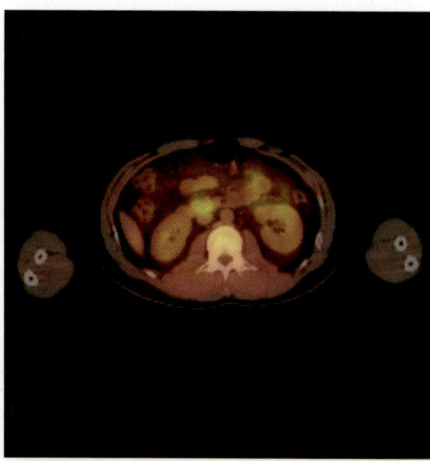

A–C

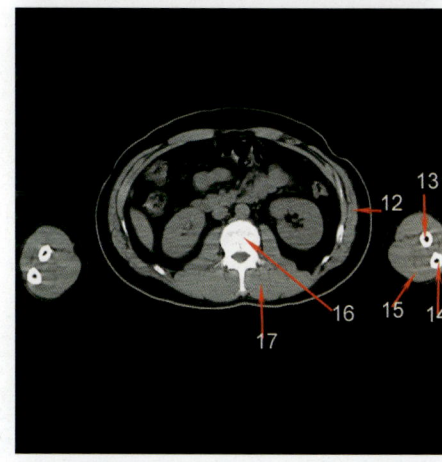
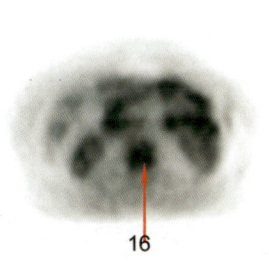
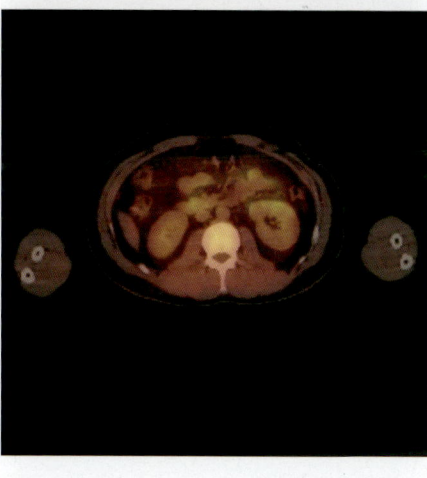

D–F

FIGURE 2-34:

1. Ascending colon
2. Pancreatic head
3. Transverse colon
4. Duodenum (third part)
5. Descending colon
6. Abdominal aorta
7. Brachioradialis muscle
8. Extensor digitorum muscle
9. Psoas muscle
10. Inferior vena cava
11. Inferior tip of liver
12. External oblique muscle
13. Radius
14. Ulna
15. Flexor digitorum profunda muscle
16. Vertebral body
17. Erector spinae muscle

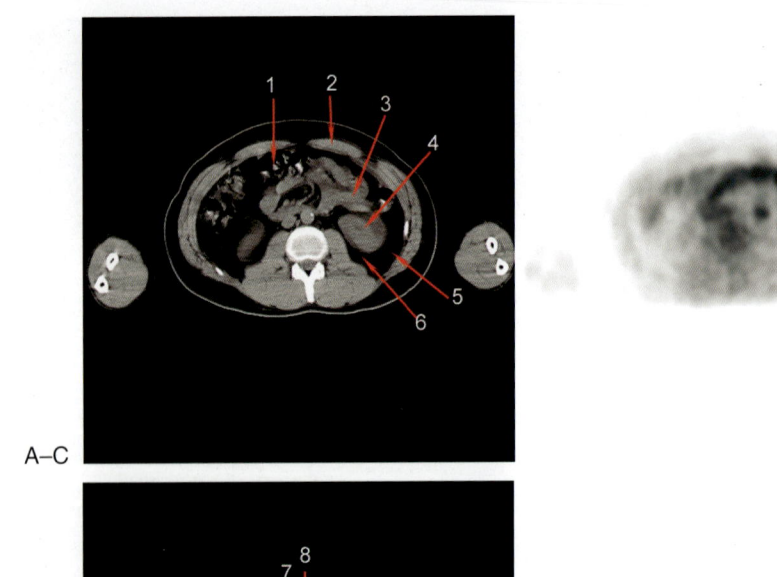
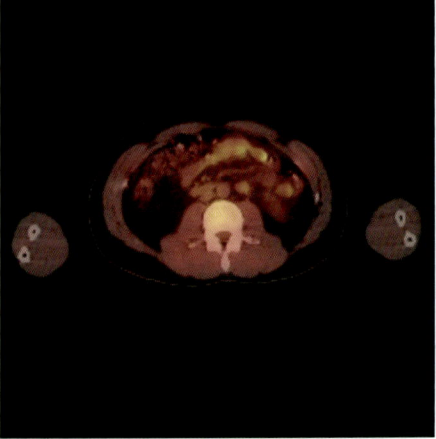

FIGURE 2-35:

1. Transverse colon
2. Rectus abdominis muscle
3. Jejunum
4. Kidney
5. Posterior pararenal space
6. Posterior perirenal space
7. Inferior vena cava
8. Abdominal aorta
9. Internal oblique muscle
10. External oblique muscle
11. Vertebral body
12. Extensor digitorum muscle
13. Flexor digitorum profunda muscle
14. Transverse process
15. Spinous process

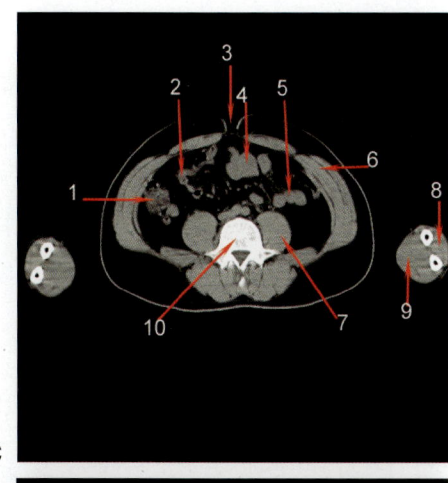
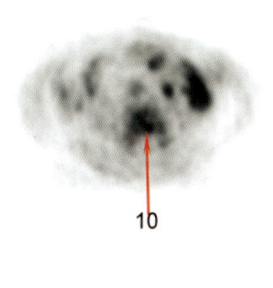
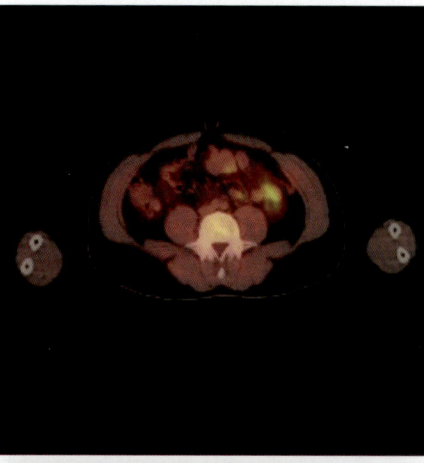

A–C

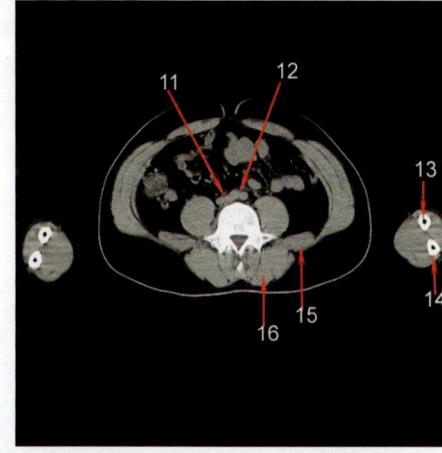

D–F

FIGURE 2-36:

1. Ascending colon
2. Ileum
3. Umbilicus
4. Transverse colon
5. Jejunum
6. Oblique muscle

7. Psoas muscle
8. Extensor digitorum muscle
9. Flexor digitorum muscle
10. Vertebral body
11. Bifurcation of inferior vena cava
12. Bifurcation of abdominal aorta

13. Radius
14. Ulna
15. Quadratus lumborum muscle
16. Erector spinae muscle

Pelvis

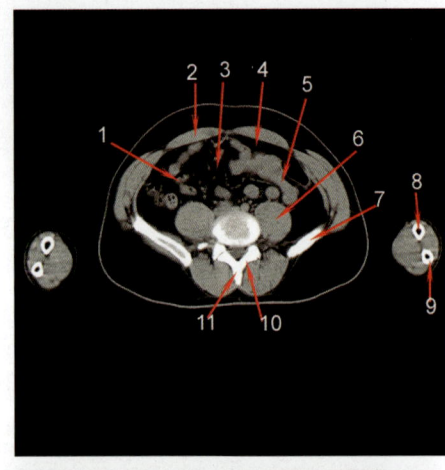
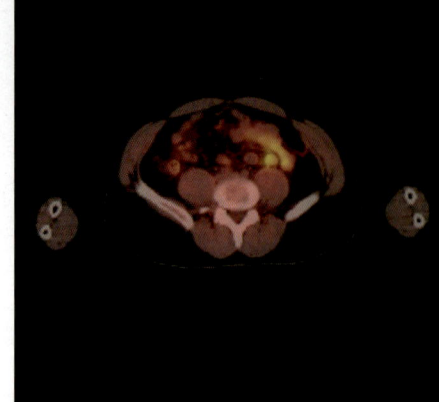

A–C

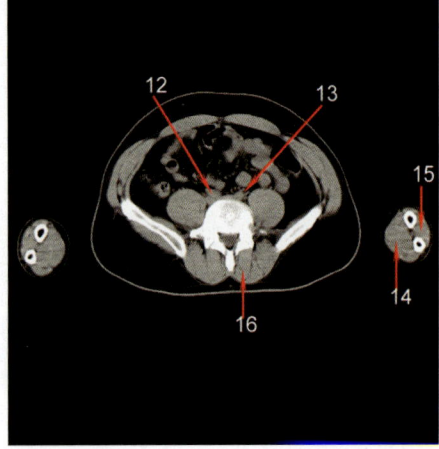
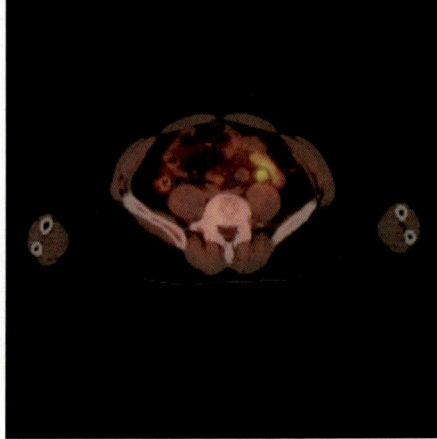

D–F

FIGURE 2-37:

1. Ileum
2. Rectus abdominis muscle
3. Mesentery (fat)
4. Omentum (fat)
5. Jejunum
6. Psoas muscle
7. Iliac crest
8. Radius
9. Ulna
10. Lamina
11. Spinous process
12. Right common iliac vessels
13. Left common iliac vessels
14. Flexor digitorum muscle
15. Extensor digitorum muscle
16. Multifidus muscle

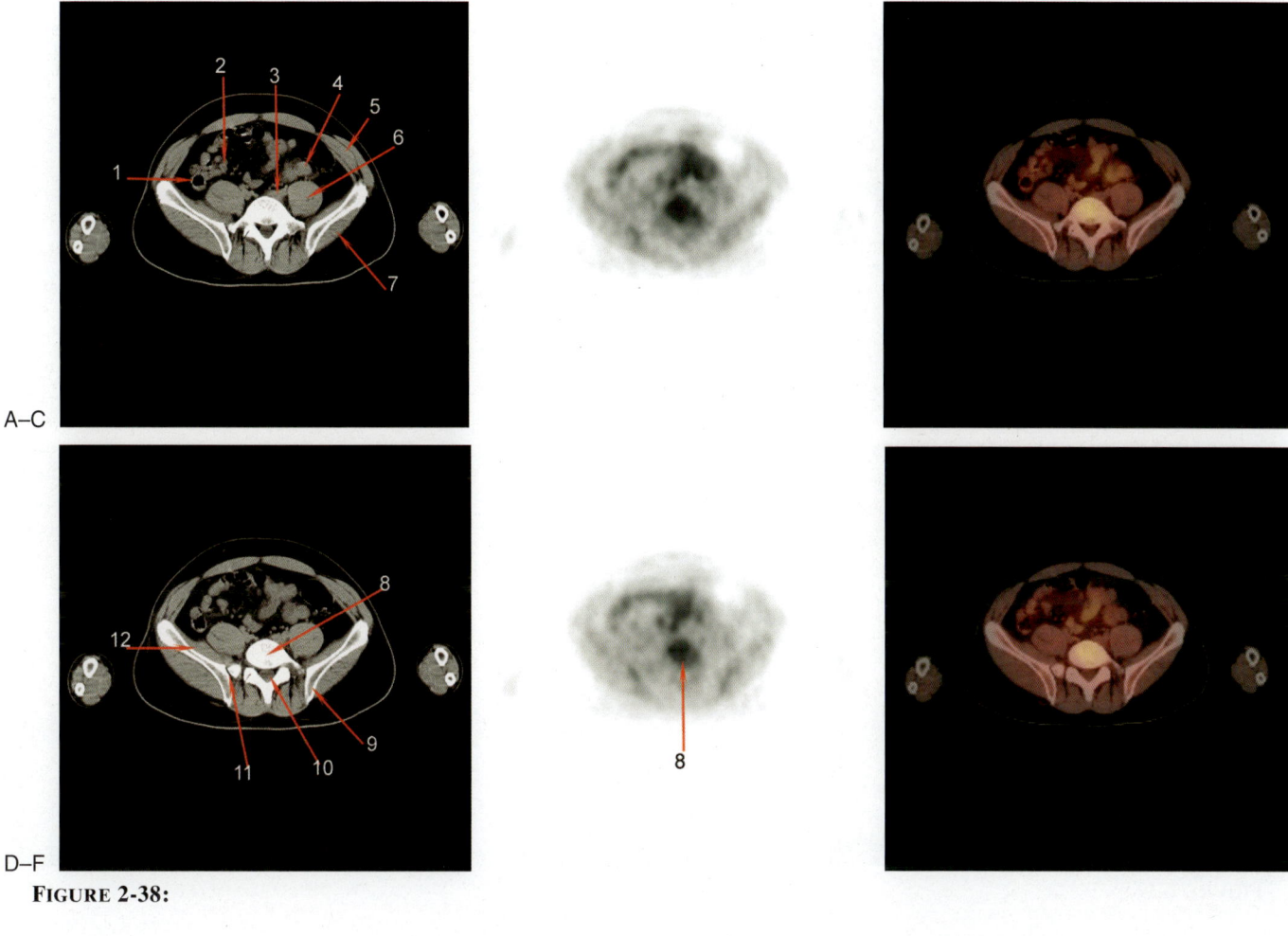

A–C

D–F

FIGURE 2-38:

1. Cecum
2. Ileum
3. Common iliac vessels
4. Jejunum

5. Oblique muscle
6. Psoas muscle
7. Gluteus medius muscle
8. Vertebral body

9. Iliac tuberosity
10. Spinal canal
11. Sacroiliac joint
12. Iliacus muscle

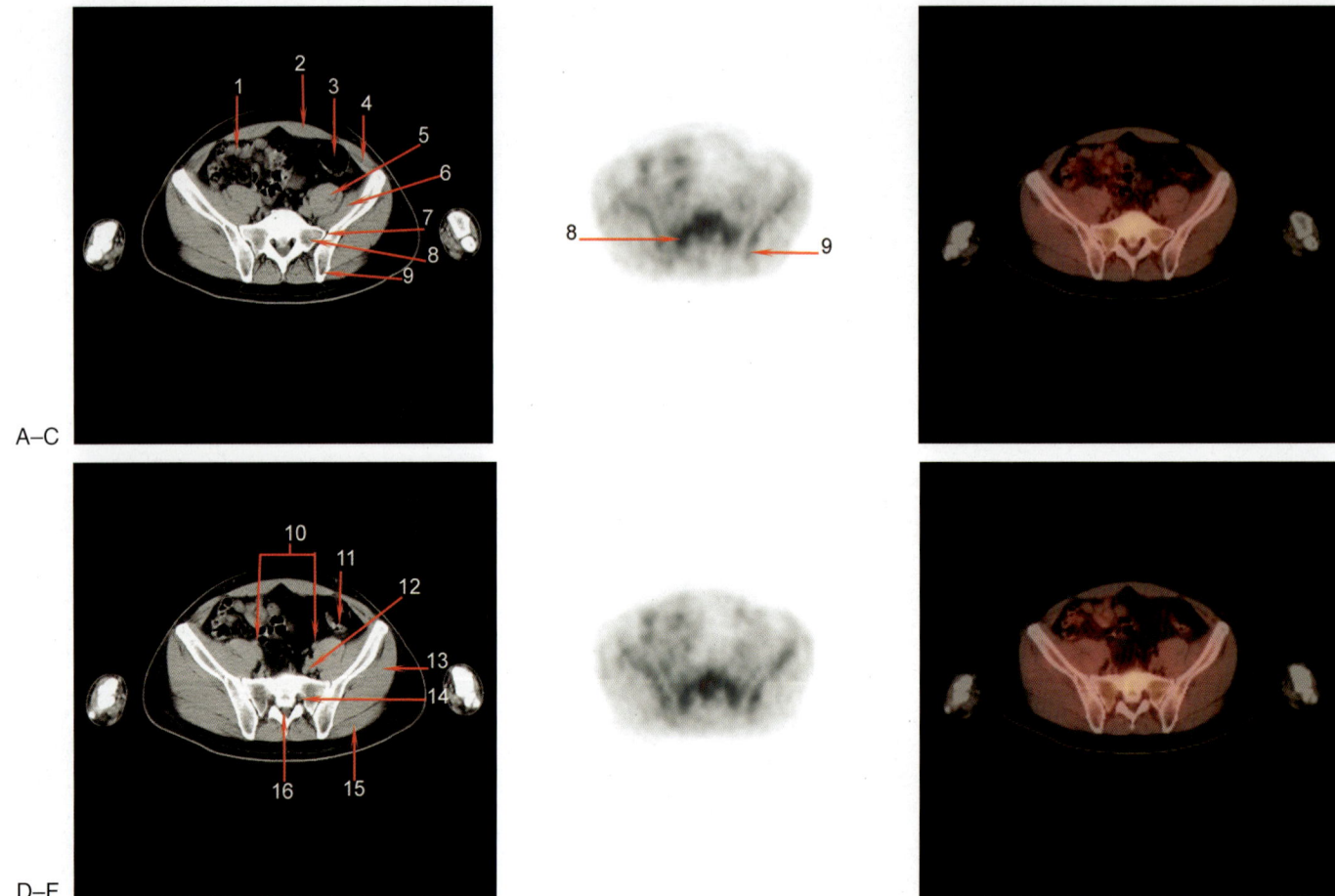

A–C

D–F

FIGURE 2-39:

1. Ileum	7. Sacroiliac joint	13. Gluteus medius
2. Rectus abdominis muscle	8. Sacral ala	14. Neural foramina of S1
3. Descending colon	9. Iliac tuberosity	15. Gluteus maximus
4. Oblique muscle	10. Ureter	16. Sacral canal
5. Psoas muscle	11. Jejunum	
6. Iliacus muscle	12. Common iliac vessels	

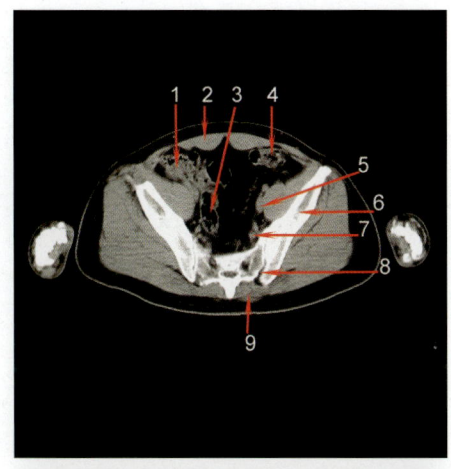

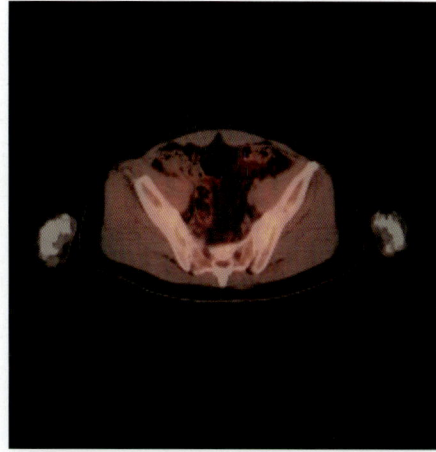

A–C

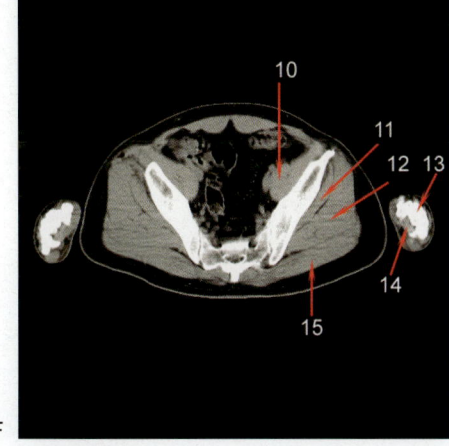

D–F

FIGURE 2-40:

1. Ileum
2. Rectus abdominis muscle
3. Sigmoid colon
4. Jejunum
5. External iliac vessels

6. Ilium
7. Internal iliac vessels
8. Sacroiliac joint
9. Multifidus muscle
10. Iliopsoas muscle

11. Gluteus minimus muscle
12. Gluteus medius muscle
13. Metacarpals
14. Flexor digitorum muscle
15. Gluteus maximus muscle

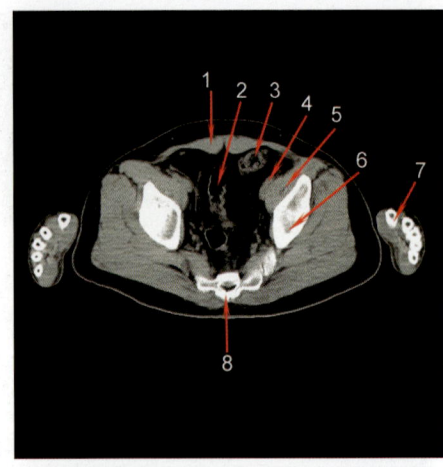
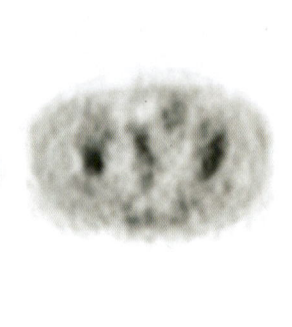
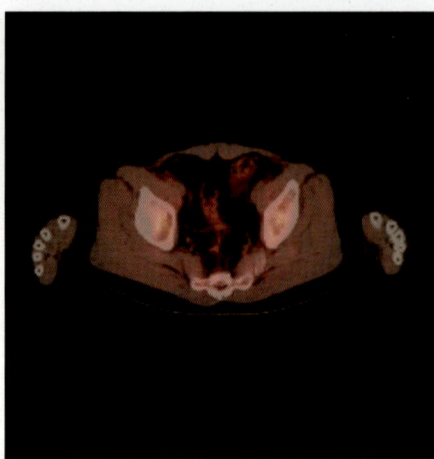

A–C

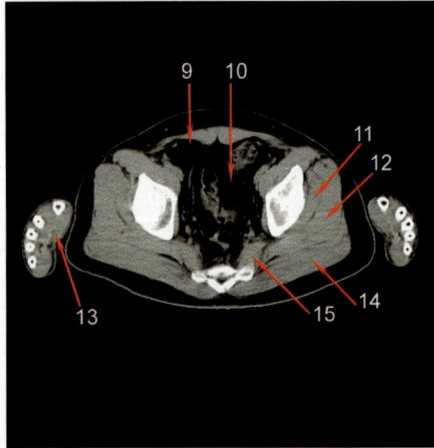
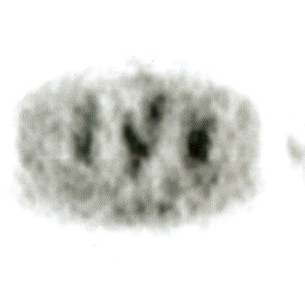
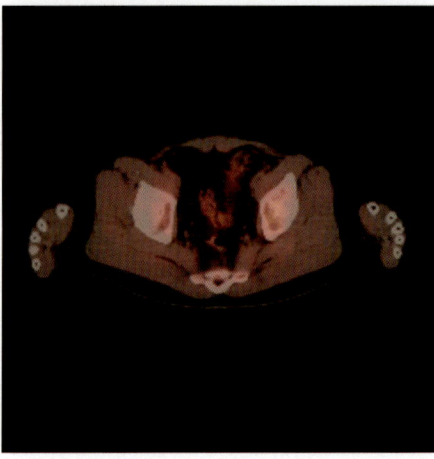

D–F

FIGURE 2-41:

1. Rectus abdominis muscle
2. Sigmoid colon
3. Jejunum
4. External iliac vessels
5. Iliopsoas muscle
6. Ilium
7. First metacarpal
8. Sacrum
9. Omentum (fat)
10. Mesentery (fat)
11. Gluteus minimus muscle
12. Gluteus medius muscle
13. Flexor digitorum muscle
14. Gluteus maximus muscle
15. Piriformis muscle

A–C

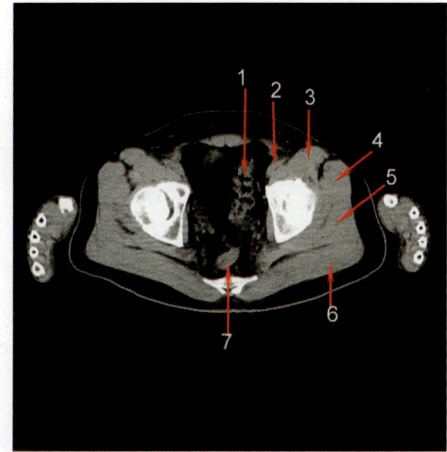
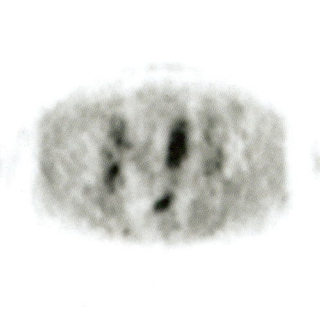
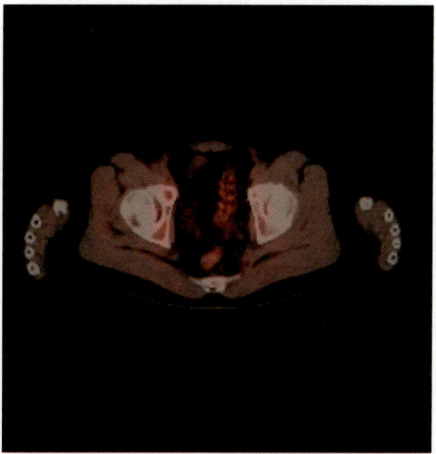

D–F

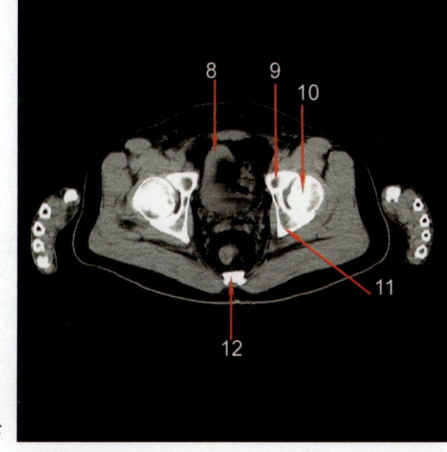
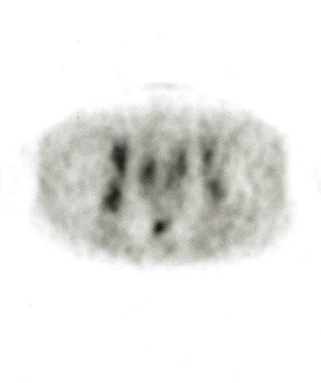
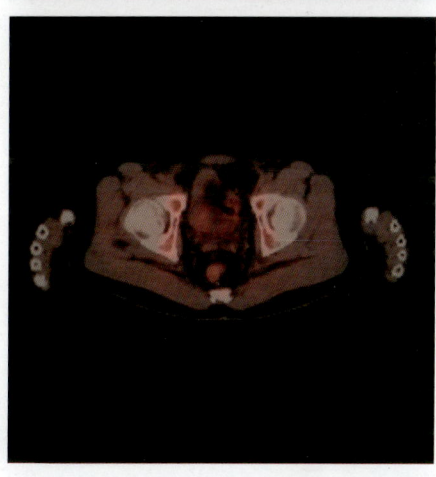

FIGURE 2-42:

1. Sigmoid colon
2. External iliac vessels
3. Iliopsoas muscle
4. Gluteus minimus muscle
5. Gluteus medius muscle
6. Gluteus maximus muscle
7. Rectum
8. Bladder dome
9. Anterior lip of acetabulum
10. Femoral head
11. Posterior lip of acetabulum
12. Coccyx

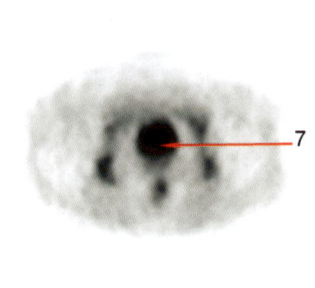
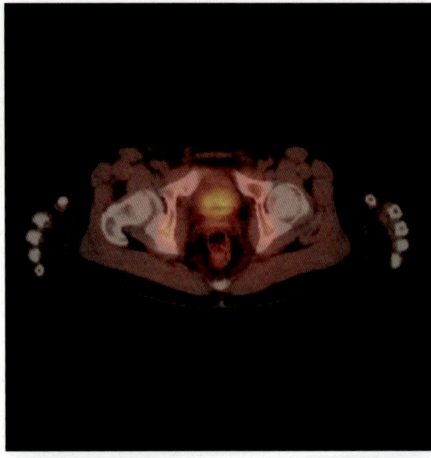

A–C

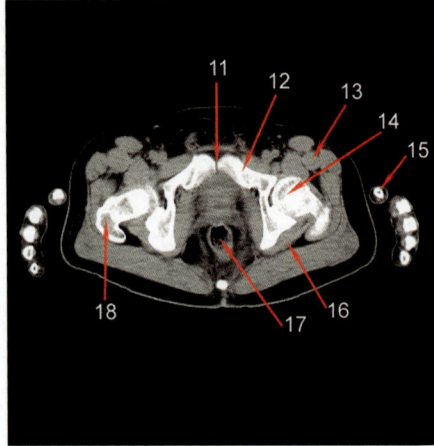
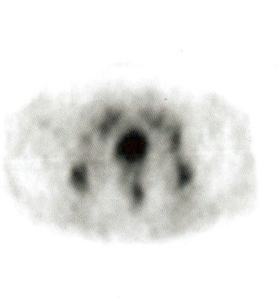
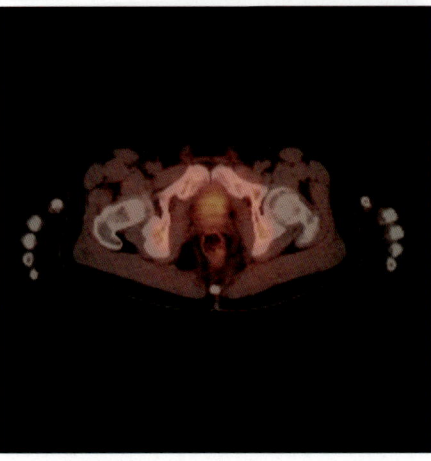

D–F

FIGURE 2-43:

1. Rectus abdominis muscle	7. Bladder	13. Rectus femoris muscle
2. Pectineus muscle	8. Gluteus maximus muscle	14. Femoral head
3. Inguinal vessels	9. Obturator internus muscle	15. First proximal phalanx
4. Sartorius muscle	10. Coccyx	16. Gemellus muscles
5. Iliopsoas muscle	11. Symphysis pubis	17. Rectum
6. Tensor fasciae latae muscle	12. Superior ramus of pubis	18. Trochanter of femur

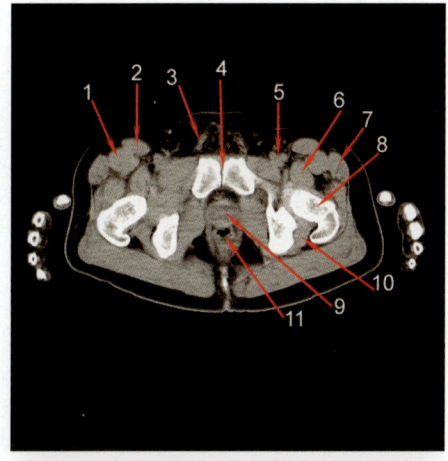
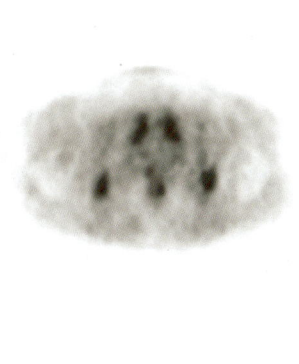
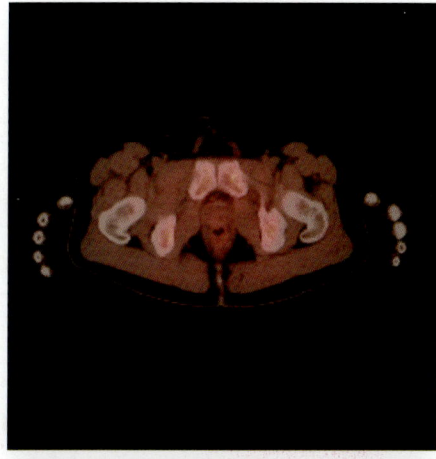

A–C

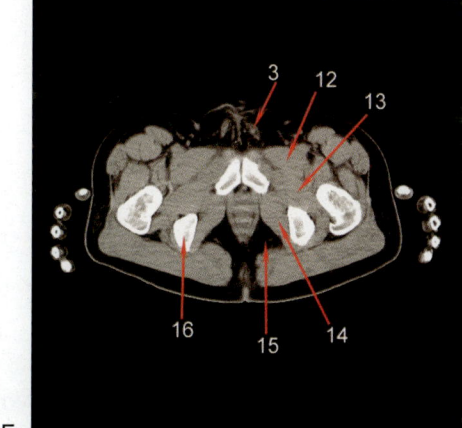
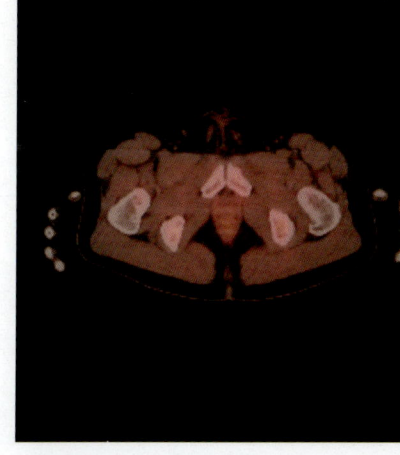

D–F

FIGURE 2-44:

1. Rectus femoris muscle
2. Sartorius muscle
3. Spermatic cord
4. Symphysis pubis
5. Femoral vessels
6. Iliopsoas muscle
7. Tensor fascia latae muscle
8. Femoral neck
9. Prostate
10. Quadratus femoris muscle
11. Rectum
12. Pectineus muscle
13. Obturator externa muscle
14. Obturator interna muscle
15. Ischiorectal fat
16. Ischial tuberosity

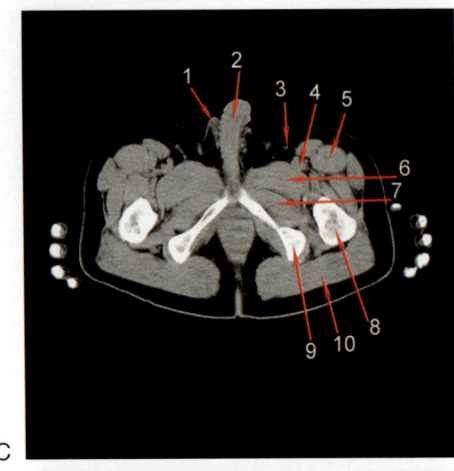
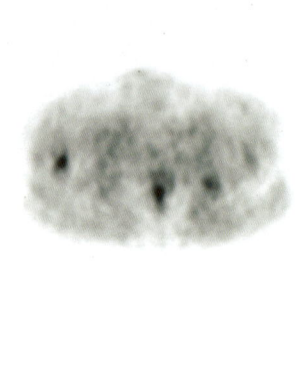
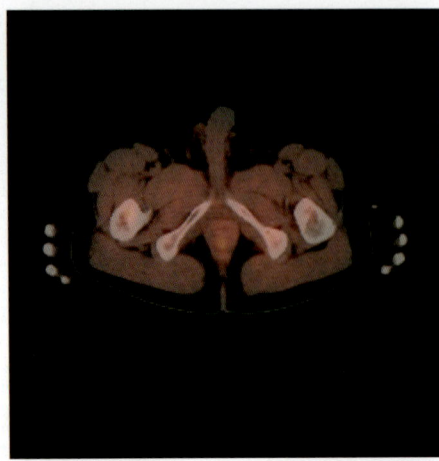

A–C

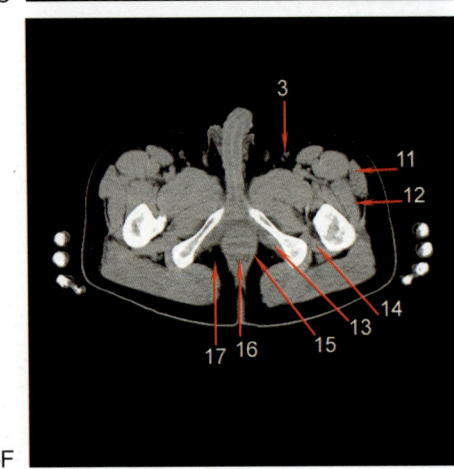
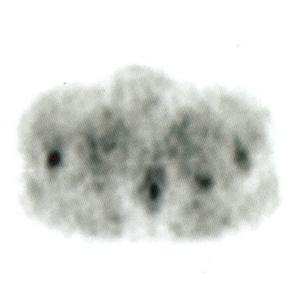
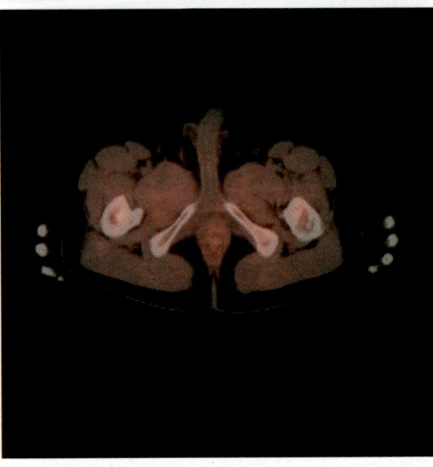

D–F

FIGURE 2-45:

1. Spermatic cord
2. Corpora muscles of penis
3. Inguinal lymph node
4. Femoral vessels
5. Rectus femoris muscle
6. Adductor brevis muscle

7. Obturator externus muscle
8. Femoral trochanter
9. Ischium
10. Gluteus maximus muscle
11. Tensor fascia latae muscle
12. Vastus lateralis muscle

13. Obturator interna muscle
14. Quadratus femoris muscle
15. Levator ani muscle
16. Anus
17. Ischiorectal fat

A–C

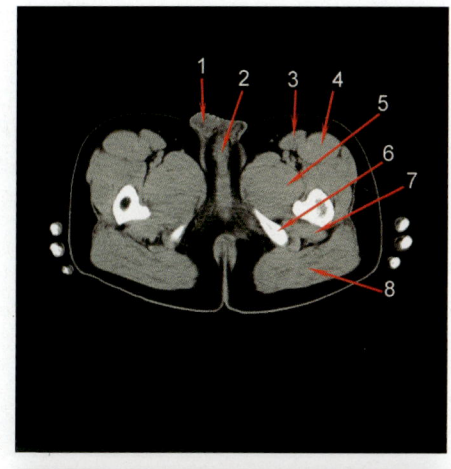 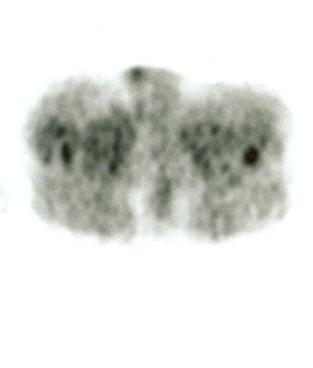 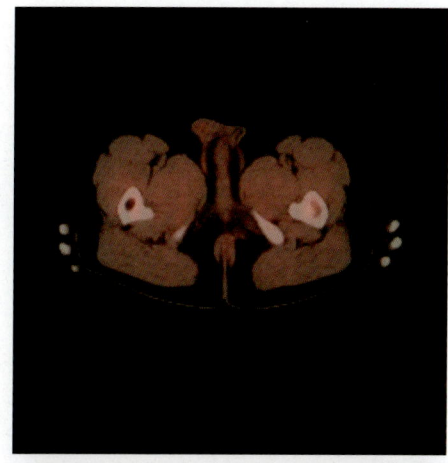

D–F

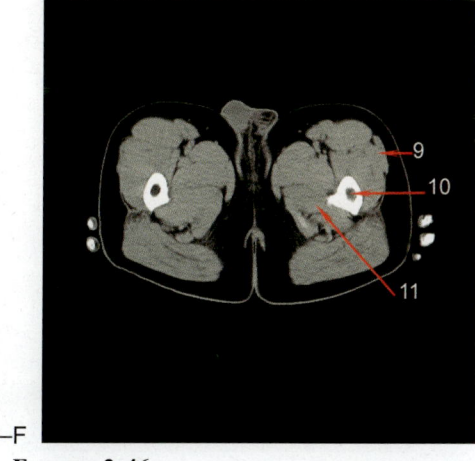 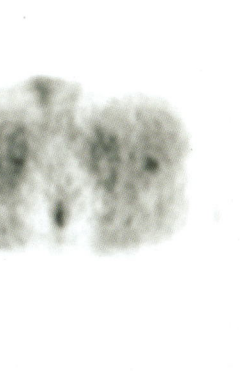 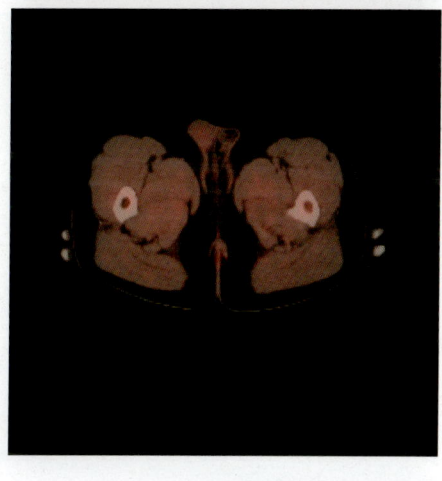

FIGURE 2-46:

1. Testis
2. Corpora muscles of penis
3. Sartorius muscle
4. Rectus femoris muscle
5. Adductor brevis muscle
6. Ischium
7. Quadratus femoris muscle
8. Gluteus maximus muscle
9. Vastus lateralis muscle
10. Femur
11. Adductor magnus muscle

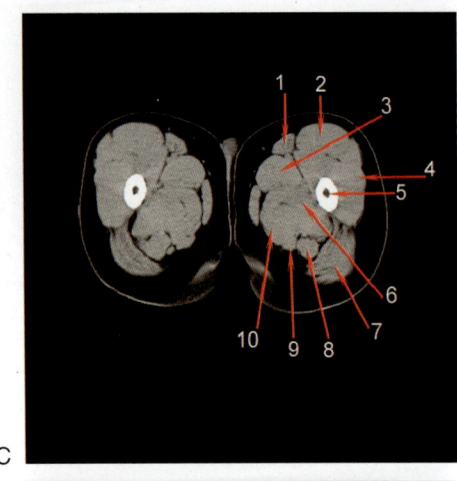
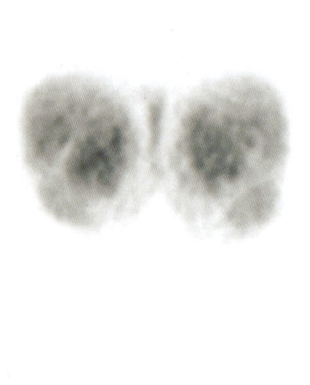
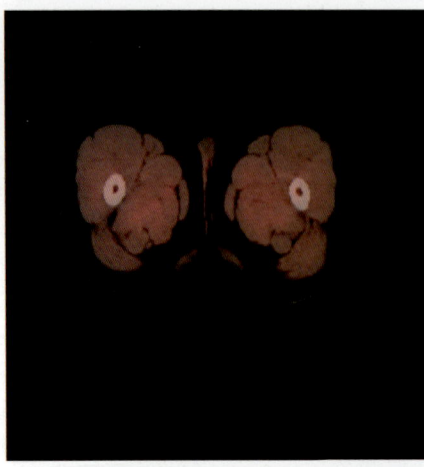

A–C

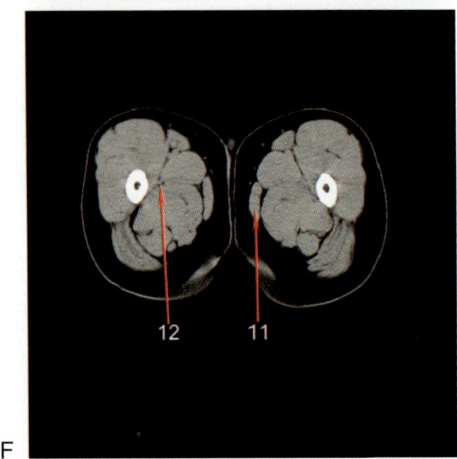
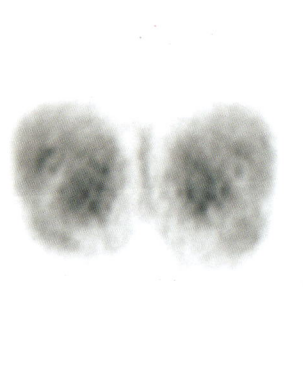
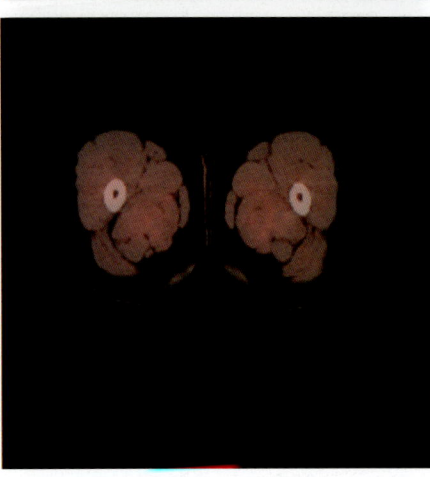

D–F

FIGURE 2-47:

1. Sartorius muscle
2. Rectus femoris muscle
3. Vastus medialis muscle
4. Vastus lateralis muscle

5. Femur
6. Adductor magnus muscle
7. Gluteus maximus muscle
8. Biceps femoris muscle

9. Semitendinosus muscle
10. Semimembranous muscle
11. Gracilis muscle
12. Femoral vessels

Section 2: PET with C-11 Acetate

Coronals

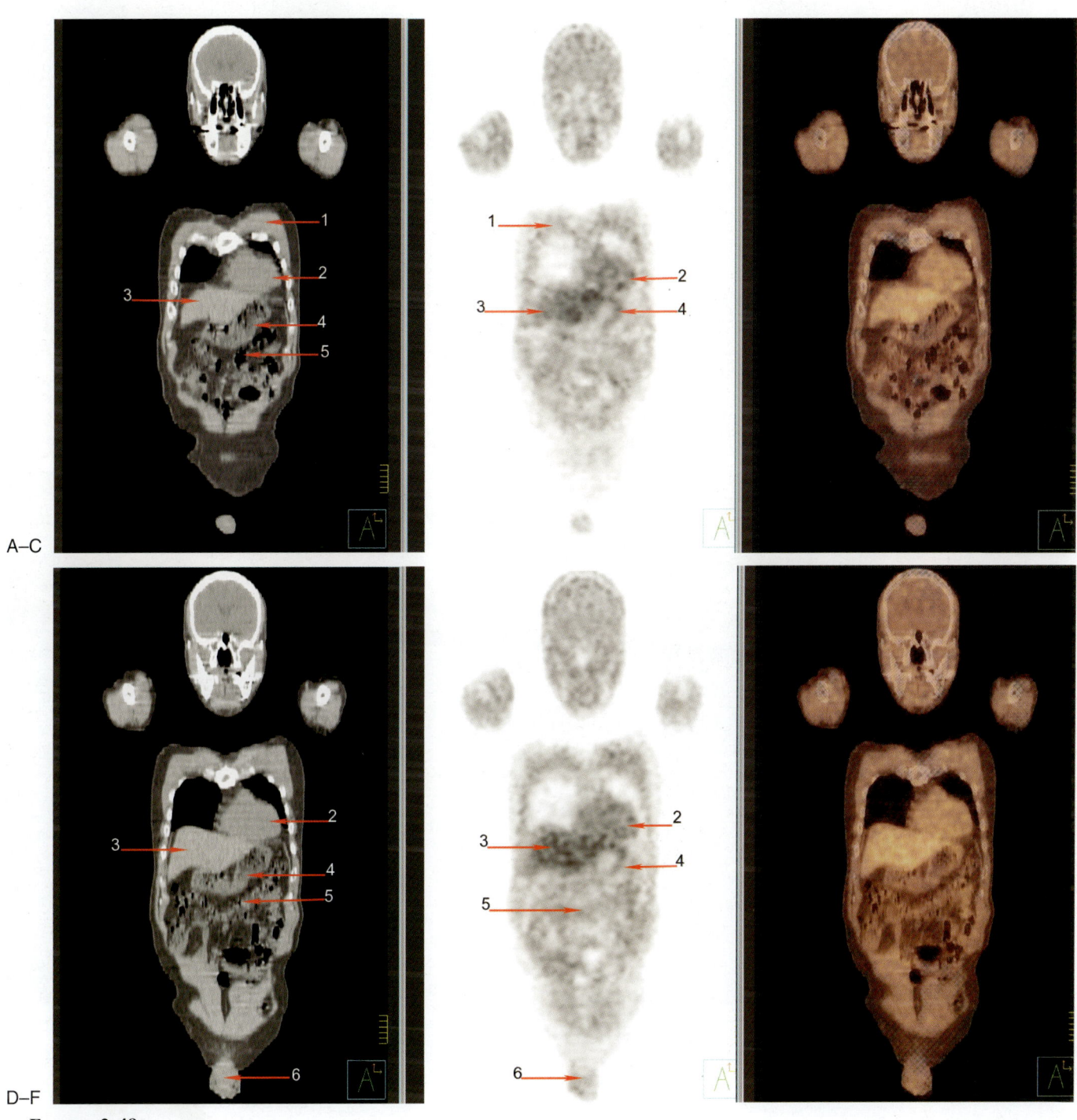

A–C

D–F

FIGURE 2-48:

1. Pectoralis major muscle
2. Left ventricle
3. Right liver
4. Stomach
5. Transverse colon
6. Penile corpora muscles

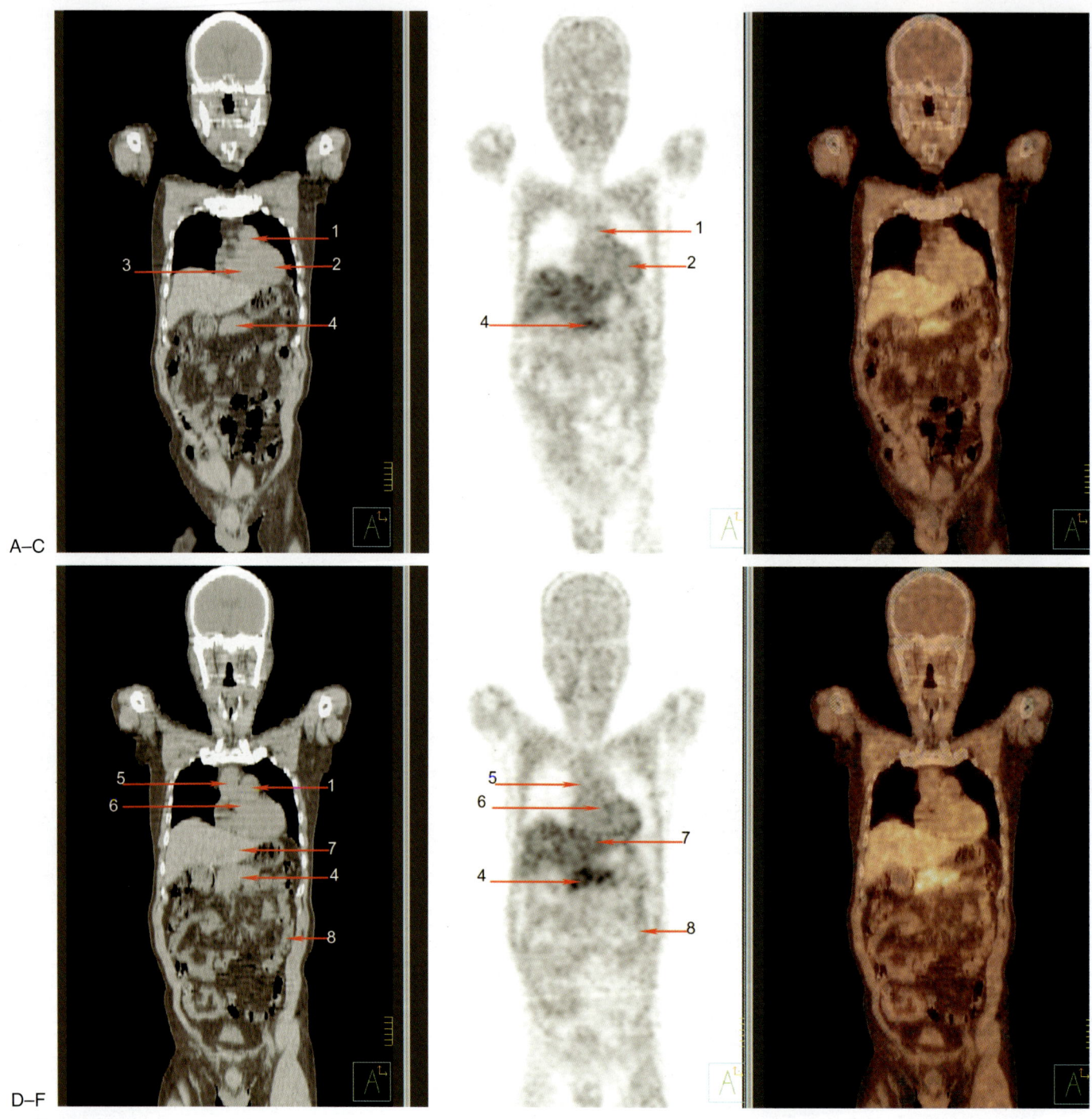

A–C

D–F

FIGURE 2-49:

1. Main pulmonary artery
2. Left ventricle
3. Right atrium

4. Pancreatic body
5. Superior vena cava
6. Ascending thoracic aorta

7. Left hepatic lobe
8. Descending colon

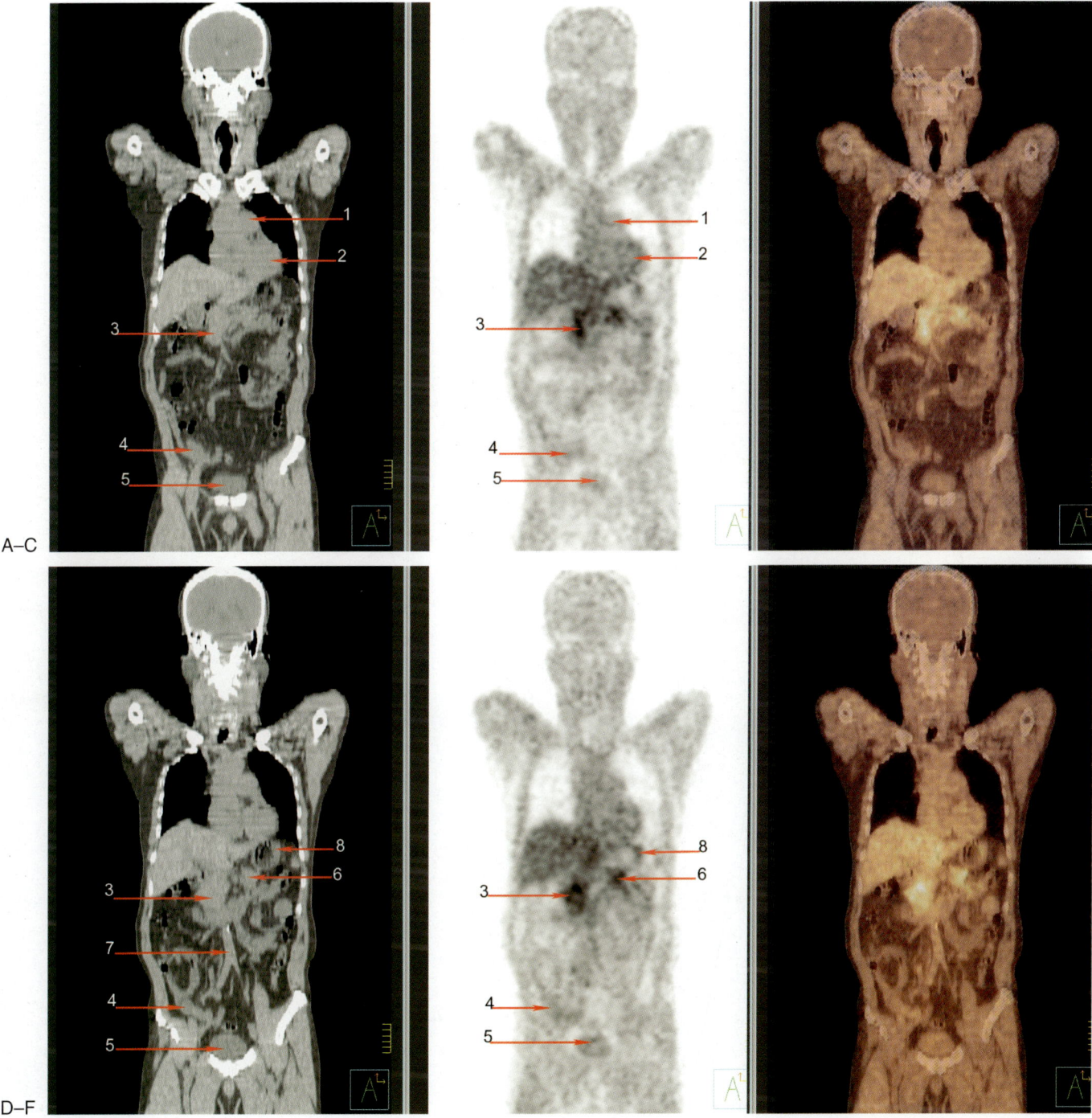

A–C

D–F

FIGURE 2-50:

1. Main pulmonary artery
2. Left ventricle
3. Pancreatic head

4. Cecum
5. Bladder
6. Pancreatic body

7. Abdominal aorta
8. Stomach

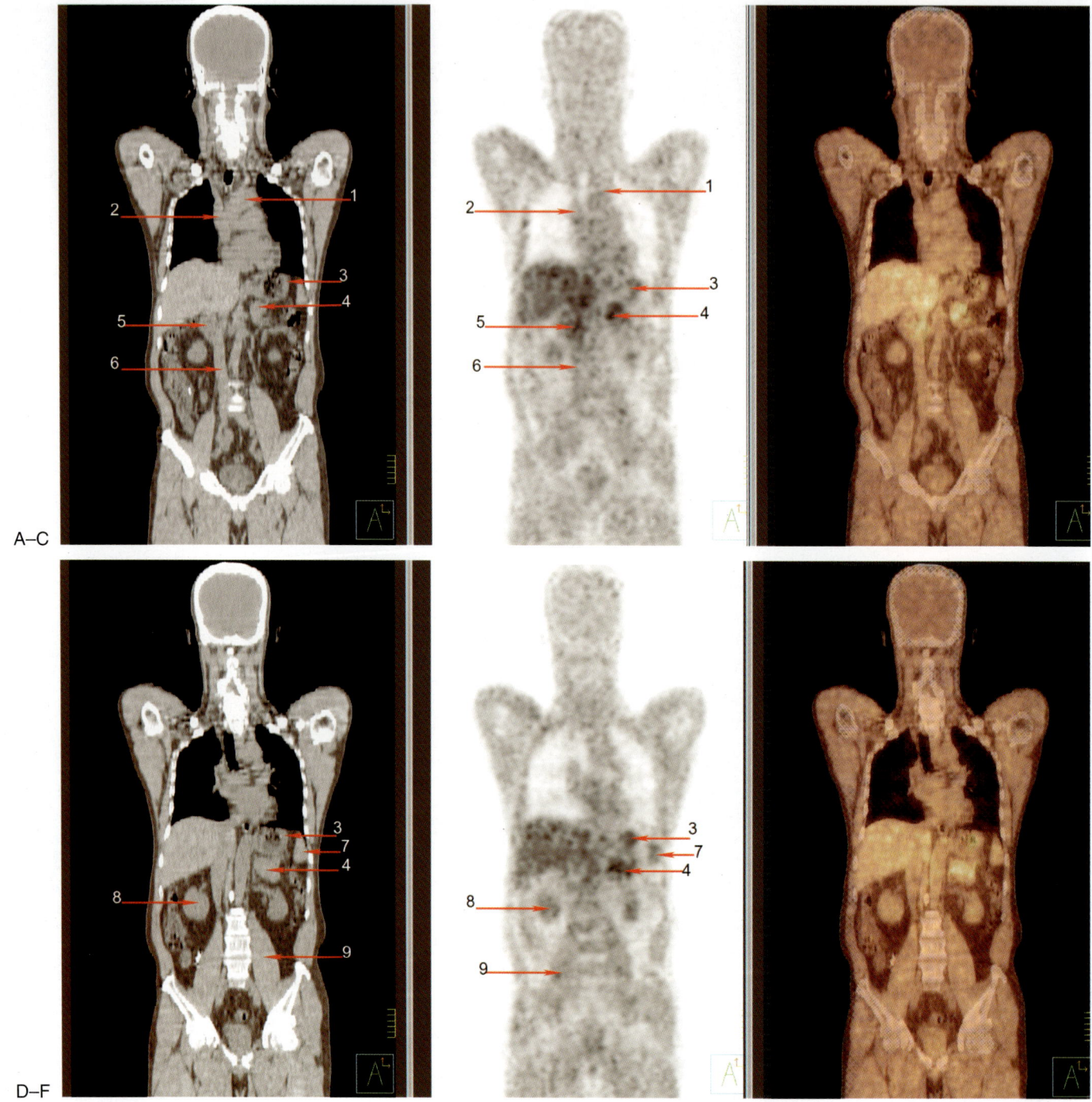

A–C

D–F

FIGURE 2-51:

1. Aortic arch
2. Superior vena cava
3. Stomach
4. Pancreatic tail
5. Pancreatic head
6. Inferior vena cava
7. Spleen
8. Kidney
9. Psoas muscle

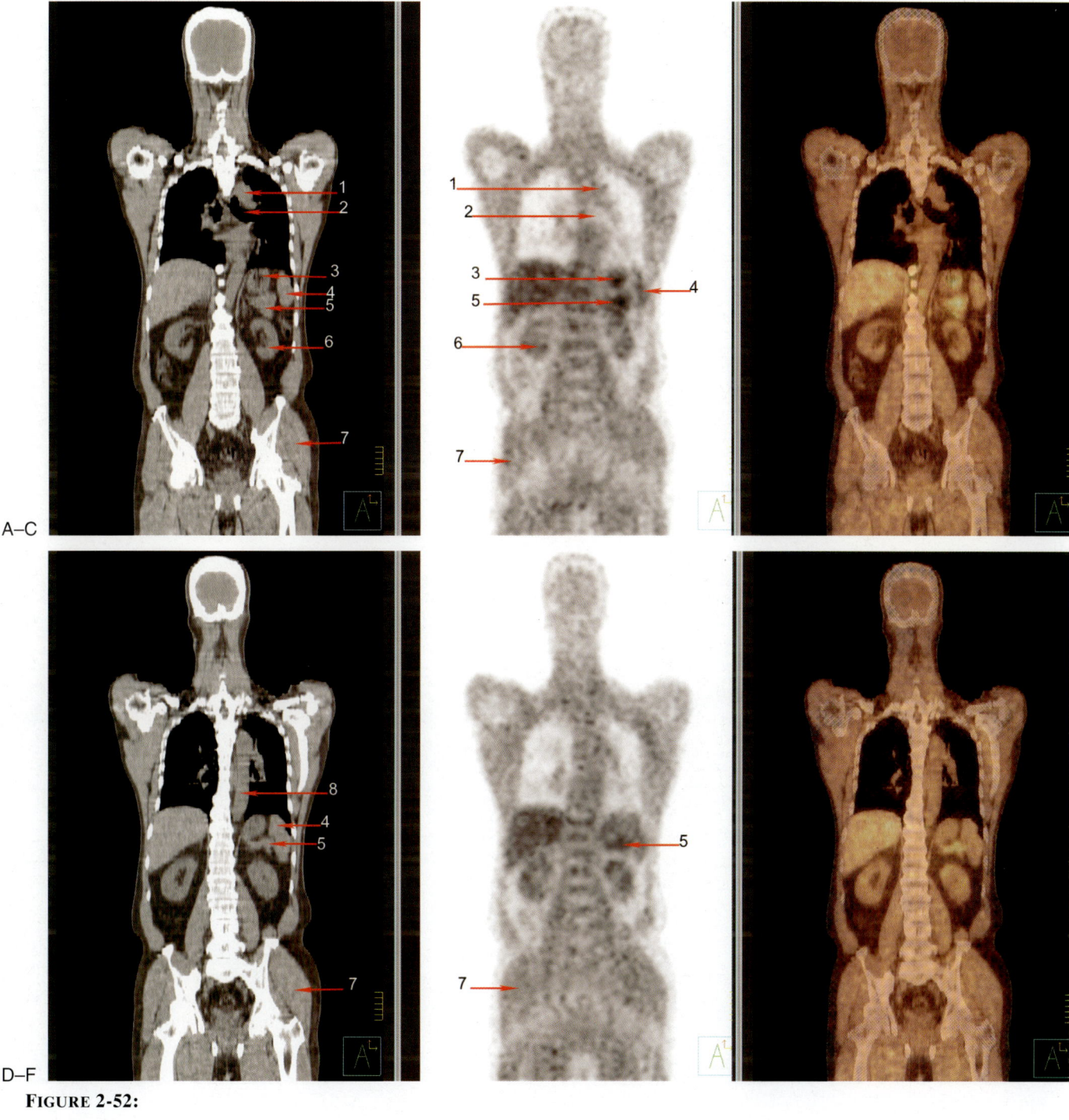

A–C

D–F

FIGURE 2-52:

1. Aortic arch
2. Left main bronchus
3. Stomach
4. Spleen
5. Pancreatic tail
6. Kidney
7. Gluteus muscle
8. Descending aorta

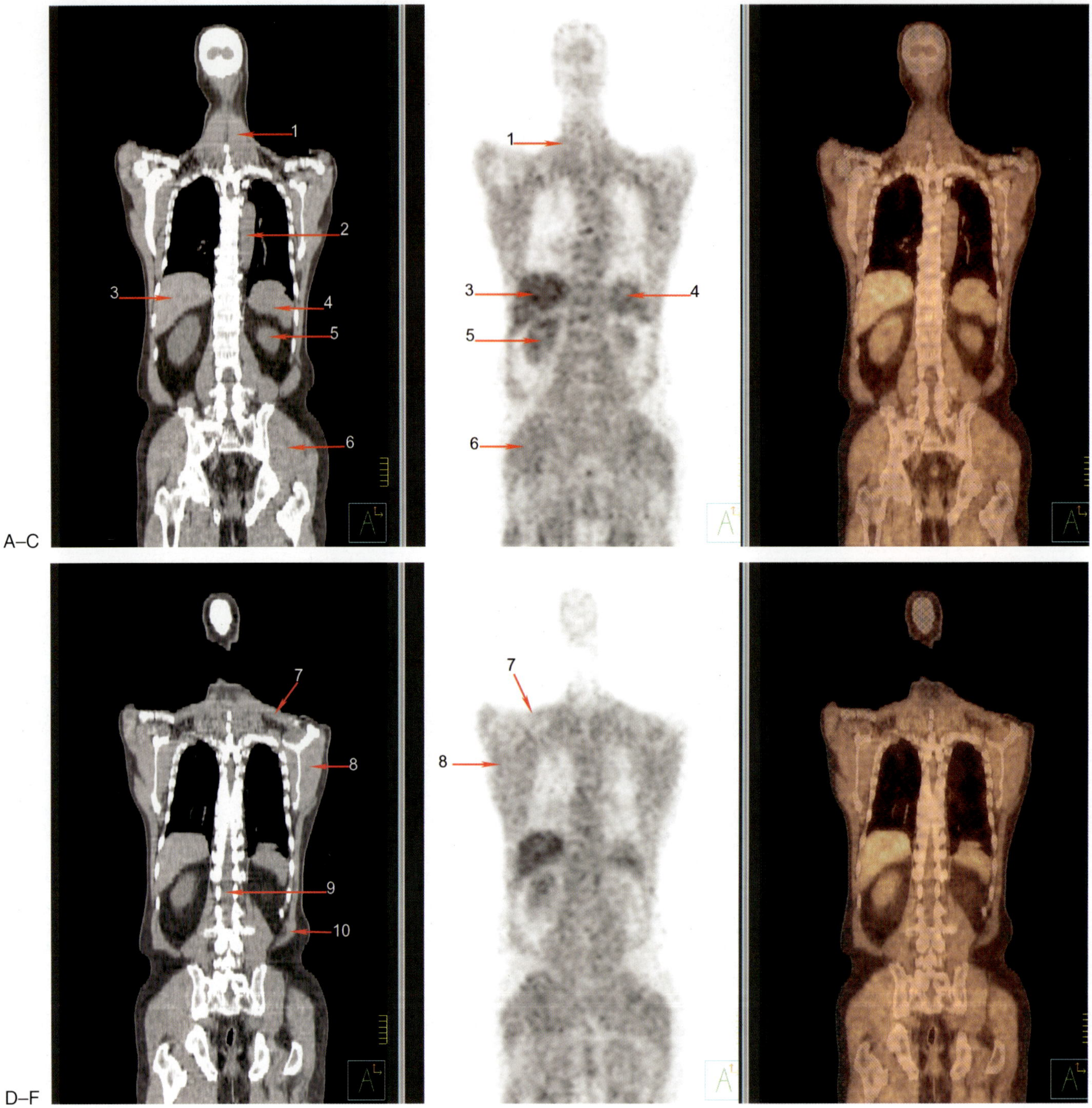

A–C

D–F

FIGURE 2-53:

1. Semispinalis cervicis muscle
2. Descending aorta
3. Right hepatic lobe
4. Spleen
5. Kidney
6. Gluteus muscle
7. Trapezius muscle
8. Infraspinatus muscle
9. Spinal canal
10. Oblique muscles

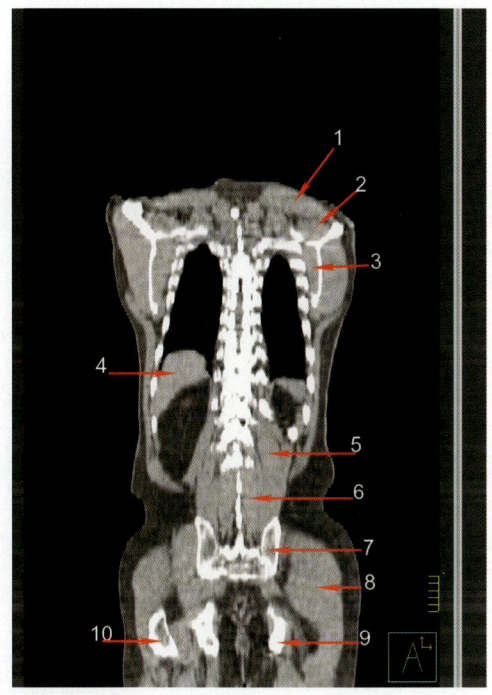
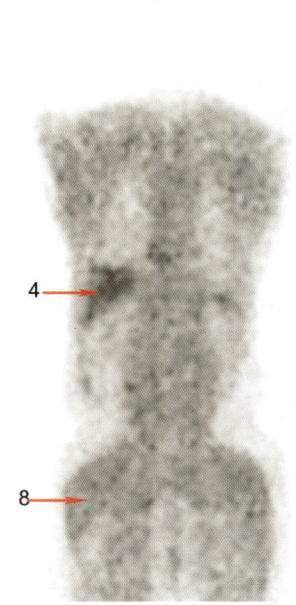
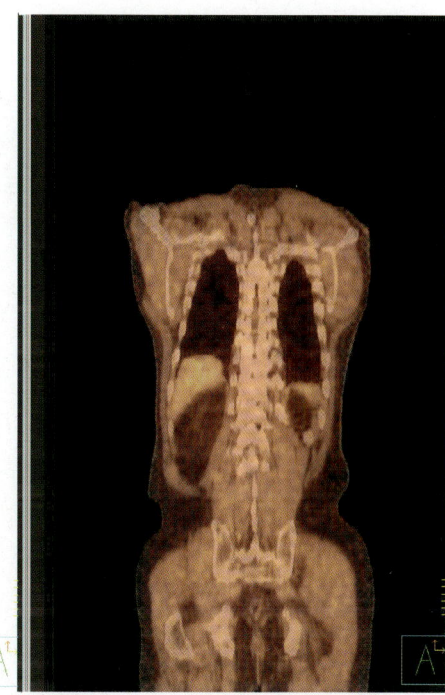

A–C

FIGURE 2-54:

1. Trapezius muscle
2. Supraspinatus muscle
3. Subscapularis muscle
4. Right hepatic lobe
5. Quadratus lumborum muscle
6. Erector spinae muscle
7. Iliac tuberosity
8. Gluteus muscles
9. Ischium
10. Trochanter of femur

Sagittals

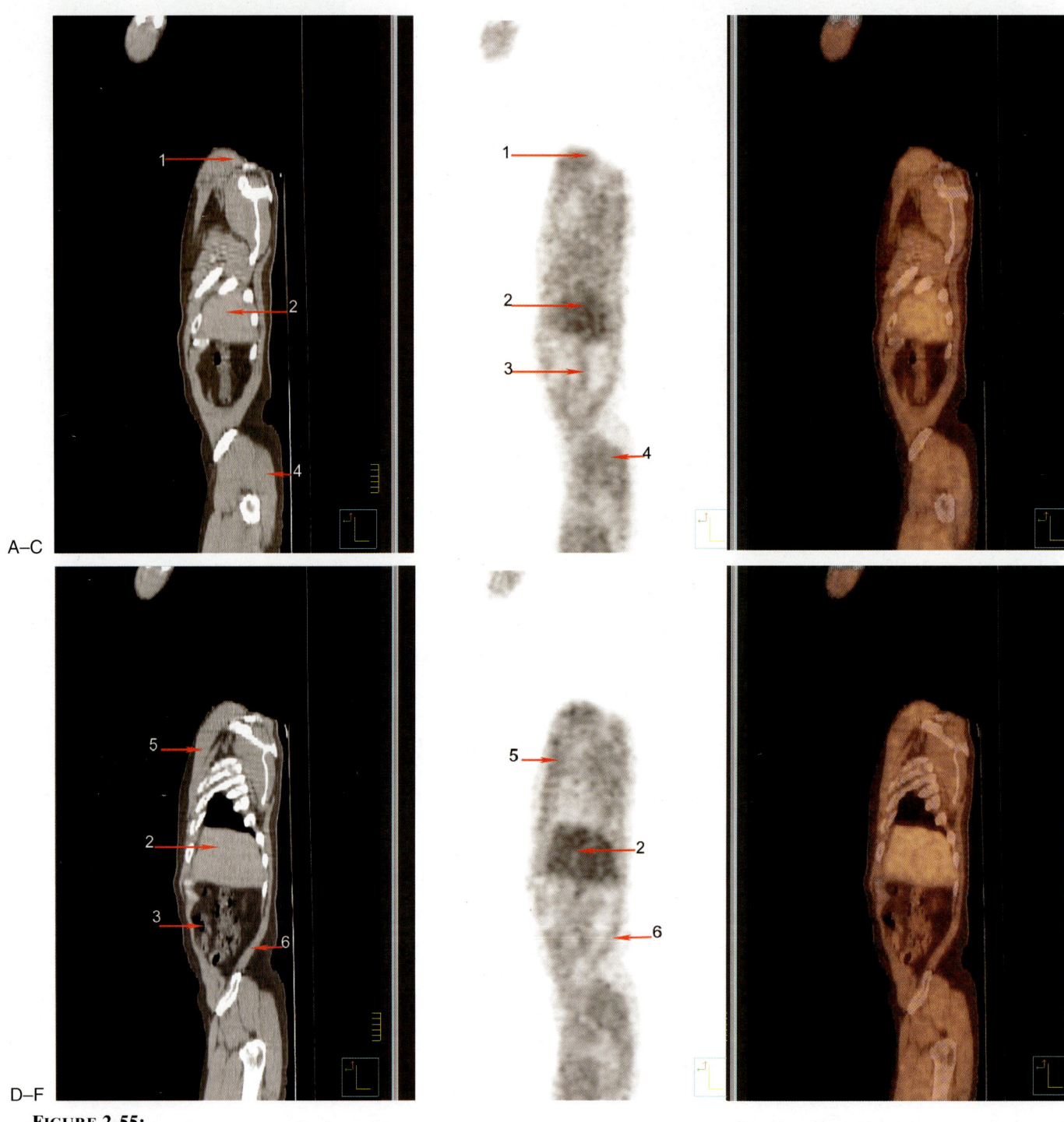

FIGURE 2-55:

1. Trapezius muscle
2. Right hepatic lobe
3. Ascending colon
4. Gluteus muscle
5. Pectoralis major muscle
6. Latissimus dorsi muscle

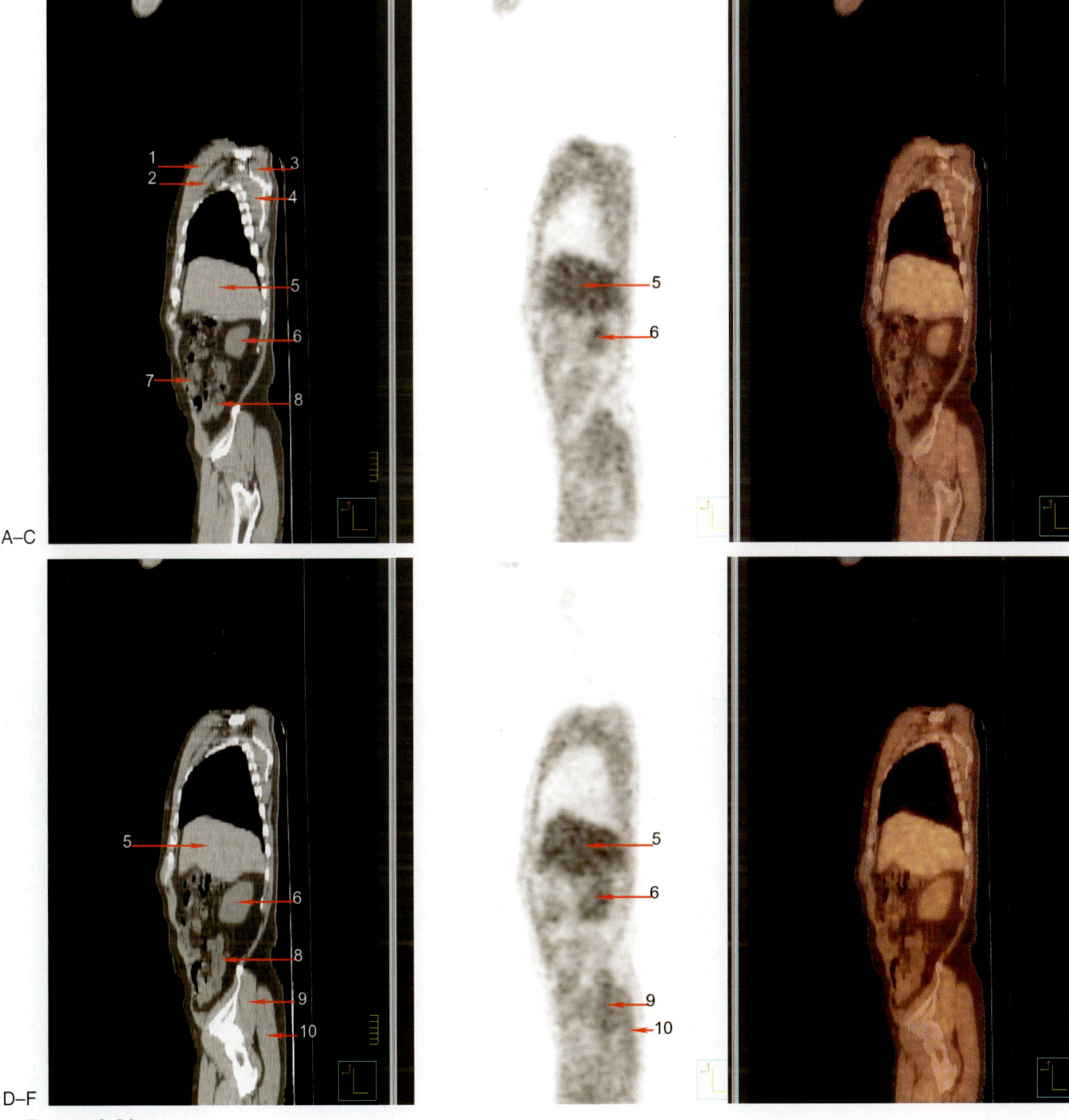

A–C

D–F

FIGURE 2-56:

1. Pectoralis major muscle
2. Pectoralis minor muscle
3. Supraspinatus muscle
4. Subscapularis muscle
5. Right hepatic lobe
6. Kidney
7. Ascending colon
8. Ileum
9. Gluteus medius muscle
10. Gluteus maximus muscle

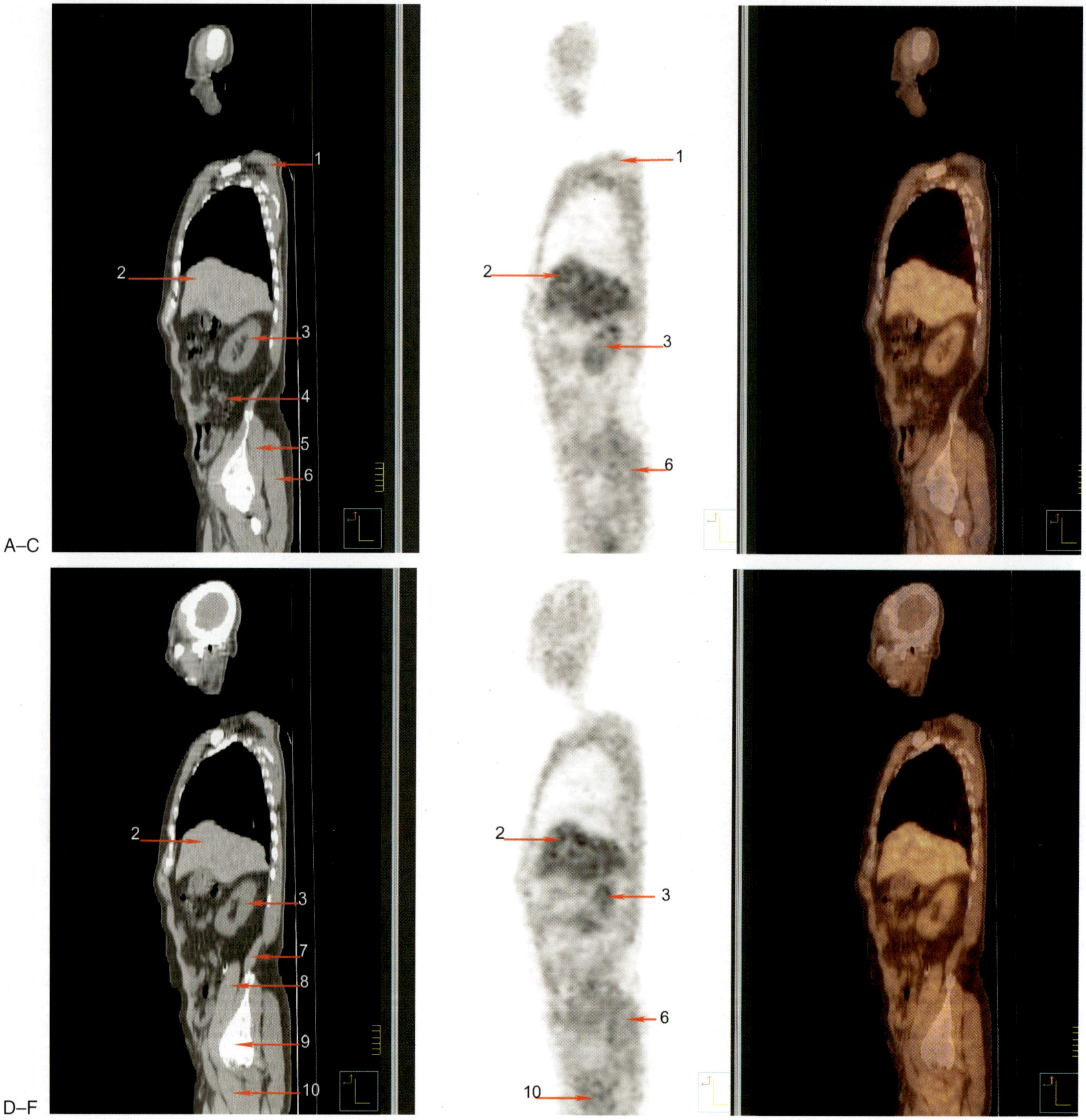

A–C

D–F

FIGURE 2-57:

1. Trapezius muscle
2. Right hepatic lobe
3. Kidney
4. Ileum
5. Gluteus medius muscle
6. Gluteus maximus muscle
7. Quadratus lumborum muscle
8. Iliacus muscle
9. Iliac acetabulum
10. Adductor muscles

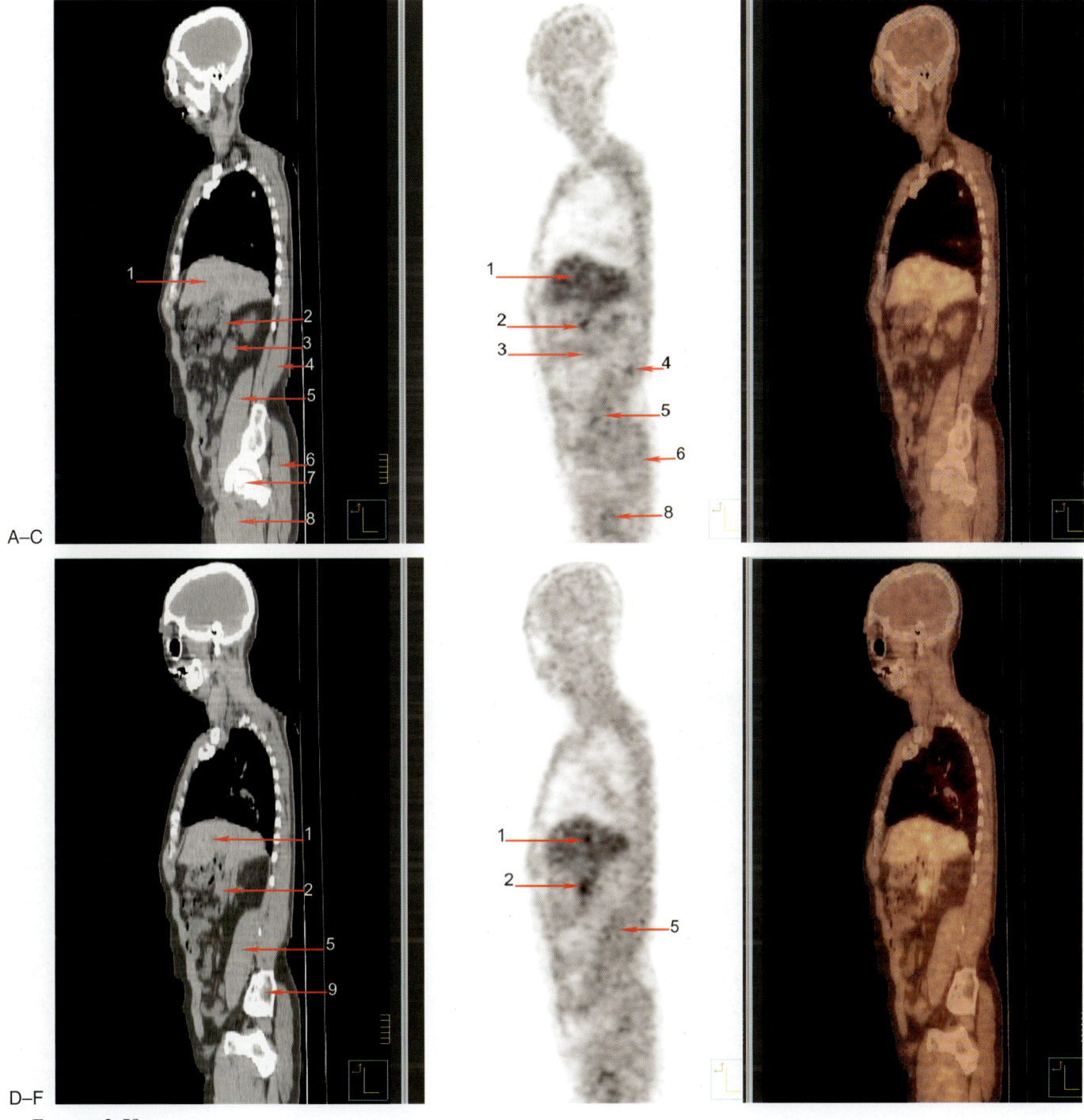

A–C

D–F

FIGURE 2-58:

1. Right hepatic lobe
2. Pancreatic head
3. Right kidney
4. Erector spinae muscle
5. Psoas muscle
6. Gluteus muscle
7. Femoral head
8. Adductor muscles
9. Iliac tuberosity

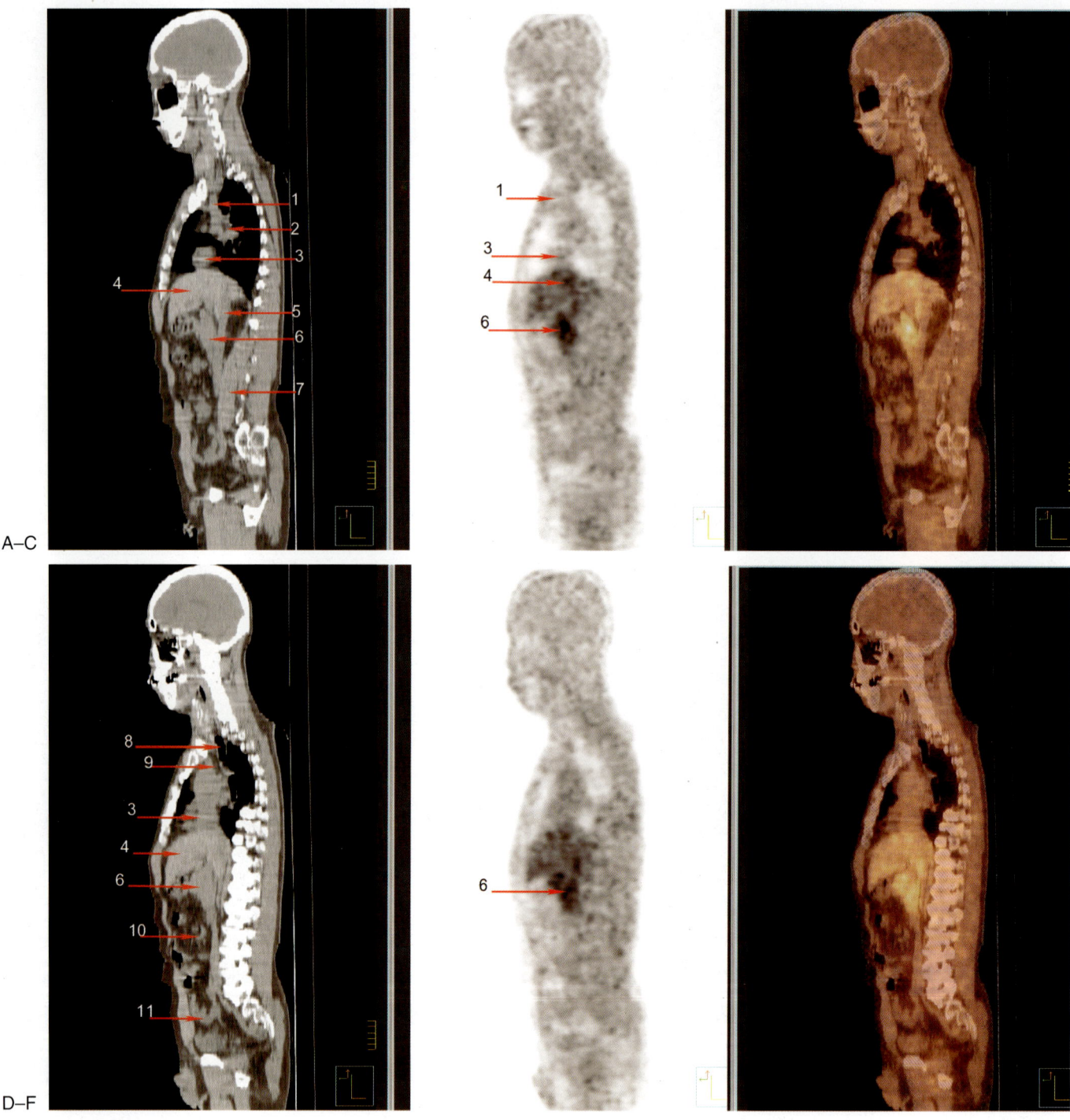

A–C

D–F

FIGURE 2-59:

1. Superior vena cava
2. Pulmonary vein
3. Right atrium
4. Right hepatic lobe
5. Inferior vena cava
6. Pancreatic head
7. Psoas muscle
8. Esophagus
9. Trachea
10. Jejunum
11. Ileum

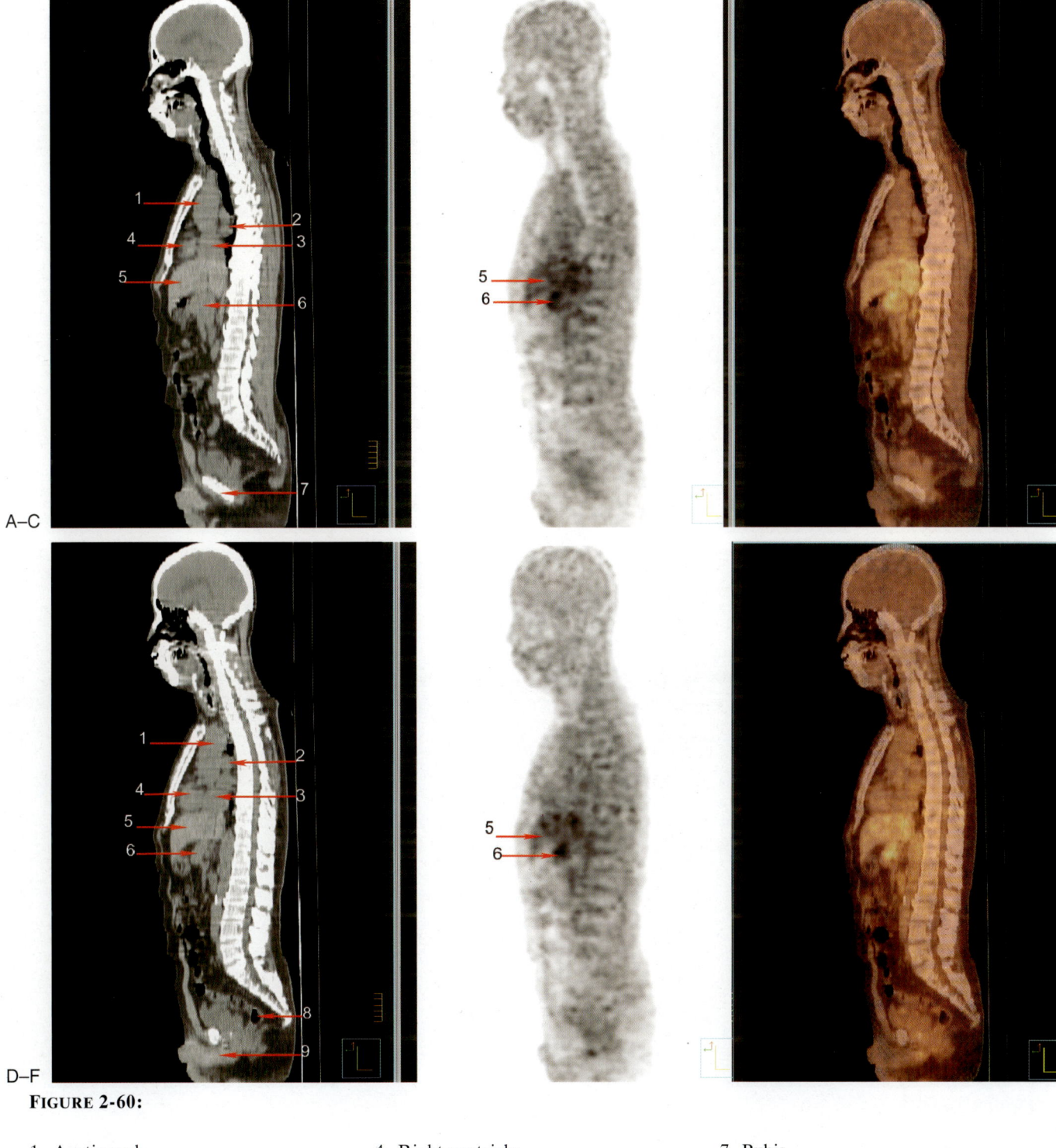

FIGURE 2-60:

1. Aortic arch
2. Pulmonary artery
3. Left ventricle
4. Right ventricle
5. Left hepatic lobe
6. Pancreatic body
7. Pubis
8. Sigmoid
9. Penile corpora muscles

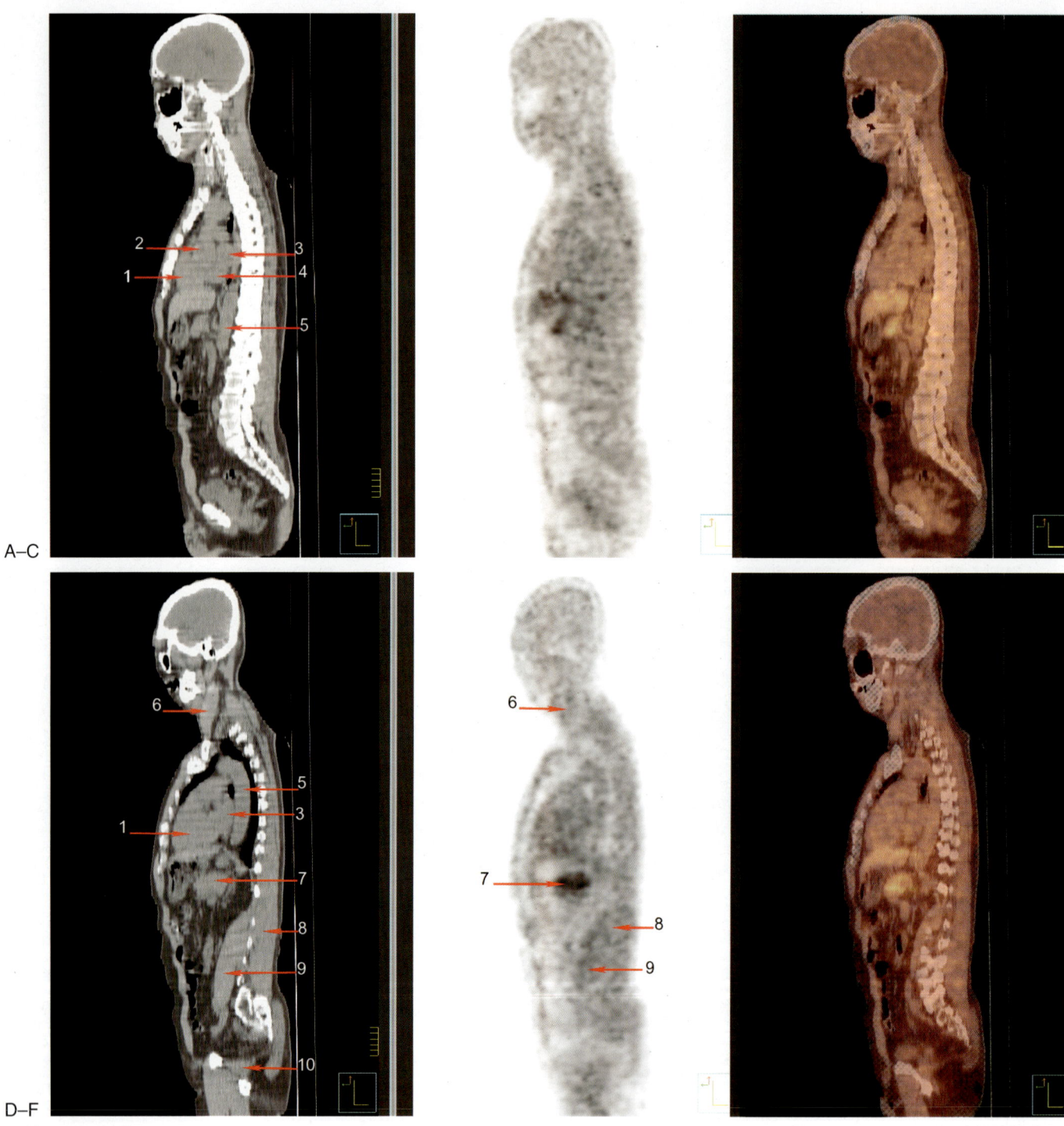

FIGURE 2-61:

1. Right ventricle
2. Ascending aorta
3. Left atrium
4. Left ventricle
5. Descending aorta
6. Sternocleidomastoid muscle
7. Pancreatic body
8. Erector spinae muscle
9. Psoas muscle
10. Bladder

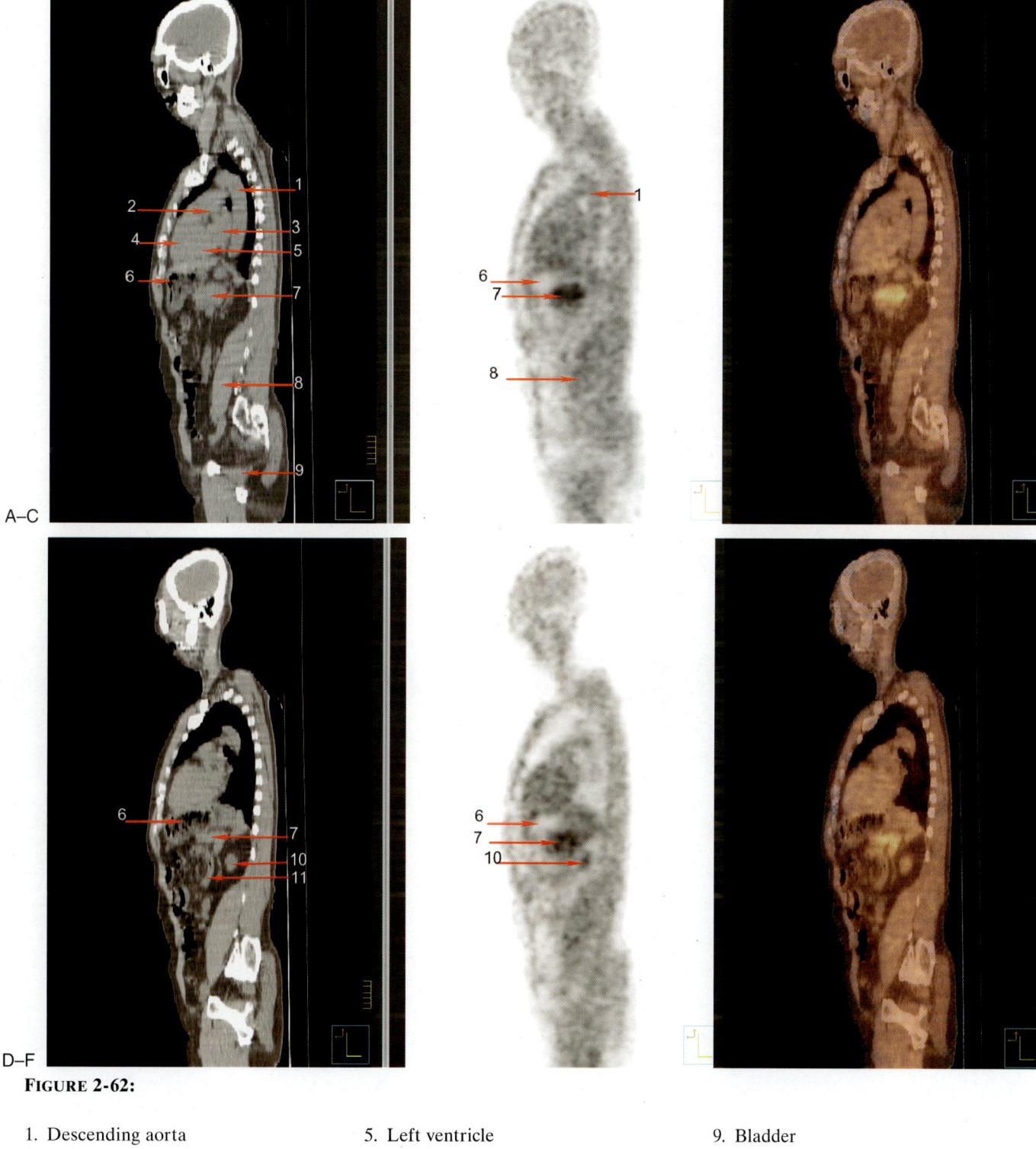

FIGURE 2-62:

1. Descending aorta
2. Left pulmonary artery
3. Left atrium
4. Right ventricle

5. Left ventricle
6. Stomach
7. Pancreatic body
8. Psoas muscle

9. Bladder
10. Left kidney
11. Jejunum

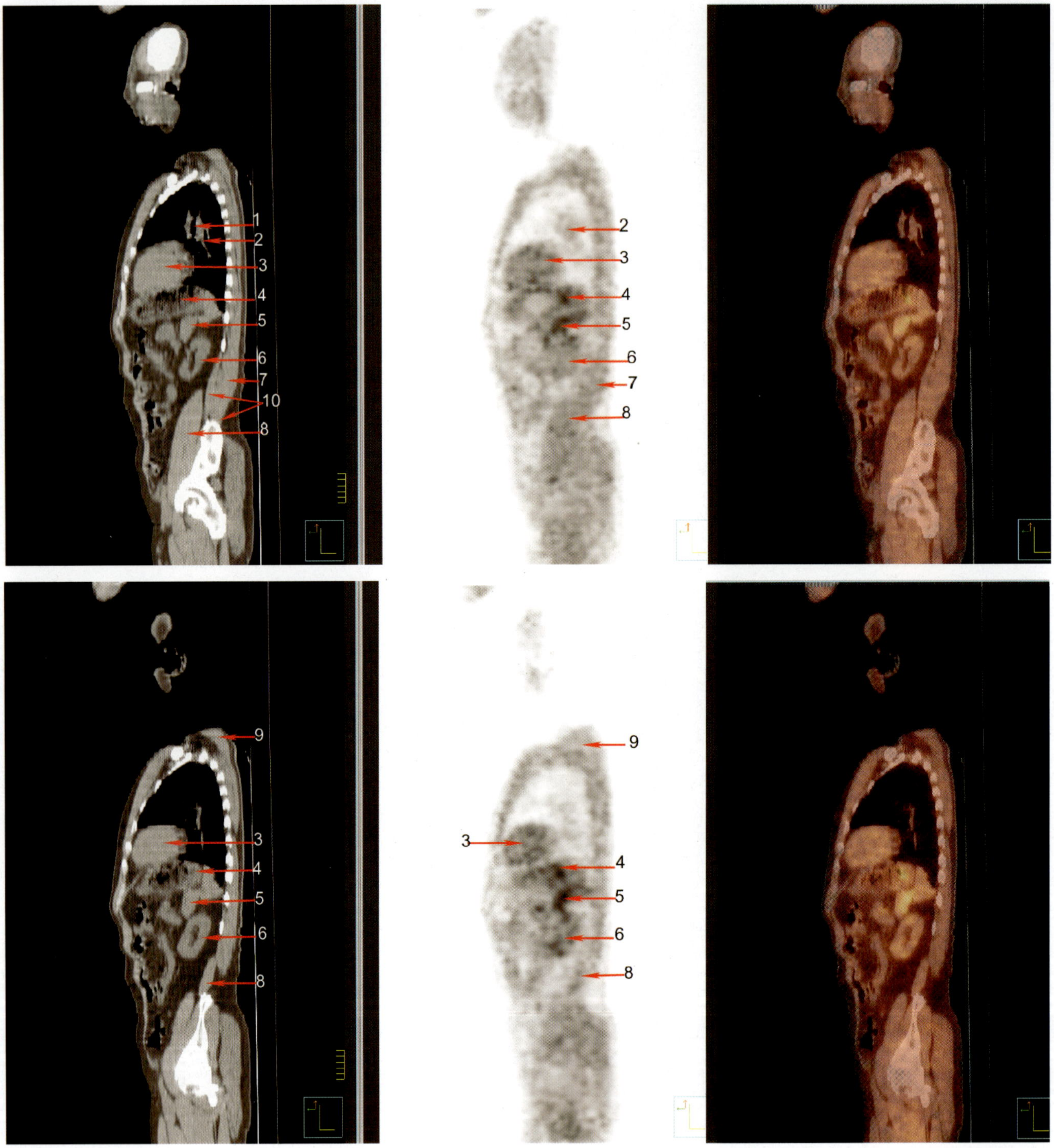

A–C

D–F

FIGURE 2-63:

1. Left main bronchus
2. Left pulmonary vein
3. Left ventricle
4. Stomach
5. Pancreatic body
6. Left kidney
7. Erector spinae muscle
8. Psoas muscle
9. Trapezius muscle
10. Quadratus lumborum muscle

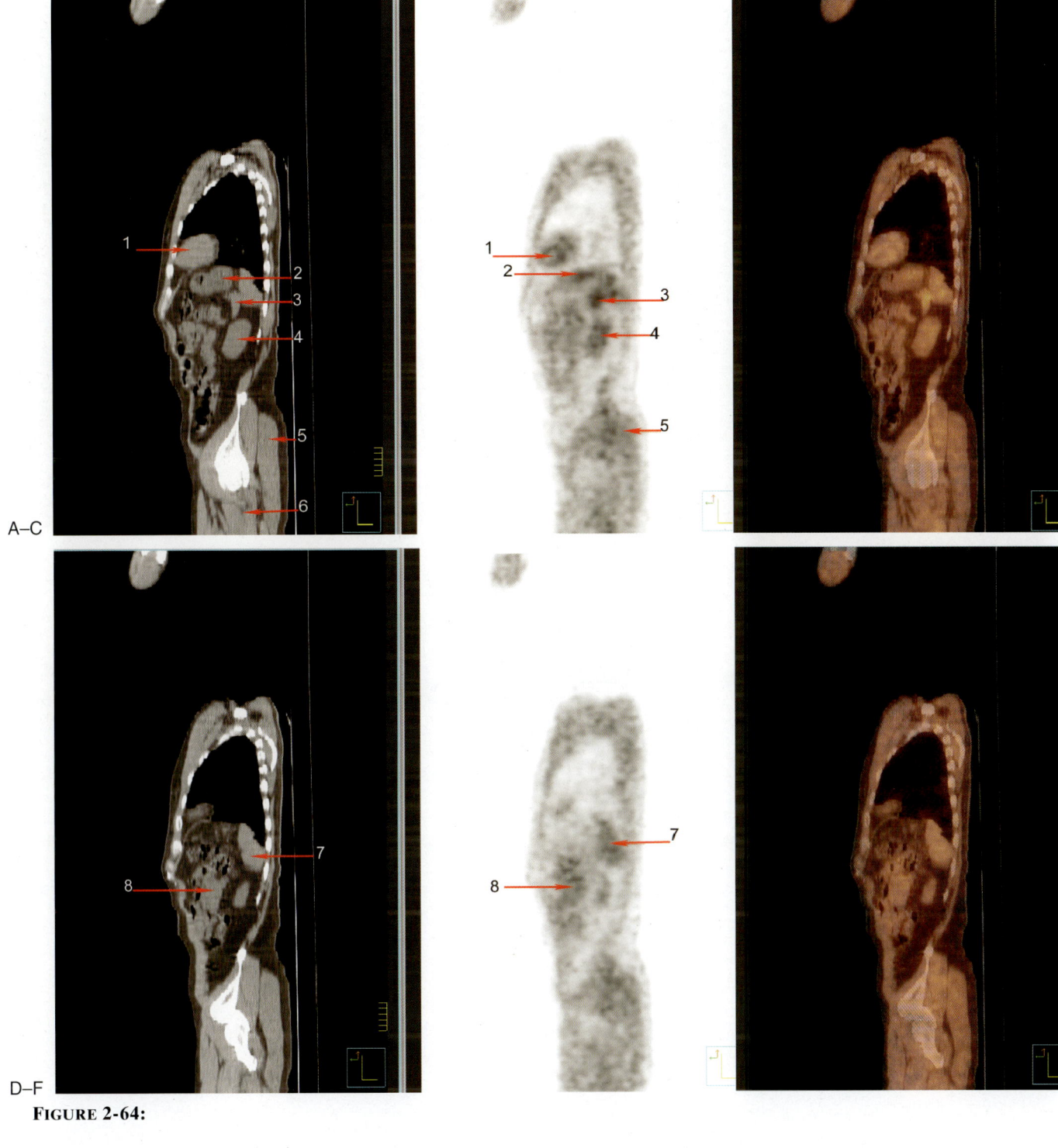

FIGURE 2-64:

1. Left ventricle
2. Stomach
3. Pancreatic tail
4. Left kidney
5. Gluteus maximus muscle
6. Adductor muscles
7. Spleen
8. Jejunum

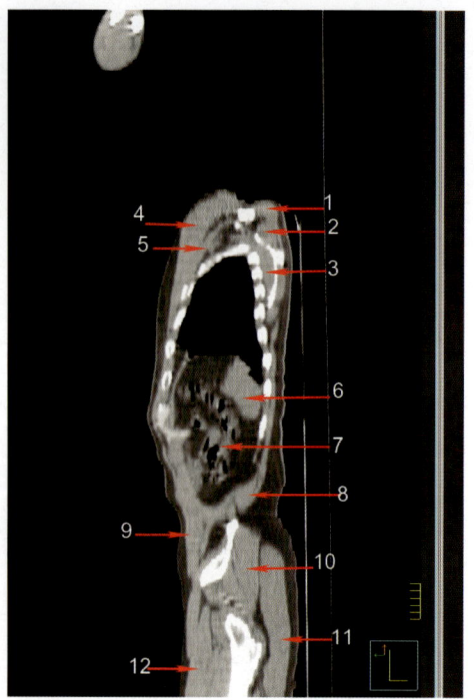
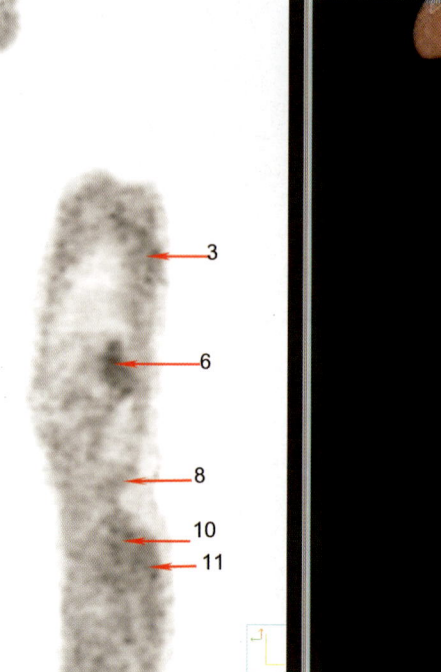
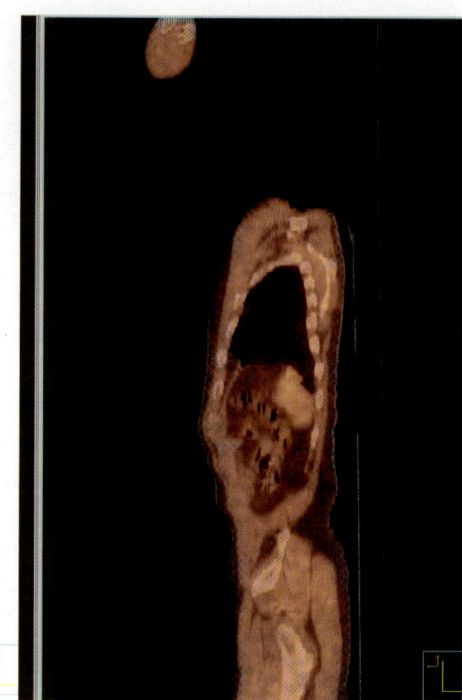

A–C

FIGURE 2-65:

1. Trapezius muscle	5. Pectoralis minor muscle	9. Oblique muscles
2. Supraspinatus muscle	6. Spleen	10. Gluteus medius muscle
3. Subscapularis muscle	7. Jejunum	11. Gluteus maximus muscle
4. Pectoralis major muscle	8. Latissimus dorsi muscle	12. Rectus femoris muscle

Transaxials

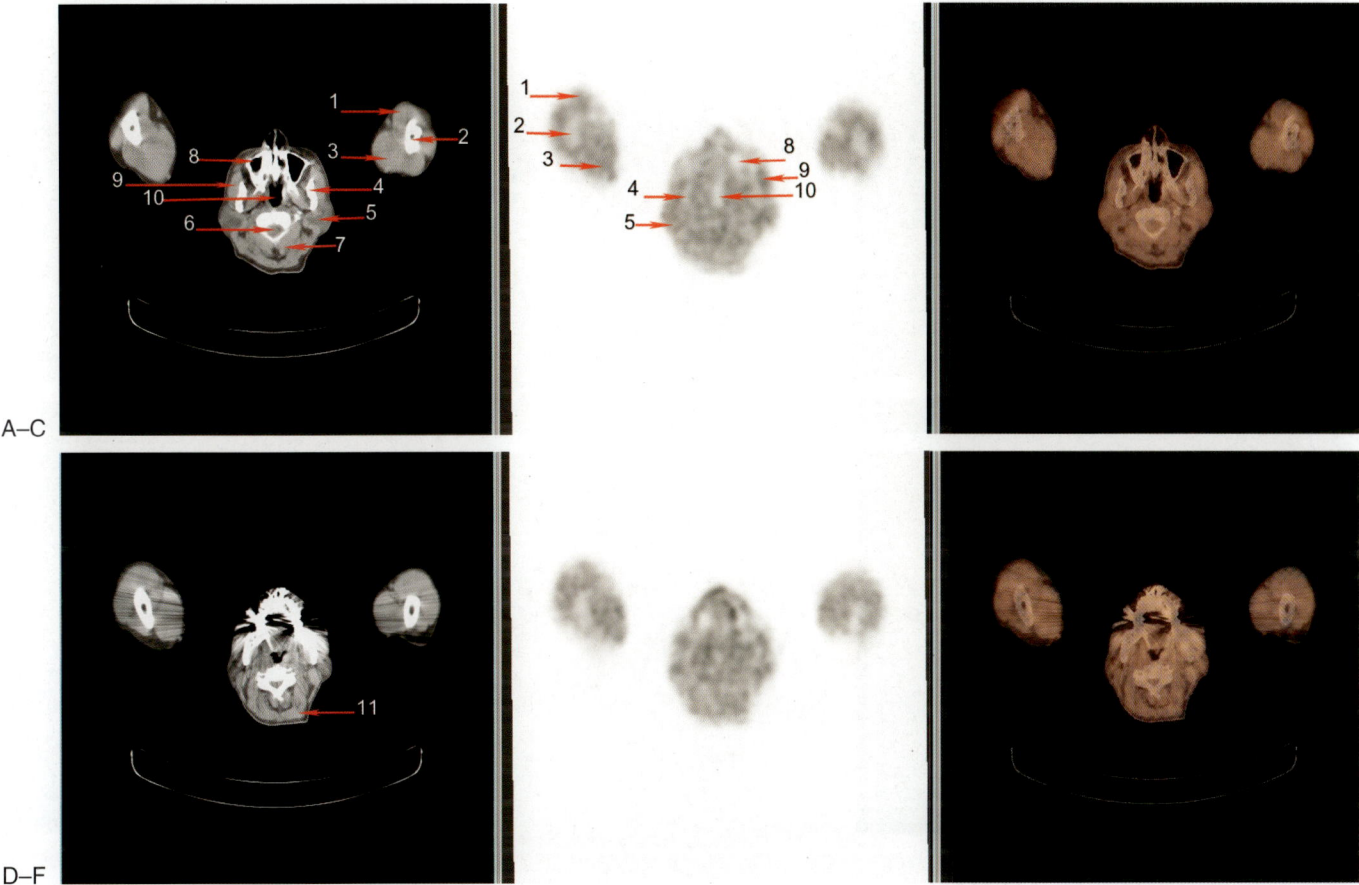

FIGURE 2-66:

1. Biceps brachii muscle
2. Humerus
3. Triceps muscle
4. Mandible
5. Parotid gland
6. Spinal cord
7. Semispinalis cervicis muscle
8. Maxillary sinus
9. Masseter muscle
10. Nasopharynx
11. Semispinalis capitis muscle

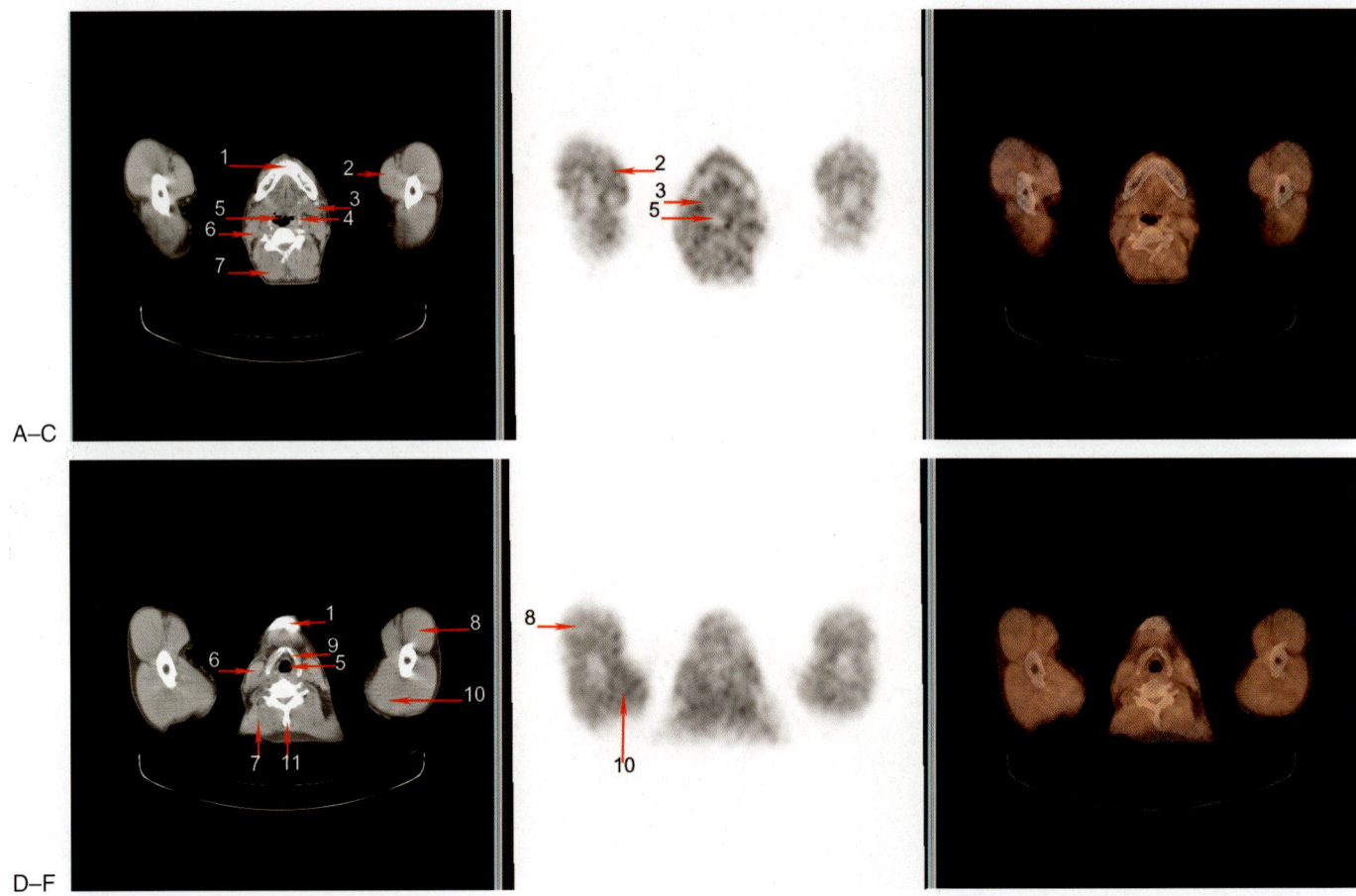

A–C

D–F

FIGURE 2-67:

1. Mandible
2. Brachialis muscle
3. Submandibular gland
4. Hyoid bone
5. Oropharynx
6. Sternocleidomastoid muscle
7. Splenius capitis muscle
8. Biceps brachii muscle
9. Thyroid cartilage
10. Triceps muscle
11. Spinous process

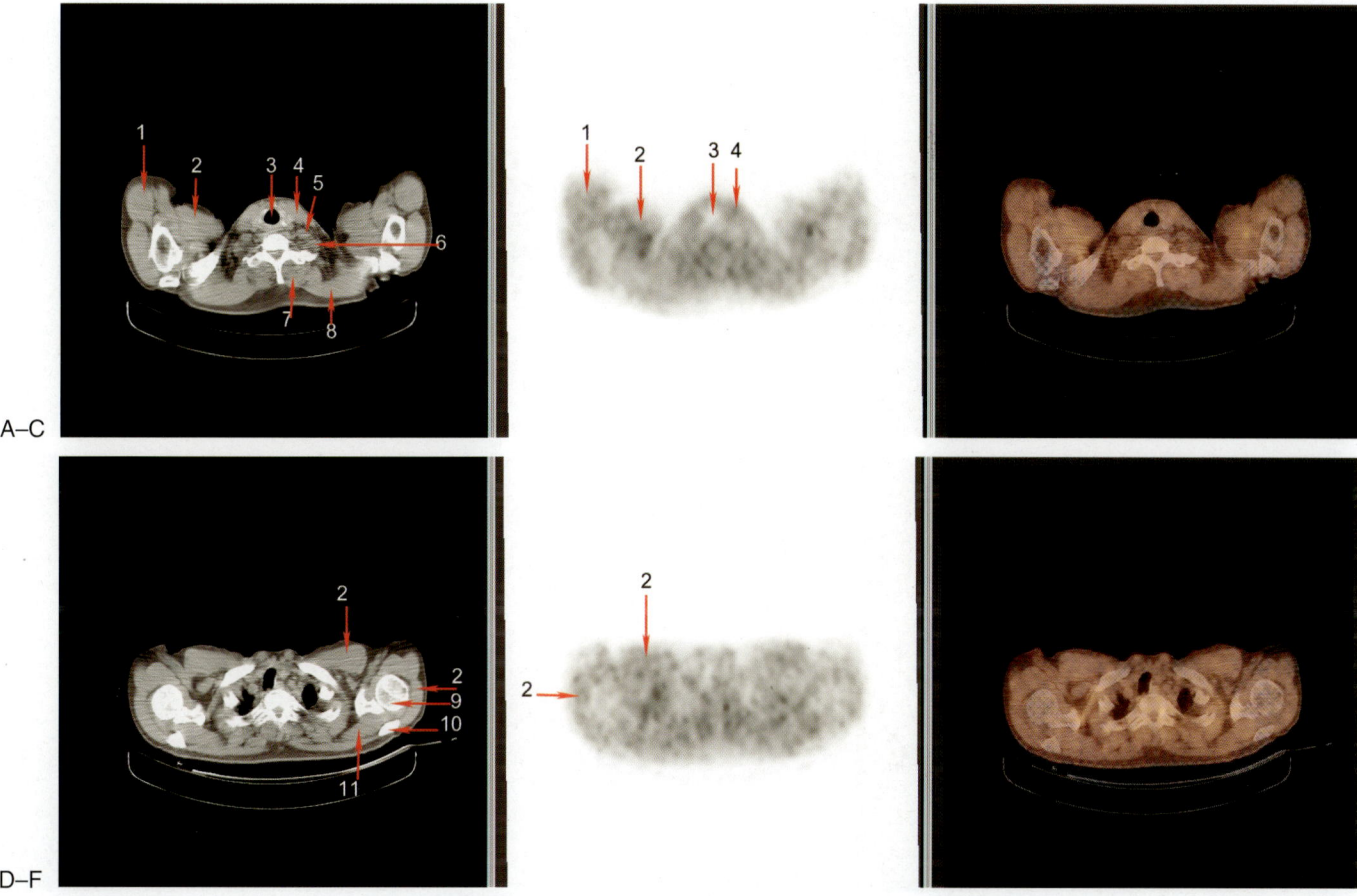

FIGURE 2-68:

1. Biceps brachii muscle
2. Deltoid muscle
3. Trachea
4. Sternocleidomastoid muscle
5. Internal jugular vein
6. Anterior scalene muscle
7. Semispinalis cervicis muscle
8. Trapezius muscle
9. Humeral head
10. Scapular spine
11. Supraspinatus muscle

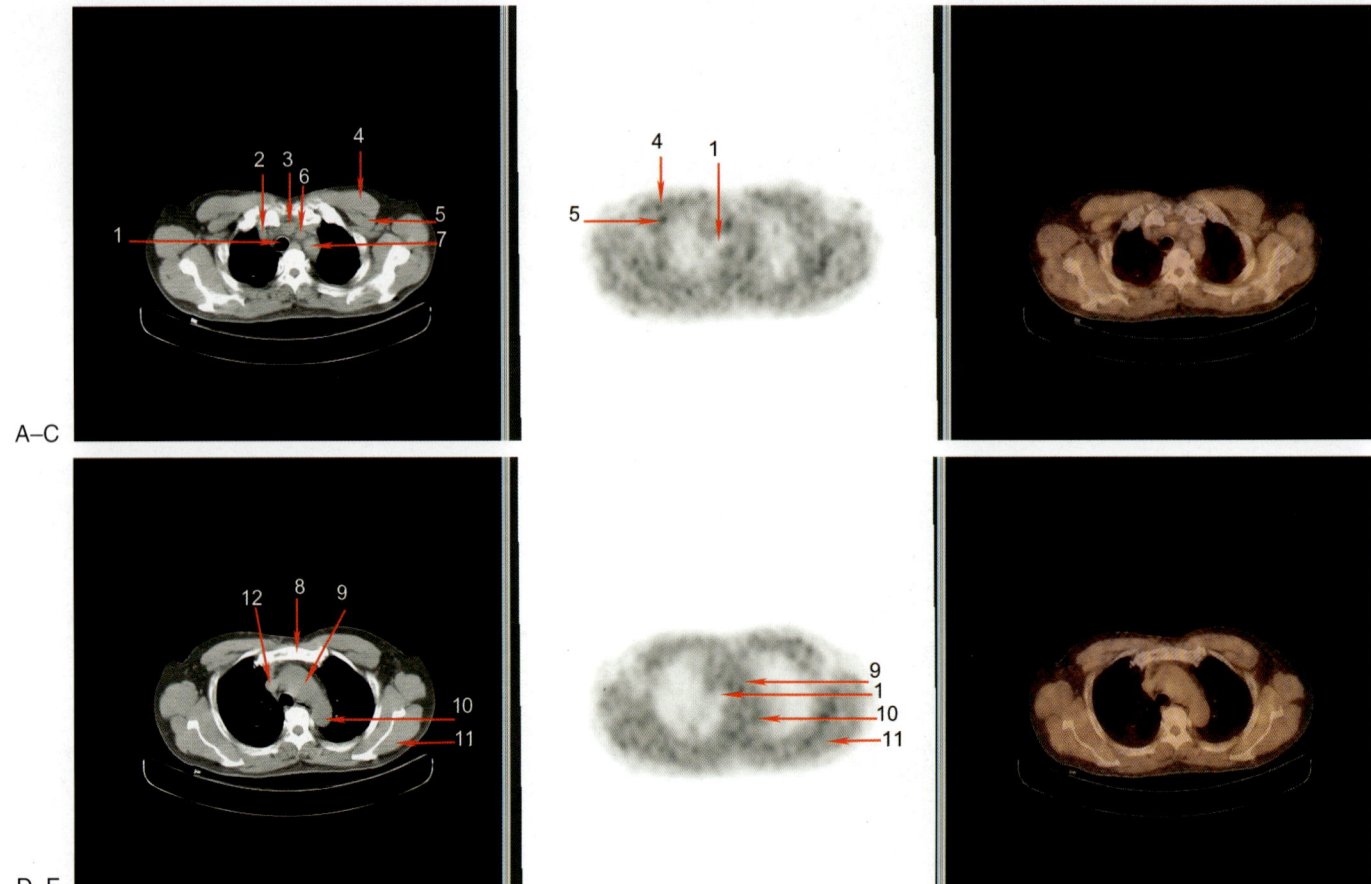

A–C

D–F

FIGURE 2-69:

1. Trachea
2. Right brachiocephalic vein
3. Brachiocephalic artery
4. Pectoralis major muscle
5. Pectoralis minor muscle
6. Left common carotid artery
7. Left subclavian artery
8. Manubrium
9. Aortic arch
10. Descending aorta
11. Teres minor muscle
12. Superior vena cava

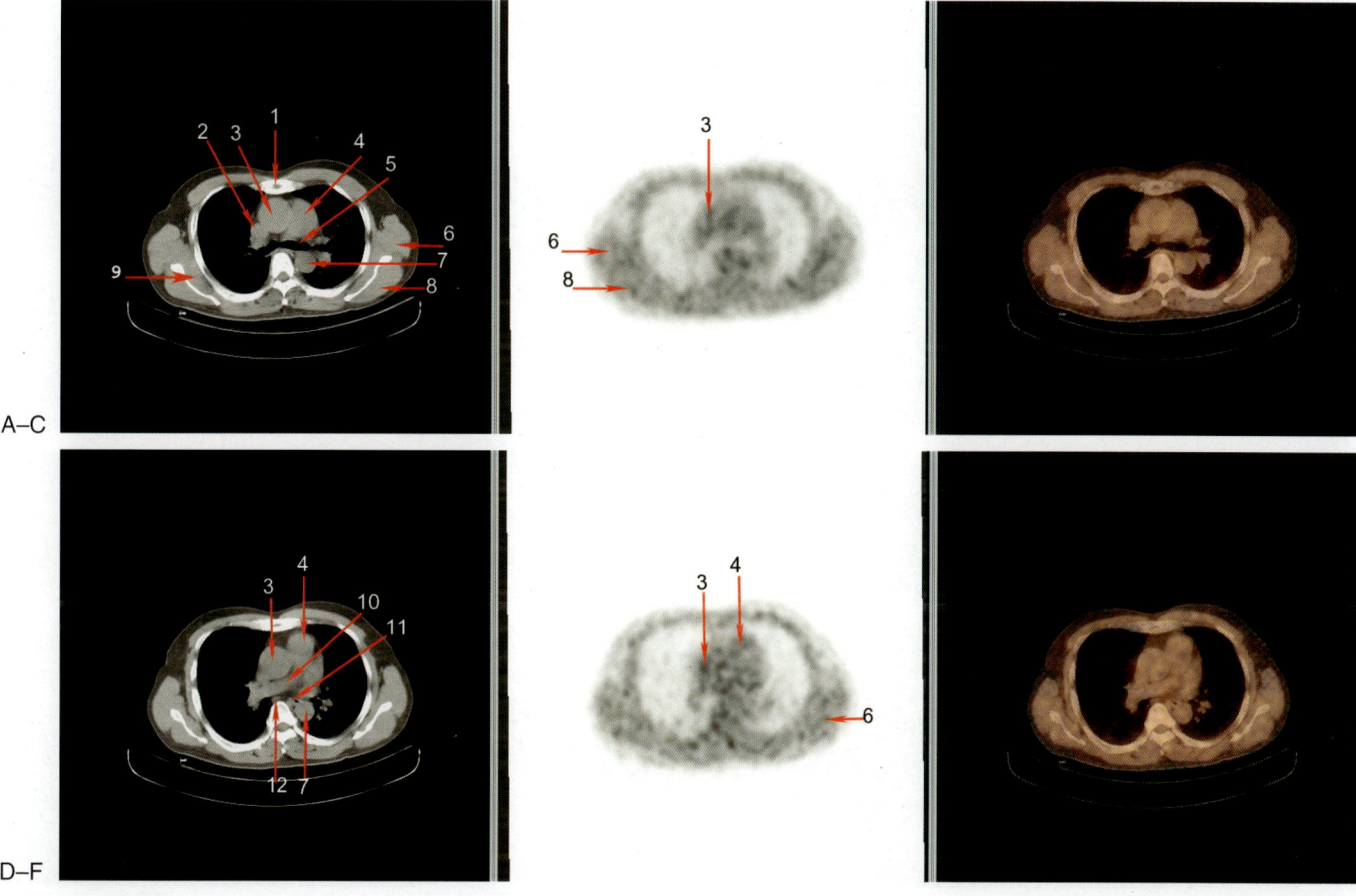

A–C

D–F

FIGURE 2-70:

1. Sternum
2. Superior vena cava
3. Ascending aorta
4. Main pulmonary artery
5. Left main bronchus
6. Teres major muscle
7. Descending aorta
8. Infraspinatus muscle
9. Subscapularis muscle
10. Right pulmonary artery
11. Esophagus
12. Azygos vein

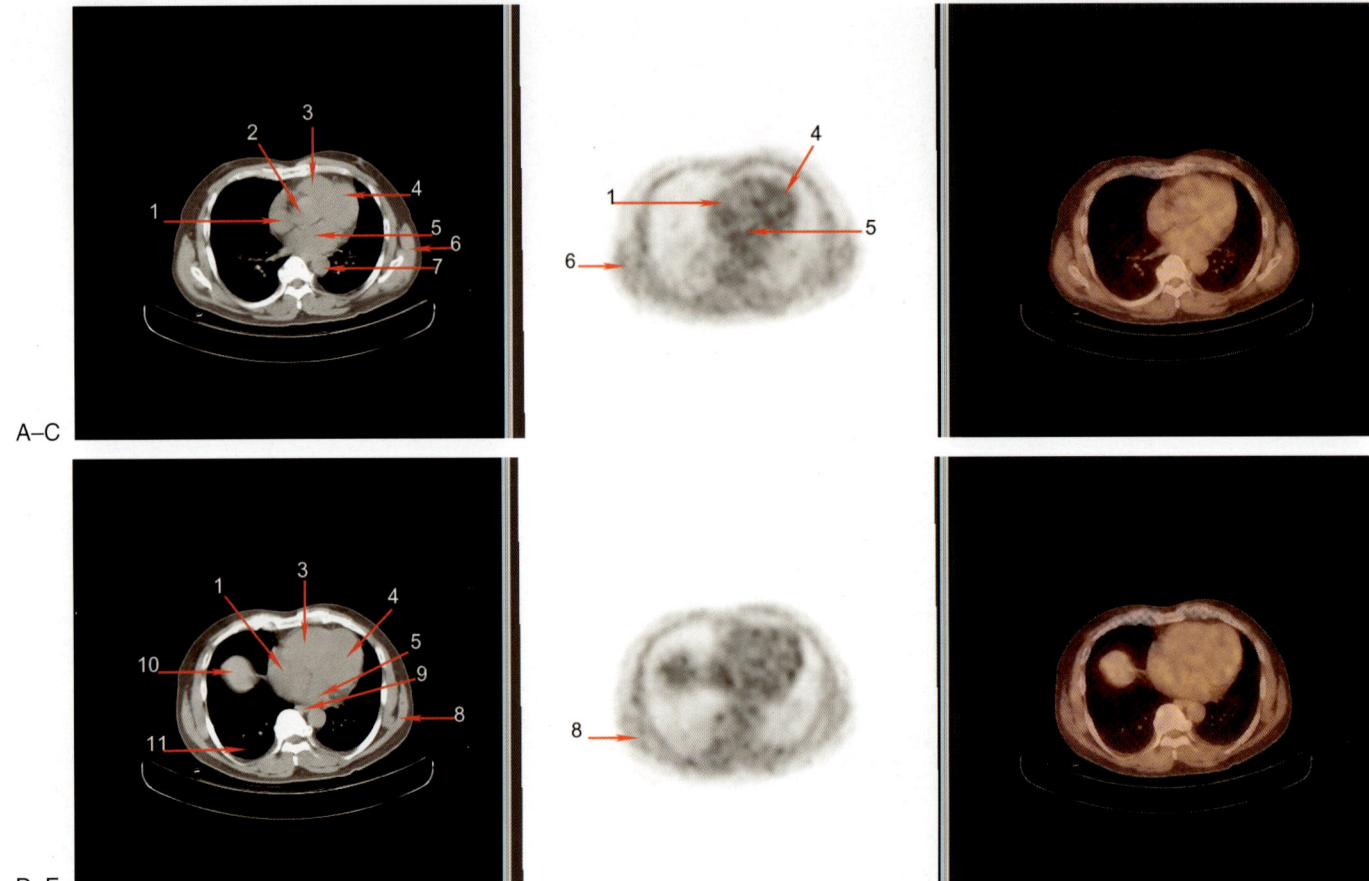

A–C

D–F

FIGURE 2-71:

1. Right atrium
2. Ascending aorta
3. Right ventricle
4. Left ventricle
5. Left atrium
6. Serratus anterior muscle
7. Descending aorta
8. Latissimus dorsi muscle
9. Esophagus
10. Cupula of right diaphragm
11. Posterior basal segment of right lower lobe of lung

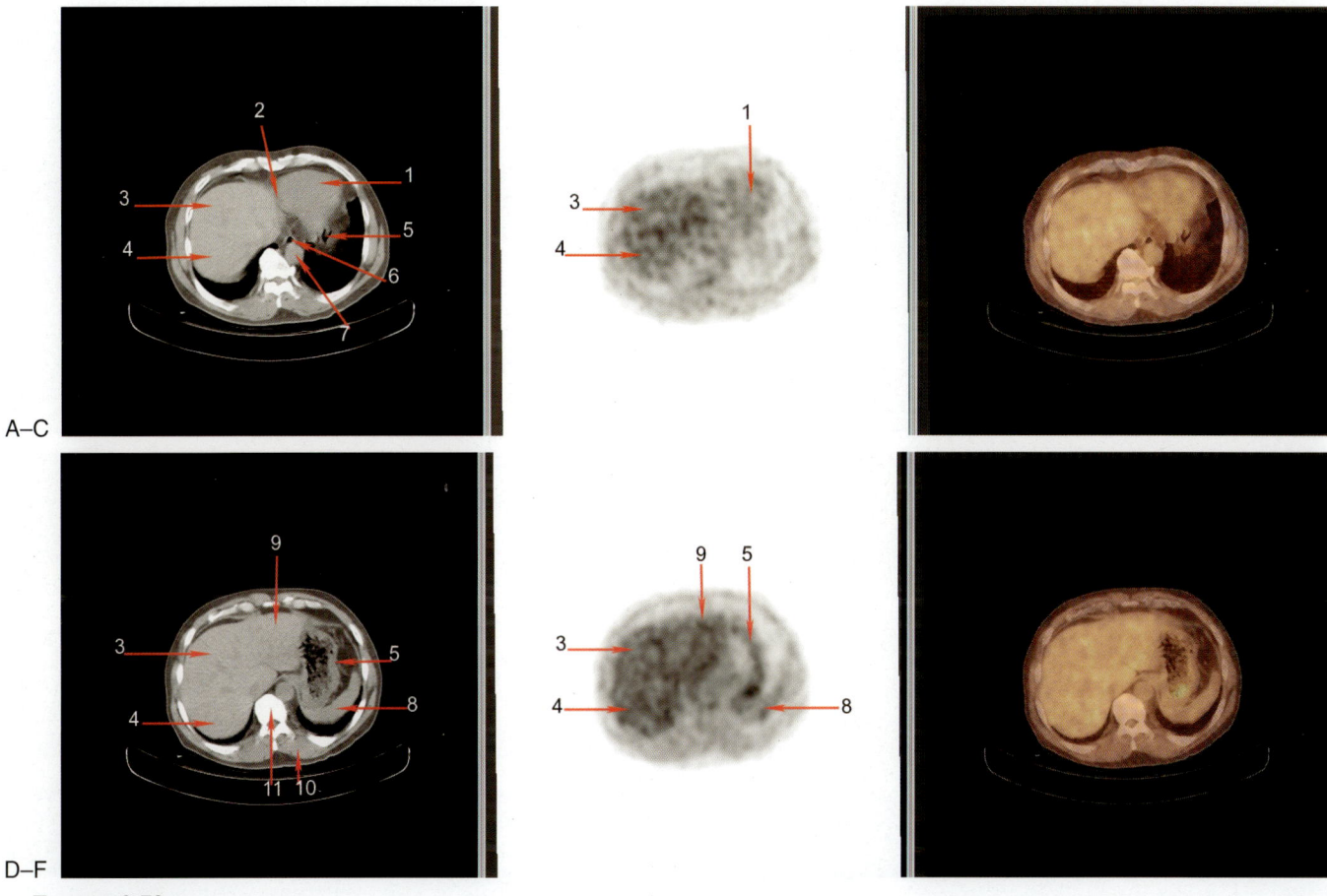

A–C

D–F

FIGURE 2-72:

1. Left ventricle
2. Pericardium
3. Superior anterior segment of right hepatic lobe (segment VIII)
4. Superior posterior segment of right hepatic lobe (segment VII)
5. Stomach
6. Esophagus
7. Descending aorta
8. Spleen
9. Lateral segment of left hepatic lobe (segment III)
10. Erector spinae muscle
11. Vertebral body

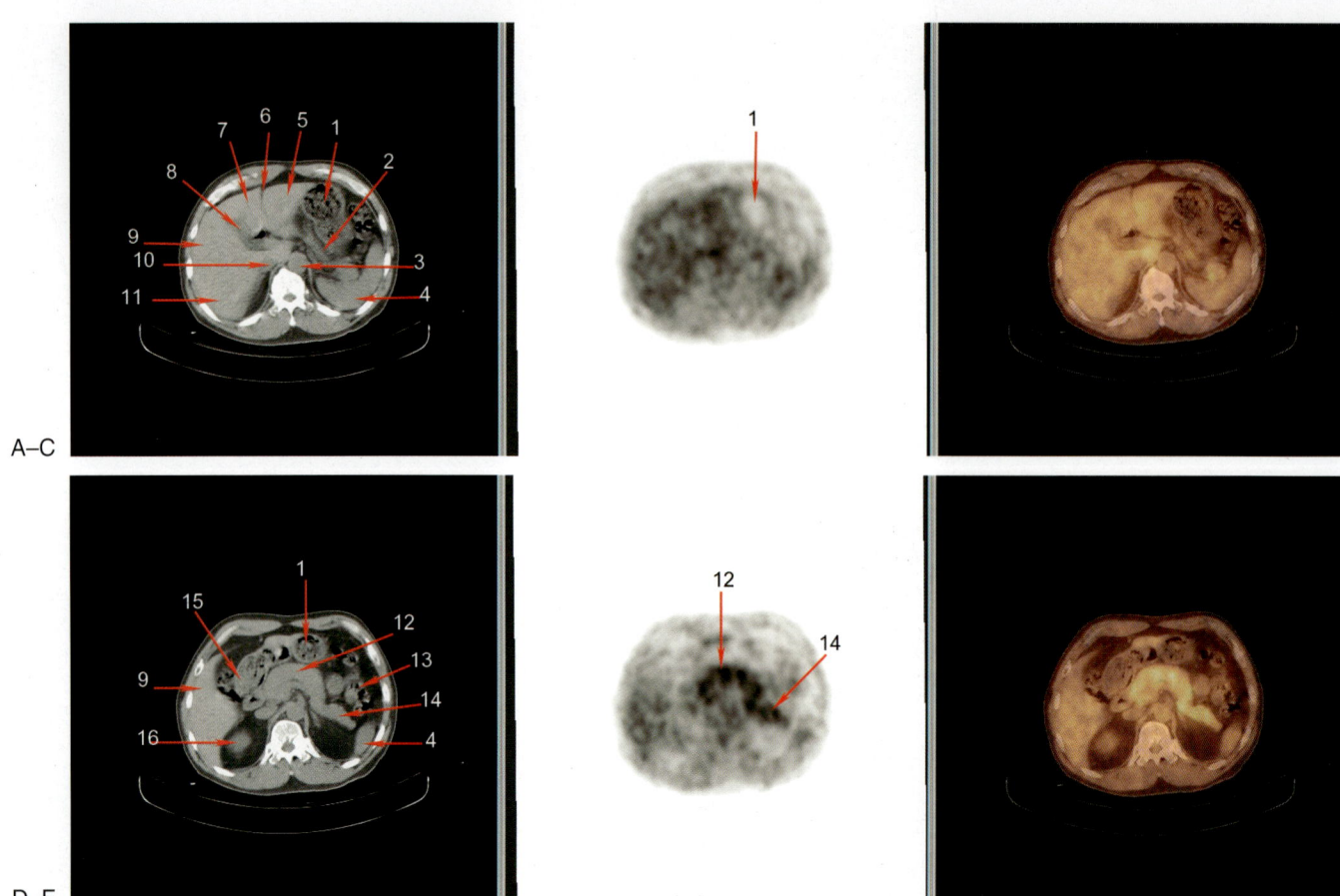

A–C

D–F

FIGURE 2-73:

1. Stomach
2. Splenic vein
3. Abdominal aorta
4. Spleen
5. Lateral segment of left hepatic lobe (segment III)
6. Falciform ligament
7. Medial segment of left hepatic lobe (segment IV)
8. Gallbladder
9. Inferior anterior segment of right hepatic lobe (segment V)
10. Inferior vena cava
11. Inferior posterior segment of right hepatic lobe (segment VI)
12. Pancreatic body
13. Jejunum
14. Pancreatic tail
15. Duodenum (second portion)
16. Superior pole of right kidney

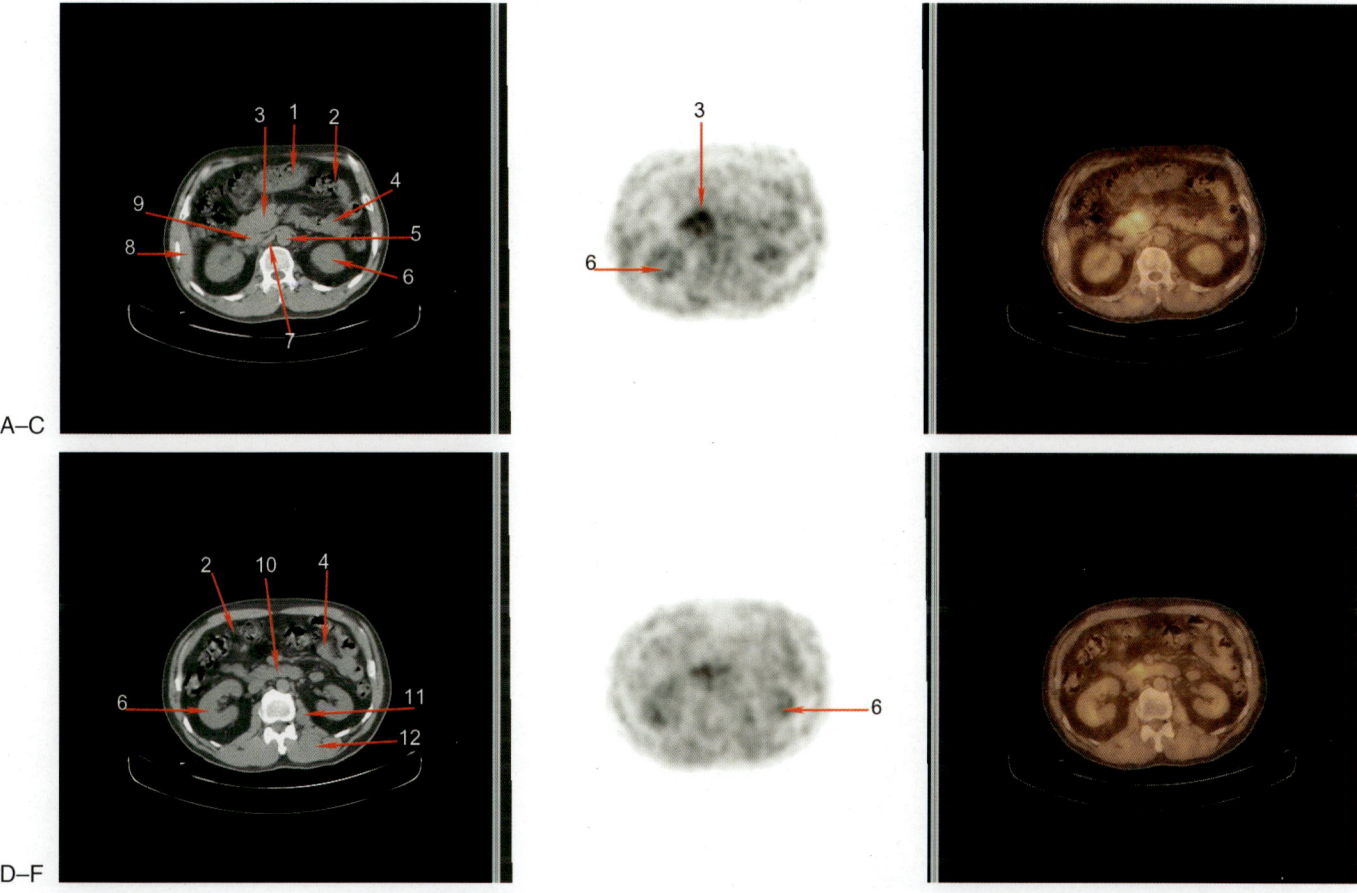

A–C

D–F

FIGURE 2-74:

1. Stomach
2. Transverse colon
3. Pancreatic head (uncinate process)
4. Jejunum
5. Abdominal aorta
6. Kidney
7. Diaphragmatic crura
8. Inferior tip of liver
9. Inferior vena cava
10. Duodenum (third portion)
11. Psoas muscle
12. Erector spinae muscle

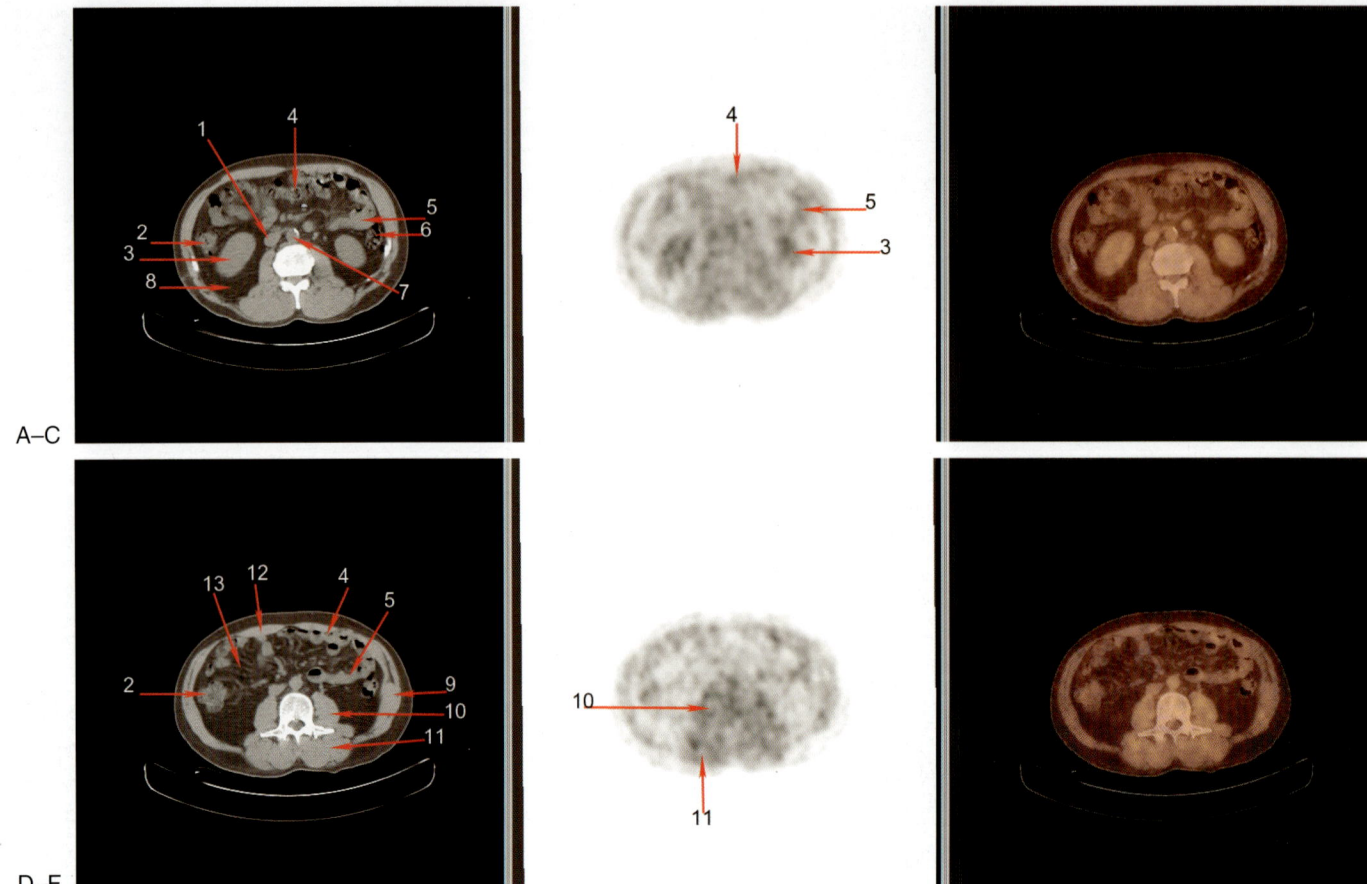

A–C

D–F

FIGURE 2-75:

1. Inferior vena cava
2. Ascending colon
3. Kidney
4. Transverse colon
5. Jejunum
6. Descending colon
7. Abdominal aorta
8. Perirenal space
9. Oblique muscles
10. Psoas muscle
11. Erector spinae muscle
12. Rectus abdominis muscle
13. Mesenteric fat

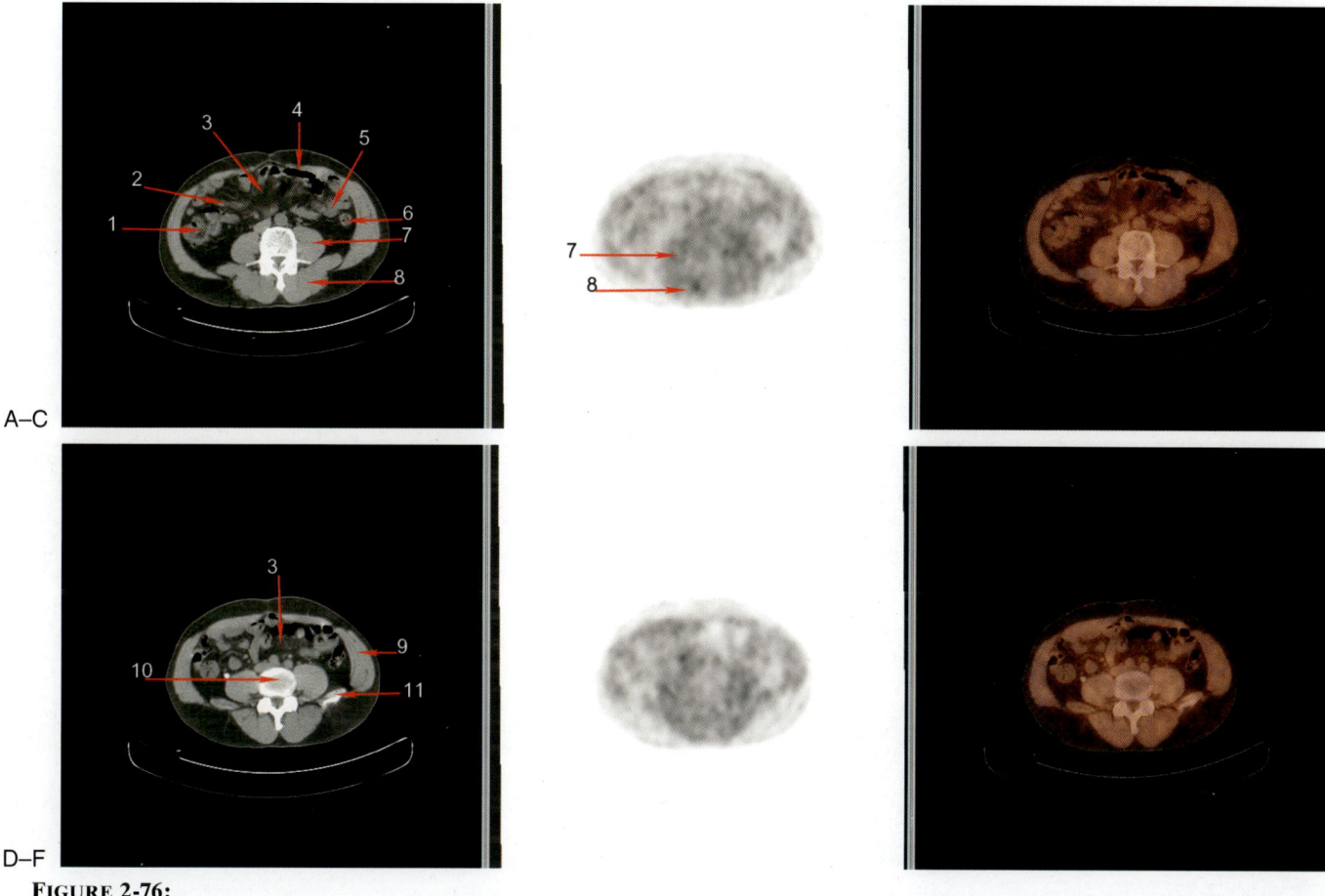

A–C

D–F

FIGURE 2-76:

1. Ascending colon
2. Ileum
3. Mesentery
4. Transverse colon

5. Jejunum
6. Descending colon
7. Psoas muscle
8. Erector spinae muscle

9. Oblique muscle
10. L5 vertebral body
11. Left iliac ala

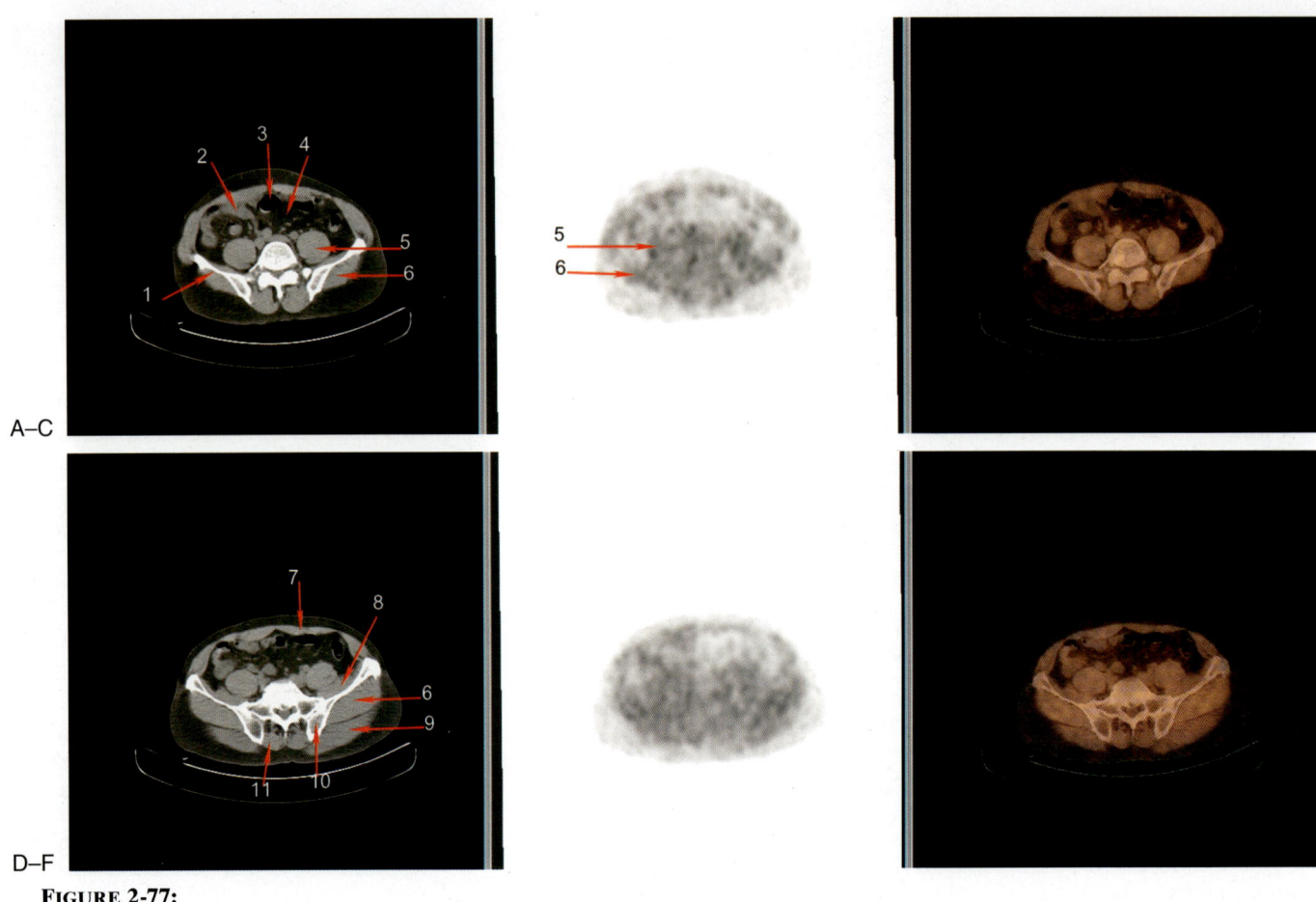

A–C

D–F

FIGURE 2-77:

1. Iliac ala
2. Ileum
3. Transverse colon
4. Mesentery

5. Psoas muscle
6. Gluteus medius muscle
7. Rectus abdominis muscle
8. Iliacus muscle

9. Gluteus maximus muscle
10. Iliac tuberosity
11. Multifidus muscle

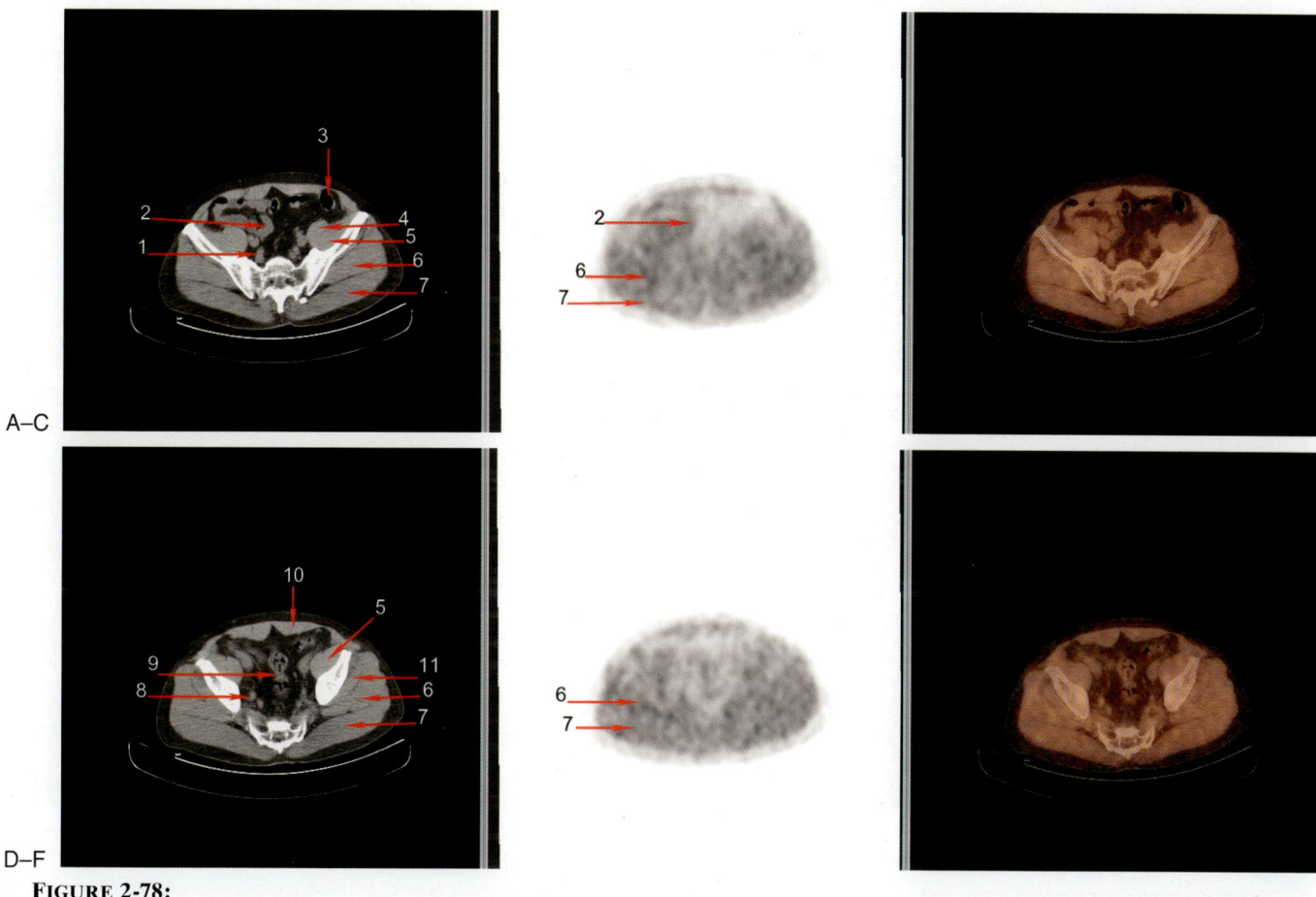

FIGURE 2-78:

1. Common iliac vessels
2. Ileum
3. Jejunum
4. Psoas muscle
5. Iliacus muscle
6. Gluteus medius muscle
7. Gluteus maximus muscle
8. Internal iliac vessels
9. Sigmoid colon
10. Rectus abdominis muscle
11. Gluteus minimus muscle

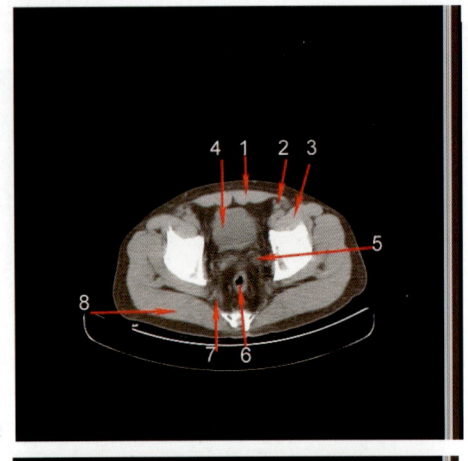
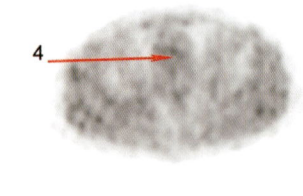
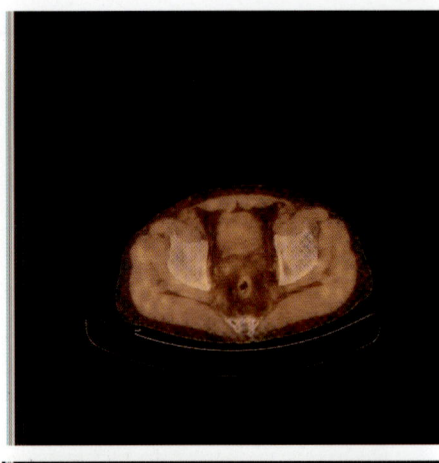

A–C

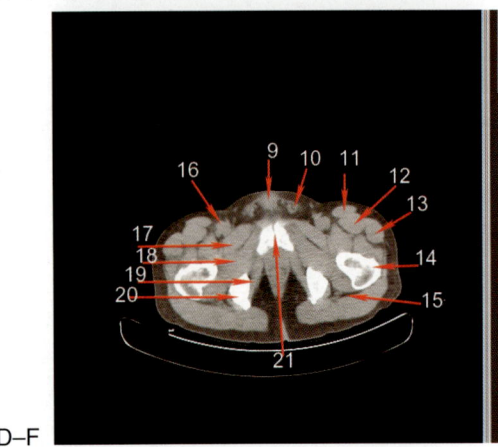

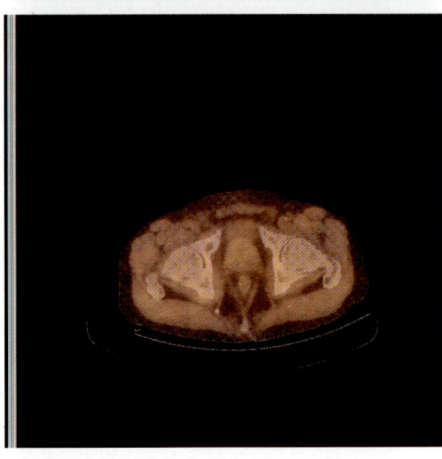

D–F

FIGURE 2-79:

1. Rectus abdominis muscle
2. External iliac vessels
3. Iliopsoas muscle
4. Bladder
5. Seminal vesicle
6. Rectum
7. Inferior gluteal vessel

8. Gluteus maximus muscle
9. Penile corpus spongiosum
10. Spermatic cord
11. Sartorius muscle
12. Iliopsoas muscle
13. Tensor fasciae latae muscle
14. Greater trochanter of femur

15. Sciatic nerve
16. Femoral vessels
17. Adductor muscle
18. External obturator muscle
19. Internal obturator muscle
20. Ischial tuberosity
21. Symphysis pubis

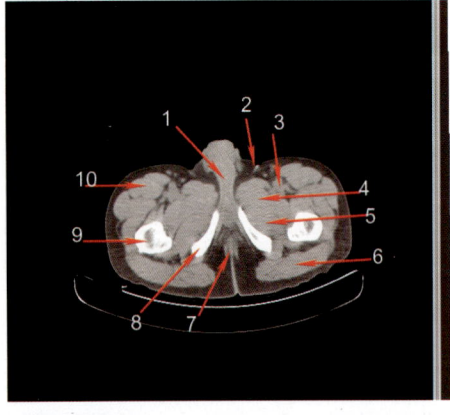
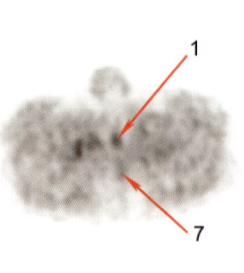
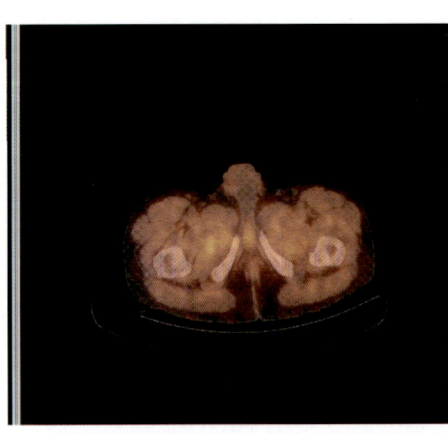

A–C

FIGURE 2-80:

1. Penile corpora cavernosum
2. Spermatic cord
3. Femoral vessels
4. Adductor muscles

5. External obturator muscle
6. Gluteus maximus muscle
7. Anus
8. Ischium

9. Femur
10. Rectus femoris muscle

Section 3: F18 Fluoride Bone PET

Coronals

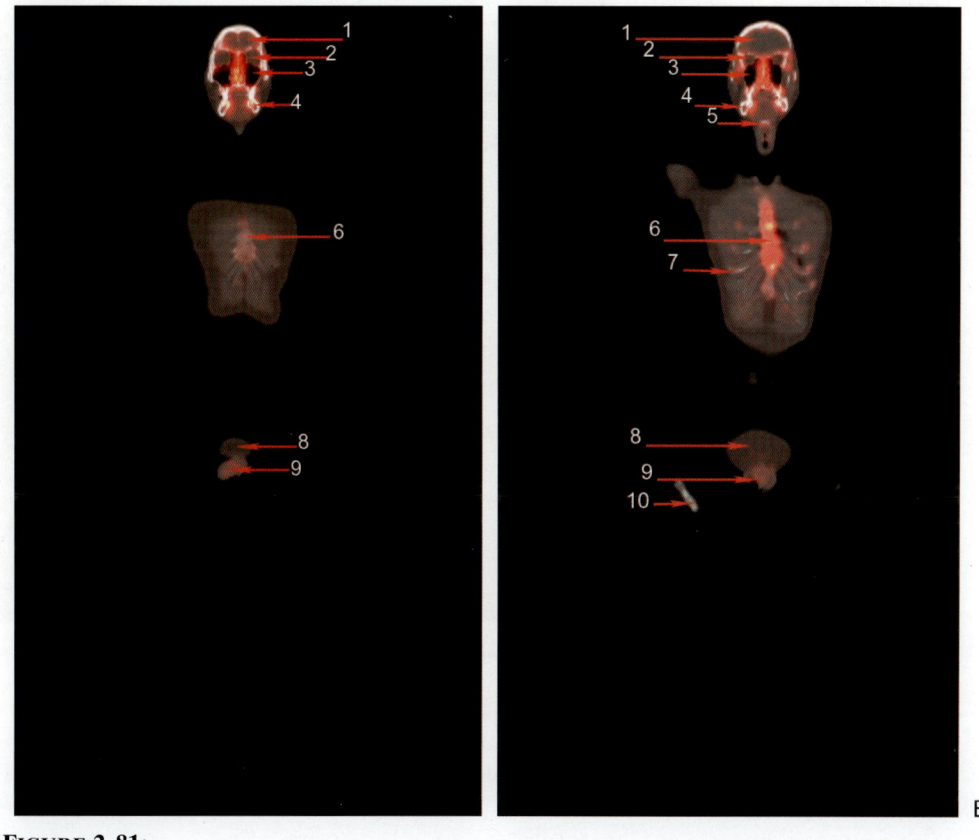

FIGURE 2-81:

1. Frontal skull	5. Hyoid bone	9. Bladder
2. Orbital globe	6. Sternum	10. Metacarpal bone
3. Maxillary sinus	7. Rib	
4. Mandible	8. Rectus abdominis muscle	

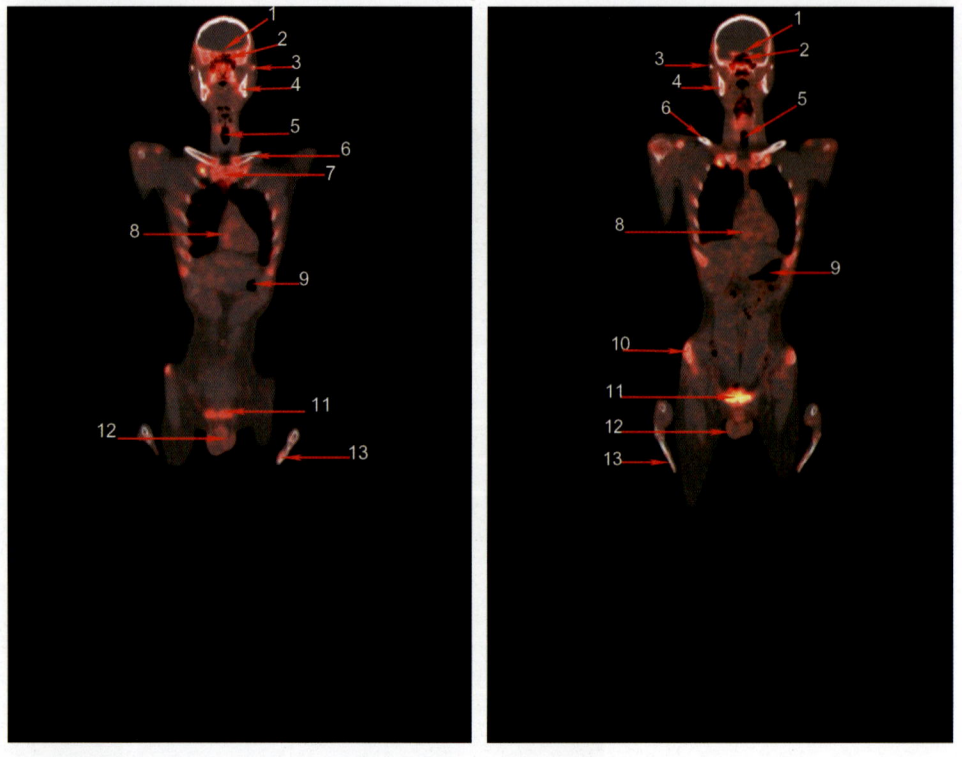

A   B

FIGURE 2-82:

1. Planum sphenoidale	6. Clavicle	10. Iliac crest
2. Sphenoidal sinus	7. Manubrium	11. Bladder
3. Zygomatic arch	8. Right ventricle	12. Scrotum
4. Mandible	9. Stomach air	13. Phalanx
5. Trachea		

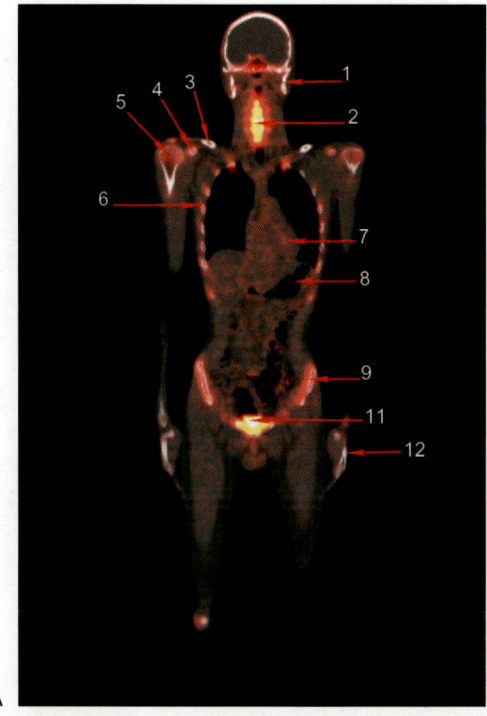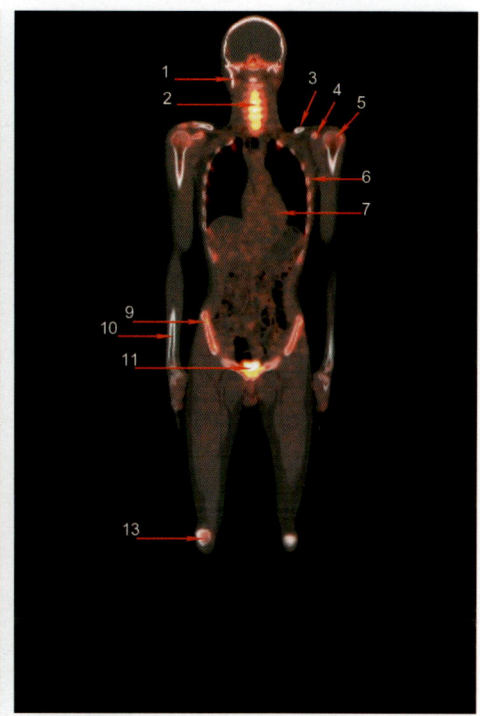

A   B

FIGURE 2-83:

1. Mandible	6. Rib	10. Radius
2. Cervical spine	7. Left ventricle	11. Bladder
3. Clavicle	8. Stomach gas	12. Metacarpal
4. Coracoid process of scapula	9. Iliac crest	13. Patella
5. Humeral head		

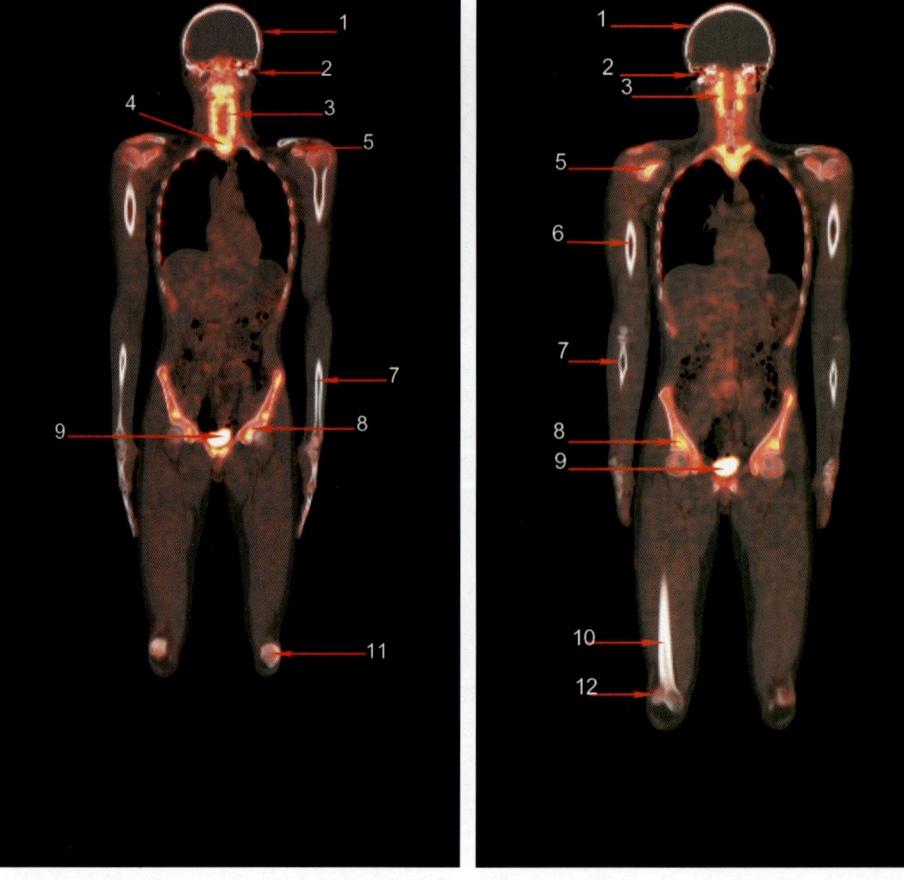

FIGURE 2-84:

1. Parietal skull
2. External auditory canal
3. Vertebral pedicle
4. Vertebral body
5. Coracoid process of scapula
6. Humerus
7. Radius
8. Acetabulum
9. Bladder
10. Femur shaft
11. Patella
12. Femoral condyle

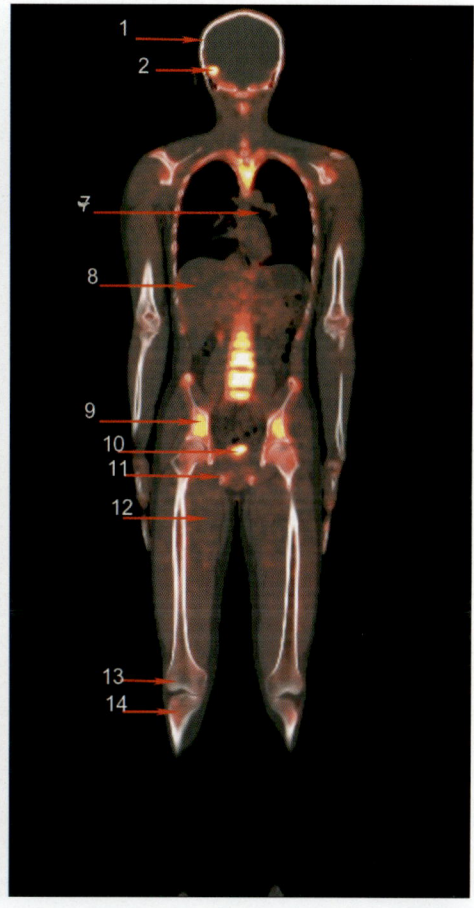

FIGURE 2-85:

1. Parietal skull	6. Carina	11. Pubic ramus
2. Probable tumor	7. Left main bronchus	12. Adductor muscles
3. Mastoid air cells	8. Right hepatic lobe	13. Femoral condyle
4. T2 vertebral body	9. Acetabulum	14. Tibial plateau
5. Left pulmonary artery	10. Bladder	

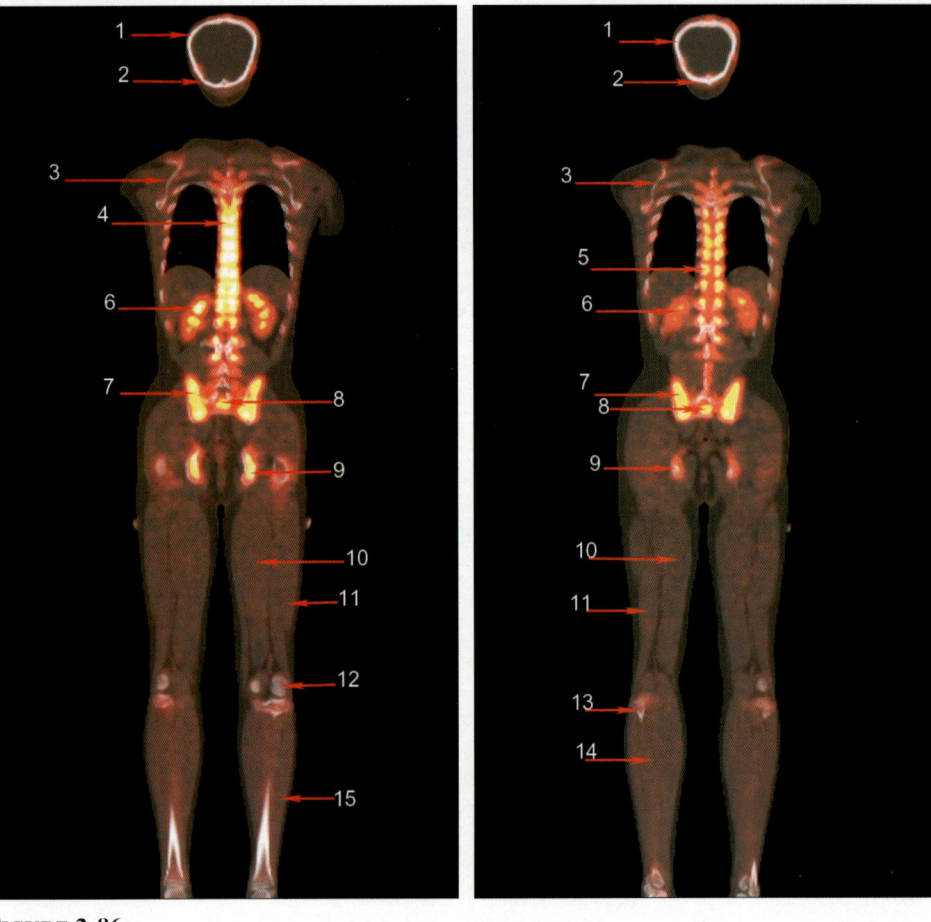

FIGURE 2-86:

1. Parietal skull
2. Occipital skull
3. Scapula
4. Vertebral body
5. Vertebral pedicle
6. Kidney
7. Iliac tuberosity
8. Sacrum
9. Ischium
10. Vastus medialis
11. Vastus lateralis
12. Femoral condyle
13. Fibular head
14. Soleus muscle
15. Gastrocnemius, lateral head

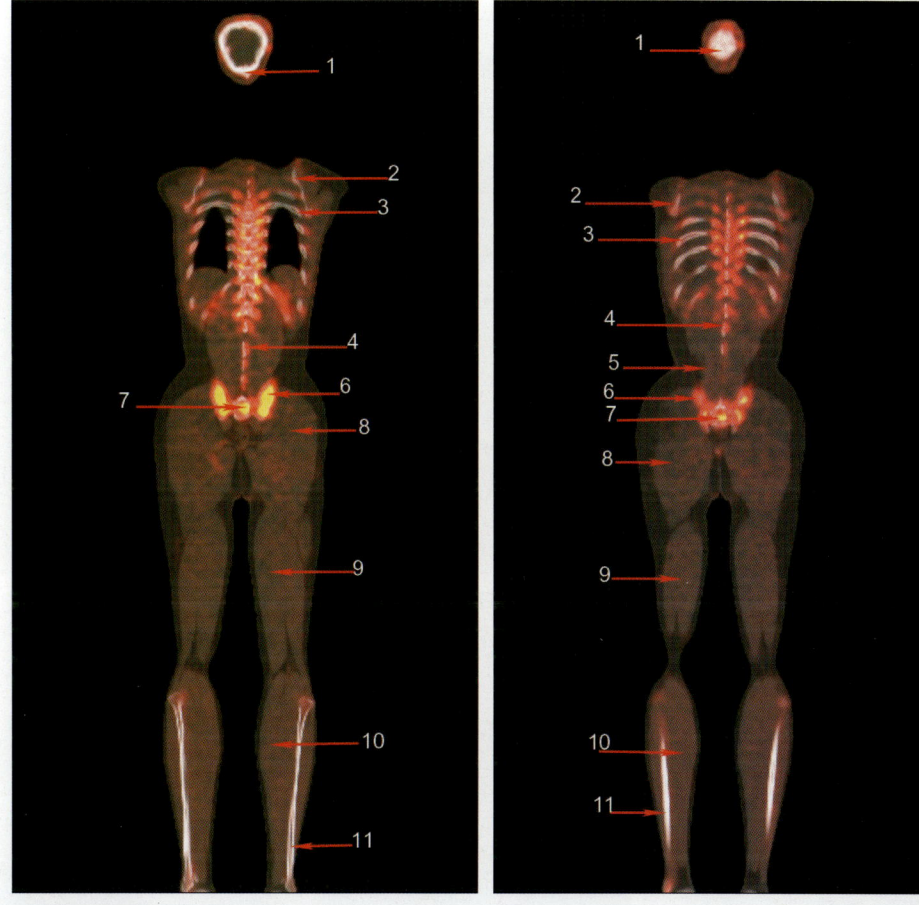

FIGURE 2-87:

1. Occipital skull
2. Scapula
3. Rib
4. Vertebral spinous process
5. Quadratus lumborum muscle
6. Iliac tuberosity
7. Sacrum
8. Gluteus muscles
9. Biceps femoris muscle
10. Gastrocnemius muscle, medial head
11. Fibula

Transaxials

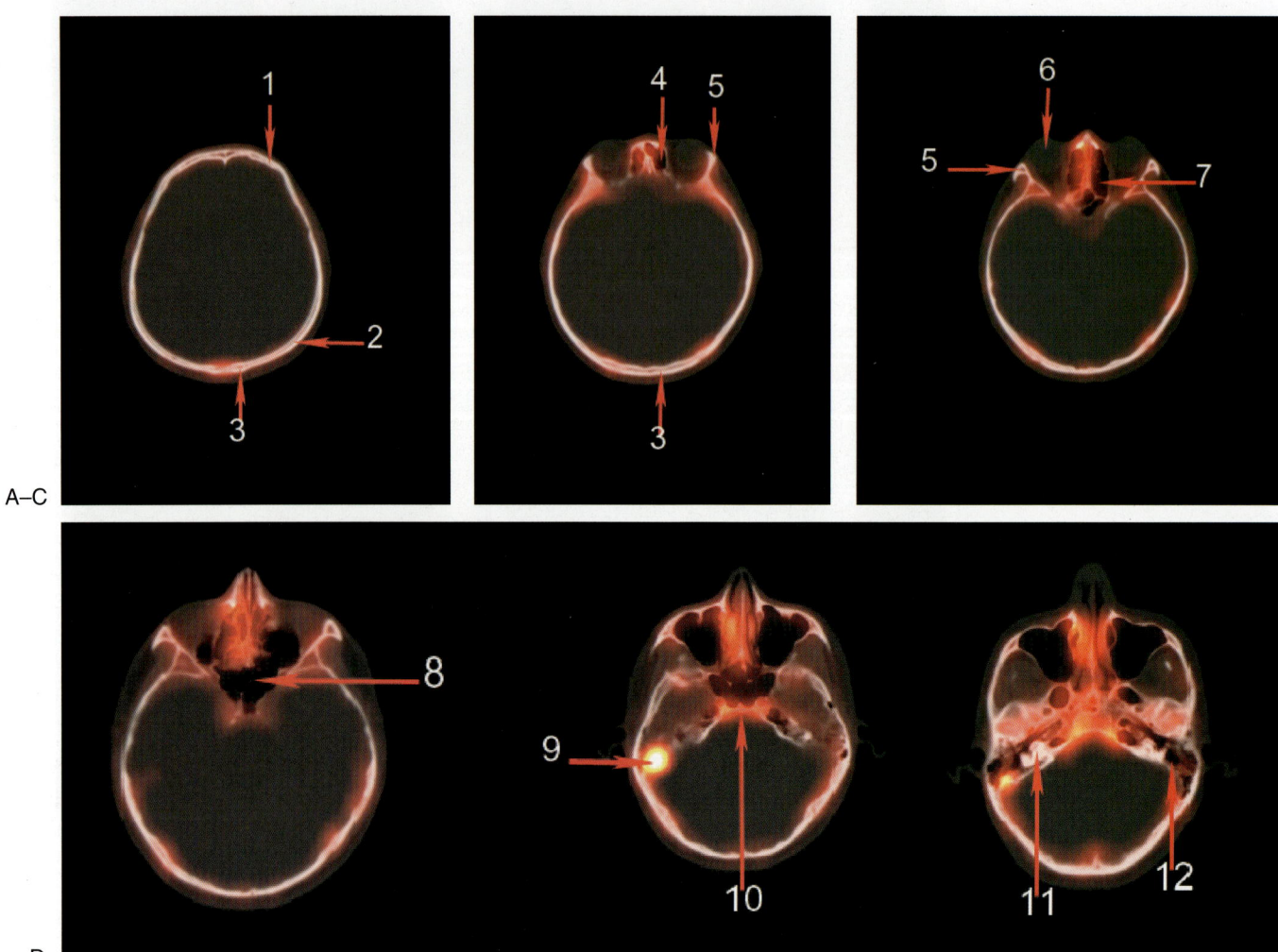

FIGURE 2-88:

1. Frontal skull	5. Zygomatic arch	9. Probable tumor
2. Parietal skull	6. Orbital globe	10. Clivus
3. Occipital skull	7. Ethmoidal sinus	11. Petrous temporal bone
4. Ethmoidal air	8. Sphenoidal sinus	12. Mastoid air cells

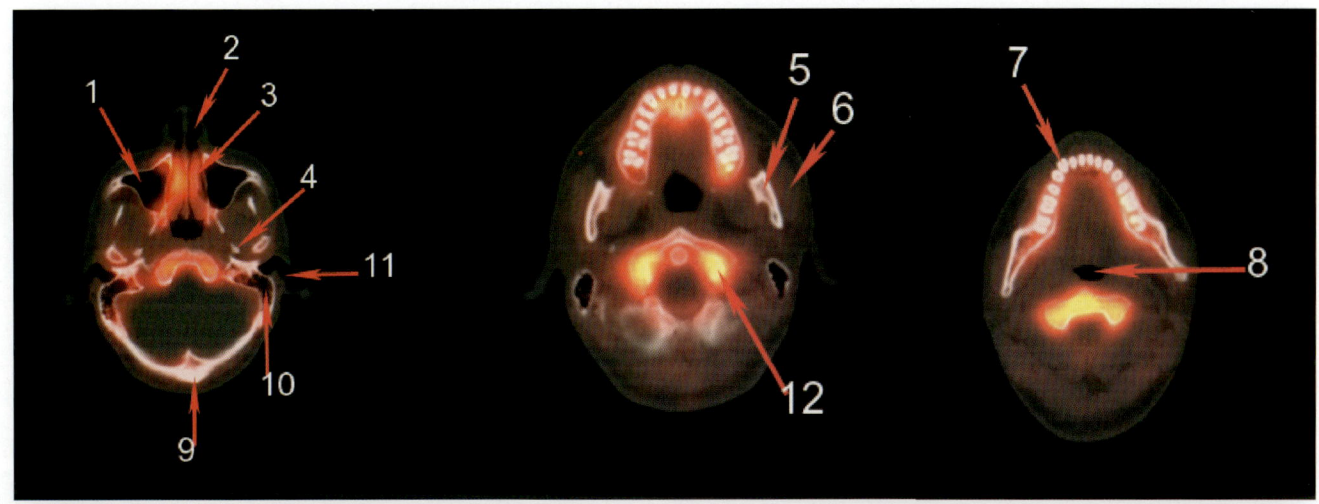

A

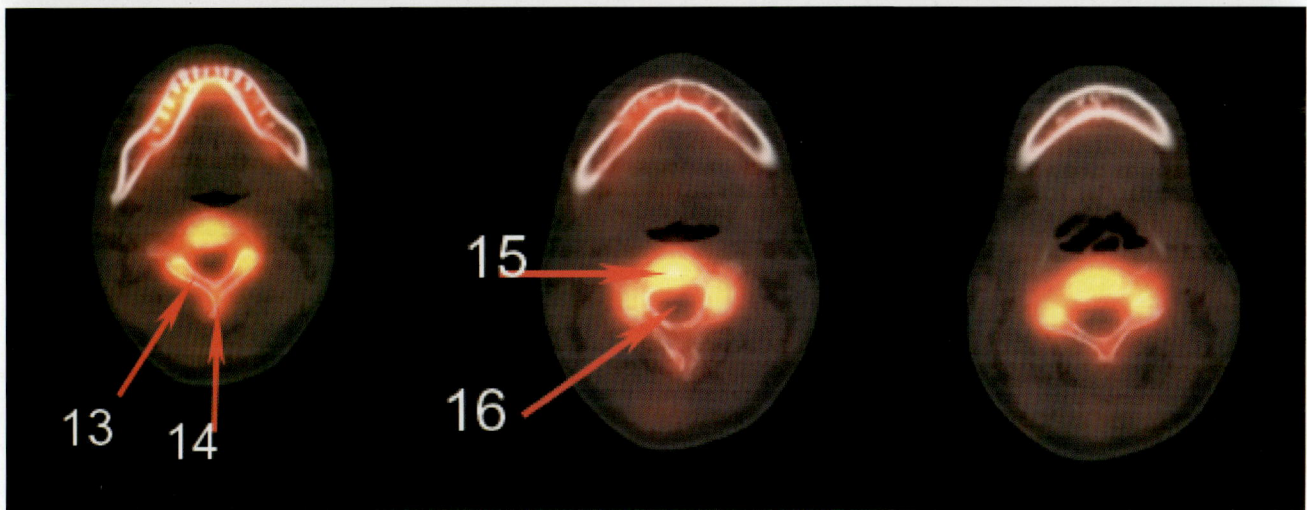

B

FIGURE 2-89:

1. Maxillary sinus
2. Nasal cavity
3. Nasal turbinate
4. Styloid process
5. Mandible
6. Parotid gland

7. Teeth
8. Oropharynx
9. Occipital skull
10. Mastoid air cells
11. External auditory canal
12. Atlas (C1)

13. Vertebral lamina
14. Vertebral spinous process
15. Vertebral body
16. Spinal canal

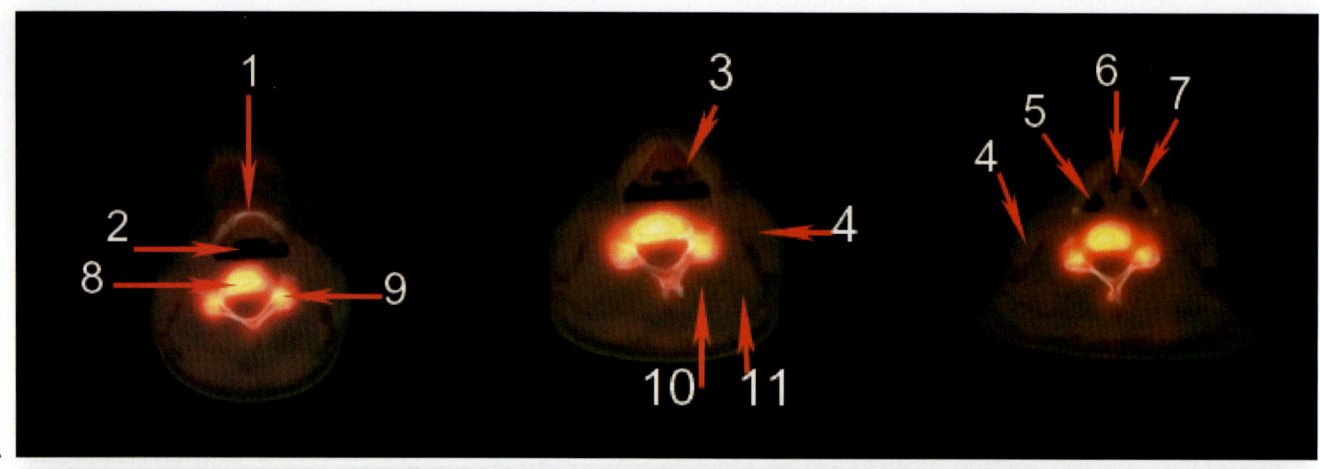

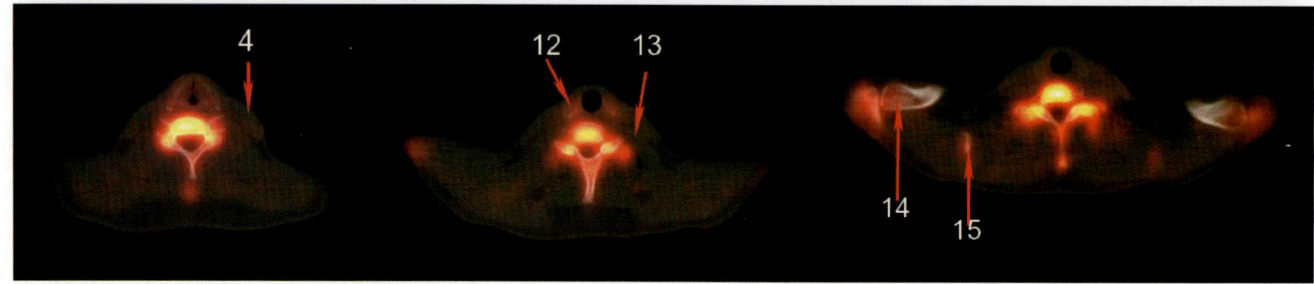

FIGURE 2-90:

1. Hyoid bone
2. Oropharynx
3. Vocal cord
4. Sternocleidomastoid muscle
5. Pyriform sinus
6. Trachea

7. Thyroid cartilage
8. Vertebral body
9. Articular pillar or process
10. Semispinalis cervicis muscle
11. Semispinalis capitis
     muscle

12. Thyroid
13. Scalene muscle
14. Clavicle
15. Scapula

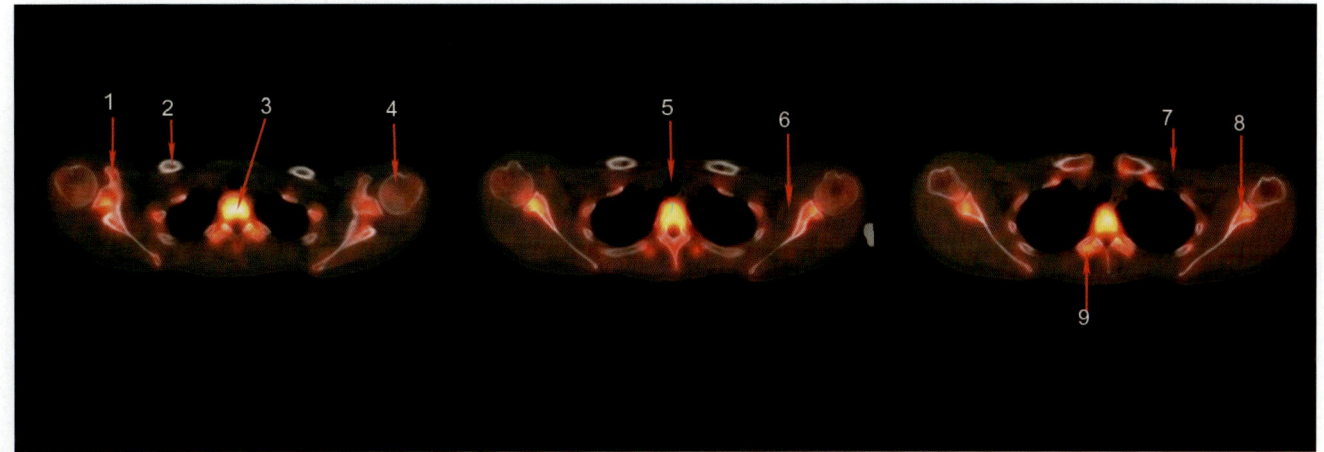

A

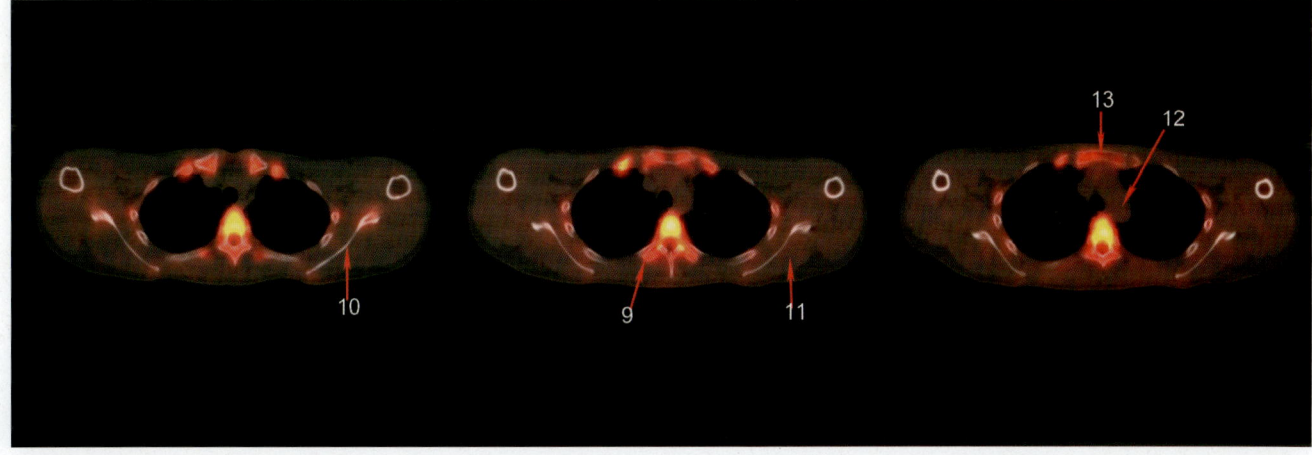

B

FIGURE 2-91:

1. Coracoid of scapula
2. Clavicle
3. Vertebral body
4. Humeral head
5. Trachea
6. Subscapularis muscle
7. Pectoralis muscles
8. Glenoid of scapula
9. Costotransverse joint
10. Scapular wing
11. Infraspinatus muscle
12. Aortic arch
13. Manubrium

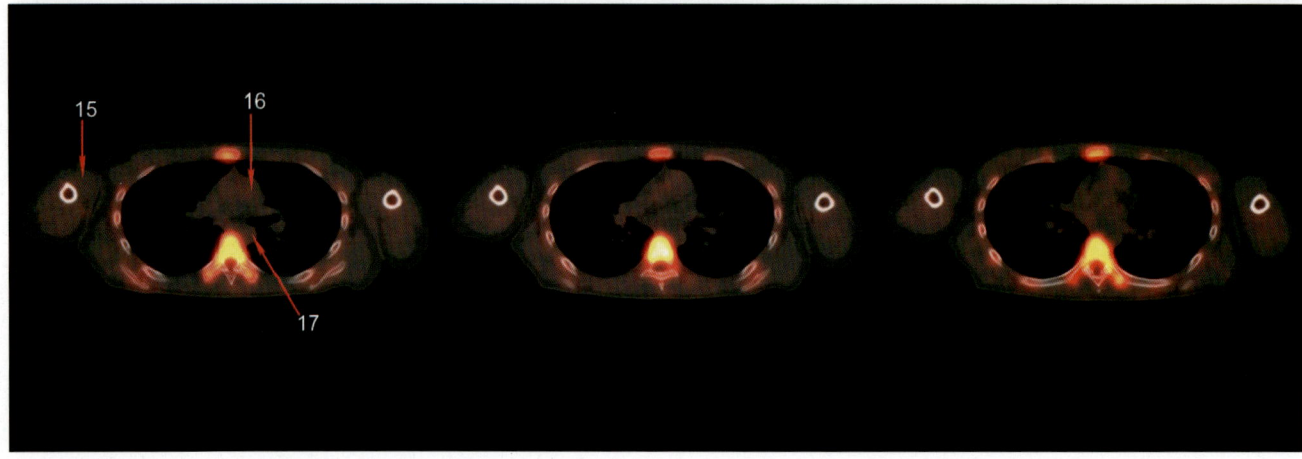

FIGURE 2-92:

1. Manubrium
2. Aortic arch
3. Pectoralis muscles
4. Deltoid muscle
5. Superior vena cava
6. Sternum
7. Left pulmonary artery
8. Subscapularis muscle
9. Right main bronchus
10. Humerus
11. Vertebral body
12. Scapula
13. Infraspinatus muscle
14. Triceps brachii muscle
15. Biceps brachii muscle
16. Right ventricle
17. Descending aorta

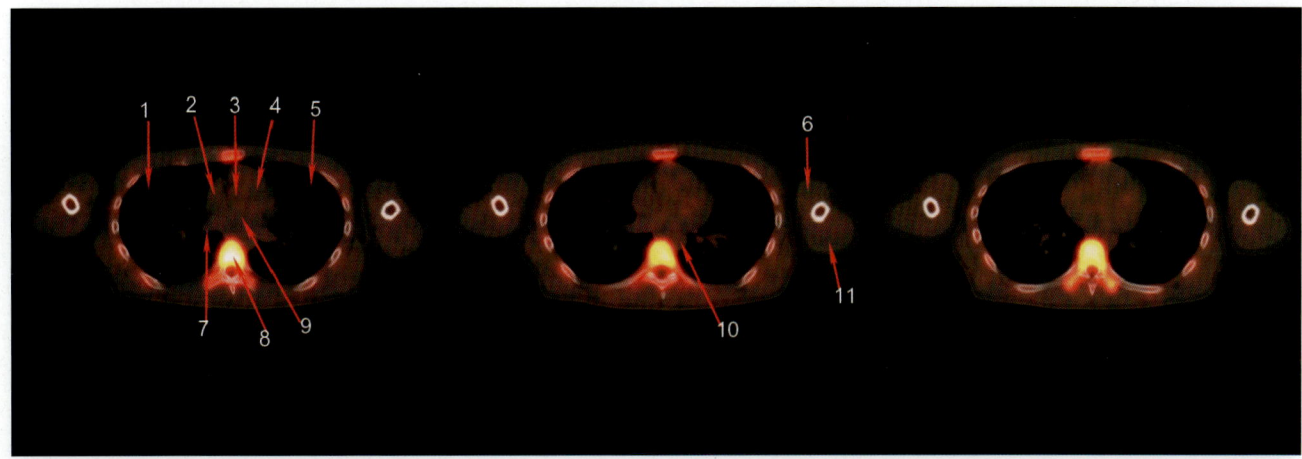

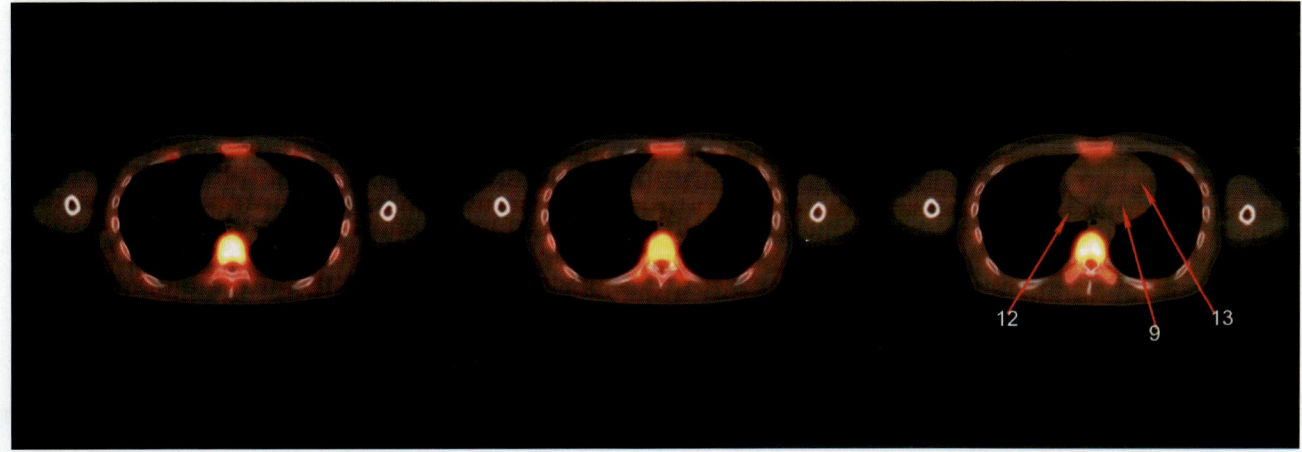

FIGURE 2-93:

1. Right upper lobe of lung
2. Right atrium
3. Ascending aorta
4. Right ventricle
5. Left upper lobe of lung
6. Biceps brachii muscle
7. Pulmonary vein
8. Vertebral body
9. Left atrium
10. Descending aorta
11. Triceps brachii muscle
12. Inferior vena cava
13. Left ventricle

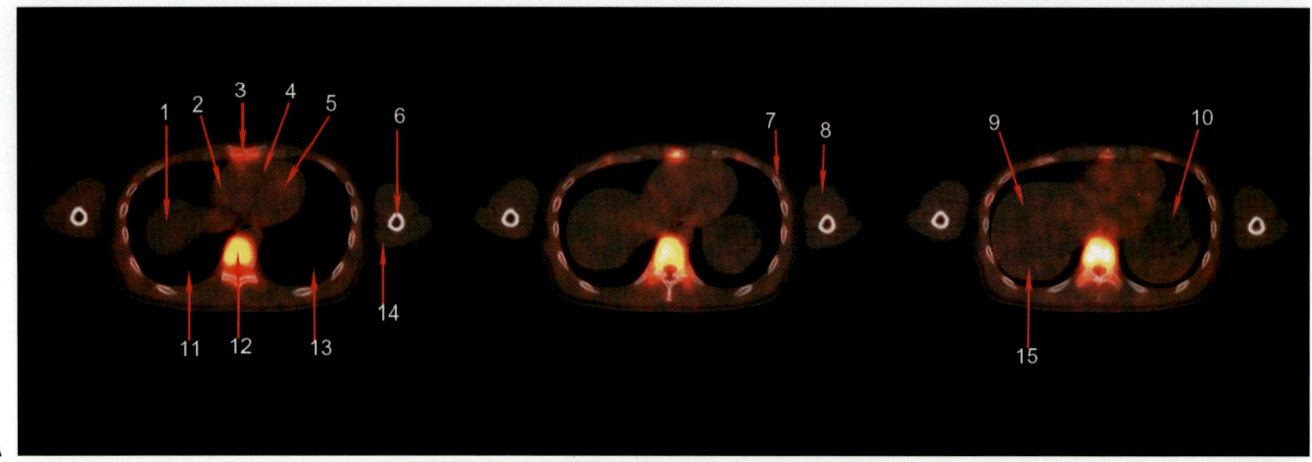

A

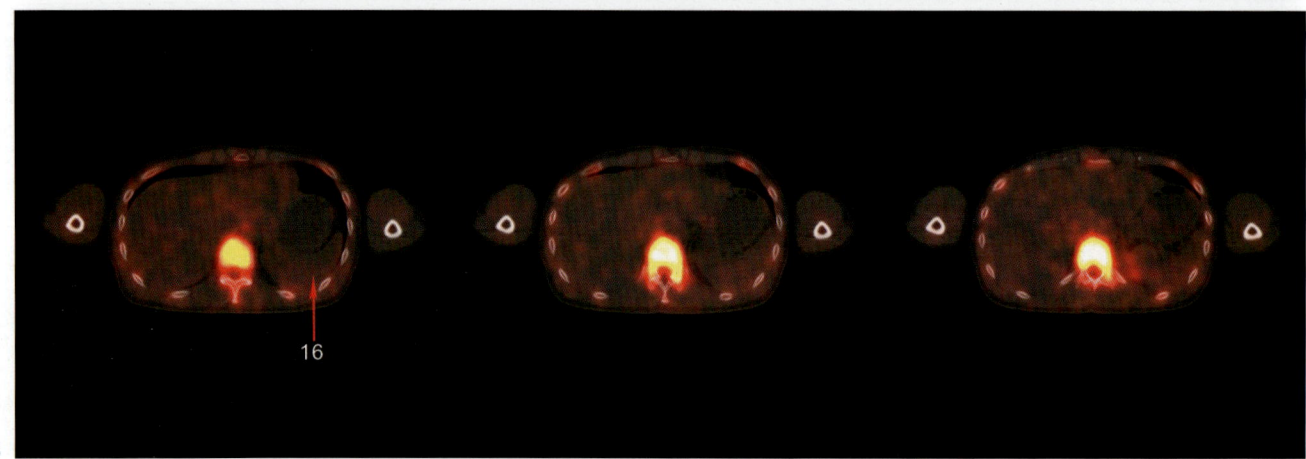

B

FIGURE 2-94:

1. Hepatic dome	7. Rib	12. Vertebral body
2. Right atrium	8. Biceps brachii muscle	13. Left lower lobe of lung
3. Sternum	9. Upper segment (#8) of right	14. Triceps brachii muscle
4. Right ventricle	hepatic lobe	15. Upper segment (#7) of right
5. Left ventricle	10. Stomach	hepatic lobe
6. Humerus	11. Right lower lobe of lung	16. Spleen

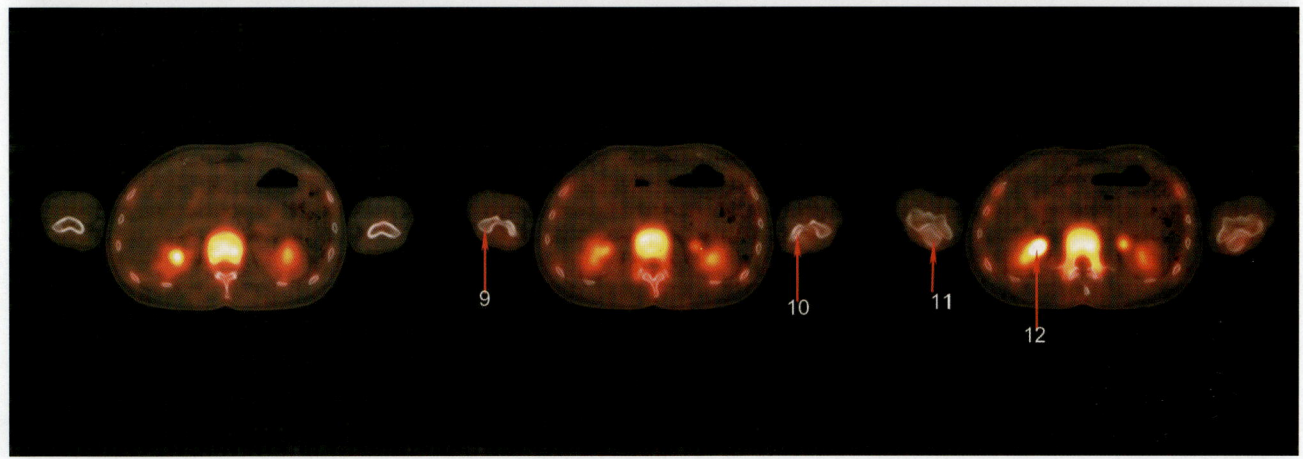

FIGURE 2-95:

1. Vertebral body
2. Lateral segment of left hepatic lobe
3. Stomach
4. Splenic flexure of colon
5. Humerus

6. Inferior segment (#5) of right hepatic lobe
7. Left renal pelvis
8. Inferior segment (#6) of right hepatic lobe

9. Lateral epicondyle of humerus
10. Medial epicondyle of humerus
11. Olecranon process of ulna
12. Right renal pelvis

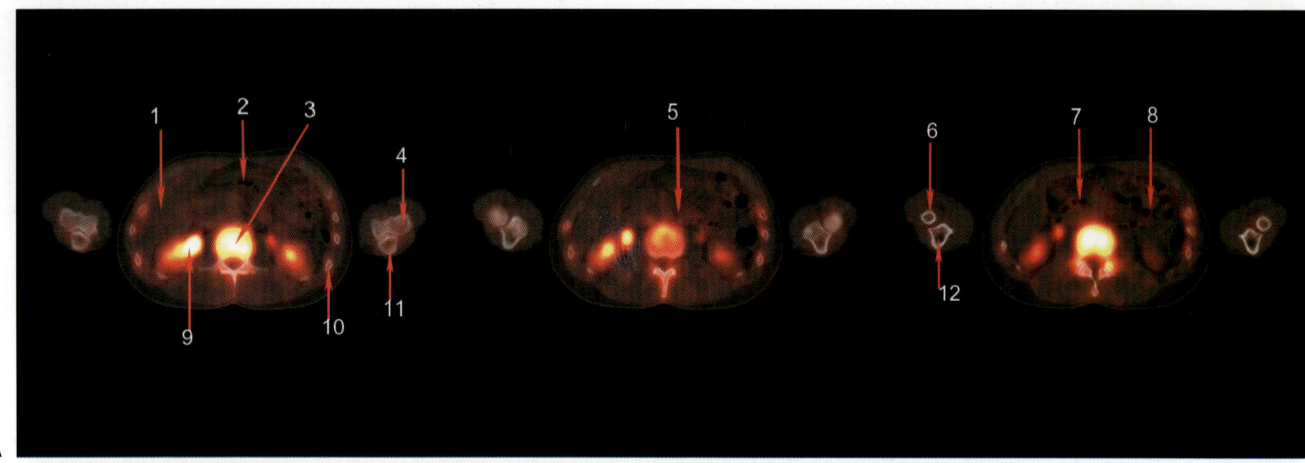

A

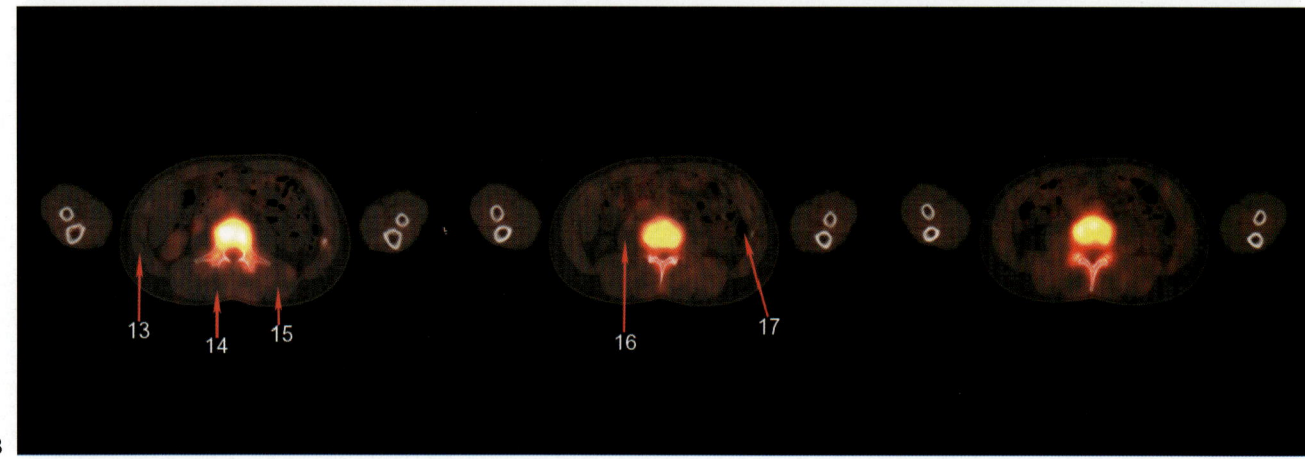

B

FIGURE 2-96:

1. Inferior segment of right hepatic
   lobe
2. Stomach
3. Vertebral body
4. Lateral epicondyle of humerus
5. Pancreas

6. Radius
7. Duodenum
8. Jejunum
9. Right renal pelvis
10. Rib
11. Olecranon process of ulna

12. Ulna
13. Oblique muscles
14. Erector spinae muscle
15. Quadratus lumborum muscle
16. Psoas muscle
17. Descending colon

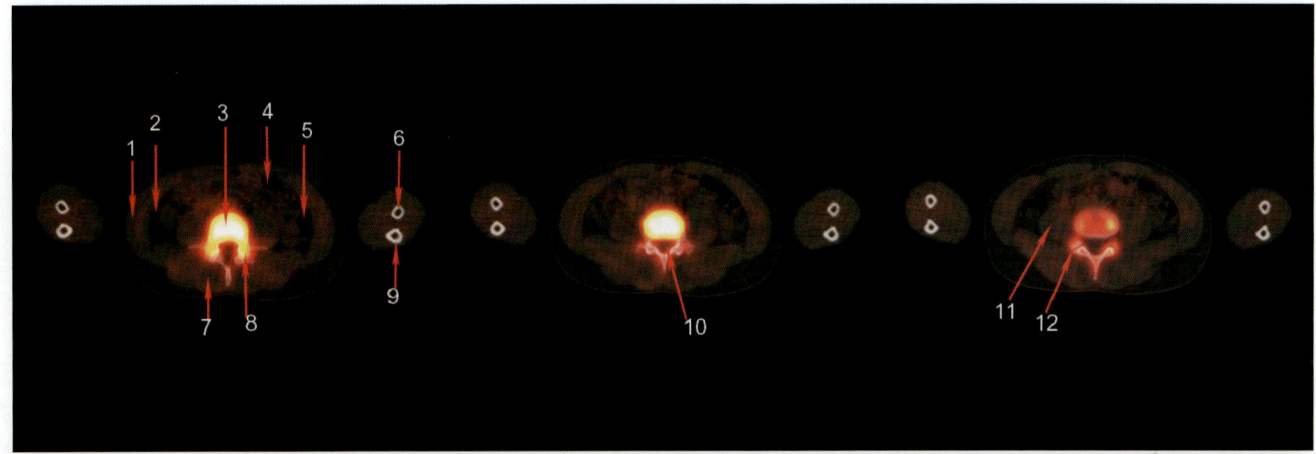

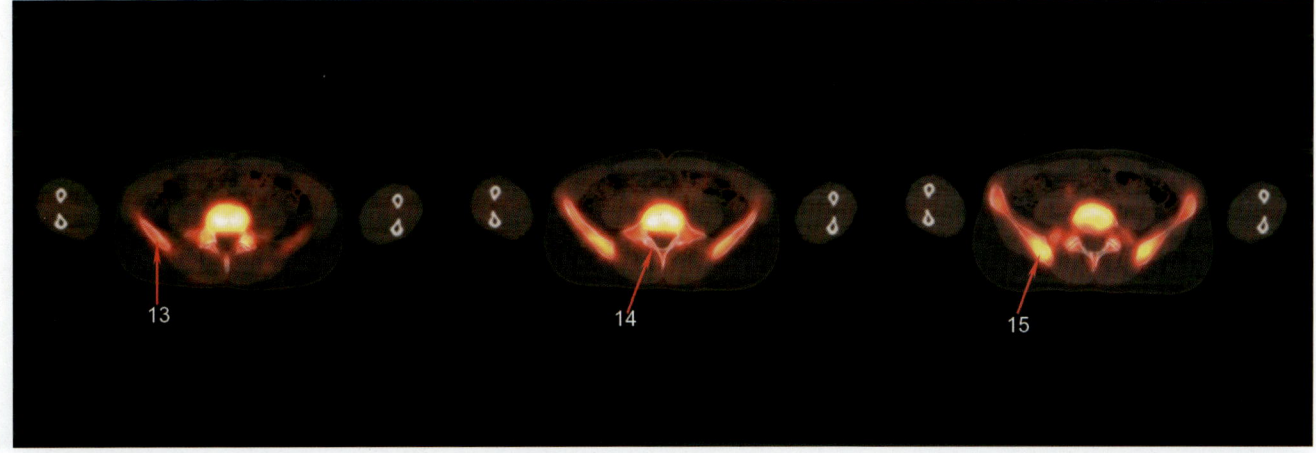

FIGURE 2-97:

1. Oblique muscles
2. Ascending colon
3. Vertebral body
4. Transverse colon
5. Descending colon

6. Radius
7. Erector spinae muscle
8. Vertebral pedicle
9. Ulna
10. Spinal canal

11. Psoas muscle
12. Facet joint
13. Iliac ala
14. Vertebral lamina
15. Iliac tuberosity

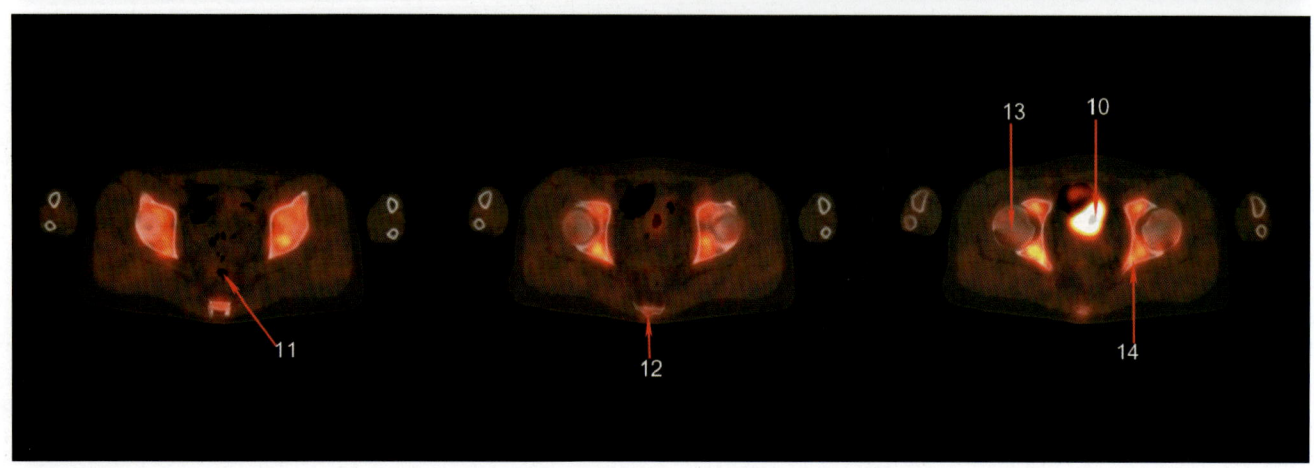

FIGURE 2-98:

1. Iliacus muscle	6. Sigmoid colon	11. Rectum
2. Iliac ala	7. Sacral ala	12. Coccyx
3. Gluteus medius muscle	8. Sacral canal	13. Femoral head
4. Cecum	9. Gluteus maximus	14. Posterior lip of acetabulum
5. Ileum	10. Bladder	

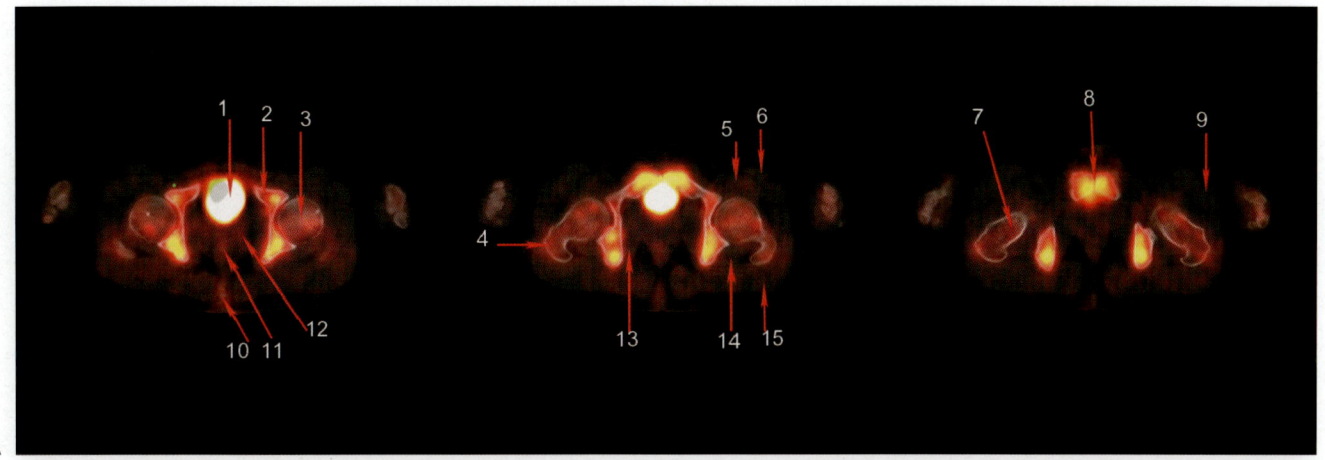

A

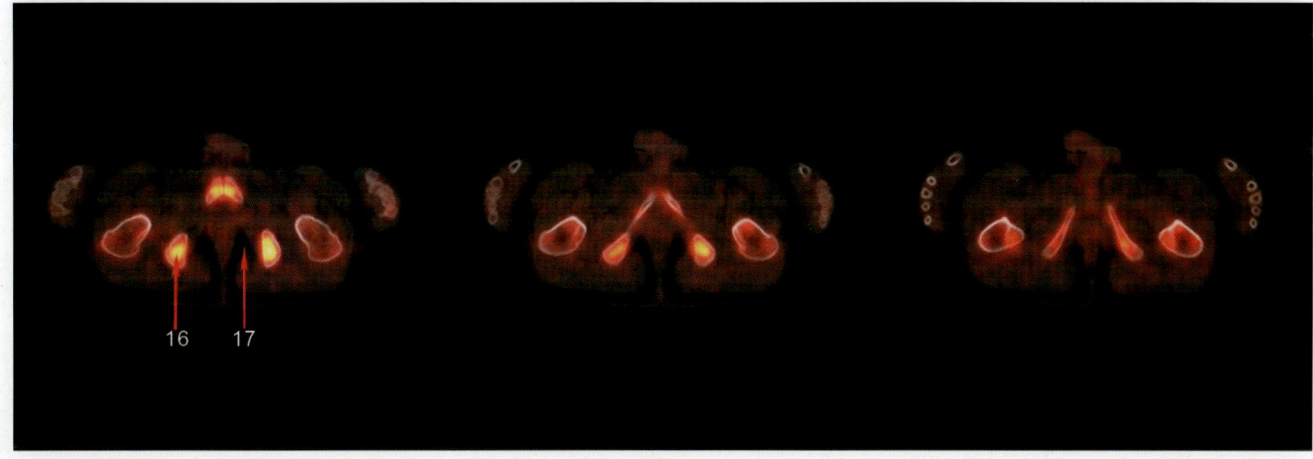

B

FIGURE 2-99:

1. Bladder
2. Anterior lip of acetabulum
3. Femoral head
4. Greater trochanter of right femur
5. Iliopsoas muscle
6. Sartorius muscle
7. Femoral neck
8. Symphysis pubis
9. Tensor fascia latae muscle
10. Coccyx
11. Anus
12. Seminal vesicle
13. Obturator internal muscle
14. Gemellus muscles
15. Gluteus maximus muscle
16. Ischial tuberosity
17. Ischiorectal fat

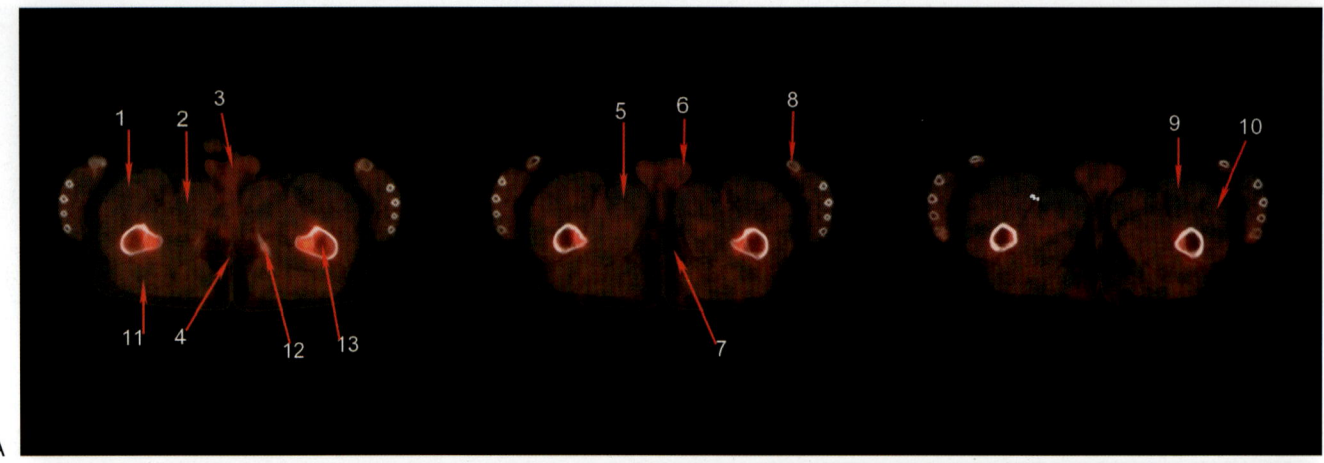

A

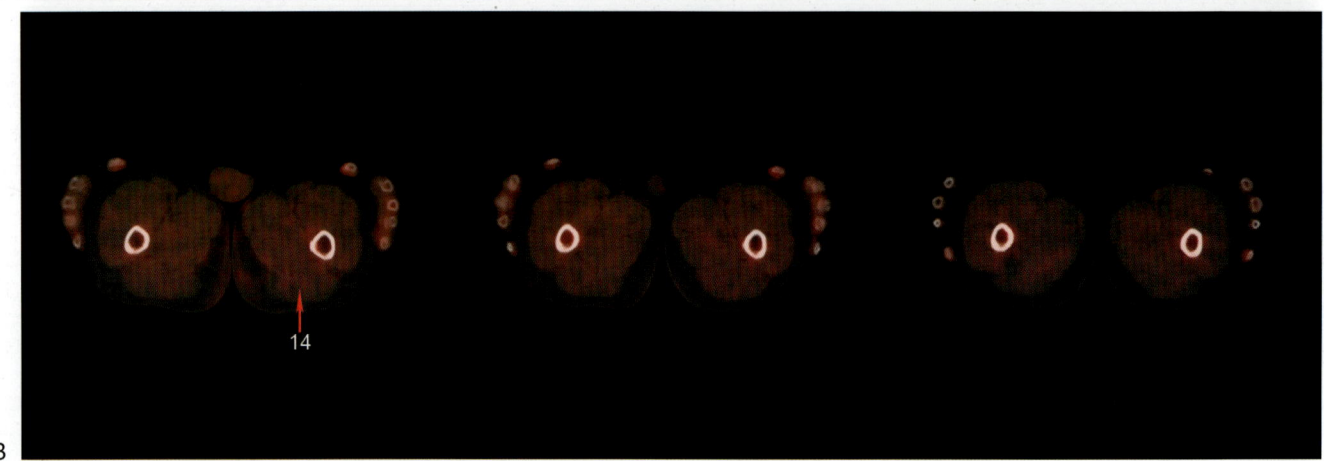

B

FIGURE 2-100:

1. Iliopsoas muscle
2. External obturator muscle
3. Corporal muscles of penis
4. Anus
5. Adductor muscles
6. Testis
7. Perianal fat
8. Thumb (metacarpal)
9. Rectus femoris muscle
10. Vastus lateralis muscle
11. Gluteus muscles
12. Ischium
13. Trochanter of femur
14. Biceps femoris muscle

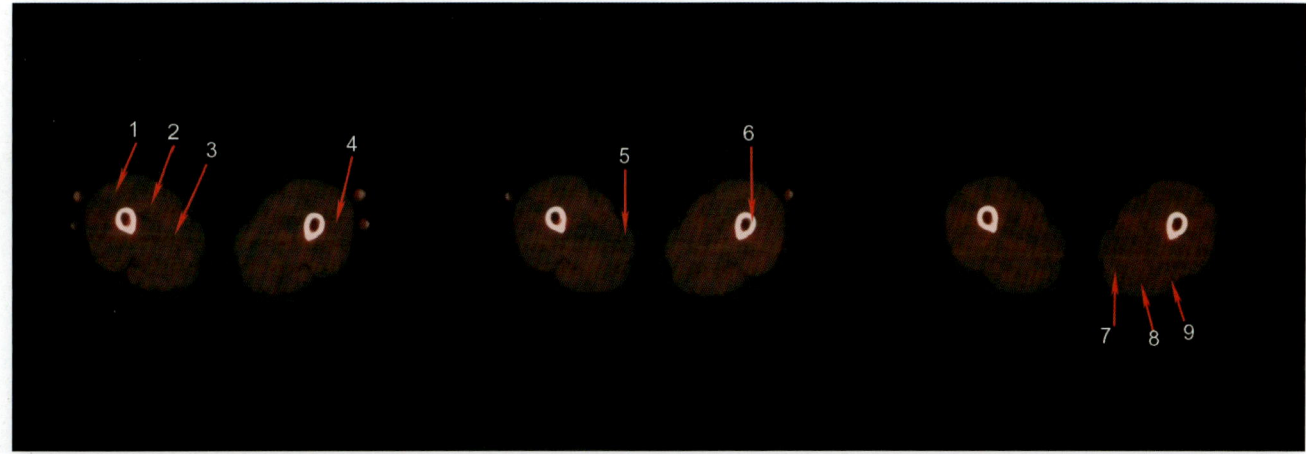

A

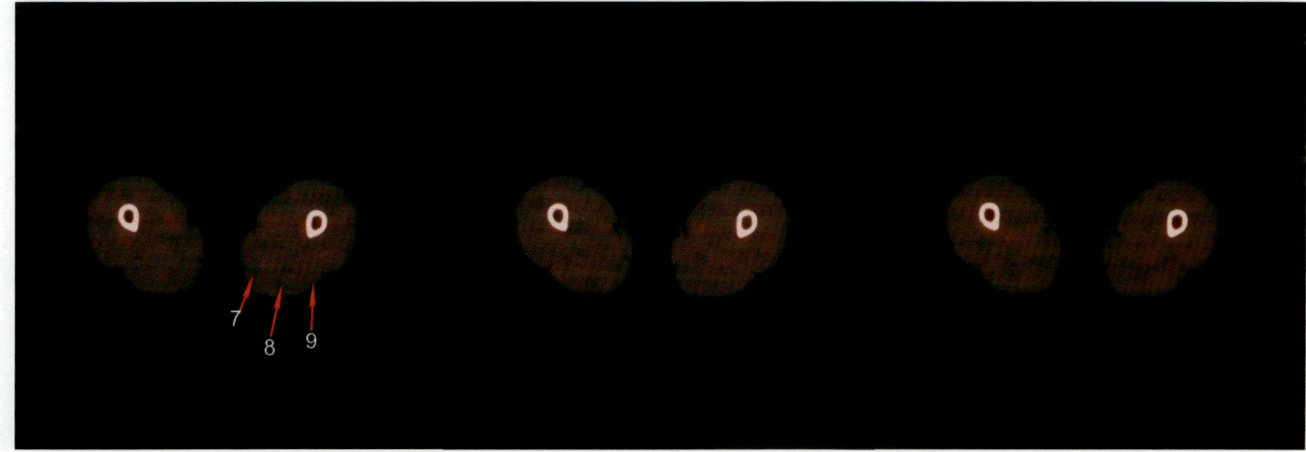

B

FIGURE 2-101:

1. Rectus femoris muscle
2. Vastus medialis muscle
3. Adductor magnus muscle
4. Vastus lateralis muscle
5. Gracilis muscle
6. Cortex of left femur
7. Semimembranous muscle
8. Semitendinosus muscle
9. Biceps femoris muscle

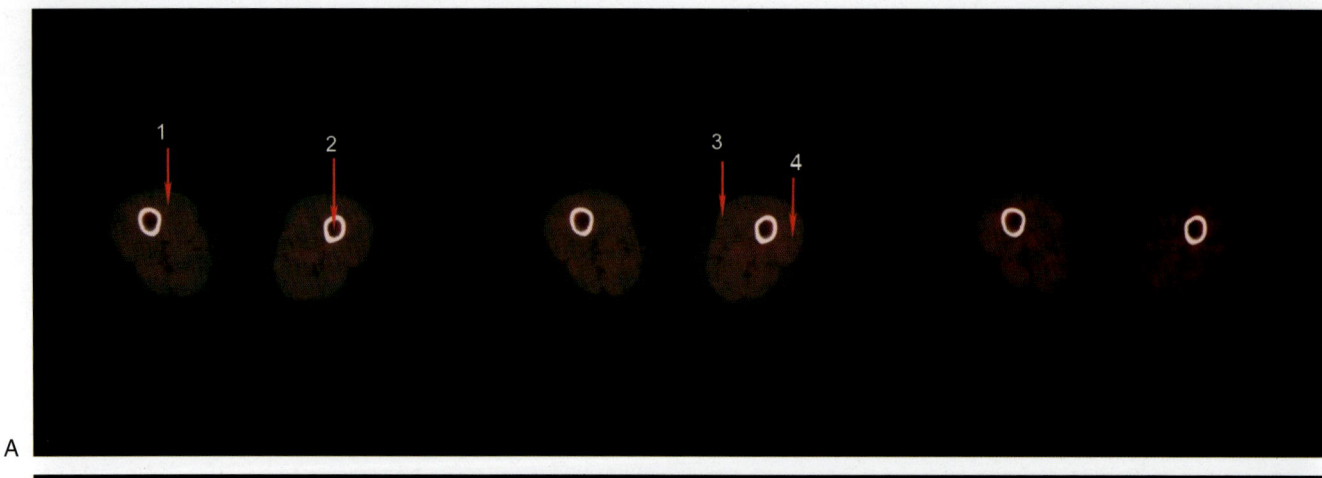

A

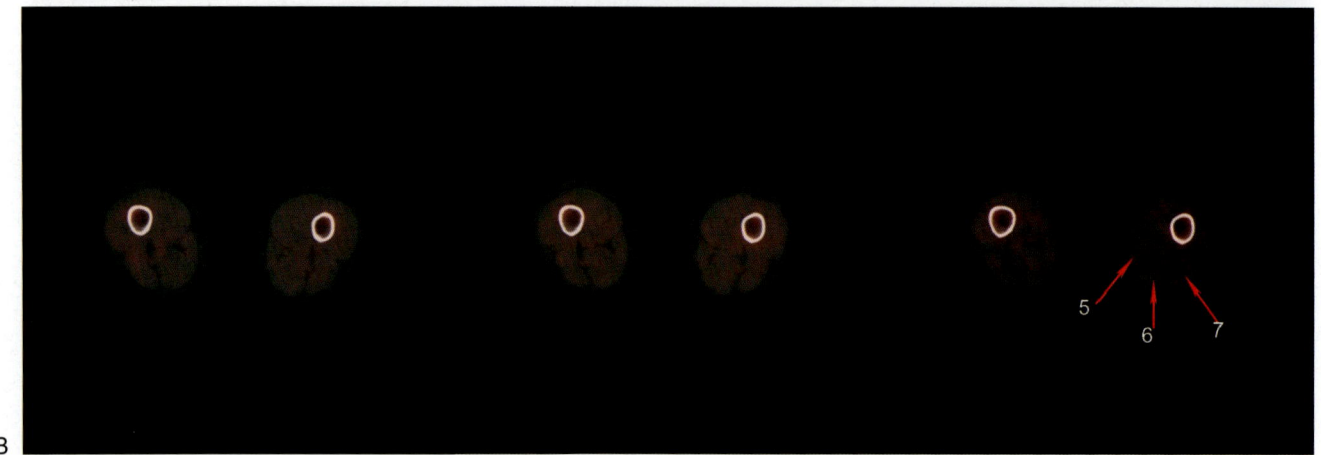

B

FIGURE 2-102:

1. Vastus medialis muscle
2. Femur
3. Sartorius muscle
4. Vastus lateralis muscle
5. Semimembranous muscle
6. Semitendinous muscle
7. Biceps femoris muscle

3 Lymphoscintigraphy SPECT/CT

Lymphoscintigraphy has been used for mapping of the sentinel lymph node (SLN) of various tumors before biopsy. It assists in tailoring the surgical field and in the determination of the surgical site.[1] It is particularly important for tumors located in regions with ambiguous nodal drainage, such as trunk and shoulders, as well as head and neck.[2] Single photon emission computed tomography (SPECT)/computed tomography (CT) identified SLNs missed on planar imaging and allowed for the precise localization in 43% of patients with the tumor in the head and neck or trunk. These SLNs were located close to the injection site and were hidden by the scattered radiation.[3] SPECT/CT also allowed for the estimation of the depth of the SLN in areas of complex anatomy in 40% of patients with melanoma, helping improve surgical planning.[4]

Section 1: Head and Neck Lymphoscintigraphy

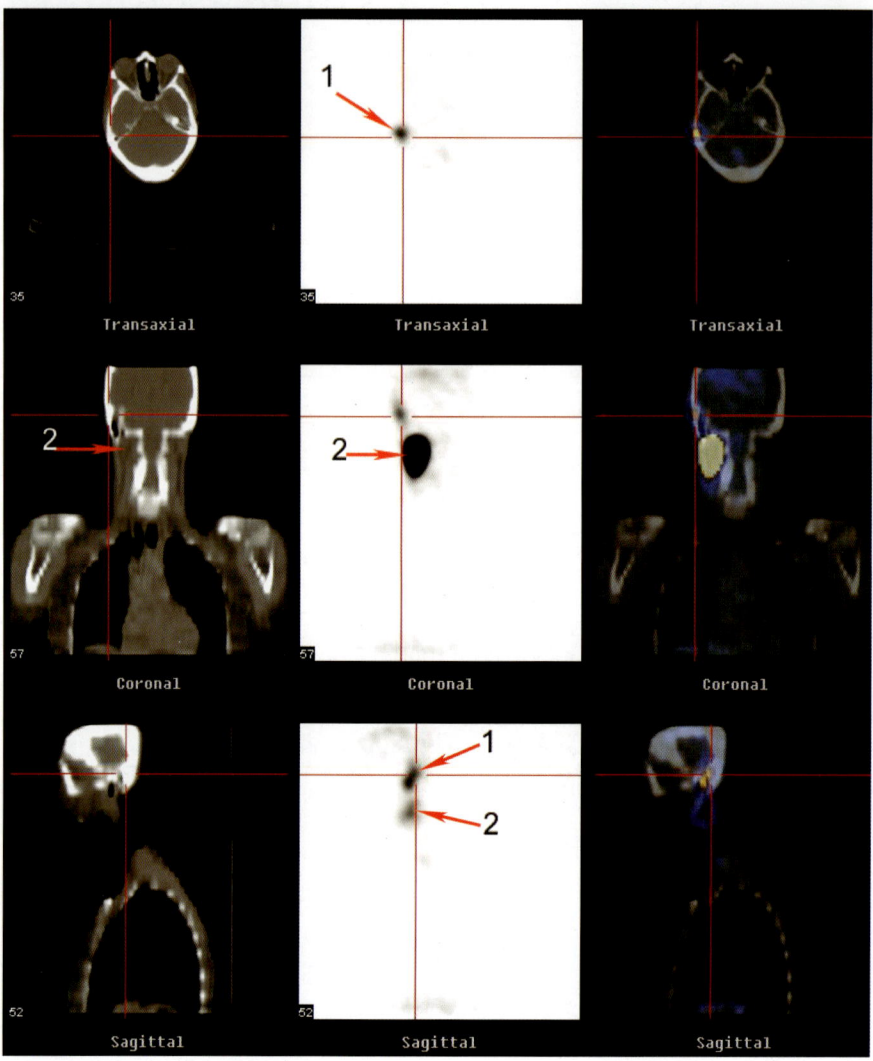

FIGURE 3-1: (Copyright 2006 The University of Texas M. D. Anderson Cancer Center. Used with Permission.)

1. Postauricular node   2. Upper jugular node (level II)

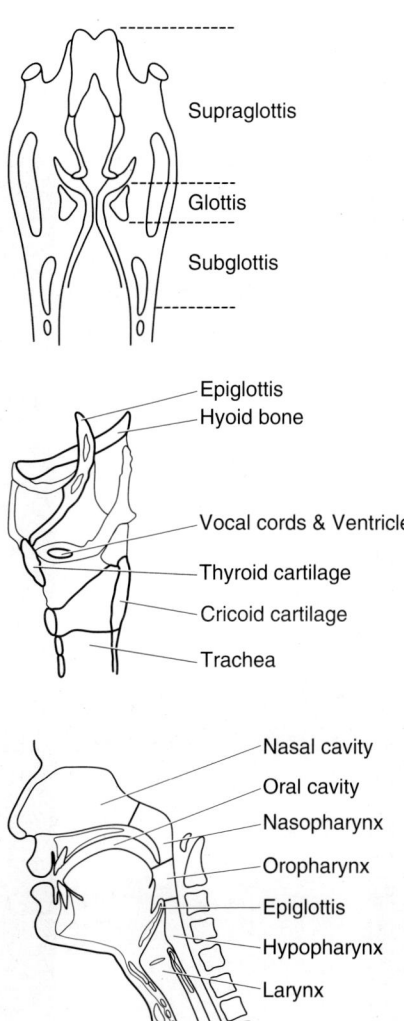

FIGURE 3-2: Anatomy of mouth and neck

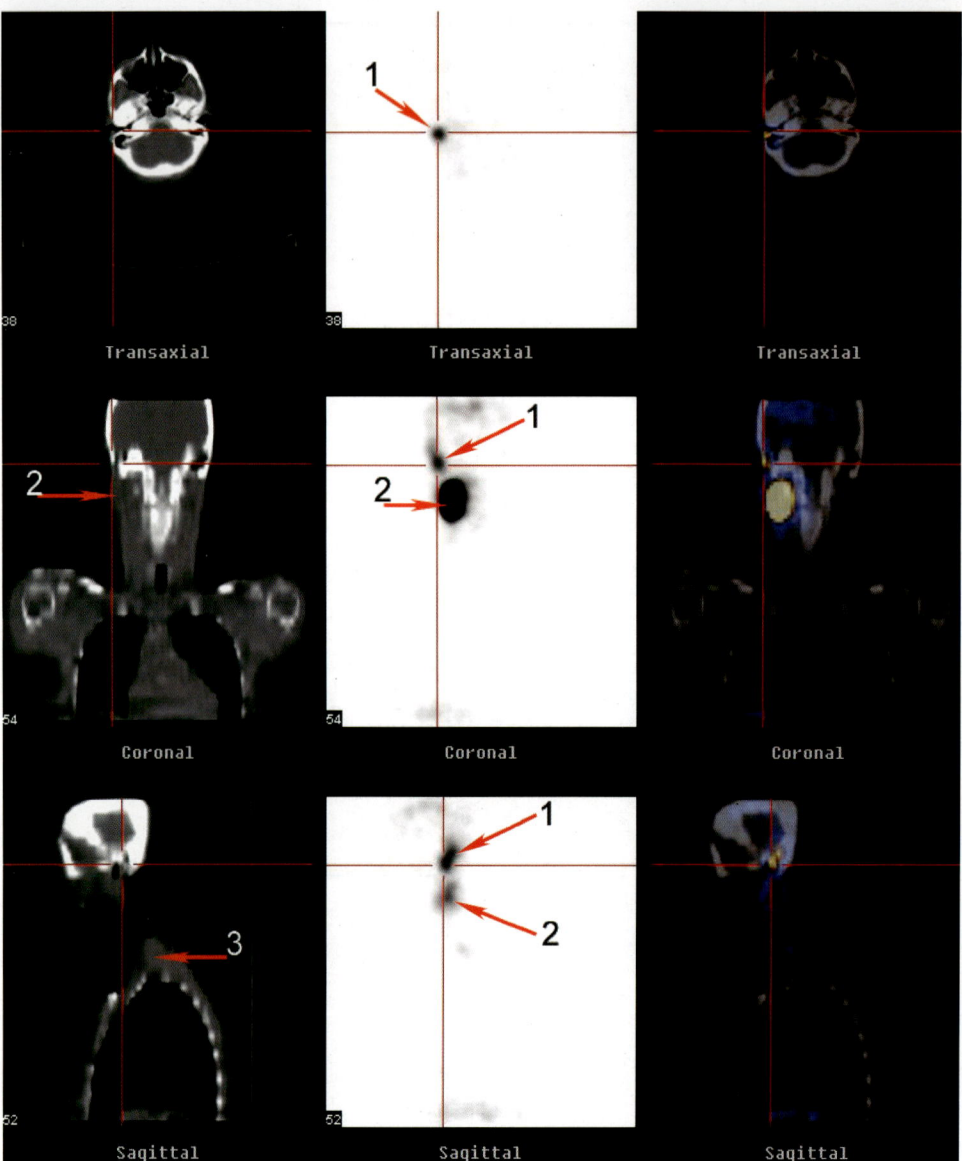

1. Postauricular node
2. Upper jugular node (level II)
3. Trapezius muscle

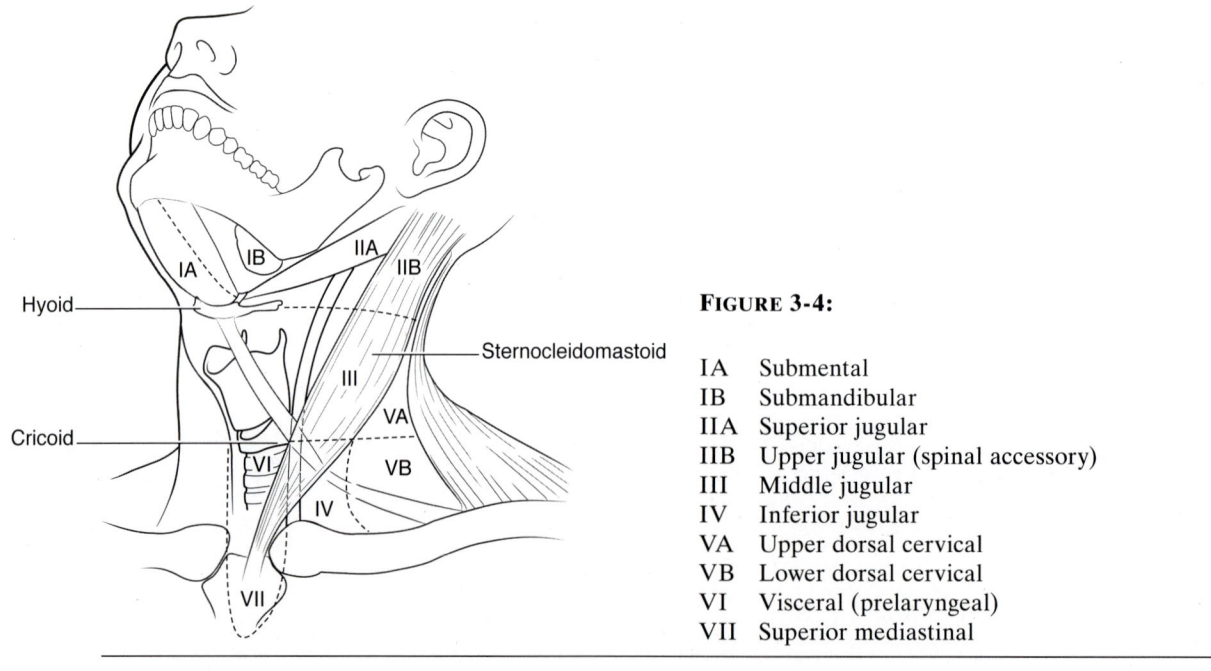

FIGURE 3-4:

IA	Submental
IB	Submandibular
IIA	Superior jugular
IIB	Upper jugular (spinal accessory)
III	Middle jugular
IV	Inferior jugular
VA	Upper dorsal cervical
VB	Lower dorsal cervical
VI	Visceral (prelaryngeal)
VII	Superior mediastinal

FIGURE 3-5: (Copyright 2006 The University of Texas M. D. Anderson Cancer Center. Used with Permission.)

1. Upper jugular node (level II)
2. Trachea
3. Right atrium
4. Left ventricle

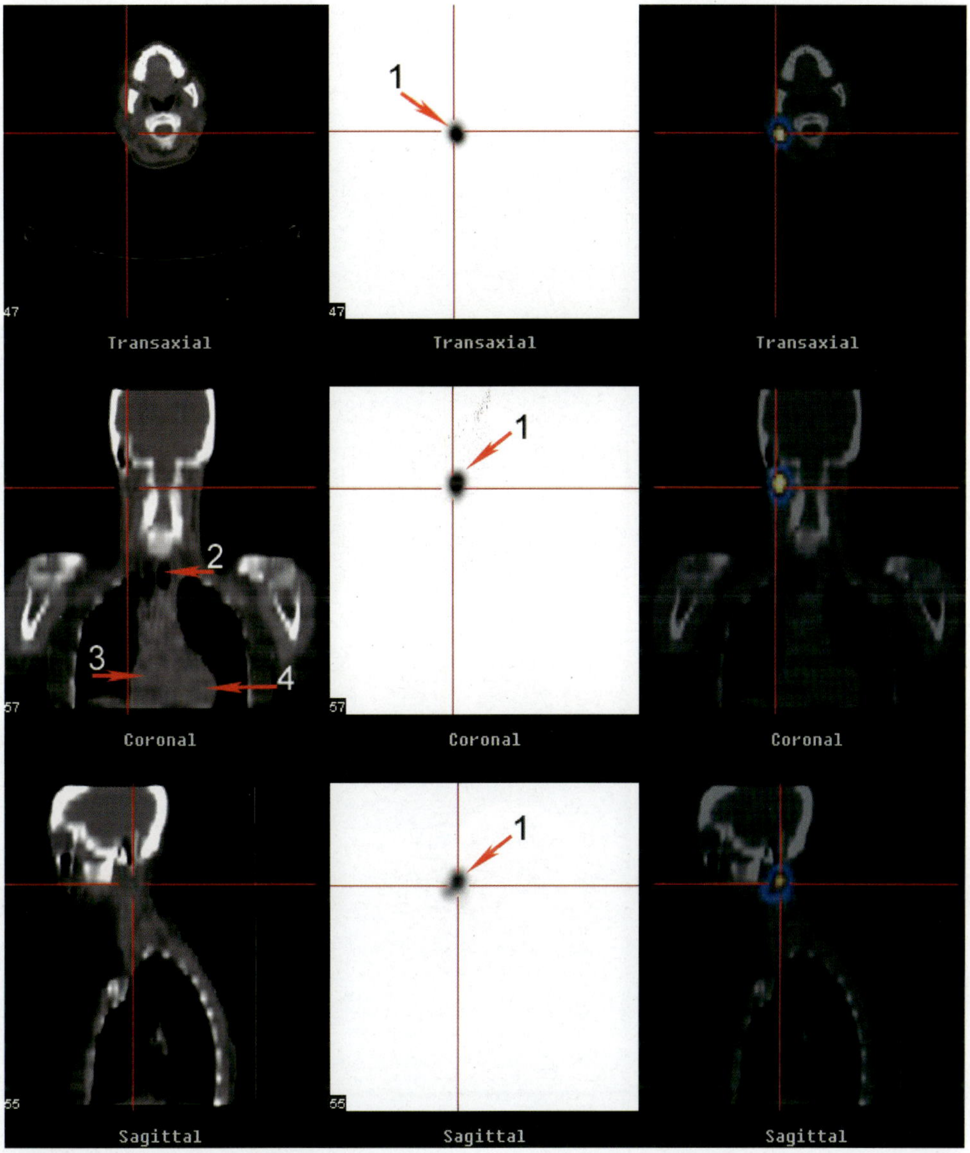

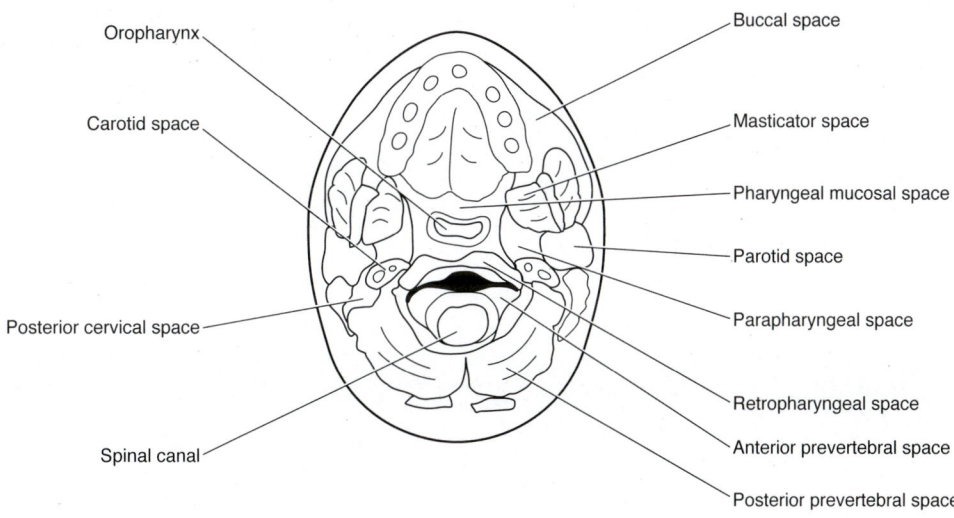

FIGURE 3-6: Transaxial view of neck (mouth level)

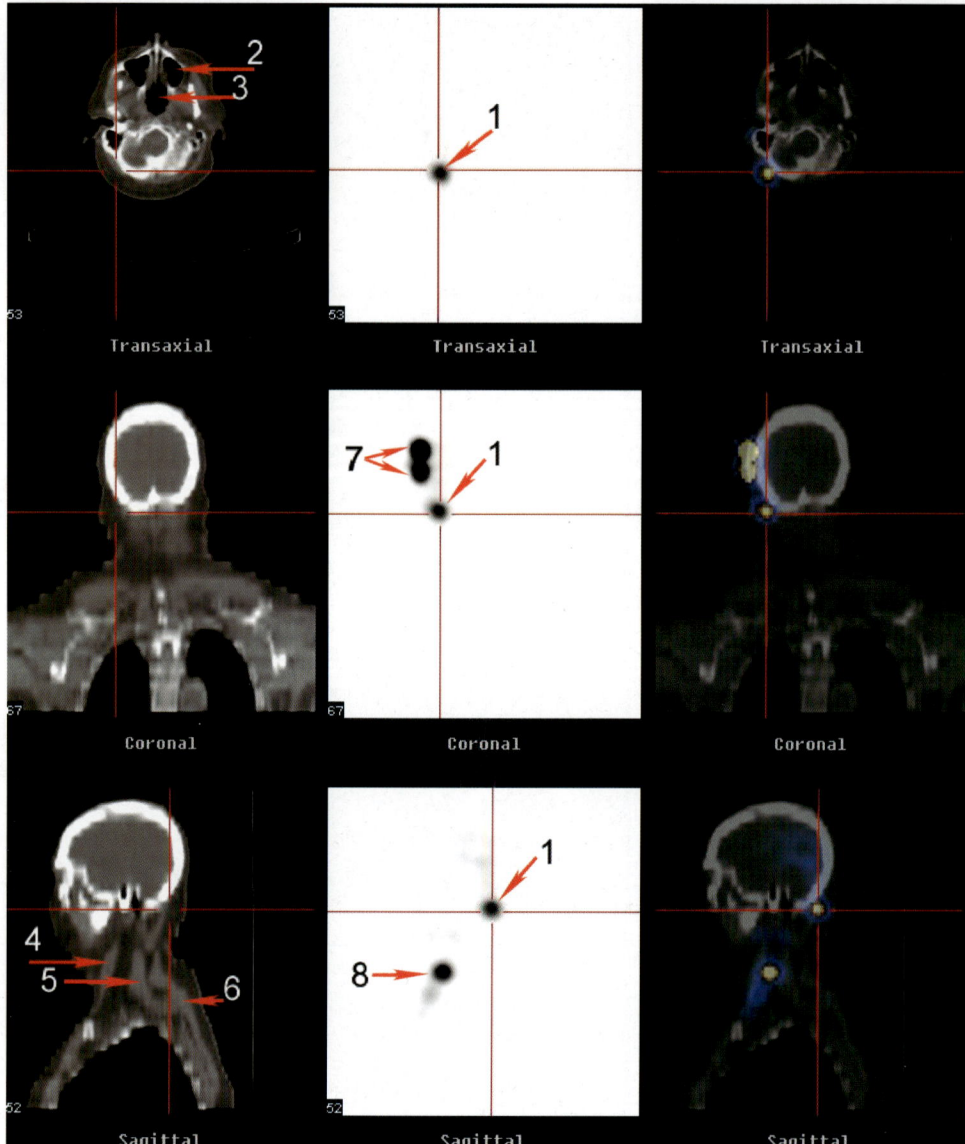

FIGURE 3-7:

1. Posterior triangle or upper dorsal cervical node (level VA)
2. Maxillary sinus
3. Nasopharynx
4. Sternocleidomastoid muscle
5. Splenius capitis muscle
6. Trapezius muscle
7. Injection site
8. Middle jugular lymph node (level III)

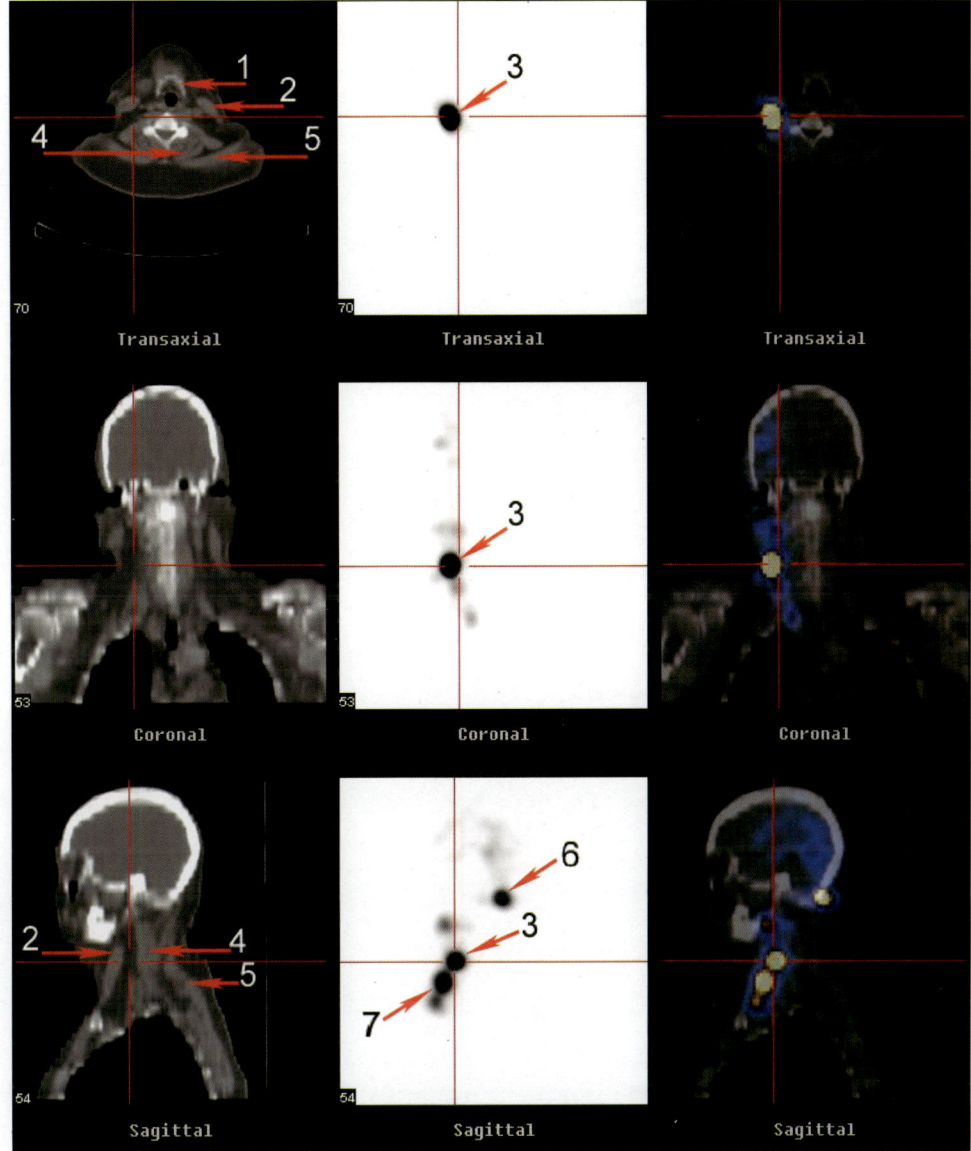

FIGURE 3-8:

1. Hyoid bone
2. Sternocleidomastoid muscle
3. Middle jugular node (level III)
4. Semispinalis cervicis muscle
5. Trapezius muscle
6. Upper cervical node (level VA)
7. Lower jugular node (level IV)

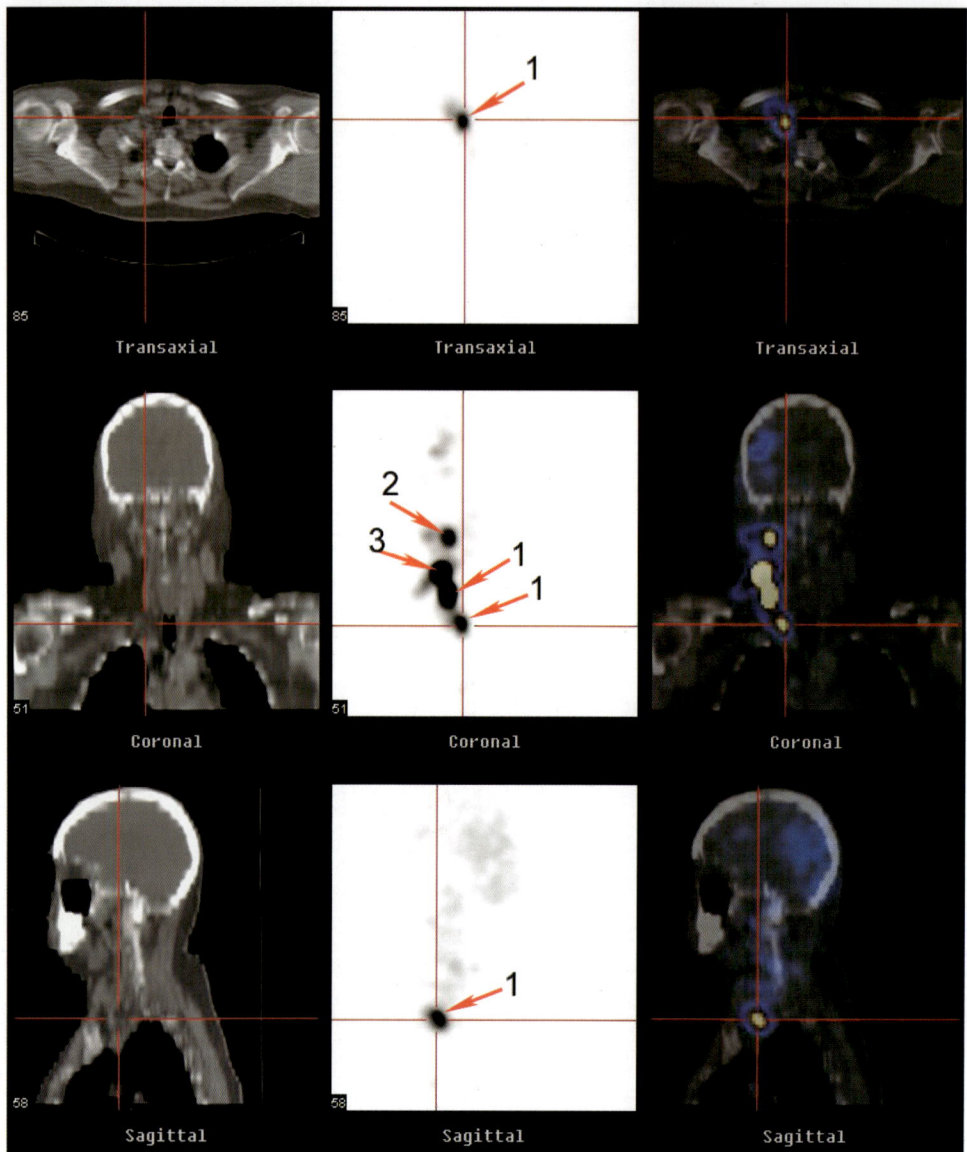

FIGURE 3-9: (Copyright 2006 The University of Texas M. D. Anderson Cancer Center. Used with Permission.)

1. Lower jugular node (level IV)
2. Upper jugular node (level II)
3. Middle jugular node (level III)

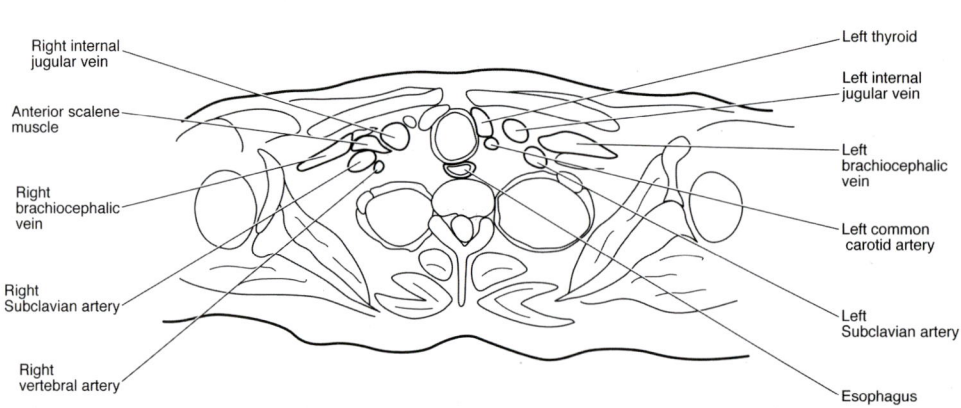

FIGURE 3-10: Transaxial view of lower neck

Right internal jugular vein

Anterior scalene muscle

Right brachiocephalic vein

Right Subclavian artery

Right vertebral artery

Left thyroid

Left internal jugular vein

Left brachiocephalic vein

Left common carotid artery

Left Subclavian artery

Esophagus

FIGURE 3-11: (Copyright 2006 The University of Texas M. D. Anderson Cancer Center. Used with Permission.)

1. Left upper jugular node (level IIA)
2. Left middle jugular node (level III)

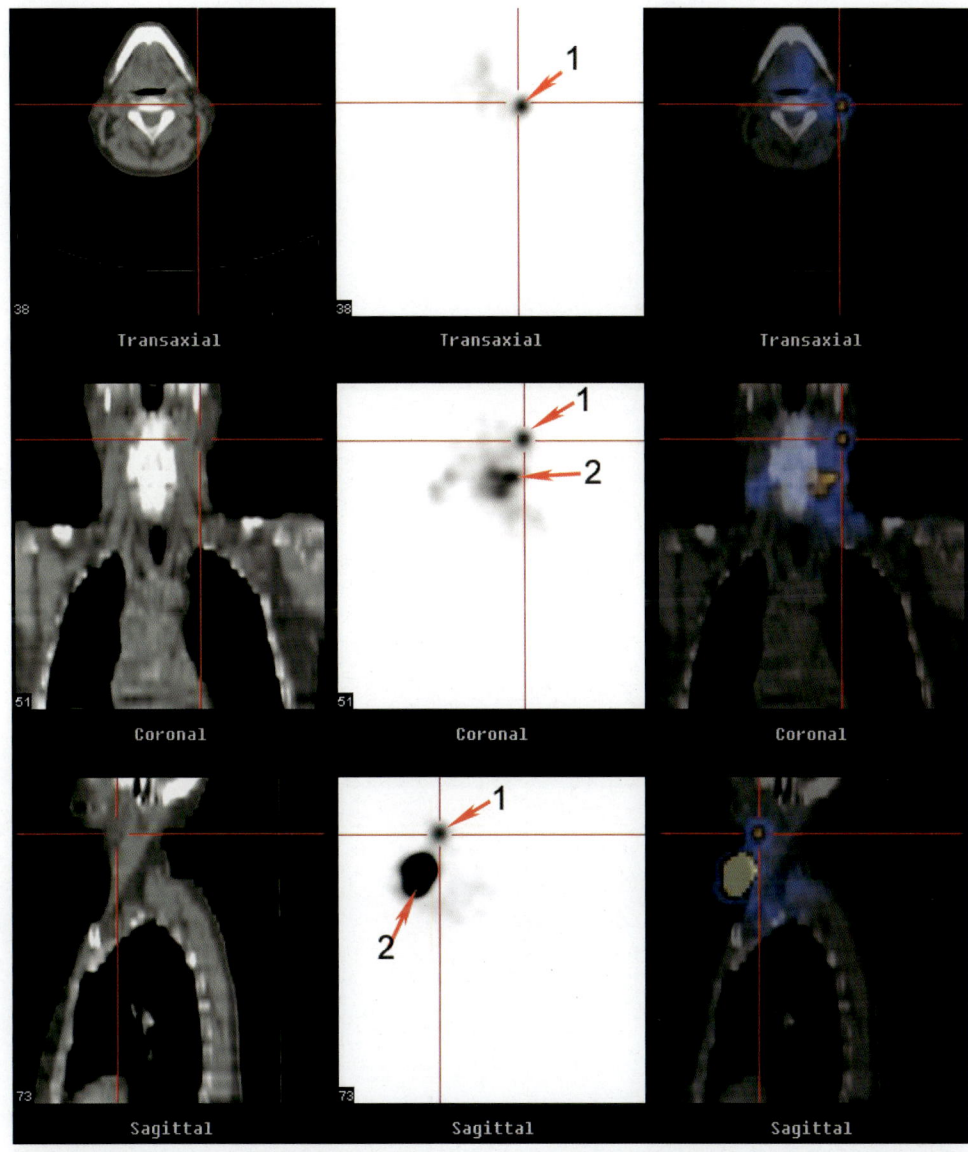

FIGURE 3-12: Coronal view of neck

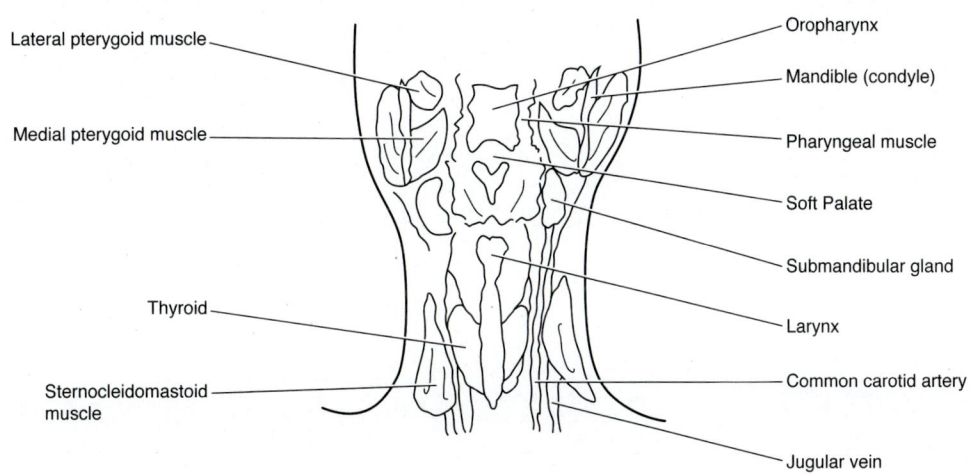

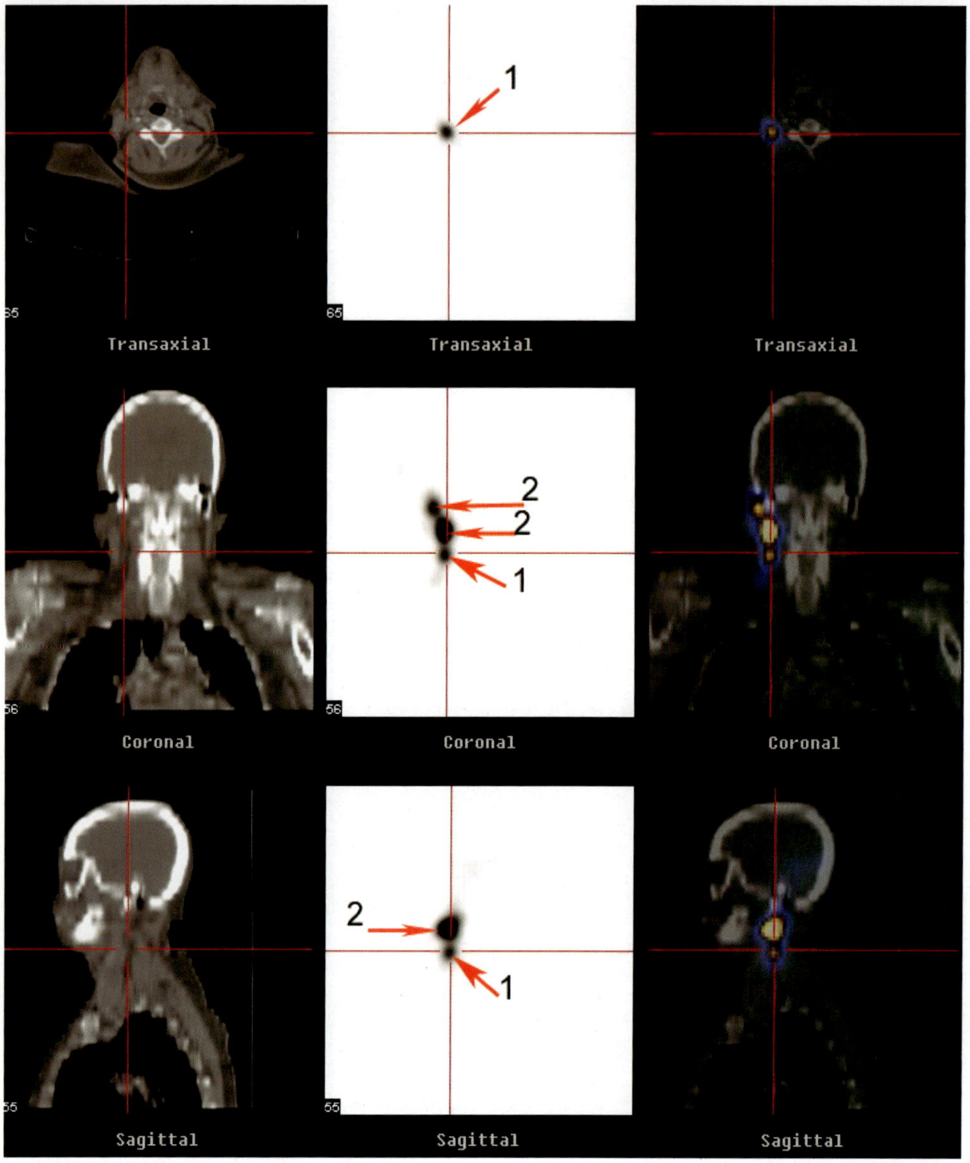

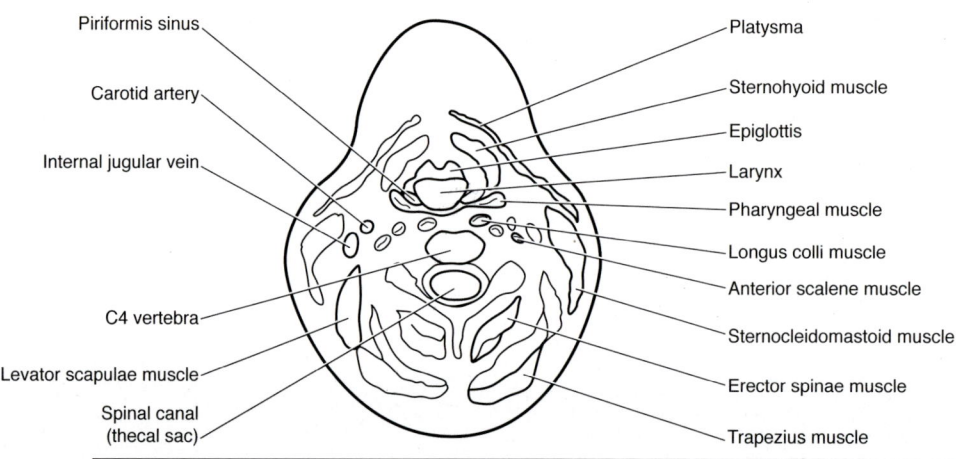

Transaxial   Transaxial   Transaxial

Coronal   Coronal   Coronal

Sagittal   Sagittal   Sagittal

FIGURE 3-13:

1. Middle jugular node (level III)
2. Upper jugular node (level IIB)

Piriformis sinus

Carotid artery

Internal jugular vein

C4 vertebra

Levator scapulae muscle

Spinal canal (thecal sac)

Platysma

Sternohyoid muscle

Epiglottis

Larynx

Pharyngeal muscle

Longus colli muscle

Anterior scalene muscle

Sternocleidomastoid muscle

Erector spinae muscle

Trapezius muscle

FIGURE 3-14: Transaxial view of upper neck

2
4
3
5
6
61
Transaxial
1
61
Transaxial
Transaxial
7
8
52
Coronal
52
Coronal
Coronal
5
3
6
75
Sagittal
9
7
8
75
Sagittal
Sagittal

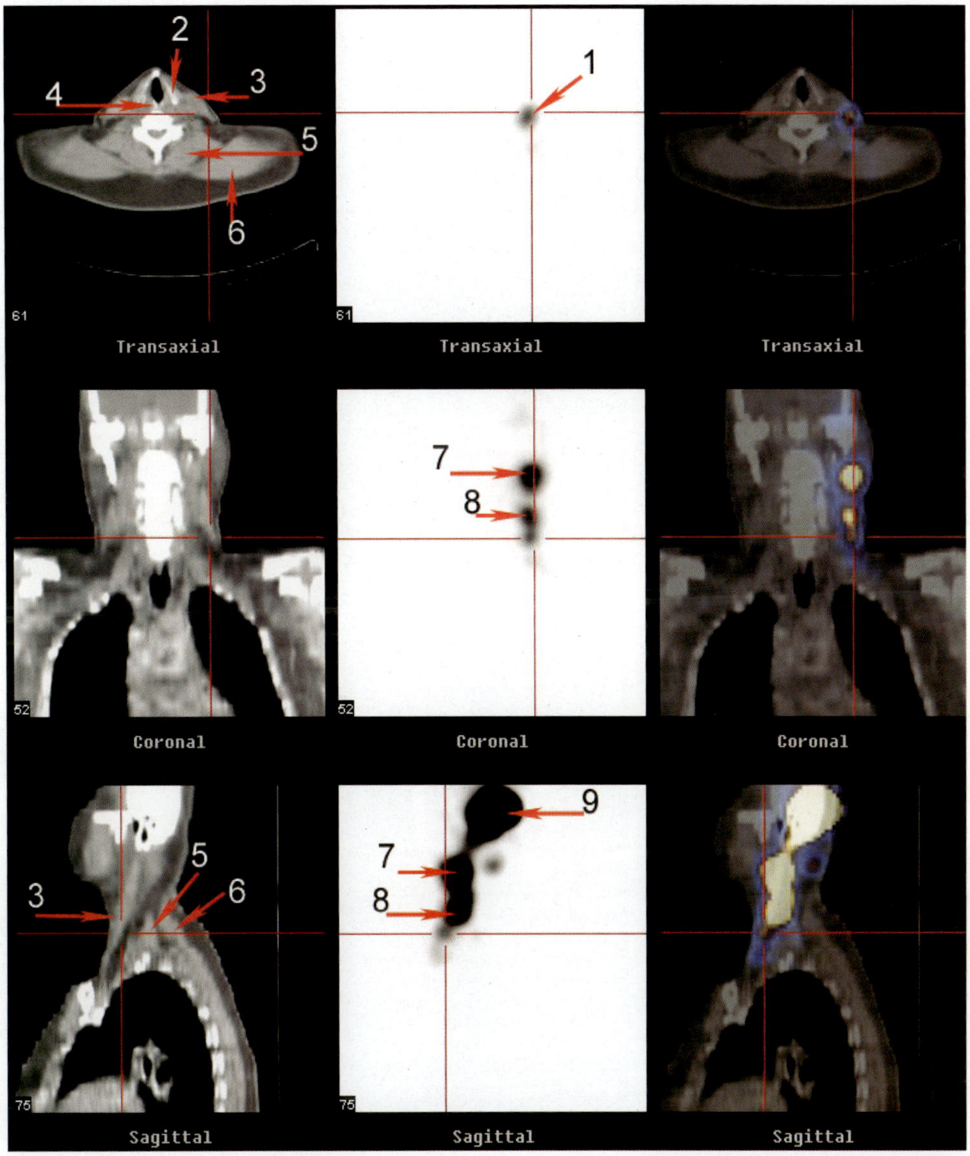

FIGURE 3-15: (Copyright 2006 The University of Texas M.D. Anderson Cancer Center. Used with Permission.)

1. Lower jugular node (level IV)
2. Thyroid cartilage
3. Sternocleidomastoid muscle
4. Cricoid cartilage
5. Semispinalis cervicis muscle
6. Trapezius muscle
7. Upper jugular node (level II)
8. Middle jugular node (level III)
9. Activity at injection site

Anterior cervical space
Visceral space
Carotid space
Anterior prevertebral space
Posterior cervical space
Posterior prevertebral space

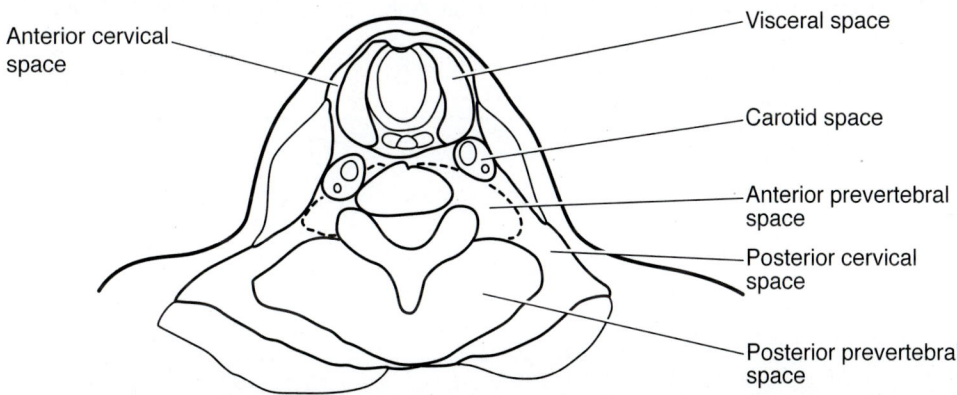

FIGURE 3-16: Transaxial view of middle neck

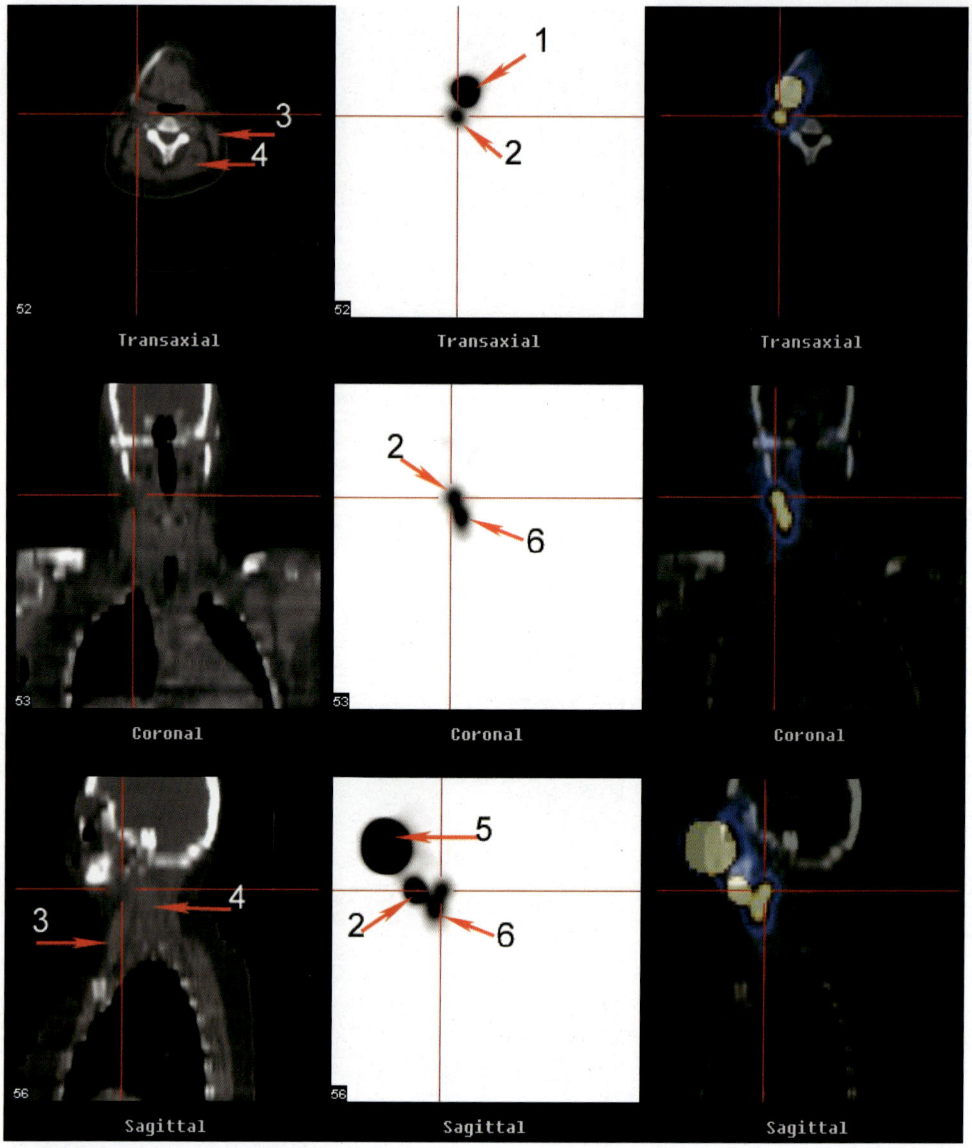

FIGURE 3-17:

1. Submandibular node (level IB)
2. Upper jugular node (level IIA)
3. Sternocleidomastoid muscle
4. Splenius capitis muscle
5. Activity at injection site
6. Middle jugular node (level III)

Section 2: Breast Lymphoscintigraphy

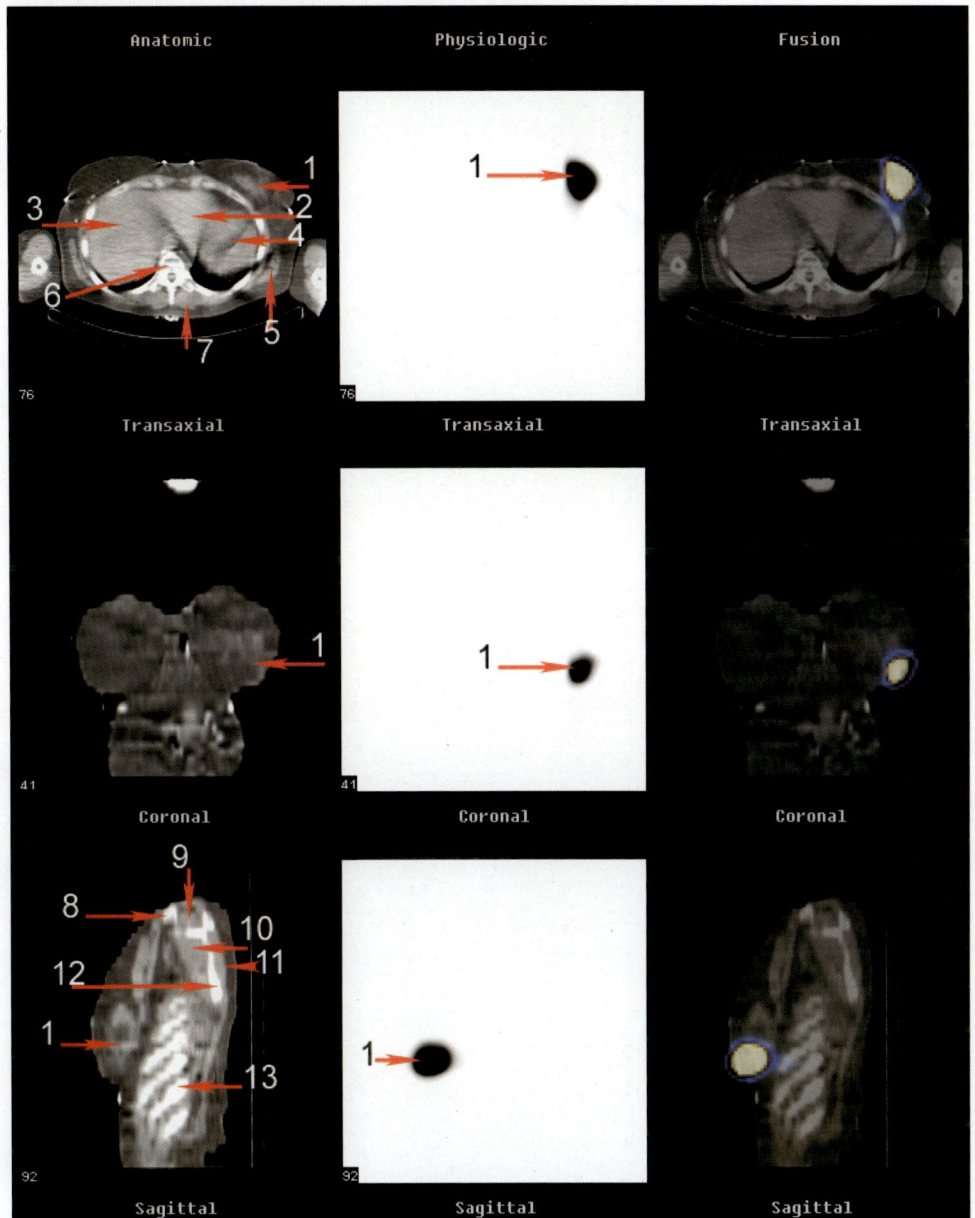

FIGURE 3-18:

1. Left breast (activity at injection site)
2. Left ventricle of heart
3. Right hepatic lobe
4. Stomach
5. Latissimus dorsi muscle
6. T12 marrow
7. Erector spinae muscle
8. Clavicle
9. Supraspinatus muscle
10. Subscapularis muscle
11. Infraspinatus muscle
12. Scapula
13. Rib

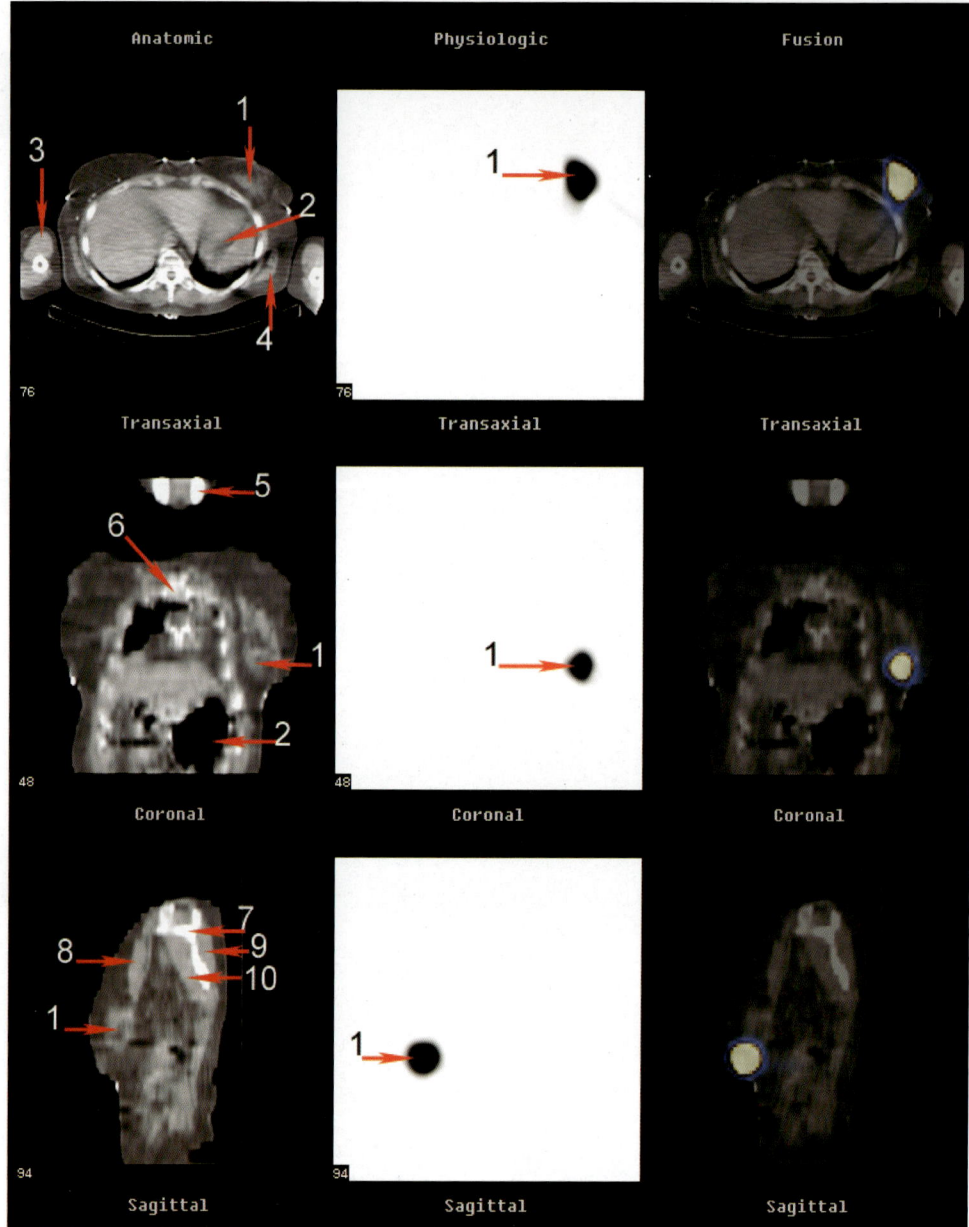

FIGURE 3-19:

1. Left breast (activity at injection site)
2. Stomach
3. Brachialis muscle
4. Latissimus dorsi muscle
5. Mandible
6. Manubrium
7. Scapula
8. Pectoralis major muscle
9. Infraspinatus muscle
10. Subscapularis muscle

Section 3: Gynecologic Lymphoscintigraphy

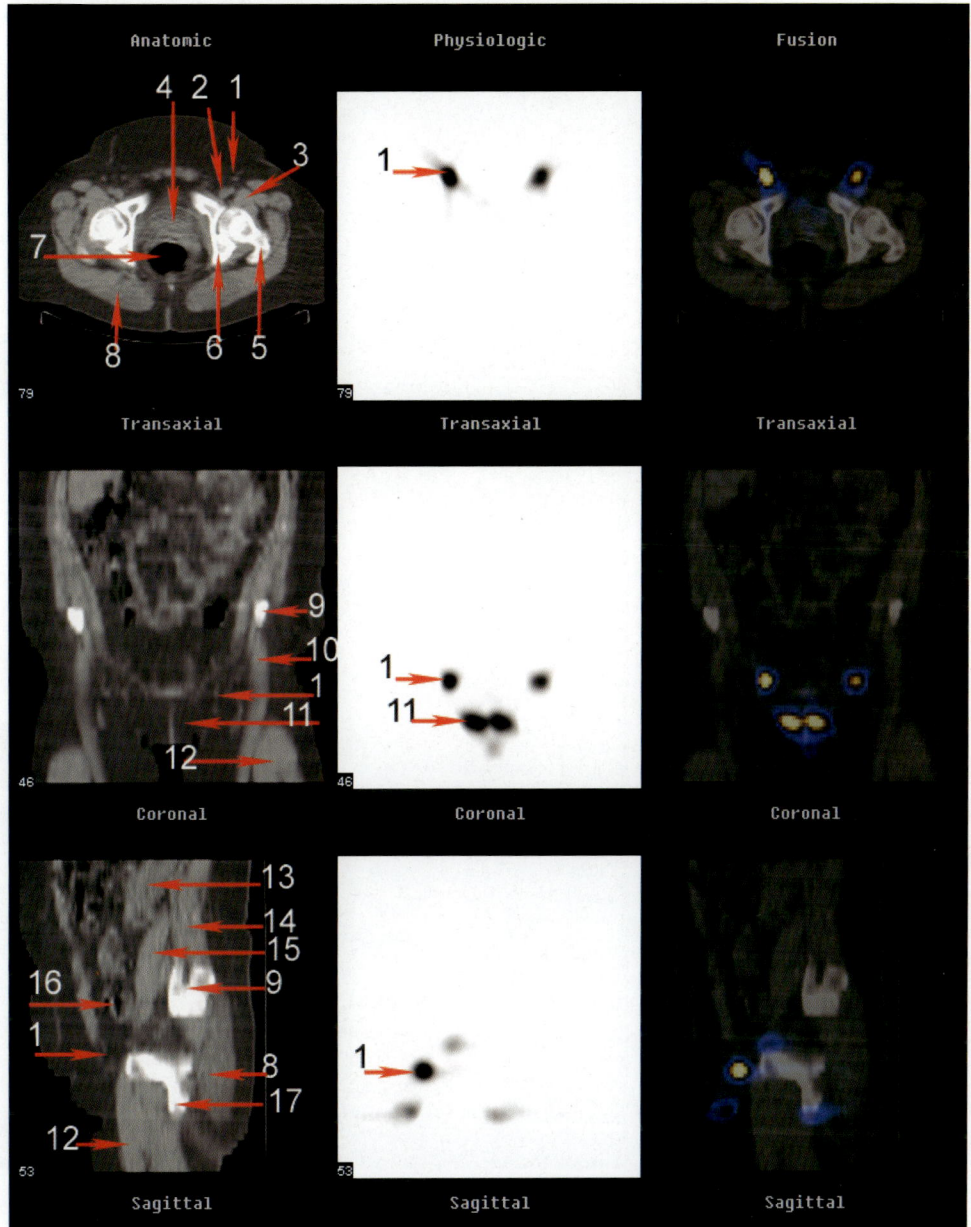

FIGURE 3-20:

1. Inguinal lymph node
2. Femoral artery and vein
3. Iliopsoas muscle
4. Bladder
5. Femoral neck
6. Posterior lip of acetabulum (ilium)
7. Rectum
8. Gluteus maximus muscle
9. Anterior superior iliac crest
10. Iliacus muscle
11. Activity at injection site of vulva
12. Rectus femoris muscle
13. Left kidney
14. Quadratus lumborum muscle
15. Psoas muscle
16. Ileum
17. Ischium

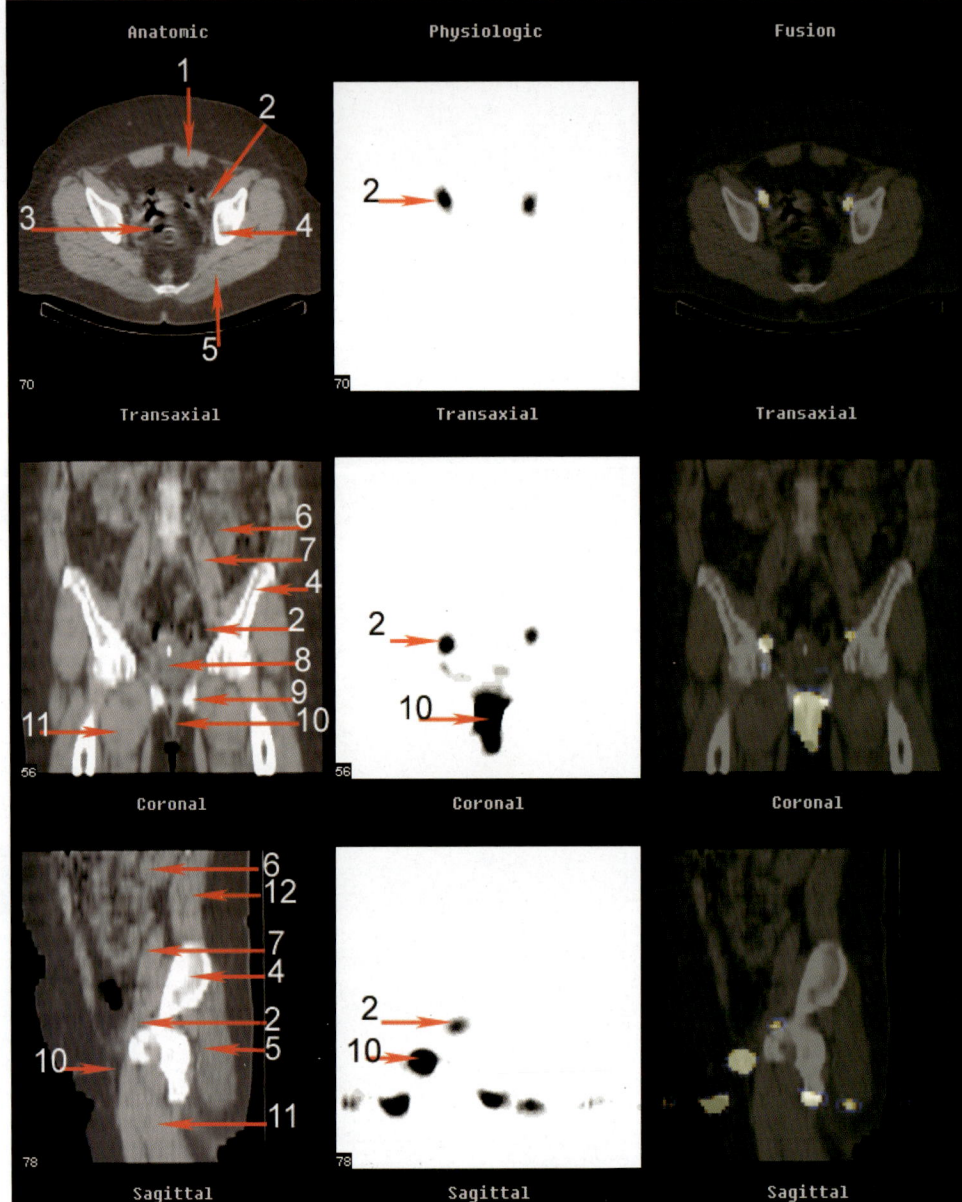

FIGURE 3-21:

1. Rectus abdominis muscle	5. Gluteus maximus muscle	10. Activity at injection site of vagina
2. External iliac (obturator) lymph node	6. Left kidney	11. Adductor brevis muscle
3. Sigmoid colon	7. Psoas muscle	12. Quadratus lumborum muscle
4. Ilium	8. Bladder	
	9. Pubic ramus	

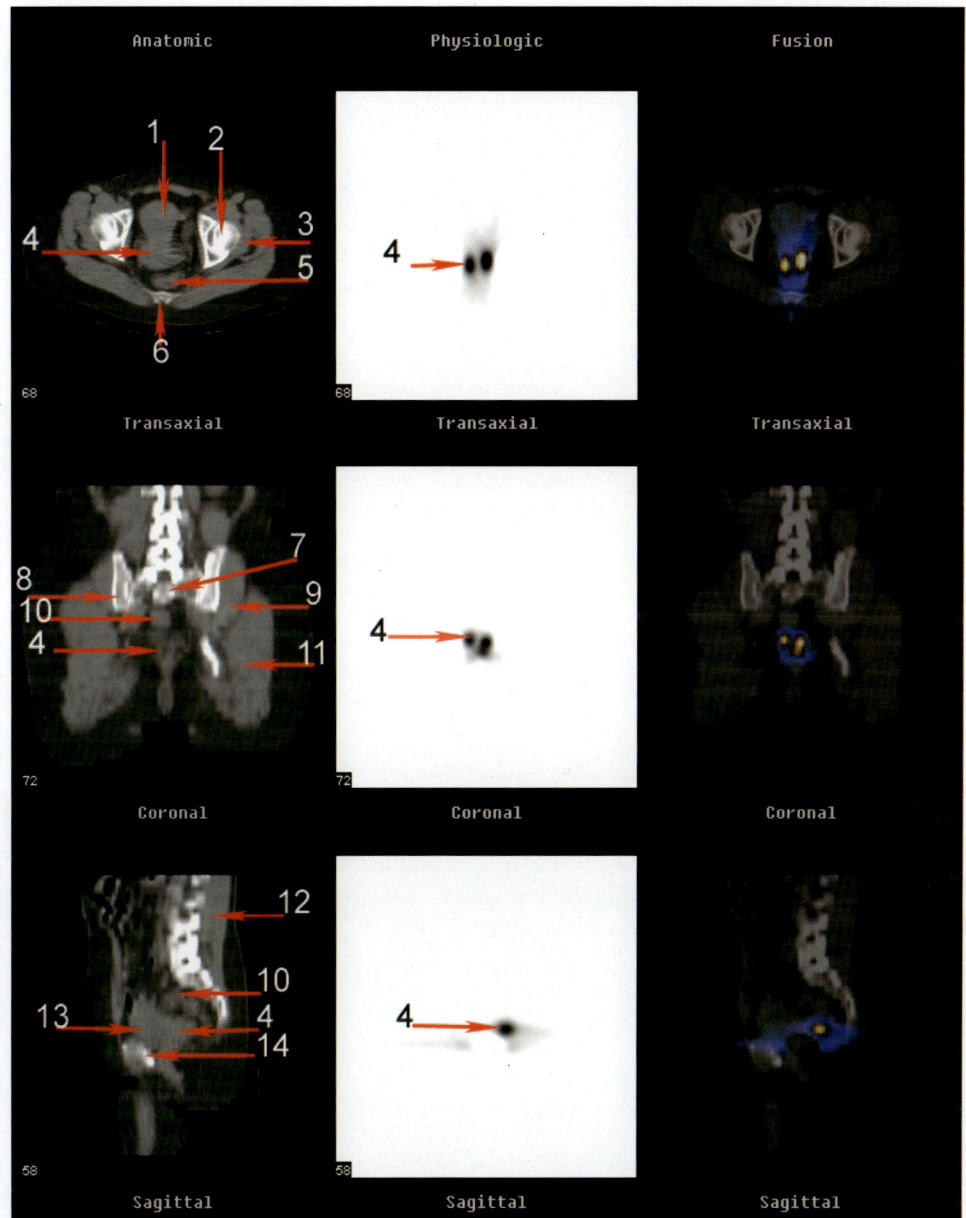

FIGURE 3-22:

1. Bladder
2. Femoral head
3. Gluteus minimus muscle
4. Uterine cervix with
   injected activity
5. Rectum
6. Coccyx
7. Sacral promontory
8. Iliac tuberosity
9. Gluteus medius muscle
10. Sigmoid colon
11. Gluteus maximus muscle
12. Erector spinae muscle
13. Uterine body
14. Symphysis pubis

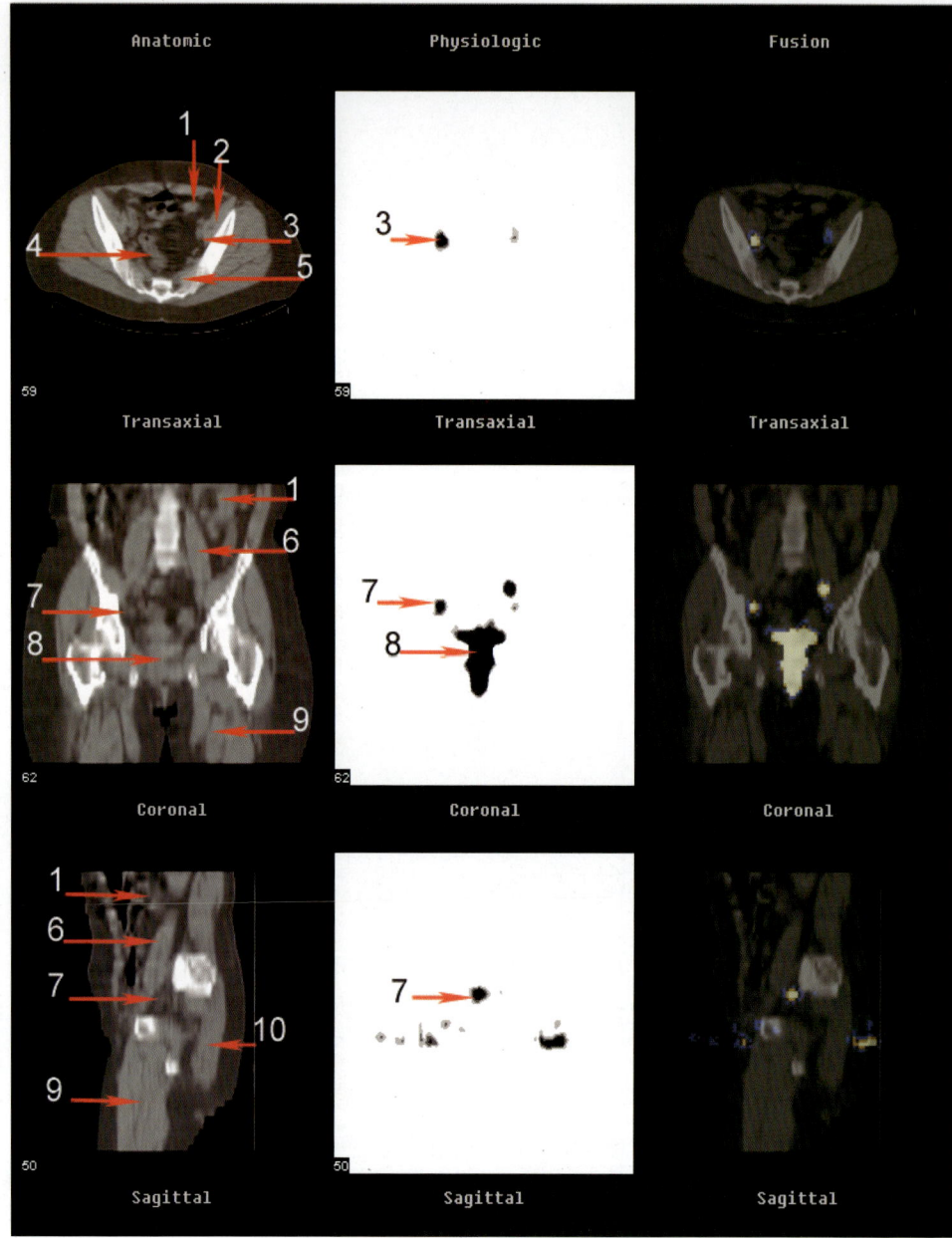

FIGURE 3-23:

1. Jejunum
2. Iliacus muscle
3. Internal iliac (hypogastric) lymph node
4. Sigmoid colon
5. Sacral ala
6. Psoas muscle
7. External iliac node
8. Vagina with injected activity
9. Adductor magnus muscle
10. Gluteus maximus muscle

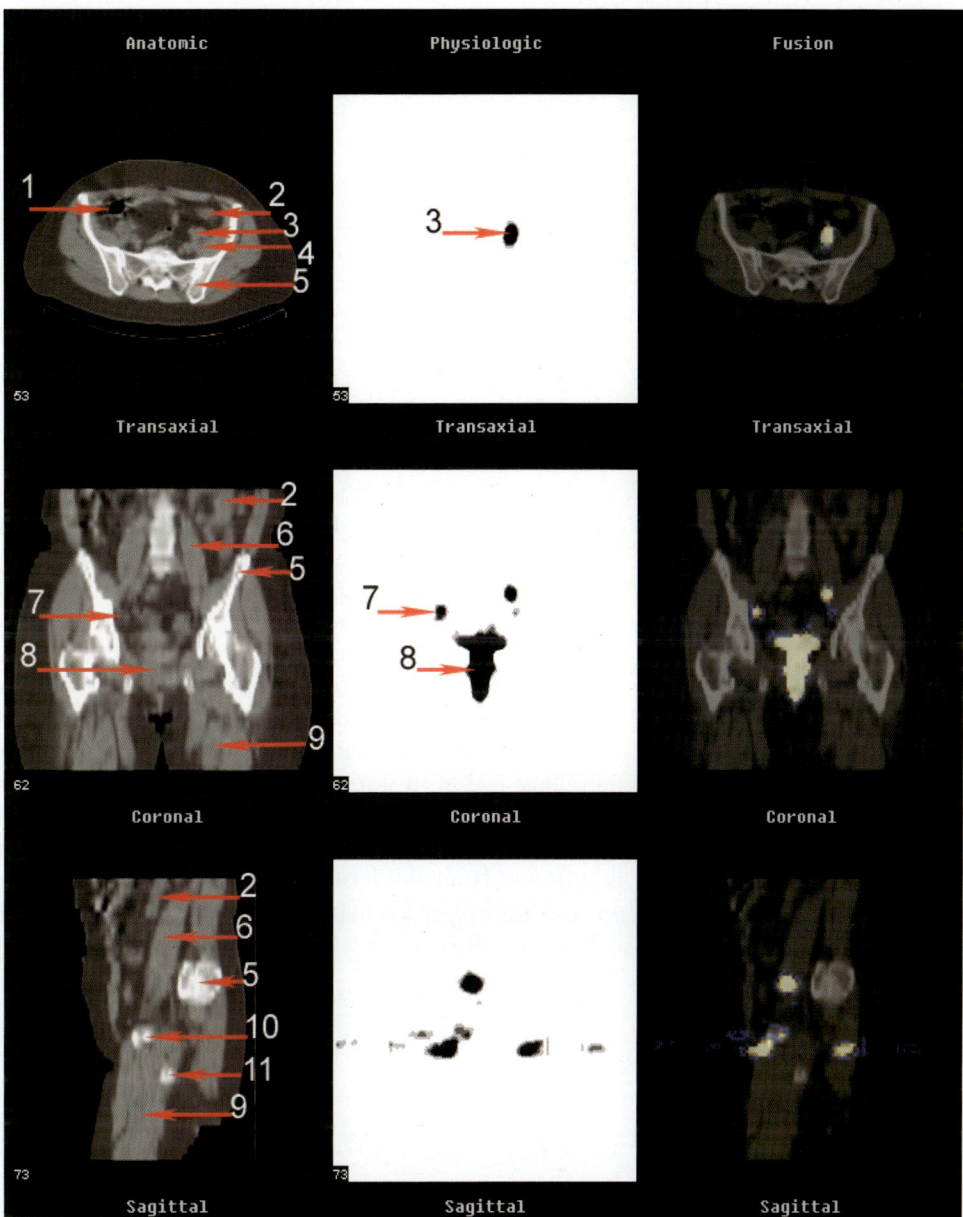

FIGURE 3-24:

1. Cecum
2. Jejunum
3. Internal iliac node
4. Iliacus muscle
5. Ilium
6. Psoas muscle
7. External iliac node
8. Vagina with injected activity
9. Adductor magnus muscle
10. Pubis
11. Ischium

4 Lung SPECT/CT

During the past decade, the radiation oncology community has entered another phase in treatment planning methodology with the use of image-guided radiotherapy (RT). A three-dimensional (3D) image of the tumor was created, as well as of the surrounding critical structures. The birth of 3D RT made it possible for treatments to be more conformal with dose escalation while maintaining the same toxicity. There is now another dimension in RT treatment planning in the form of functional imaging. The biologic target volume can be identified by using functional imaging that is then fused to the anatomic image derived from computed tomography (CT).[1] Functional imaging may have a role in normal tissue sparing when irradiating adjacent tumors. Single photon emission computed tomography (SPECT) of lung perfusion using technetium (Tc)-99m macroaggregated albumin (MAA) particles has been used to avoid perfused lung tissue from the irritated volumes.[2] SPECT/CT fusion was useful in detecting 48% of patients with hypoperfused regions of the lung. In 11% of patients, the RT field angles were altered to avoid highly functional lung tissue.[3]

Section 1: Coronals

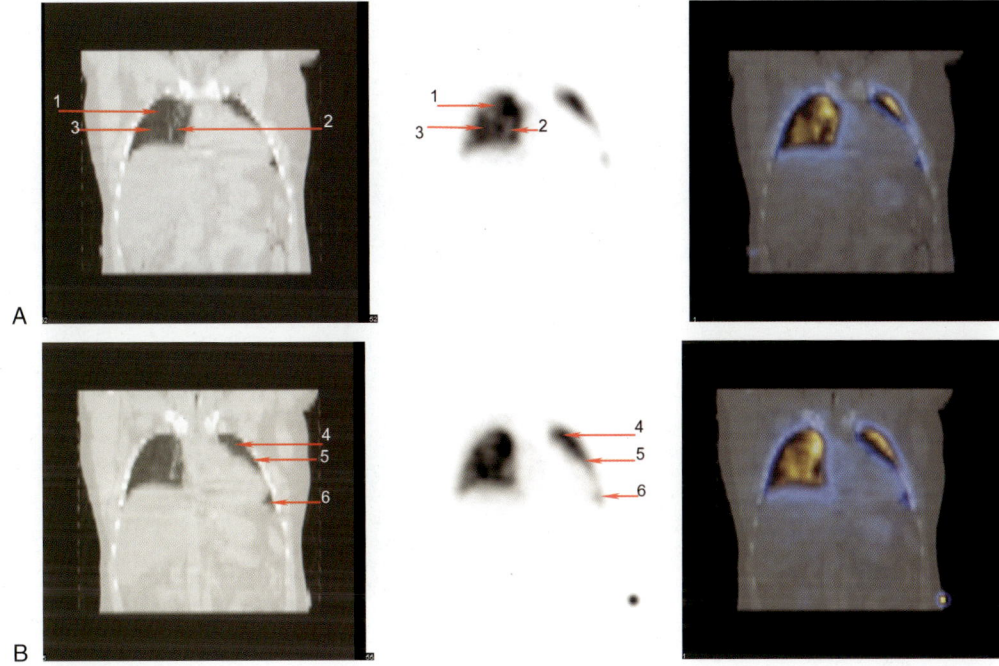

FIGURE 4-1:

1. Right upper lung (anterior segment)
2. Right middle lung (medial segment)
3. Right middle lung (lateral segment)
4. Left upper lung (anterior segment)
5. Left upper lung (superior lingular segment)
6. Left upper lung (inferior lingular segment)

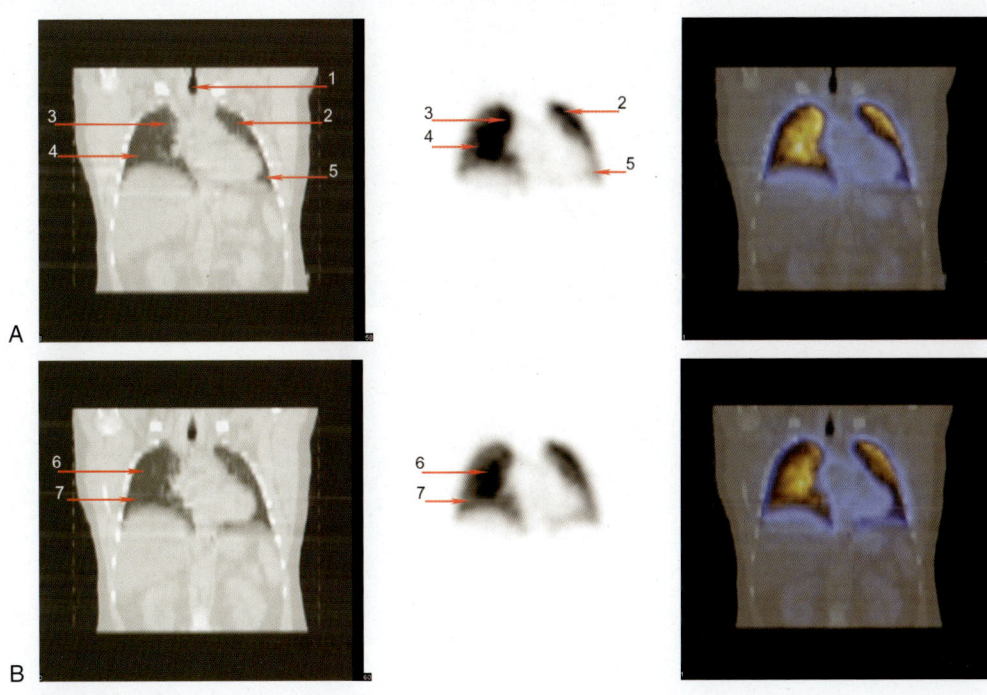

FIGURE 4-2:

1. Trachea
2. Left upper lung (apicoposterior segment)
3. Right upper lung (apical segment)
4. Right lower lung (anterior basal segment)
5. Left lower lung (lateral basal segment)
6. Right upper lung (anterior segment)
7. Right lower lung (lateral basal segment)

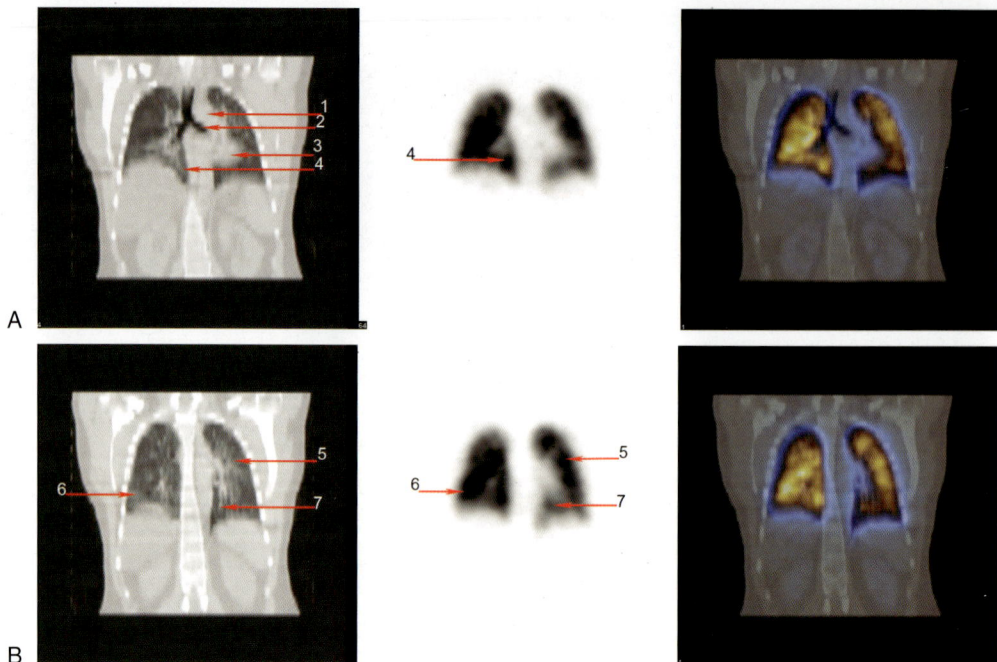

FIGURE 4-3:

1. Aortic arch
2. Left main bronchus
3. Left ventricle
4. Right lower lung
   (posterior basal segment)

5. Left lower lung (superior
   segment)
6. Right lower lung (lateral
   basal segment)

7. Left lower lung (posterior
   basal segment)

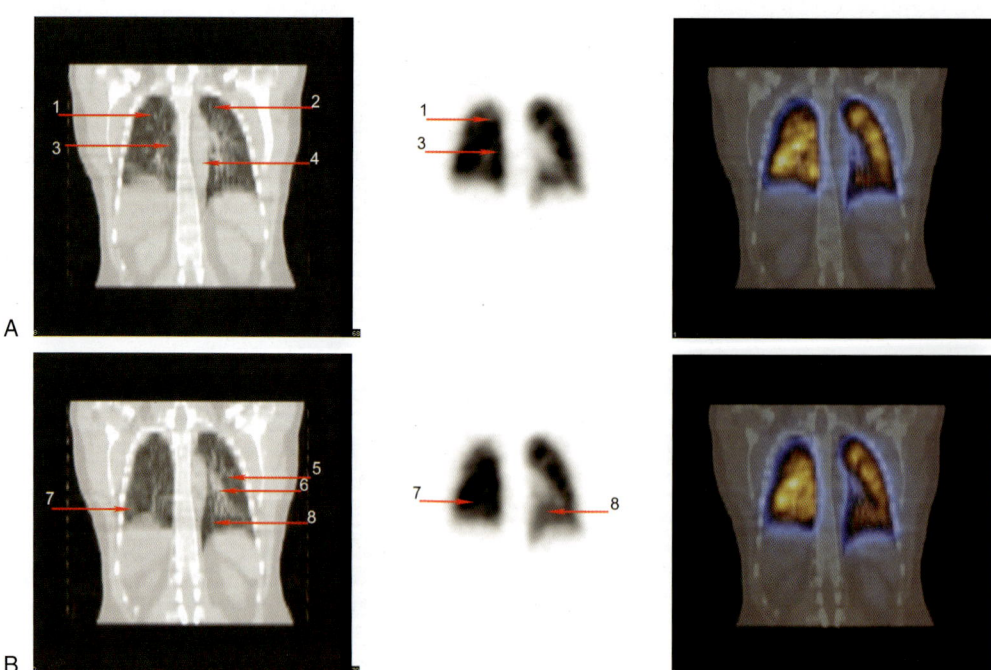

FIGURE 4-4:

1. Right upper lung
   (posterior segment)
2. Left upper lung
   (apicoposterior segment)
3. Right lower lung
   (superior segment)

4. Descending aorta
5. Left lower lung (superior
   segment)
6. Left pulmonary vein

7. Right lower lung (lateral
   basal segment)
8. Left lower lung (posterior
   basal segment)

Section 2: Sagittals

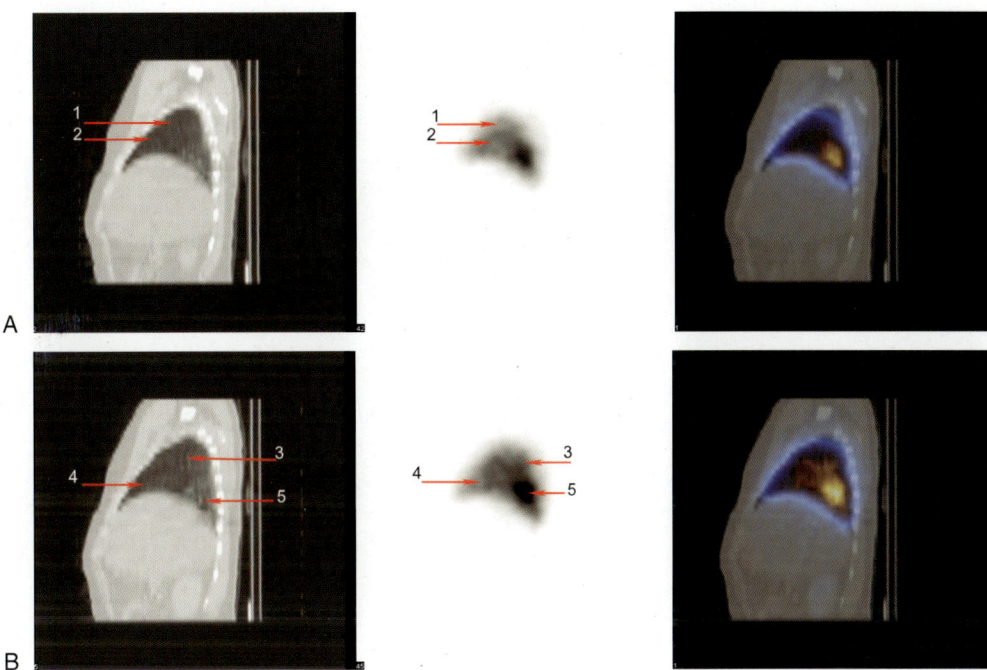

FIGURE 4-5:

1. Right upper lung (anterior segment)
2. Right middle lung (lateral segment)
3. Right upper lung (posterior segment)
4. Right lower lung (anterior basal segment)
5. Right lower lung (posterior basal segment)

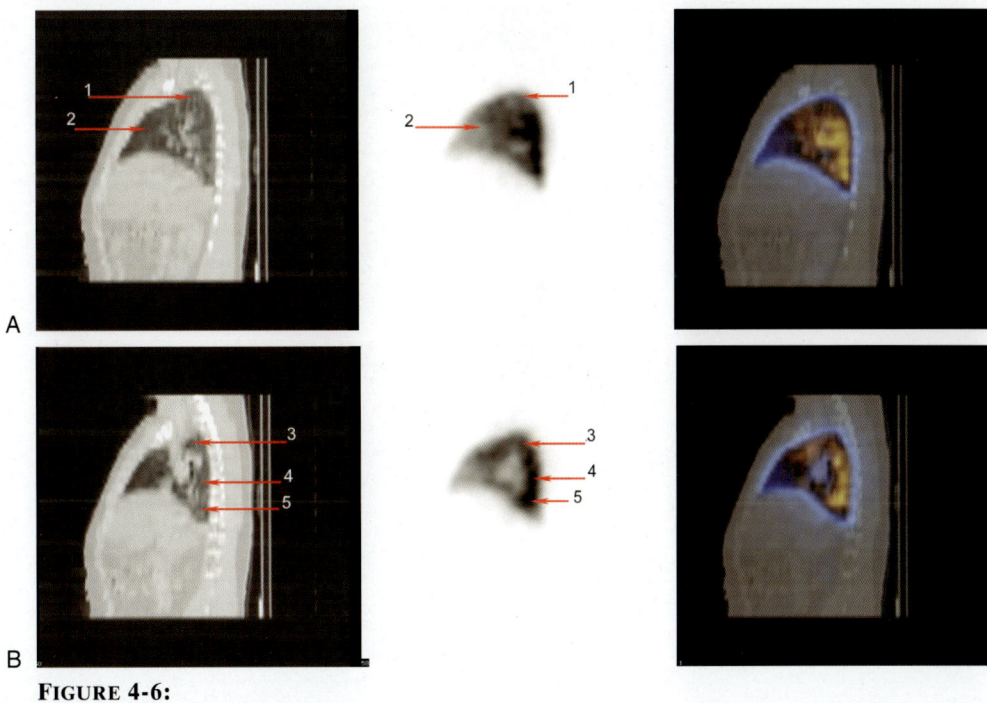

FIGURE 4-6:

1. Right upper lung (apical segment)
2. Right middle lung (medial segment)
3. Right upper lung (posterior segment)
4. Right lower lung (superior segment)
5. Right lower lung (posterior basal segment)

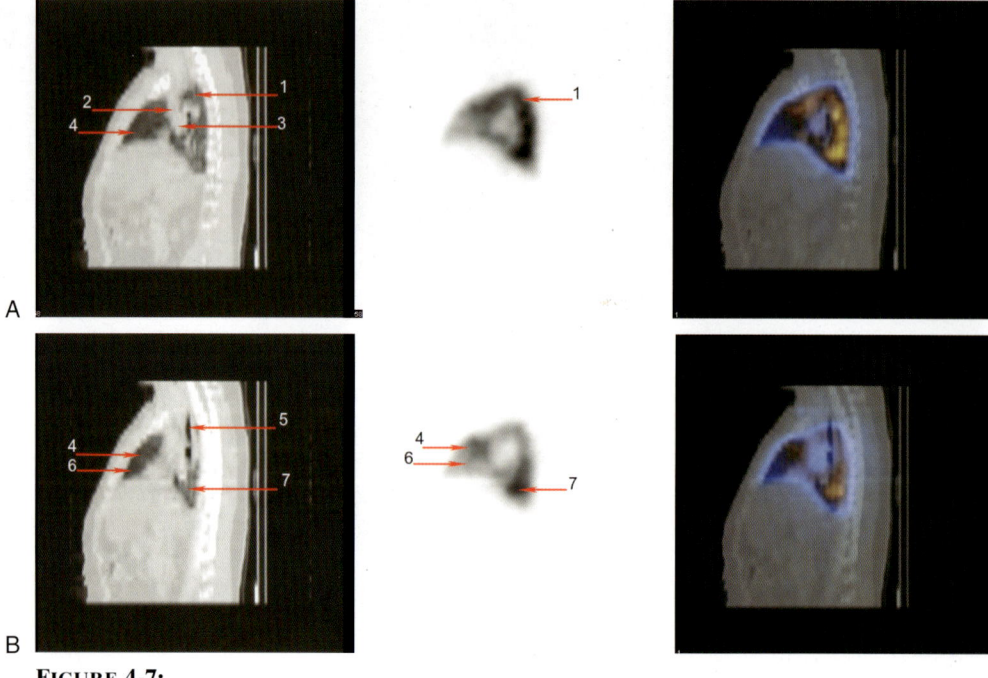

FIGURE 4-7:

1. Right upper lung   4. Right middle lung   6. Right lower lung (medial
   (posterior segment)   (medial segment)   basal segment)
2. Superior vena cava   5. Trachea   7. Right lower lung
3. Main pulmonary vein   (posterior basal segment)

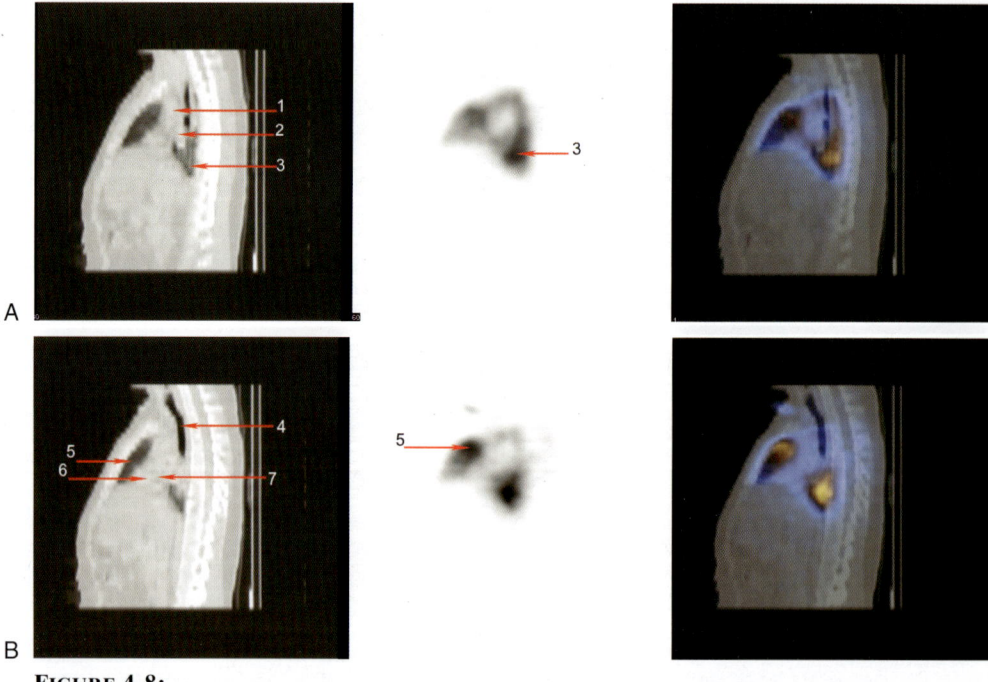

FIGURE 4-8:

1. Aortic arch   4. Trachea   6. Right ventricle
2. Right pulmonary artery   5. Right middle lung   7. Left atrium
3. Right lower lobe   (medial segment)
   (posterior basal segment)

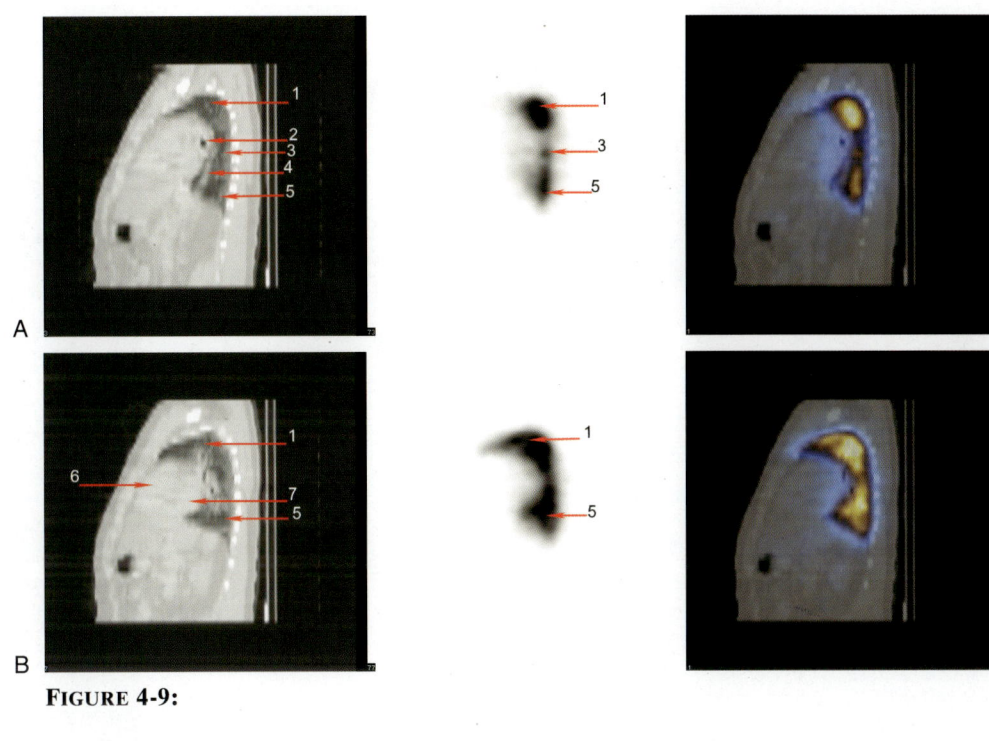

FIGURE 4-9:

1. Left upper lung (apicoposterior segment)
2. Left bronchus
3. Left lower lung (superior segment)
4. Descending aorta
5. Left lower lung (posterior basal segment)
6. Right ventricle
7. Left ventricle

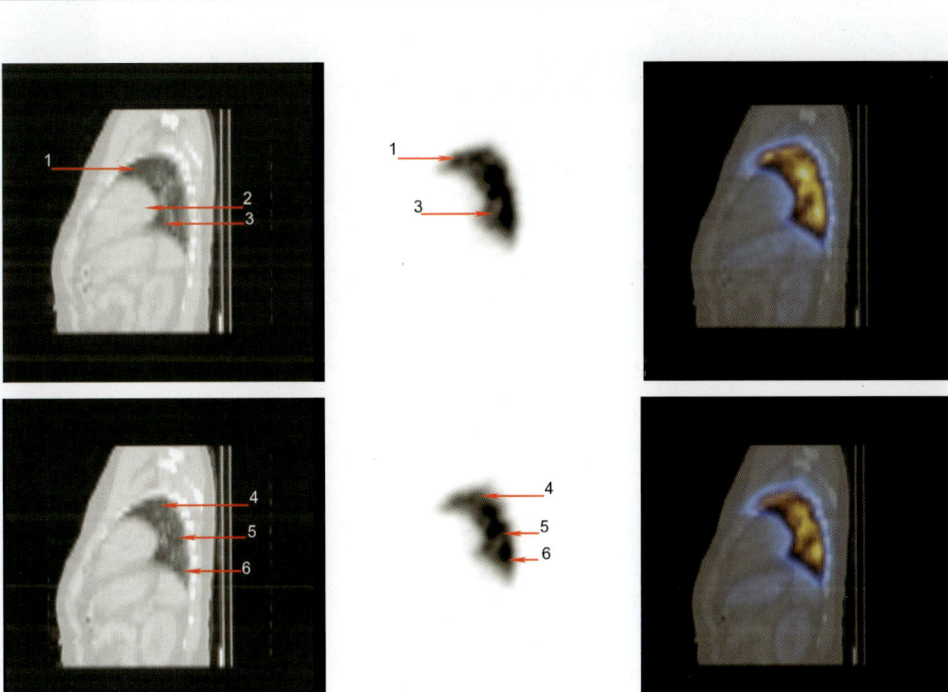

FIGURE 4-10:

1. Left upper lung (anterior segment)
2. Left ventricle
3. Left lower lung (anteromedial basal segment)
4. Left upper lung (apicoposterior segment)
5. Left lower lung (superior segment)
6. Left lower lung (posterior basal segment)

Section 3: Transaxials

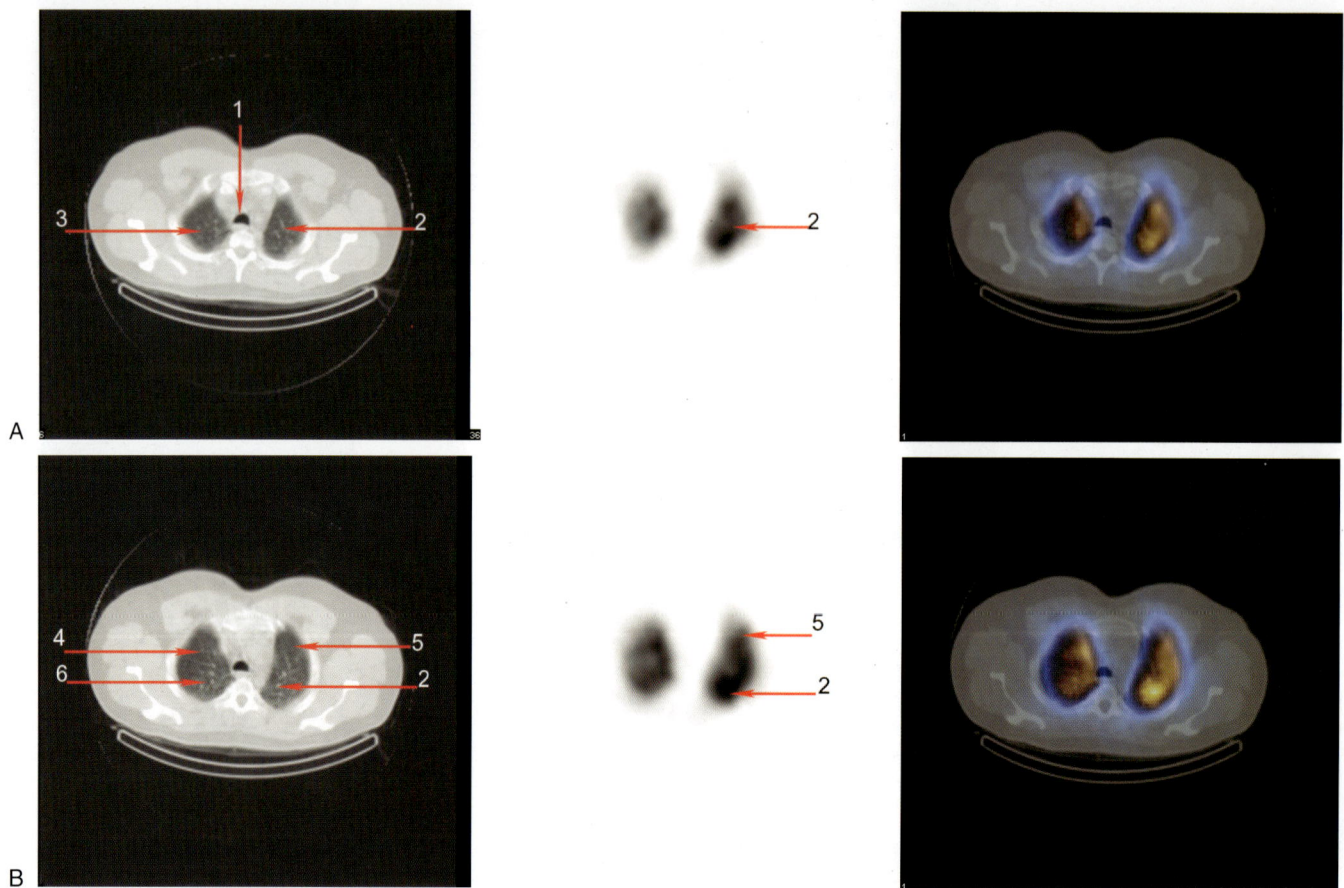

FIGURE 4-11:

1. Trachea
2. Left upper lung (apicoposterior segment)
3. Right upper lung (apical segment)
4. Right upper lung (anterior segment)
5. Left upper lung (anterior segment)
6. Right upper lung (posterior segment)

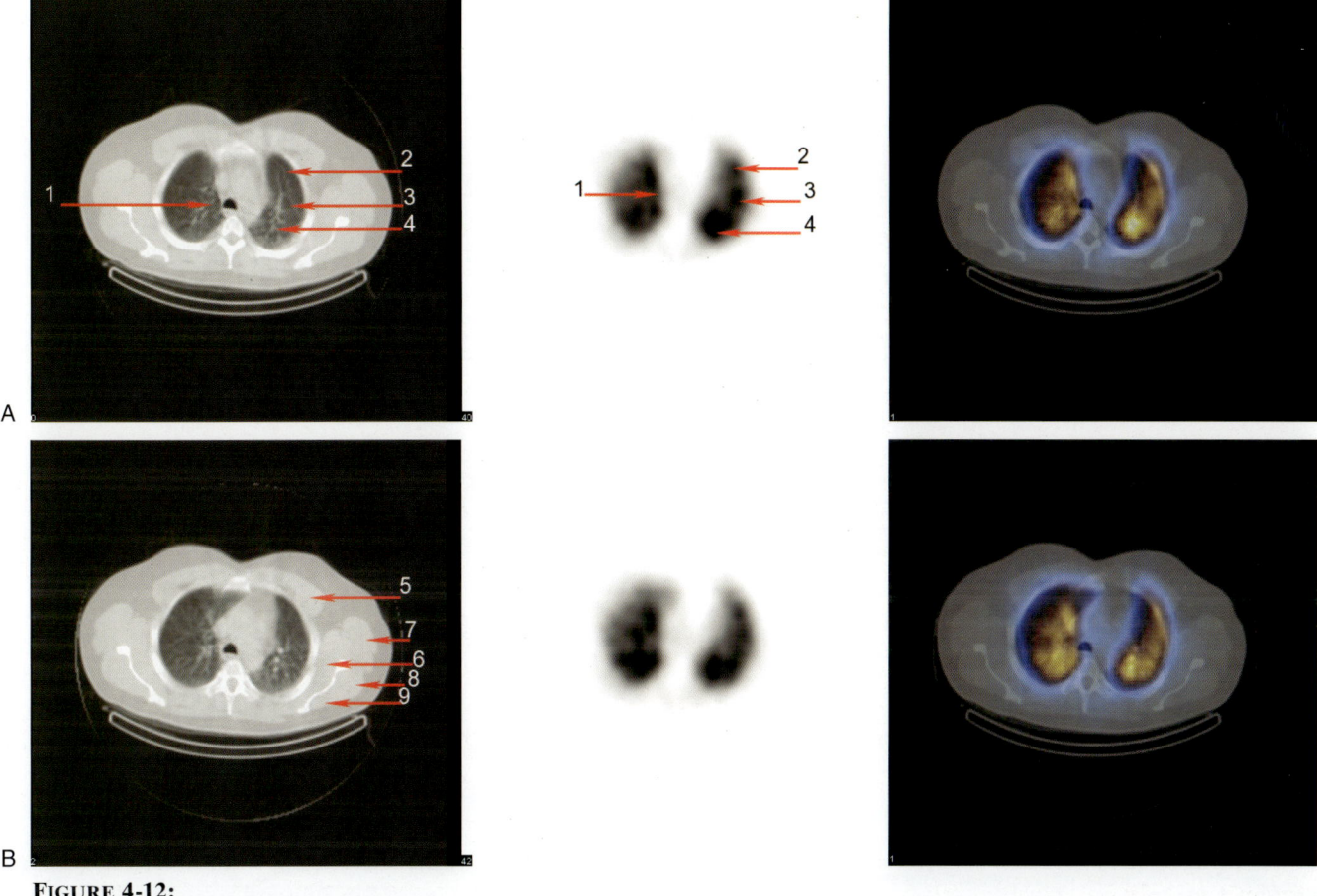

FIGURE 4-12:

1. Right upper lung (apical segment)
2. Left upper lung (anterior segment)
3. Left upper lung (apicoposterior segment)
4. Left lower lung (superior segment)
5. Pectoralis major muscle
6. Subscapularis muscle
7. Teres major muscle
8. Teres minor muscle
9. Infraspinatus muscle

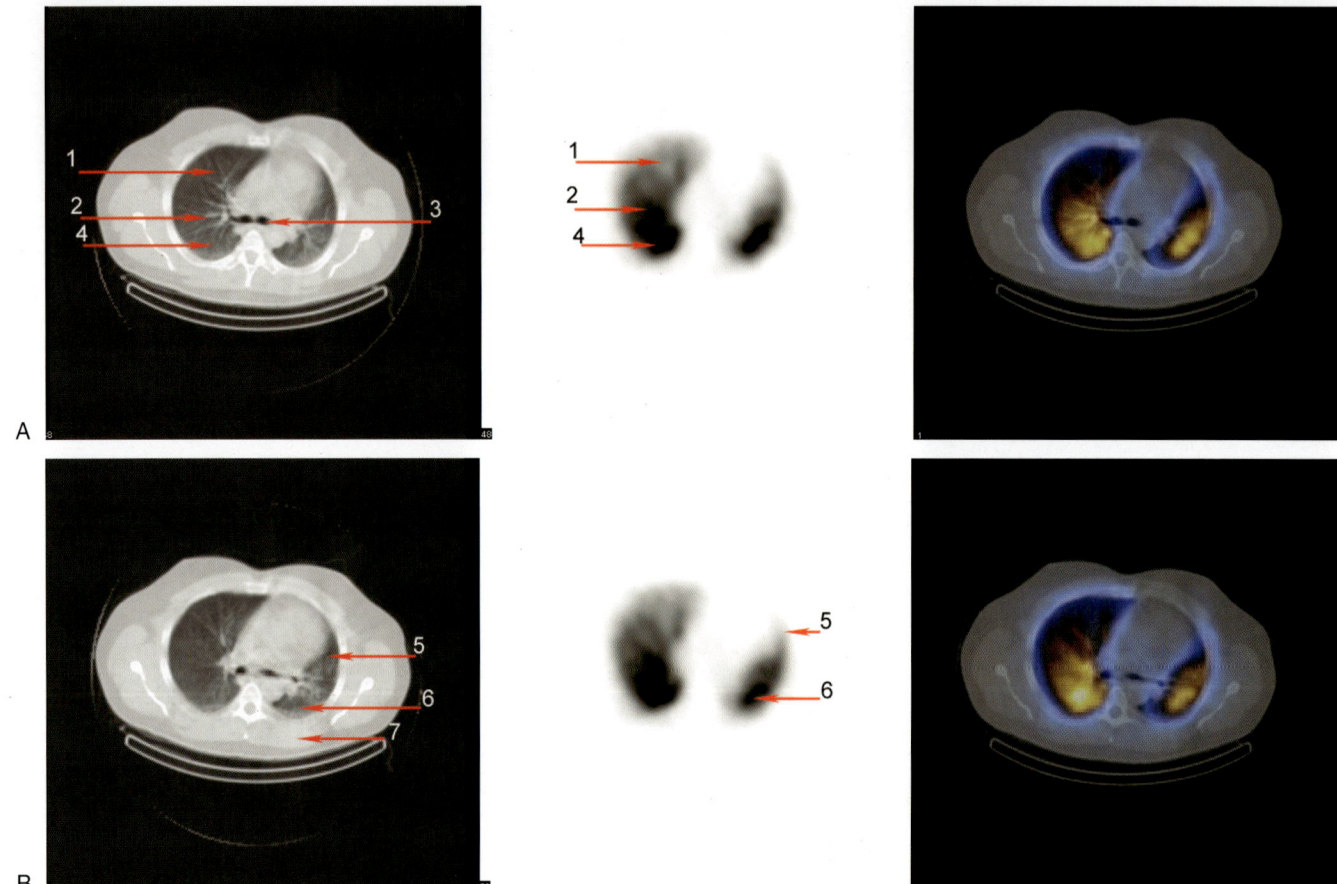

FIGURE 4-13:

1. Right upper lung (anterior segment)
2. Right upper lung (posterior segment)
3. Left main bronchus
4. Right lower lung (superior segment)
5. Left upper lung (superior lingular segment)
6. Left lower lung (superior segment)
7. Trapezius muscle

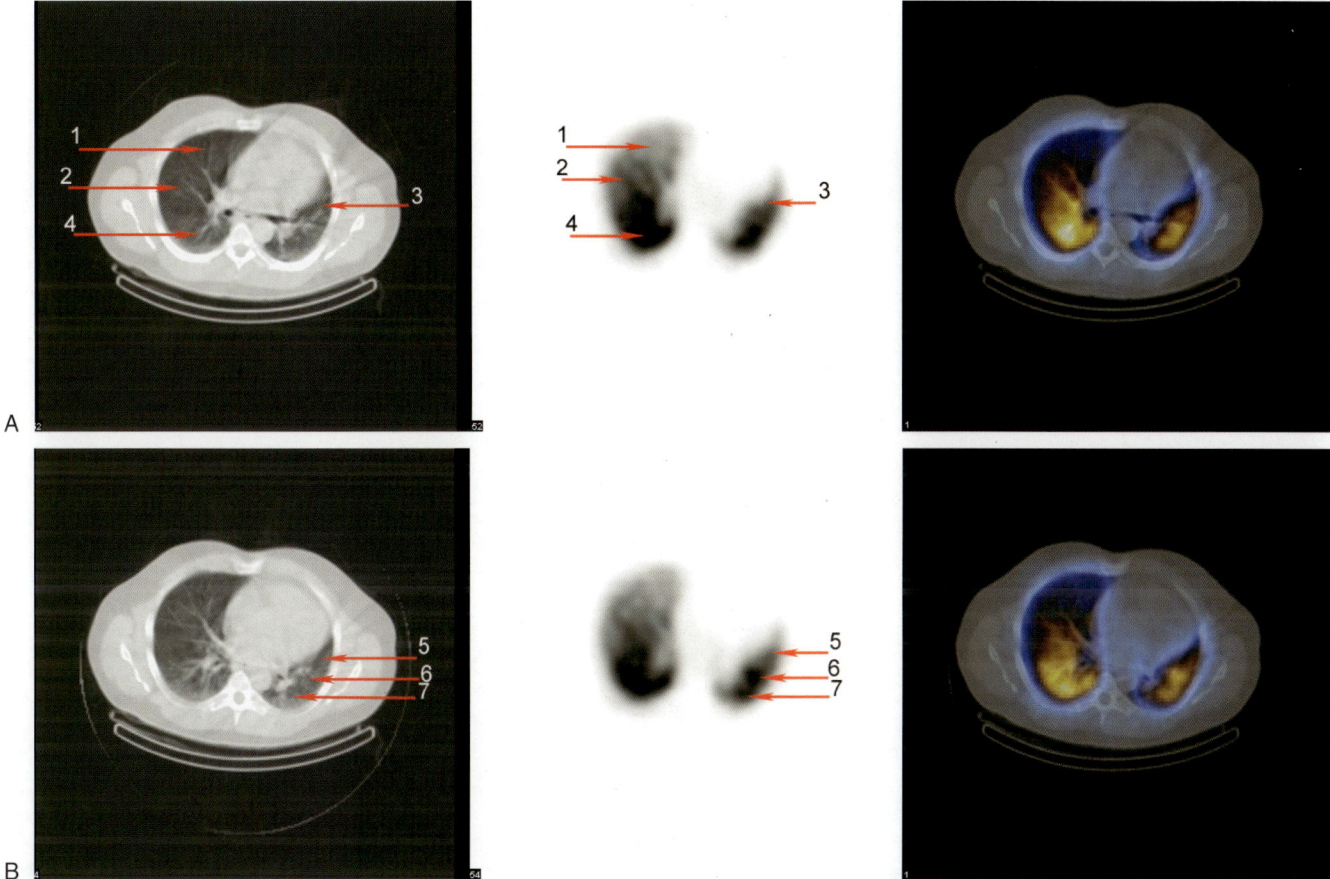

FIGURE 4-14:

1. Right middle lung (medial segment)
2. Right middle lung (lateral segment)
3. Left upper lung (inferior lingular segment)
4. Right lower lung (superior segment)
5. Left lower lung (anteromedial basal segment)
6. Left lower lung (lateral basal segment)
7. Left lower lung (posterior basal segment)

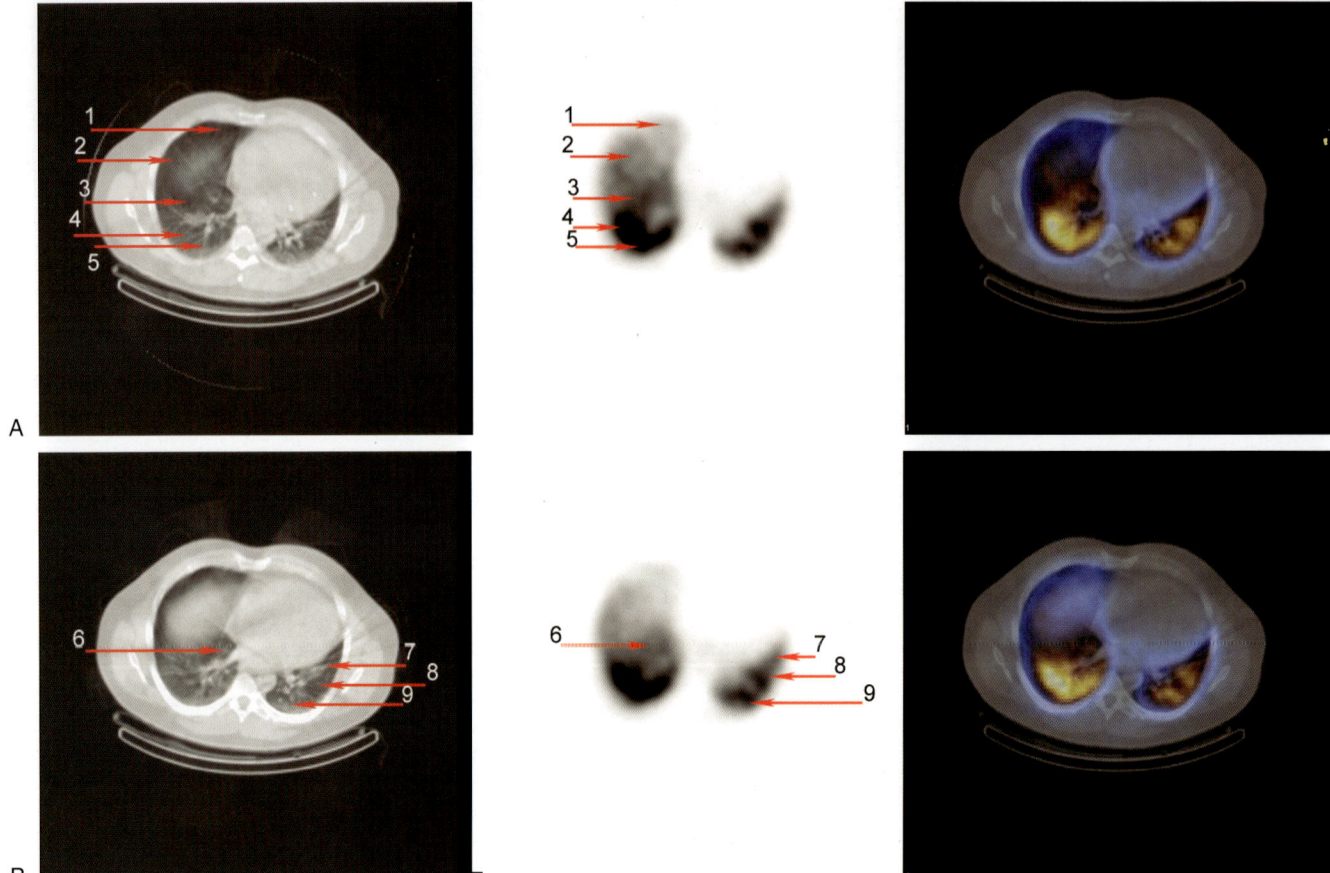

FIGURE 4-15:

1. Right middle lung (medial segment)
2. Right middle lung (lateral segment)
3. Right lower lung (anterior basal segment)
4. Right lower lung (lateral basal segment)
5. Right lower lung (superior segment)
6. Right lower lung (medial basal segment)
7. Left lower lung (anteromedial basal segment)
8. Left lower lung (lateral basal segment)
9. Left lower lung (posterior basal segment)

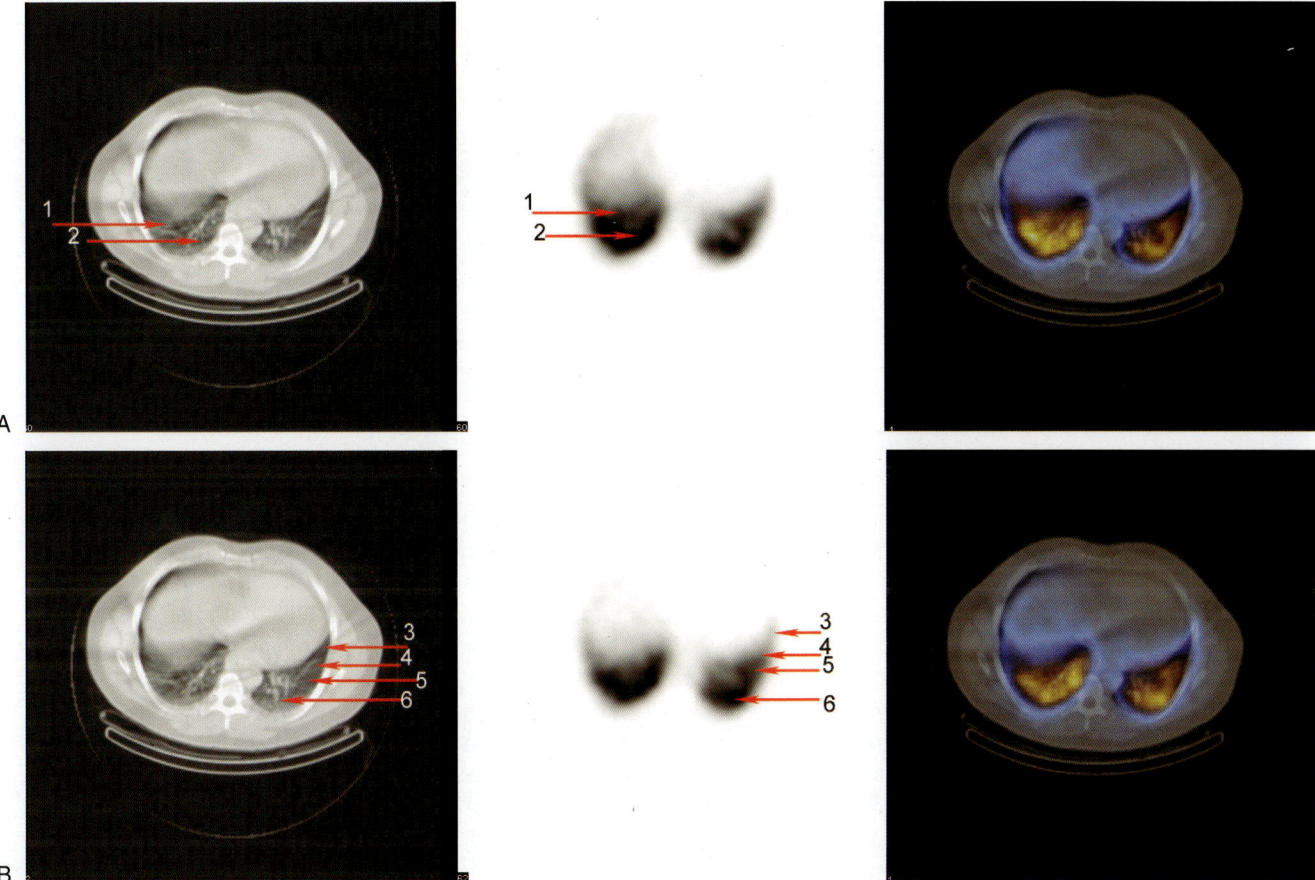

FIGURE 4-16:

1. Right lower lung (lateral segment)
2. Right lower lung (posterior basal segment)
3. Left upper lung (inferior lingular segment)
4. Left lower lung (anteromedial basal segment)
5. Left lower lung (lateral basal segment)
6. Left lower lung (posterior basal segment)

5 Parathyroid SPECT/CT

Parathyroid scintigraphy has been used for localization of a parathyroid adenoma in the evaluation of hyperparathyroidism. Sestamibi-SPECT (single photon emission computed tomography) and computed tomography (CT) fusion has improved the localization of a parathyroid adenoma.[1] In two of eight patients with ectopic thoracic adenoma studied with MIBI-SPECT/CT, the fused imaging was the only procedure to provide precise localization of the adenoma for the surgical planning.[2] An incremental value of hybrid SPECT/CT also was documented in all nine patients with hyperparathyroidism, with a major benefit in four patients with an ectopic parathyroid. In addition, SPECT/CT facilitated successful surgery, even in a minimally invasive approach, and excluded the need for further or additional imaging studies.[3]

Section 1: Coronals

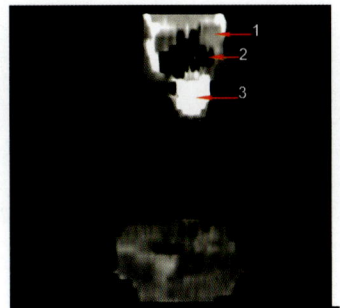 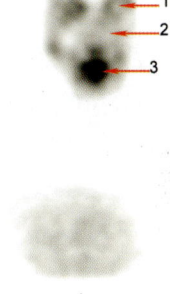 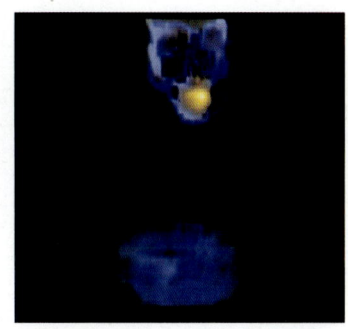

FIGURE 5-1:

1. Orbital globe   2. Maxillary sinus   3. Tongue or oropharynx

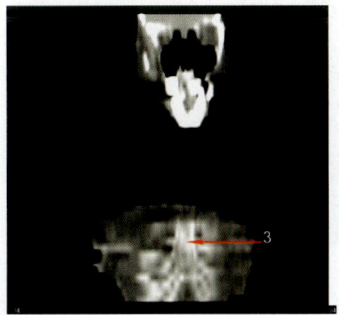 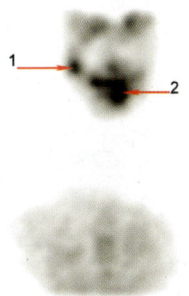 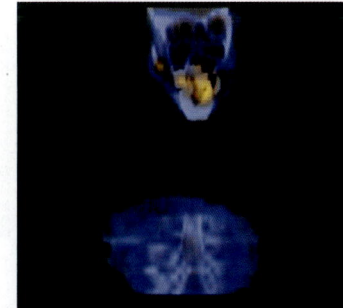

FIGURE 5-2:

1. Parotid gland   2. Tongue or oropharynx   3. Sternum

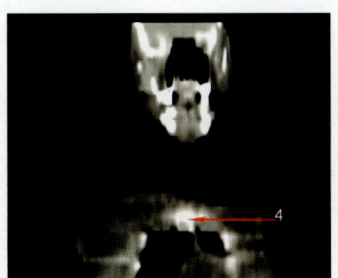 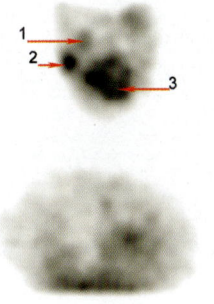 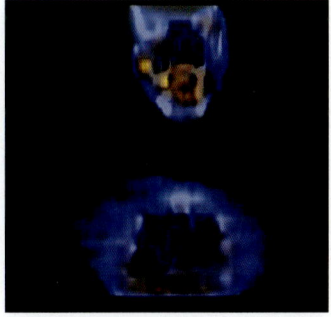

FIGURE 5-3:

1. Masseter muscle   3. Tongue   4. Manubrium
2. Parotid gland

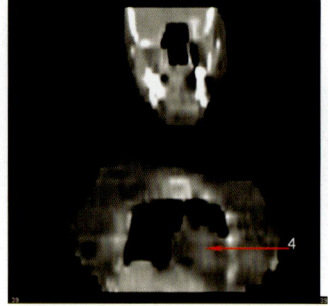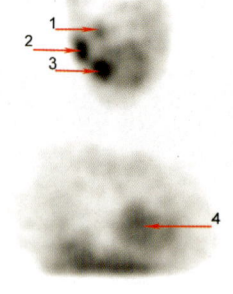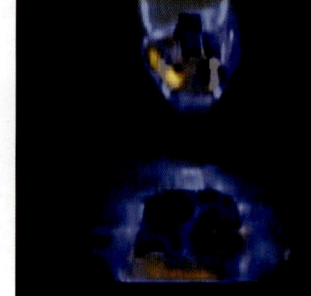

FIGURE 5-4:

1. Masseter muscle
2. Parotid gland

3. Submandibular gland

4. Heart

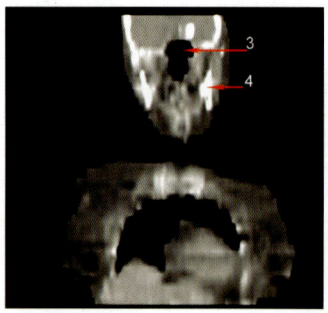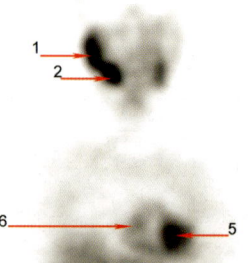

FIGURE 5-5:

1. Parotid gland
2. Submandibular gland

3. Sphenoid sinus
4. Mandible

5. Left ventricle
6. Right ventricle

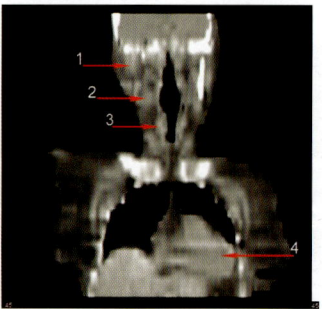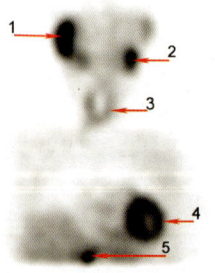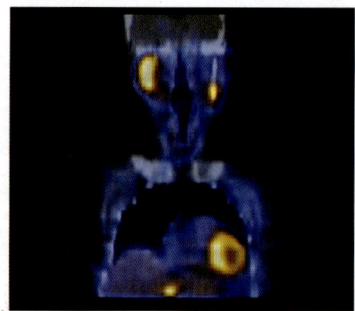

FIGURE 5-6:

1. Parotid gland
2. Submandibular gland
3. Thyroid

4. Left ventricular
   myocardium

5. Hepatic metastasis of
   carcinoid

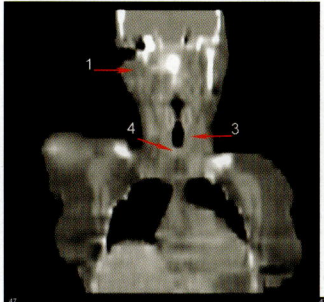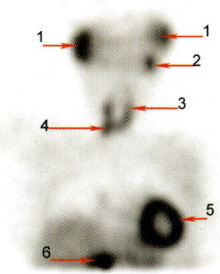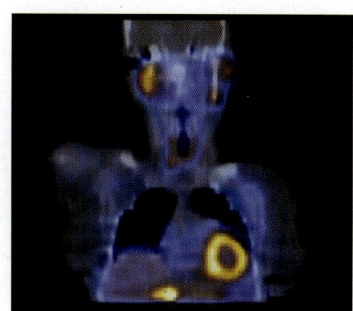

FIGURE 5-7:

1. Parotid gland
2. Submandibular gland
3. Thyroid
4. Parathyroid adenoma
5. Left ventricular
   myocardium
6. Hepatic metastasis of
   carcinoid

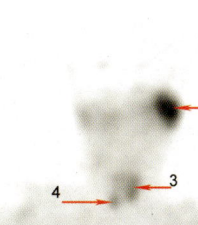

FIGURE 5-8:

1. Parotid gland
2. Sternocleidomastoid
   muscle
3. Thyroid
4. Parathyroid adenoma
5. Right atrial appendage
6. Breast
7. Right axilla

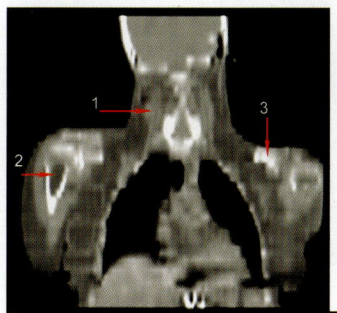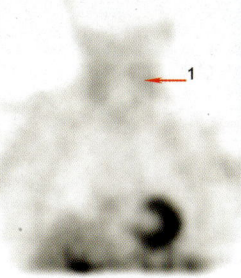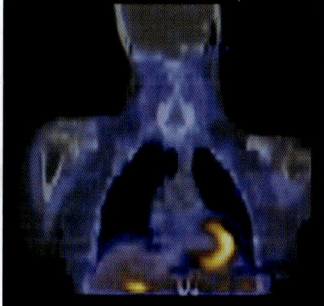

FIGURE 5-9:

1. Sternocleidomastoid
   muscle
2. Humerus
3. Left scapula

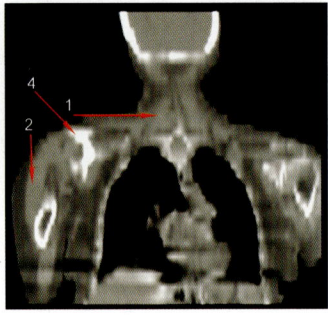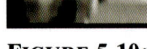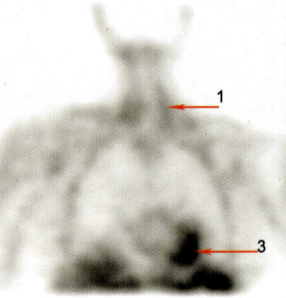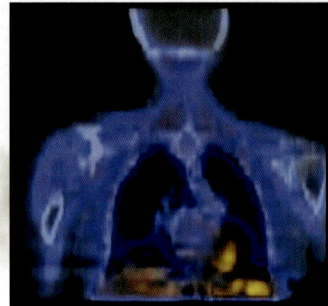

FIGURE 5-10:

1. Scalene muscles
2. Deltoid muscle

3. Apex of left ventricle

4. Right scapula

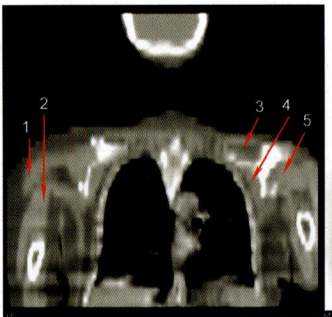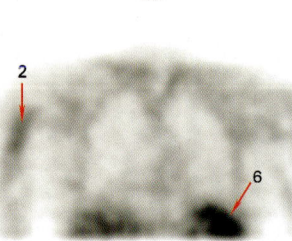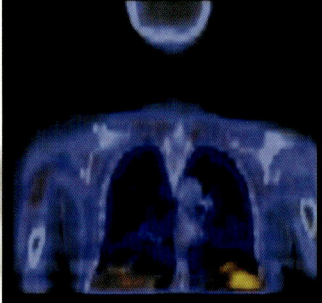

FIGURE 5-11:

1. Deltoid muscle
2. Biceps brachii muscle

3. Supraspinatus muscle
4. Subscapularis muscle

5. Infraspinatus muscle
6. Spleen

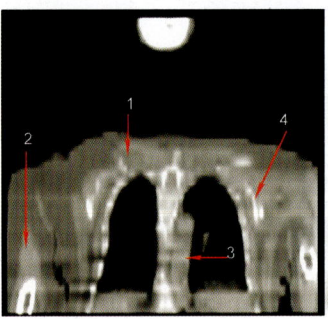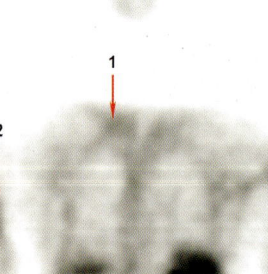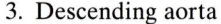

FIGURE 5-12:

1. Trapezius muscle
2. Biceps brachii muscle

3. Descending aorta

4. Left scapula

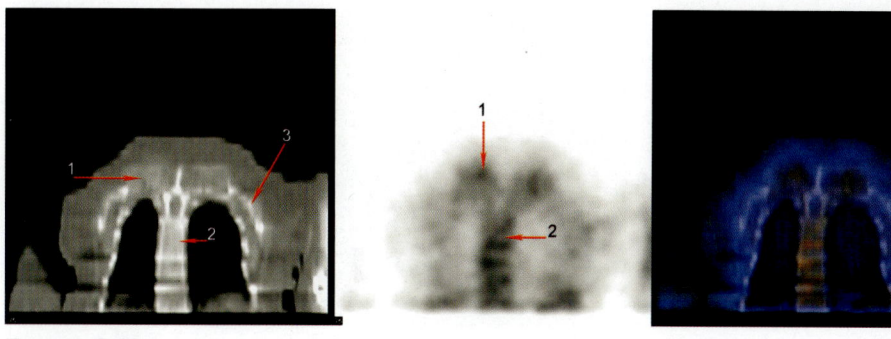

FIGURE 5-13:

1. Trapezius muscle   2. T4 marrow   3. Left scapula

Section 2: Sagittals

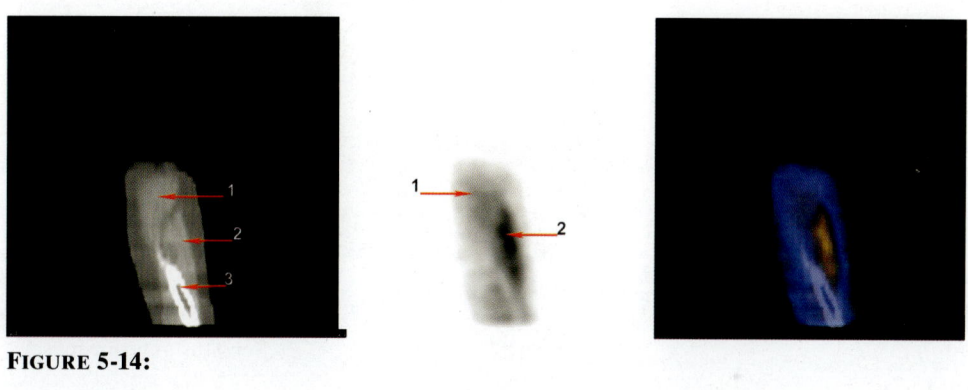

FIGURE 5-14:

1. Deltoid muscle   2. Triceps brachii muscle   3. Humerus

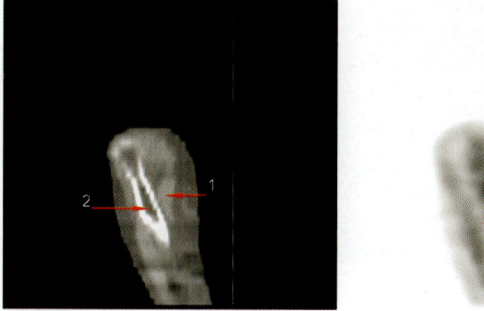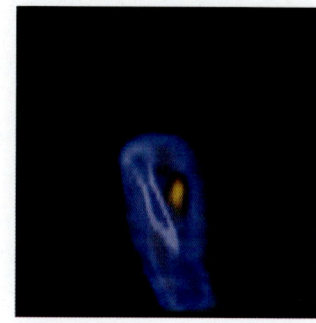

FIGURE 5-15:

1. Triceps brachii muscle   2. Humerus

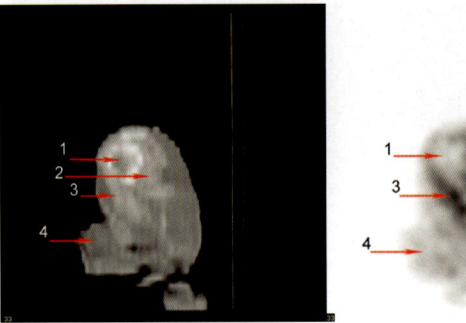

FIGURE 5-16:

1. Humeral head
2. Triceps brachii muscle
3. Biceps brachii muscle
4. Breast

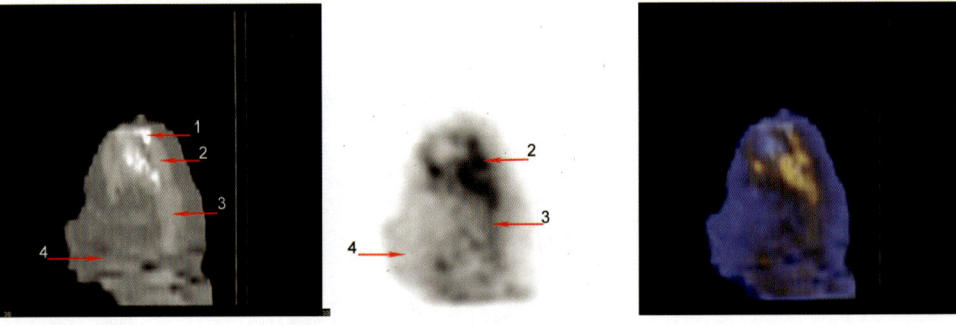

FIGURE 5-17:

1. Scapula
2. Subscapularis muscle
3. Serratus anterior muscle
4. Breast

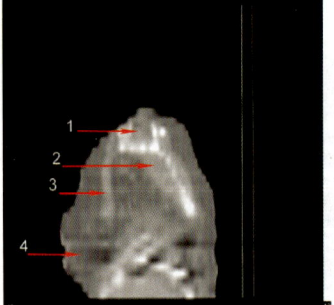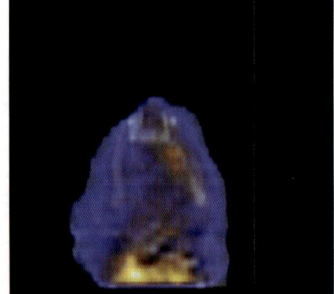

FIGURE 5-18:

1. Supraspinatus muscle
2. Subscapularis muscle
3. Pectoralis major muscle
4. Breast
5. Liver

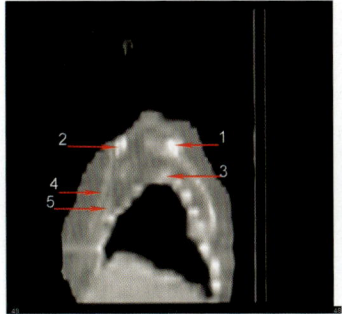
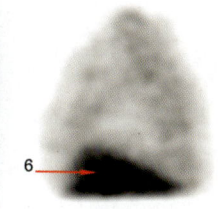
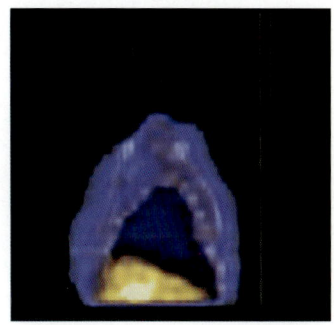

FIGURE 5-19:

1. Scapula
2. Clavicle
3. Rib
4. Pectoralis major muscle
5. Pectoralis minor muscle
6. Liver

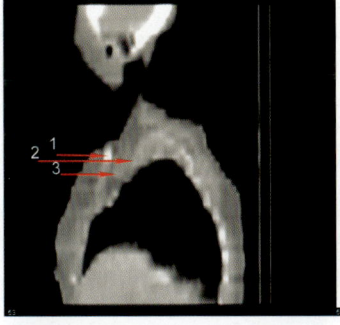
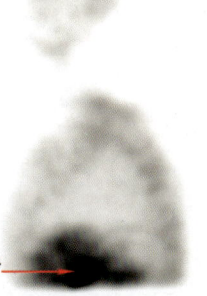
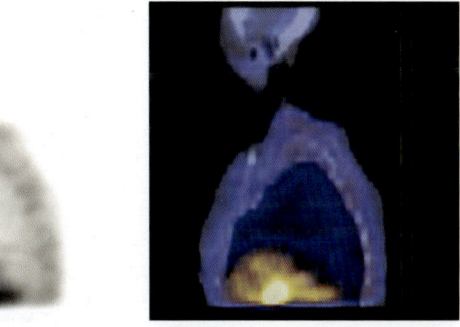

FIGURE 5-20:

1. Clavicle
2. Brachial plexus
3. Subclavian vein
4. Hepatic metastasis of carcinoid

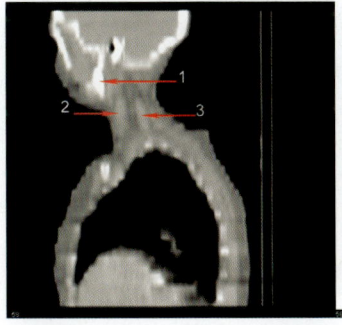
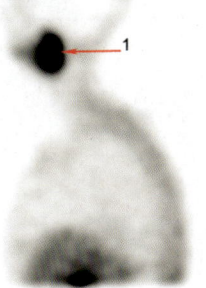
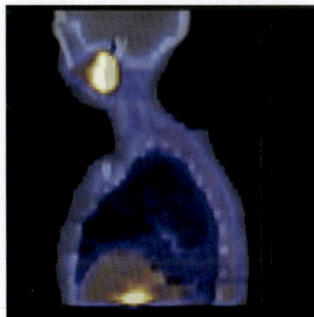

FIGURE 5-21:

1. Submandibular gland
2. Sternocleidomastoid muscle
3. Scalene muscle

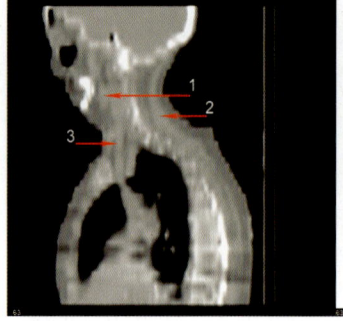
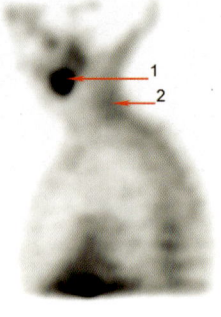
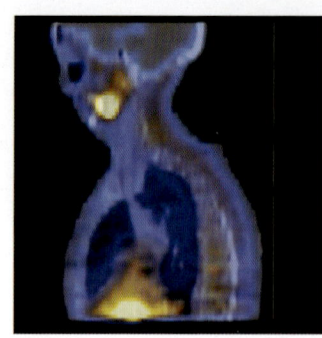

FIGURE 5-22:

1. Submandibular gland
2. Trapezius muscle
3. Sternocleidomastoid muscle

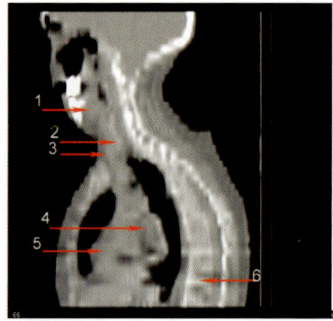
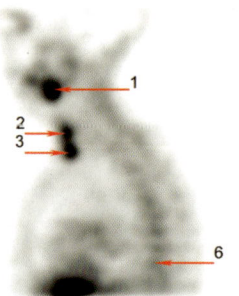
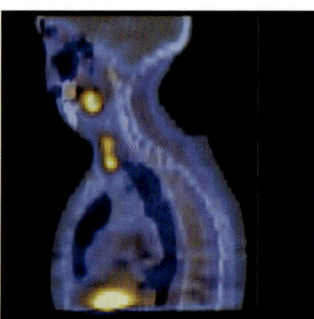

FIGURE 5-23:

1. Submandibular gland
2. Right thyroid
3. Parathyroid adenoma
4. Aortic arch
5. Right ventricle
6. Thoracic spine

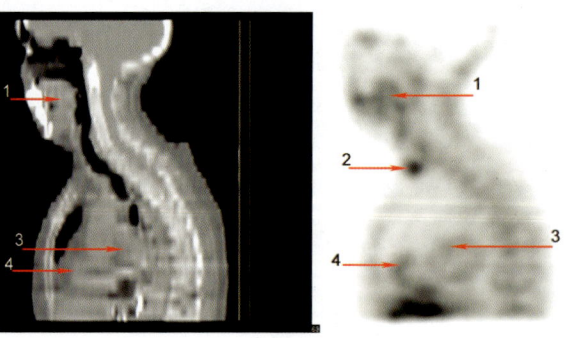
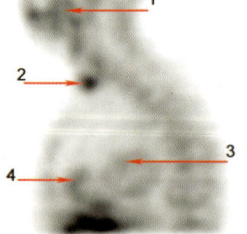
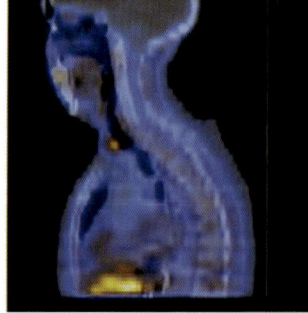

FIGURE 5-24:

1. Tongue
2. Parathyroid adenoma
3. Left atrium
4. Right ventricle

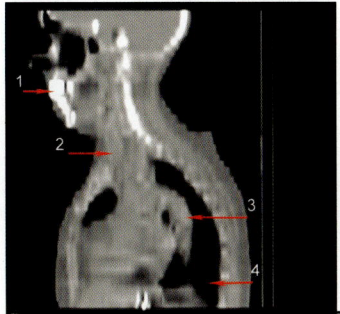
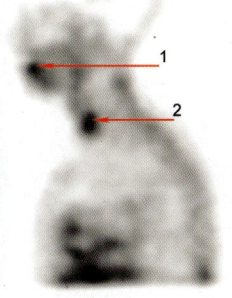
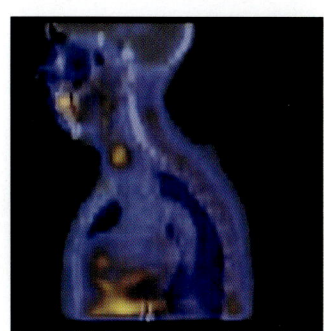

FIGURE 5-25:

1. Mandible or hyoglossus muscle
2. Left thyroid
3. Descending aorta
4. Left lower lung (posterior basal segment)

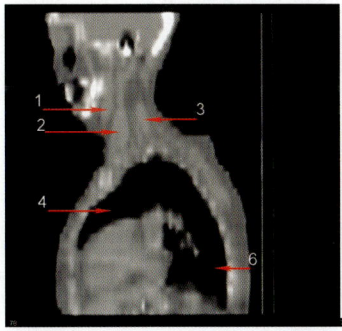
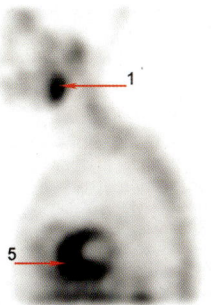
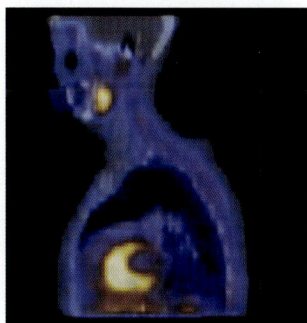

FIGURE 5-26:

1. Submandibular gland
2. Sternocleidomastoid muscle
3. Trapezius muscle
4. Left upper lung
5. Left ventricular myocardium
6. Left lower lung

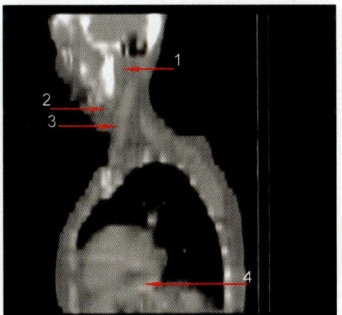
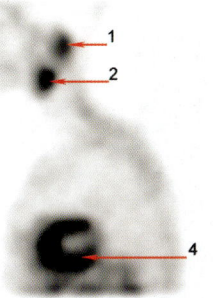
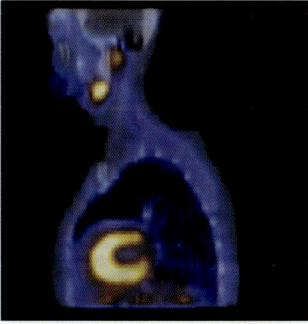

FIGURE 5-27:

1. Parotid gland
2. Submandibular gland
3. Sternocleidomastoid muscle
4. Left ventricular myocardium

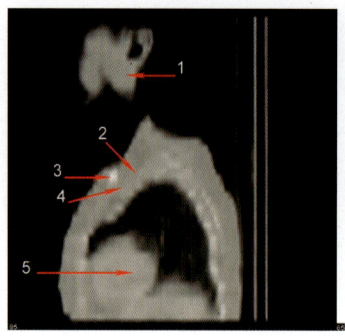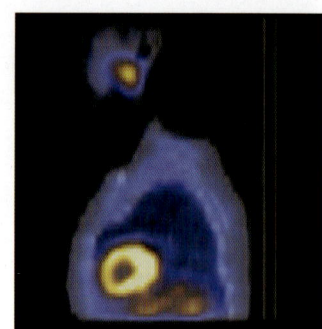

FIGURE 5-28:

1. Parotid gland	3. Clavicle	5. Left ventricular
2. Brachial plexus	4. Subclavian vein	myocardium

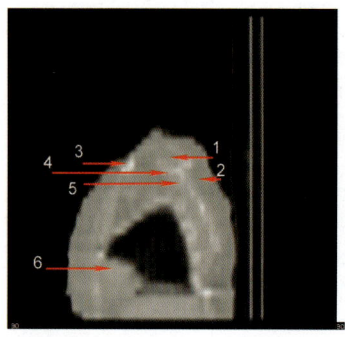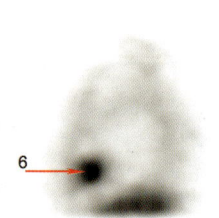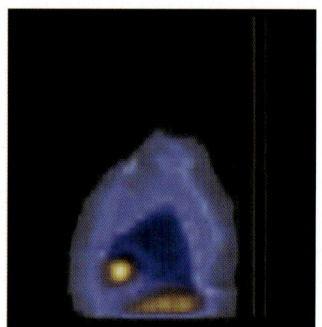

FIGURE 5-29:

1. Supraspinatus muscle	3. Clavicle	5. Subscapularis muscle
2. Infraspinatus muscle	4. Scapula	6. Left ventricle

Section 3: Transaxials

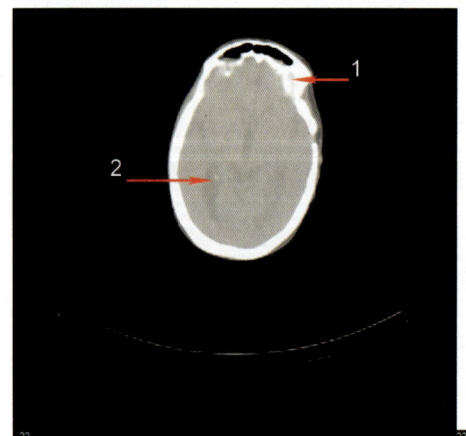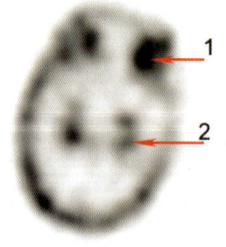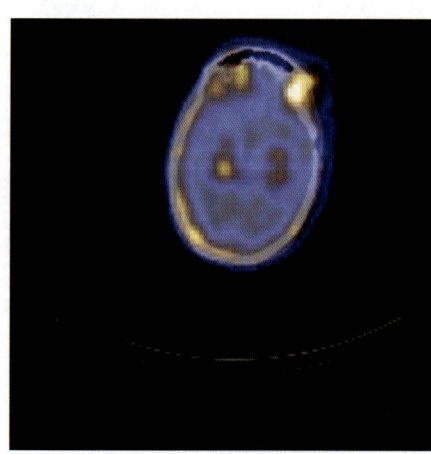

FIGURE 5-30:

1. Orbital fat   2. Choroidal plexus

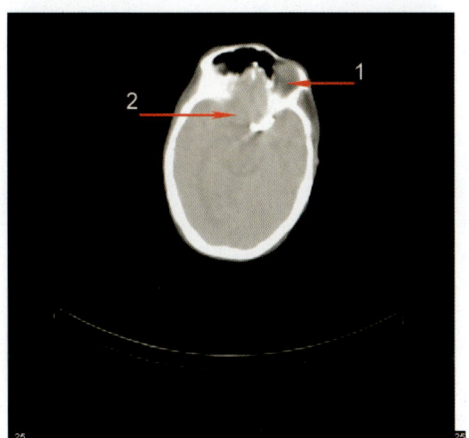
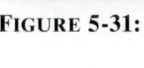
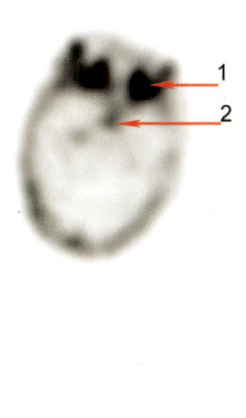

FIGURE 5-31:

1. Orbital fat   2. Pituitary gland

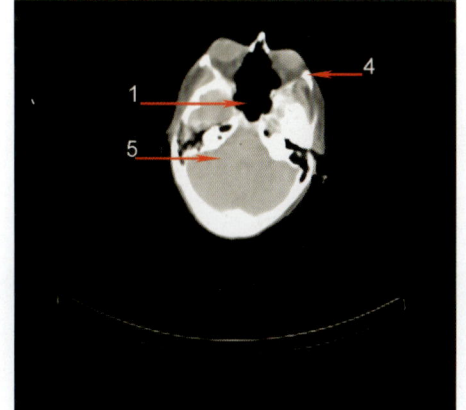
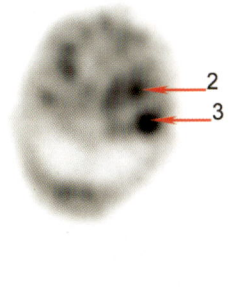
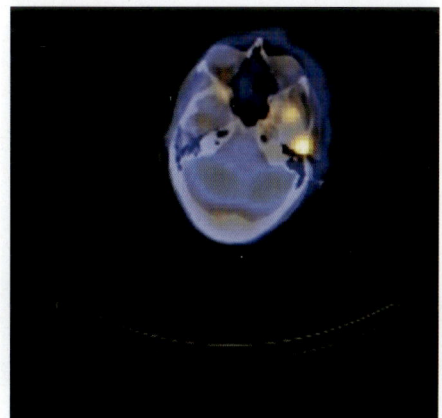

FIGURE 5-32:

1. Pharynx   3. Parotid gland   5. Cerebellopontine angle
2. Pterygoid muscle   4. Zygomatic arch

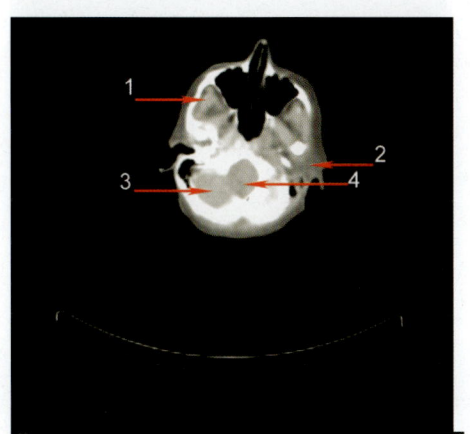
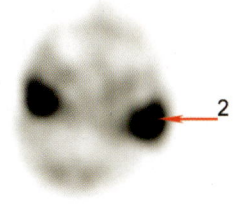
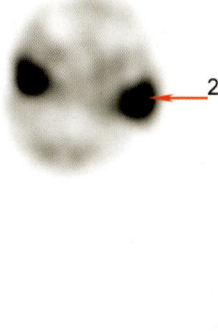
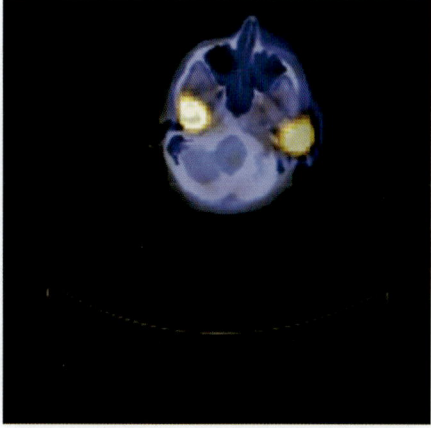

FIGURE 5-33:

1. Pterygoid muscle   3. Cerebellum   4. Pons
2. Parotid gland

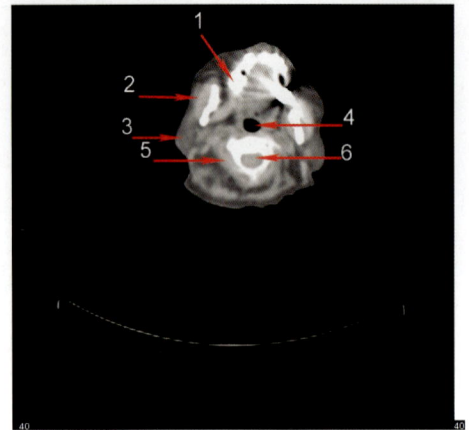 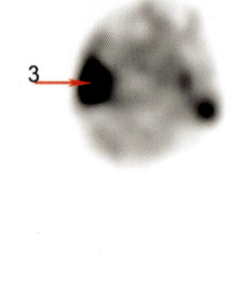 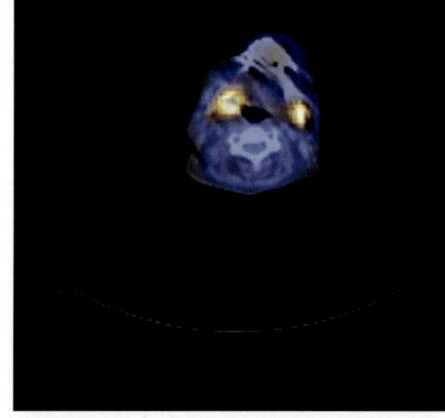

FIGURE 5-34:

1. Mandible
2. Masseter muscle
3. Parotid gland
4. Oropharynx
5. Semispinalis cervicis muscle
6. Spinal cord

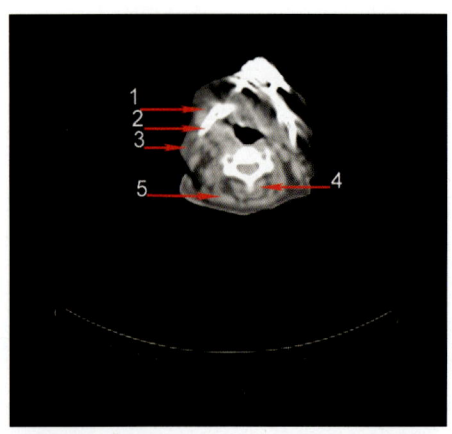 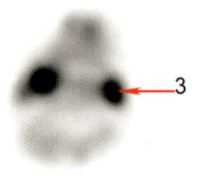 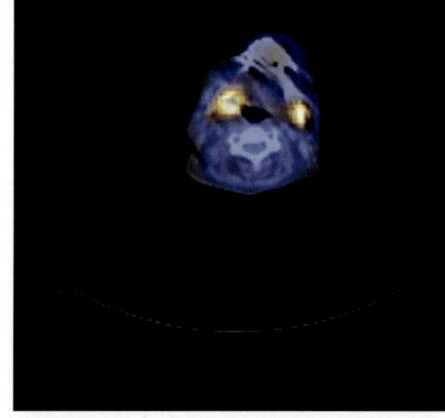

FIGURE 5-35:

1. Masseter muscle
2. Mandible
3. Parotid gland
4. Semispinalis cervicis muscle
5. Splenius capitis muscle

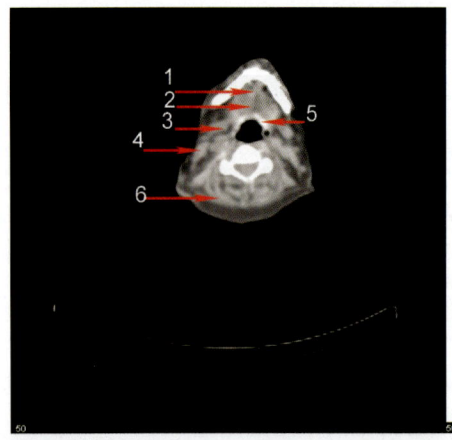 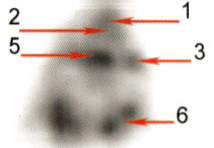 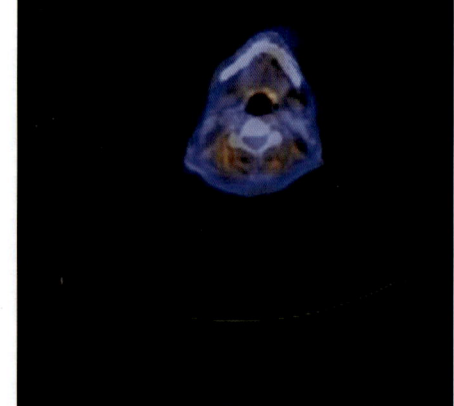

FIGURE 5-36:

1. Hyoglossus muscle
2. Genioglossus muscle
3. Palatine tonsil
4. Sternocleidomastoid muscle
5. Lingual tonsil
6. Semispinalis capitis muscle

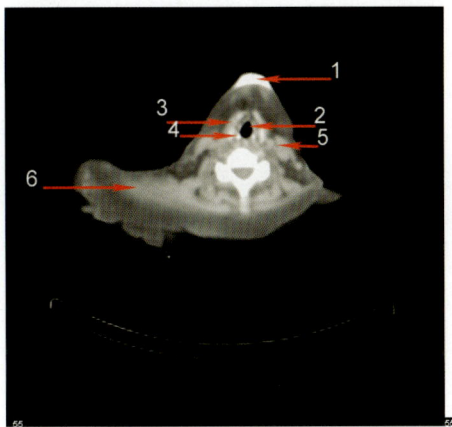 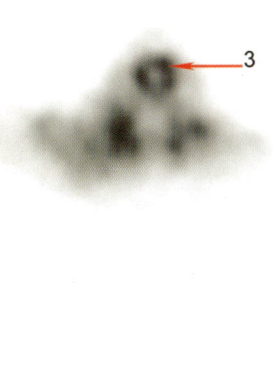

FIGURE 5-37:

1. Mandible	3. Thyroid cartilage	5. Sternocleidomastoid muscle
2. Oropharynx	4. Cricoid cartilage	6. Trapezius muscle

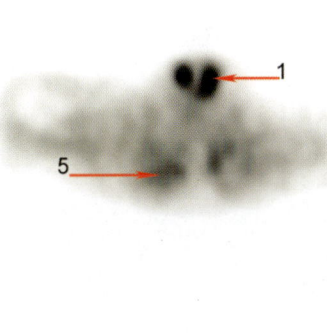 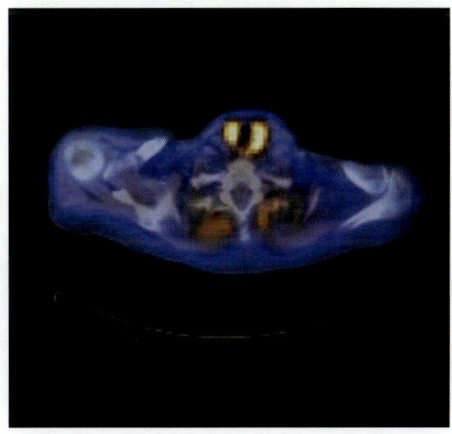

FIGURE 5-38:

1. Thyroid	3. Supraspinatus muscle	5. Trapezius muscle
2. Deltoid muscle	4. Infraspinatus muscle	

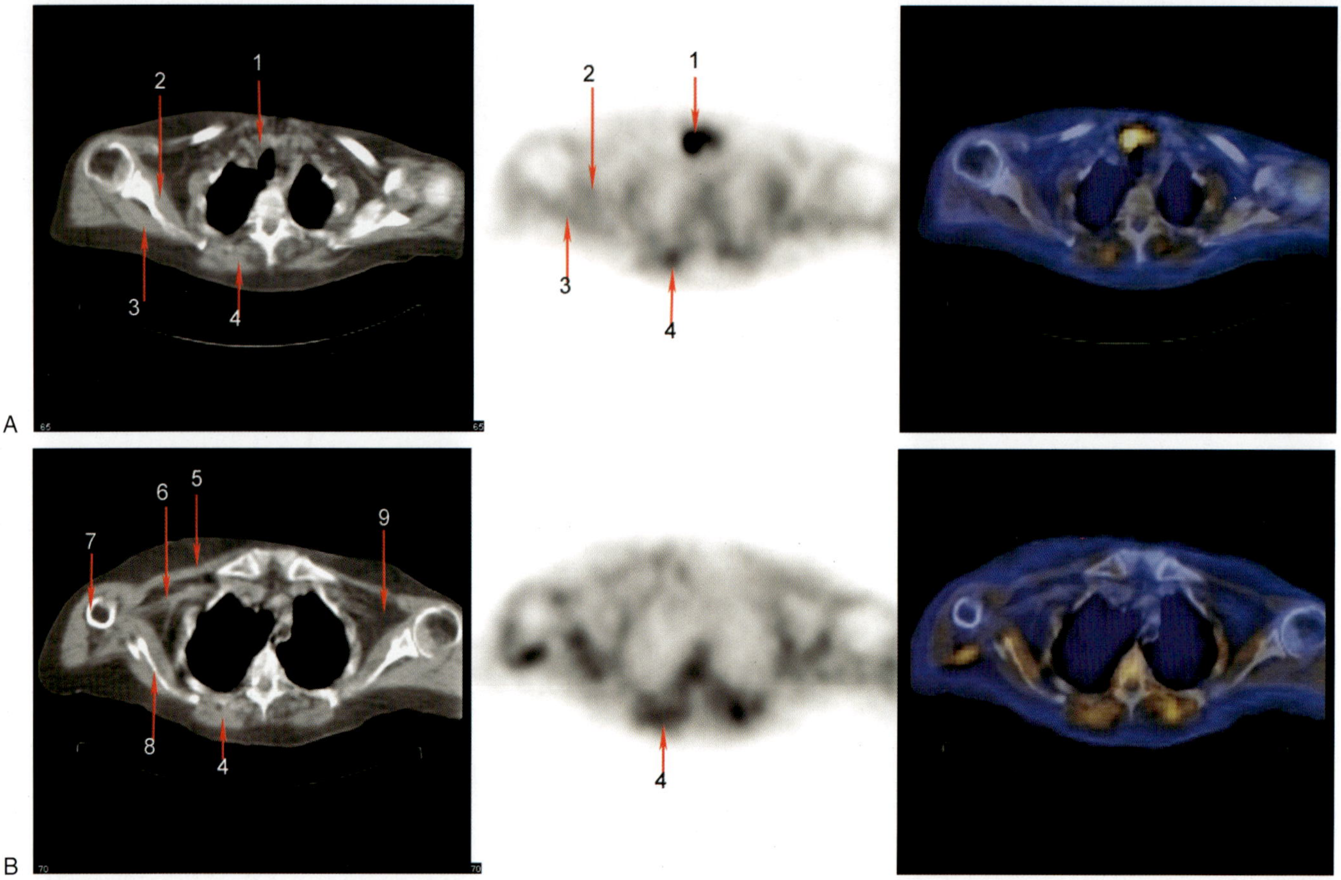

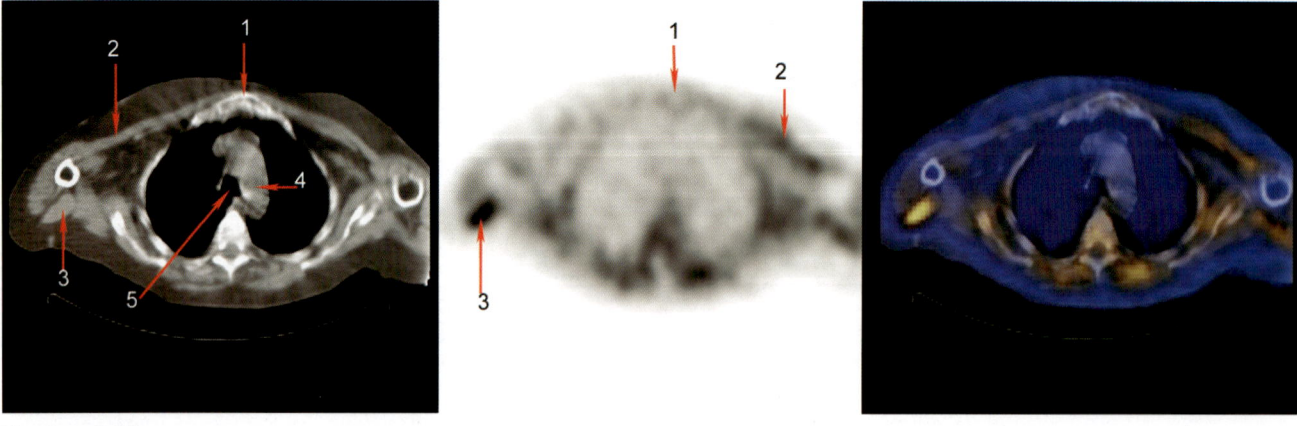

FIGURE 5-39:

1. Parathyroid adenoma
2. Subscapularis muscle
3. Infraspinatus muscle
4. Trapezius muscle
5. Pectoralis major muscle
6. Pectoralis minor muscle
7. Humerus
8. Scapula
9. Left axilla

FIGURE 5-40:

1. Sternum
2. Pectoralis major muscle
3. Triceps brachii muscle
4. Aortic arch
5. Trachea

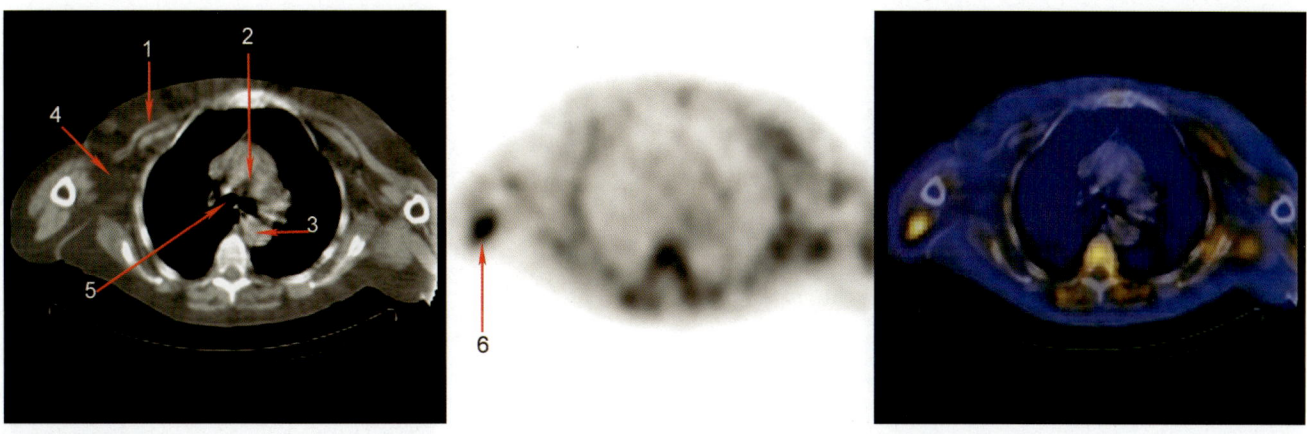

FIGURE 5-41:

1. Pectoralis major muscle
2. Aorta pulmonic window
3. Descending aorta
4. Axilla
5. Right main bronchus
6. Triceps brachii muscle

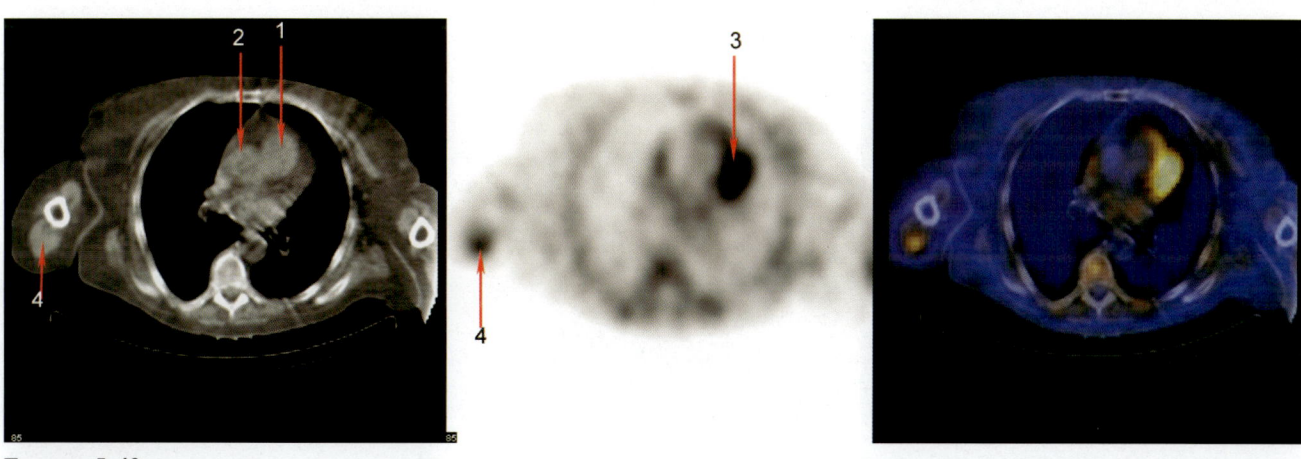

FIGURE 5-42:

1. Pulmonary artery
2. Ascending aorta
3. Left ventricular myocardium
4. Triceps brachii muscle

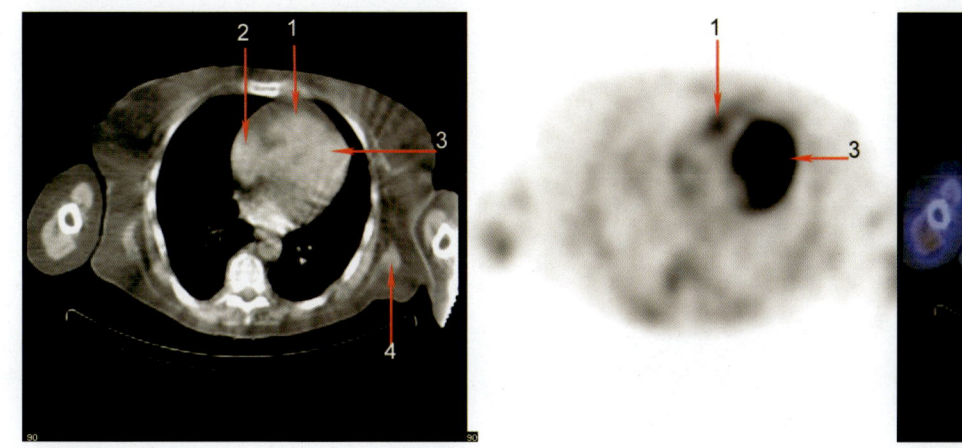
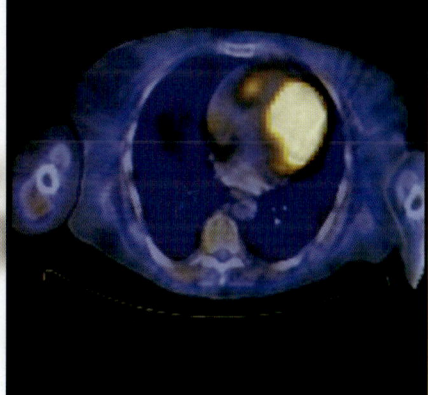

FIGURE 5-43:

1. Right ventricle
2. Right atrium
3. Left ventricle
4. Subscapularis muscle

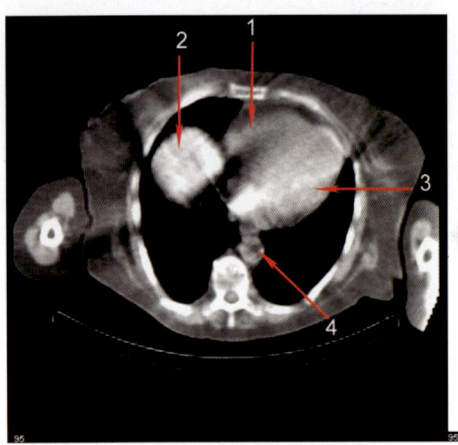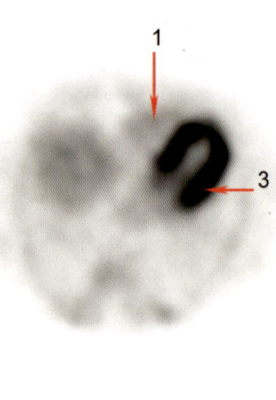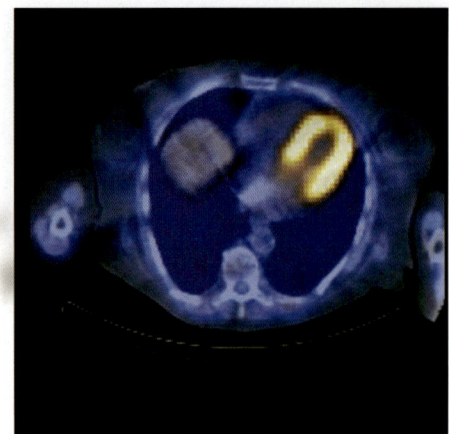

FIGURE 5-44:

1. Right ventricle   3. Left ventricular myocardium   4. Descending aorta
2. Right hepatic lobe (segment VIII)

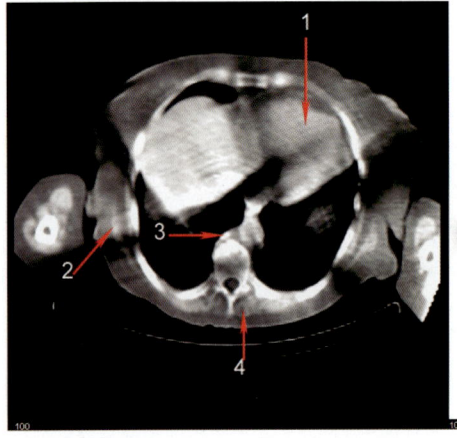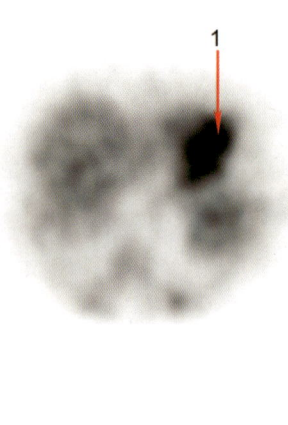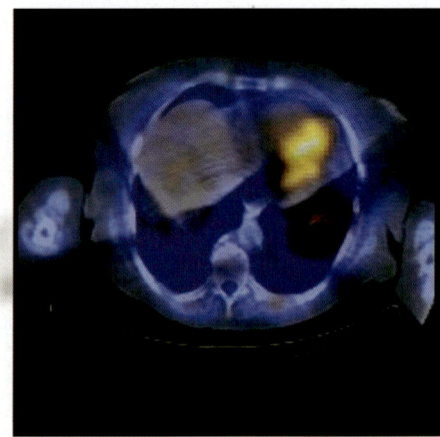

FIGURE 5-45:

1. Left ventricle   3. Azygos vein   4. Erector spinae muscle
2. Serratus anterior muscle

6 Bone SPECT/CT

Radionuclide bone scan using technetium-99m methylene diphosphonate is simple, sensitive, and efficient for the evaluation of bony metastasis, but it lacks specificity with a wide range of differential diagnoses. In a study designed to assess the clinical value of hybrid imaging in cancer patients for suspected bone metastasis, 78 patients with equivocal findings on routine nuclear bone scan underwent single photon emission computed tomography (SPECT)/computed tomography (CT) studies. Fusion of SPECT and CT images was found to improve the specificity of radionuclide bone scan and the diagnosis of the suspicious sites in about 80% of equivocal lesions on routine/ bone scans.[1]

Section 1: Coronals

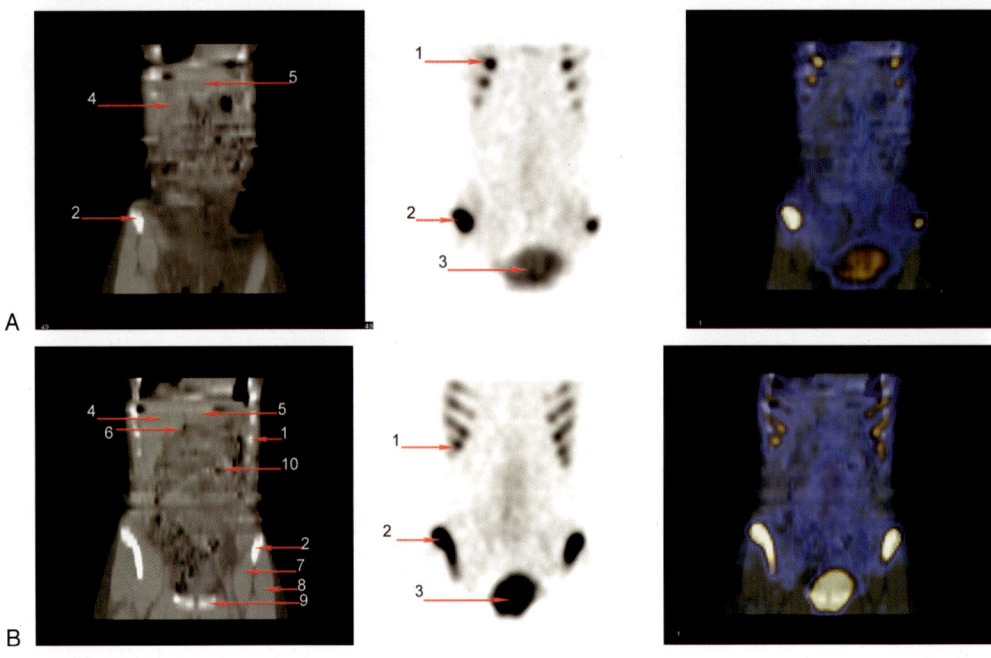

FIGURE 6-1:

1. Rib end
2. Anterior iliac crest
3. Bladder
4. Right hepatic lobe
5. Left hepatic lobe
6. Falciform ligament
7. Iliopsoas muscle
8. Rectus femoris muscle
9. Pubis
10. Jejunum

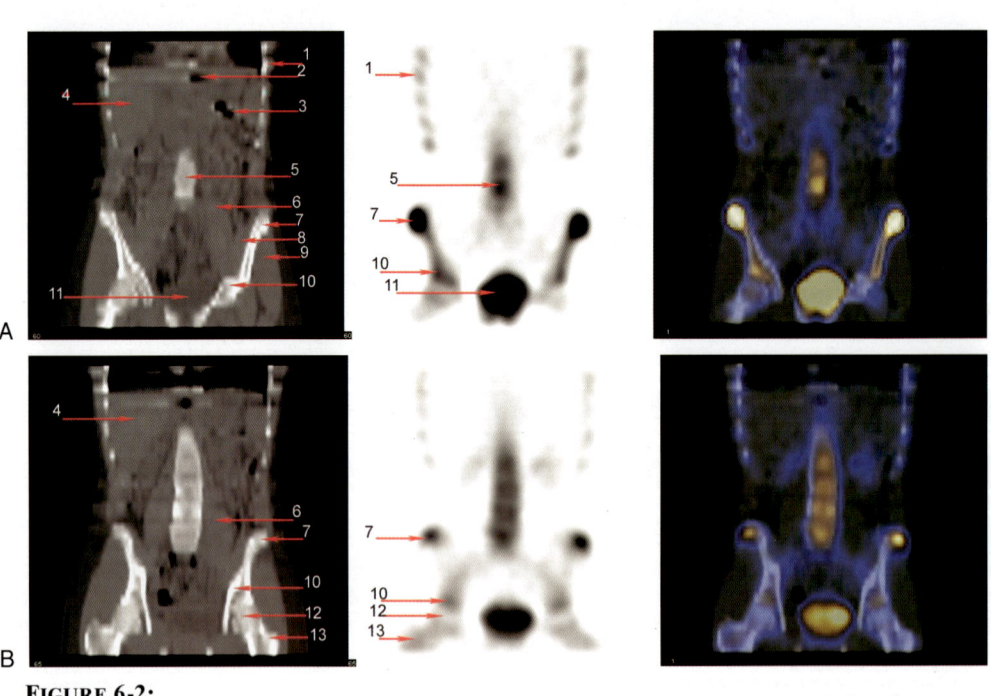

FIGURE 6-2:

1. Rib end
2. Stomach
3. Jejunum
4. Right hepatic lobe
5. L4 marrow
6. Psoas muscle
7. Anterior superior iliac crest
8. Iliacus muscle
9. Gluteus muscle
10. Acetabulum
11. Bladder
12. Femoral head
13. Greater trochanter of proximal femur

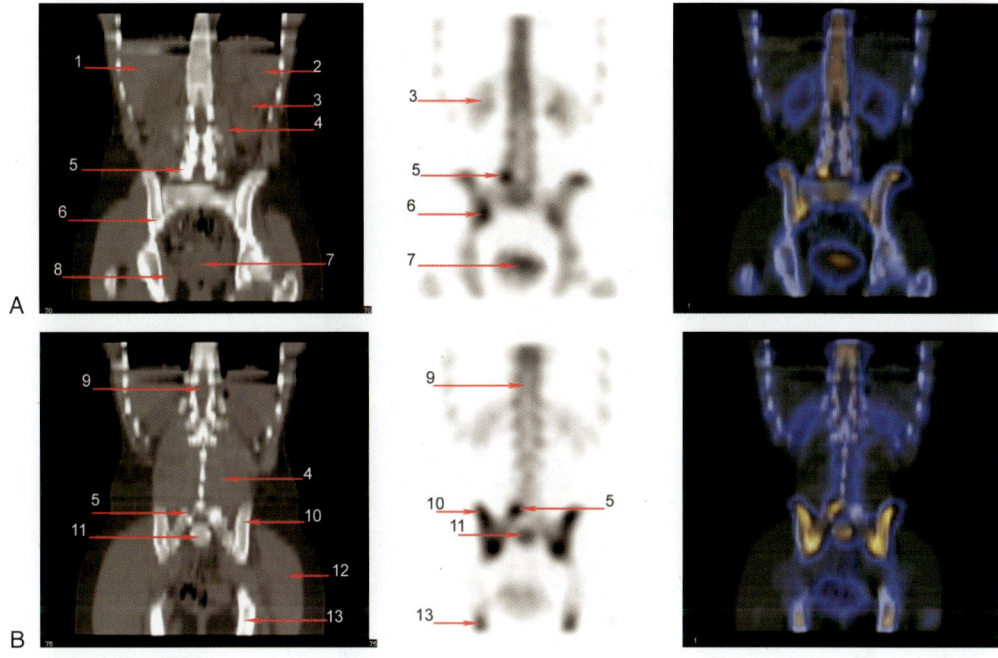

FIGURE 6-3:

1. Right hepatic lobe
2. Spleen
3. Kidney
4. Psoas muscle
5. Spur or facet joint of L5
6. Inferior iliac tuberosity
7. Bladder
8. Obturator internus muscle
9. Spinal canal
10. Superior iliac tuberosity
11. Sacral promontory
12. Gluteus muscle
13. Ischium

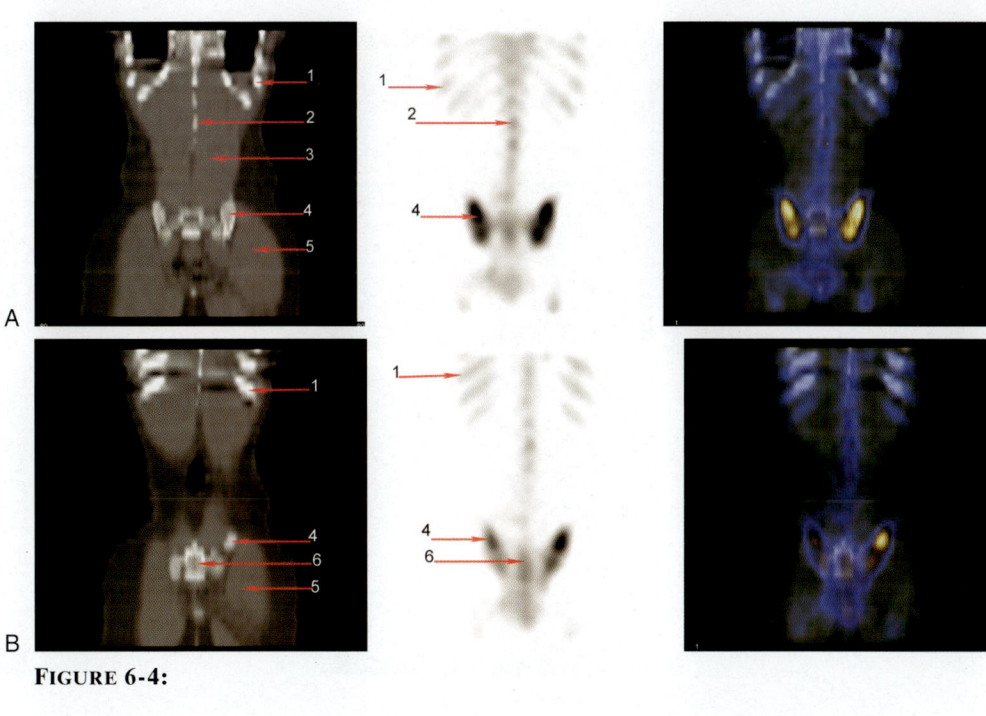

FIGURE 6-4:

1. Rib
2. Spinous process of L1
3. Erector spinae muscle
4. Iliac tuberosity
5. Gluteus muscle
6. Sacrum

Section 2: Sagittals

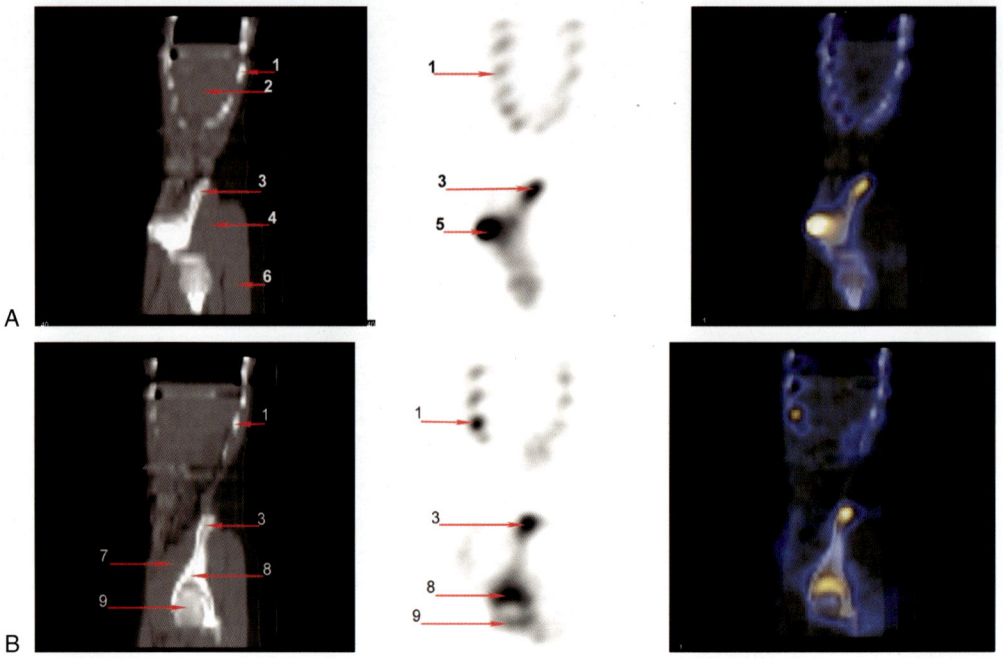

FIGURE 6-5:

1. Rib
2. Right hepatic lobe
3. Ilium
4. Gluteus medius muscle
5. Bladder
6. Gluteus maximus muscle
7. Iliacus muscle
8. Ischium (acetabulum)
9. Femoral head

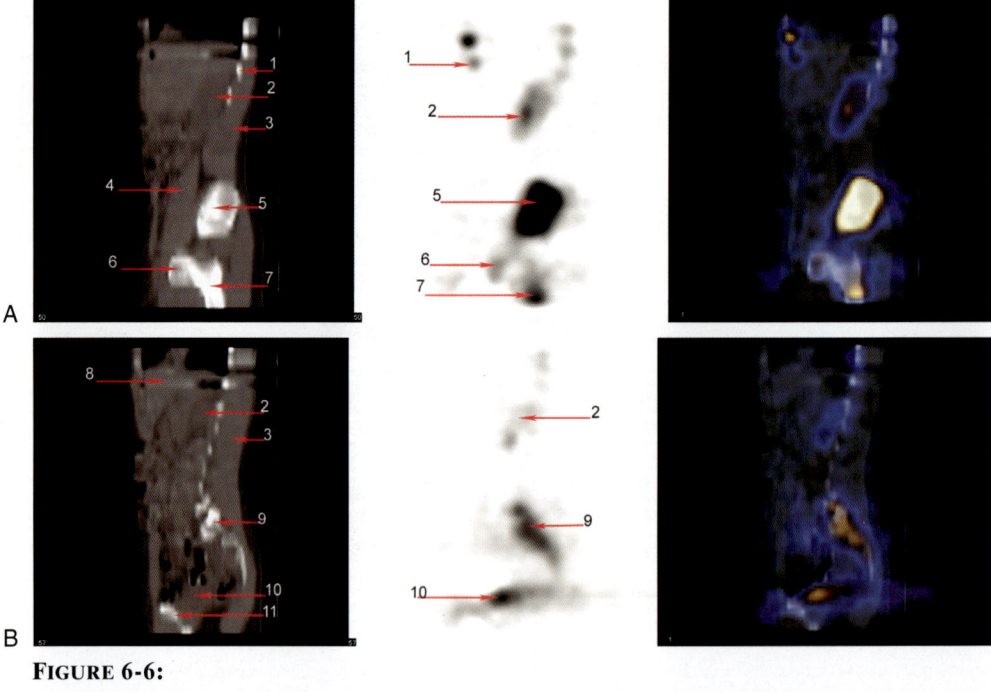

FIGURE 6-6:

1. Rib
2. Kidney
3. Erector spinae muscle
4. Psoas muscle
5. Iliac tuberosity
6. Symphysis pubis
7. Ischial tuberosity
8. Right hepatic lobe
9. Sacrum
10. Bladder
11. Symphysis pubis

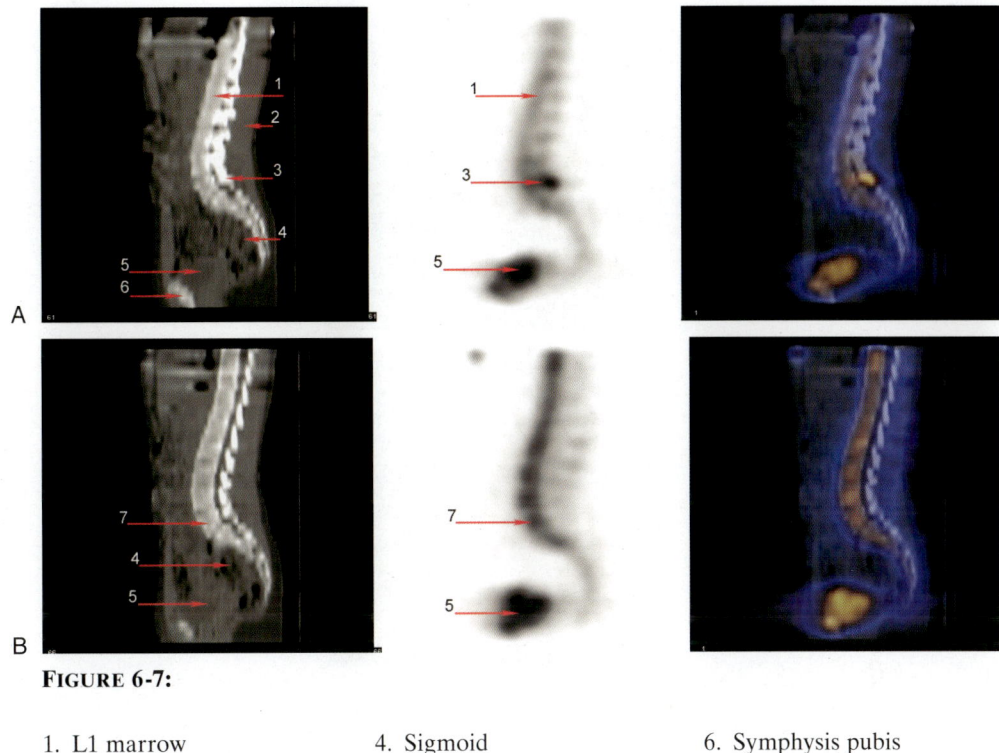

FIGURE 6-7:

1. L1 marrow
2. Erector spinae muscle
3. Facet joint or spur of L5
4. Sigmoid
5. Bladder
6. Symphysis pubis
7. L5 marrow

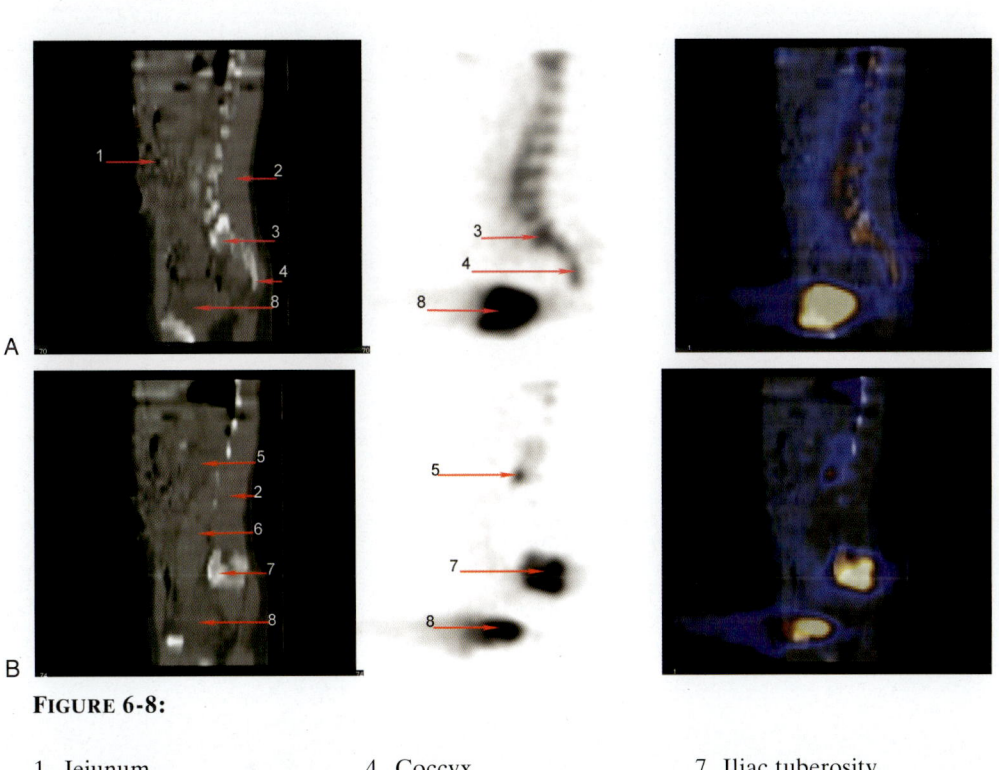

FIGURE 6-8:

1. Jejunum
2. Erector spinae muscle
3. S1 marrow
4. Coccyx
5. Kidney
6. Psoas muscle
7. Iliac tuberosity
8. Bladder

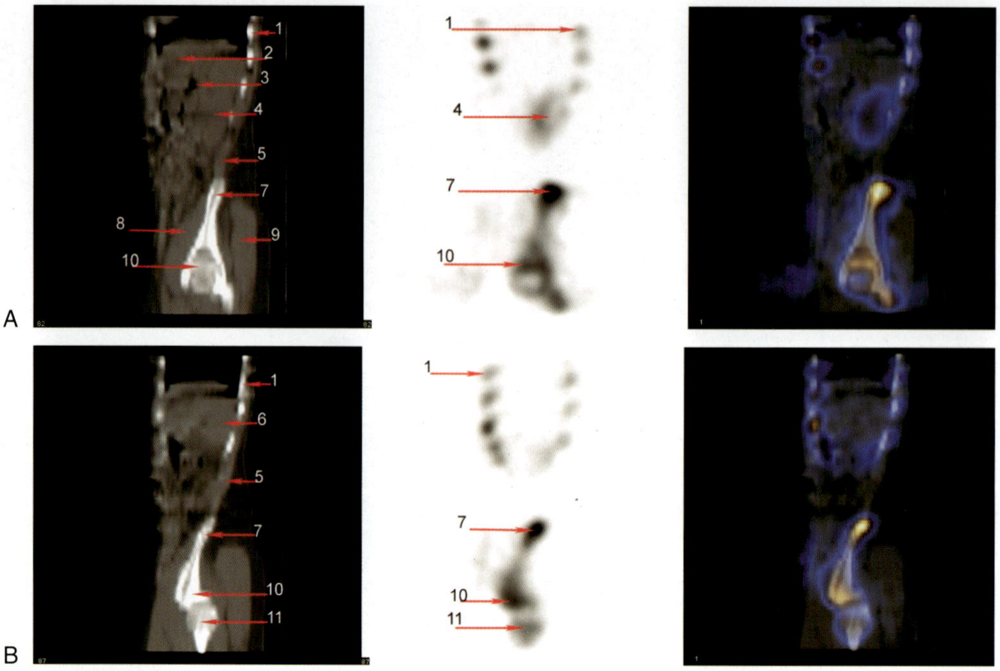

FIGURE 6-9:

1. Rib
2. Left hepatic lobe
3. Stomach
4. Kidney
5. Latissimus dorsi muscle
6. Spleen
7. Anterior superior iliac crest
8. Iliacus muscle
9. Gluteus maximus muscle
10. Acetabulum
11. Femoral head

Section 3: Transaxials

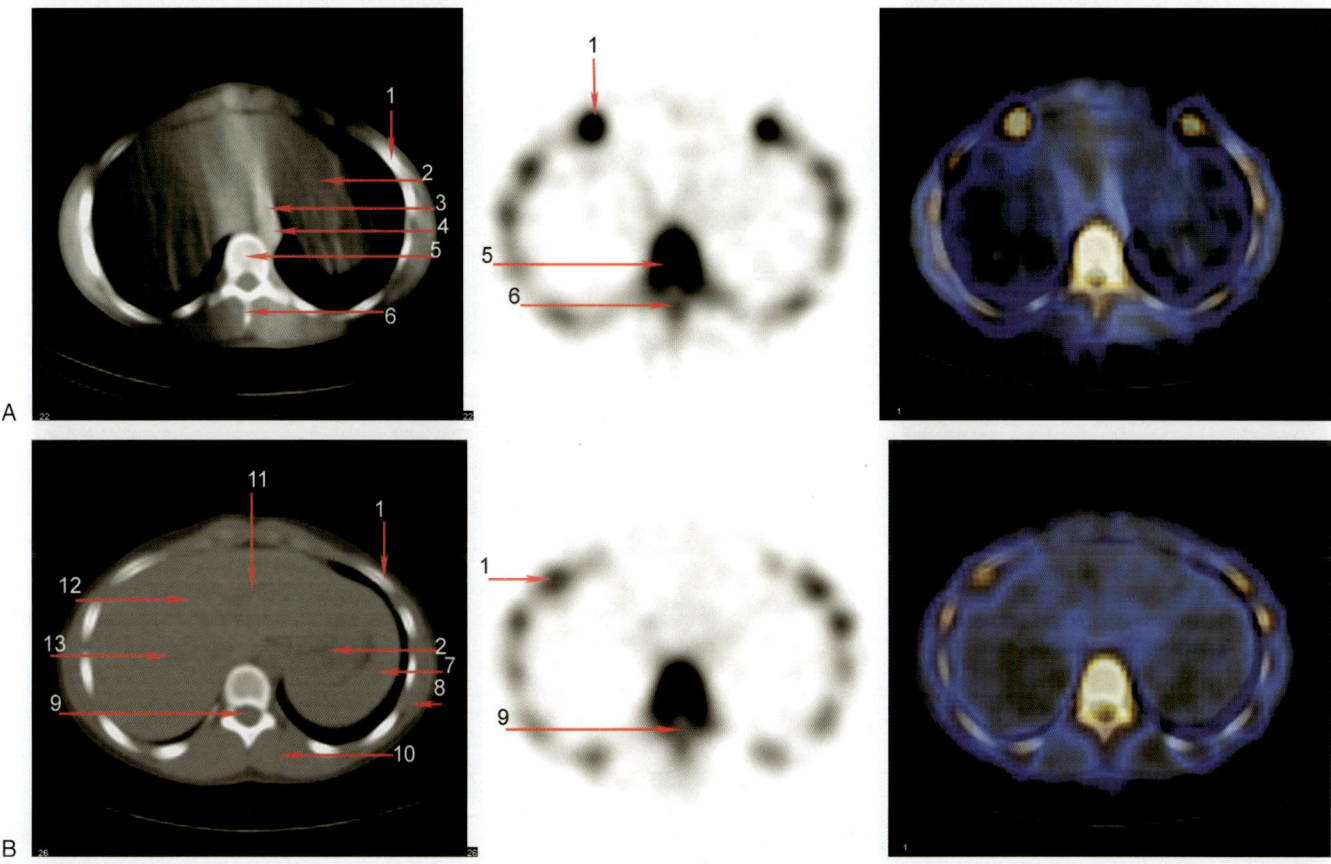

FIGURE 6-10:

1. Rib (costochondral joint)
2. Stomach
3. Esophagus
4. Abdominal aorta
5. T11 marrow

6. Spinous process of T11
7. Spleen
8. Latissimus dorsi muscle
9. Spinal canal

10. Erector spinae muscle
11. Left hepatic vein
12. Middle hepatic vein
13. Right hepatic vein

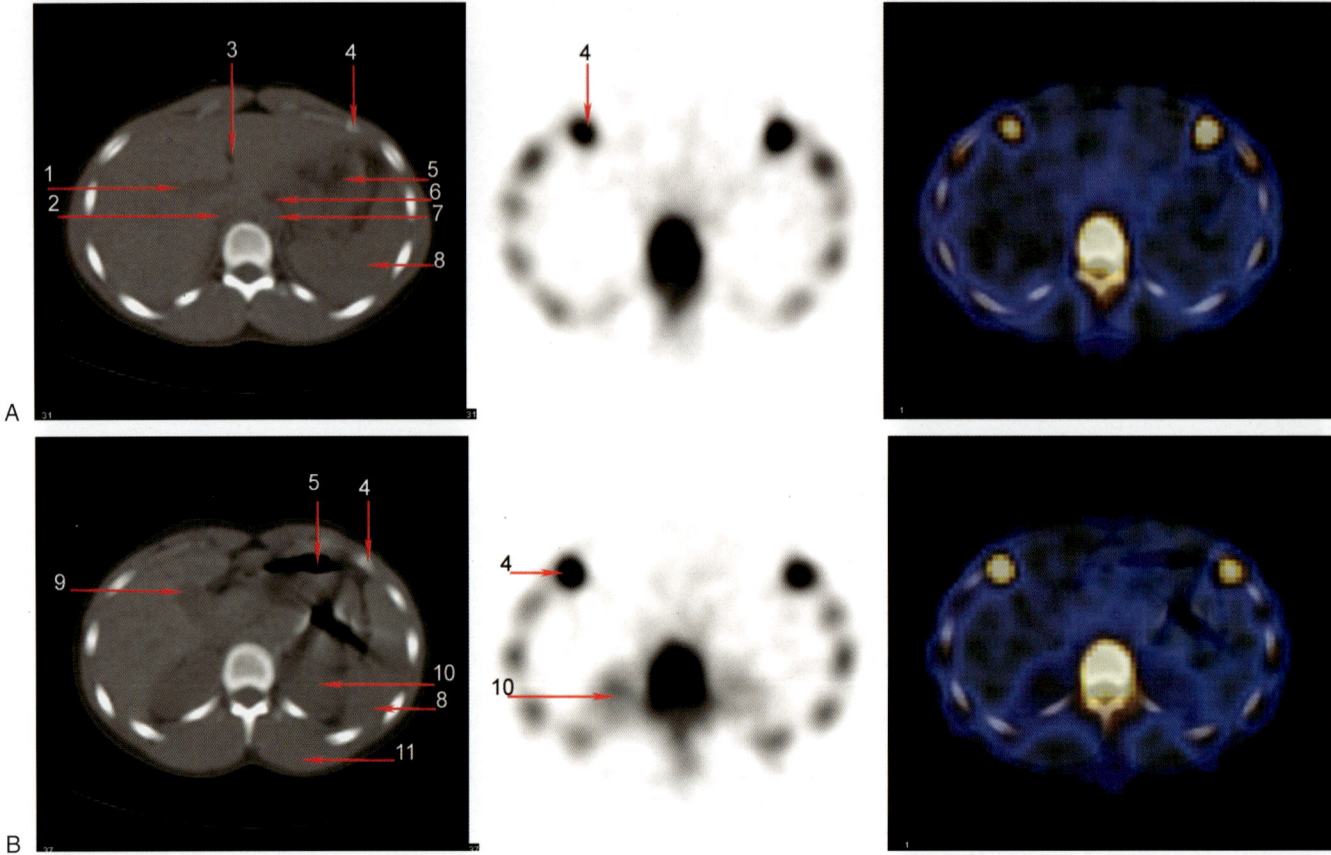

FIGURE 6-11:

1. Portal vein
2. Inferior vena cava
3. Falciform ligament
4. Rib end
5. Stomach
6. Esophagus
7. Abdominal aorta
8. Spleen
9. Gallbladder
10. Kidney
11. Erector spinae muscle

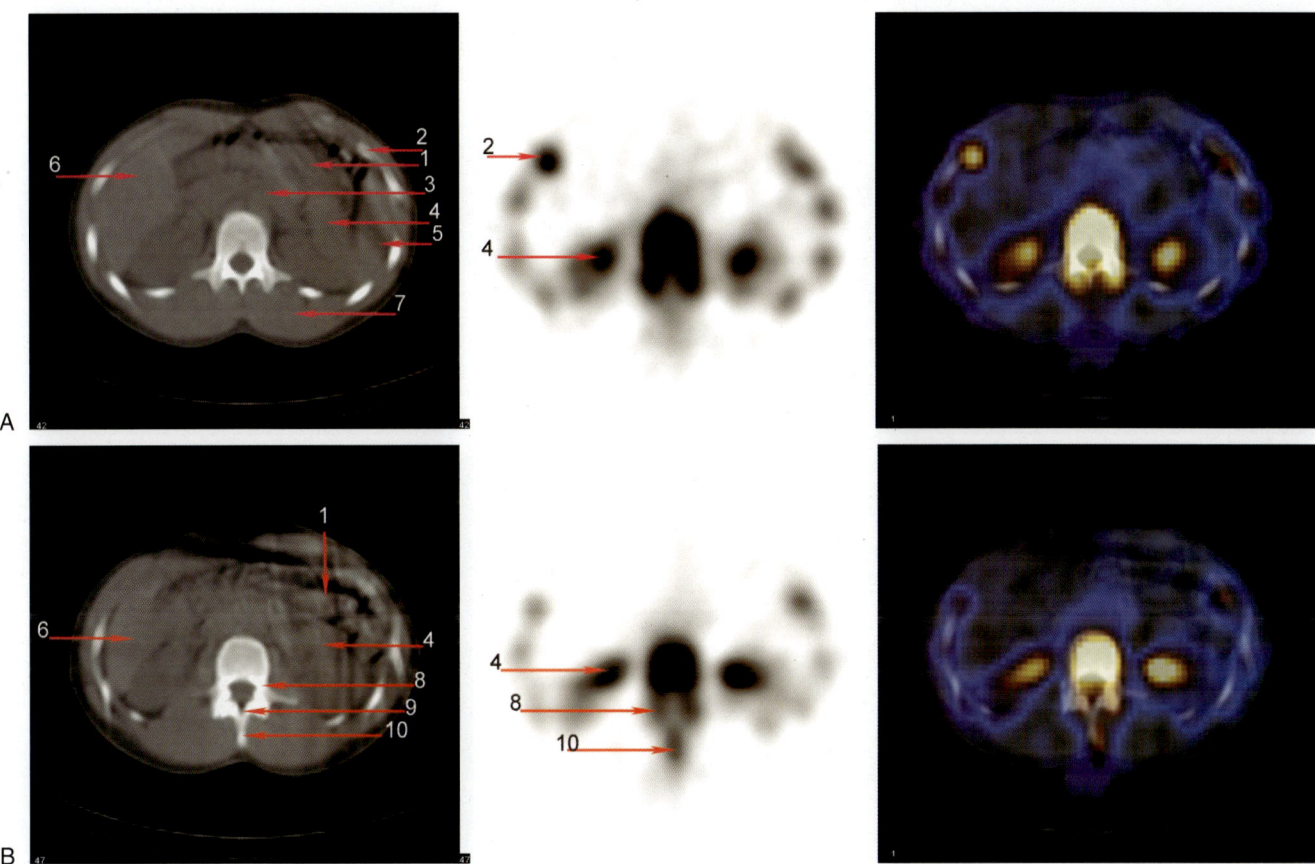

A

B

FIGURE 6-12:

1. Jejunum
2. Rib end
3. Pancreas
4. Kidney

5. Spleen
6. Right hepatic lobe (segment V)
7. Erector spinae muscle
8. Pedicle of L2

9. Lamina of L2
10. Spinous process of L2

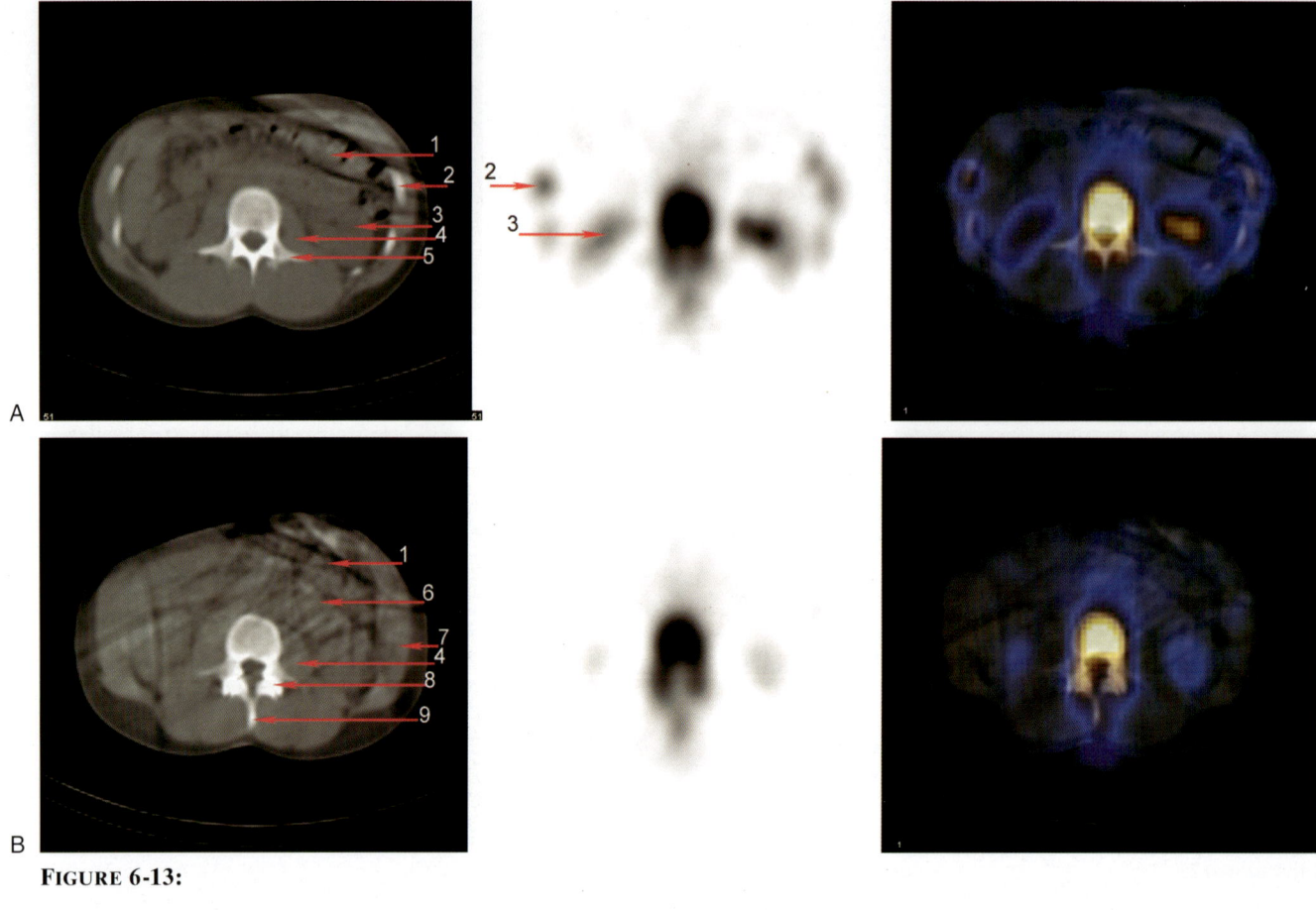

FIGURE 6-13:

1. Transverse colon	4. Psoas muscle	7. Oblique muscle
2. Rib end	5. Transverse process of L2	8. Facet joint of L2 and L3
3. Kidney	6. Jejunum	9. Spinous process of L2

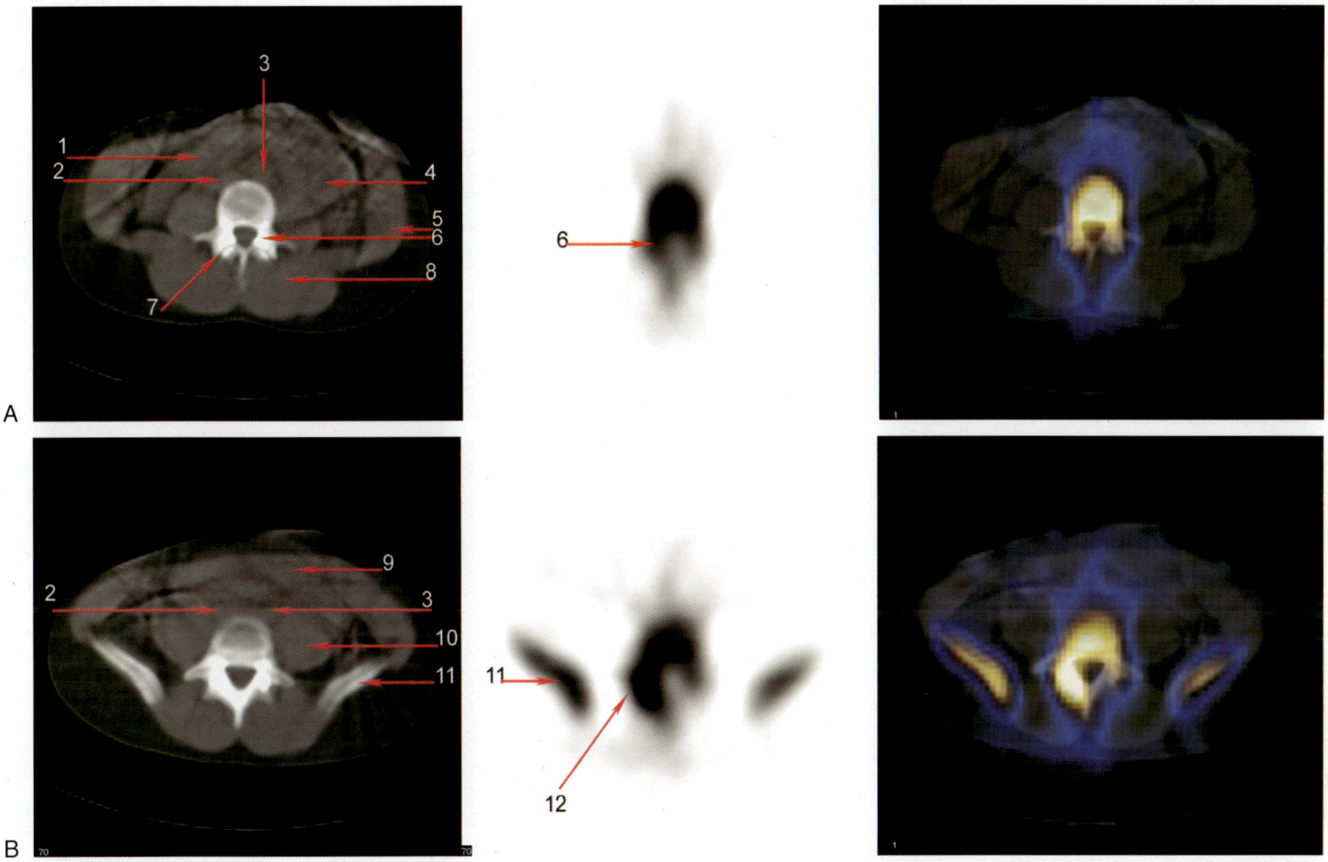

FIGURE 6-14:

1. Ileum
2. Inferior vena cava
3. Abdominal aorta
4. Jejunum

5. Oblique muscles
6. Pedicle of L4
7. Facet joint of L4 and L5
8. Erector spinae muscle

9. Rectus abdominis muscle
10. Psoas muscle
11. Iliac crest
12. Facet joint of L5 and S1

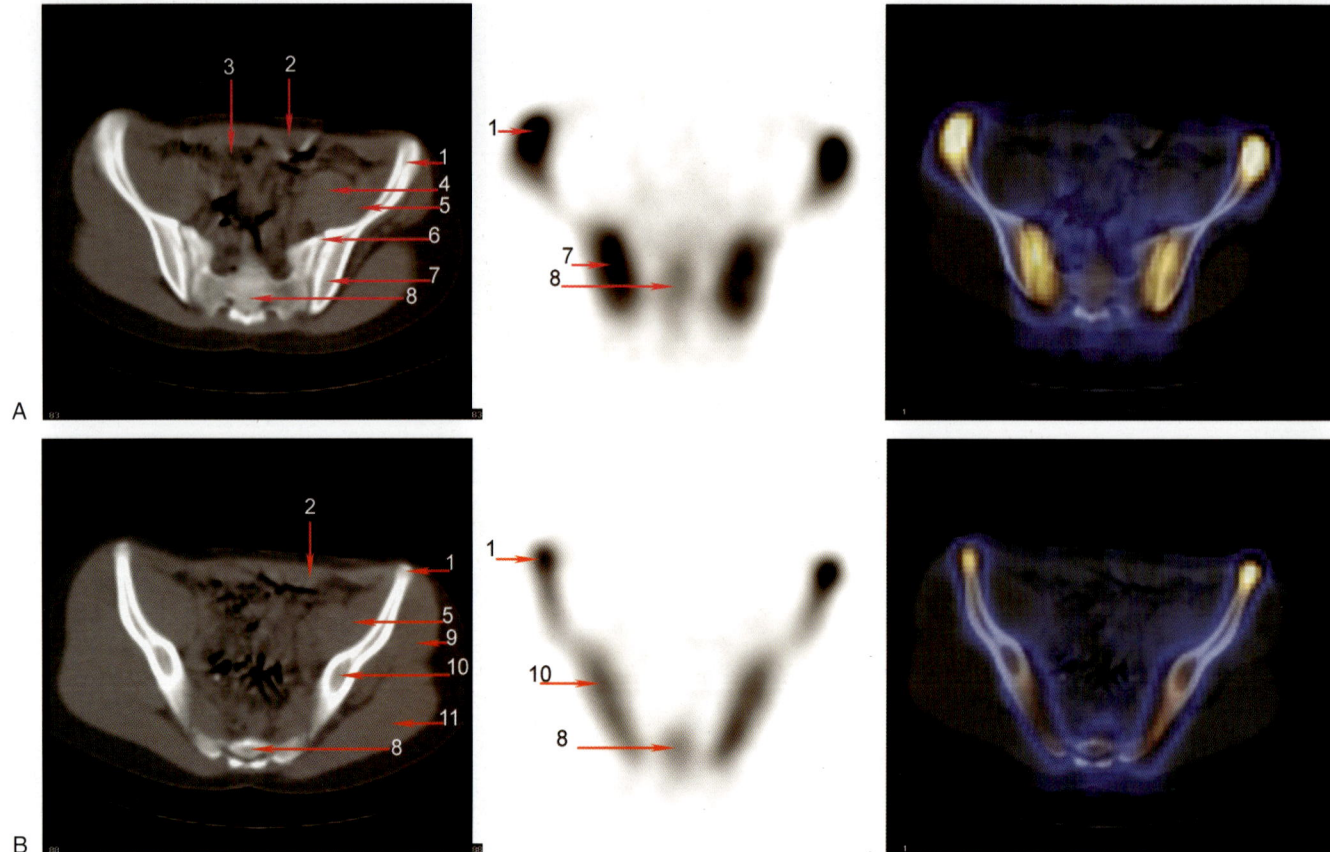

FIGURE 6-15:

1. Anterior superior iliac crest
2. Rectus abdominalis muscle
3. Ileum
4. Psoas muscle
5. Iliacus muscle
6. Sacroiliac joint
7. Iliac tuberosity
8. Sacrum
9. Gluteus medius muscle
10. Iliac crest
11. Gluteus maximus muscle

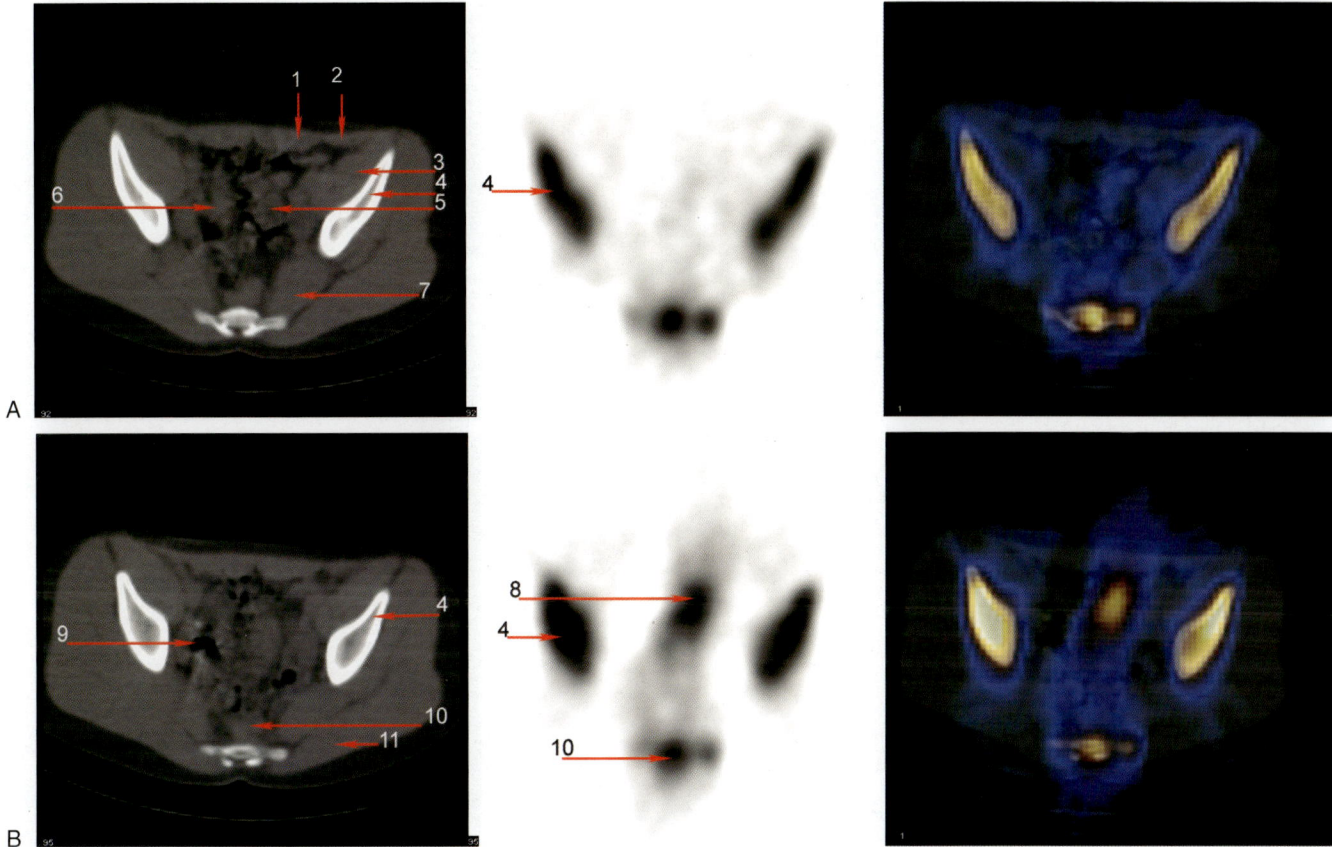

FIGURE 6-16:

1. Rectus abdominalis muscle
2. External oblique muscle
3. Iliacus muscle
4. Iliac crest
5. Jejunum
6. Ileum
7. Piriformis muscle
8. Sigmoid colon
9. Cecum
10. Rectosigmoid colon and coccyx
11. Gluteus maximus muscle

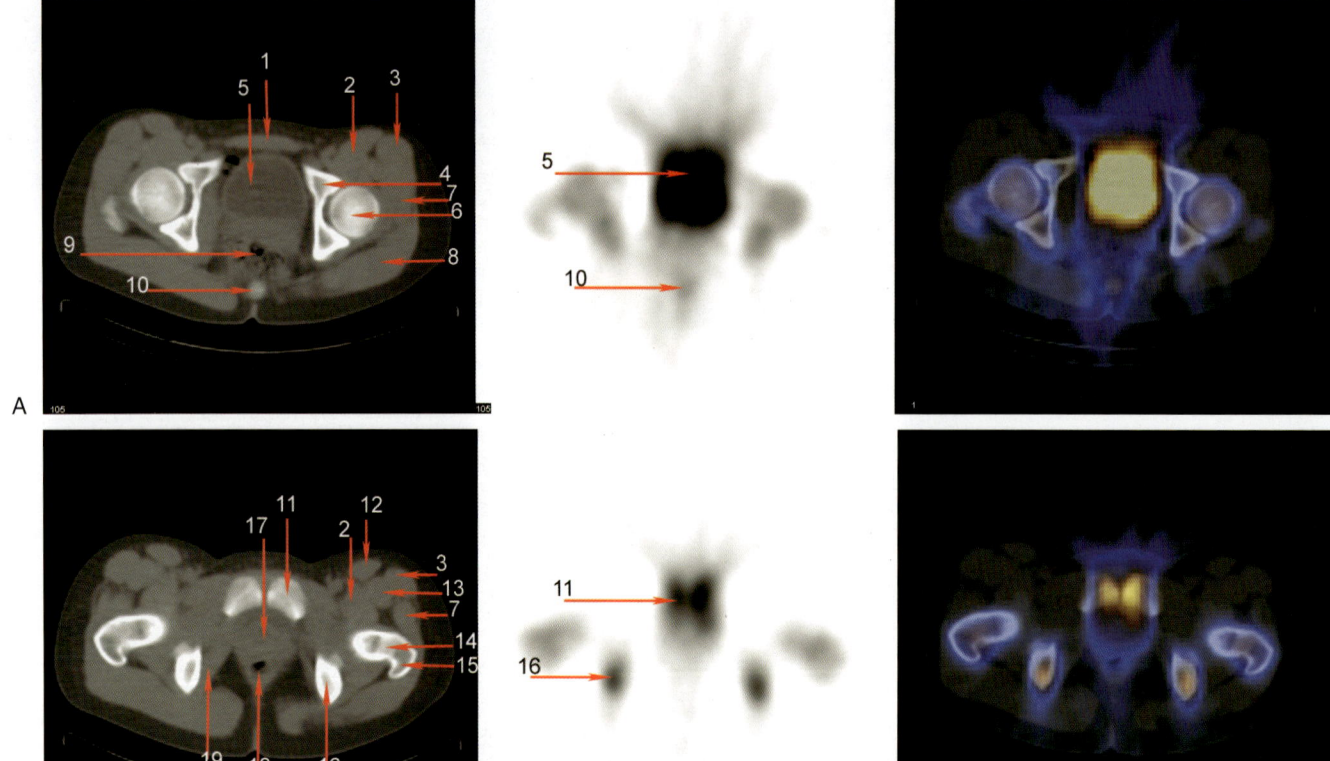

FIGURE 6-17:

1. Rectus abdominis muscle
2. Iliopsoas muscle
3. Tensor fascia latae muscle
4. Anterior lip of acetabulum (ilium)
5. Bladder
6. Femoral head
7. Gluteus medius muscle
8. Gluteus maximus muscle
9. Rectum
10. Coccyx
11. Pubis
12. Sartorius muscle
13. Rectus femoris muscle
14. Femoral neck
15. Trochanter of proximal femur
16. Ischial tuberosity
17. Prostate
18. Anus
19. Internal obturator muscle

7   131-I SPECT/CT

The risk of local recurrence and distant metastasis of thyroid cancer demands a continuous monitoring that is performed by using radioiodine whole-body scan and measurement of serum thyroglobulin. The uptake of radioiodine is based on sodium iodide symporter and organification in the thyroid tissue and cancer cells.[1] The scan data make an impact on patient management, with surgical removal of localized cancer or treatment with a high dose of I-131 sodium iodide. Precise localization of the focal lesions may benefit from fusion of single photon emission computed tomography (SPECT) and computed tomography (CT) images.[2] This is of critical significance because diagnosis of limited disease suggests the need for curative surgery.[3] SPECT/CT also improves the specificity of whole-body scan by differentiating pathologic uptake from physiologic uptake of radioiodine.[4] It helps to analyze the significance of iodine uptake in light of avidity, site, and invasion into surrounding tissues.

Section 1: Coronals

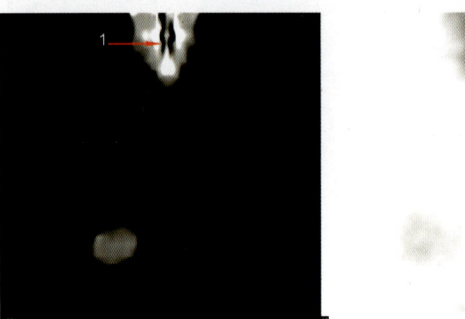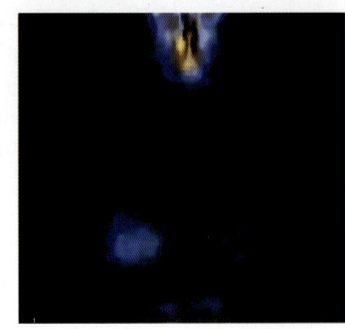

FIGURE 7-1: 131-I Coronal 1

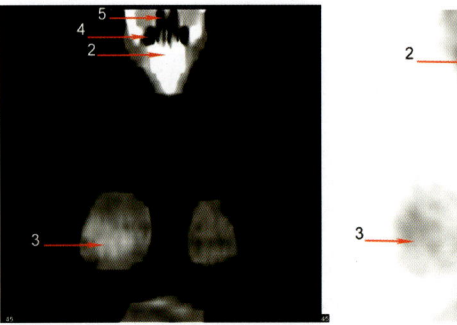

FIGURE 7-2: 131-I Coronal 2

1. Nasopharynx   3. Breast   5. Frontal sinus
2. Oral cavity   4. Maxillary sinus

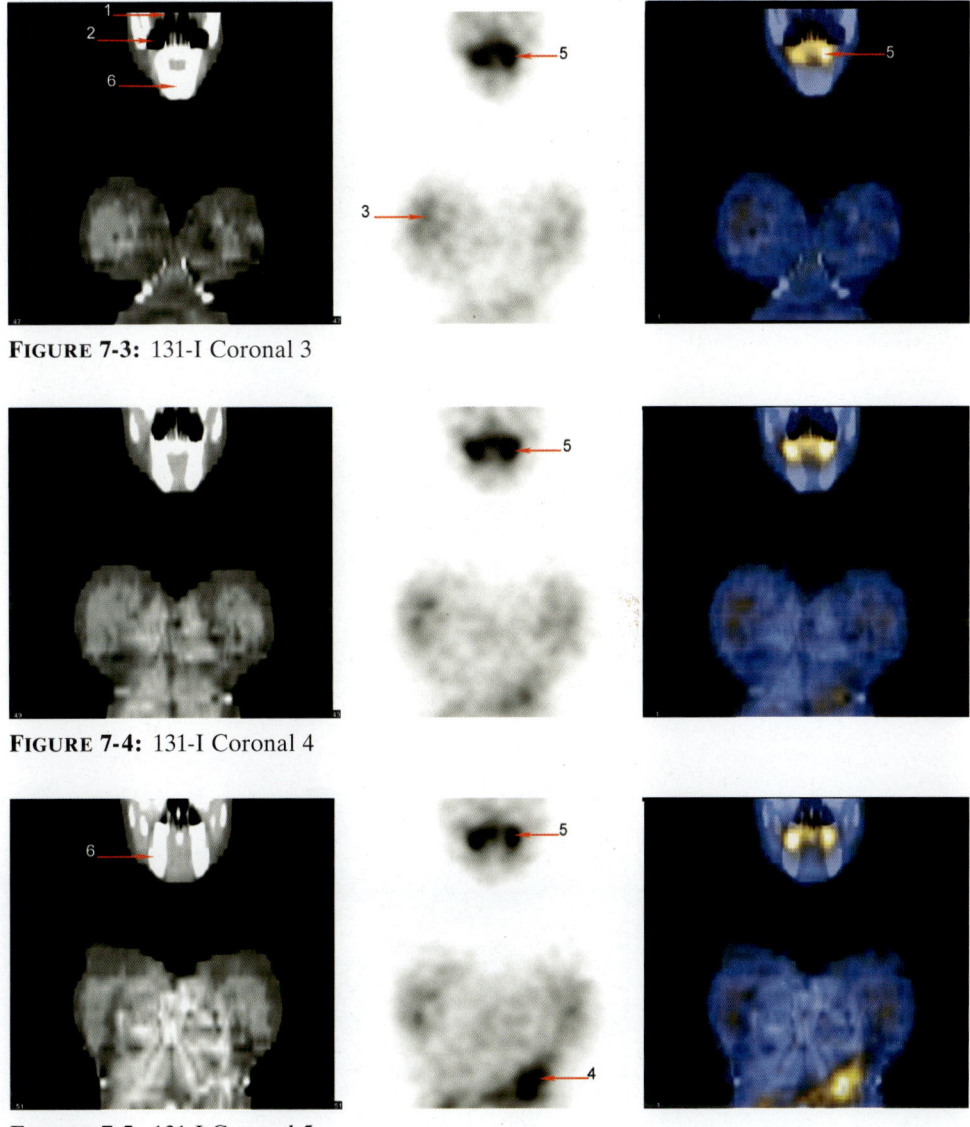

FIGURE 7-3: 131-I Coronal 3

FIGURE 7-4: 131-I Coronal 4

FIGURE 7-5: 131-I Coronal 5

1. Frontal sinus
2. Maxillary sinus
3. Breast
4. Stomach
5. Misregistered maxillary activity
6. Mandible

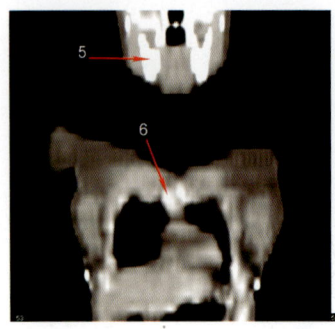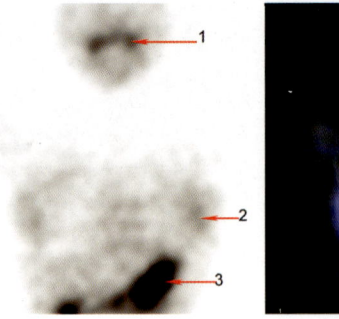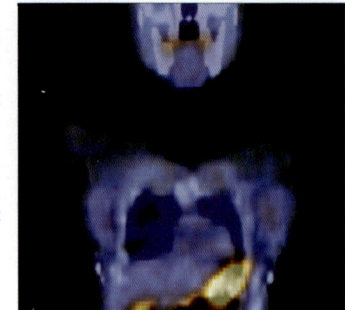

FIGURE 7-6: 131-I Coronal 6

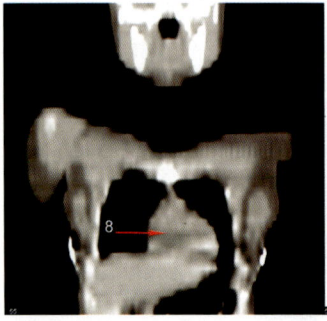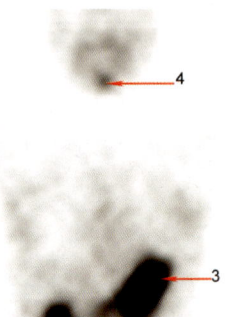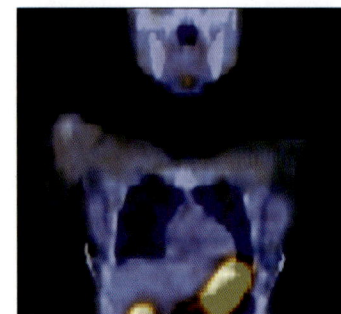

FIGURE 7-7: 131-I Coronal 7

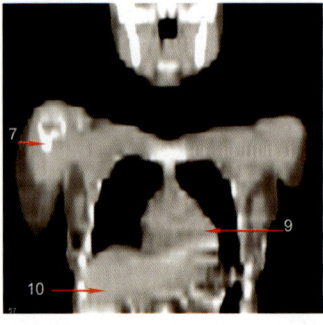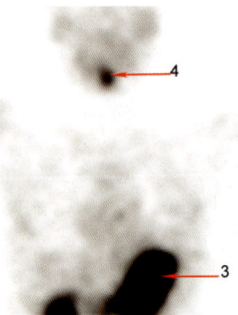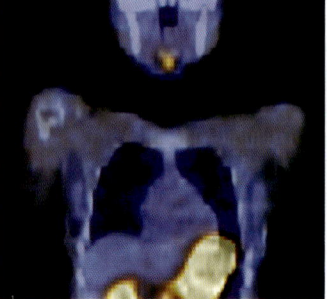

FIGURE 7-8: 131-I Coronal 8

1. Maxillary sinus	5. Mandible	9. Left ventricle
2. Breast	6. Sternum	10. Liver
3. Stomach	7. Humerus	
4. Thyroglossal duct	8. Right ventricle	

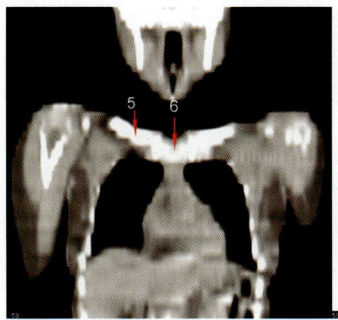
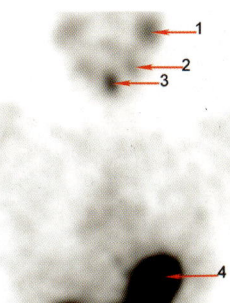
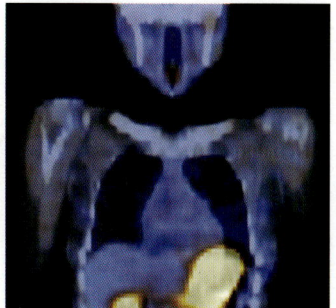

FIGURE 7-9: 131-I Coronal 9

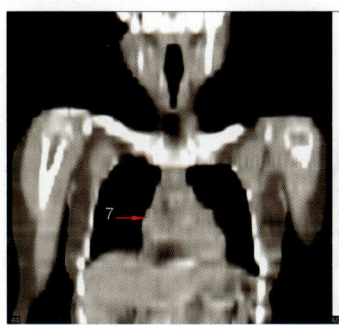

FIGURE 7-10: 131-I Coronal 10

1. Parotid gland
2. Submandibular gland
3. Thyroglossal duct
4. Stomach
5. Clavicle
6. Sternum
7. Hilum

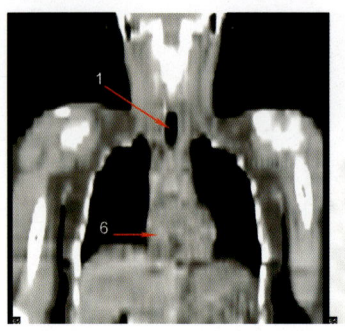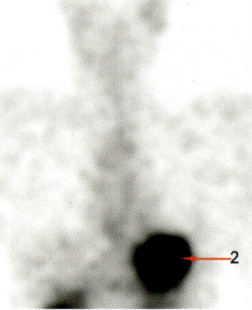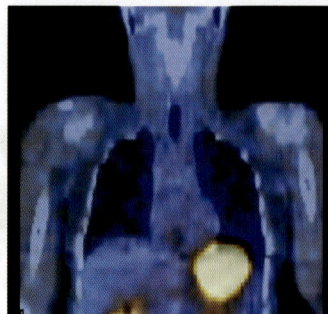

FIGURE 7-11: 131-I Coronal 11

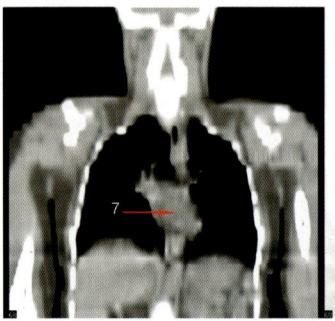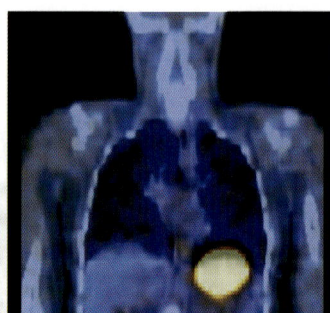

FIGURE 7-12: 131-I Coronal 12

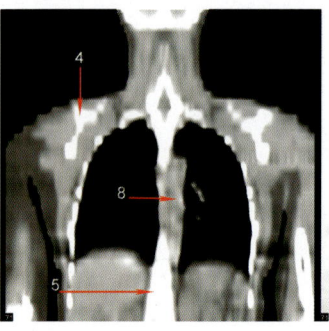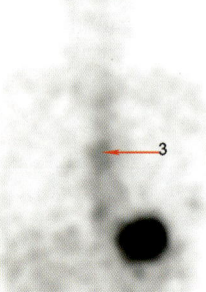

FIGURE 7-13: 131-I Coronal 13

1. Trachea	4. Scapula	7. Ascending aorta
2. Stomach	5. Spine	8. Descending aorta
3. Esophagus	6. Right ventricle	

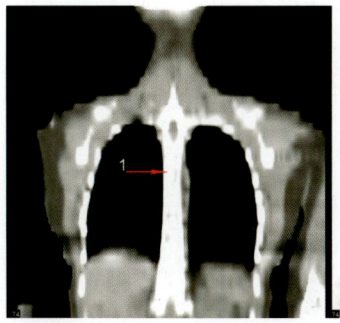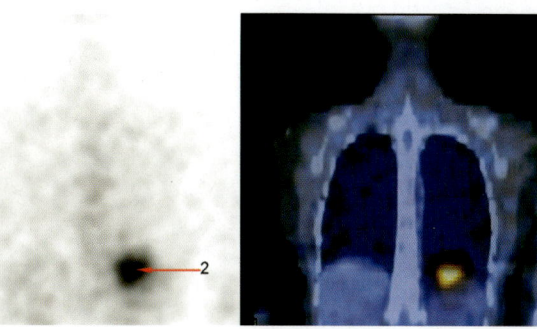

FIGURE 7-14: 131-I Coronal 14

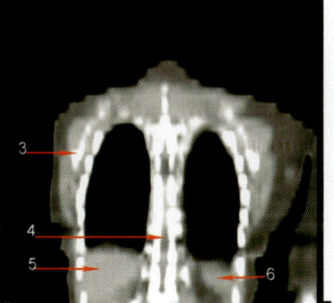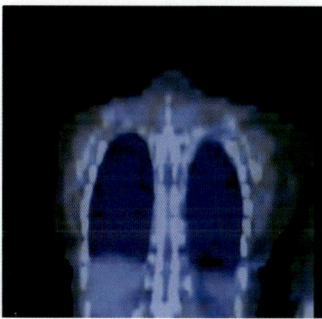

FIGURE 7-15: 131-I Coronal 15

1. Thoracic spine
2. Stomach (fundus)
3. Scapula
4. Spinal canal
5. Right liver
6. Spleen

Section 2: Sagittals

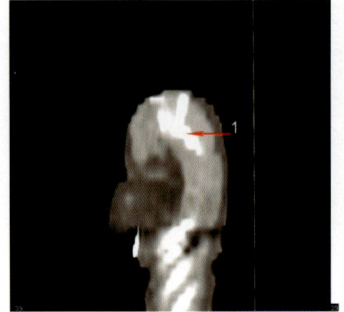  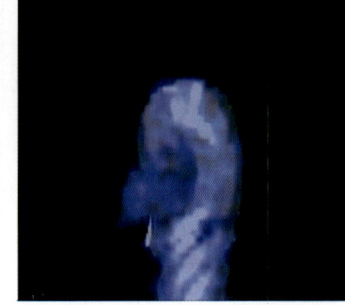

FIGURE 7-16: 131-I Sagittal 1

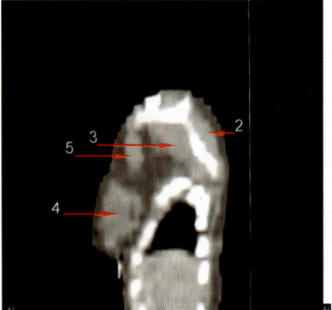 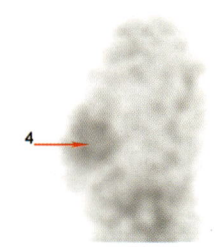 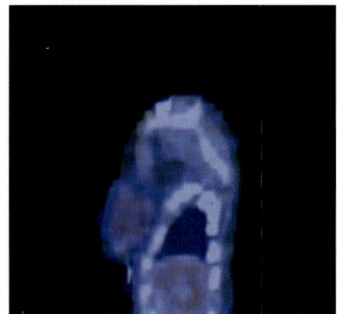

FIGURE 7-17: 131-I Sagittal 2

1. Scapula	3. Subscapularis	5. Pectoralis major muscle
2. Infraspinatus muscle	4. Breast	

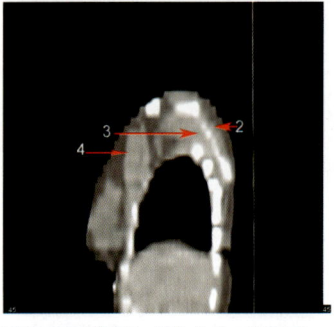 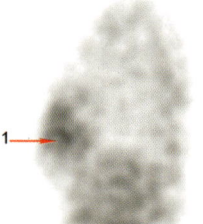 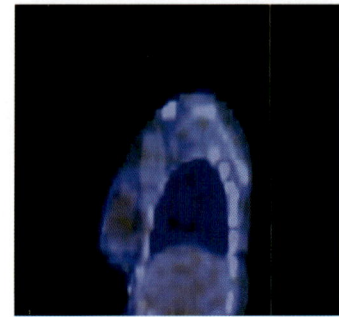

FIGURE 7-18: 131-I Sagittal 3

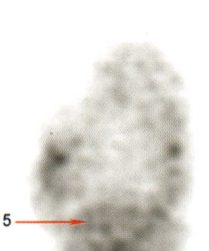 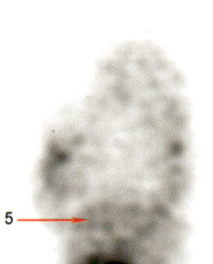 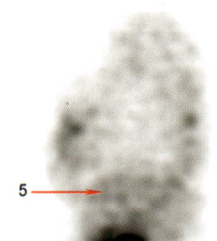 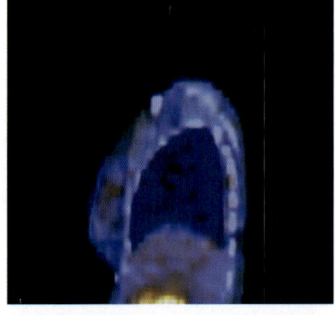

FIGURE 7-19: 131-I Sagittal 4

1. Breast	3. Subscapularis	5. Liver
2. Infraspinatus muscle	4. Pectoralis major muscle	

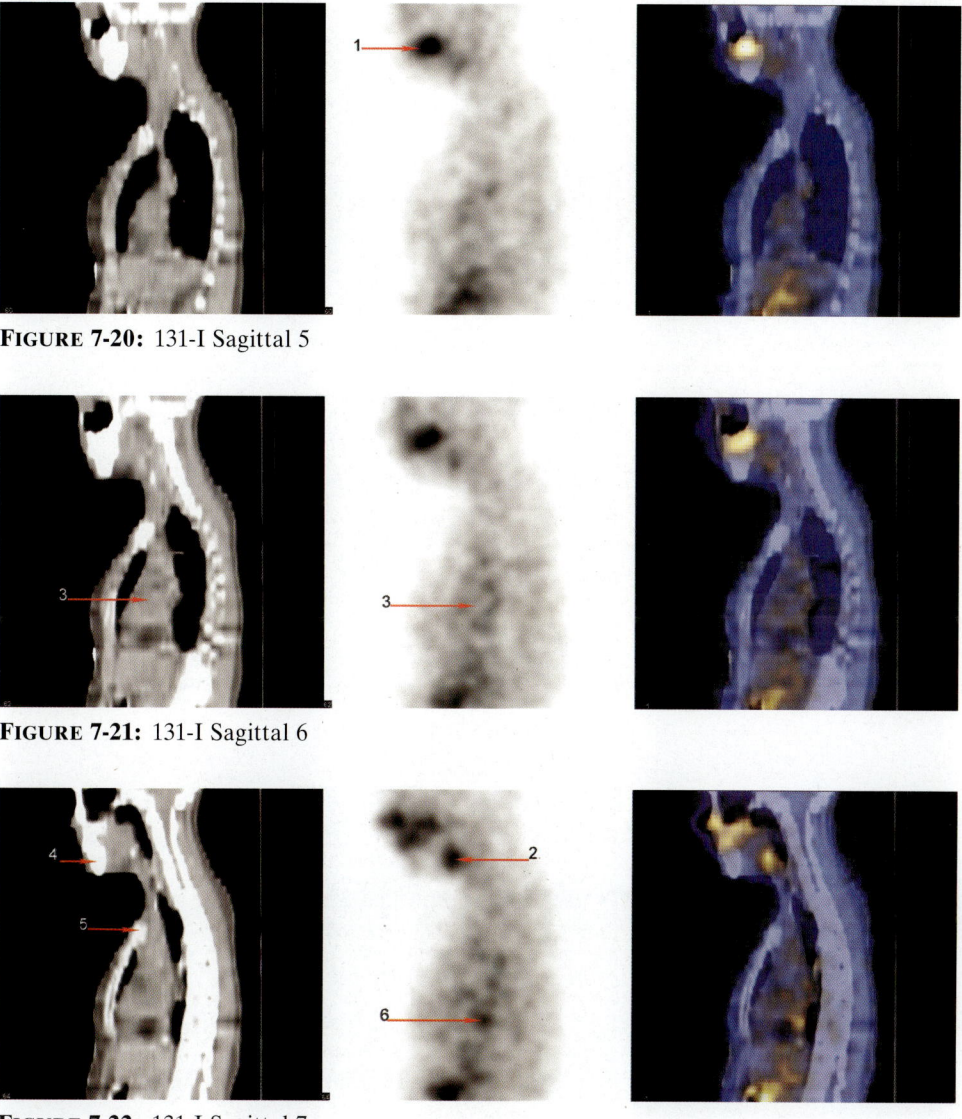

FIGURE 7-20: 131-I Sagittal 5

FIGURE 7-21: 131-I Sagittal 6

FIGURE 7-22: 131-I Sagittal 7

1. Saliva activity in oral cavity
2. Thyroglossal duct
3. Heart
4. Mandible
5. Manubrium
6. Distal esophagus

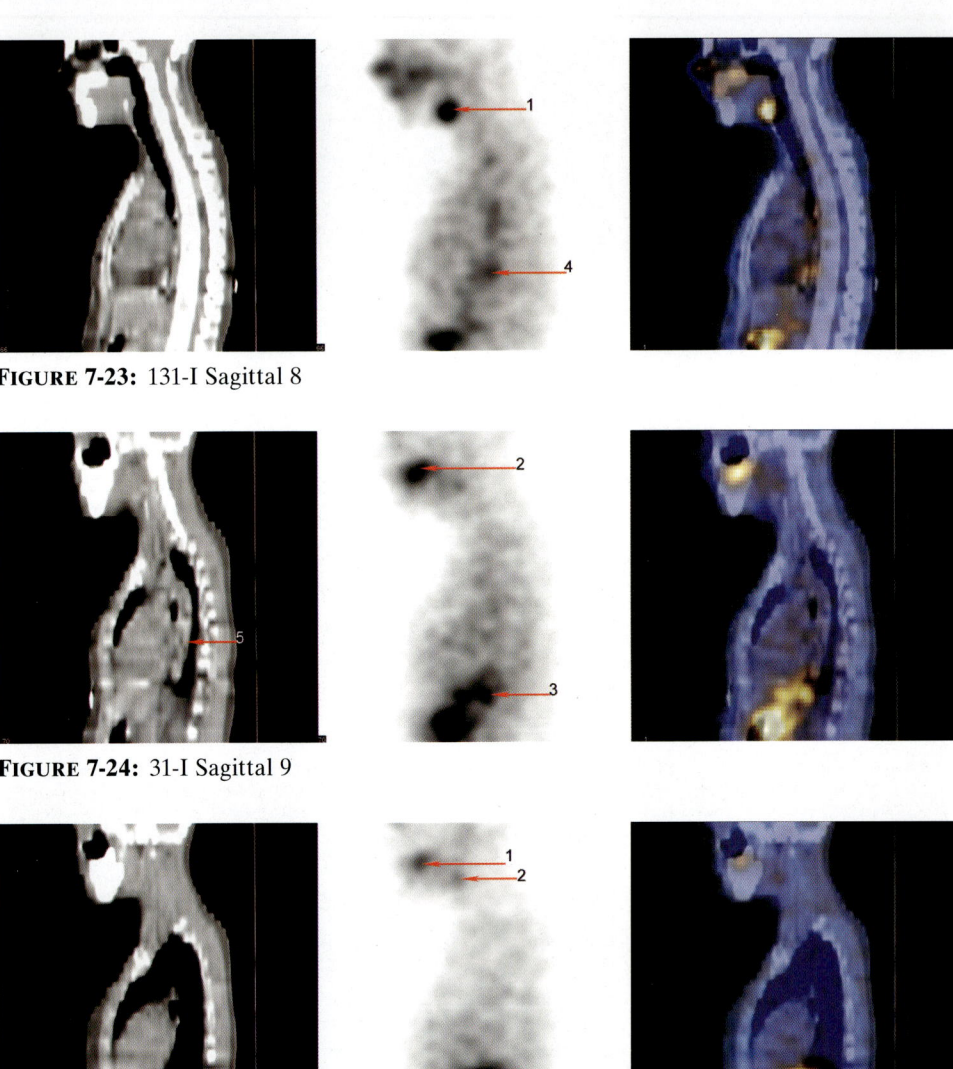

FIGURE 7-23: 131-I Sagittal 8

FIGURE 7-24: 31-I Sagittal 9

FIGURE 7-25: 131-I Sagittal 10

1. Thyroglossal duct
2. Saliva activity in oral cavity
3. Stomach
4. Distal esophagus
5. Descending aorta

Section 3: Transaxials

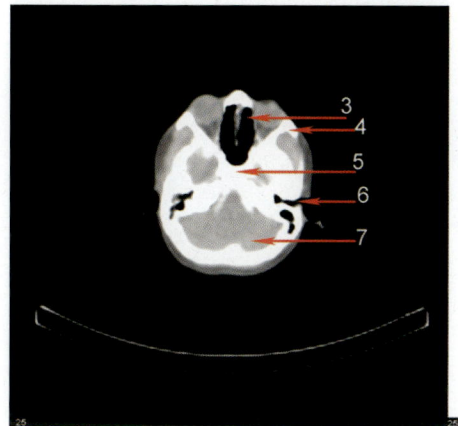
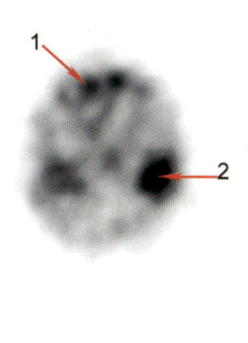
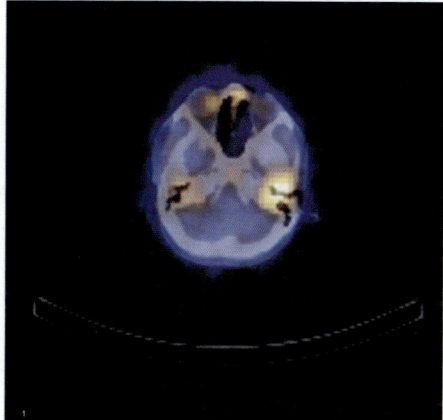

FIGURE 7-26: 131-I Transaxial 1

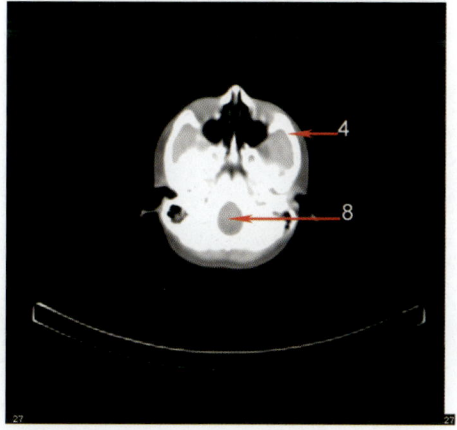
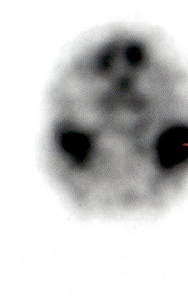

FIGURE 7-27: 131-I Transaxial 2

1. Orbit
2. Parotid gland
3. Nasopharynx
4. Zygomatic arch
5. Dorsum sella
6. External auditory canal
7. Cerebellum
8. Spinal canal

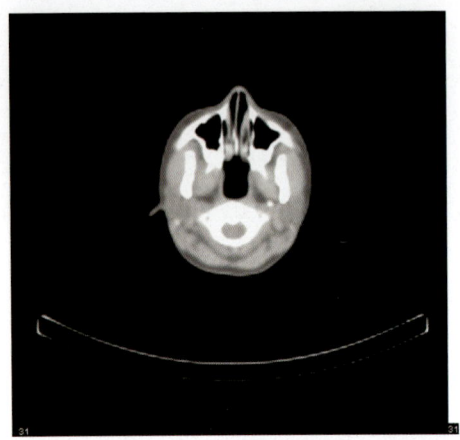
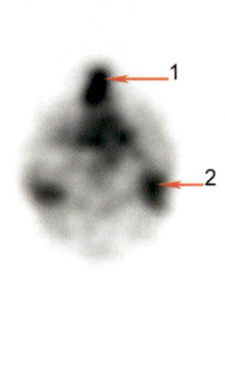
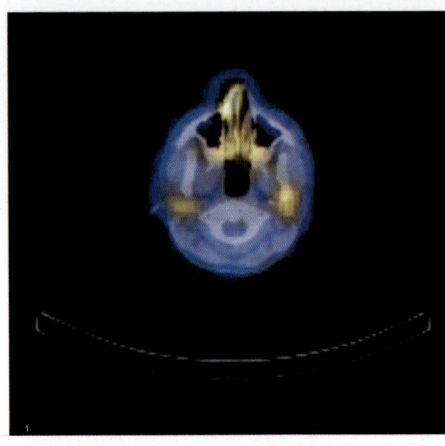

FIGURE 7-28: 131-I Transaxial 3

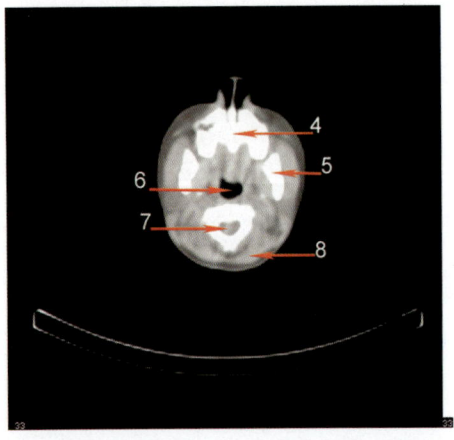
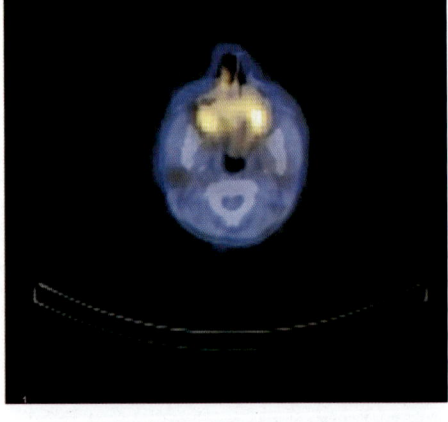

FIGURE 7-29: 131-I Transaxial 4

1. Nasal cavity
2. Parotid gland
3. Maxillary sinus activity
4. Maxilla
5. Mandible
6. Oropharynx
7. Spinal cord
8. Splenius capitis muscle

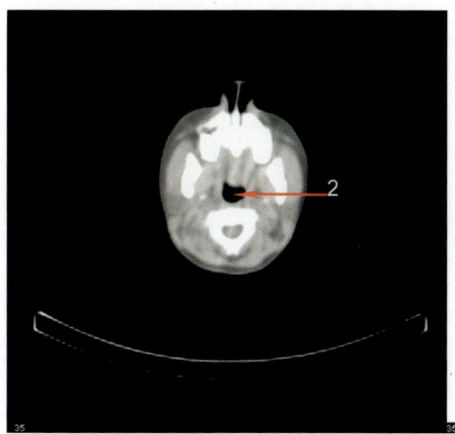
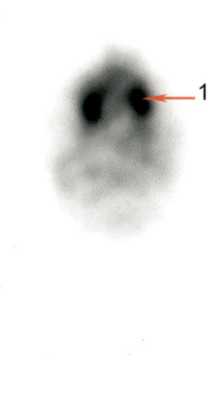
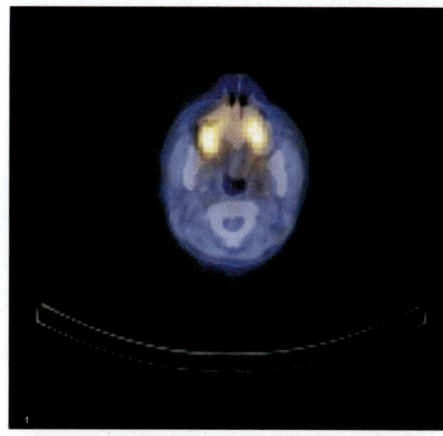

FIGURE 7-30: 131-I Transaxial 5

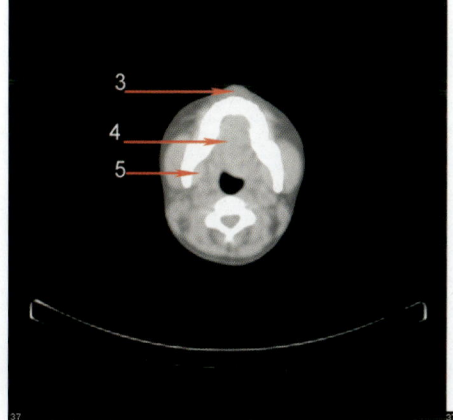
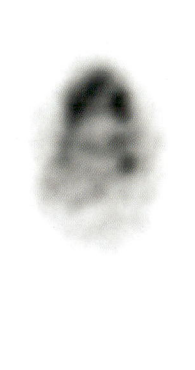
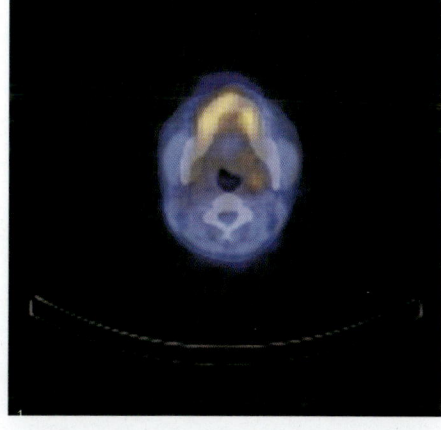

FIGURE 7-31: 131-I Transaxial 6

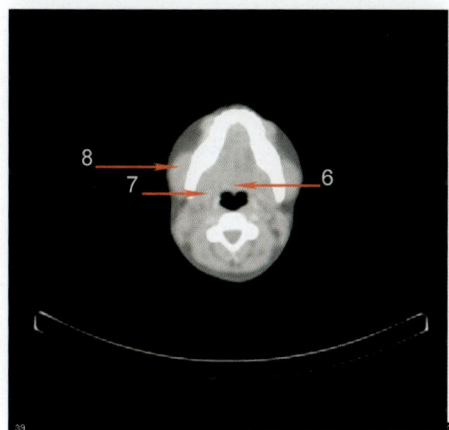
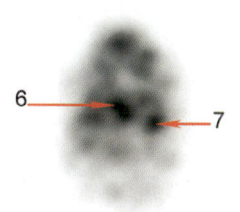
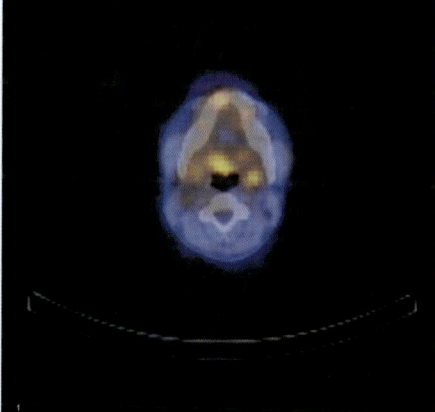

FIGURE 7-32: 131-I Transaxial 7

1. Maxillary sinus activity
2. Oropharynx
3. Periodontal muscle
4. Sublingual gland
5. Submandibular gland
6. Sublingual tonsil
7. Palatine tonsil
8. Masseter muscle

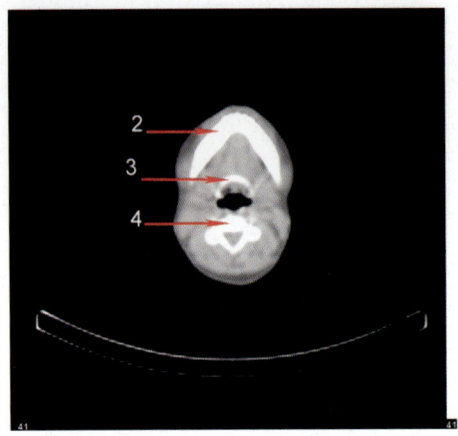
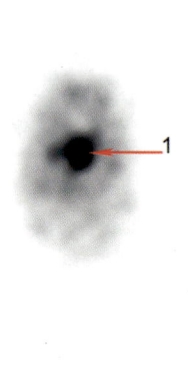
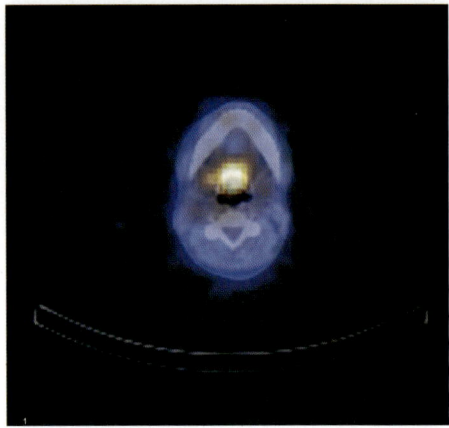

FIGURE 7-33: 131-I Transaxial 8

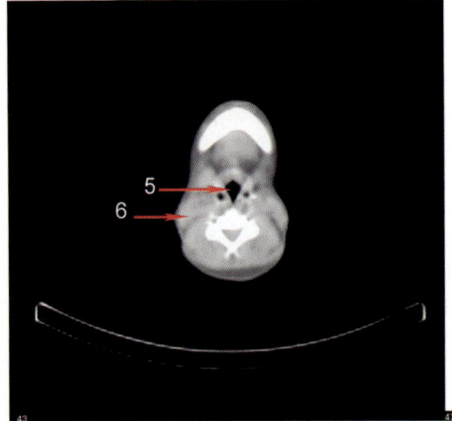
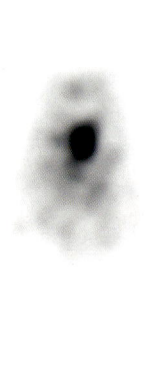
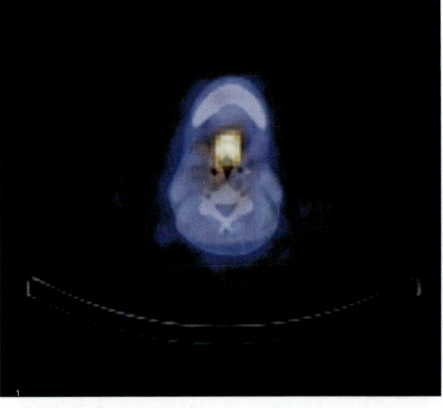

FIGURE 7-34: 131-I Transaxial 9

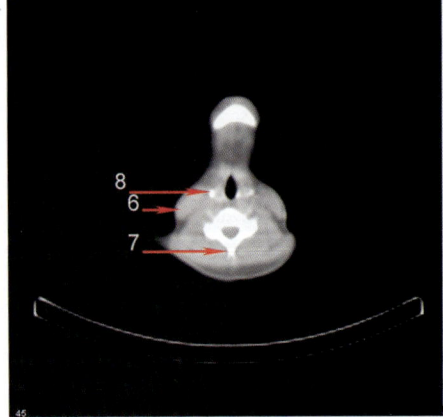
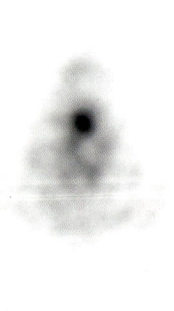

FIGURE 7-35: 131-I Transaxial 10

1. Thyroglossal duct
2. Mandible
3. Hyoid
4. Cervical vertebral body
5. Trachea
6. Sternocleidomastoid muscle
7. Spinous process of cervical spine
8. Cricoid cartilage

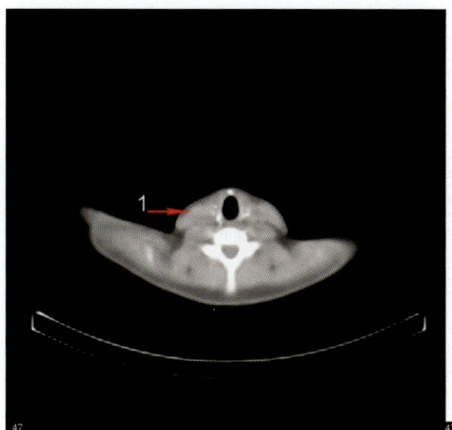
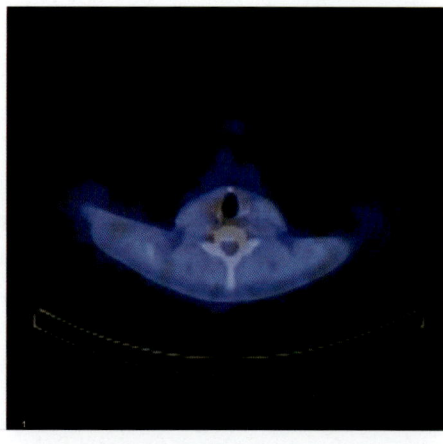

FIGURE 7-36: 131-I Transaxial 11

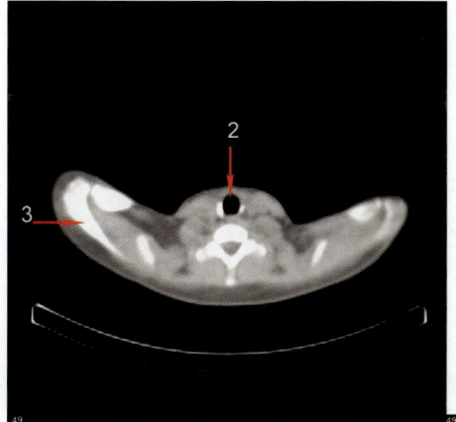
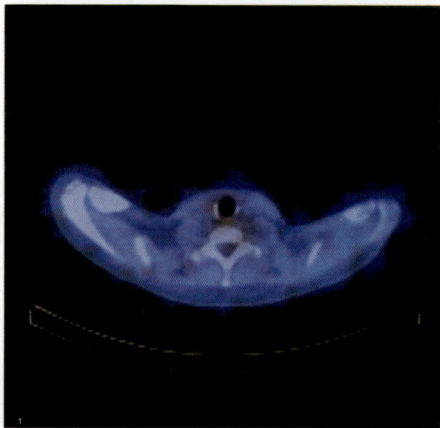

FIGURE 7-37: 131-I Transaxial 12

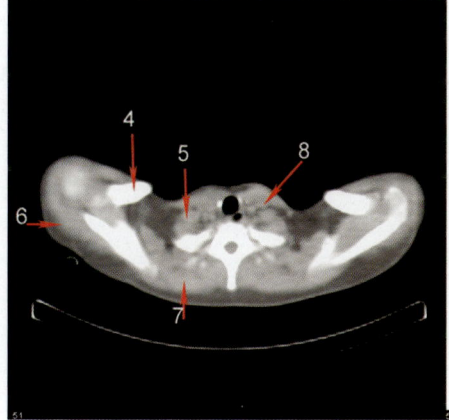
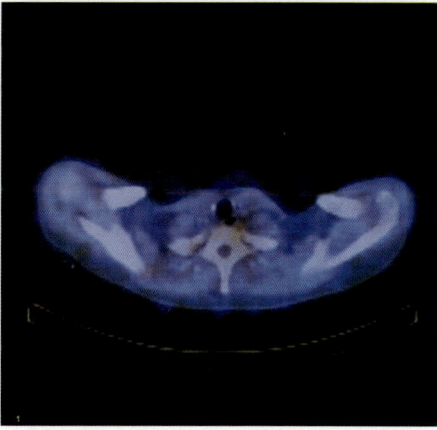

FIGURE 7-38: 131-I Transaxial 13

1. Sternocleidomastoid muscle
2. Trachea
3. Scapula
4. Clavicle
5. Scalene muscle
6. Deltoid muscle
7. Trapezius muscle
8. Internal jugular vein

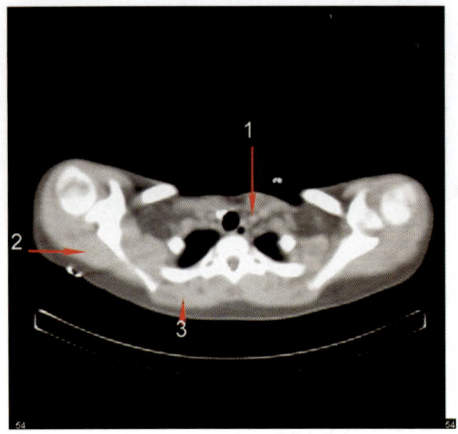 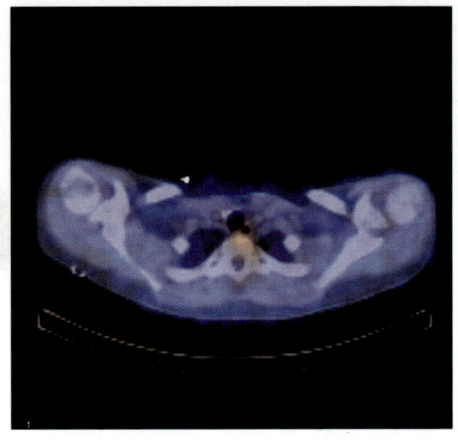

FIGURE 7-39: 131-I Transaxial 14

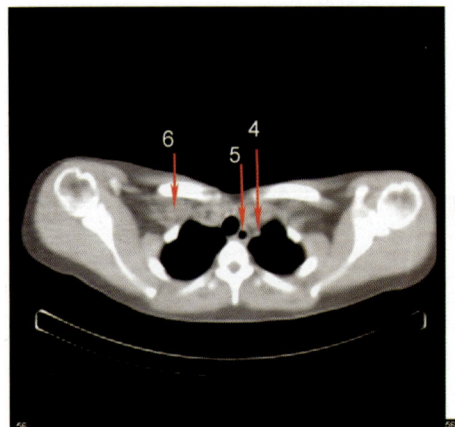 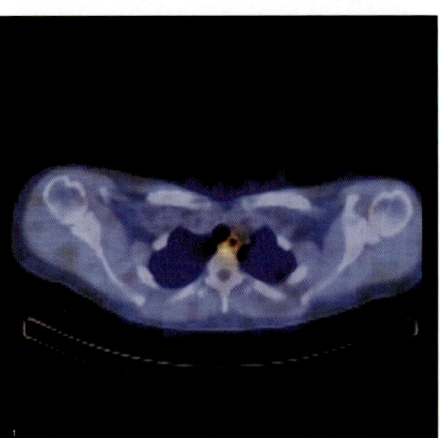

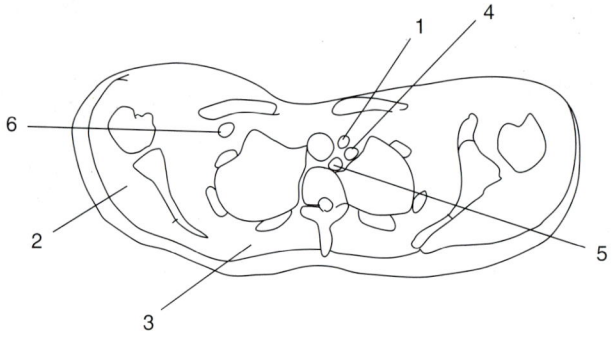

FIGURE 7-40: 131-I Transaxial 15

1. Left common carotid artery
2. Deltoid muscle
3. Trapezius muscle
4. Left subclavian artery
5. Esophagus
6. Right jugular vein

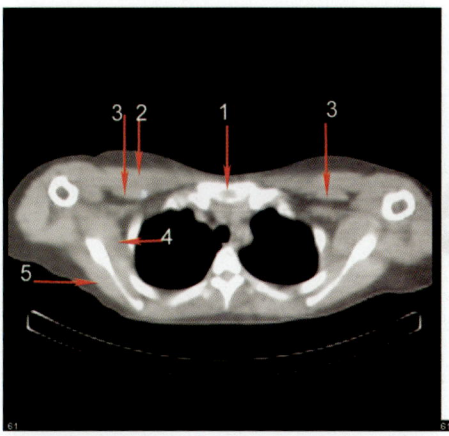
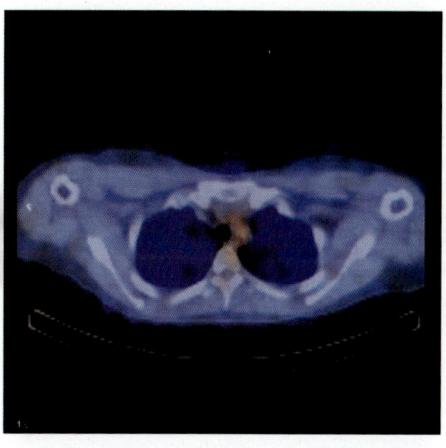

FIGURE 7-41: 131-I Transaxial 16

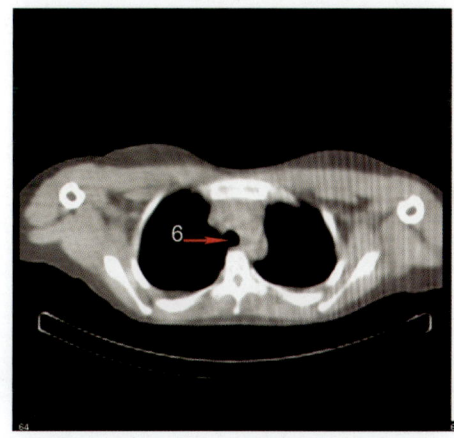
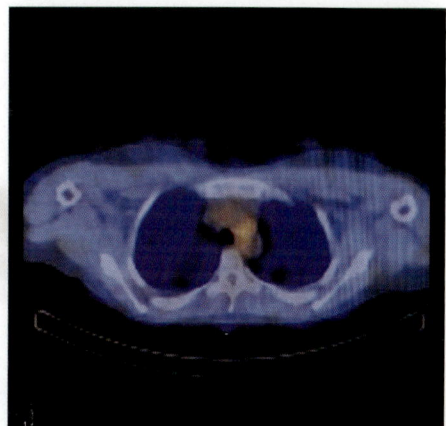
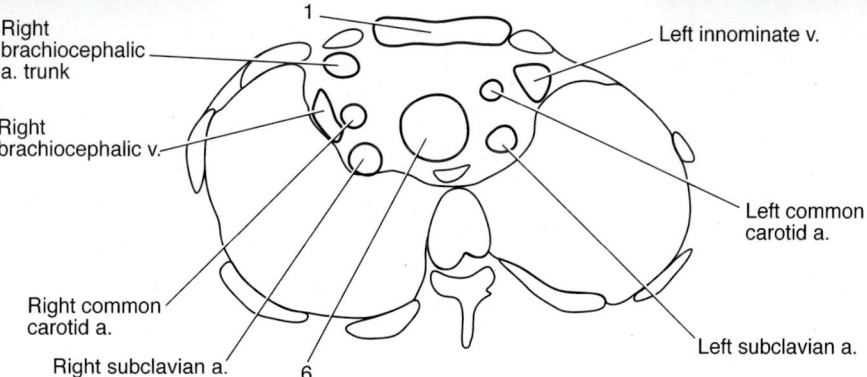

FIGURE 7-42: 131-I Transaxial 17

1. Manubrium	3. Pectoralis minor muscle	5. Infraspinatus muscle
2. Pectoralis major muscle	4. Subscapularis muscle	6. Trachea

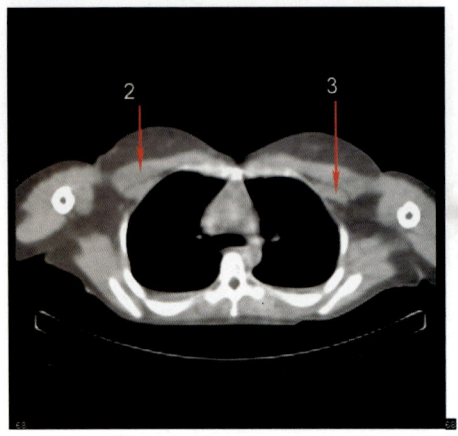
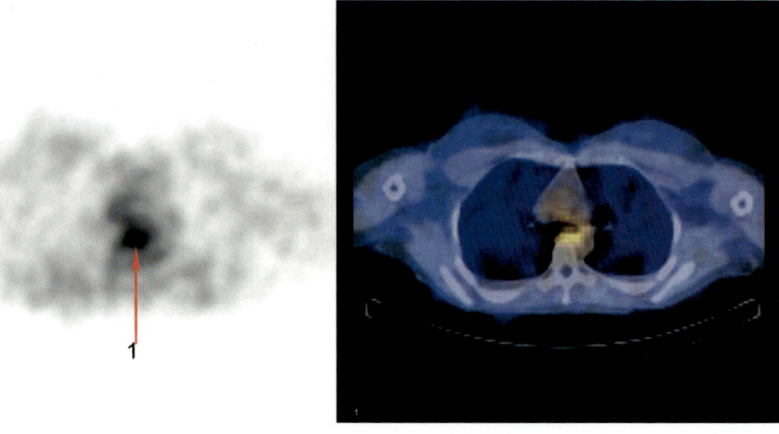

FIGURE 7-43: 131-I Transaxial 18

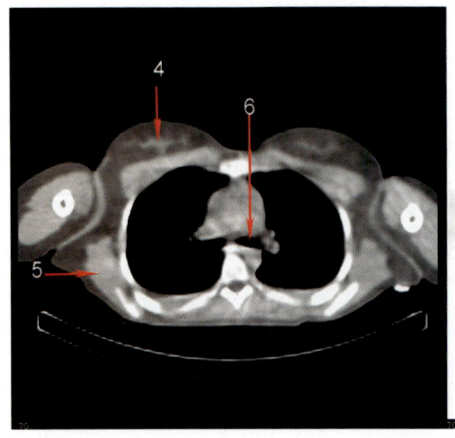
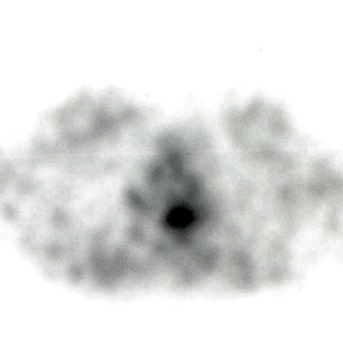
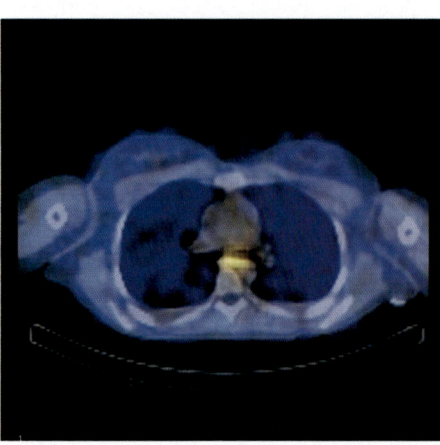

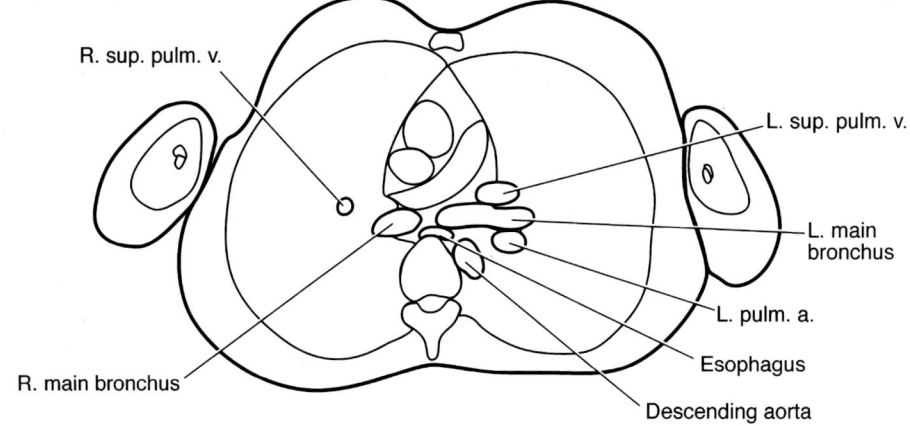

R. sup. pulm. v.

L. sup. pulm. v.

L. main bronchus

L. pulm. a.

Esophagus

Descending aorta

R. main bronchus

FIGURE 7-44: 131-I Transaxial 19

1. Esophagus
2. Pectoralis major muscle
3. Pectoralis minor muscle
4. Breast
5. Subscapularis muscle
6. Left main bronchus

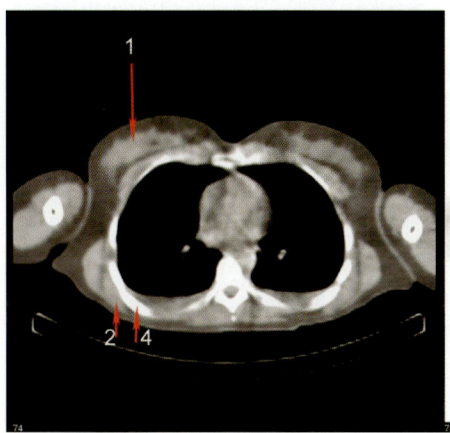
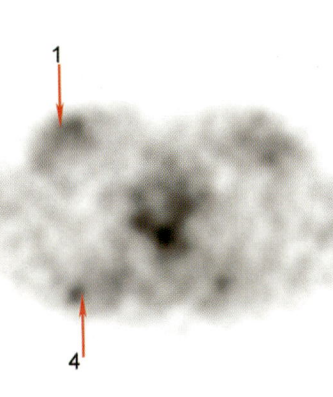
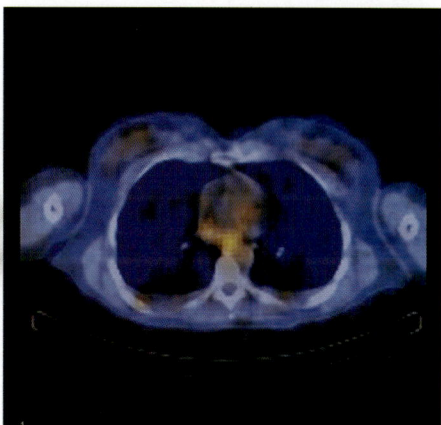

FIGURE 7-45: 131-I Transaxial 20

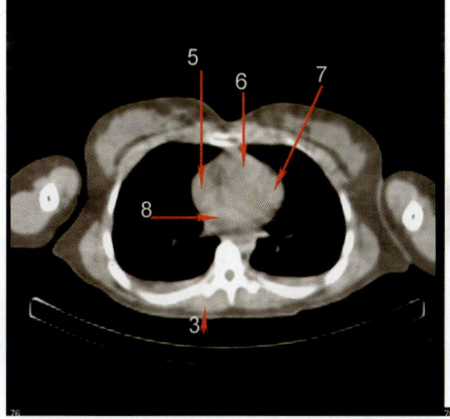
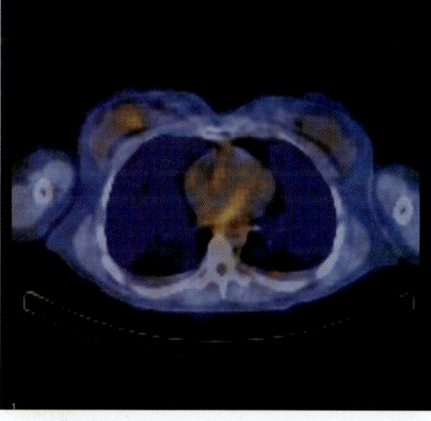

FIGURE 7-46: 131-I Transaxial 21

1. Breast
2. Rhomboideus major muscle
3. Erector spinae muscle
4. Rib
5. Right atrium
6. Right ventricle
7. Left ventricle
8. Left atrium

8 MIBG SPECT/CT

Pheochromocytoma and neuroblastoma are the most common tumors arising from the adrenergic nervous system. Ten percent of pheochromocytoma are malignant, 10% of cases are bilateral, and 10% of adult cases and 30% of pediatric cases have extraadrenal lesions.[1] Adrenal pheochromocytomas are easily visualized on computed tomography (CT) or magnetic resonance imaging (MRI) at diagnosis. Radiolabeled metaiodobenzylguanidine (MIBG) scan provides the functional characterization of the tumor and the detection of extraadrenal spread with 80% sensitivity and 96% specificity.[2] A detailed anatomic evaluation of MIBG-avid foci can be performed by hybrid single photon emission computed tomography (SPECT)/CT or fusion of MIBG-SPECT with MRI.[3] Hybrid imaging provides a better localization of tumors and the detection of bone and bone marrow involvement. It improves the delineation of physiologic activities, thus alleviating the need for additional imaging. It may help to characterize a residual viable mass seen on CT or MRI and also differentiate MIBG uptake in hyperplastic adrenal gland following contralateral adrenalectomy from retroperitoneal recurrence.[4]

Section 1: Coronals

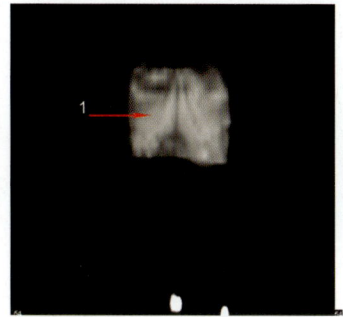 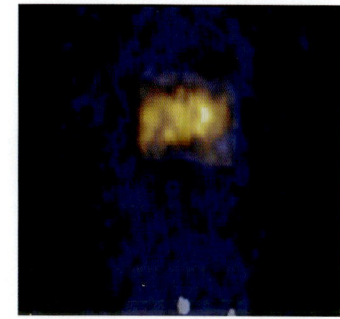

FIGURE 8-1: MIBG Coronal 1

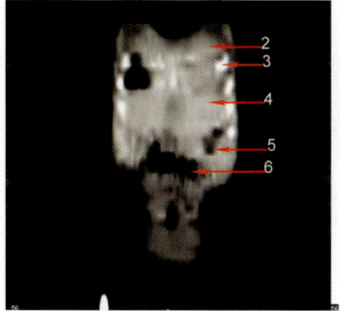 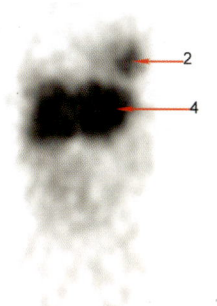

FIGURE 8-2: MIBG Coronal 2

1. Right hepatic lobe
2. Apex of left ventricle
3. Rib
4. Left hepatic lobe
5. Stomach
6. Transverse colon

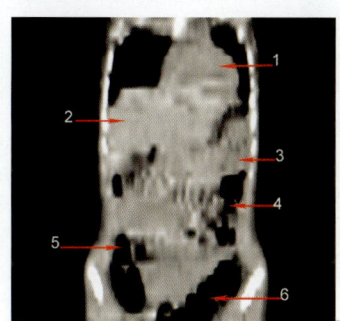 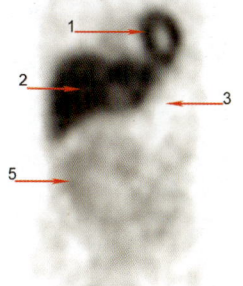 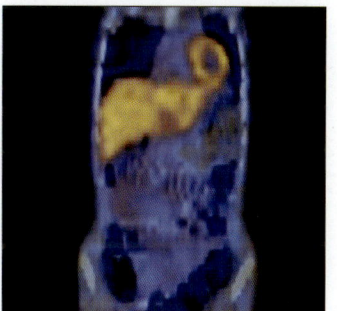

FIGURE 8-3: MIBG Coronal 3

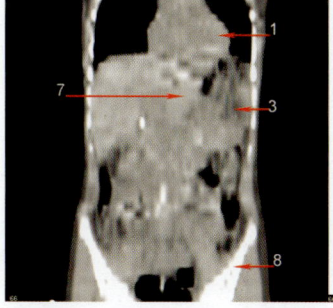 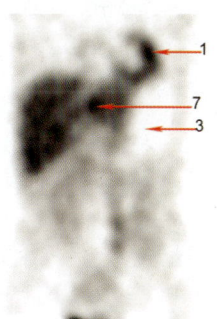 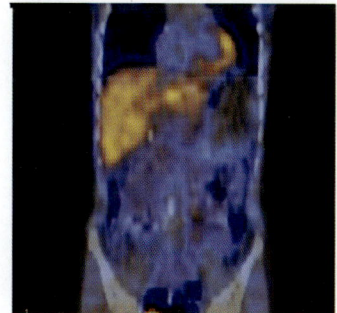

FIGURE 8-4: MIBG Coronal 4

1. Left ventricle
2. Right hepatic lobe
3. Stomach
4. Descending colon
5. Ascending colon
6. Sigmoid colon
7. Left hepatic lobe
8. Ilium

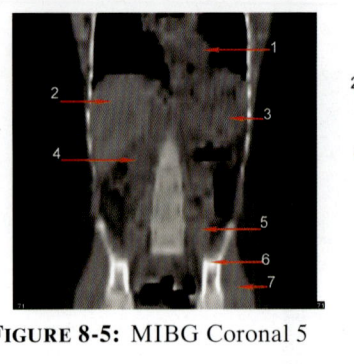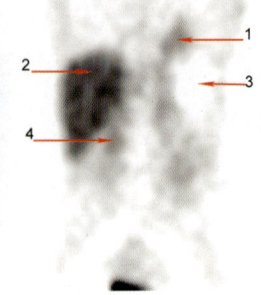

FIGURE 8-5: MIBG Coronal 5

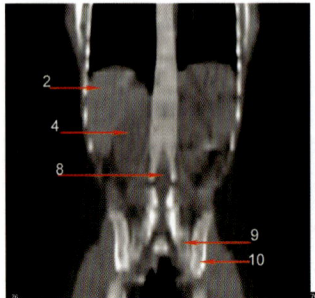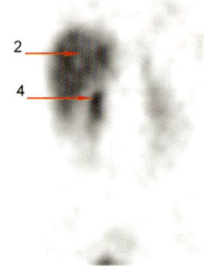

FIGURE 8-6: MIBG Coronal 6

1. Left ventricle
2. Right hepatic lobe
3. Stomach
4. Right kidney
5. Psoas muscle
6. Ilium
7. Gluteus muscle
8. Spinal canal
9. Sacral ala
10. Iliac tuberosity

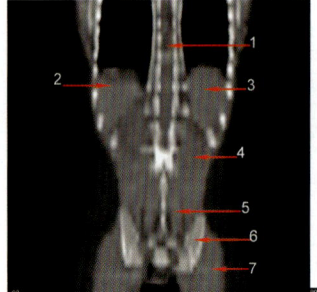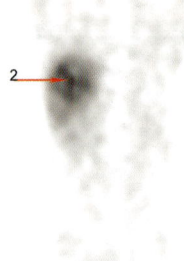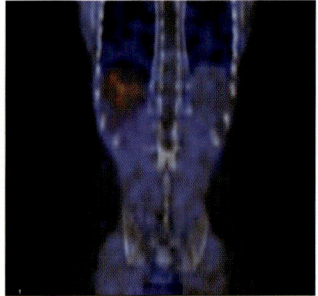

FIGURE 8-7: MIBG Coronal 7

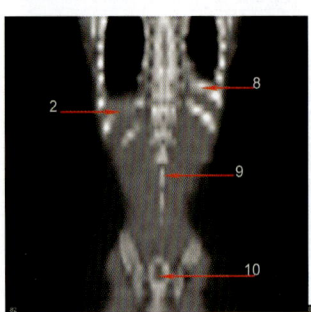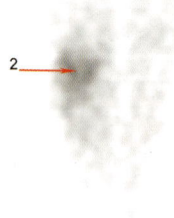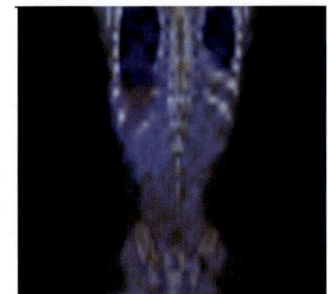

FIGURE 8-8: MIBG Coronal 8

1. Thoracic spinal canal
2. Right hepatic lobe
3. Spleen
4. Quadratus lumborum muscle
5. Erector spinae muscle
6. Iliac tuberosity
7. Gluteus muscle
8. Rib
9. Spinous process of L2
10. Sacral foramen

Section 2: Sagittals

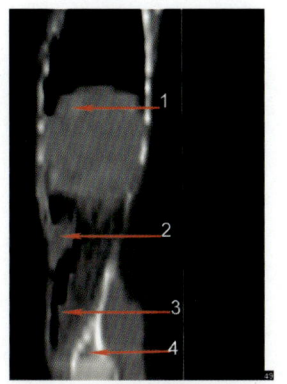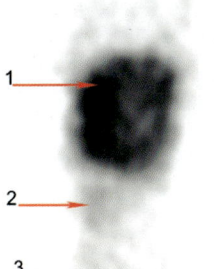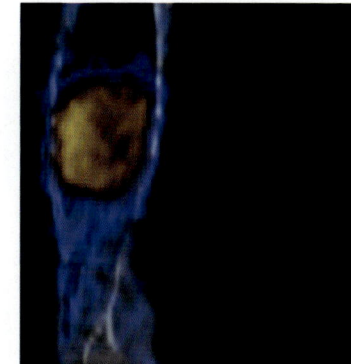

FIGURE 8-9: MIBG Sagittal 1

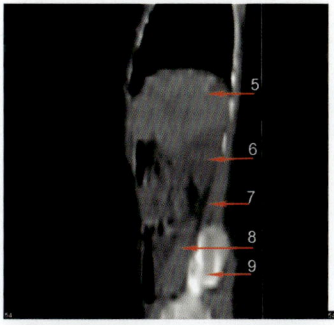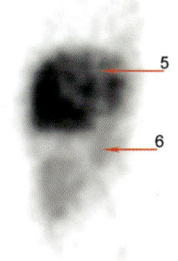

FIGURE 8-10: MIBG Sagittal 2

1. Right hepatic lobe (segment VIII)
2. Transverse colon
3. Ascending colon
4. Acetabulum (ilium)
5. Right hepatic lobe (segment VII)
6. Kidney
7. Quadratus lumborum muscle
8. Iliopsoas muscle
9. Iliac tuberosity

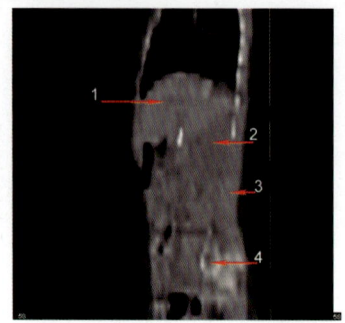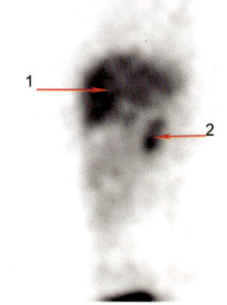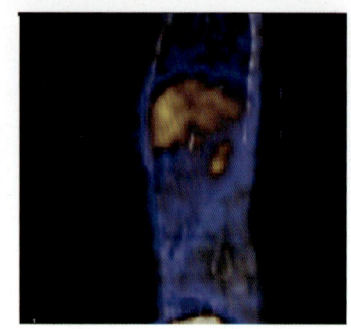

FIGURE 8-11: MIBG Sagittal 3

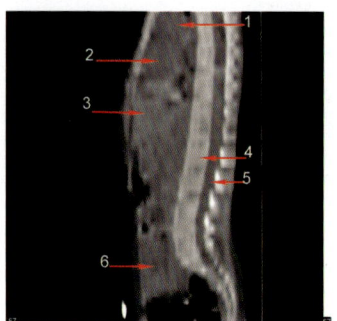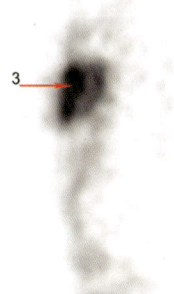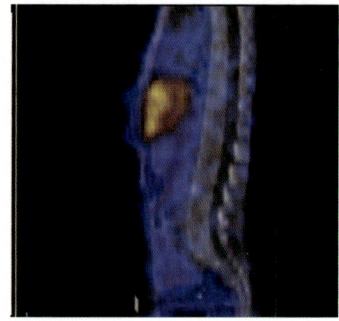

FIGURE 8-12: MIBG Sagittal 4

1. Right hepatic lobe	4. Iliac tuberosity	7. Ileum
2. Kidney	5. Left hepatic lobe	8. Sigmoid colon
3. Erector spinae muscle	6. L1 vertebral body	

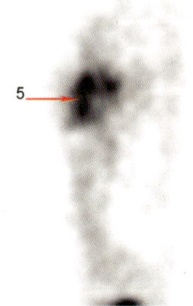

FIGURE 8-13: MIBG Sagittal 5

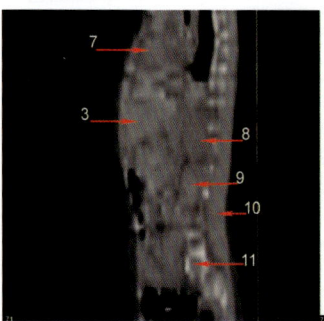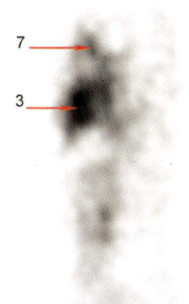

FIGURE 8-14: MIBG Sagittal 6

1. Aortic arch	5. Spinal canal	9. Psoas muscle
2. Right ventricle	6. Ileum	10. Erector spinae muscle
3. Left hepatic lobe	7. Left ventricle	11. Iliac tuberosity
4. L1 vertebral body	8. Kidney	

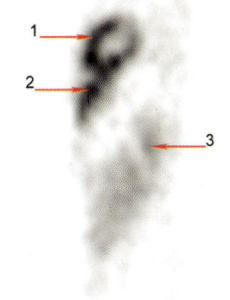 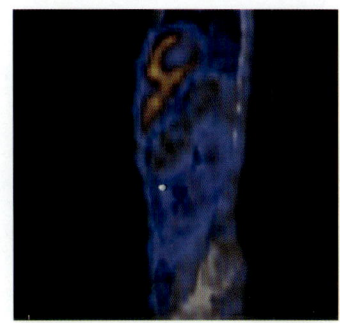

FIGURE 8-15: MIBG Sagittal 7

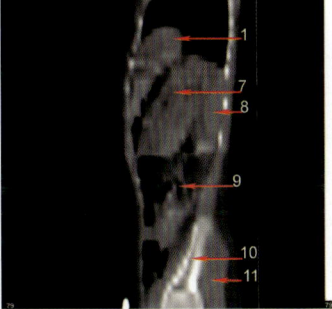 

FIGURE 8-16: MIBG Sagittal 8

1. Left ventricle
2. Left hepatic lobe
3. Kidney
4. Quadratus lumborum muscle
5. Iliacus muscle
6. Iliac tuberosity
7. Stomach
8. Spleen
9. Jejunum
10. Iliac crest
11. Gluteus muscle

Section 3: Transaxials

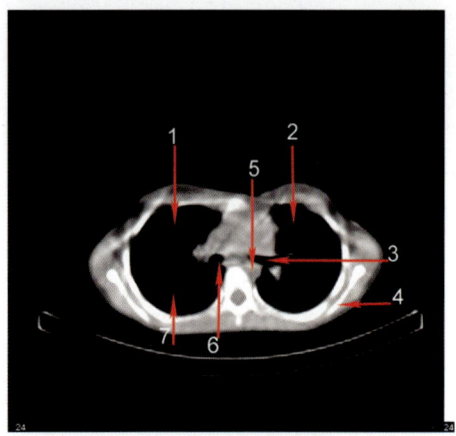 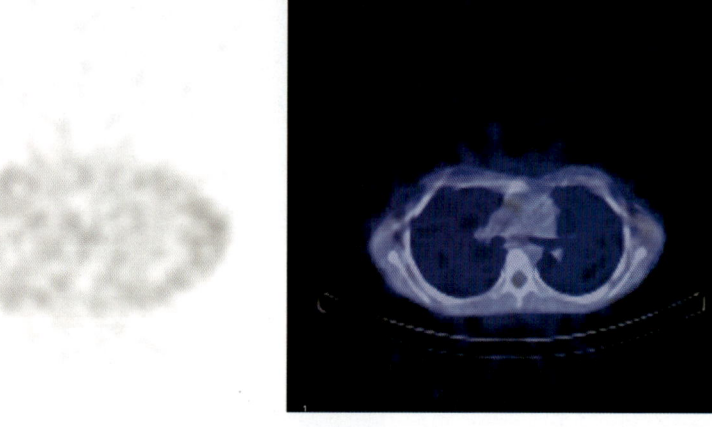

FIGURE 8-17: MIBG Transaxial 1

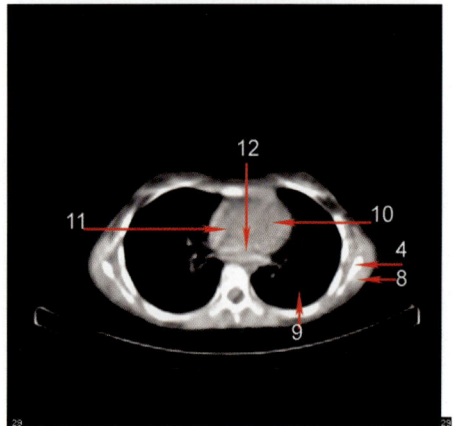 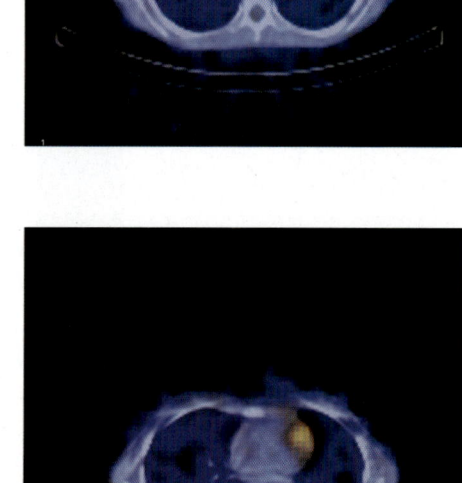

FIGURE 8-18: MIBG Transaxial 2

1. Right upper lobe
2. Left upper lobe
3. Left main bronchus
4. Left scapula
5. Descending aorta
6. Right main bronchus
7. Right lower lobe
8. Teres major muscle
9. Left lower lobe
10. Left ventricle
11. Right atrium
12. Left atrium

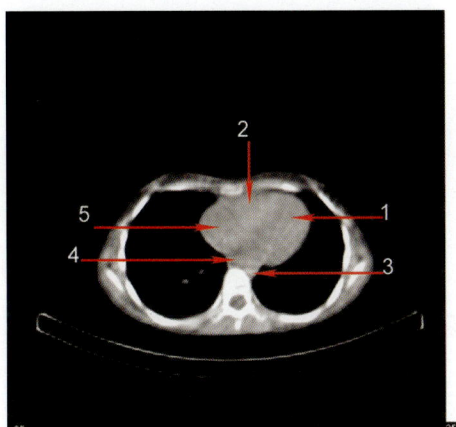 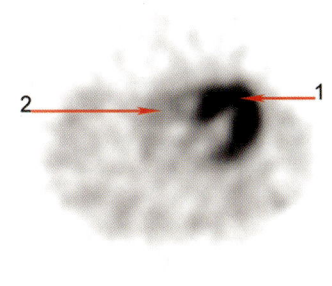 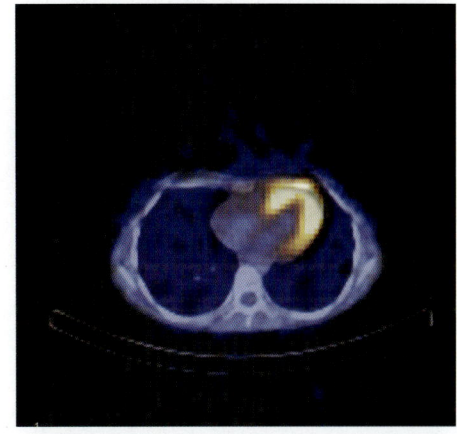

FIGURE 8-19: MIBG Transaxial 3

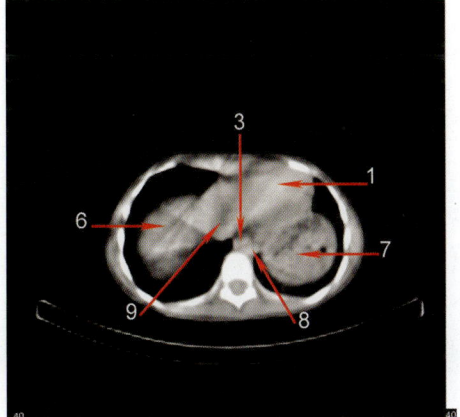 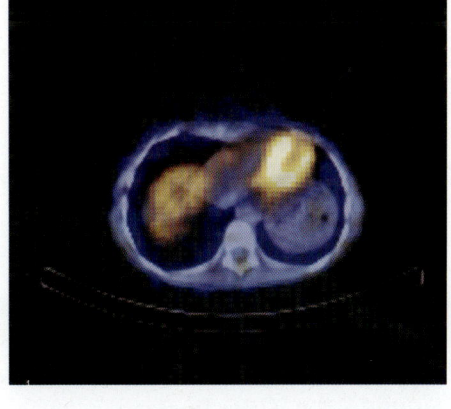

FIGURE 8-20: MIBG Transaxial 4

1. Left ventricle
2. Right ventricle
3. Descending aorta
4. Esophagus
5. Right atrium
6. Hepatic dome (cupula of right diaphragm)
7. Stomach
8. Hemiazygos vein
9. Inferior vena cava

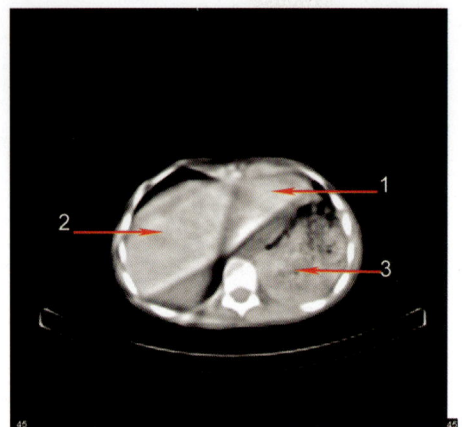
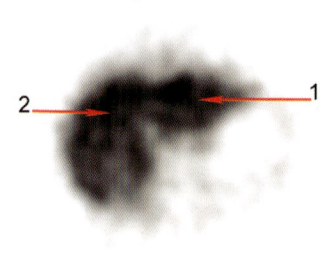
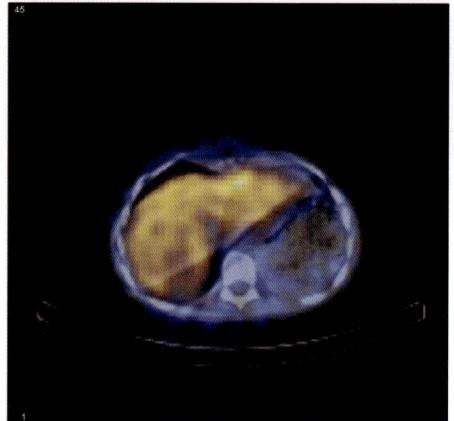

FIGURE 8-21: MIBG Transaxial 5

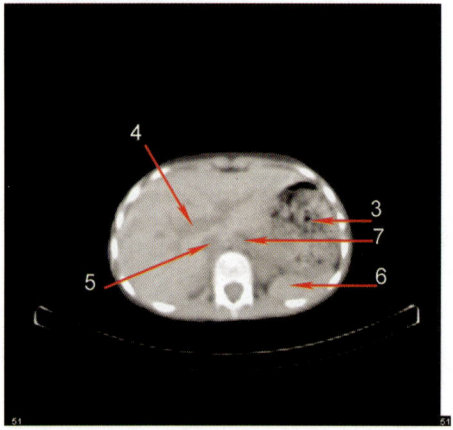
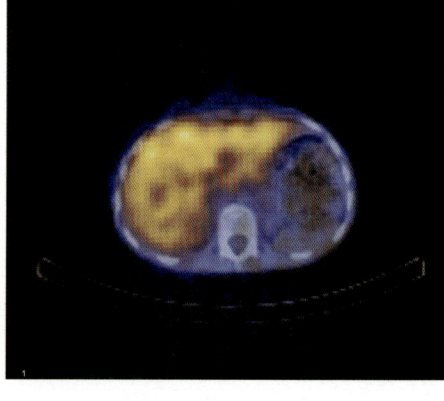

FIGURE 8-22: MIBG Transaxial 6

1. Left hepatic lobe
2. Right hepatic lobe
3. Stomach
4. Portal vein
5. Inferior vena cava
6. Spleen
7. Abdominal aorta

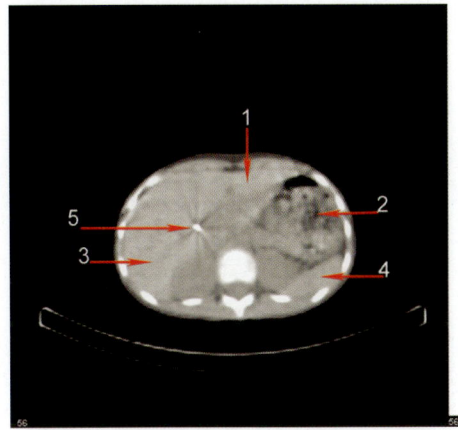
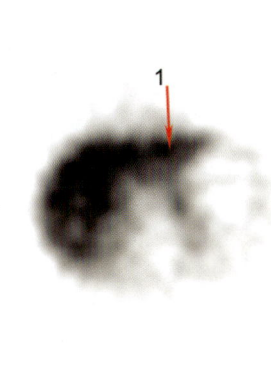
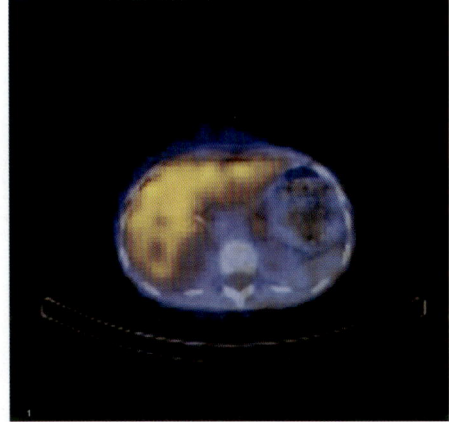

FIGURE 8-23: MIBG Transaxial 7

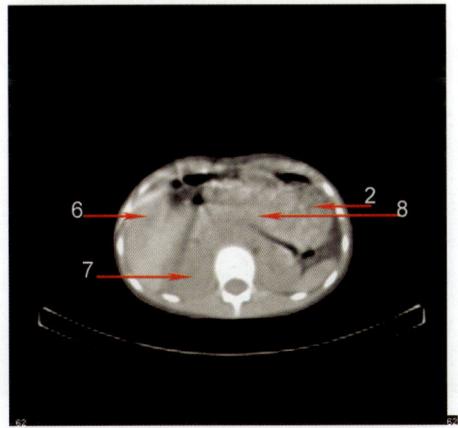
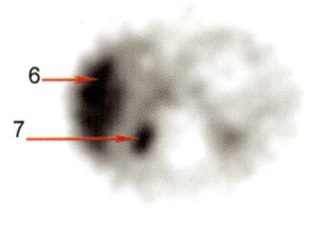
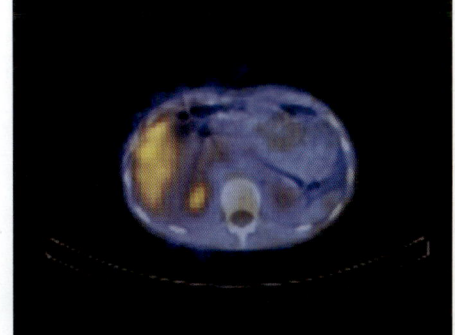

FIGURE 8-24: MIBG Transaxial 8

1. Left hepatic lobe (segment III)
2. Stomach
3. Right hepatic lobe (segment VI)
4. Spleen
5. Surgical clip
6. Right hepatic lobe (segment V)
7. Right kidney
8. Pancreas

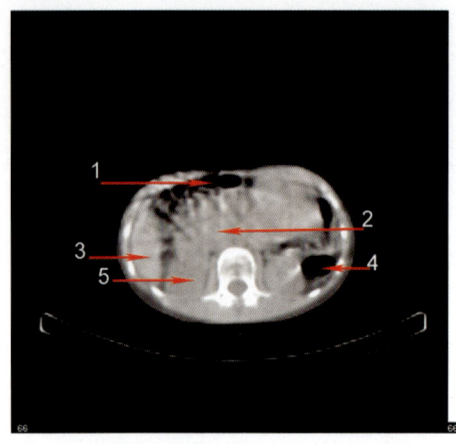
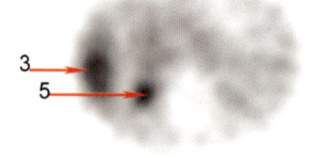
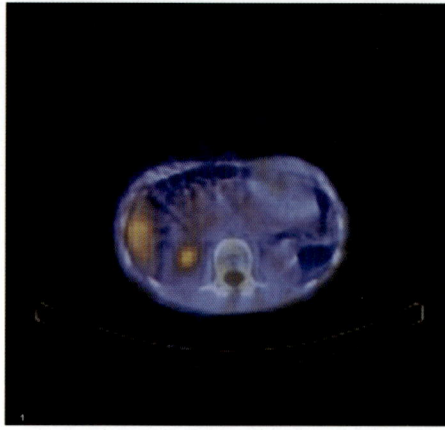

FIGURE 8-25: MIBG Transaxial 9

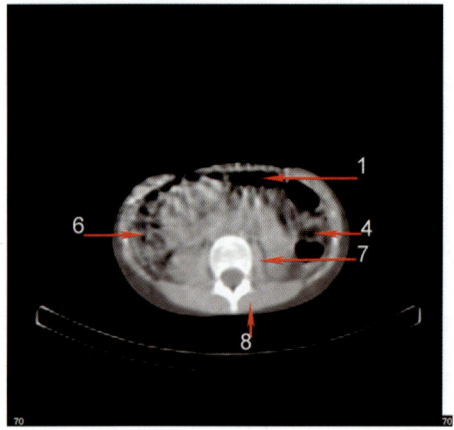
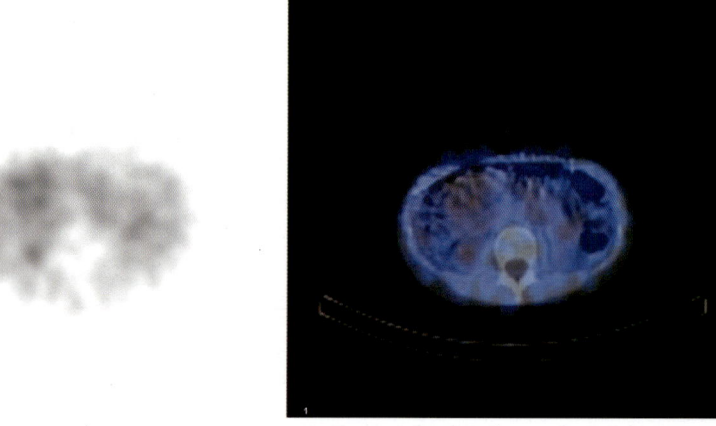

FIGURE 8-26: MIBG Transaxial 10

1. Transverse colon
2. Pancreas
3. Right hepatic lobe
4. Descending colon
5. Right kidney
6. Ascending colon
7. Psoas muscle
8. Erector spinae muscle

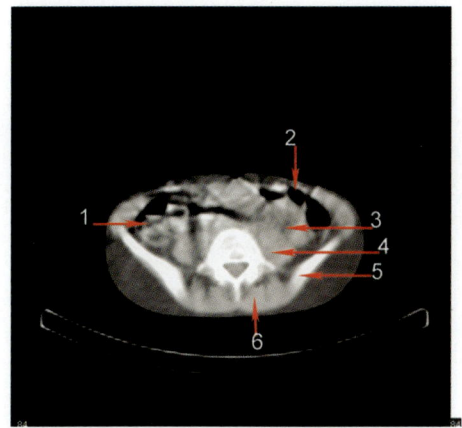 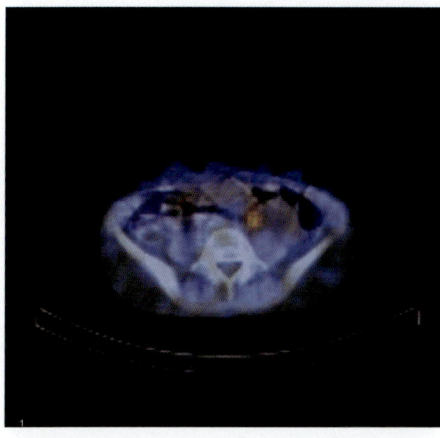

FIGURE 8-27: MIBG Transaxial 11

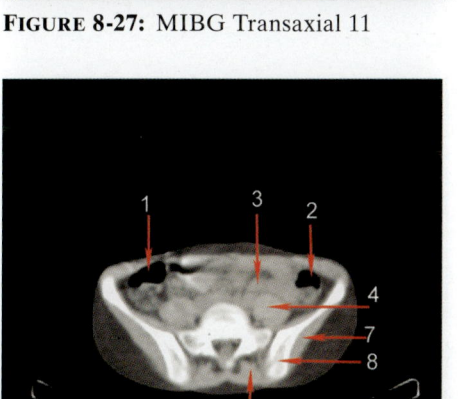 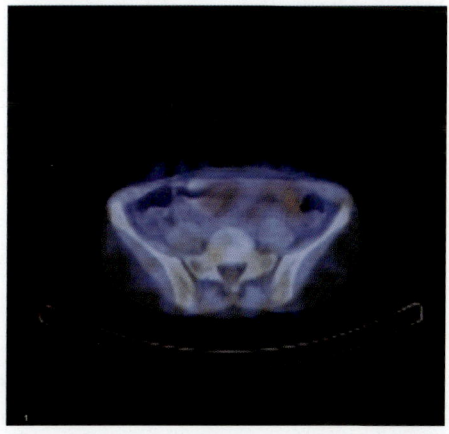

FIGURE 8-28: MIBG Transaxial 12

1. Cecum
2. Transverse colon
3. Jejunum

4. Psoas muscle
5. Ilium
6. Multifidus muscle

7. Gluteus minimus muscle
8. Iliac tuberosity

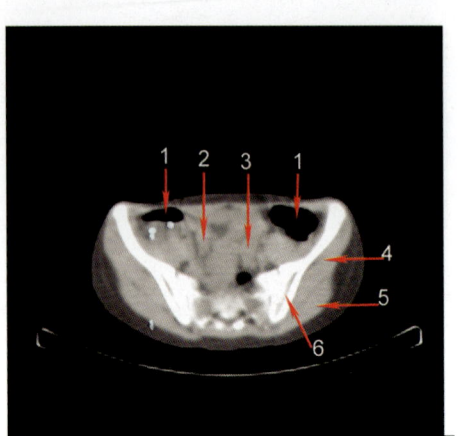
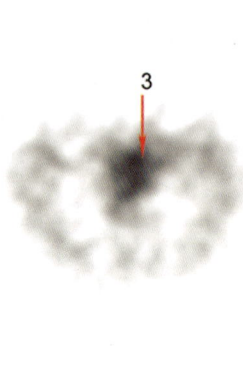
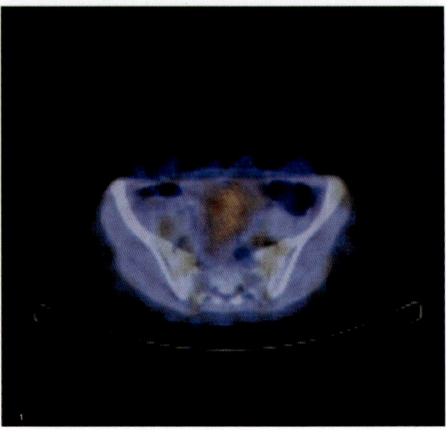

FIGURE 8-29: MIBG Transaxial 13

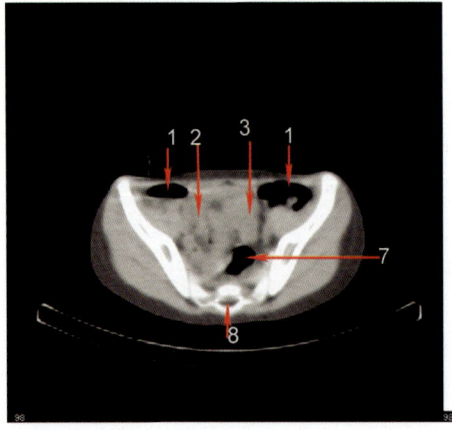
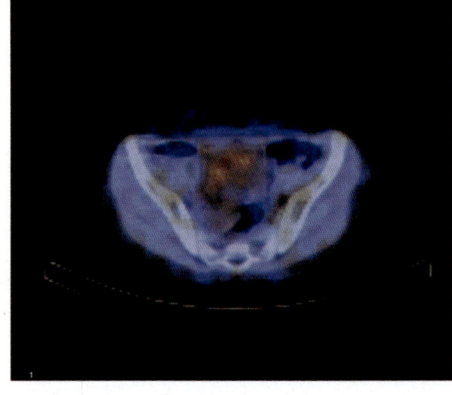

FIGURE 8-30: MIBG Transaxial 14

1. Colon
2. Ileum
3. Jejunum

4. Gluteus medius muscle
5. Gluteus maximus muscle
6. Iliac tuberosity

7. Sigmoid colon
8. Sacral foramen

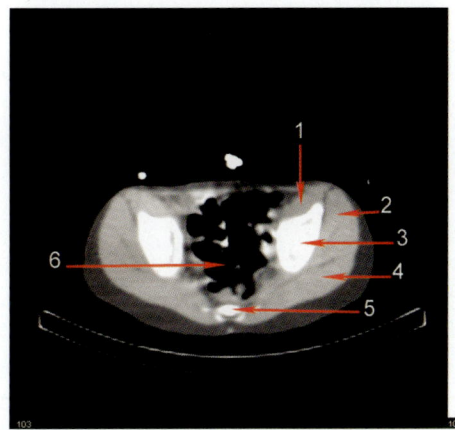
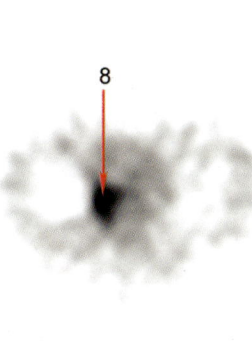
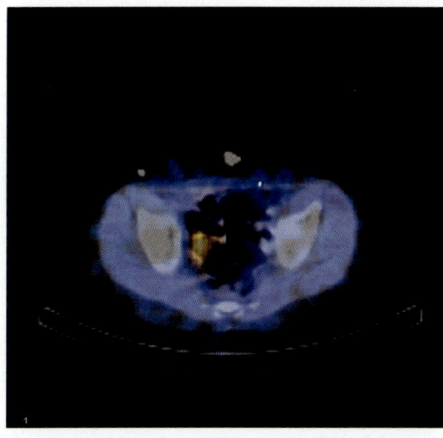

FIGURE 8-31: MIBG Transaxial 15

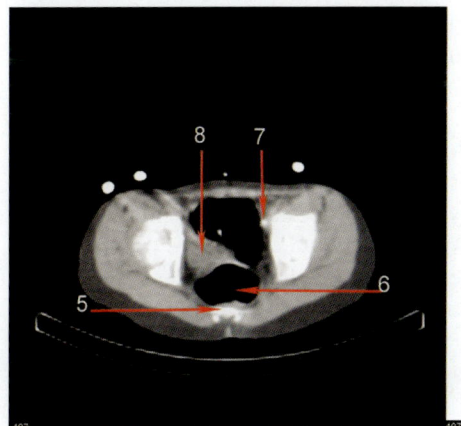
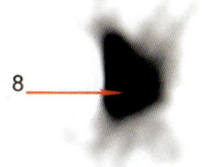

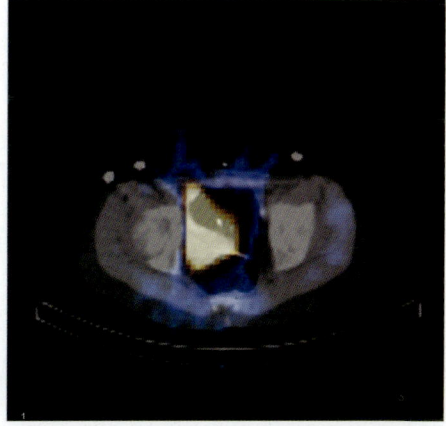

FIGURE 8-32: MIBG Transaxial 16

1. Iliopsoas muscle
2. Gluteus medius muscle
3. Ilium
4. Gluteus maximus muscle
5. Coccyx
6. Sigmoid colon gas
7. External iliac vessels
8. Bladder

9 Gallium SPECT/CT

Gallium 67-citrate scan (GS) has been considered the modality of choice until recently for functional assessment of lymphoma. It is unique for early detection and differentiation of viable lymphoma from a residual fibrositic or necrotic nonmalignant mass, although inferior to computed tomography (CT) in the initial staging.[1] It is of clinical value after initial therapy in defining complete response and in early detection of relapse.[2] It also is a good predictor of long-term prognosis.[3] Interpretation of GS often is hampered by physiologic activities, as well as benign processes such as inflammation and reactive hyperplasias of nodes.[4] Hybrid imaging using Ga-67 single photon emission CT (SPECT)/CT enables the correct localization of lymphoma lesions, defines areas of Ga-67 activity as physiologic uptake or pathology other than lymphoma, and leads to the diagnosis of additional, previously unidentified sites of disease.[5] Increased uptake of a tracer dose of the radiolabeled antibody suggests a benefit from radioimmunotherapy. SPECT and CT coregistration has been shown to improve radiation dosimetry in lymphoma patients.[6]

Section 1: Coronals

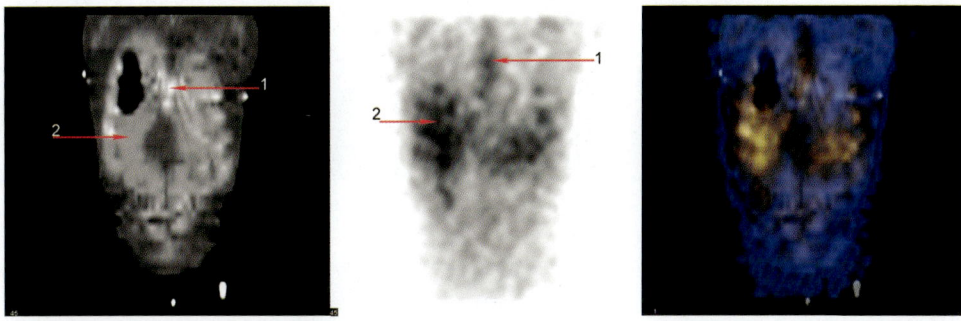

FIGURE 9-1: Gallium Abdominal Coronal 1

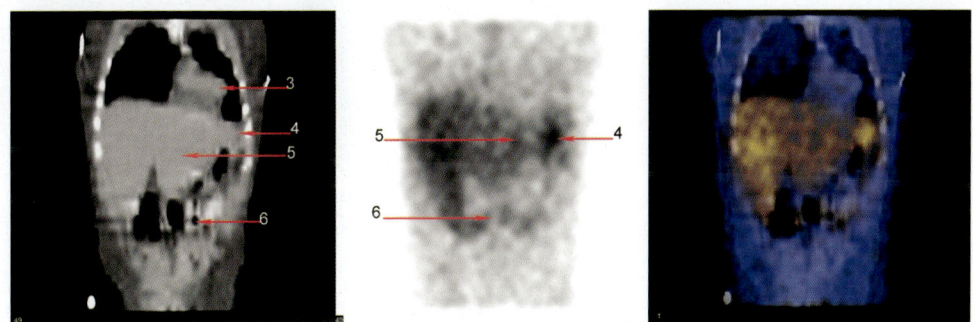

FIGURE 9-2: Gallium Abdominal Coronal 2

1. Sternum
2. Right hepatic lobe
   (segment VIII)
3. Right ventricle
4. Splenic flexure
5. Left hepatic lobe
6. Transverse colon

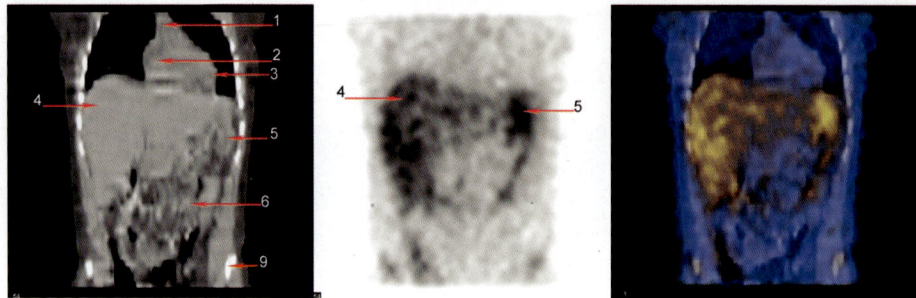

FIGURE 9-3: Gallium Abdominal Coronal 3

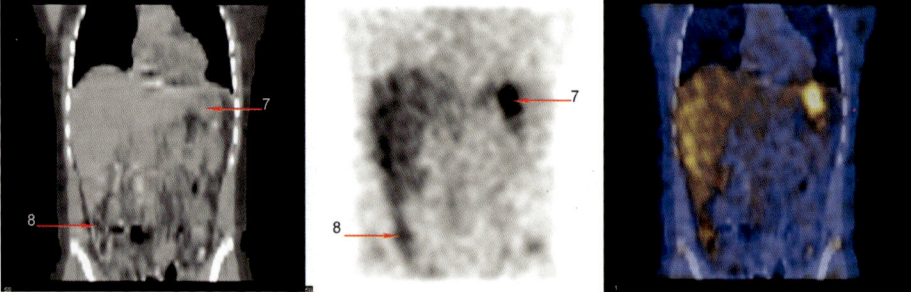

FIGURE 9-4: Gallium Abdominal Coronal 4

1. Superior vena cava
2. Right atrium
3. Left ventricle
4. Right hepatic lobe
   (segment VIII)

5. Splenic flexure
6. Transverse colon
7. Stomach

8. Ascending colon
9. Iliac crest

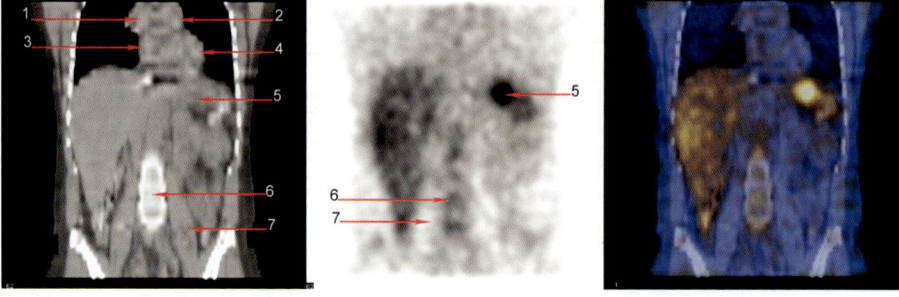

FIGURE 9-5: Gallium Abdominal Coronal 5

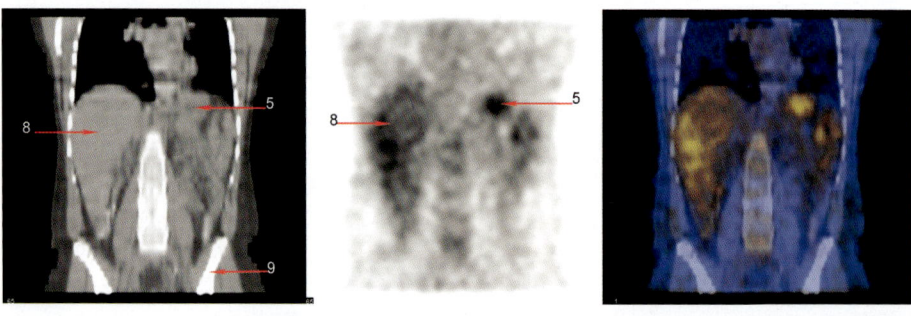

FIGURE 9-6: Gallium Abdominal Coronal 6

1. Ascending aorta
2. Main pulmonary artery
3. Right atrium
4. Left ventricle

5. Stomach
6. L3 marrow
7. Psoas muscle

8. Right hepatic lobe
   (segment VII)
9. Left iliac crest

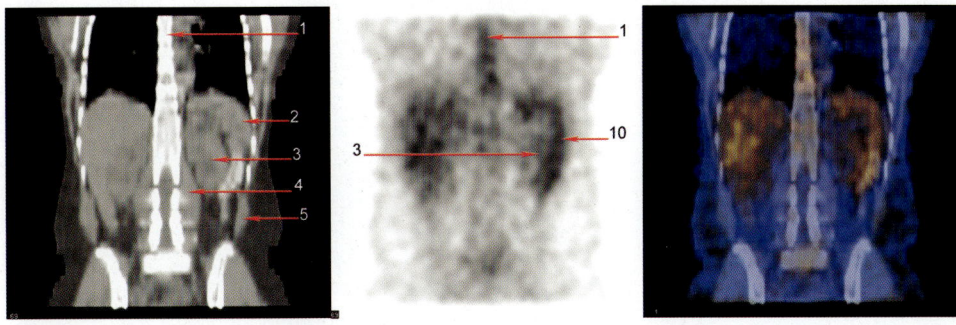

FIGURE 9-7: Gallium Abdominal Coronal 7

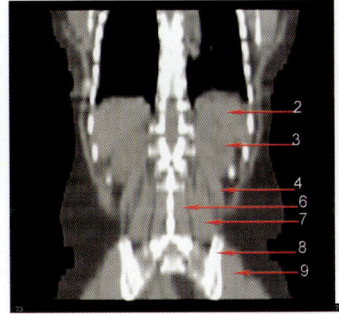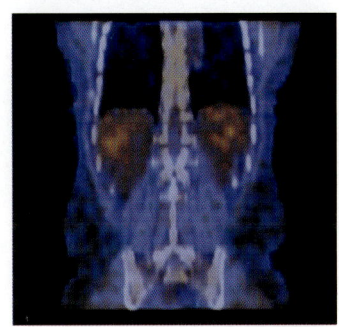

FIGURE 9-8: Gallium Abdominal Coronal 8

1. T5 marrow
2. Spleen
3. Left kidney
4. Psoas muscle
5. Internal oblique muscle
6. Erector spinae muscle
7. Quadratus lumborum muscle
8. Iliac tuberosity
9. Gluteus maximus muscle
10. Descending colon

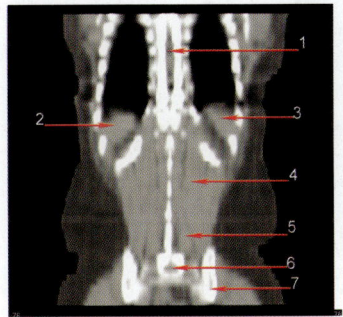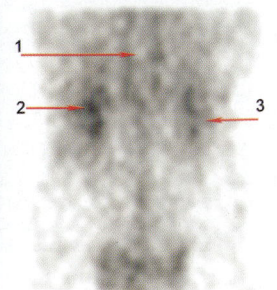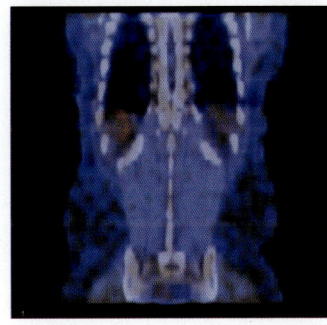

FIGURE 9-9: Gallium Abdominal Coronal 9

1. Thoracic spinal canal
2. Right hepatic lobe (segment VII)
3. Spleen
4. Quadratus lumborum muscle
5. Erector spinae muscle
6. Sacral foramen
7. Iliac tuberosity

Section 2: Sagittals

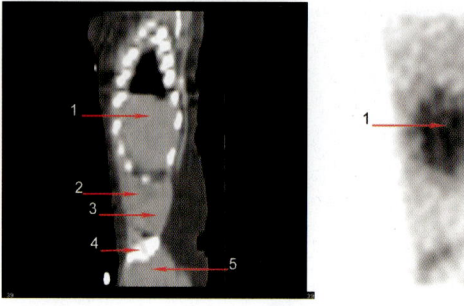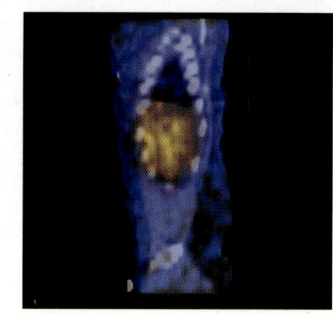

FIGURE 9-10: Gallium Abdominal Sagittal 1

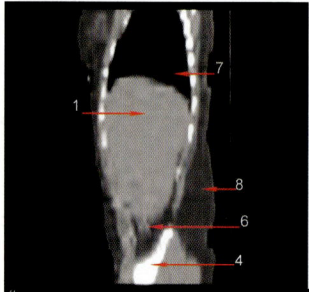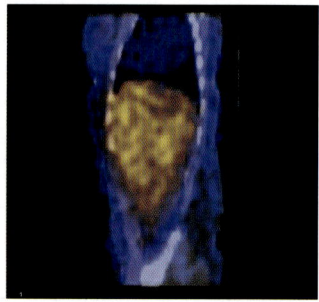

FIGURE 9-11: Gallium Abdominal Sagittal 2

1. Right hepatic lobe
2. Oblique muscles
3. Transverse abdominis muscle
4. Ilium
5. Gluteus medius muscle
6. Ascending colon
7. Posterior basal segment of right lower lobe of lung
8. Fat tissue

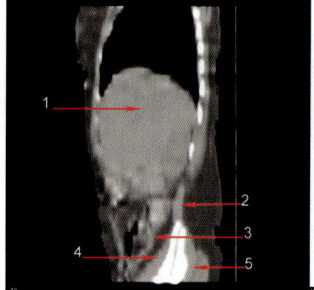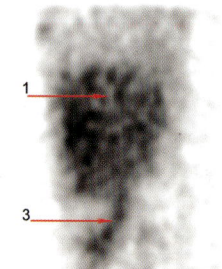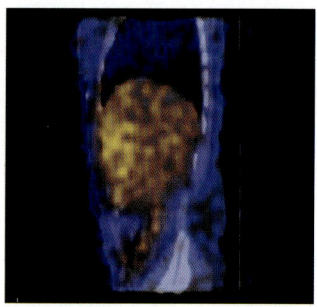

FIGURE 9-12: Gallium Abdominal Sagittal 3

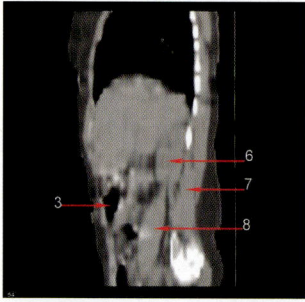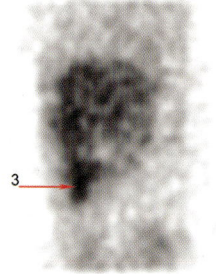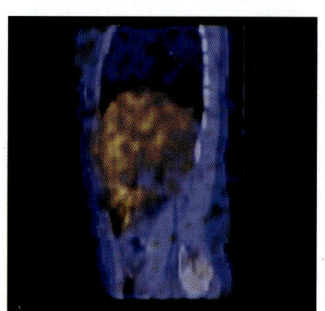

FIGURE 9-13: Gallium Abdominal Sagittal 4

1. Right hepatic lobe
2. Latissimus dorsi muscle
3. Ascending colon
4. Iliacus muscle
5. Gluteus medius muscle
6. Right kidney
7. Quadratus lumborum muscle
8. Psoas muscle

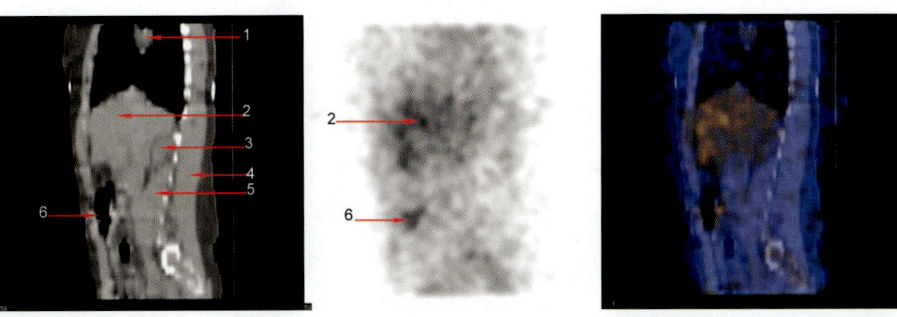

FIGURE 9-14: Gallium Abdominal Sagittal 5

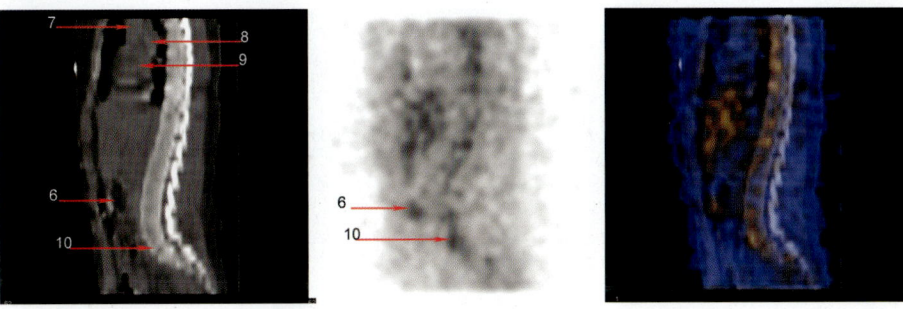

FIGURE 9-15: Gallium Abdominal Sagittal 6

1. Right hilum
2. Right hepatic lobe
3. Right kidney
4. Erector spinae muscle
5. Psoas muscle
6. Transverse colon
7. Superior vena cava
8. Right pulmonary artery
9. Right atrium
10. L5 marrow

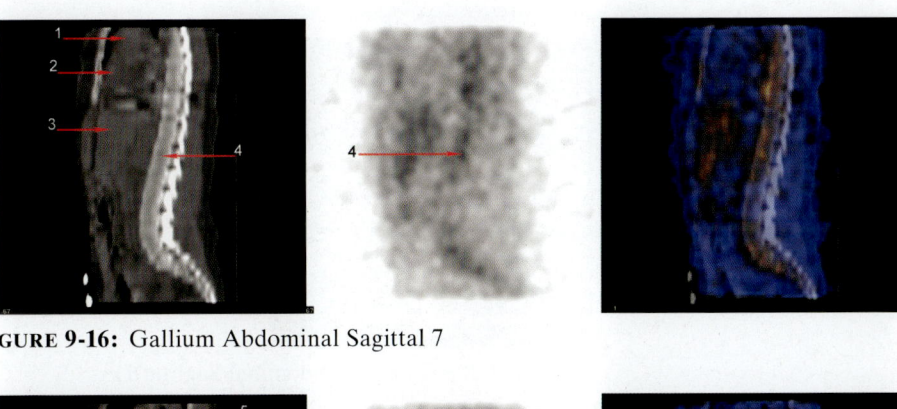

FIGURE 9-16: Gallium Abdominal Sagittal 7

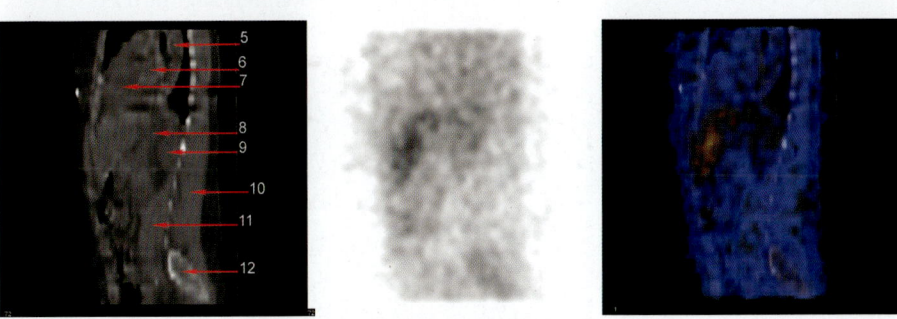

FIGURE 9-17: Gallium Abdominal Sagittal 8

1. Aortic arch
2. Right ventricle
3. Left hepatic lobe
4. T11 marrow
5. Descending aorta
6. Left atrium
7. Right ventricle
8. Stomach
9. Left kidney
10. Erector spinae muscle
11. Psoas muscle
12. Iliac tuberosity

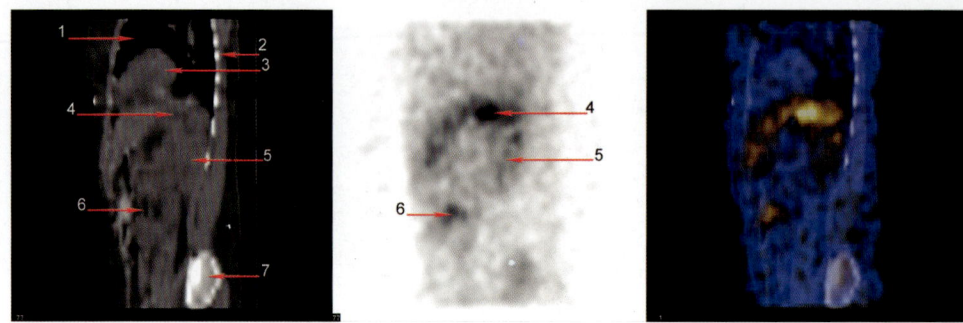

FIGURE 9-18: Gallium Abdominal Sagittal 9

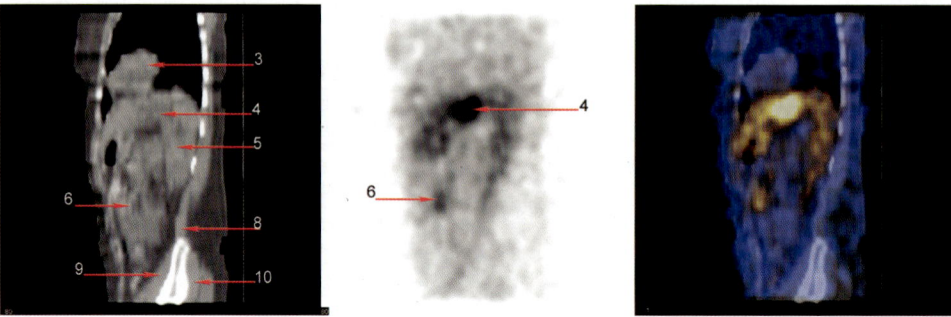

FIGURE 9-19: Gallium Abdominal Sagittal 10

1. Left upper lobe of lung
2. Rib
3. Left ventricle
4. Stomach
5. Left kidney
6. Jejunum
7. Iliac tuberosity
8. Latissimus dorsi muscle
9. Iliacus muscle
10. Gluteus medius muscle

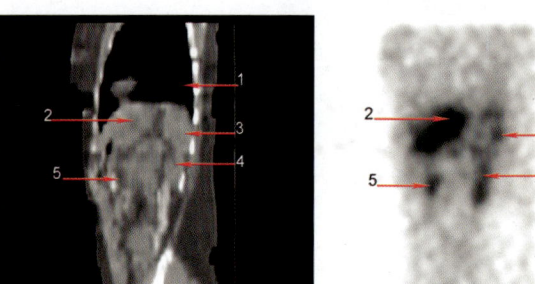
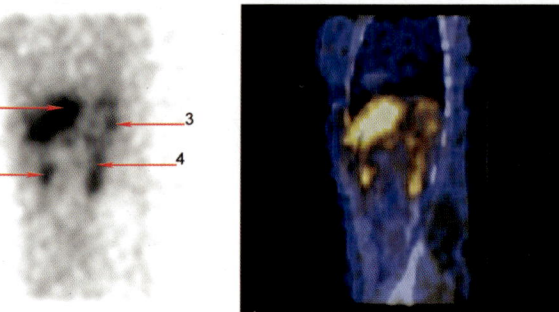

FIGURE 9-20: Gallium Abdominal Sagittal 11

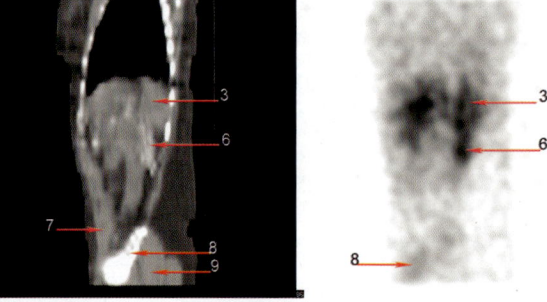
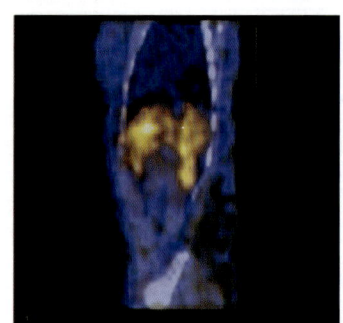

FIGURE 9-21: Gallium Abdominal Sagittal 12

1. Posterior basal segment of left lower lobe
2. Stomach
3. Spleen
4. Splenic flexure
5. Jejunum
6. Descending colon
7. Rectus abdominalis muscle
8. Ilium
9. Gluteus medius muscle

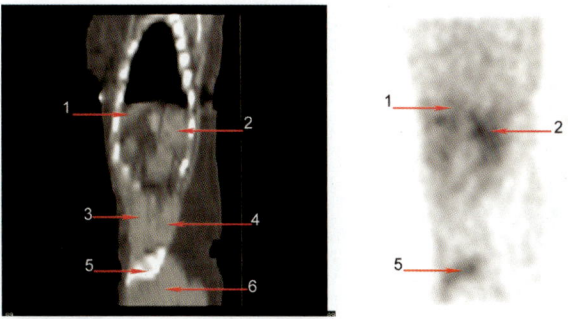
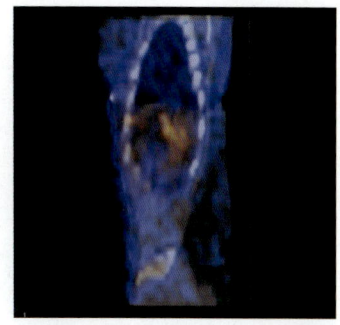

FIGURE 9-22: Gallium Abdominal Sagittal 13

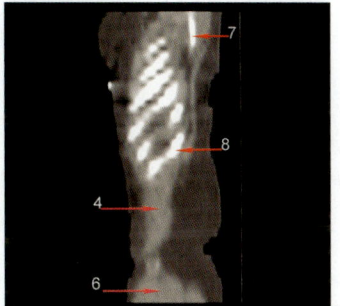

FIGURE 9-23: Gallium Abdominal Sagittal 14

1. Left hepatic lobe
2. Jejunum
3. Oblique muscles
4. Transverse abdominis muscle
5. Ilium
6. Gluteus medius muscle
7. Scapula
8. Rib

Section 3: Transaxials

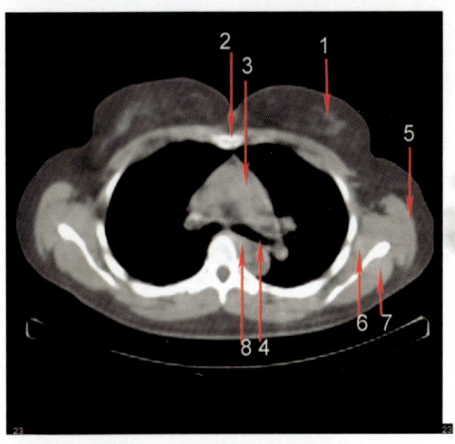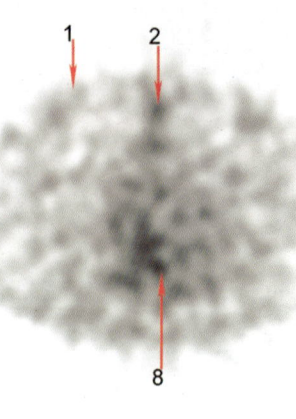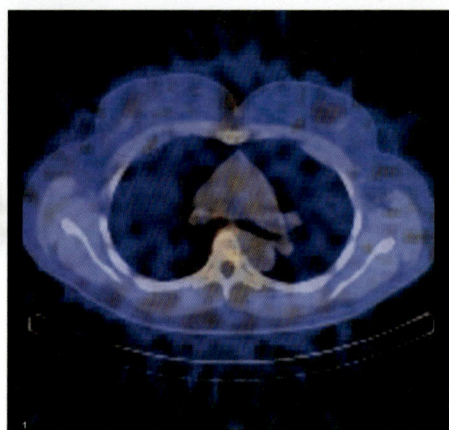

FIGURE 9-24: Gallium Abdominal Transaxial 1

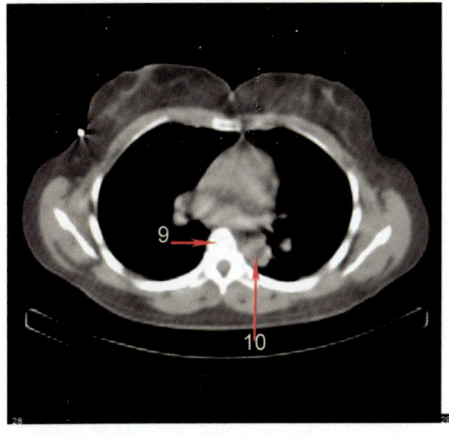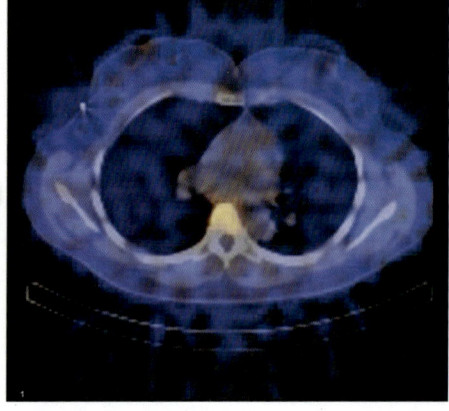

FIGURE 9-25: Gallium Abdominal Transaxial 2

1. Breast
2. Sternum
3. Main pulmonary artery
4. Left main bronchus

5. Teres major muscle
6. Subscapularis muscle
7. Infraspinatus muscle
8. Esophagus

9. T6 marrow
10. Descending aorta

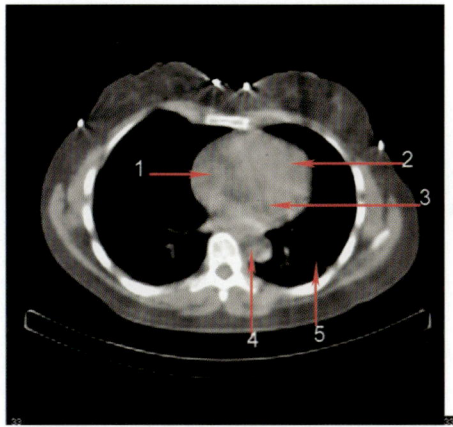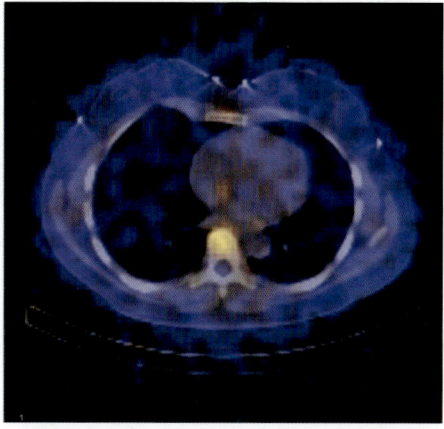

FIGURE 9-26: Gallium Abdominal Transaxial 3

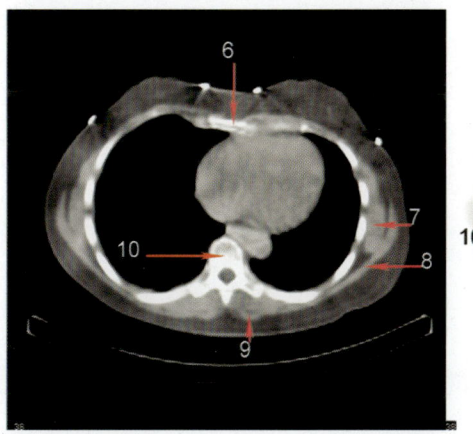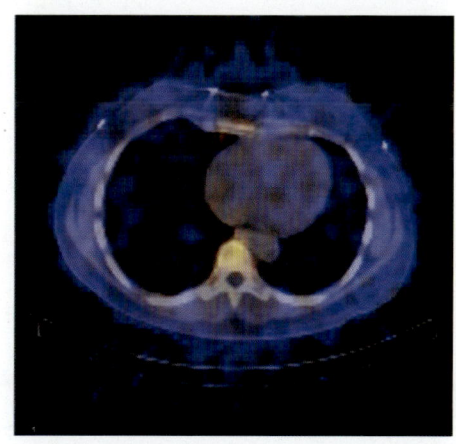

FIGURE 9-27: Gallium Abdominal Transaxial 4

1. Right atrium
2. Left ventricle
3. Left atrium
4. Descending aorta
5. Left lower lobe of lung (superior segment)
6. Sternum
7. Serratus anterior muscle
8. Latissimus dorsi muscle
9. Erector spinae muscle
10. T8 marrow

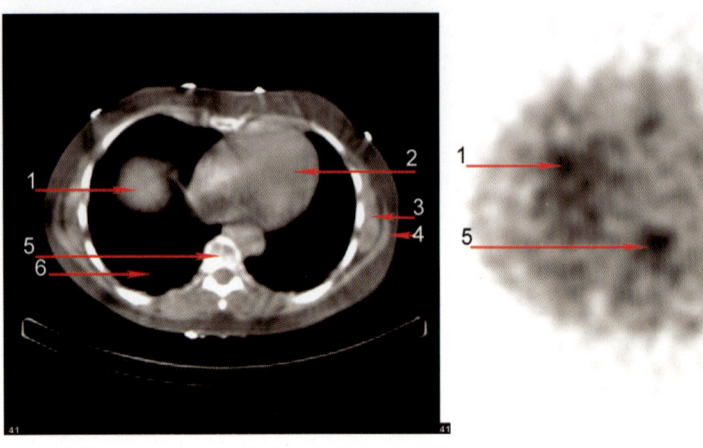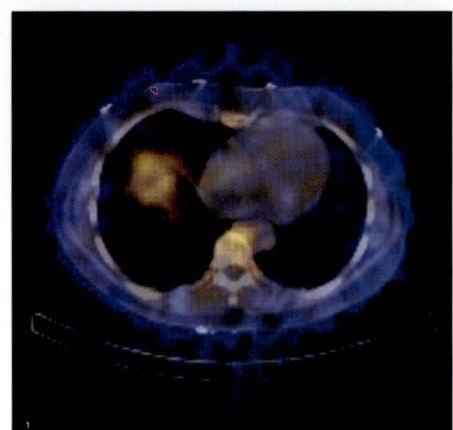

FIGURE 9-28: Gallium Abdominal Transaxial 5

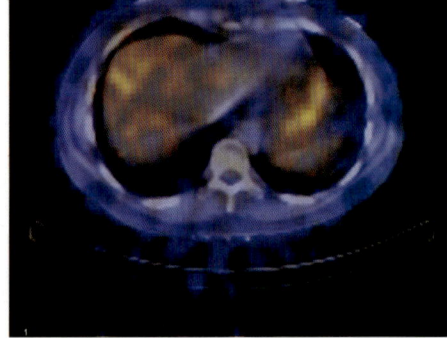

FIGURE 9-29: Gallium Abdominal Transaxial 6

1. Right hepatic lobe (segment VIII)
2. Left ventricle
3. Serratus anterior muscle
4. Latissimus dorsi muscle
5. T10 marrow
6. Right lower lobe of lung (posterior basal segment)
7. Stomach
8. Hemiazygos vein
9. Abdominal aorta

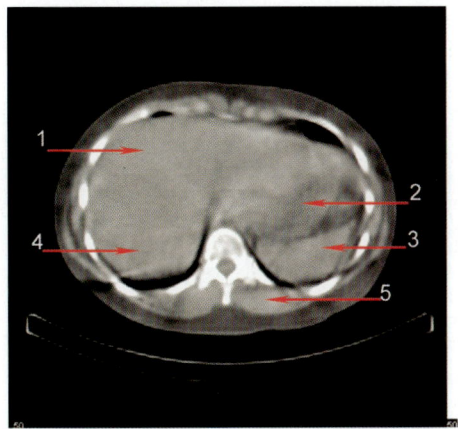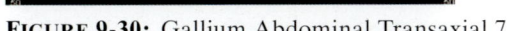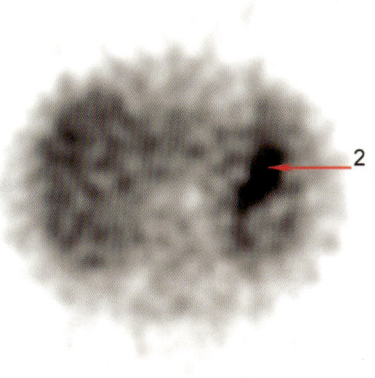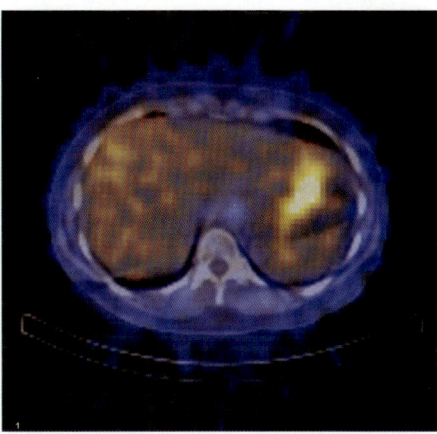

FIGURE 9-30: Gallium Abdominal Transaxial 7

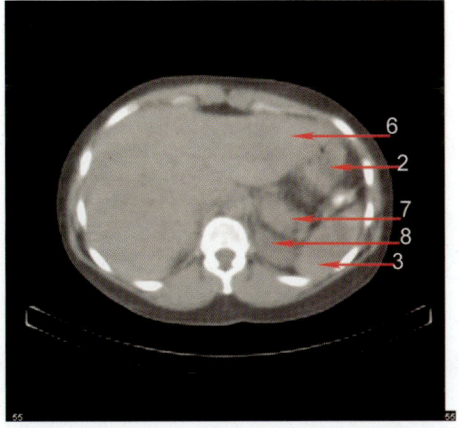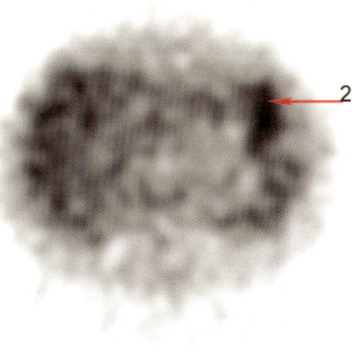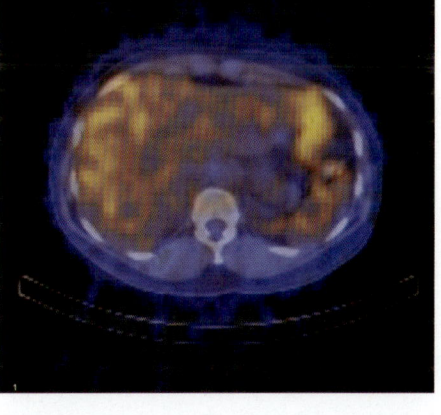

FIGURE 9-31: Gallium Abdominal Transaxial 8

1. Right hepatic lobe (segment VIII)
2. Stomach
3. Spleen
4. Right hepatic lobe (segment VII)
5. Erector spinae muscle
6. Left hepatic lobe
7. Pancreatic tail
8. Superior pole of left kidney

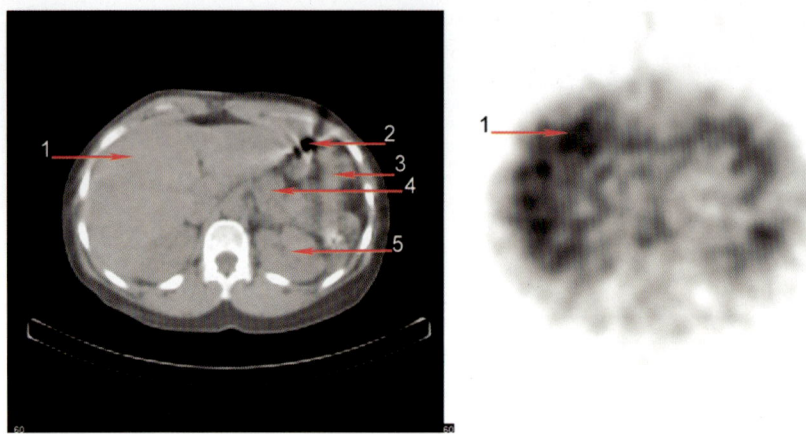
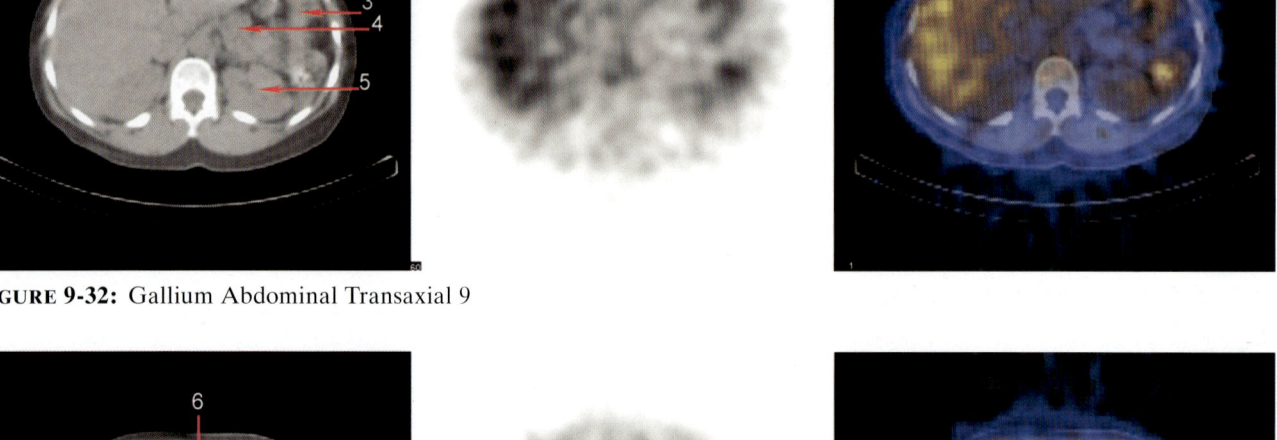

FIGURE 9-32: Gallium Abdominal Transaxial 9

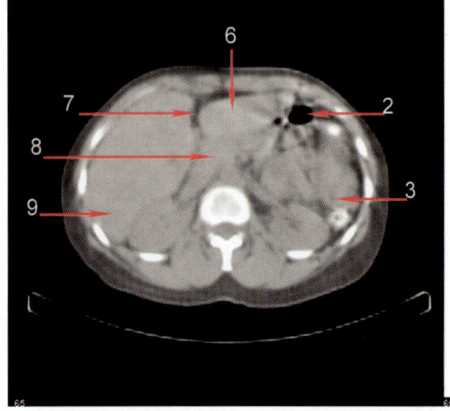
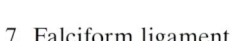

FIGURE 9-33: Gallium Abdominal Transaxial 10

1. Right hepatic lobe (segment V)
2. Stomach
3. Colon
4. Pancreatic body
5. Left kidney
6. Left hepatic lobe (segment III)
7. Falciform ligament
8. Duodenum
9. Right hepatic lobe (segment VI)

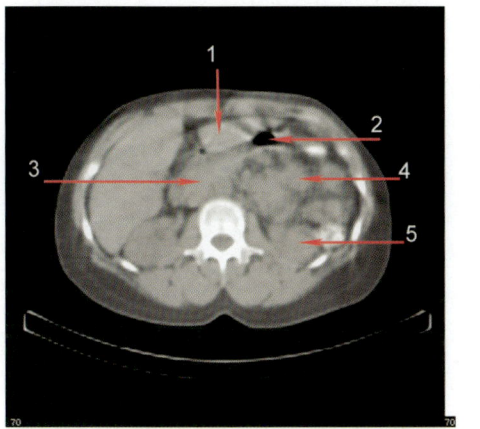 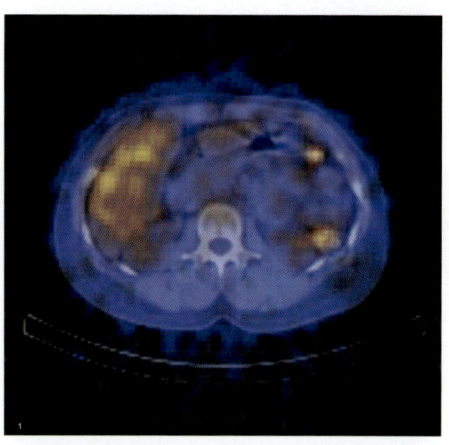

FIGURE 9-34: Gallium Abdominal Transaxial 11

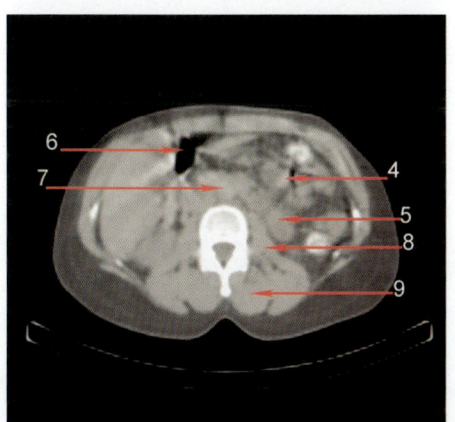 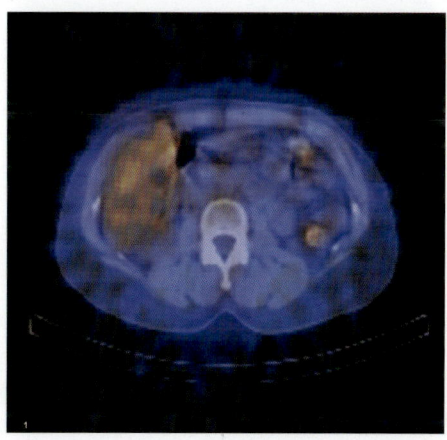

FIGURE 9-35: Gallium Abdominal Transaxial 12

1. Left hepatic lobe (segment III)
2. Stomach
3. Pancreatic head

4. Jejunum
5. Left kidney
6. Colon

7. Aorta
8. Psoas muscle
9. Erector spinae muscle

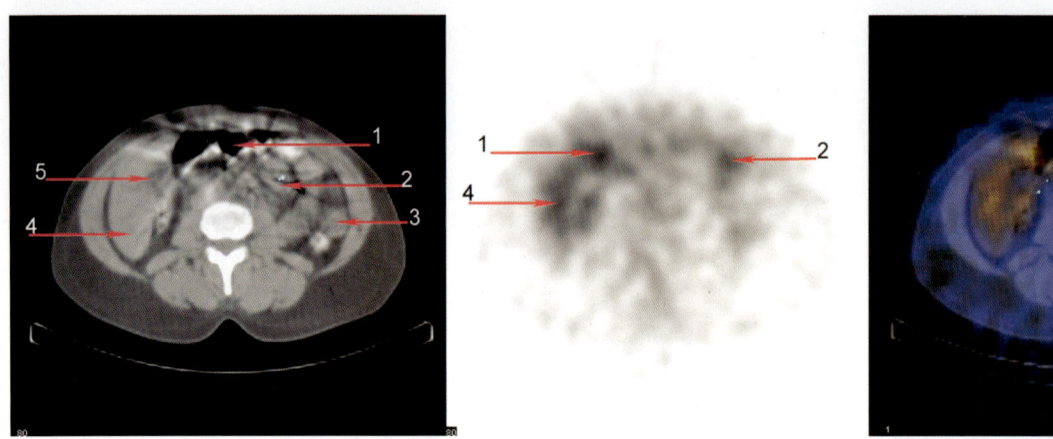

FIGURE 9-36: Gallium Abdominal Transaxial 13

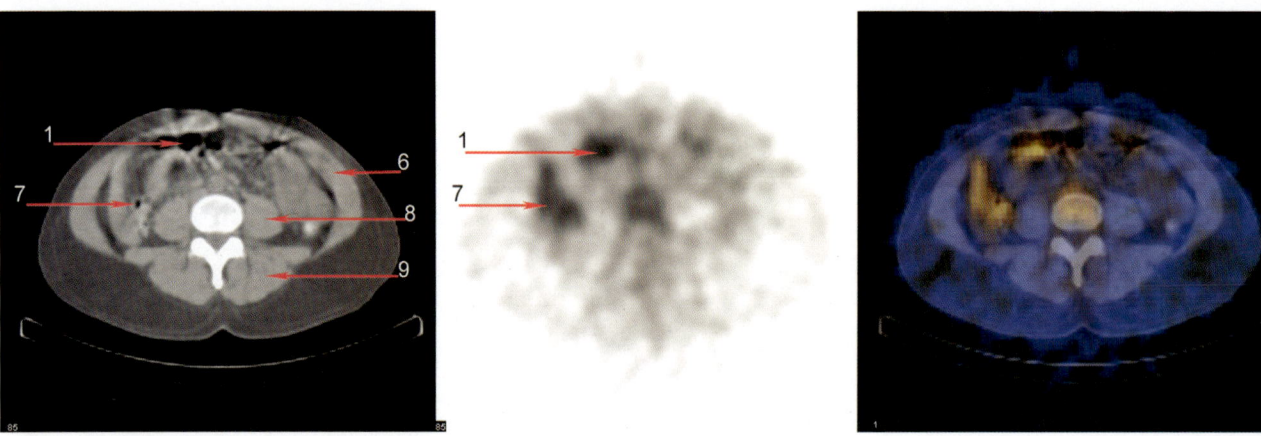

FIGURE 9-37: Gallium Abdominal Transaxial 14

1. Transverse colon
2. Jejunum
3. Descending colon
4. Right hepatic lobe (segment VI)
5. Gallbladder fossa
6. Oblique muscles
7. Duodenum
8. Psoas muscle
9. Iliocostalis lumborum muscle

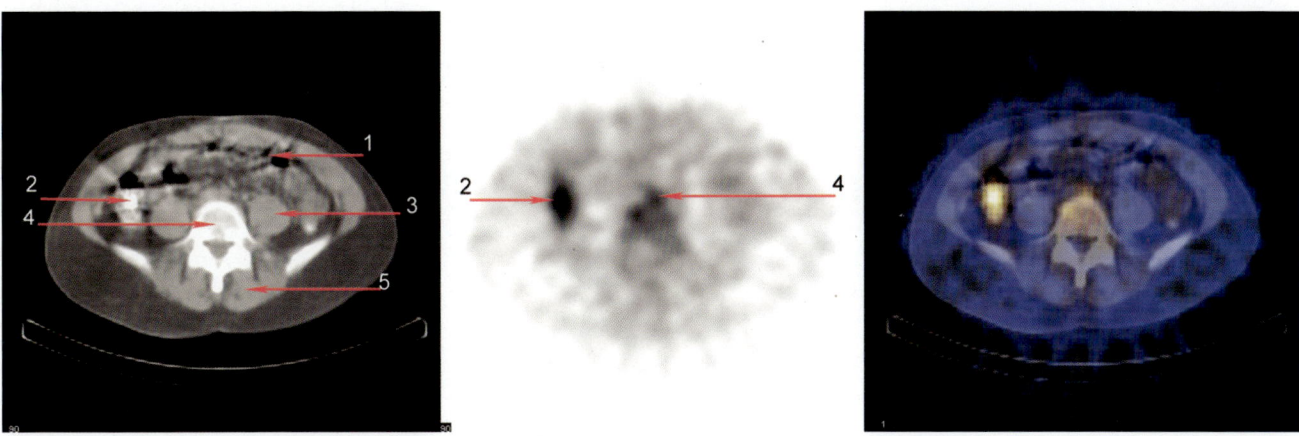

FIGURE 9-38: Gallium Abdominal Transaxial 15

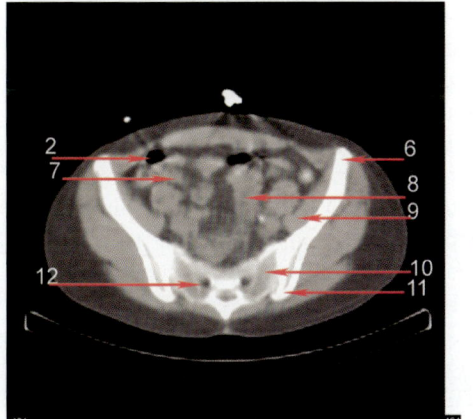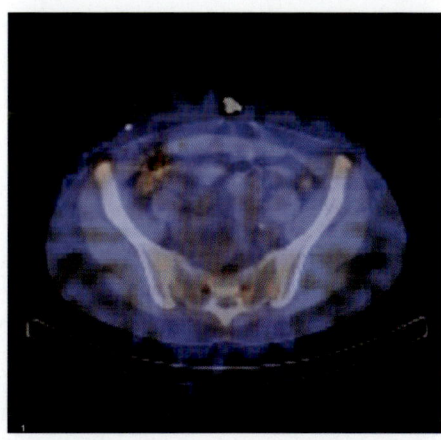

FIGURE 9-39: Gallium Abdominal Transaxial 16

1. Transverse colon
2. Cecum
3. Psoas muscle
4. L5 marrow

5. Multifidus muscle
6. Anterior superior iliac crest
7. Ileum
8. Jejunum

9. Iliacus muscle
10. Sacral ala
11. Iliac tuberosity
12. Sacral neural canal

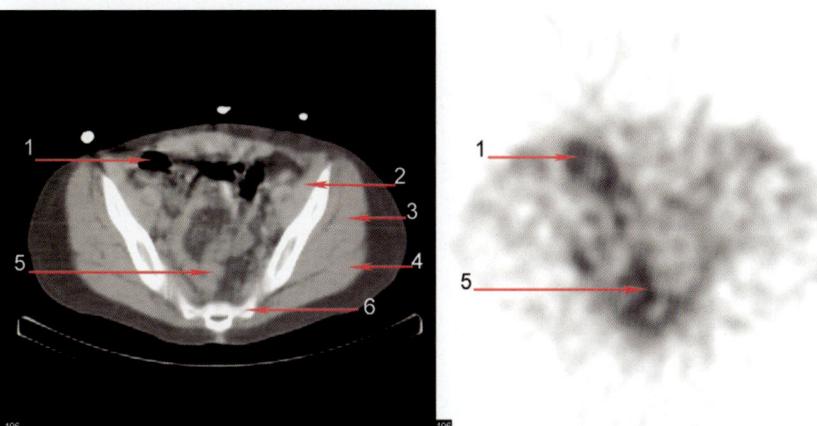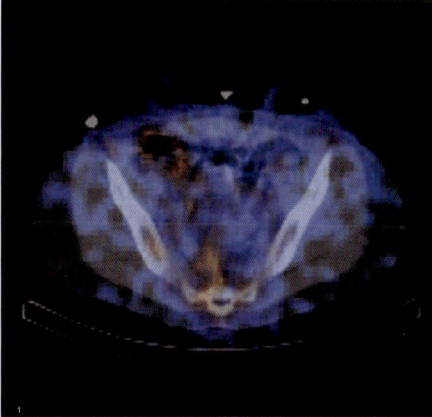

FIGURE 9-40: Gallium Abdominal Transaxial 17

1. Cecum
2. Iliacus muscle

3. Gluteus medius muscle
4. Gluteus maximus muscle

5. Sigmoid colon
6. Sacrococcygeal junction

10 Octreotide SPECT/CT

Detection of neuroendocrine tumor sites is critical for optimal surgical treatment planning, but localization of tumors may be difficult because of their small size and lack of anatomic delineation.[1] Diagnosis, staging, and follow-up have advanced considerably with I-111 octreotide, which is accumulated in tumors with somatostatin receptor subtype 2 or 5. Octreotide scan facilitates the detection of receptor-dense microscopic foci during radio-guided surgery and determines the completeness of the surgical procedure. It also identifies the receptor-status of metastases for somatostatin therapy.[2] Hybrid imaging using single photon emission computed tomography (SPECT)/computed tomography (CT) can define the precise organ involved, determine the presence or absence of invasion into surrounding tissue, and has been reported to have an impact on patient management.[3] It also may help in the choice of the appropriate treatment. When disease is confined to a single organ, a localized node of organ-specific therapy is suggested. However, surgery is inadvisable when soft tissue tumor has invaded an adjacent bone.[4]

Section 1: Coronals

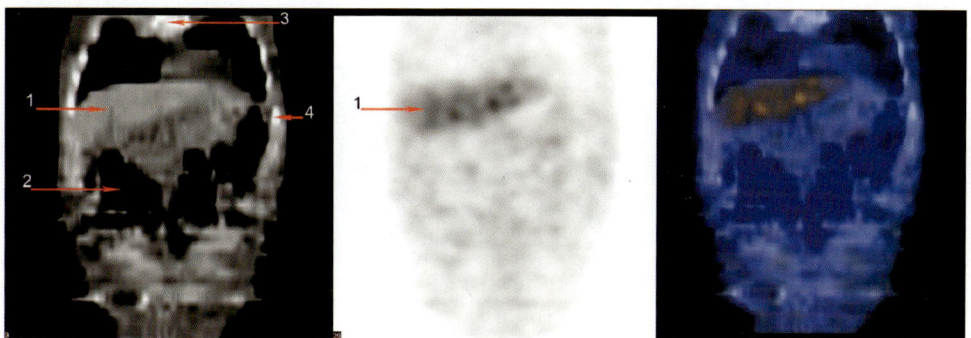

FIGURE 10-1: Octreotide Coronal 1

1. Liver (right lobe)   3. Manubrium   4. Rib
2. Bowel gas (transverse
   colon)

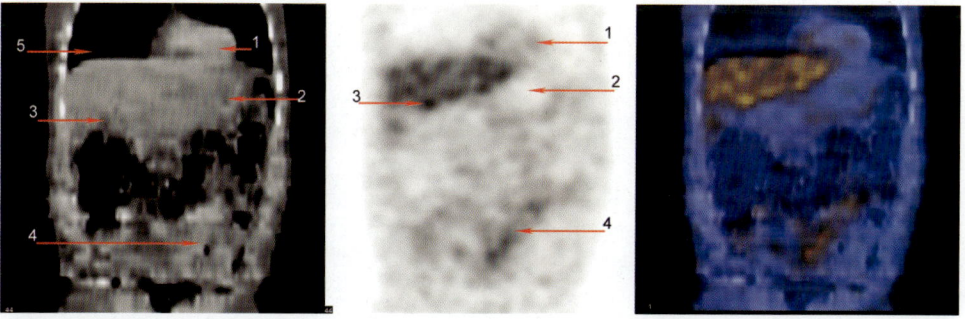

FIGURE 10-2: Octreotide Coronal 2

1. Heart   3. Gallbladder   5. Right lower lobe of lung
2. Stomach   4. Intestine (jejunum)

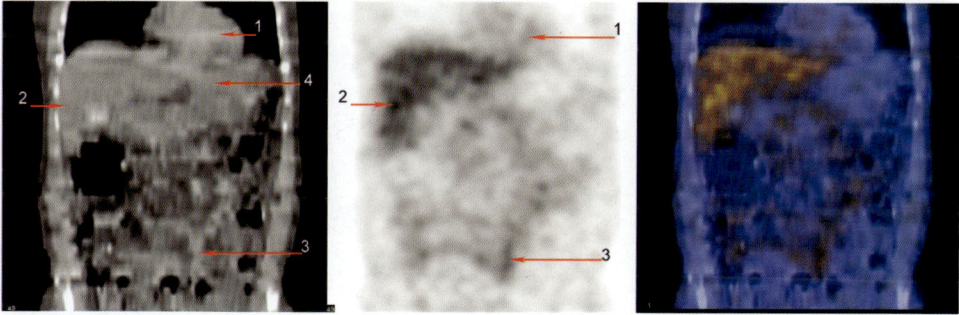

FIGURE 10-3: Octreotide Coronal 3

1. Heart   3. Jejunum   4. Stomach
2. Right hepatic lobe

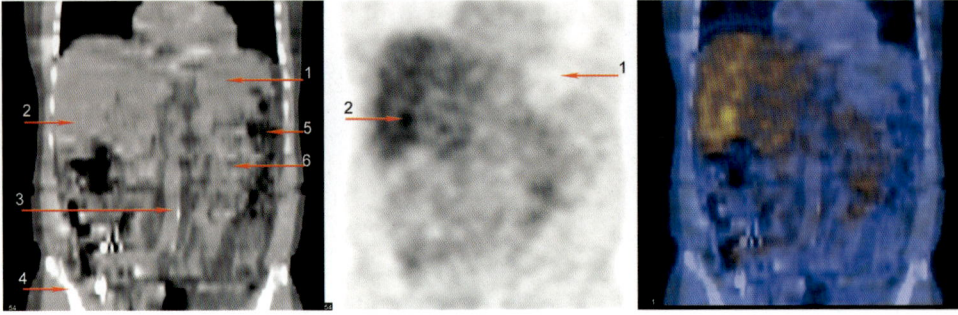

FIGURE 10-4: Octreotide Coronal 4

1. Stomach	3. Inferior vena cava	5. Splenic flexure of colon
2. Right hepatic lobe	4. Iliac crest	6. Jejunum

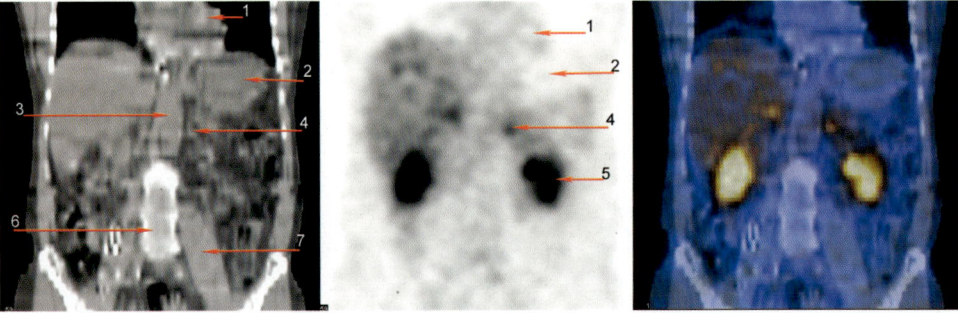

FIGURE 10-5: Octreotide Coronal 5

1. Heart	4. Adrenal gland	6. L5 body
2. Stomach	5. Kidney	7. Psoas muscle
3. Abdominal aorta		

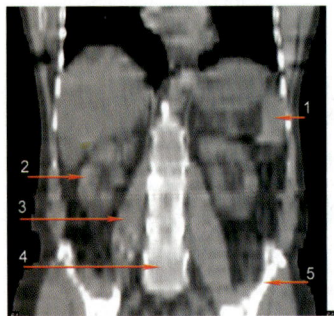
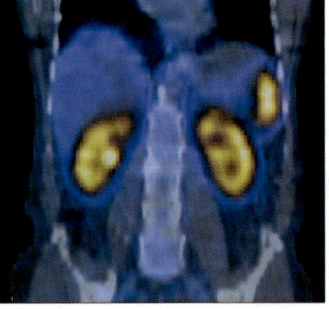

FIGURE 10-6: Octreotide Coronal 6

1. Spleen	3. Psoas muscle	5. Iliac crest
2. Kidney	4. Sacrum	

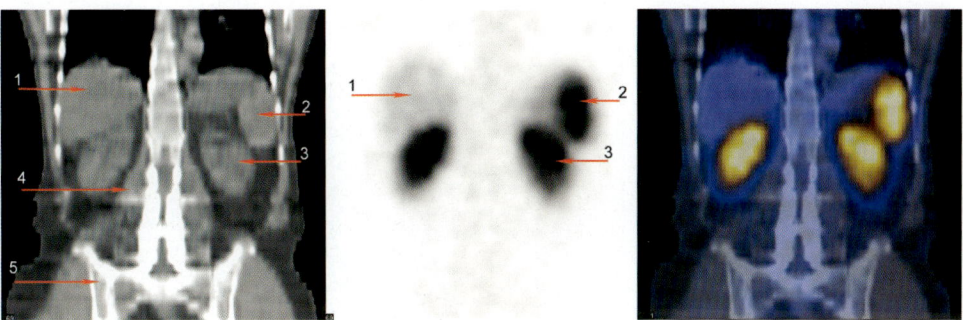

FIGURE 10-7: Octreotide Coronal 7

1. Right hepatic lobe   3. Kidney   5. Iliac tuberosity
2. Spleen   4. Psoas muscle

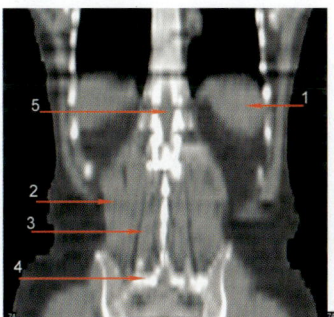

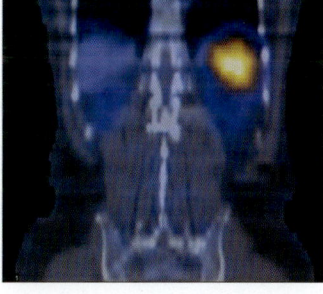

FIGURE 10-8: Octreotide Coronal 8

1. Spleen   3. Erector spinae muscle   5. Spinal canal
2. Psoas muscle   4. Sacrum

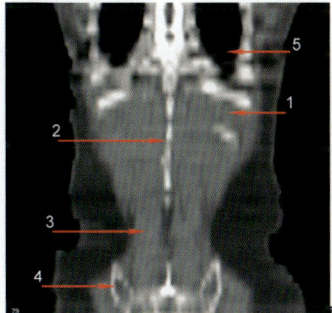
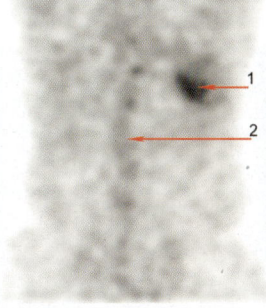

FIGURE 10-9: Octreotide Coronal 9

1. Spleen   3. Erector spinae muscle   5. Left lower lung
2. Spinous process   4. Iliac tuberosity

Section 2: Sagittals

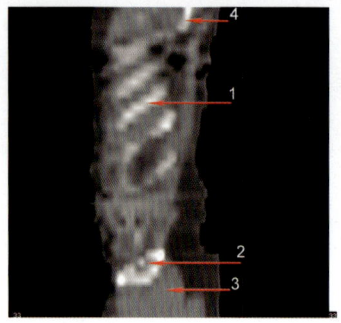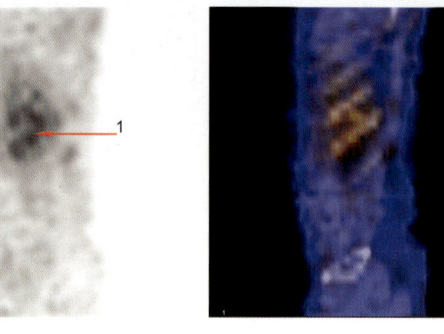

FIGURE 10-10: Octreotide Sagittal 1

1. Rib
2. Acetabulum

3. Obturator externus muscle

4. Scapula

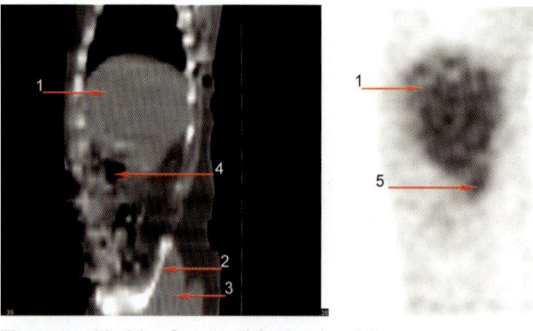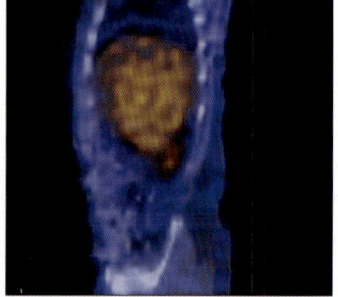

FIGURE 10-11: Octreotide Sagittal 2

1. Right hepatic lobe (segment VIII)

2. Iliac crest
3. Gluteus muscle

4. Hepatic flexure of colon
5. Right kidney

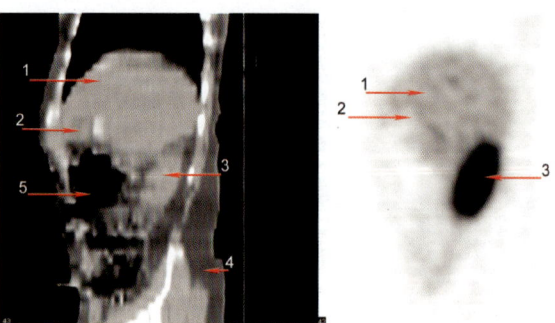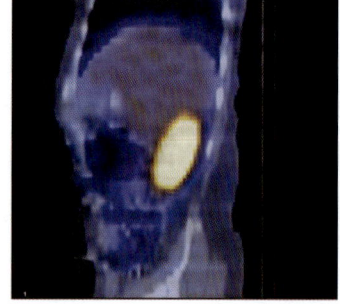

FIGURE 10-12: Octreotide Sagittal 3

1. Right hepatic lobe (segment VIII)

2. Gallbladder fossa
3. Right kidney

4. Subcutaneous fat
5. Transverse colon

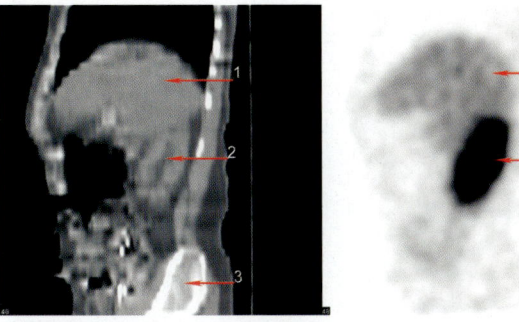
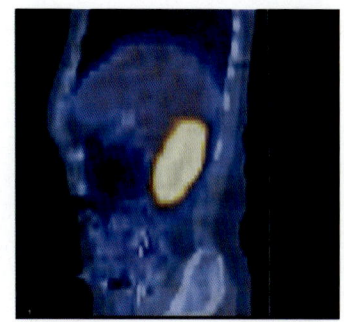

FIGURE 10-13: Octreotide Sagittal 4

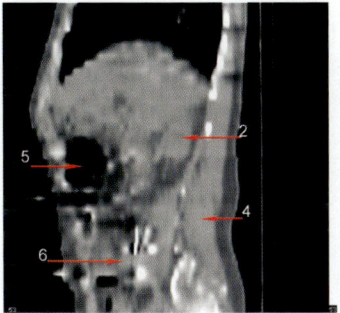

FIGURE 10-14: Octreotide Sagittal 5

1. Right hepatic lobe (segment VII)
2. Right kidney
3. Ilium
4. Erector spinae muscle
5. Transverse colon
6. Ileum

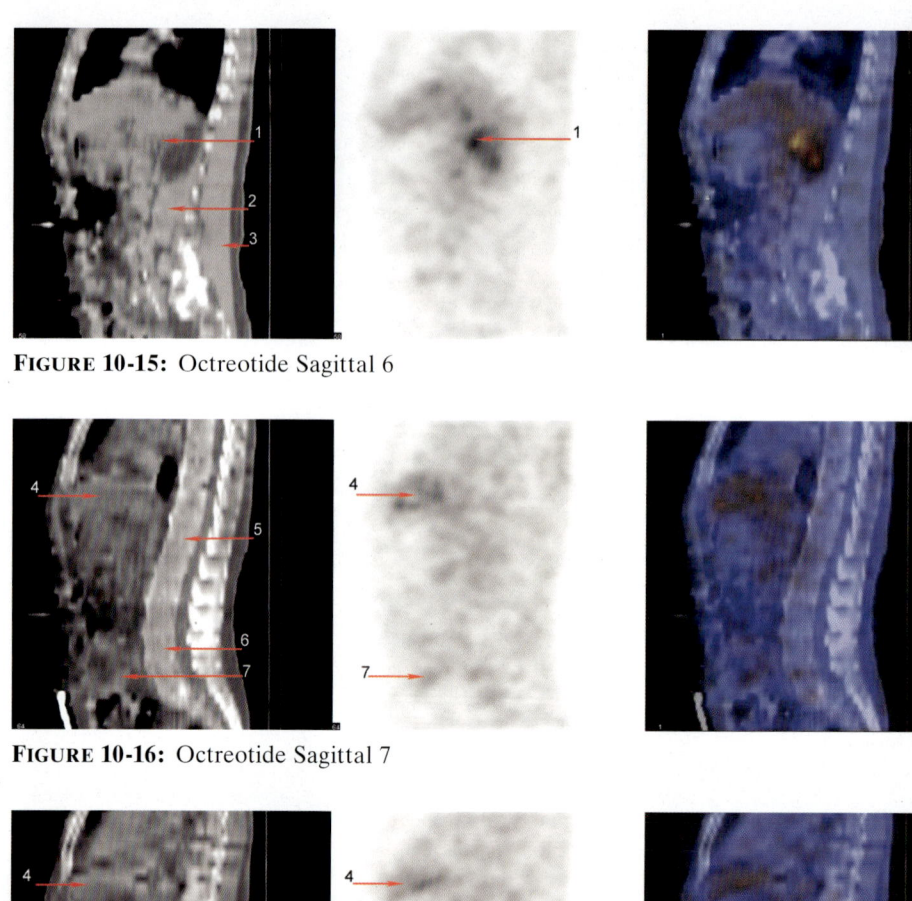

FIGURE 10-15: Octreotide Sagittal 6

FIGURE 10-16: Octreotide Sagittal 7

FIGURE 10-17: Octreotide Sagittal 8

1. Kidney
2. Psoas muscle
3. Erector spinae muscle
4. Left hepatic lobe (segment II)
5. T12
6. L4
7. Ileum

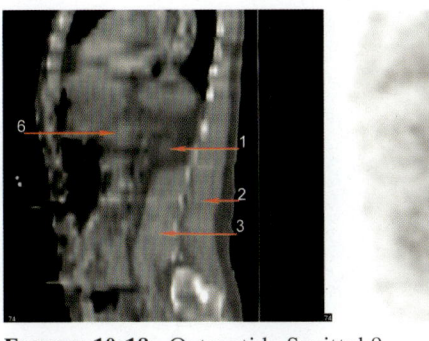
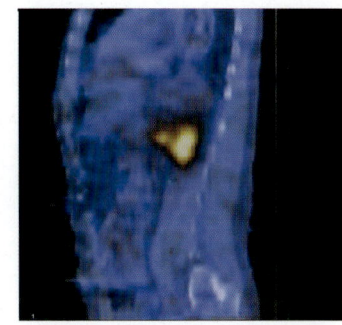

FIGURE 10-18: Octreotide Sagittal 9

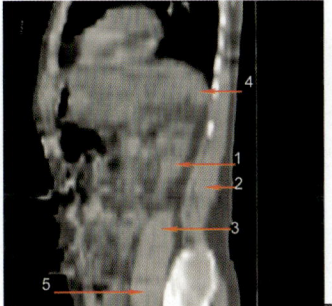
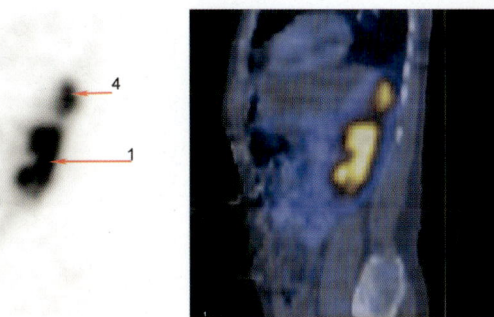

FIGURE 10-19: Octreotide Sagittal 10

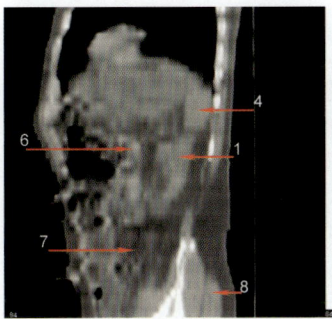
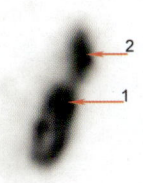
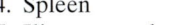

FIGURE 10-20: Octreotide Sagittal 11

1. Left kidney
2. Erector spinae muscle
3. Psoas muscle
4. Spleen
5. Iliacus muscle
6. Pancreas
7. Mesenteric fat
8. Gluteus muscle

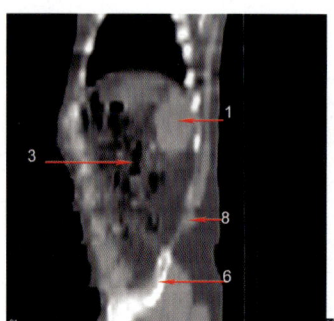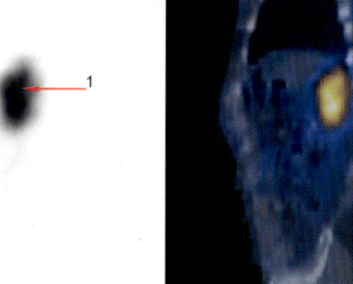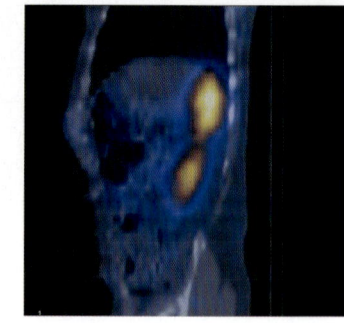

FIGURE 10-21: Octreotide Sagittal 12

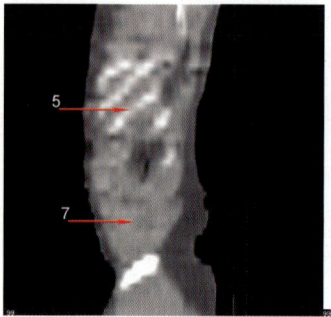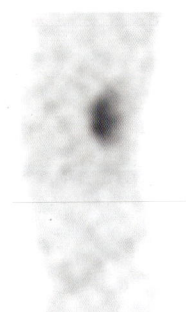

FIGURE 10-22: Octreotide Sagittal 13

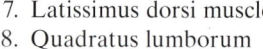

FIGURE 10-23: Octreotide Sagittal 14

1. Spleen	4. Ileum	7. Latissimus dorsi muscle
2. Left kidney	5. Rib	8. Quadratus lumborum
3. Jejunum	6. Ilium	muscle

Section 3: Transaxials

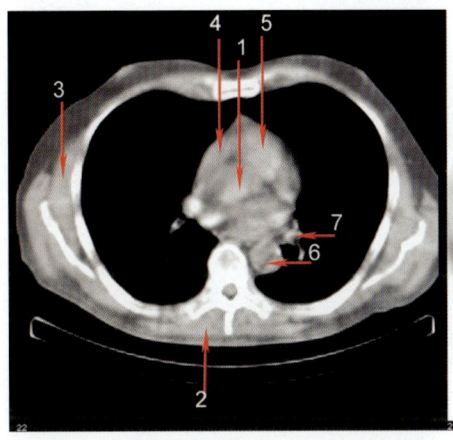 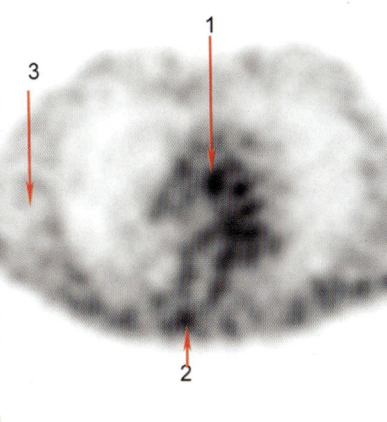 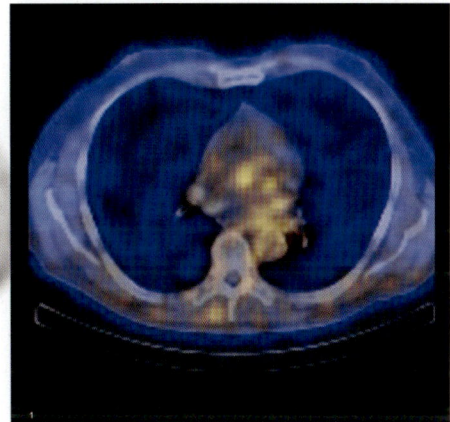

FIGURE 10-24: Octreotide Transaxial 1

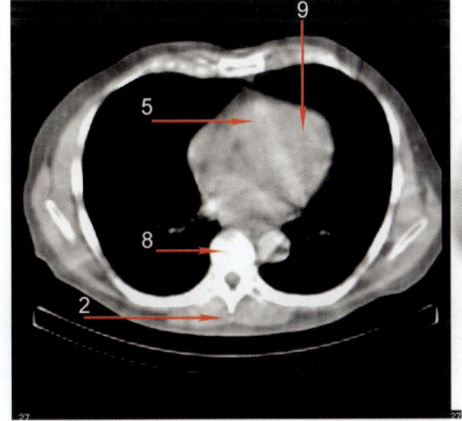 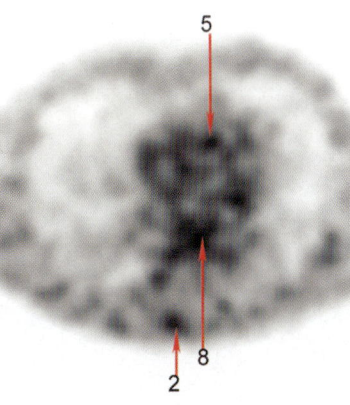 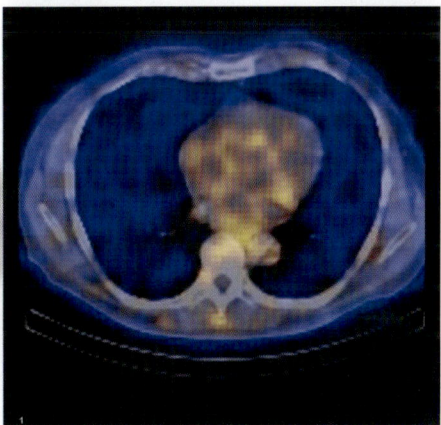

FIGURE 10-25: Octreotide Transaxial 2

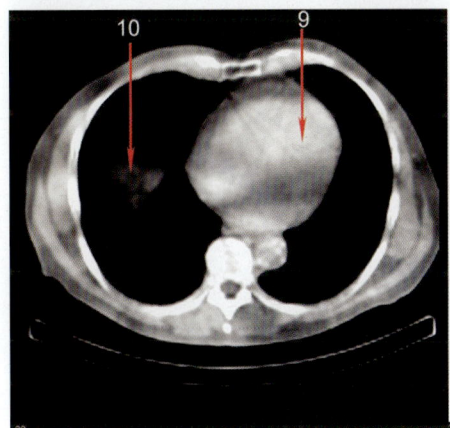 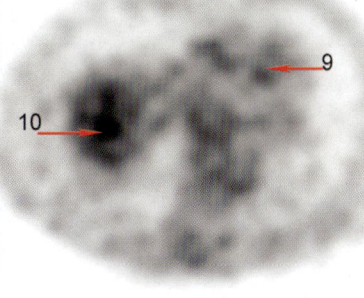

FIGURE 10-26: Octreotide Transaxial 3

1. Ascending aorta
2. Semispinalis muscle
3. Serratus anterior muscle
4. Right atrium
5. Right ventricle
6. Descending aorta
7. Pulmonary vein
8. T8
9. Left ventricle
10. Diaphragm

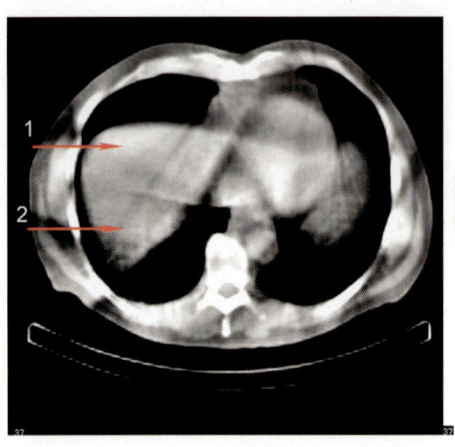
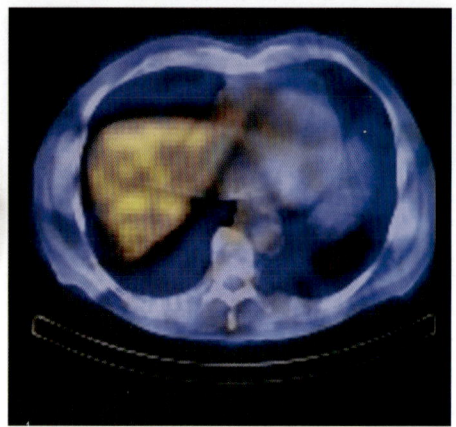

FIGURE 10-27: Octreotide Transaxial 4

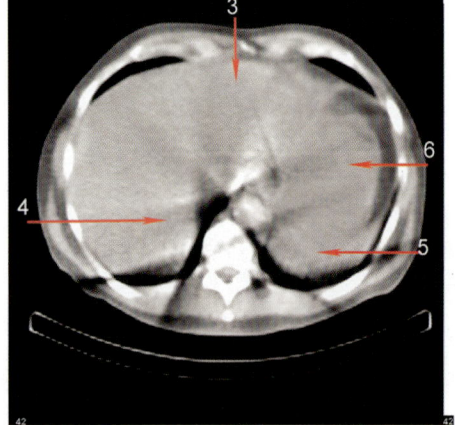
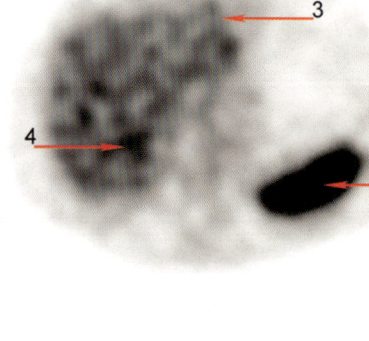
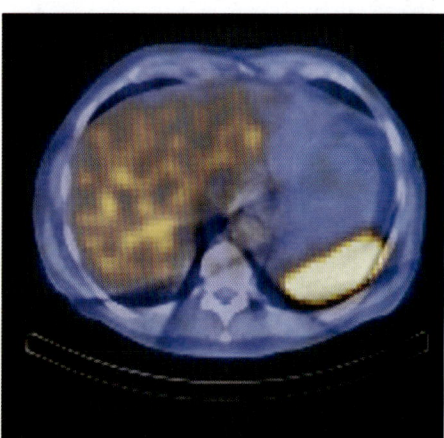

FIGURE 10-28: Octreotide Transaxial 5

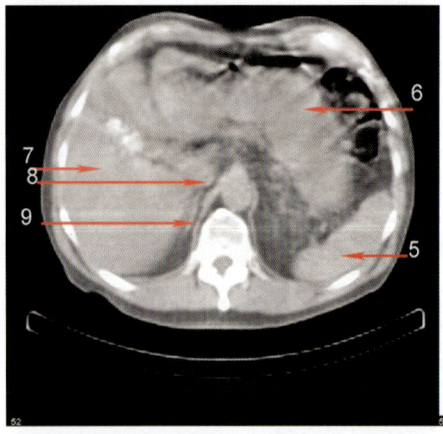
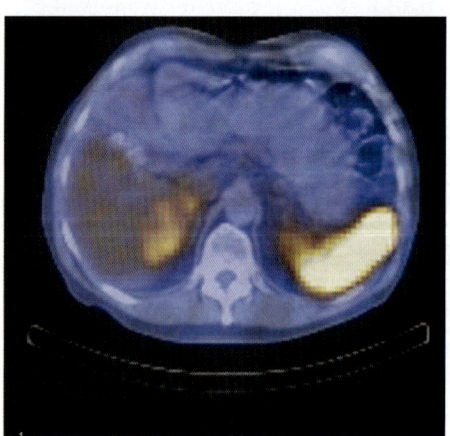

FIGURE 10-29: Octreotide Transaxial 6

1. Right hepatic lobe (segment VIII)
2. Right hepatic lobe (segment VII)
3. Left hepatic lobe (segment III)

4. Right hepatic lobe (segment VI)
5. Spleen
6. Stomach

7. Right hepatic lobe (segment V)
8. Inferior vena cava
9. Diaphragmatic crura

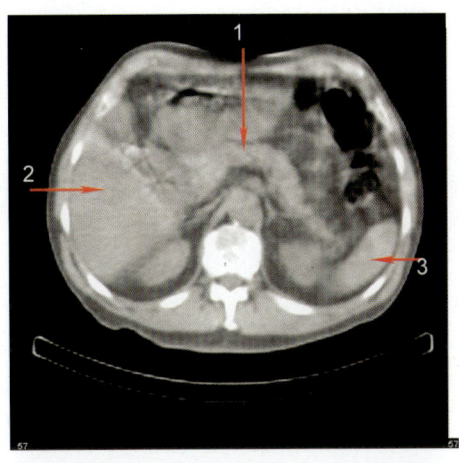
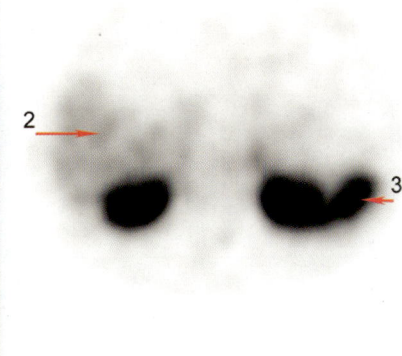
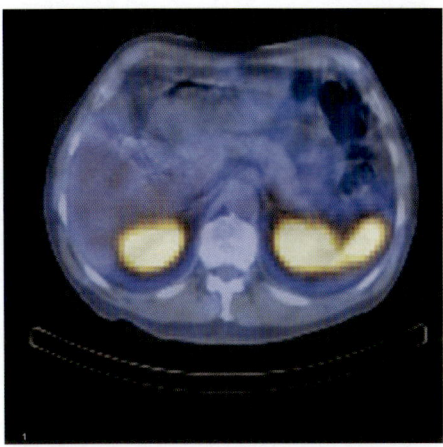

FIGURE 10-30: Octreotide Transaxial 7

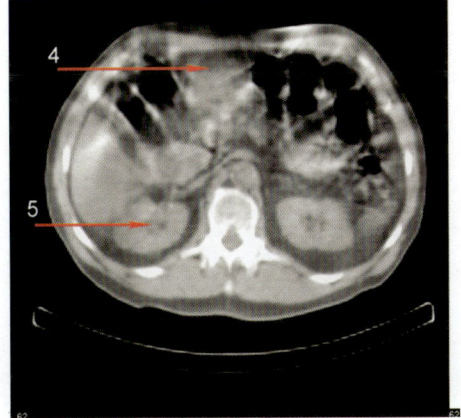
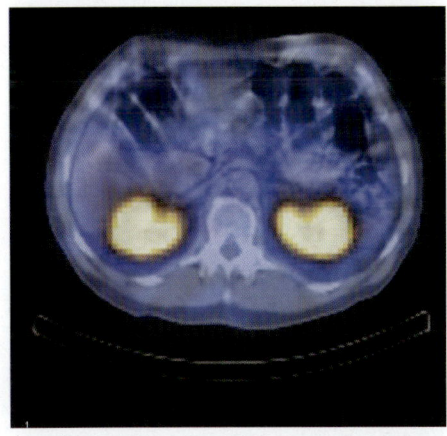

FIGURE 10-31: Octreotide Transaxial 8

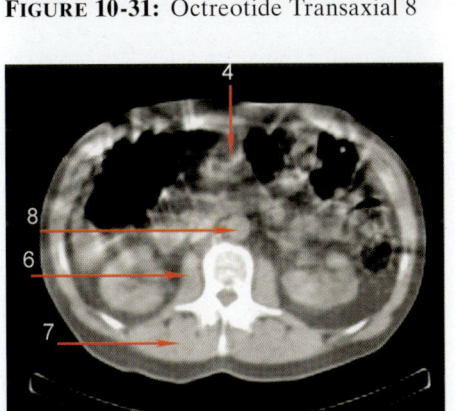
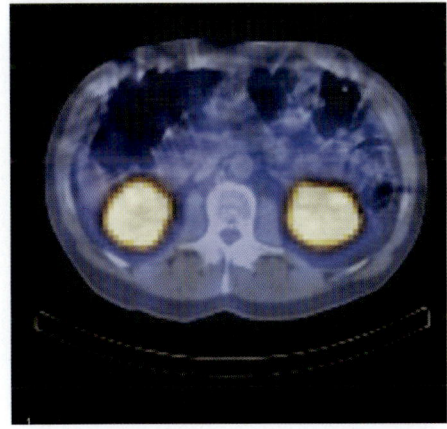

FIGURE 10-32: Octreotide Transaxial 9

1. Pancreas
2. Right hepatic lobe (segment V)
3. Spleen
4. Transverse colon
5. Kidney
6. Psoas muscle
7. Erector spinae muscle
8. Abdominal aorta

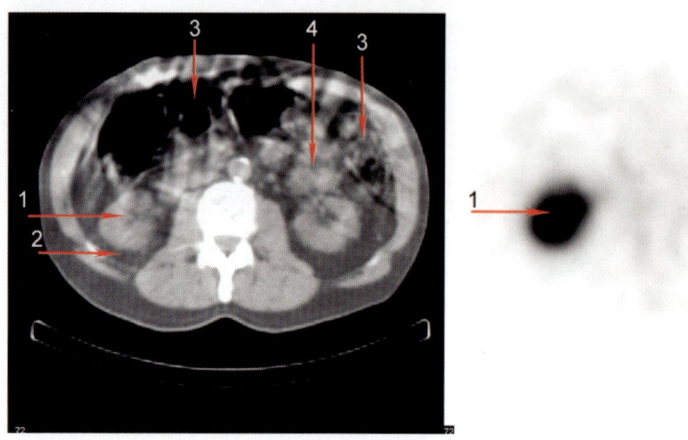 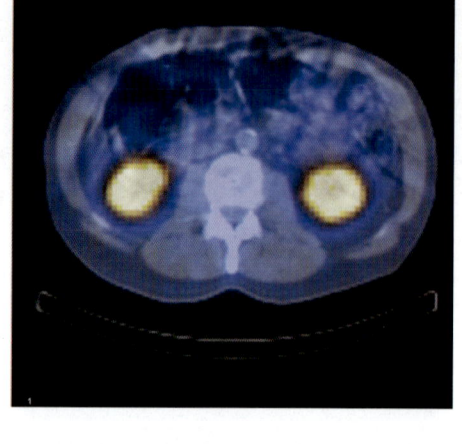

FIGURE 10-33: Octreotide Transaxial 10

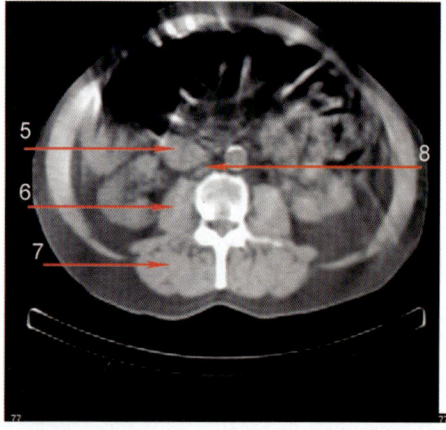 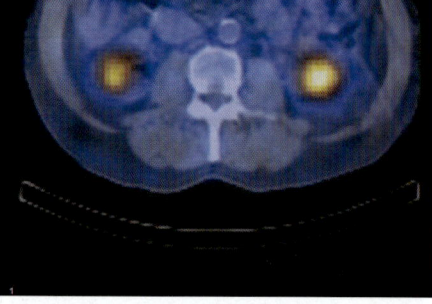

FIGURE 10-34: Octreotide Transaxial 11

1. Kidney
2. Posterior perirenal space
3. Transverse colon

4. Jejunum
5. Duodenum
6. Psoas muscle

7. Erector spinae muscle
8. Inferior vena cava

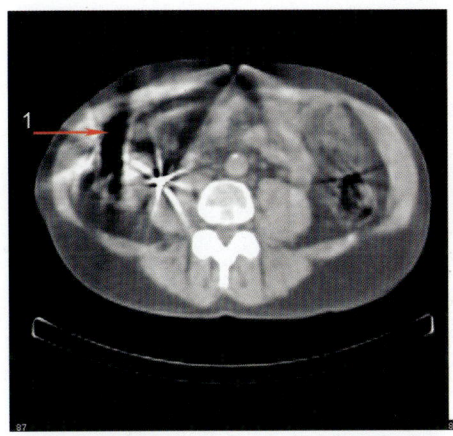
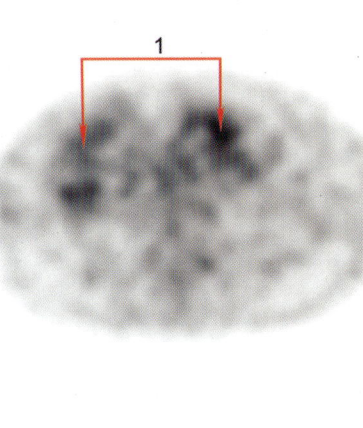
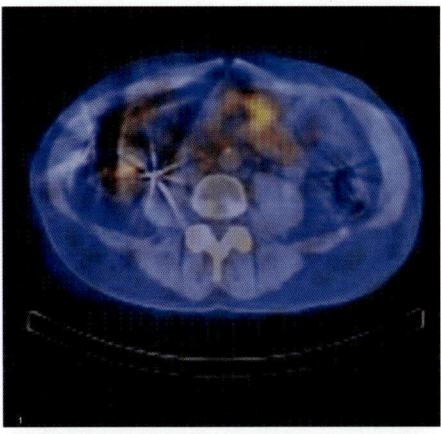

FIGURE 10-35: Octreotide Transaxial 12

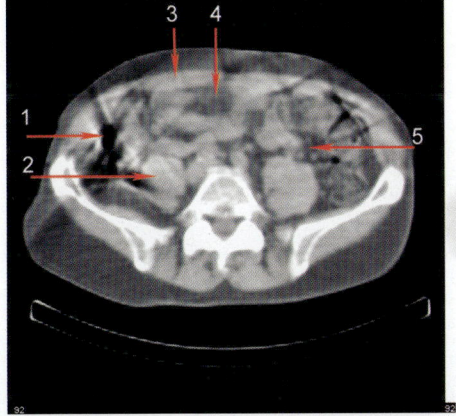
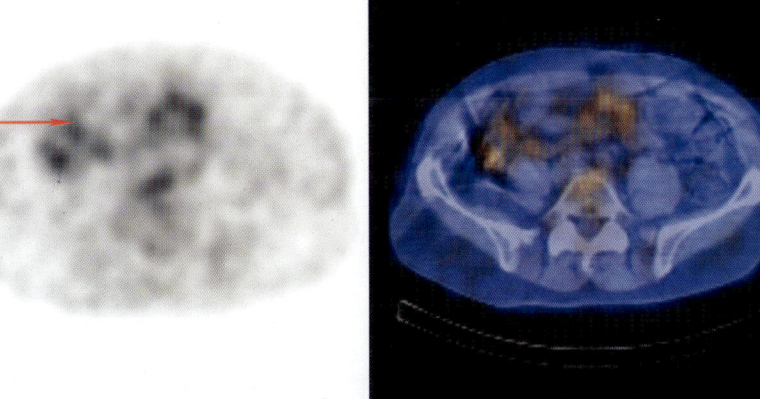

FIGURE 10-36: Octreotide Transaxial 13

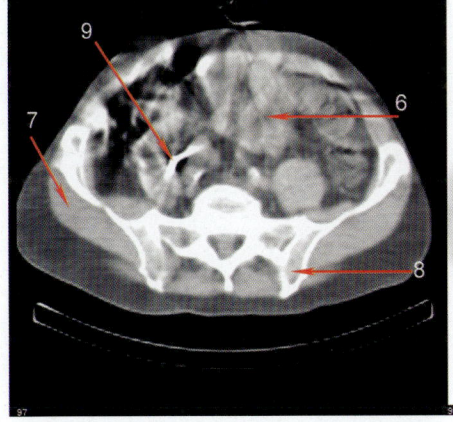
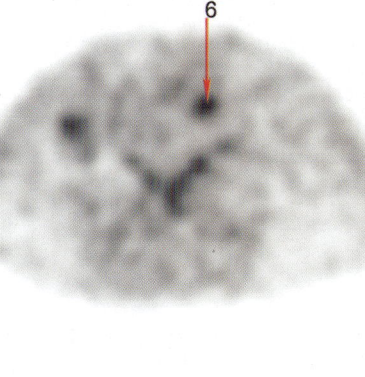
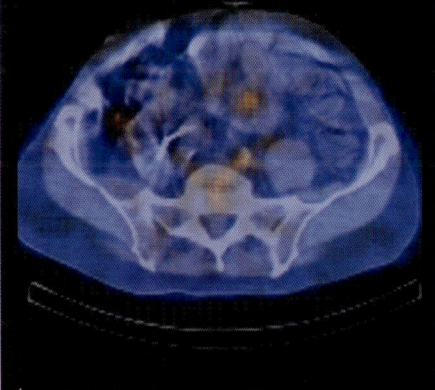

FIGURE 10-37: Octreotide Transaxial 14

1. Colon
2. Psoas muscle
3. Rectus abdominalis muscle
4. Omentum
5. Mesentery
6. Ileum
7. Gluteus muscle
8. Iliac tuberosity
9. Metallic clip

Part II
Anatomic Variations and Artifacts of PET/CT and SPECT/CT

As with all imaging techniques, thorough knowledge of normal distribution and anatomic, physiologic, and pathologic variants is required to avoid misinterpretation. Lymphoid tissues may demonstrate the significant uptake of F-18 fluorodeoxyglucose (FDG), and the increased activity is often seen in the tonsils and adenoids, particularly marked in children.[1] Pharyngeal lymphoid activity may be asymmetric, making it more difficult to differentiate from physiologic activity. After exercise or if contraction takes place during the uptake period after injection of F-18 FDG, there is an increased uptake in active skeletal muscles, which relates to increased aerobic glycolysis of active muscles. Laryngeal muscle activity may be related to speech, and swallowing may cause the tongue base activity. Increased uptake of FDG in stressed sternocleidomastoids, scalene, and trapezius muscles is one of the common causes of positron emission tomography (PET) interpretation problems. Symmetric FDG uptakes in the neck, shoulders, and thoracic paraspinal areas is probably related to activated brown fatty tissues in underweight patients during increased sympathetic nerve activity, attributable to cold stress.[2] Therefore, exercise should be prohibited on the day of scanning to minimize muscle activities, and benzodiazepines may be used to abolish the muscle uptakes often seen in tense patients.

Anxiety-related increased muscular tension may cause symmetric or asymmetric uptakes in neck and paraspinal muscles. Hyperventilation also may produce a diaphragmatic activity. Problems may result from involuntary muscle spasm, such as that seen with torticollis, which may lead to asymmetric activity in the neck muscle. Trauma and inflammation may also cause enhanced muscle activity.[3]

A number of areas in the gastrointestinal system may show the uptake of FDG, which may be partly attributable to smooth muscle activity. FDG activity is most noticeable in the large intestine and to a lesser extent in the stomach (particularly gastric fundus) and small bowl. Sagittal images may be helpful in defining the oblique position of the stomach as it passes caudally. Marked activity may be seen in the cecum and rectosigmoid at times. Cecal activity may be attributable to lymphoid tissues.[4] Hepatic and splenic activities on an attenuation-corrected image are slightly nonuniform. Hepatic activity on uncorrected images is similar to lung uptakes with slightly increased activity in the periphery, and there is no gallbladder activity.

Significant and variable urinary activity is seen in all patients because FDG is not totally reabsorbed in the renal tubules.[4] If there is a significant activity in the renal collecting system of an obstructed kidney, reconstruction artifacts may interfere with visualization of the upper abdomen. The skin contamination from urinary activity may be limited at the superficial areas. Glandular breast tissue often demonstrates moderate uptake of FDG in premenopausal women. There is little breast activity after menopause, but women taking estrogen for hormone replacement therapy may show enhanced uptake of FDG. Focal increased activity may be noted at the nipple, and asymmetric breast activity may be seen in a woman who fed her baby on one side. Thymic activity with an inverted V shape may be identified in children or those in their late teens, because thymus harbors many white cells. The activities in the bone marrow and spleen are markedly enhanced by the growth factors or cytokine therapy. Increased marrow activity is also noted in patients with acute infection because of increased production of white cells.[5] There may be higher activity in superficial structures, such as skin, if attenuation correction is not performed on whole-body FDG PET. A common artifact resulting from this phenomenon is caused by the axillary skinfold. Photon-deficient areas may result from metallic hip prostheses, breast implants, medallions, coins, and keys.[6] Ring artifacts may occur if there is a misregistration between transmission and emission data, and they are particularly apparent at borders where there are sudden large changes in activity, especially at a metal prosthesis. A partly infiltrated injection not only may cause reconstruction artifacts across the trunk, but also may result in a low count study.[7] Focal axillary FDG activity may occur following subcutaneous extravasation. FDG whole-body imaging can lead to unusual appearance if the patient moves between bed-scan positions. The upper part of an arm may be visible in the higher scanning positions with absent or amputated lower down when moved out of the field of view of the scanning positions. The lacrimal glands may show moderately intense uptake of C-11 methionine, but may be easily recognized by the symmetric distribution and the anterior location below the frontal lobes of the brain. There is a normal bone marrow uptake of methionine within the sphenoid and clivus. There are also activities within a recent biopsy site, and within the pons following radiotherapy to the surrounding region. Increased salivary gland activity is symmetrically demonstrated by methionine PET, but asymmetric activity after unilateral surgery or radiotherapy of head and neck tumors may be problematic. The majority of patients show no thyroid uptake of methionine. However, diffusely increased activity has been shown in patients with thyroiditis and thyrotoxicosis. Focal uptake may be noted in benign thyroid nodules. In the abdomen, physiologic uptake of methionine within the bowel may limit the use of methionine PET in the study of colon and pelvic tumors. High pancreatic uptake is normally identified, and pancreatic tumors may obscure the uptake of methionine. The urinary methionine activity is variable, but occurs in the minority of patients. There is minimal renal cortical activity, which would not interfere with the interpretation of renal PET.[8]

Anatomic variants of normal structures scanned on an unusual computed tomography (CT) plane should not be misinterpreted as mass lesions. Scars can be quite difficult to distinguish from tumors if the scar tissue forms a mass lesion. Scar tissue does not enhance until the interstitial phase after injection of contrast agent. A pseudo thrombus effect occurs most often at the confluence of veins, particularly pulmonary and hepatic veins.[9] Bowel loops after biliary surgery often collapse and may easily be mistaken for local mass. Bladder wall edema following urethral resection can mimic a residual tumor.[10]

In studies using relatively thick CT sections and nonoverlapping reconstructions, partial volume effects can reduce the contrast of small pulmonary or hepatic lesions. Scans that cut a portion of an adjacent structure may simulate lesions where none exist. Examples are posterior attachment of the first rib to the sternum, upper pole of the right kidney, gallbladder, and air-filled bowel segments. The lower part of hepatic caudate lobe may be mistaken for a lymph node at the porta hepatis on CT. If the CT window is set too wide, image contrast is reduced. Conversely, too narrow a window setting can significantly increase image noise and abolish gray-scale differentiation in fatty tissue areas. The low-energy spectral components are absorbed more than the higher-energy components, as the thickness of the scanned object increases. The increased beam hardening would cause a decline in CT numbers, because absorption is reduced at higher energies. If the local atomic composition of the object differs markedly from that of water (bone, metal, contrast medium), beam-hardening artifact will continue to occur.[11] Image noise represents random fluctuations of the measured

CT number and can be reduced by increasing the section thickness or using smoothing filter kernel. Image noise is increased in thin-section multislice CT images because of the narrower collimation and the reduced detector dose. Noise and spatial resolution are not uniformly distributed throughout a set of spiral CT images. Motion in the scanned section during one rotation of the X-ray tube will result in nonconsistent projection data because of different configurations of the scanned objects in the various X-ray projections. As a result, varying degrees of motion artifact will appear throughout the reconstructed image, being most pronounced in the region of the moving structure.[12] CT scanners can cause a variety of equipment-related artifacts, most of which are attributable to errors of application, adjustment problems, or scanner defects. Spiral artifacts may occur in spiral or multislice CT at interfaces that are slightly angulated relative to the scan plane. Some beam artifacts are attributable to the geometry of the X-ray beam and are more pronounced the more detector rows present.[13]

11 PET/CT Anatomy: Variations and Artifacts

The individual component of integrated positron emission tomography (PET)/computed tomography (CT) systems can introduce artifacts when images are combined and fused. Because the two systems are integrated and heavily dependent on each other, specific problems arise in addition to the normal variants of the individual system. Artifacts and variants are related to the technology, patient, operator, and algorithm.[1] The technology of detectors and electronics determines spatial and temporal resolution of the modalities. The hardware electronics determine temporal resolution of the PET acquisition. Bed motion and pitch have an effect on the longitudinal axis sampling and determine the z-axis resolution in CT. This combination together with the final reconstruction software determines the three-dimensional spatial resolution of the system.[2] Motion and movement artifacts are difficult to resolve and are attributable to involuntary and uncontrollable movement of internal organs, as well as preventable shift and sliding because of patient discomfort during scanning. Artifacts induced by respiration are inevitable and cannot be circumvented when patients are imaged over an extended period of time.[3] Several interesting physiologic variants have been identified with PET/CT. Foremost among these is the identification of brown fat as fluorodeoxyglucose-avid tissue in the supraclavicular area and paramediastinum.

Section 1: PET/CT Anatomy Variations

Esophageal: Variations

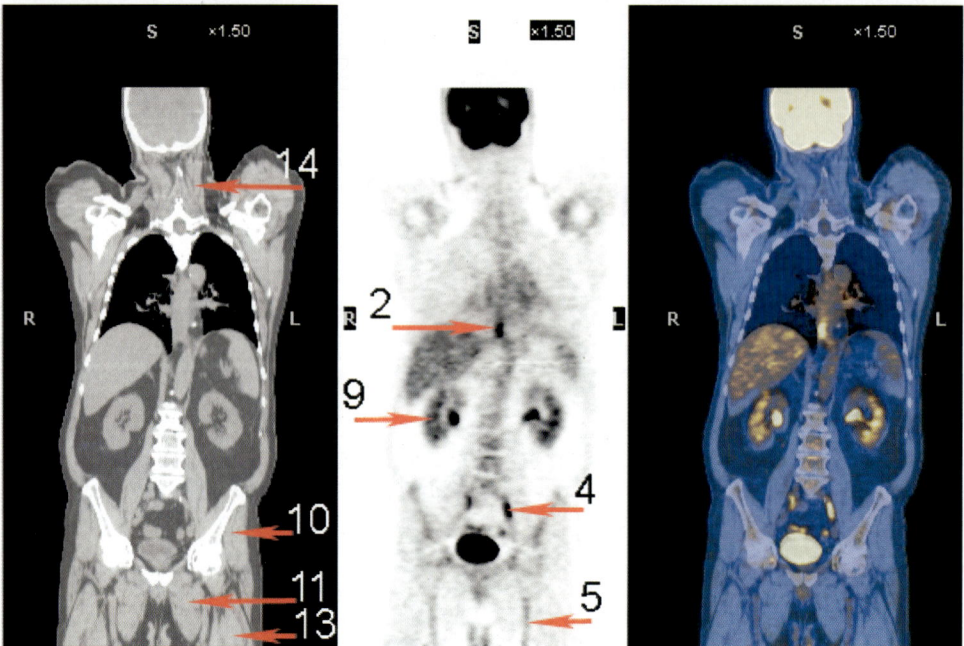

FIGURE 11-1: Esophageal Coronal 1

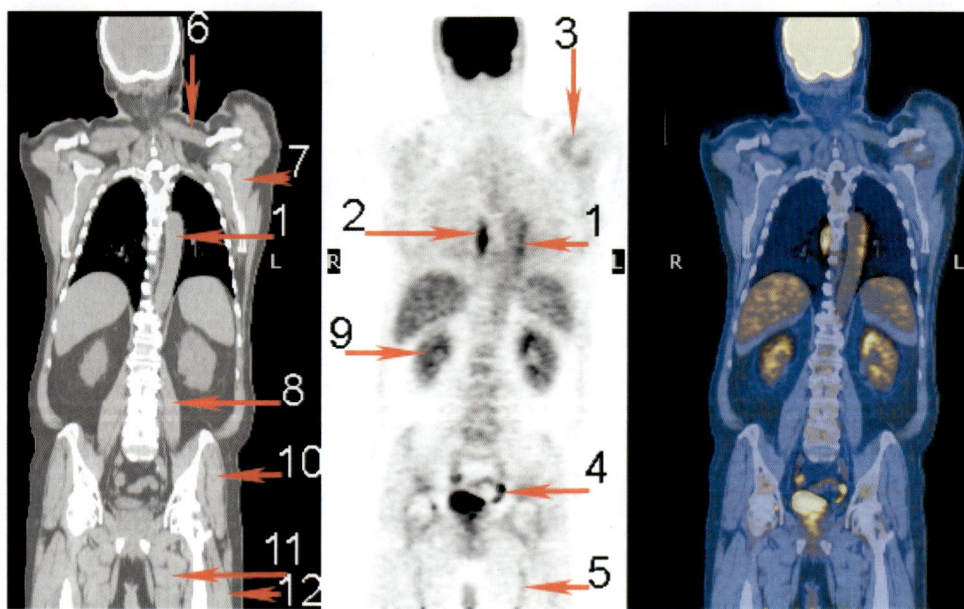

FIGURE 11-2: Esophageal Coronal 2

1. Descending aorta
2. Esophagus
3. Shoulder joint capsule
4. Distal ureter
5. Femoral artery
6. Trapezius muscle
7. Infraspinatus muscle
8. Psoas muscle
9. Kidney
10. Gluteus muscle
11. Adductor muscle
12. Vastus lateralis muscle
13. Rectus femoris muscle
14. Scalene muscle

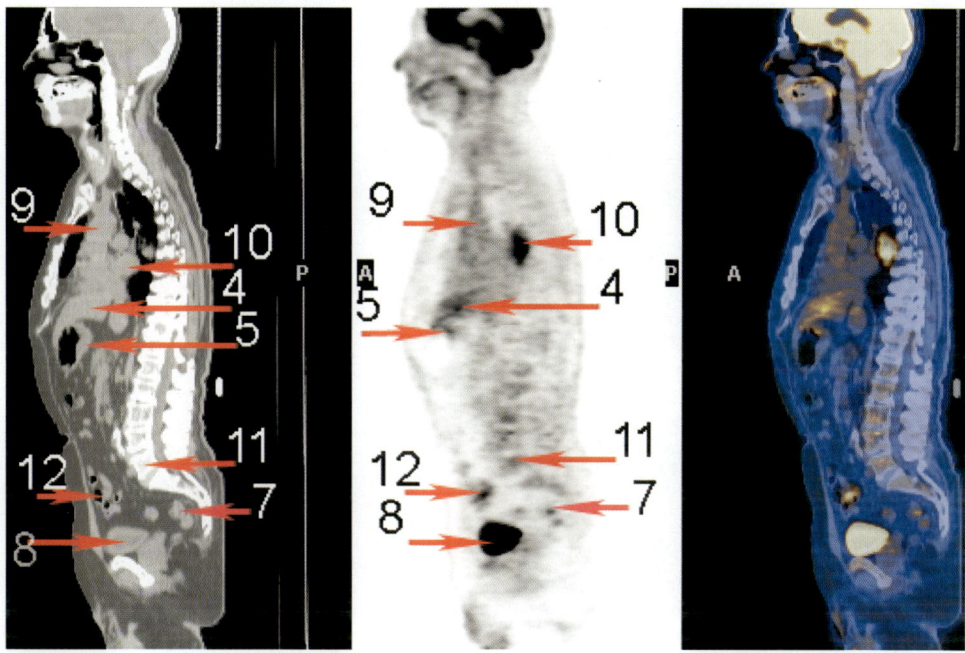

FIGURE 11-3: Esophageal Sagittal 1

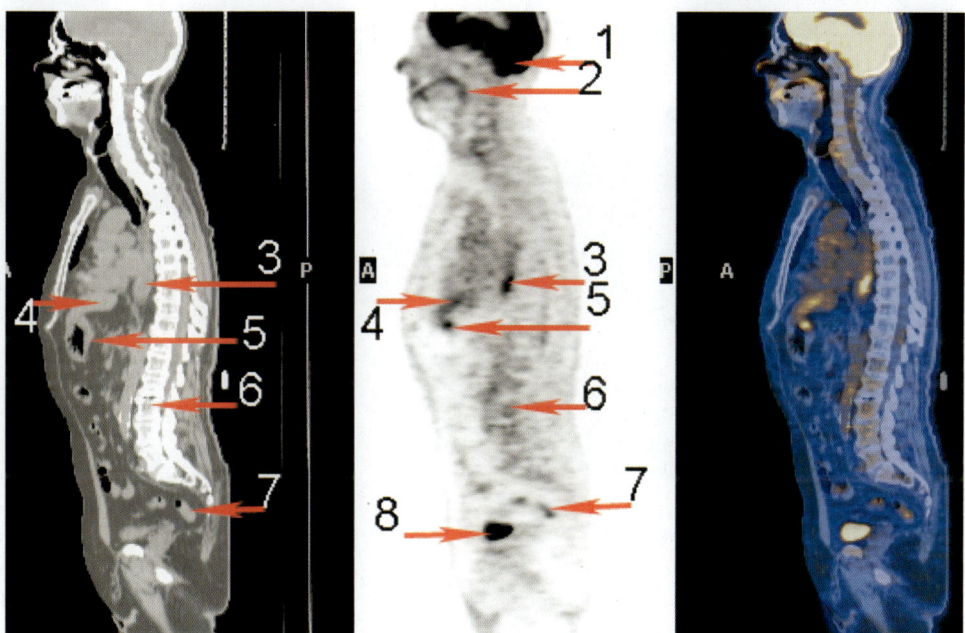

FIGURE 11-4: Esophageal Sagittal 2

1. Cerebellum
2. Soft palate
3. Distal esophagus
4. Left hepatic lobe
5. Transverse colon
6. L3 marrow
7. Sigmoid colon
8. Bladder
9. Aortic arch
10. Pulmonary artery
11. L5 marrow
12. Ileum

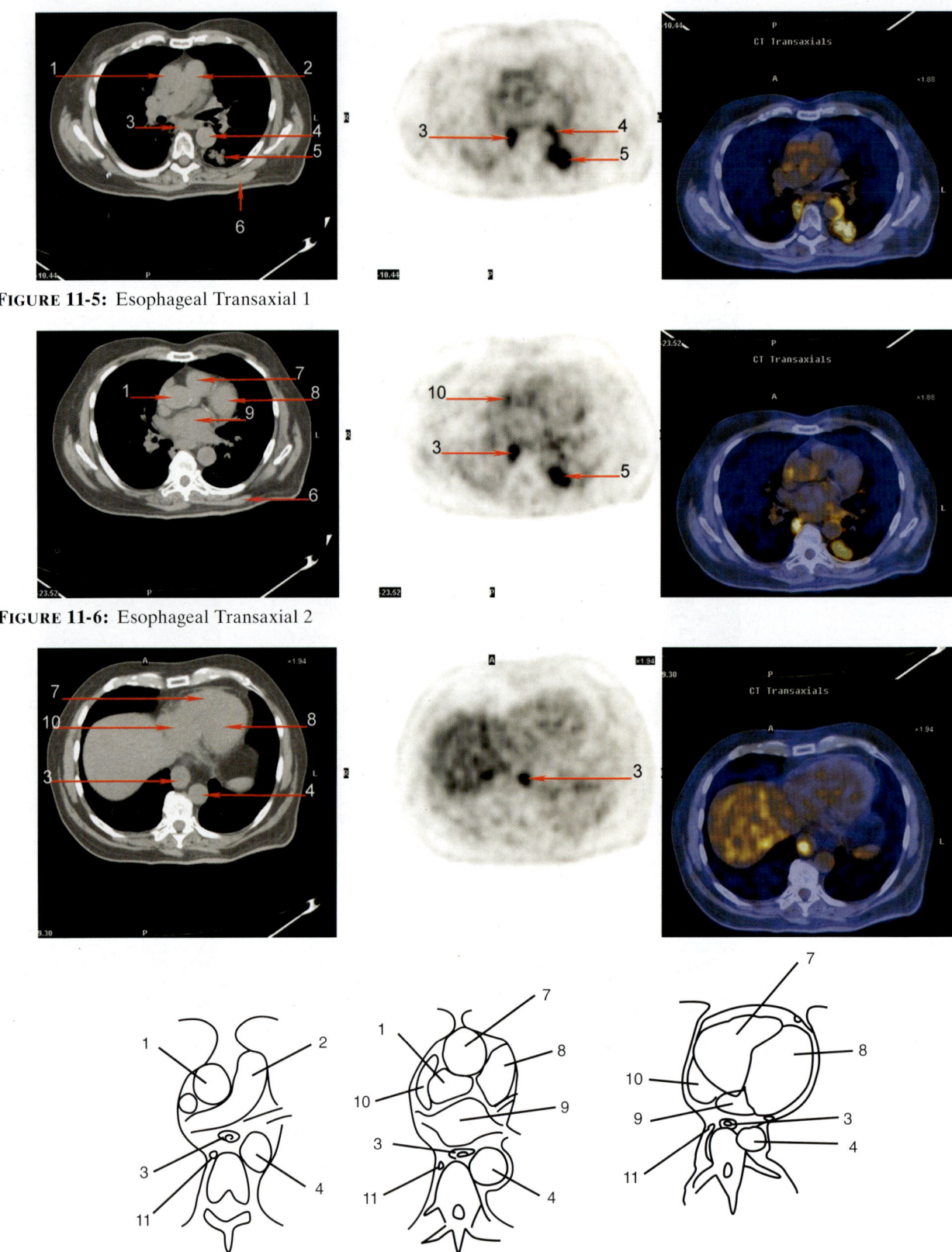

FIGURE 11-5: Esophageal Transaxial 1

FIGURE 11-6: Esophageal Transaxial 2

FIGURE 11-7: Esophageal Transaxial 3

1. Aorta
2. Main pulmonary artery
3. Esophagus
4. Descending aorta
5. Tumor in superior segment of left lower lung
6. Trapezius muscle
7. Right ventricle
8. Left ventricle
9. Left atrium
10. Right atrium
11. Azygos vein

Liver: Variations

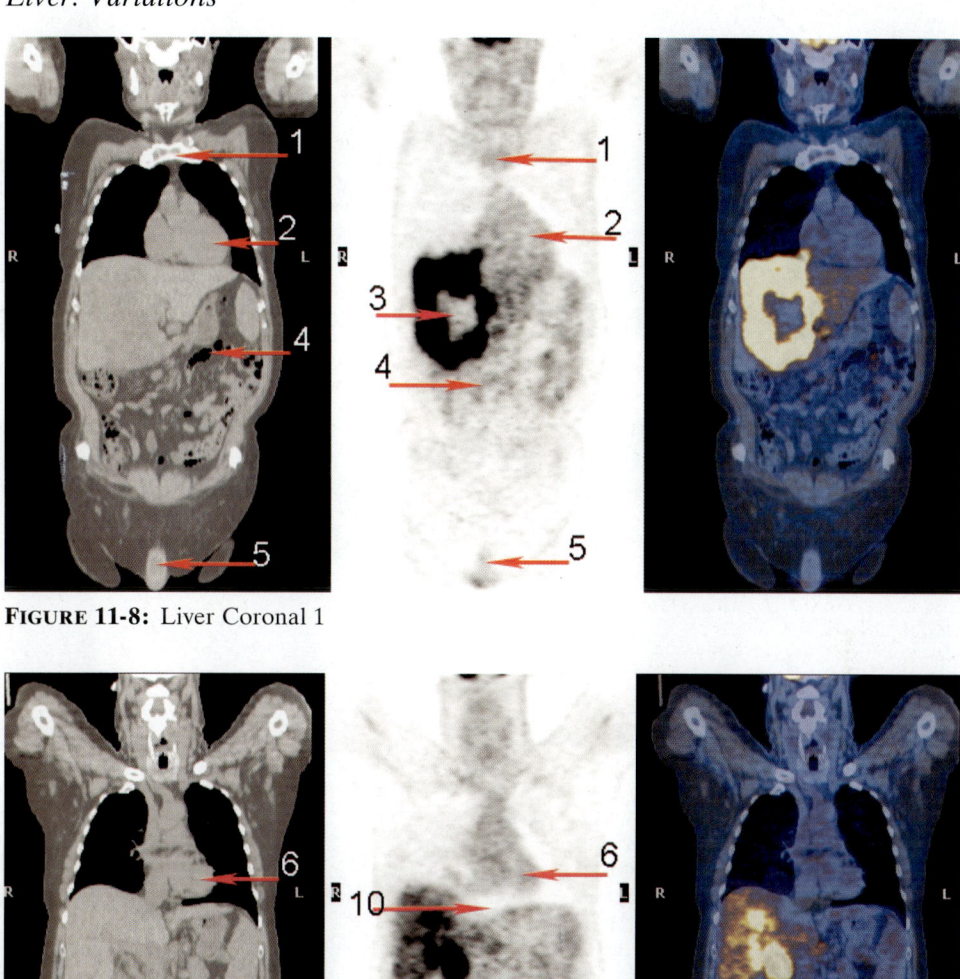

FIGURE 11-8: Liver Coronal 1

FIGURE 11-9: Liver Coronal 2

1. Manubrium
2. Right ventricle
3. Metastatic tumor in right hepatic lobe with central necrosis
4. Transverse colon
5. Penis
6. Left ventricle
7. Ascending colon
8. Bladder
9. Scrotum
10. Reconstruction artifacts (linear photon deficiency)

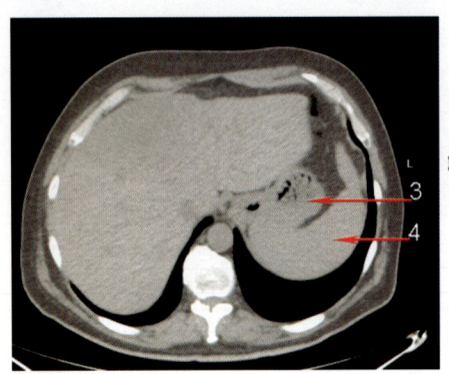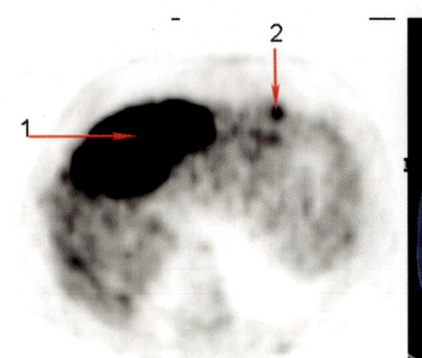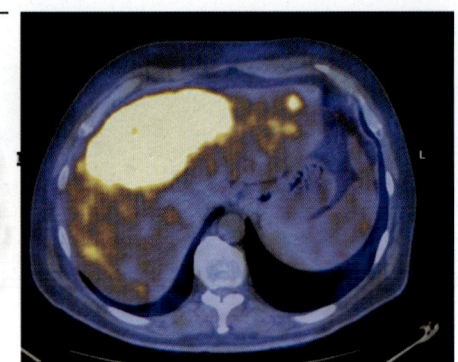

FIGURE 11-10: Liver Transaxial 1

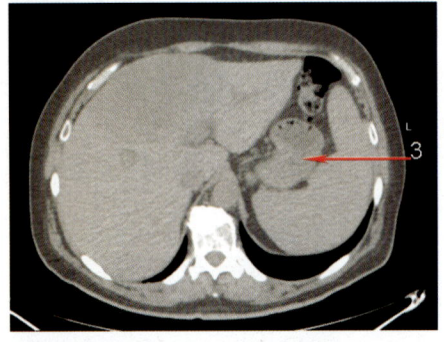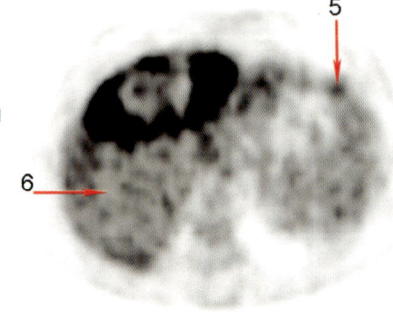

FIGURE 11-11: Liver Transaxial 2

1. Metastatic tumor in upper segment of right hepatic lobe (segment VIII)
2. Metastatic tumor in lateral segment of left hepatic lobe (segment III)
3. Stomach
4. Spleen
5. Metastatic tumor in the anterior tip of spleen
6. Upper posterior segment of right hepatic lobe (segment VII)

Lung: Variations

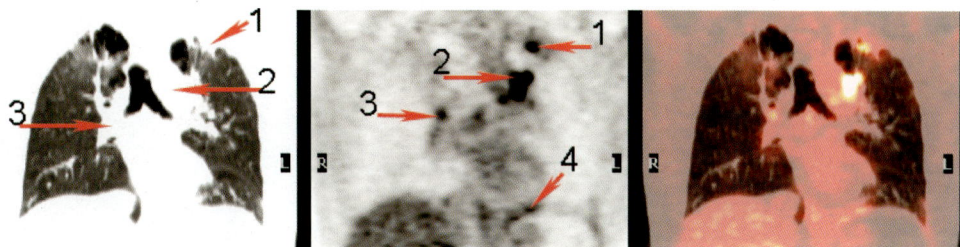

FIGURE 11-12: Lung Coronal 1

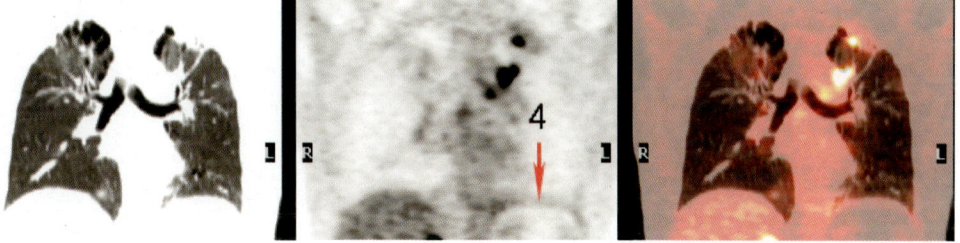

FIGURE 11-13: Lung Coronal 2

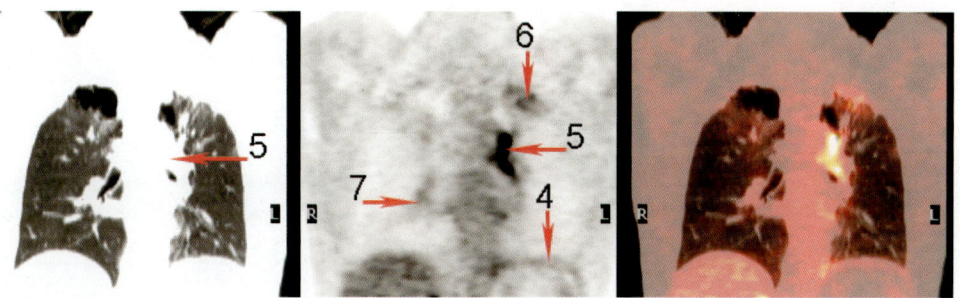

FIGURE 11-14: Lung Coronal 3

1. Metastatic tumor in left lung apex
2. Metastatic tumor in aortopulmonic window
3. Metastatic tumor in right hilum
4. Pleural inflammatory reaction or pleuritis
5. Metastatic tumor in left perihilum
6. Inflammatory lesion in left lung apex
7. Inflammatory lesion in right infrahilum

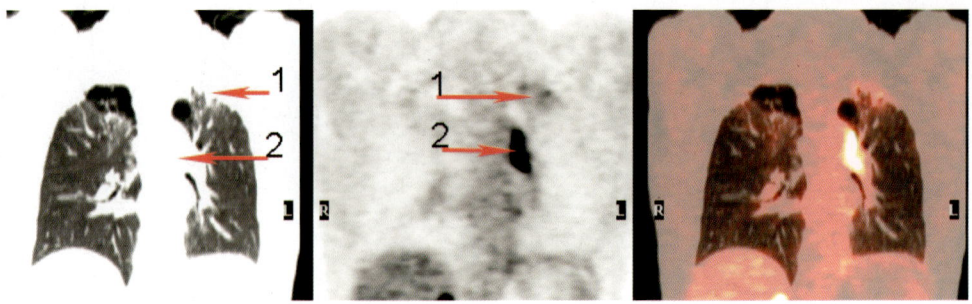

FIGURE 11-15: Lung Coronal 4

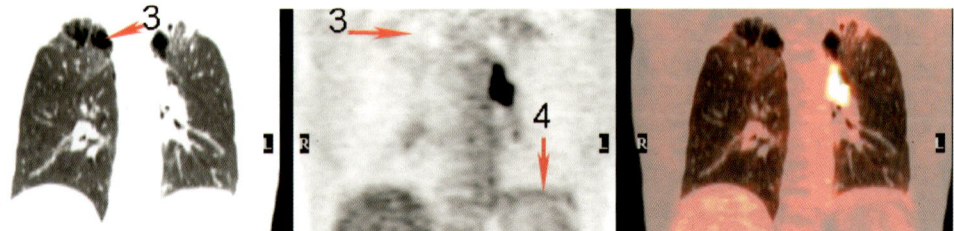

FIGURE 11-16: Lung Coronal 5

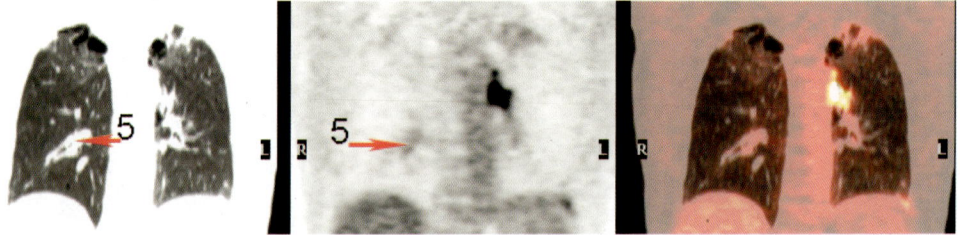

FIGURE 11-17: Lung Coronal 6

1. Inflammatory lesion in
   left lung apex
2. Metastatic tumor in left
   perihilum

3. Bullae in right lung apex
4. Inflammatory reaction or
   pleuritis

5. Inflammatory lesion in
   right infrahilum

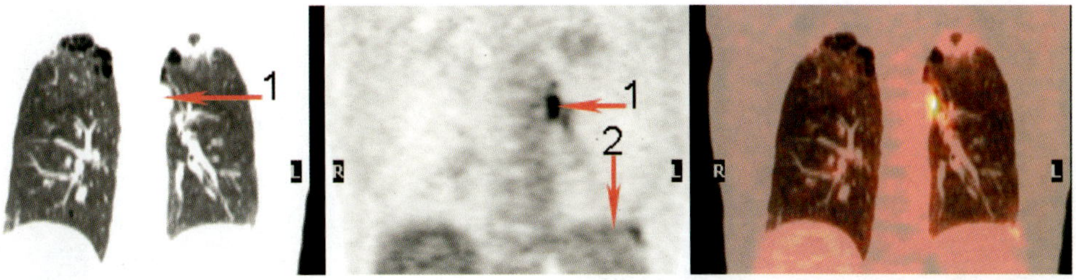

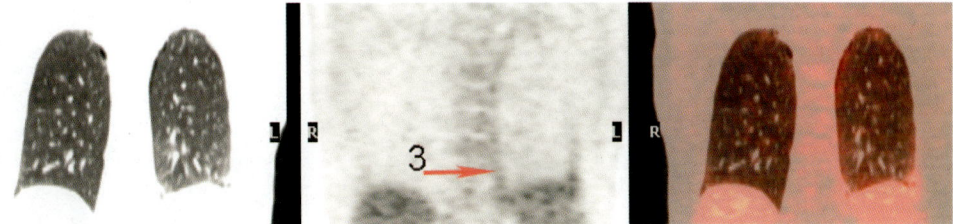

FIGURE 11-18: Lung Coronal 7

FIGURE 11-19: Lung Coronal 8

1. Metastatic tumor in left perihilum

2. Inflammatory reaction or pleuritis in left lung base

3. Inflammatory reaction or pleuritis in left paramediastinum

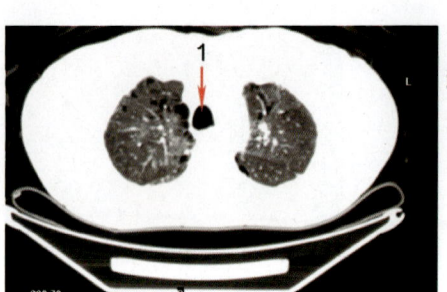
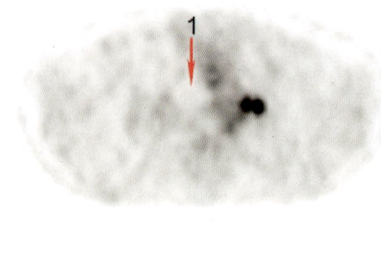
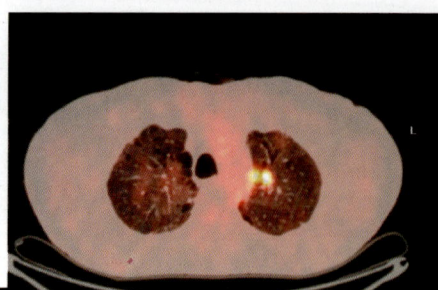

FIGURE 11-20: Lung Transaxial 1

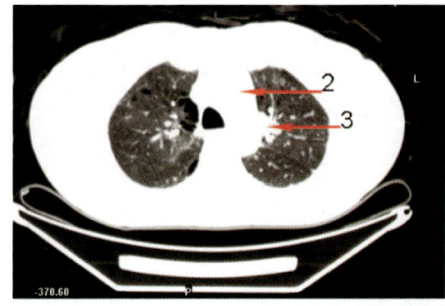
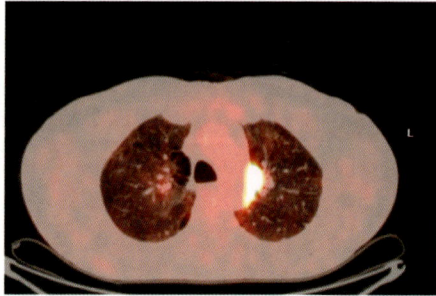

FIGURE 11-21: Lung Transaxial 2

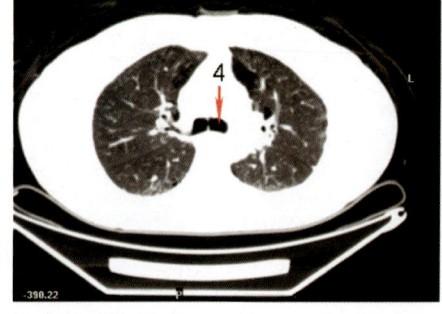
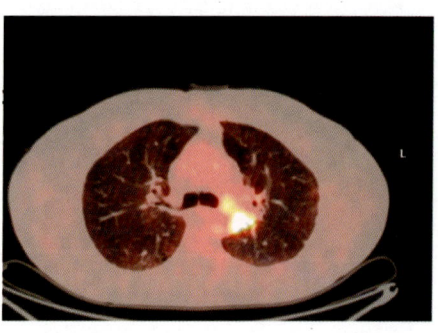

FIGURE 11-22: Lung Transaxial 3

1. Trachea
2. Aortic arch

3. Metastatic tumor in left perihilum

4. Left main bronchus

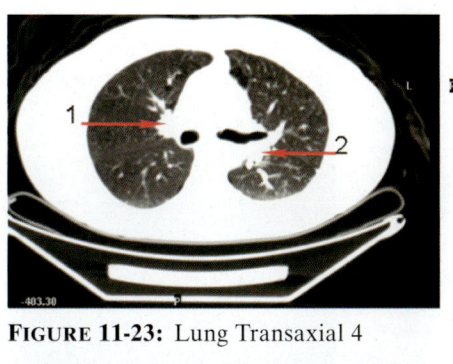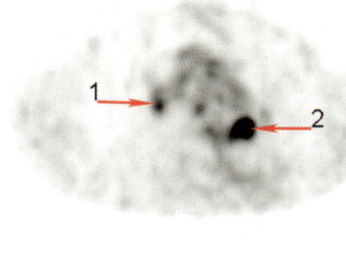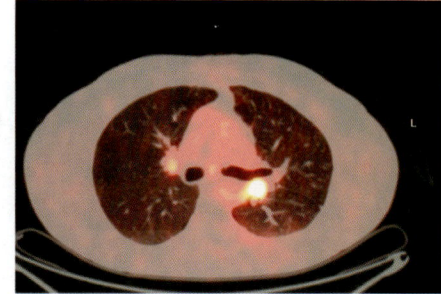

FIGURE 11-23: Lung Transaxial 4

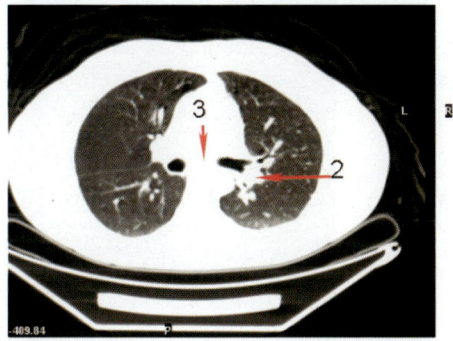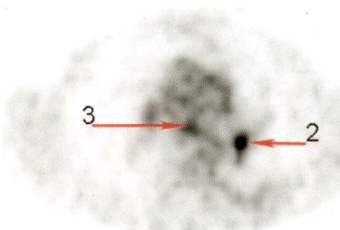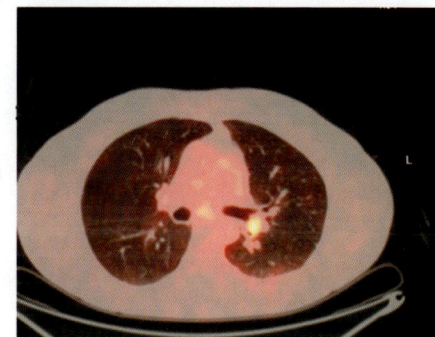

FIGURE 11-24: Lung Transaxial 5

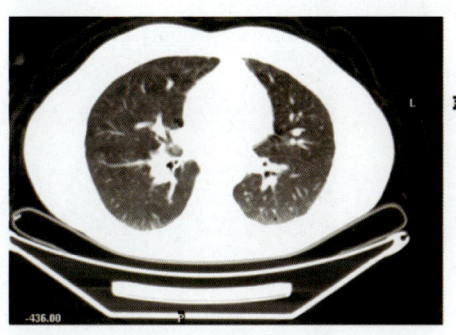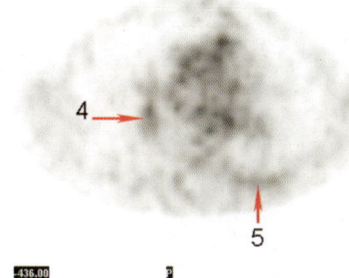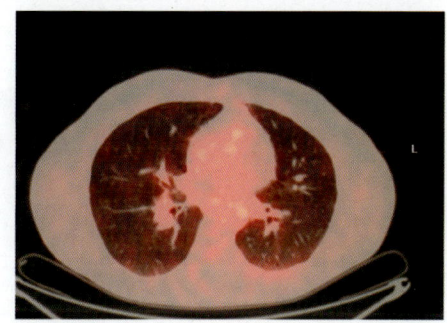

FIGURE 11-25: Lung Transaxial 6

1. Metastatic tumor in right hilum
2. Metastatic tumor in left hilum
3. Metastatic tumor in carina

4. Inflammatory lesion in right infrahilum

5. Inflammatory reaction or pleuritis

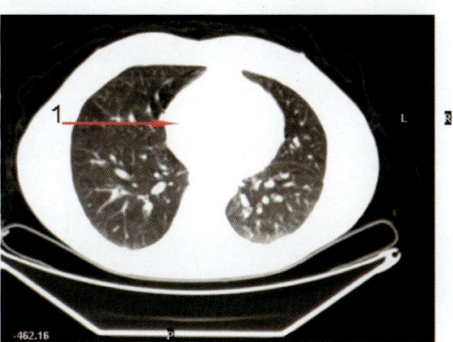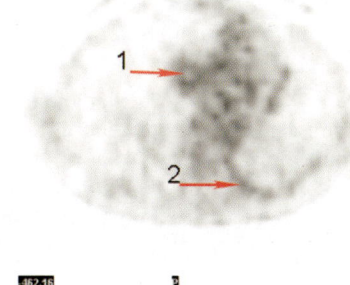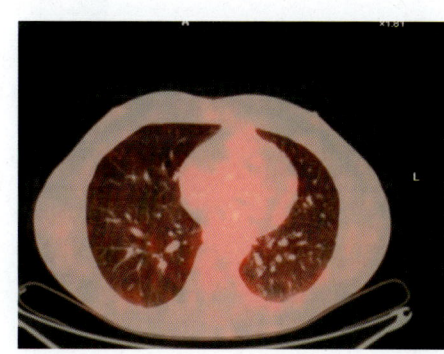

FIGURE 11-26: Lung Transaxial 7

1. Right atrium

2. Inflammatory reaction or pleuritis

Fluid Accumulation: Variations

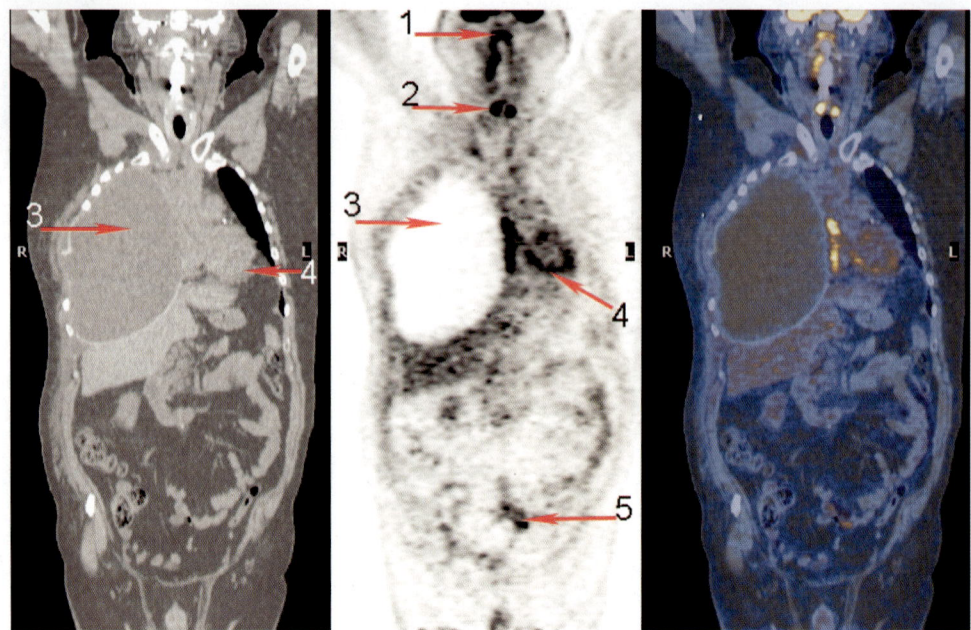

FIGURE 11-27: Fluid Accumulation Coronal 1

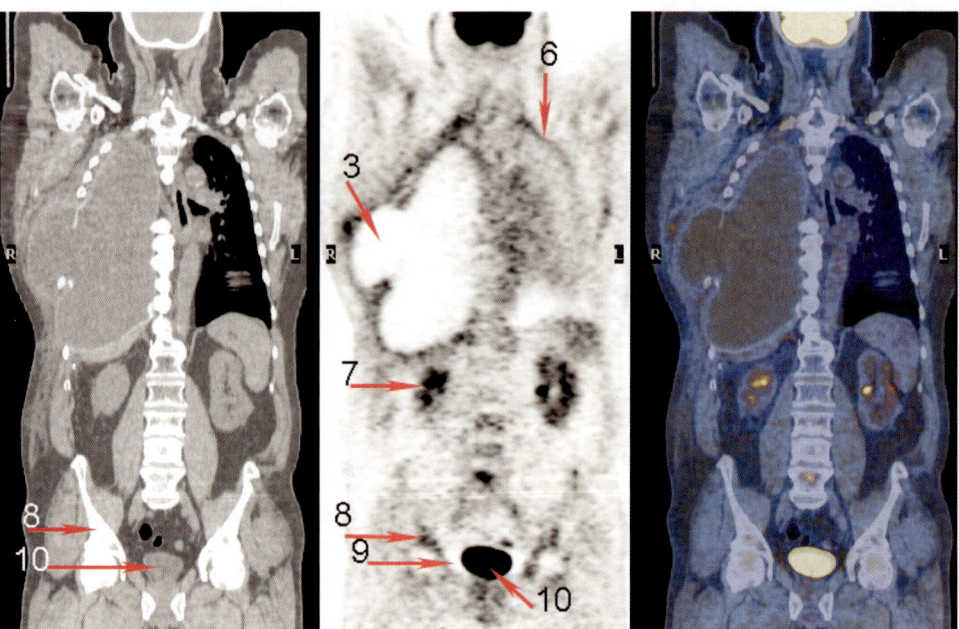

FIGURE 11-28: Fluid Accumulation Coronal 2

1. Tongue base	5. Sigmoid colon	9. Reconstruction artifact (linear photon deficiency)
2. Vocal cord muscle	6. Pleural reaction or pleuritis	
3. Fluid in right thorax	7. Kidney	10. Bladder
4. Left ventricular myocardium	8. Acetabulum	

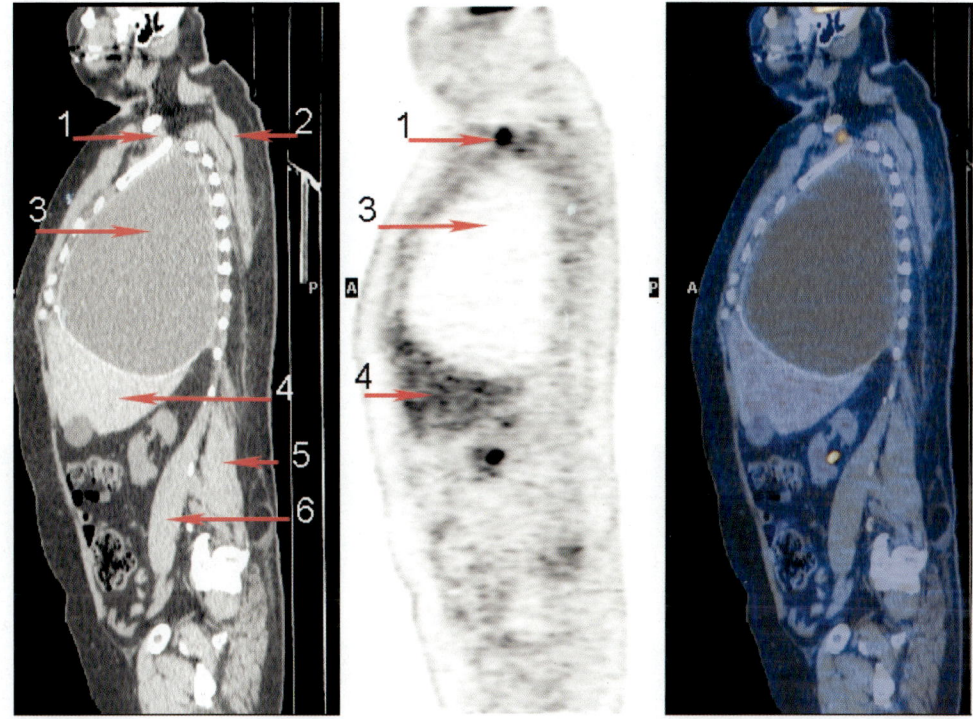

FIGURE 11-29: Fluid Accumulation Sagittal 1

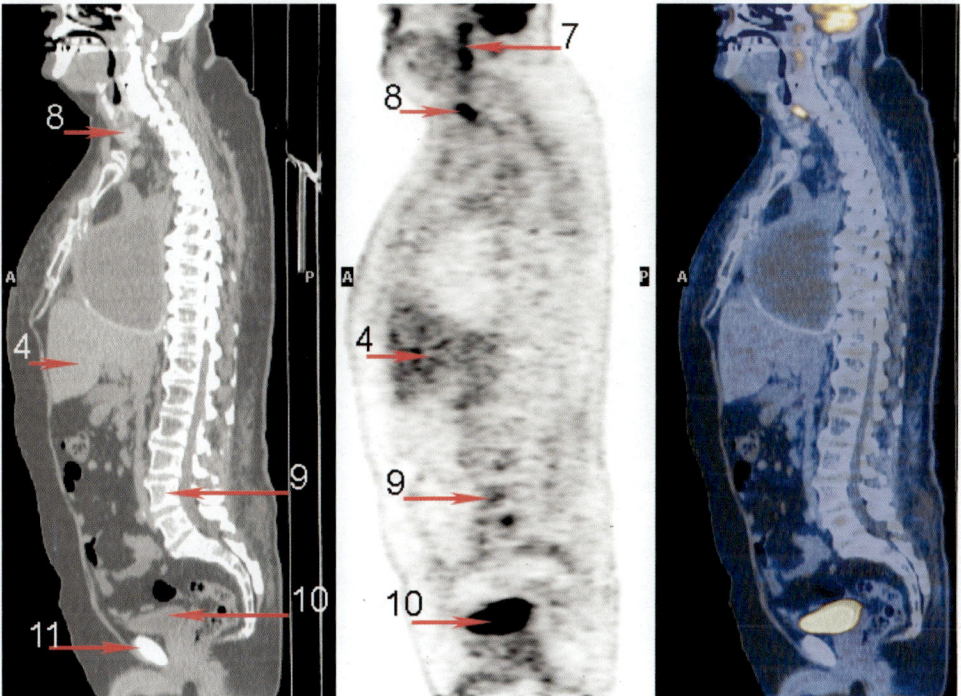

FIGURE 11-30: Fluid Accumulation Sagittal 2

1. Right subclavian vein
2. Trapezius muscle
3. Fluid in right thorax
4. Liver

5. Erector spinae muscle
6. Psoas muscle
7. Parotid gland
8. Vocal cord muscle

9. L4 marrow
10. Bladder
11. Symphysis pubis

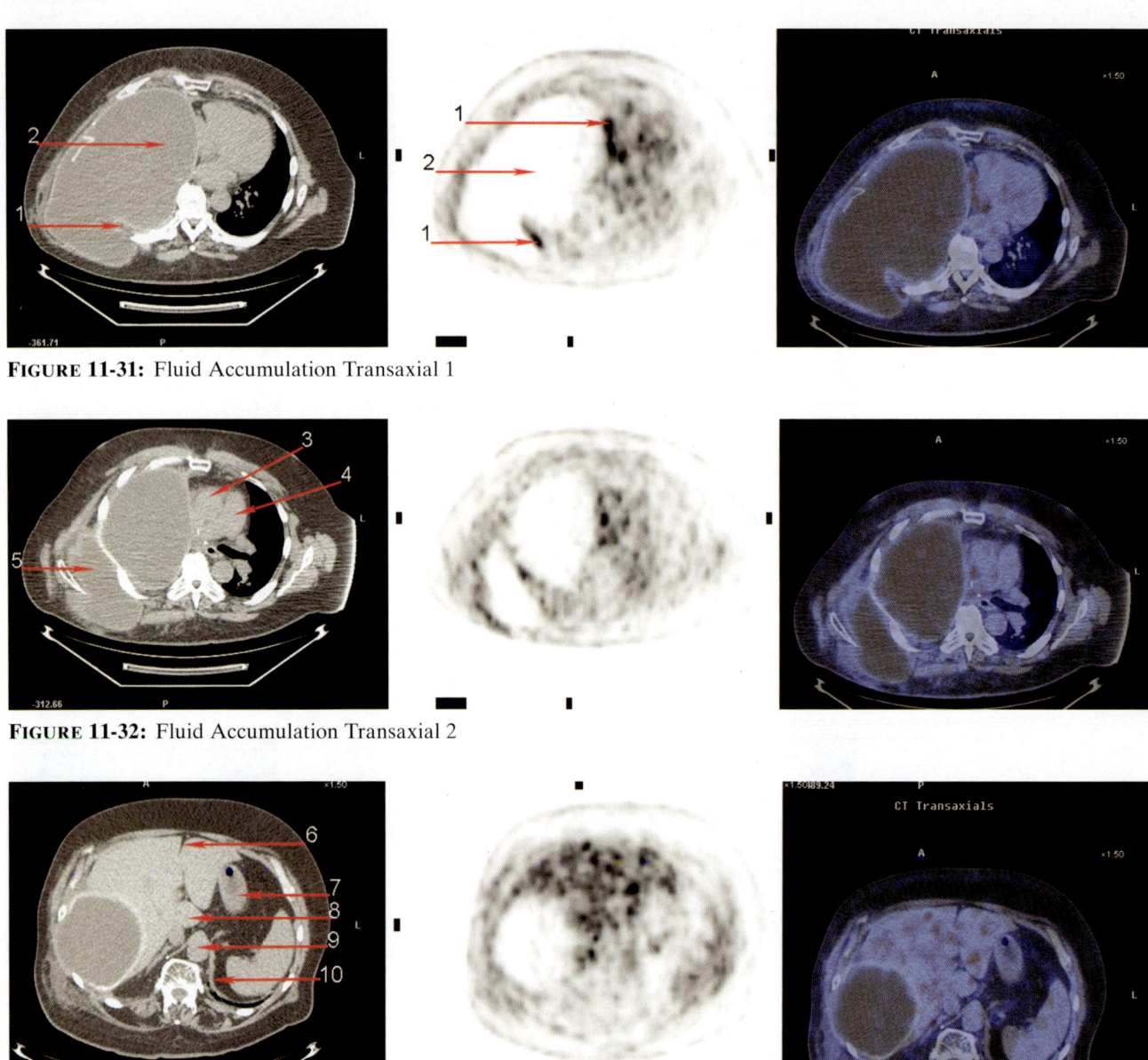

FIGURE 11-31: Fluid Accumulation Transaxial 1

FIGURE 11-32: Fluid Accumulation Transaxial 2

FIGURE 11-33: Fluid Accumulation Transaxial 3

1. Inflammatory reaction or lesion
2. Fluid in right thorax
3. Aorta
4. Main pulmonary artery
5. Fluid in subscapularis space
6. Falciform ligament
7. Stomach
8. Caudate hepatic lobe (segment I)
9. Aorta
10. Diaphragmatic crura

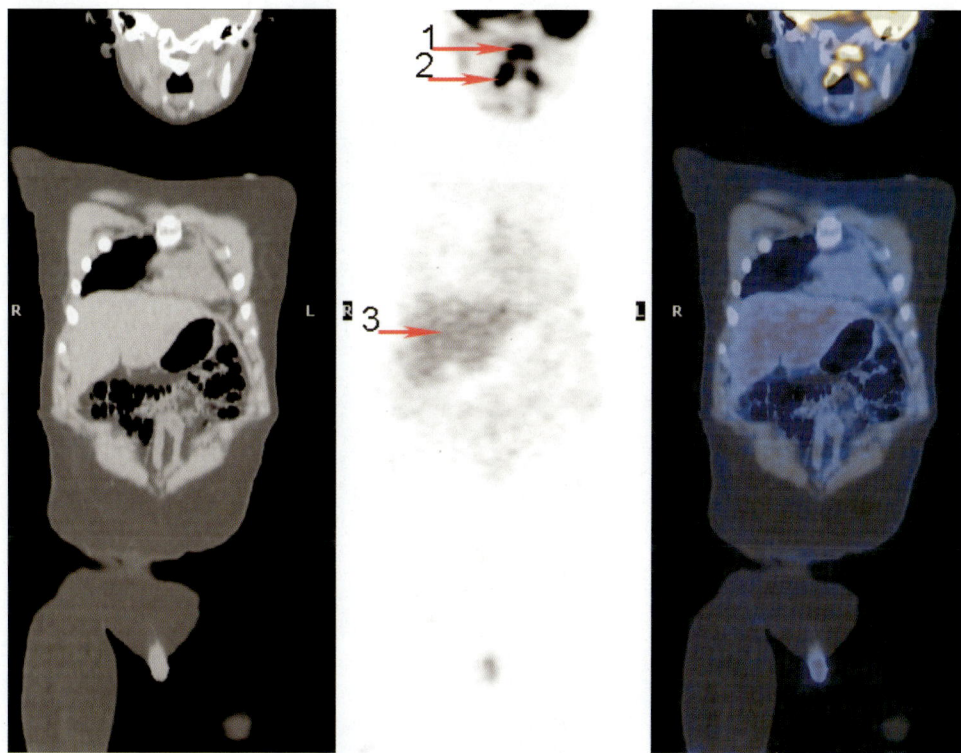

FIGURE 11-34: Pediatrics Coronal 1

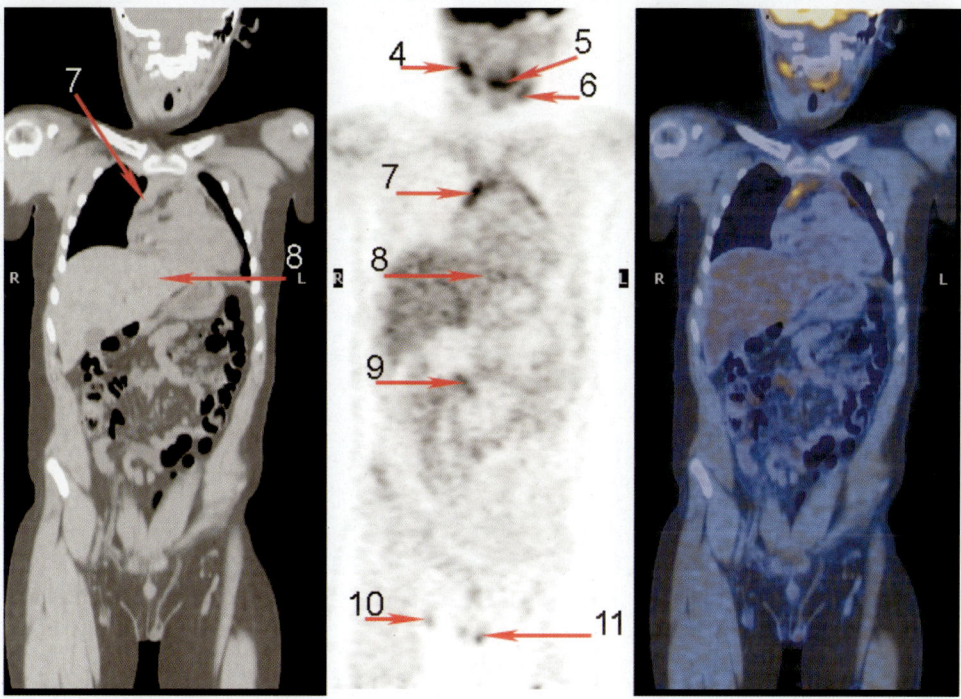

FIGURE 11-35: Pediatrics Coronal 2

1. Soft palate
2. Mylohyoid muscles
3. Right hepatic lobe
4. Right parotid gland
5. Tongue base
6. Left submandibular
7. Thymus
8. Left hepatic lobe
9. Transverse colon
10. Right femoral node
11. Scrotum

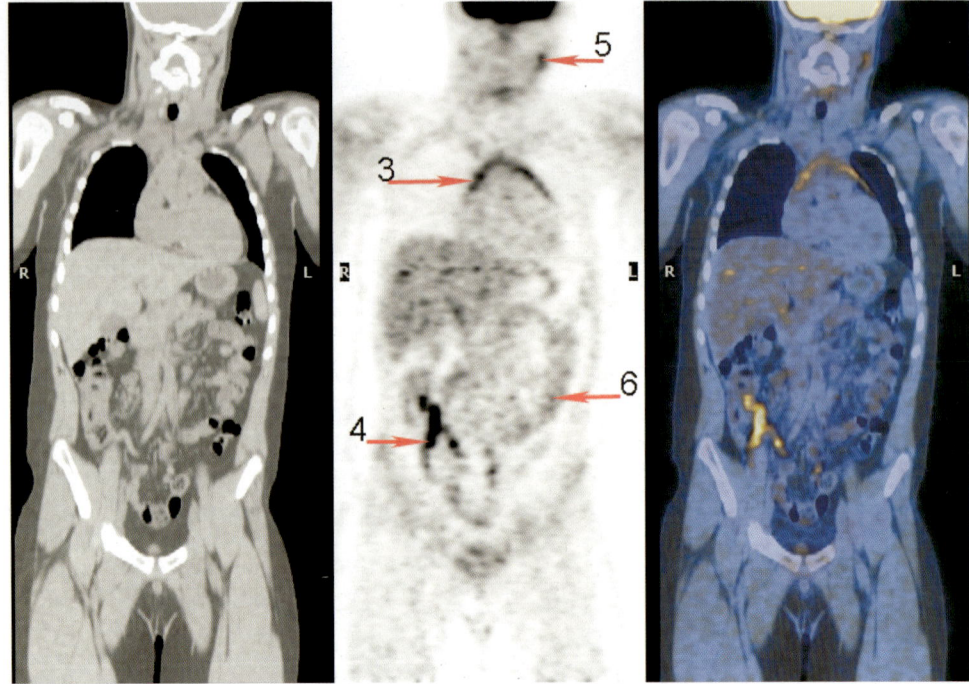

FIGURE 11-36: Pediatrics Coronal 3

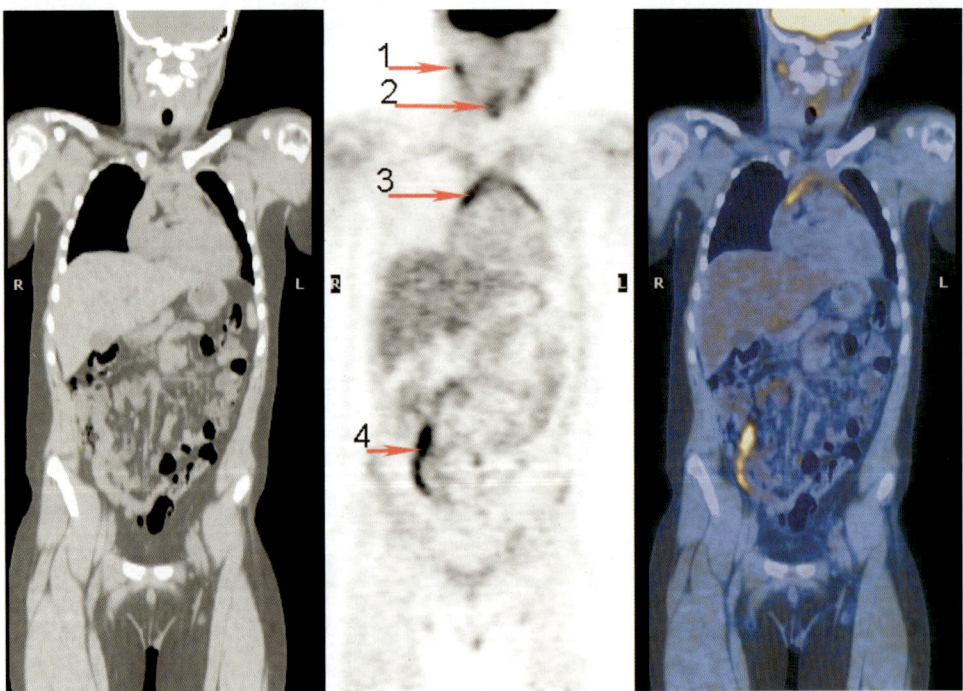

FIGURE 11-37: Pediatrics Coronal 4

1. Right parotid gland	3. Thymus	5. Left parotid gland
2. Vocal cord muscle	4. Ileum	6. Descending colon

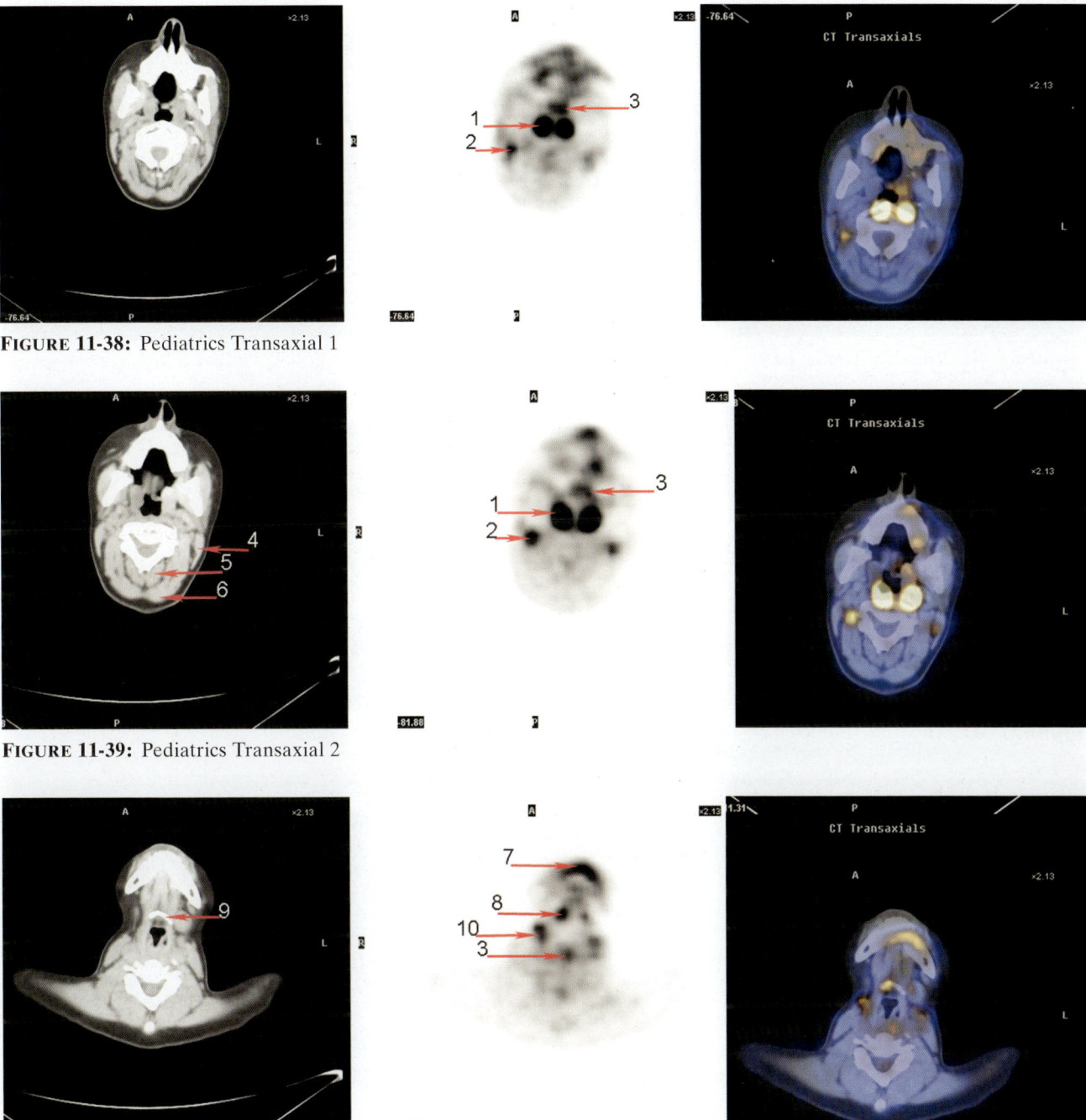

FIGURE 11-38: Pediatrics Transaxial 1

FIGURE 11-39: Pediatrics Transaxial 2

FIGURE 11-40: Pediatrics Transaxial 3

1. Lingual tonsil
2. Internal jugular vein
3. Longus capitis muscle
4. Sternocleidomastoid muscle
5. Semispinalis cervicis muscle
6. Splenius capitis muscle
7. Periodontal geniohyoid muscle
8. Genioglossus muscle
9. Hyoid bone
10. Palatine tonsil

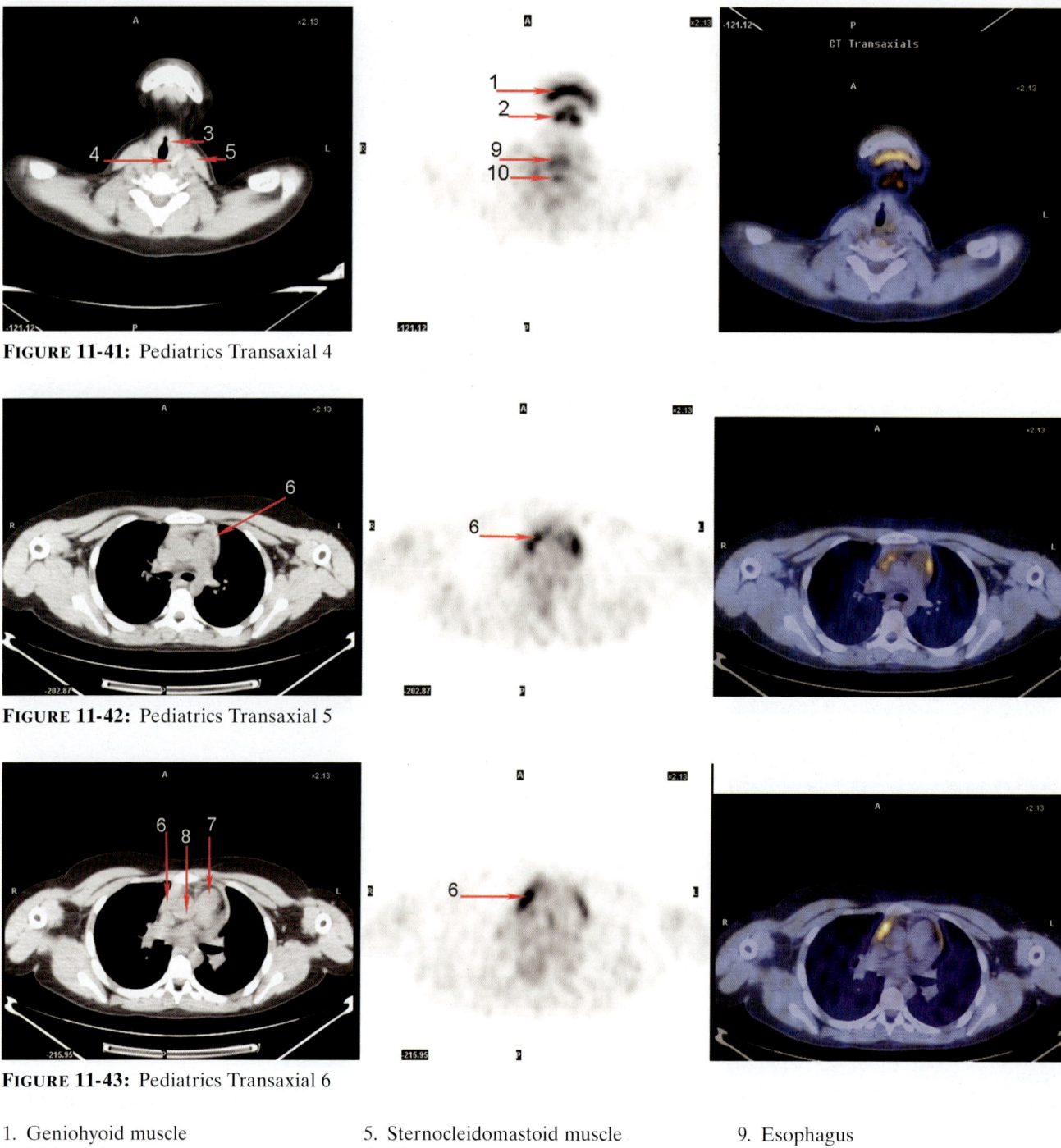

FIGURE 11-41: Pediatrics Transaxial 4

FIGURE 11-42: Pediatrics Transaxial 5

FIGURE 11-43: Pediatrics Transaxial 6

1. Geniohyoid muscle
2. Mylohyoid muscle
3. Thyroid cartilage
4. Cricoid cartilage
5. Sternocleidomastoid muscle
6. Thymus
7. Main pulmonary artery
8. Aorta
9. Esophagus
10. Longus colli muscle

Heart: Variations

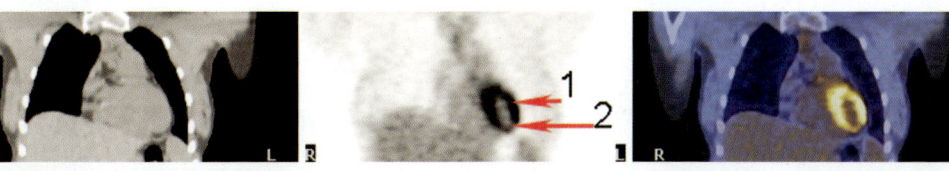

FIGURE 11-44: Heart Coronal 1

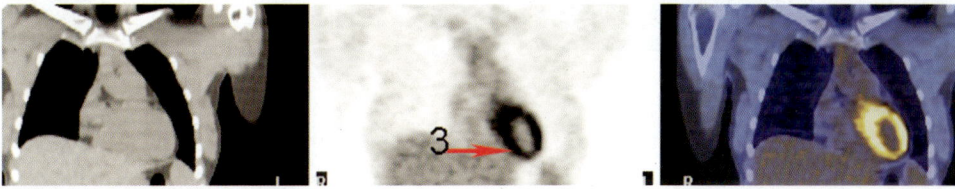

FIGURE 11-45: Heart Coronal 2

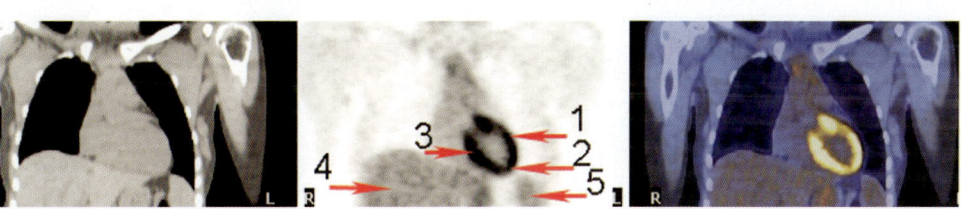

FIGURE 11-46: Heart Coronal 3

1. Lateral segment of left
   ventricular myocardium
2. Apex of left ventricle

3. Septal segment of left
   ventricle
4. Liver

5. Spleen

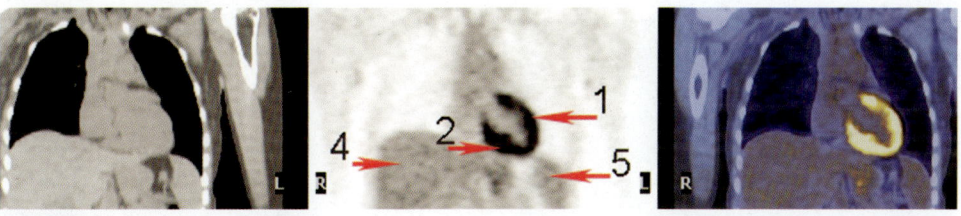

FIGURE 11-47: Heart Coronal 4

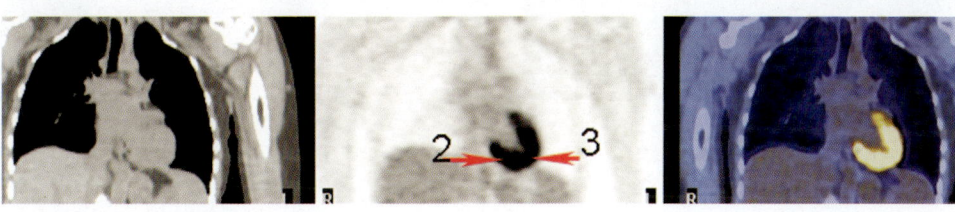

FIGURE 11-48: Heart Coronal 5

1. Lateral segment of left
   ventricular myocardium
2. Inferior segment of left
   ventricular myocardium

3. Apex of left ventricle
4. Liver

5. Spleen

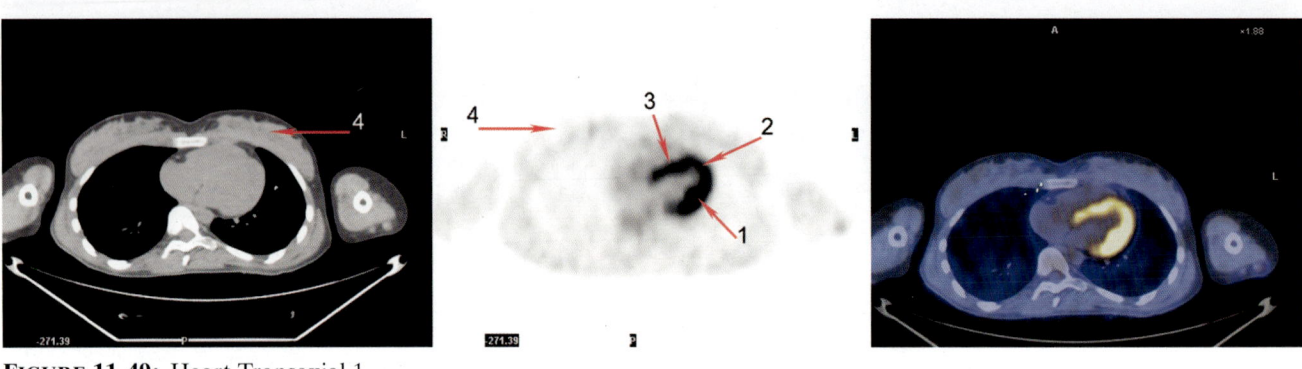

FIGURE 11-49: Heart Transaxial 1

1. Lateral segment of left ventricular myocardium
2. Apex of left ventricle
3. Septal segment of left ventricle
4. Breast

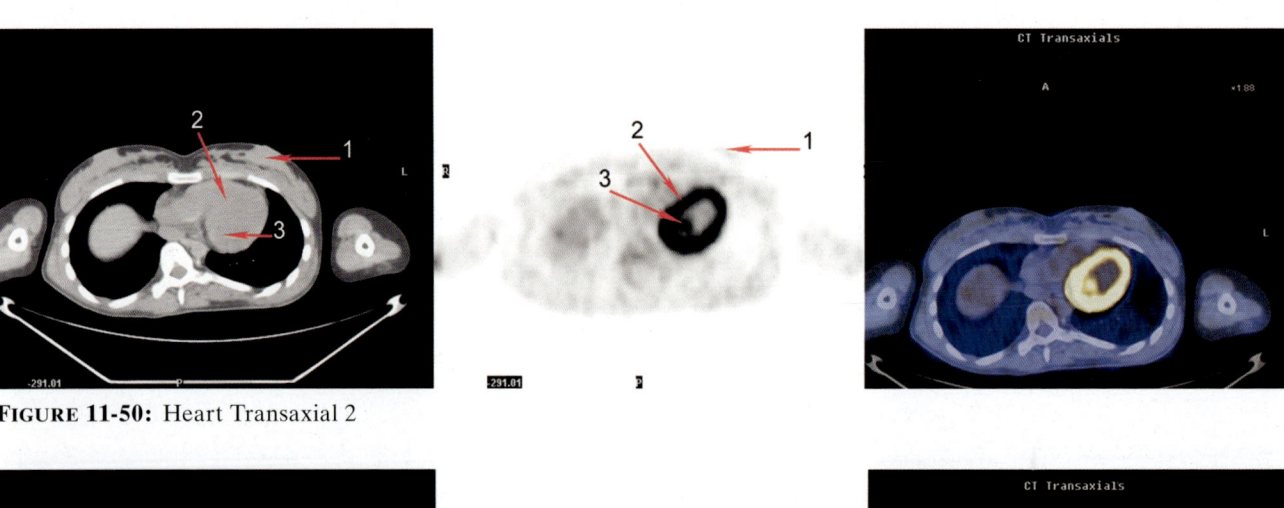

FIGURE 11-50: Heart Transaxial 2

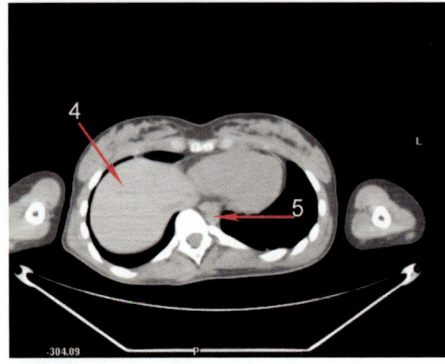
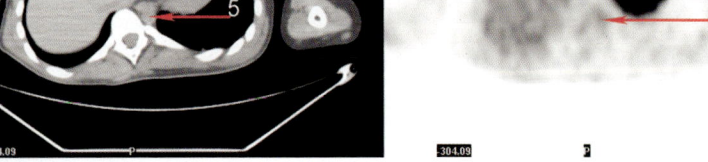
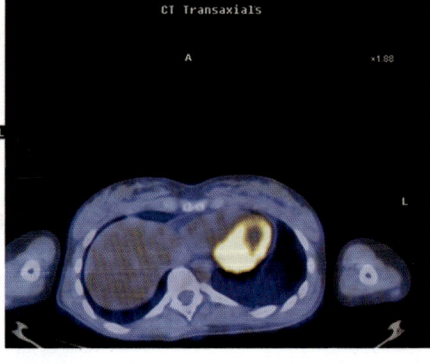

FIGURE 11-51: Heart Transaxial 3

1. Breast
2. Left ventricular myocardium
3. Left atrial appendage
4. Right hepatic lobe
5. Esophagus

Lymphoma: Variations

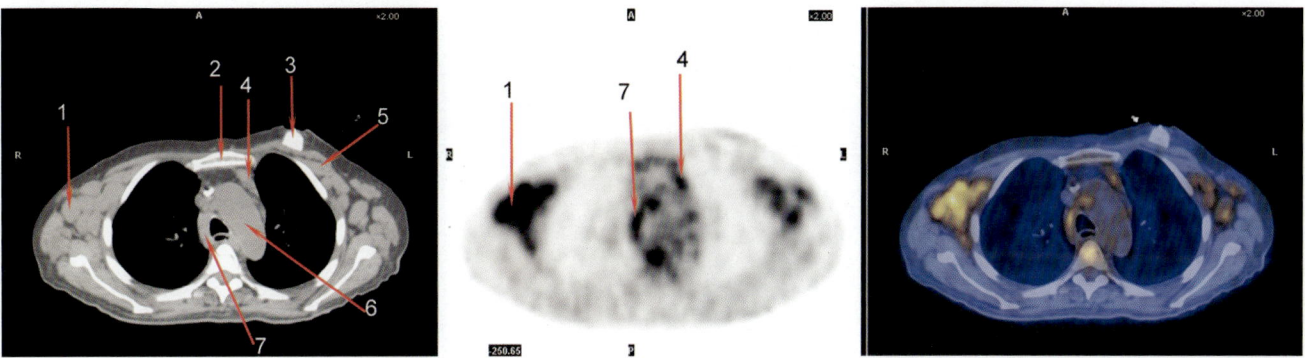

FIGURE 11-52: Lymphoma Transaxial 1

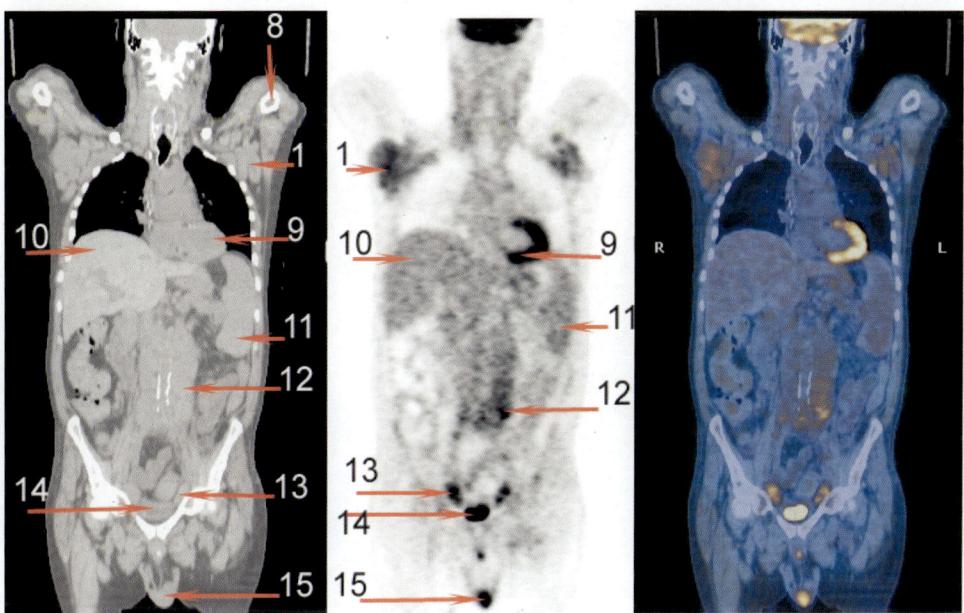

FIGURE 11-53: Lymphoma Coronal 1

1. Enlarged axillary nodes (level I)
2. Manubrium
3. Porta cath
4. Prevascular node
5. Pectoralis minor muscle
6. Aortic arch
7. Paratracheal node
8. Humerus
9. Left ventricular myocardium
10. Right liver
11. Spleen
12. Paraaortic node
13. External iliac node
14. Bladder
15. Testis

Necrosis: Variations

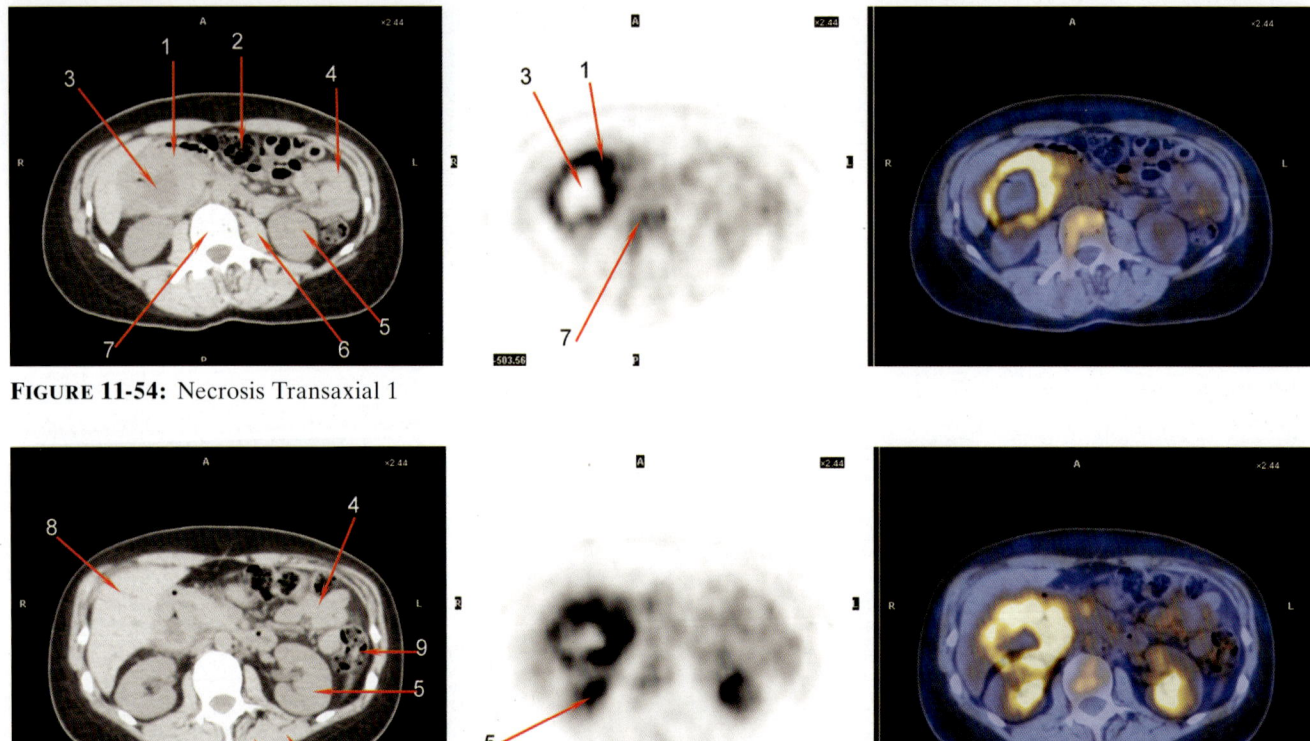

FIGURE 11-54: Necrosis Transaxial 1

FIGURE 11-55: Necrosis Transaxial 2

1. Active tumor
2. Transverse colon
3. Central necrosis
4. Jejunum

5. Kidney
6. Psoas muscle
7. Vertebral body
8. Right hepatic lobe (V)

9. Descending colon
10. Quadratus lumborum muscle
11. Erector spinae muscle

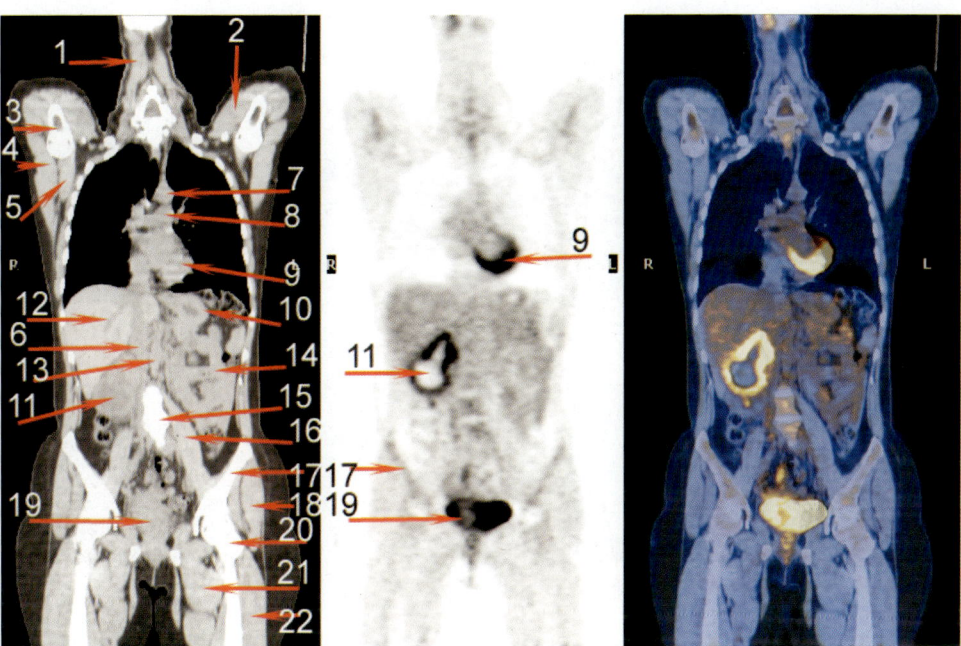

FIGURE 11-56: Necrotic mass Coronal 1

1. Longus colli muscle
2. Deltoid muscle
3. Humerus
4. Serratus anterior muscle
5. Subscapularis muscle
6. Inferior vena cava
7. Aortic arch
8. Left main pulmonary artery
9. Left ventricle
10. Stomach
11. Necrotic tumor
12. Hepatic vein
13. Abdominal aorta
14. Jejunum
15. L5 body
16. Psoas muscle
17. Ilium
18. Gluteus muscle
19. Bladder
20. Trochanter of femur
21. Adductor muscle
22. Vastus lateralis muscle

Head Neck Mouth: Variations

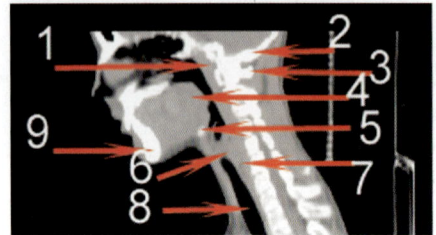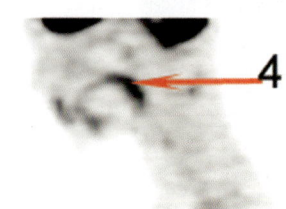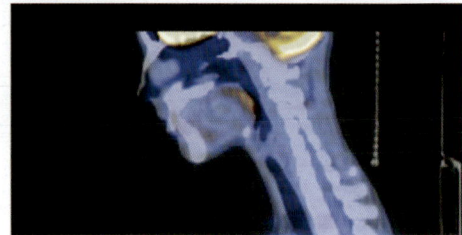

FIGURE 11-57: Head Neck Mouth Sagittal 1

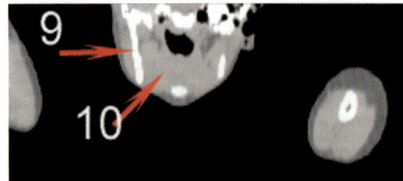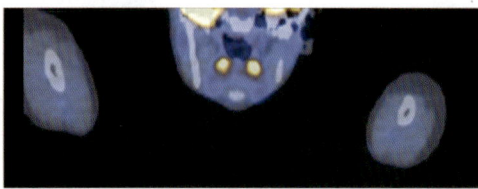

FIGURE 11-58: Head Neck Mouth Coronal 1

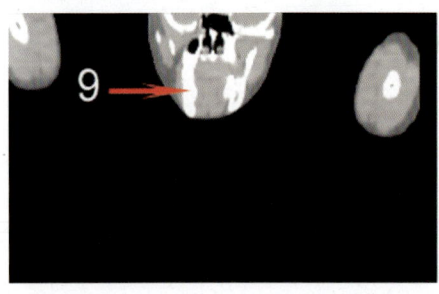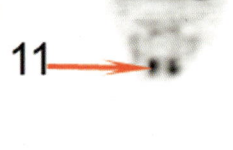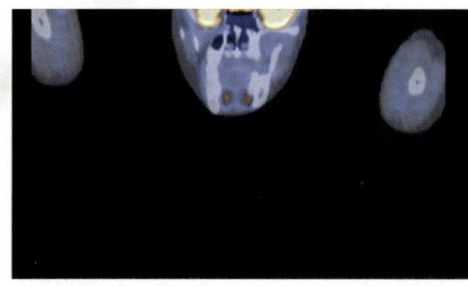

FIGURE 11-59: Head Neck Mouth Coronal 2

1. Adenoid
2. Occipital skull
3. Spinous process of C2
4. Tongue base
5. Hyoid bone
6. Vocal cord
7. Hypopharynx
8. Larynx
9. Mandible
10. Tongue
11. Lingual tonsil

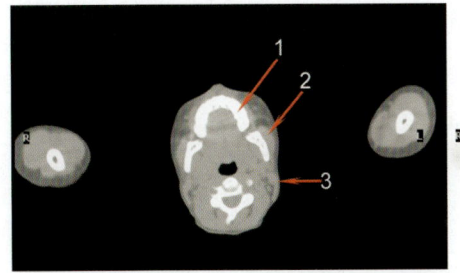
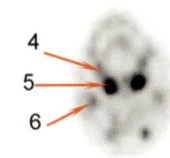
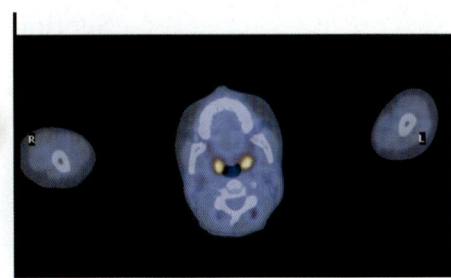

FIGURE 11-60: Head Neck Mouth Transaxial 1

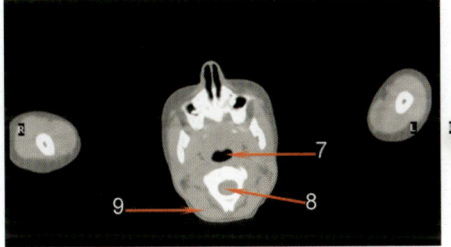
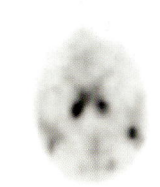
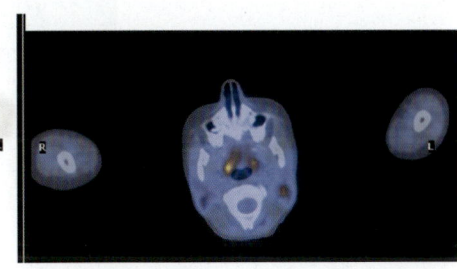

FIGURE 11-61: Head Neck Mouth Transaxial 2

1. Mandible
2. Masseter muscle
3. Sternocleidomastoid muscle
4. Submandibular gland
5. Lingual tonsil
6. Parotid gland
7. Oropharynx
8. Spinal cord
9. Splenius capitis muscle

Section 2: PET/CT Anatomy Artifacts

Breast Implants: Artifacts

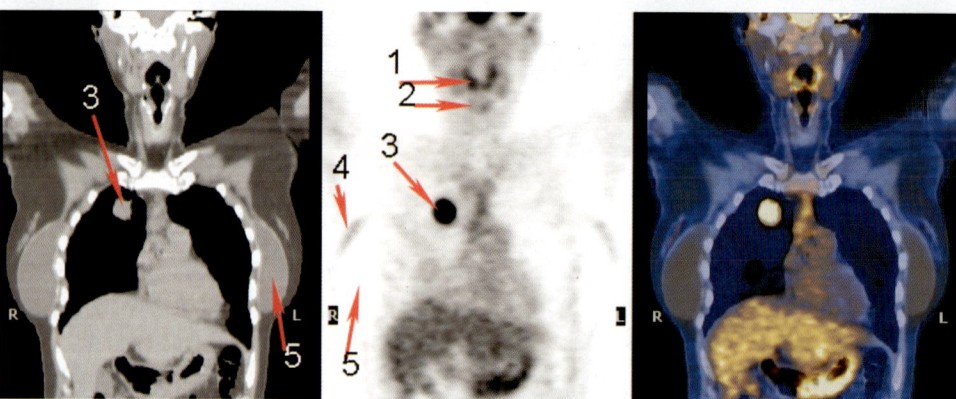

FIGURE 11-62: Breast Implants Coronal 1

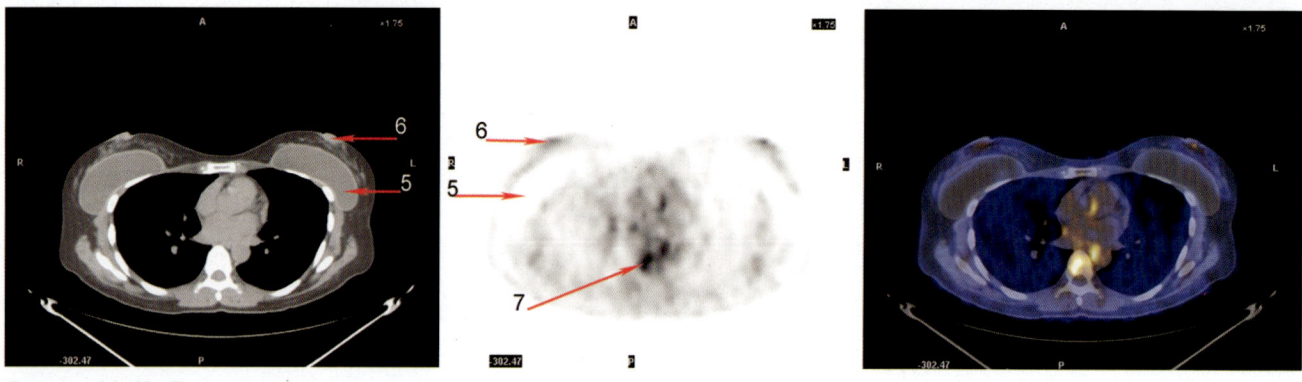

FIGURE 11-63: Breast Implants Transaxial 1

1. Lingual tonsil
2. Vocal cord muscle
3. Metastatic tumor in right upper lung
4. Inflammatory reaction
5. Implant in right breast
6. Nipple of right breast
7. Metastatic tumor in T7 vertebral body

Muscle Arm Uptake: Artifacts

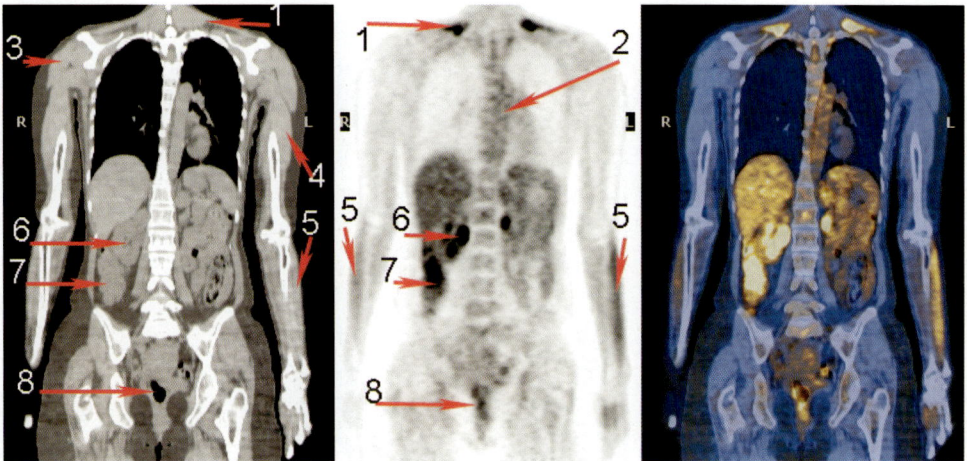

FIGURE 11-64: Muscle Arm Uptake Coronal 1

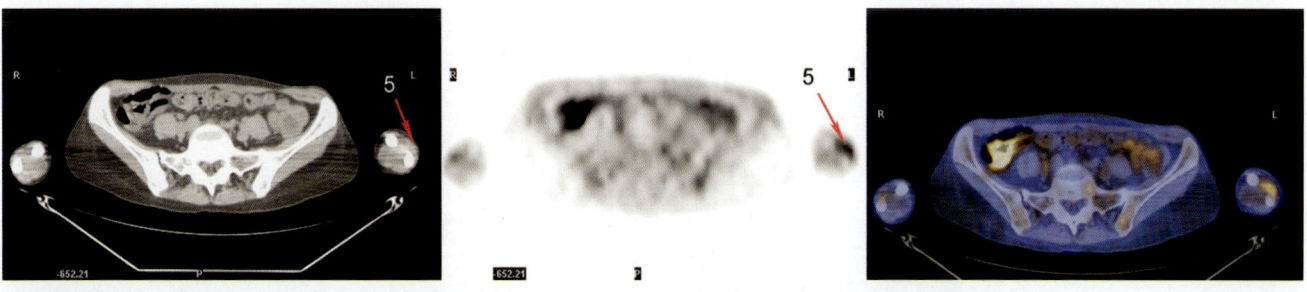

FIGURE 11-65: Muscle Arm Uptake Transaxial 1

1. Trapezius muscle
2. Descending aorta
3. Deltoid muscle
4. Biceps brachii muscle
5. Flexor carpi ulnaris muscle
6. Right kidney
7. Ascending colon
8. Rectosigmoid colon

Brown Fat: Artifacts

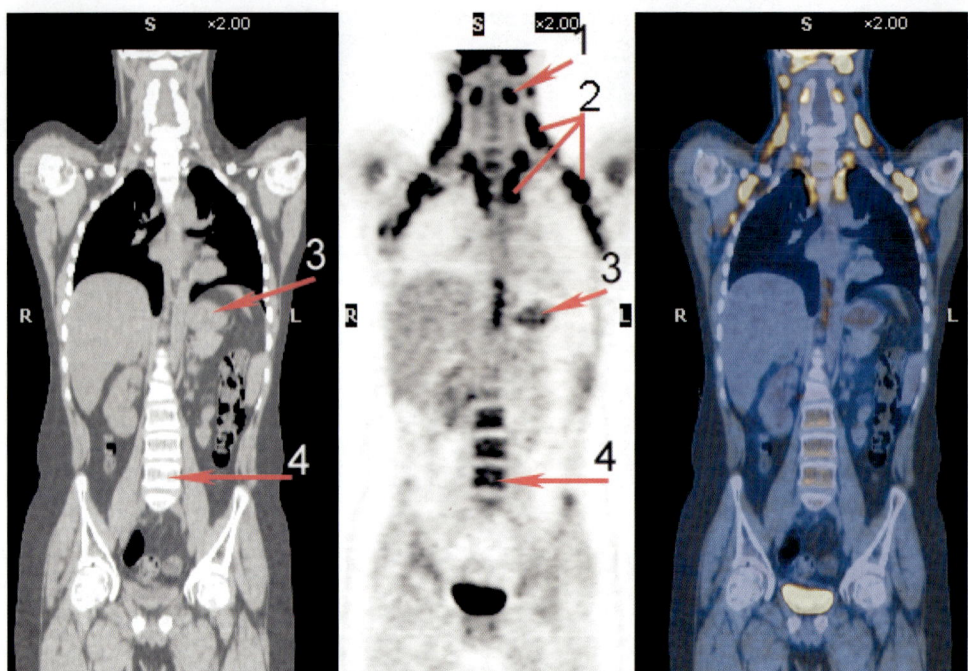

FIGURE 11-66: Brown Fat Coronal 1

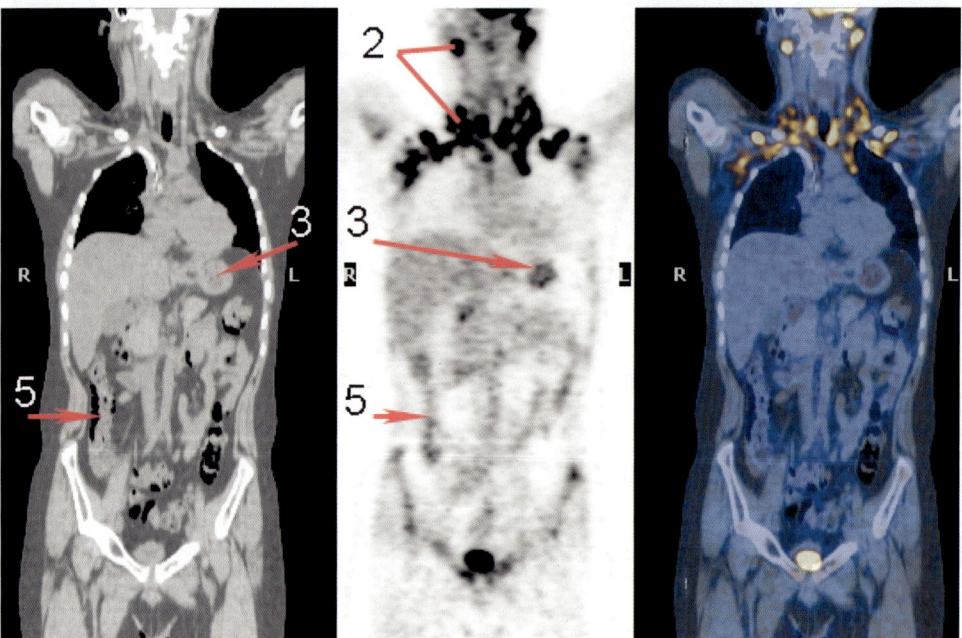

FIGURE 11-67: Brown Fat Coronal 2

1. Lingual tonsil	3. Gastric fundus	5. Ascending colon
2. Brown fat tissues	4. L5 marrow	

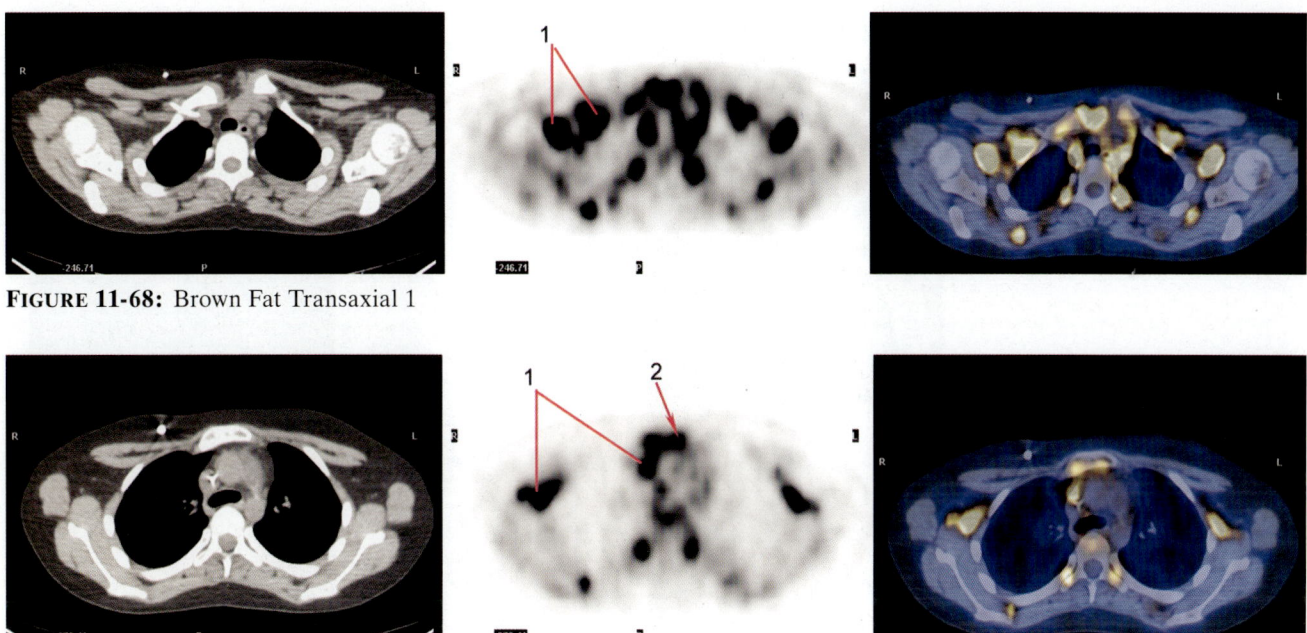

FIGURE 11-68: Brown Fat Transaxial 1

FIGURE 11-69: Brown Fat Transaxial 2

1. Brown fat tissues   2. Sternum

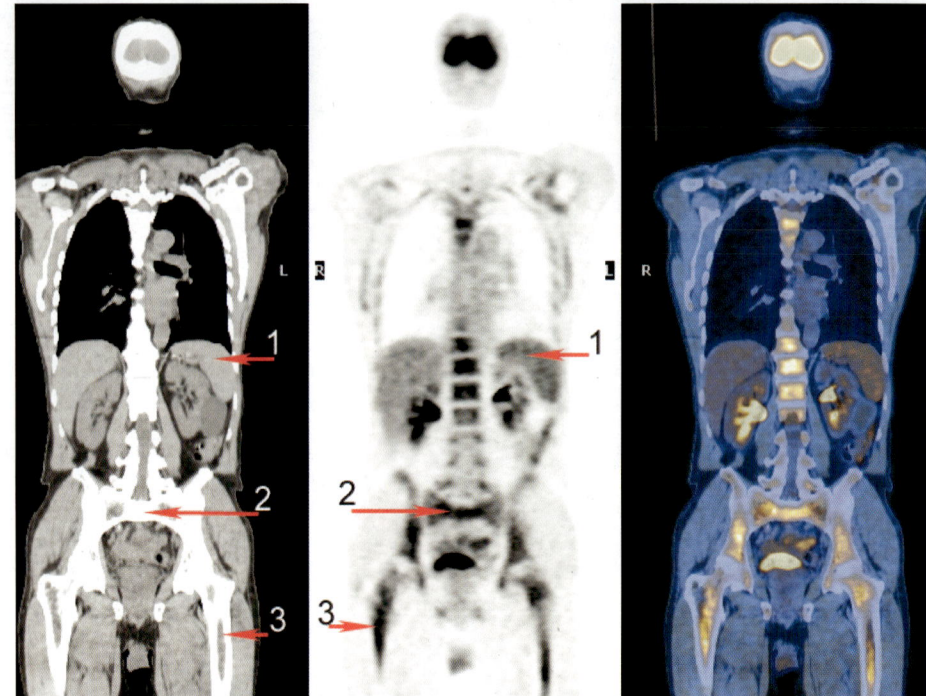

FIGURE 11-70: Chemotherapy Coronal 1

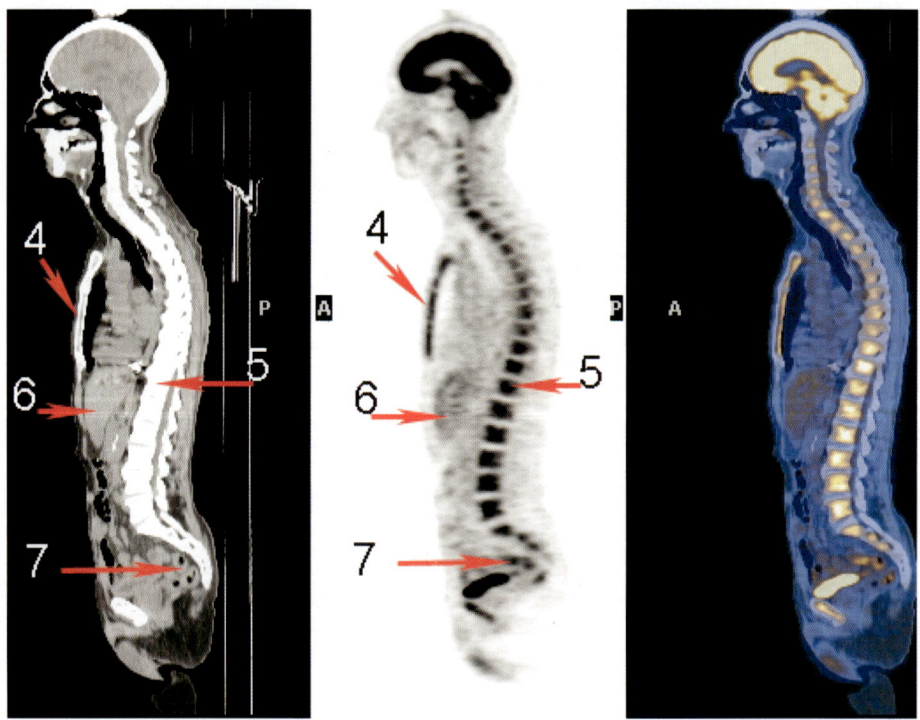

FIGURE 11-71: Chemotherapy Sagittal 1

1. Stimulated spleen
2. Stimulated sacral marrow
3. Stimulated femoral marrow
4. Stimulated sternal marrow
5. Stimulated T12 marrow
6. Liver
7. Sigmoid colon

Denture: Artifacts

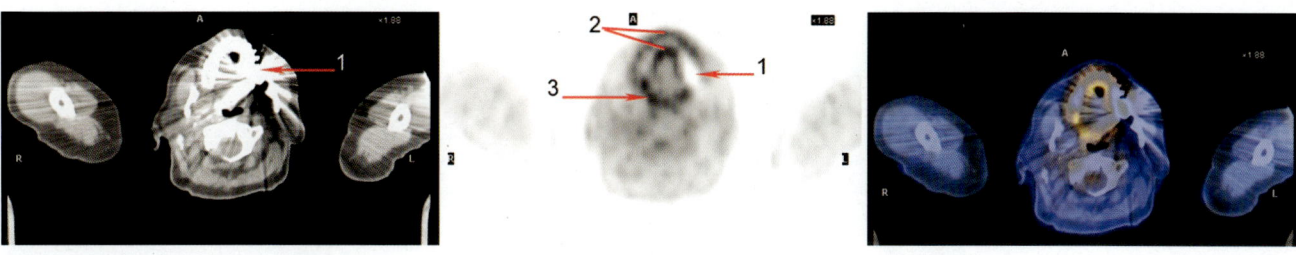

FIGURE 11-72: Denture Transaxial 1

1. Artifacts by dental crown (beam hardening on CT, decreased activity on PET)

2. Periodontal muscles

3. Lymphoid tissues in tongue base

Bladder Streak: Artifacts

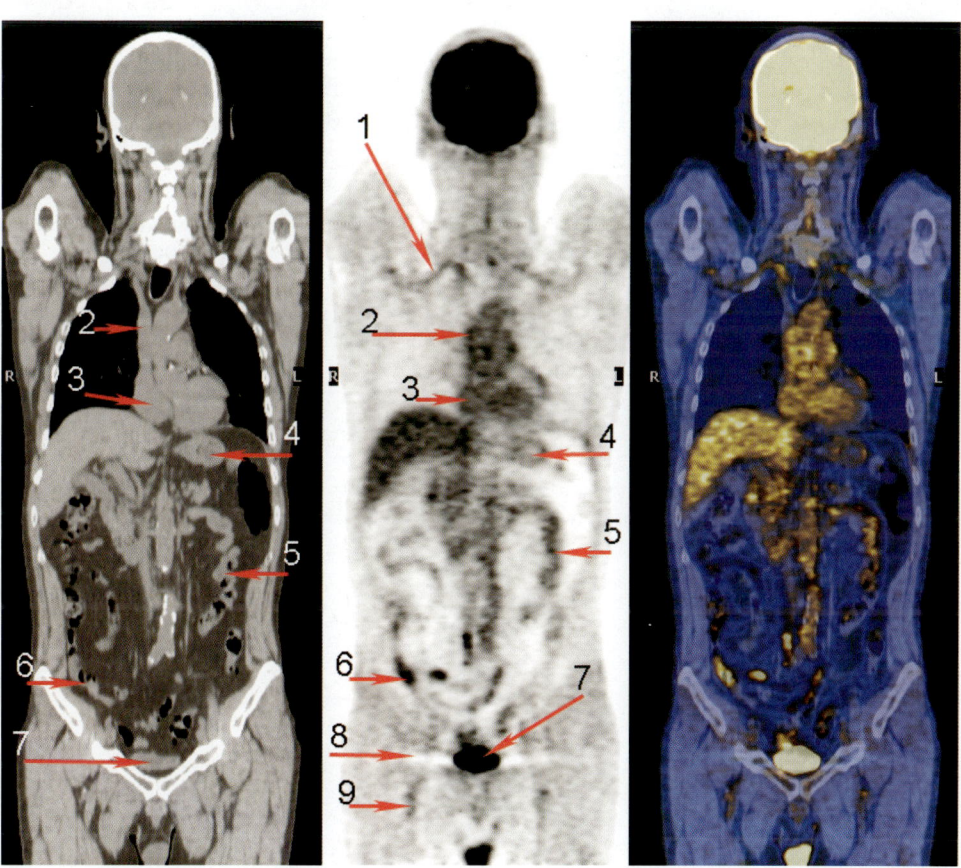

FIGURE 11-73: Bladder Streak Coronal 1

1. Right subclavian vein
2. Superior vena cava
3. Right atrium
4. Gastric fundus

5. Jejunum
6. Ileum
7. Bladder

8. Reconstruction artifact (linear photon deficiency)
9. Femoral artery

Respiratory Motion: Artifacts

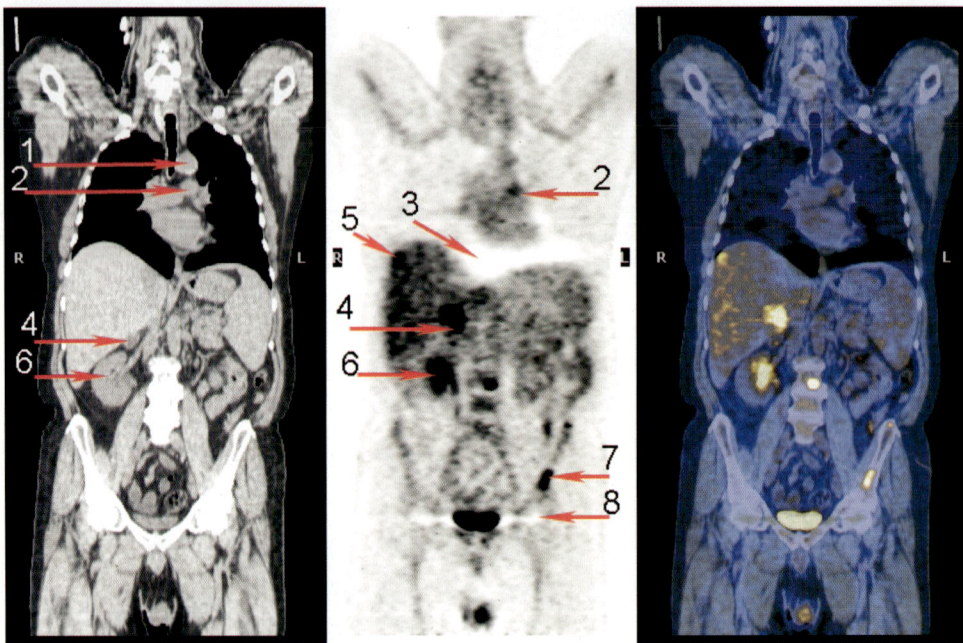

FIGURE 11-74: Respiratory Motion Coronal 1

1. Aortic arch
2. Left pulmonary artery
3. Reconstruction artifact caused by breathing motion (linear photon deficiency)
4. Gallbladder
5. Metastatic tumor in upper segment of right hepatic lobe (segment VIII)
6. Right kidney
7. Metastasis in left iliac crest
8. Reconstruction artifact caused by intense bladder activity (linear photon deficiency)

Misregistration: Artifacts

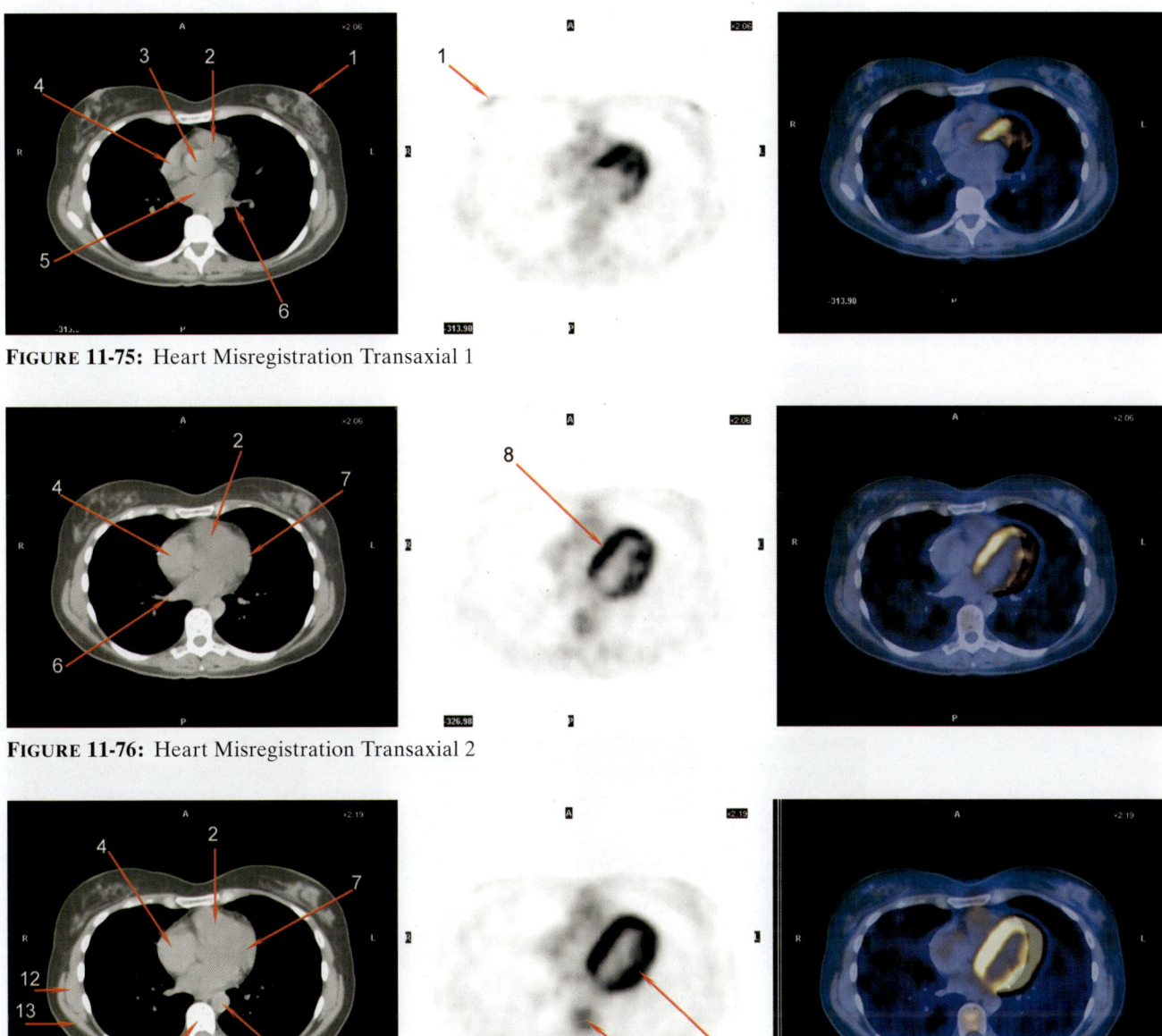

FIGURE 11-75: Heart Misregistration Transaxial 1

FIGURE 11-76: Heart Misregistration Transaxial 2

FIGURE 11-77: Heart Misregistration Transaxial 3

1. Breast nipple
2. Right ventricle
3. Ascending aorta
4. Right atrium
5. Left atrium
6. Segmented pulmonary vein

7. Left ventricle
8. Septal segment of left ventricular myocardium
9. Lateral segment of left ventricular myocardium
10. Vertebral body

11. Descending aorta
12. Serratus anterior muscle
13. Latissimus dorsi muscle

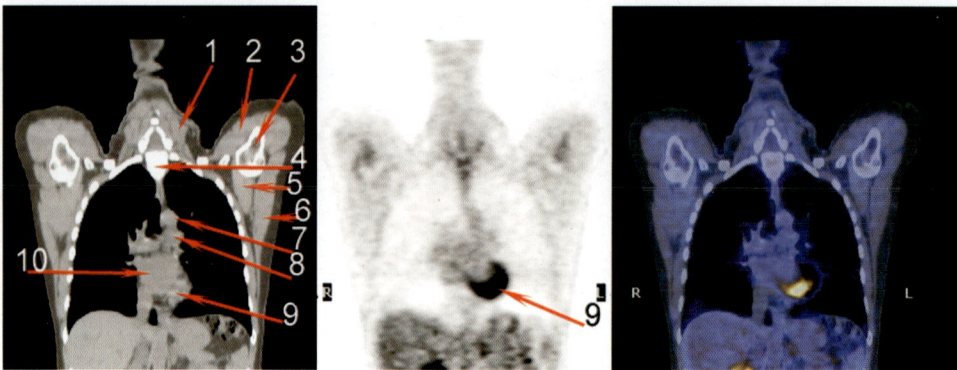

FIGURE 11-78: Heart Misregistration Coronal 1

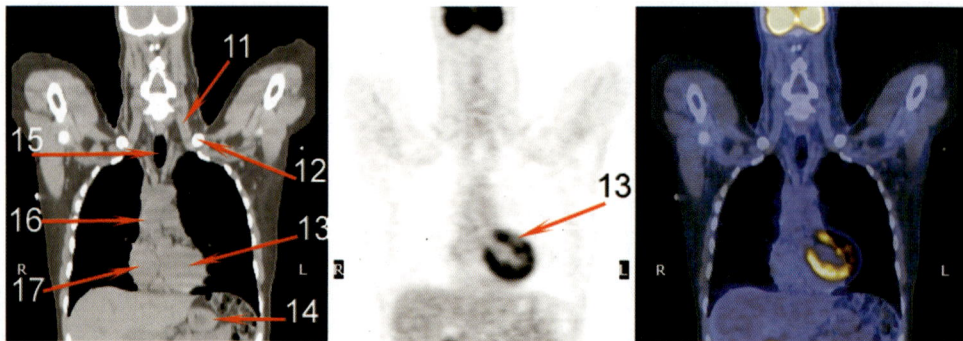

FIGURE 11-79: Heart Misregistration Coronal 2

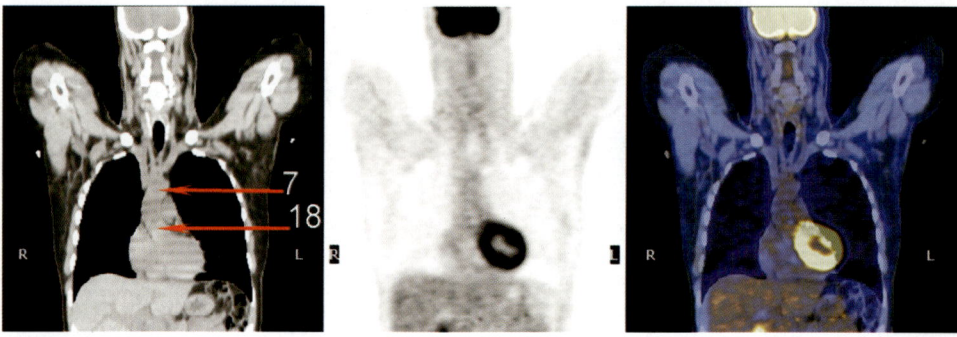

FIGURE 11-80: Heart Misregistration Coronal 3

1. Scalene muscle
2. Deltoid muscle
3. Humerus
4. Vertebral body
5. Subscapularis muscle
6. Serratus anterior muscle
7. Aortic arch
8. Left pulmonary artery

9. Apex of left ventricle
10. Left atrium
11. Sternocleidomastoid muscle
12. Clavicle
13. Lateral segment of left ventricular myocardium
14. Stomach

15. Trachea
16. Superior vena cava
17. Right atrium
18. Ascending aorta

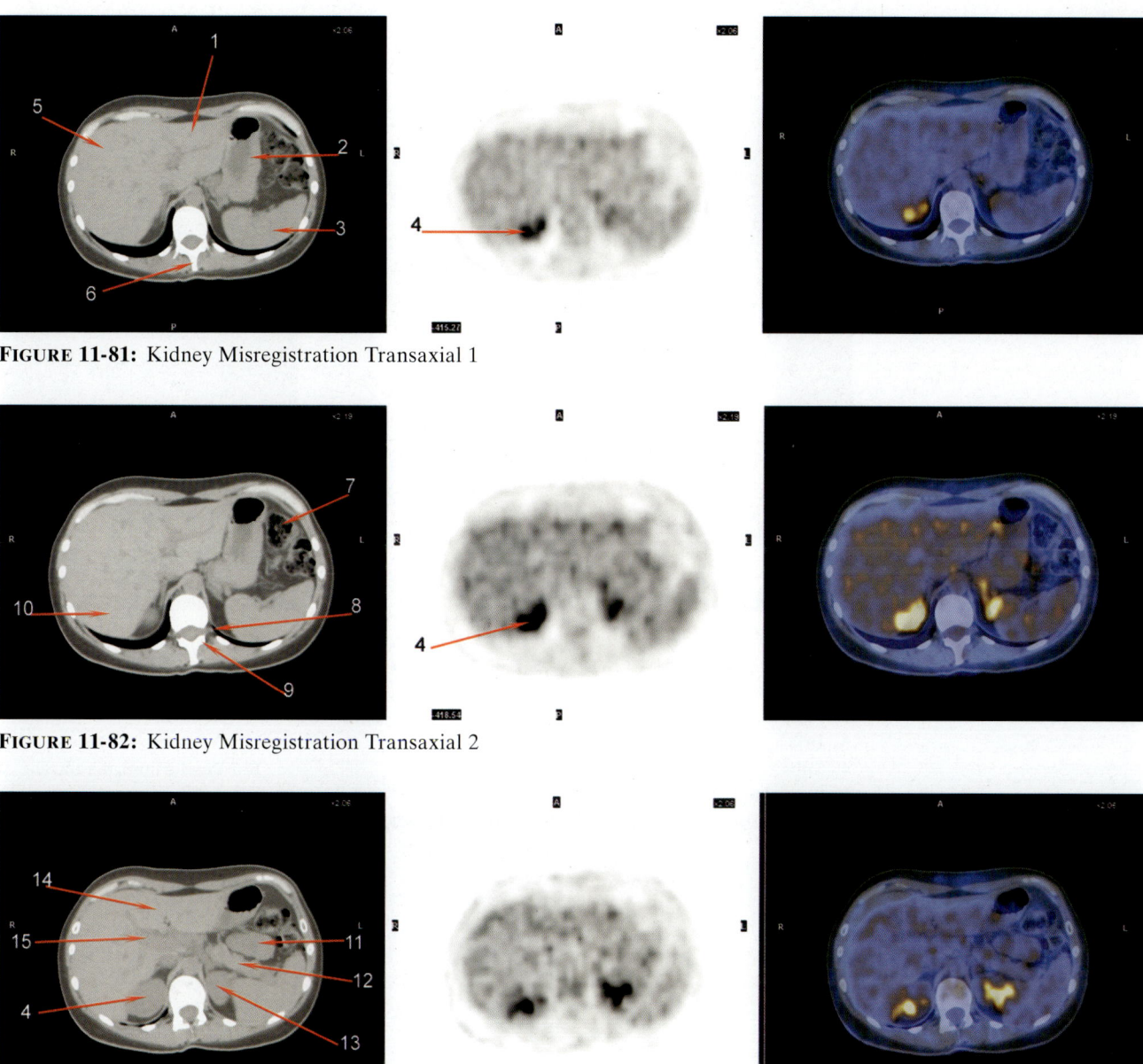

FIGURE 11-81: Kidney Misregistration Transaxial 1

FIGURE 11-82: Kidney Misregistration Transaxial 2

FIGURE 11-83: Kidney Misregistration Transaxial 3

1. Left hepatic lobe (segment III)
2. Stomach
3. Spleen
4. Right kidney
5. Right hepatic lobe (segment VIII)
6. Vertebral spinous process
7. Splenic flexure of colon
8. Neural canal
9. Vertebral lamina
10. Right hepatic lobe (segment VII)
11. Jejunum
12. Pancreas
13. Left kidney
14. Falciform ligament
15. Porta hepatic

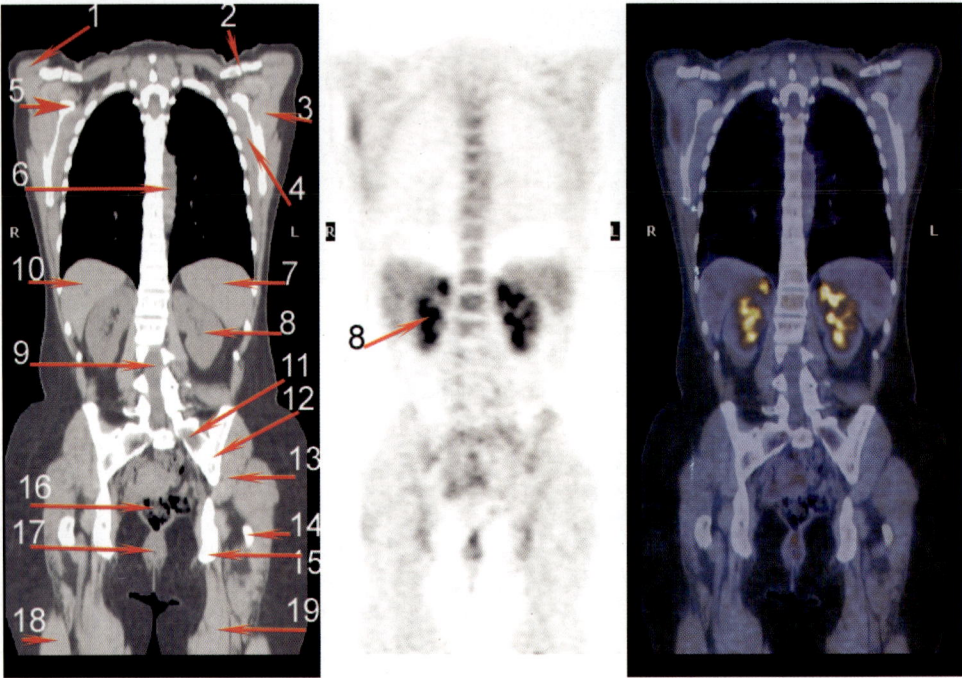

FIGURE 11-84: Kidney Misregistration Coronal 1

1. Deltoid muscle
2. Acromioclavicular joint
3. Infraspinatus muscle
4. Subscapularis muscle
5. Scapula
6. Descending aorta
7. Spleen
8. Kidney
9. Lumbar spinal canal
10. Right liver
11. Sacral ala
12. Iliac tuberosity
13. Gluteus muscle
14. Trochanter of femur
15. Ischial tuberosity
16. Rectum
17. Anus
18. Vastus lateralis muscle
19. Adductor muscle

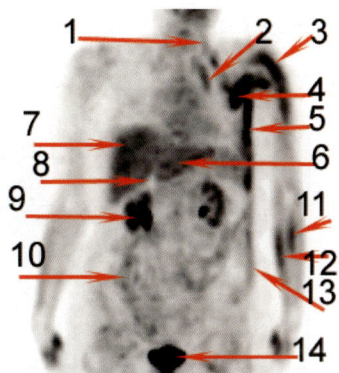

FIGURE 11-85: Muscle tension by crutch Coronal 1

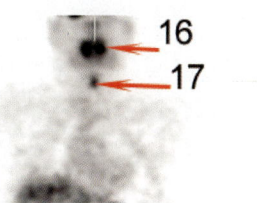

FIGURE 11-86: Snoring Coronal 1

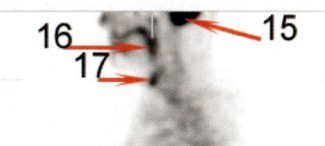

FIGURE 11-87: Snoring Sagittal 1

FIGURE 11-88: Snoring Transaxial 1

1. Scalene muscle
2. Sternocleidomastoid muscle
3. Deltoid muscle
4. Infraspinatus muscle
5. Serratus anterior muscle
6. Left liver
7. Right liver
8. Hepatic hilus
9. Right kidney
10. Ascending colon
11. Flexor digitorum profundus muscle
12. Flexor carpi radialis muscle
13. Oblique muscle
14. Bladder
15. Cerebellar gray matter
16. Lingual tonsil
17. Vocal muscle

Section 3: Non-FDG PET/CT Variants: C11-Methionine

Coronals

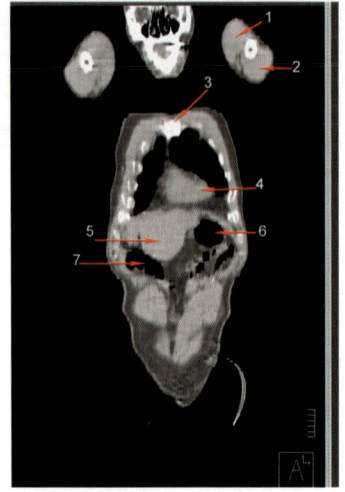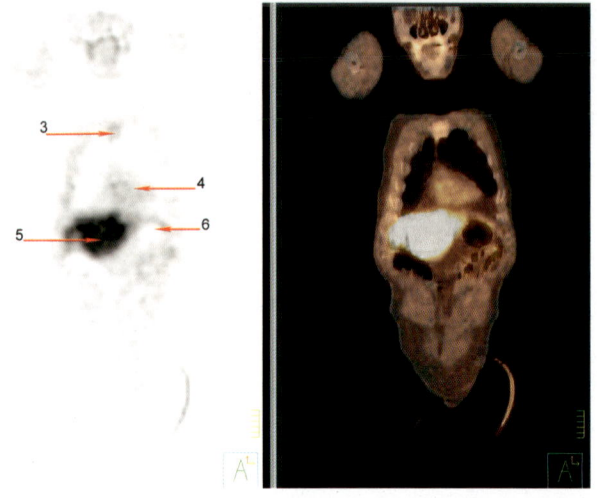

FIGURE 11-89: Coronal 1

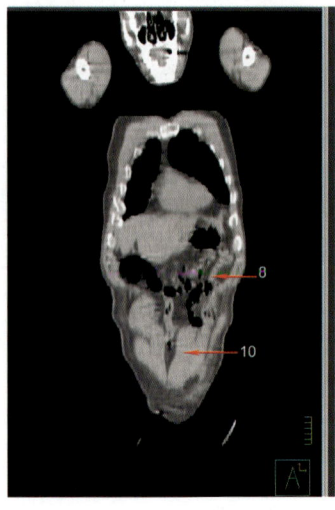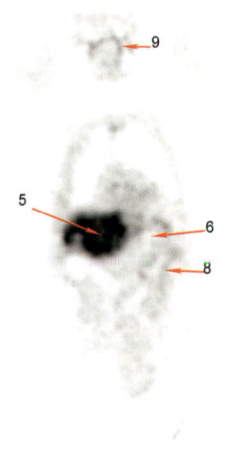

FIGURE 11-90: Coronal 2

1. Biceps brachii muscle	5. Left hepatic lobe	9. Periodontal muscle
2. Triceps brachii muscle	6. Stomach	(omohyoid muscle)
3. Manubrium	7. Transverse colon	10. Rectus abdominis
4. Right ventricle	8. Descending colon	muscle

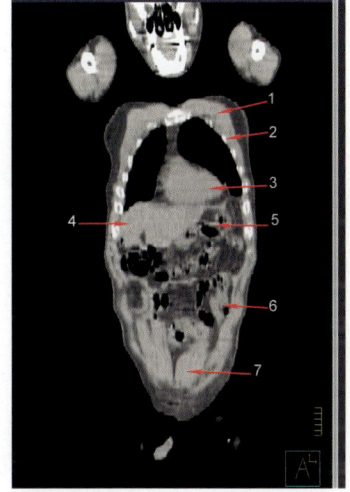
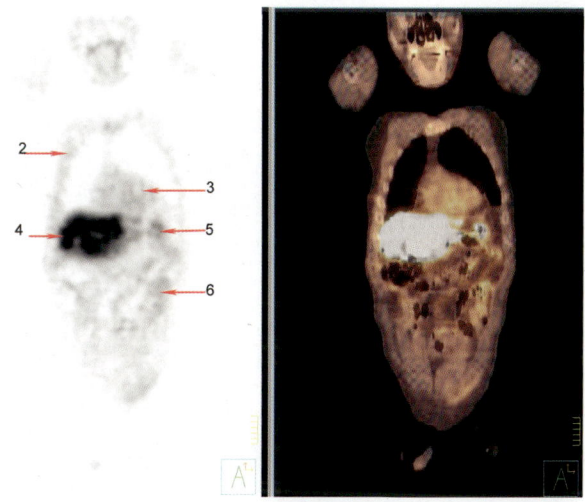

FIGURE 11-91: Coronal 3

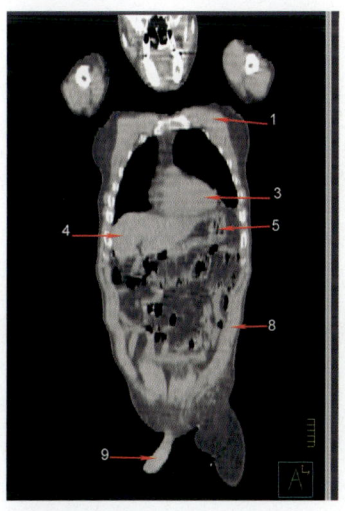
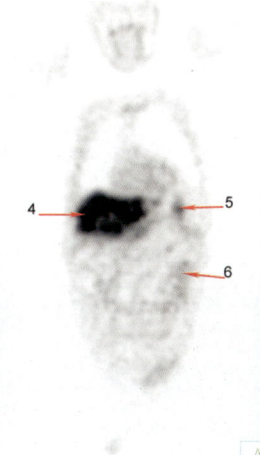

FIGURE 11-92: Coronal 4

1. Pectoralis major muscle
2. Rib
3. Left ventricle
4. Right hepatic lobe
5. Stomach
6. Descending colon
7. Rectus abdominis muscle
8. Oblique muscles
9. Penis

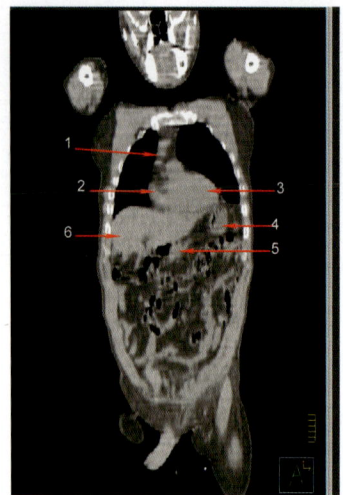 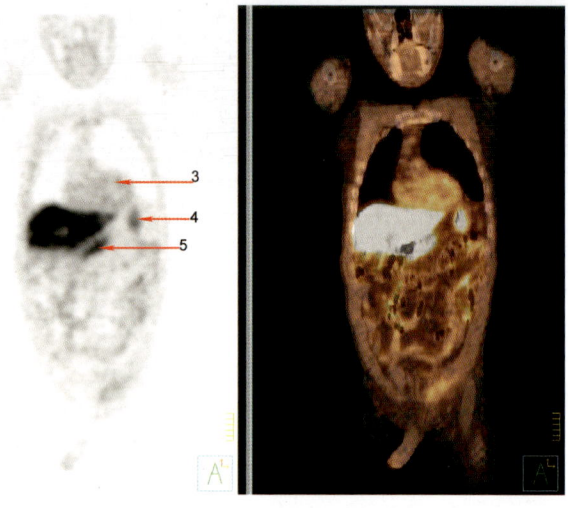

FIGURE 11-93: Coronal 5

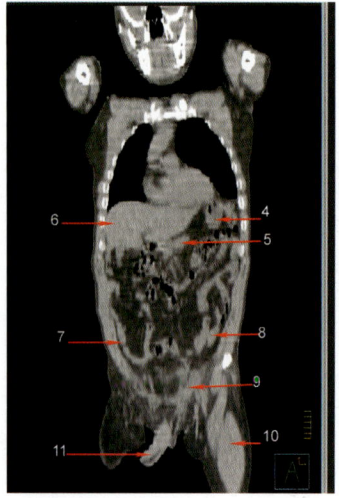 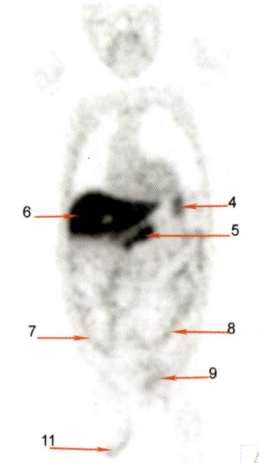 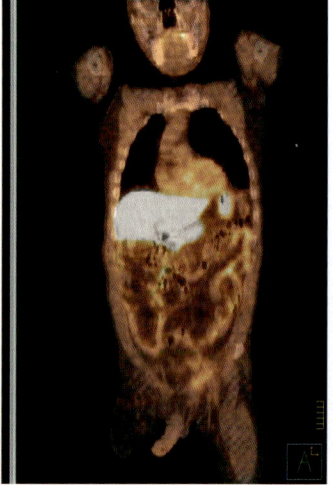

FIGURE 11-94: Coronal 6

1. Right brachiocephalic
   vein
2. Right atrium
3. Left ventricle
4. Stomach
5. Pancreas (body)
6. Right hepatic lobe
7. Ascending colon
8. Descending colon
9. Sigmoid
10. Rectus femoris muscle
11. Penis

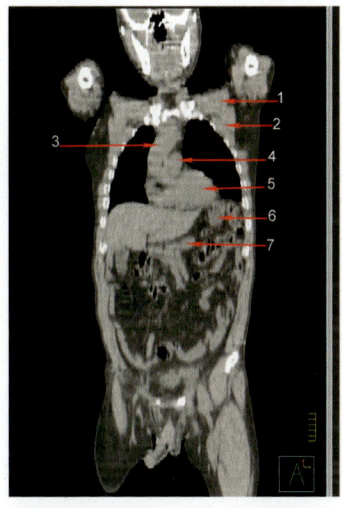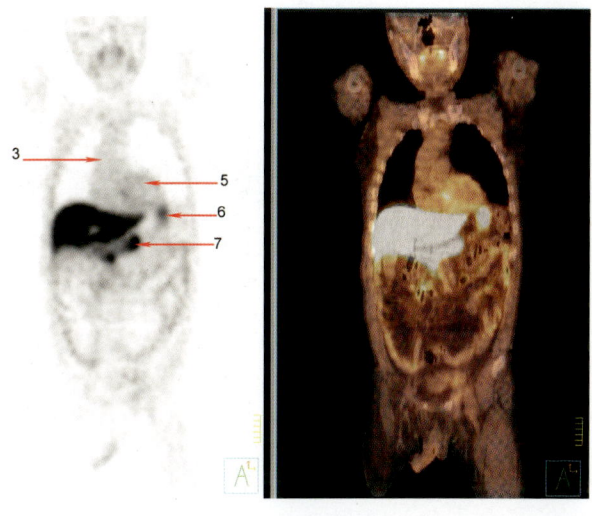

FIGURE 11-95: Coronal 7

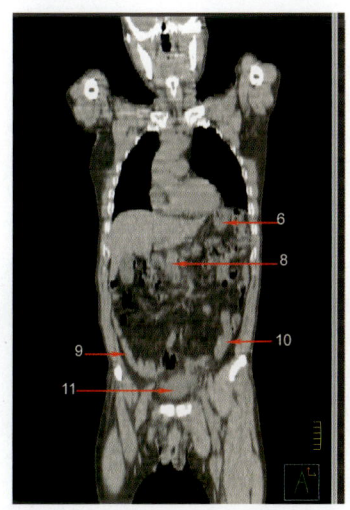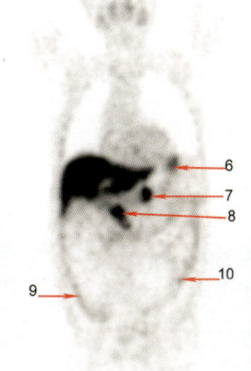

FIGURE 11-96: Coronal 8

1. Pectoralis major muscle
2. Pectoralis minor muscle
3. Ascending aorta
4. Main pulmonary artery
5. Left ventricle
6. Stomach
7. Pancreas (body)
8. Pancreas (head)
9. Ascending colon
10. Descending colon
11. Bladder

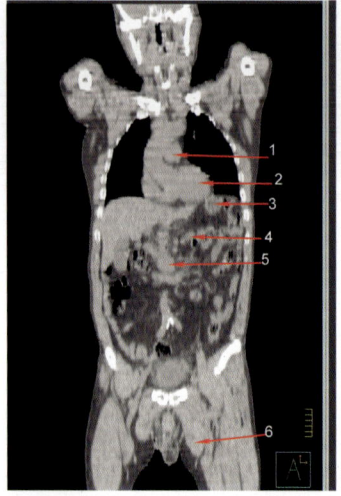
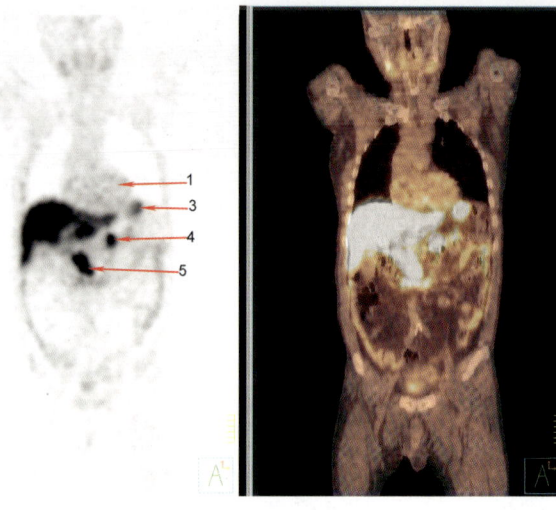

FIGURE 11-97: Coronal 9

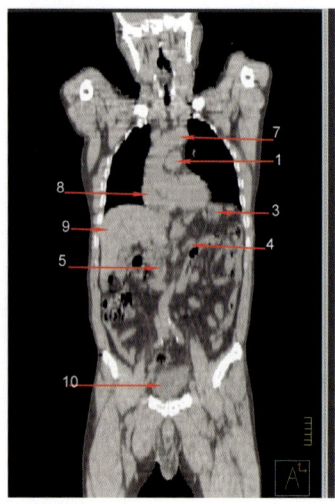
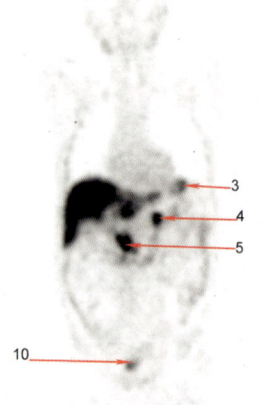
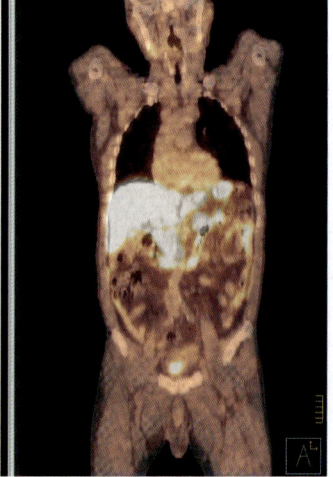

FIGURE 11-98: Coronal 10

1. Main pulmonary artery
2. Left ventricle
3. Stomach
4. Pancreas (tail)
5. Pancreas (head)
6. Adductor muscles
7. Aortic arch
8. Right atrium
9. Right hepatic lobe
10. Bladder

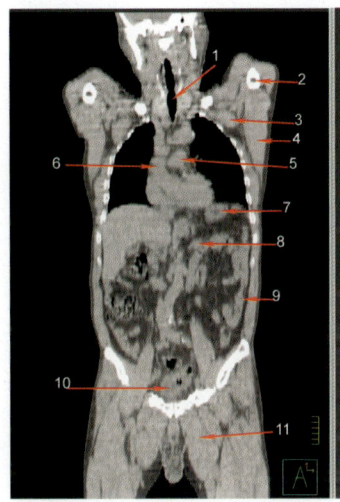
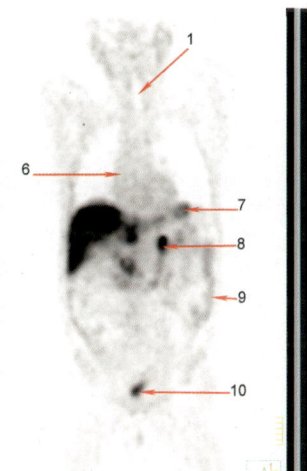
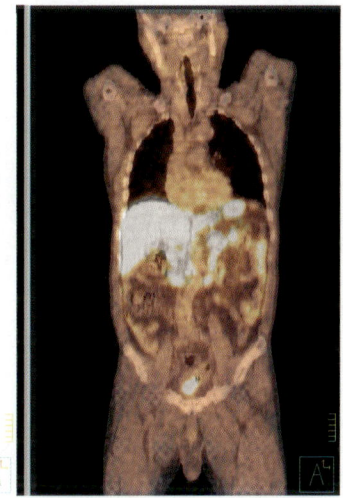

FIGURE 11-99: Coronal 11

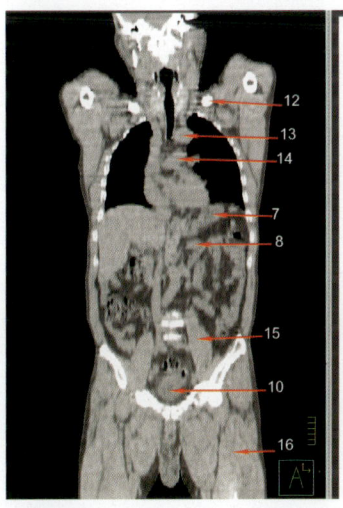
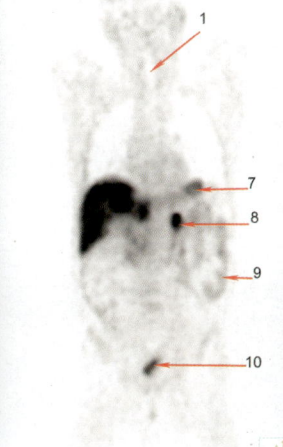
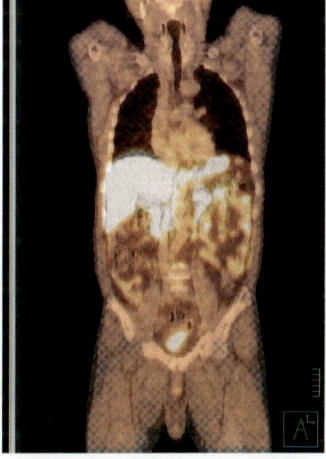

FIGURE 11-100: Coronal 12

1. Trachea	7. Stomach	13. Aortic arch
2. Humerus	8. Pancreas (tail)	14. Pulmonary trunk
3. Pectoralis minor muscle	9. Descending colon	15. Psoas muscle
4. Subscapularis muscle	10. Bladder	16. Rectus femoris muscle
5. Main pulmonary artery	11. Adductor muscles	
6. Superior vena cava	12. Clavicle	

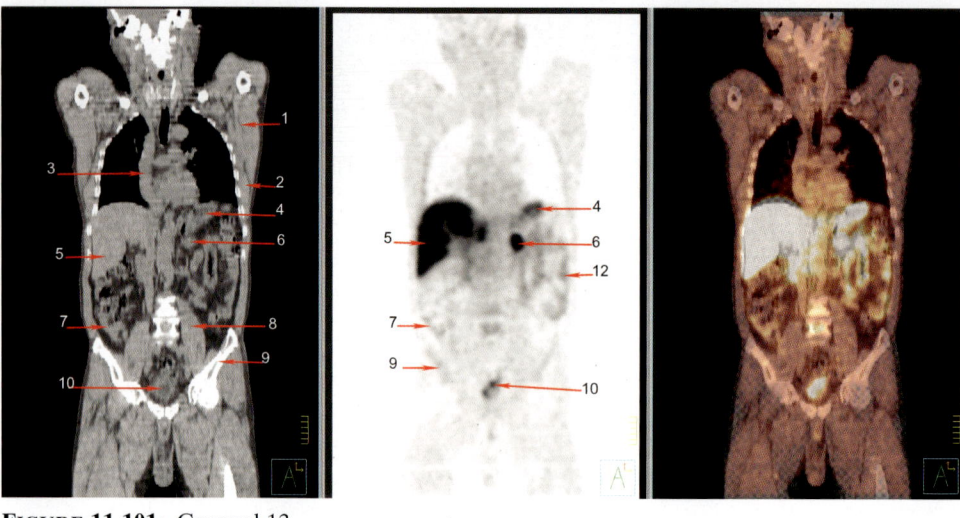

FIGURE 11-101: Coronal 13

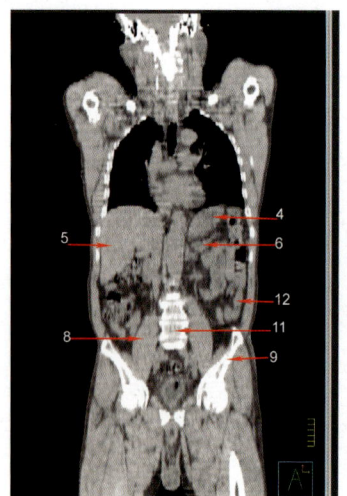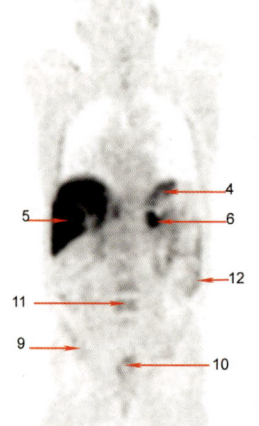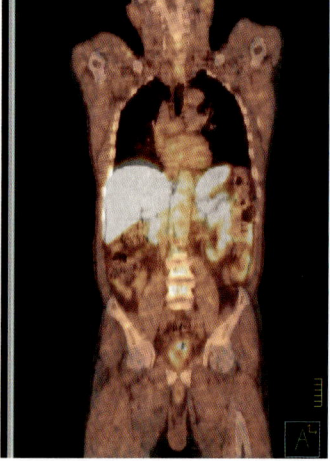

FIGURE 11-102: Coronal 14

1. Subscapularis muscle
2. Serratus anterior muscle
3. Superior vena cava
4. Stomach
5. Right hepatic lobe
6. Pancreas (tail)
7. Cecum
8. Psoas muscle
9. Iliac crest
10. Bladder
11. L5 marrow
12. Descending colon

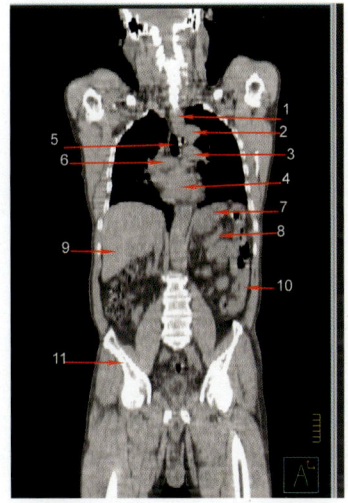
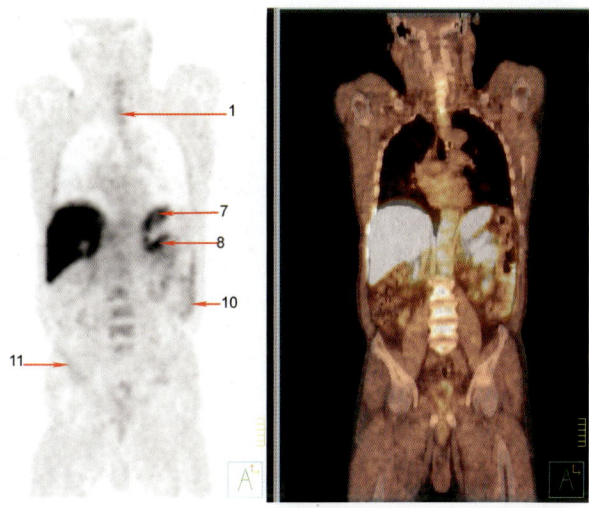

FIGURE 11-103: Coronal 15

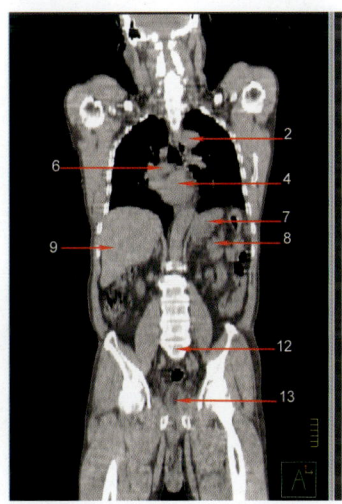
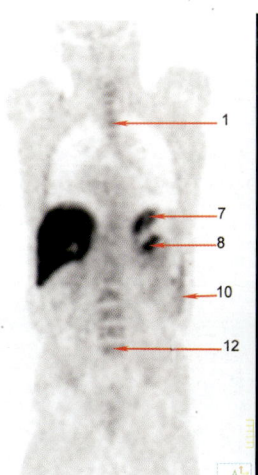
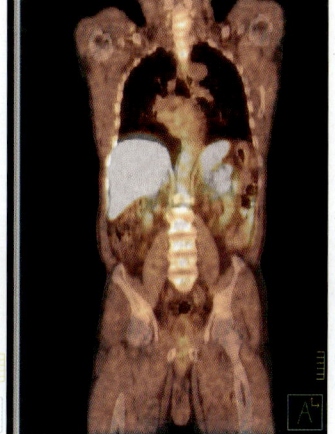

FIGURE 11-104: Coronal 16

1. Esophagus
2. Aortic arch
3. Left pulmonary artery
4. Left ventricle
5. Trachea
6. Left atrium
7. Stomach
8. Pancreas (tail)
9. Right hepatic lobe
10. Descending colon
11. Ilium
12. Sacrum
13. Rectum

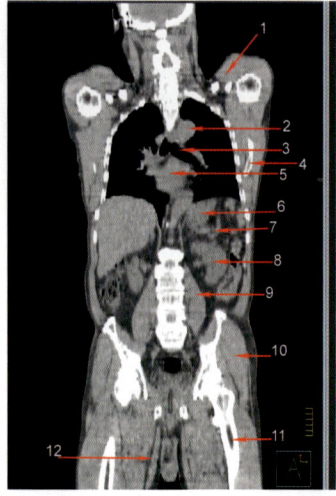 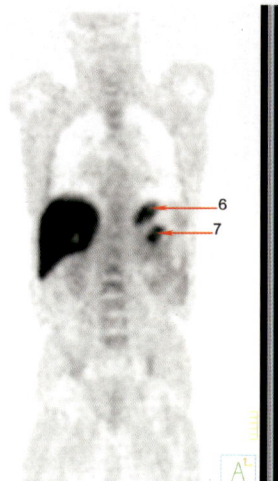 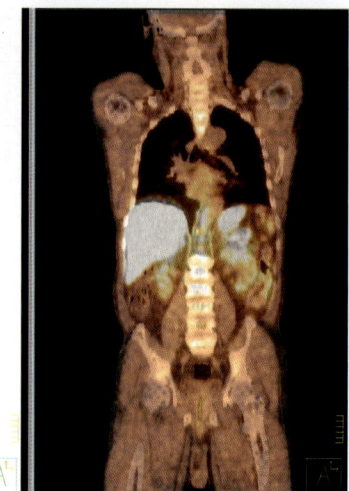

FIGURE 11-105: Coronal 17

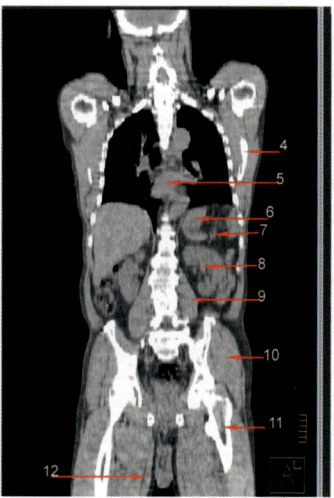 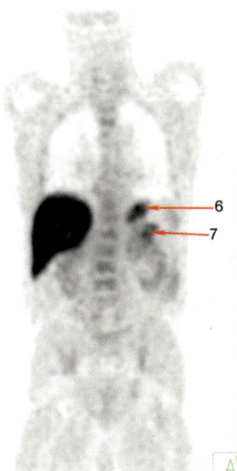 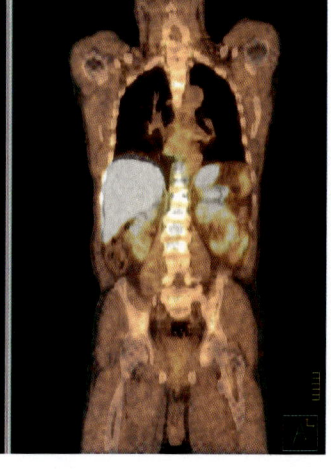

FIGURE 11-106: Coronal 18

1. Supraspinatus muscle
2. Aortic arch
3. Left main bronchus
4. Scapula
5. Left atrium
6. Stomach (fundus)
7. Pancreas (tail)
8. Left kidney
9. Psoas muscle
10. Gluteus muscle
11. Femur
12. Adductor muscles

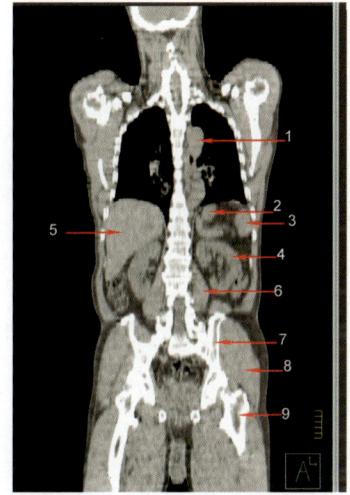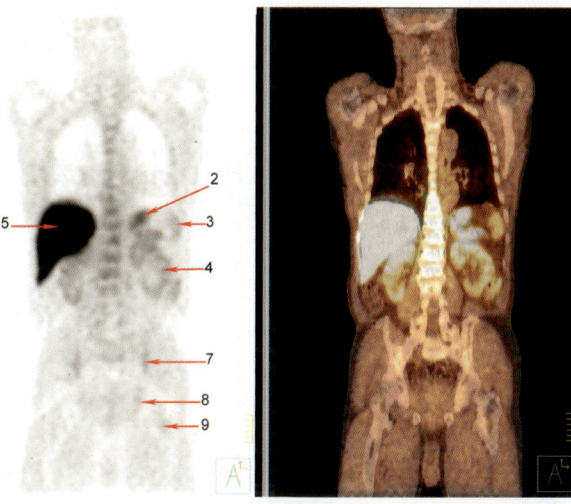

FIGURE 11-107: Coronal 19

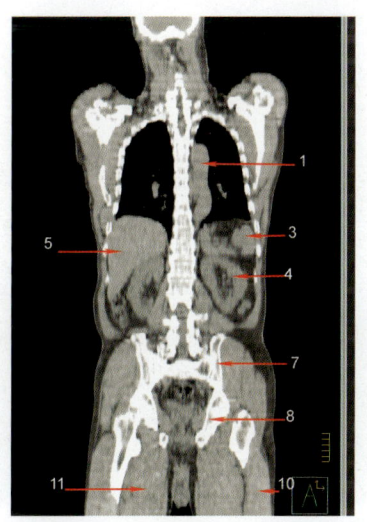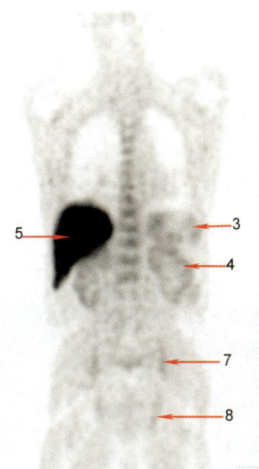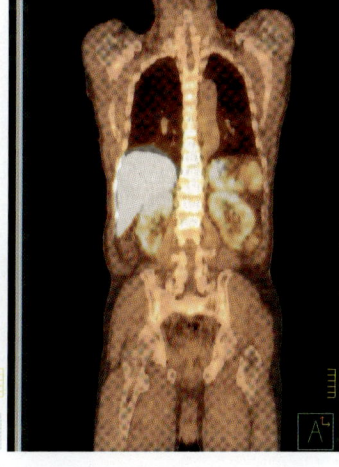

FIGURE 11-108: Coronal 20

1. Descending aorta
2. Stomach (fundus)
3. Spleen
4. Kidney
5. Right hepatic lobe
6. Psoas muscle
7. Iliac tuberosity
8. Gluteus muscle
9. Trochanter of femur
10. Vastus lateralis muscle
11. Adductor muscle
12. Ischium

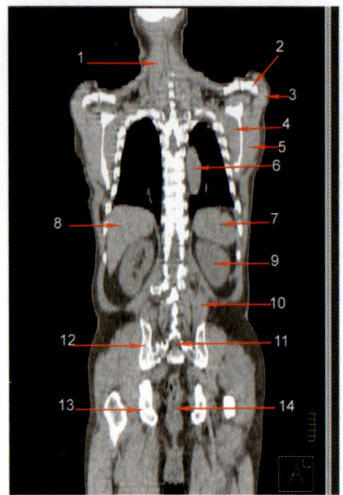
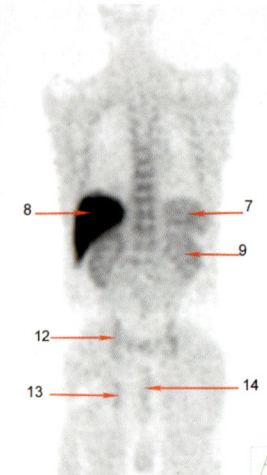
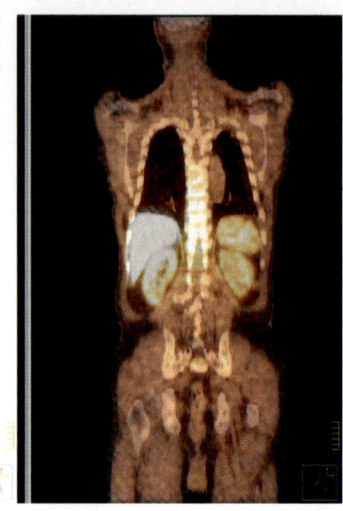

FIGURE 11-109: Coronal 21

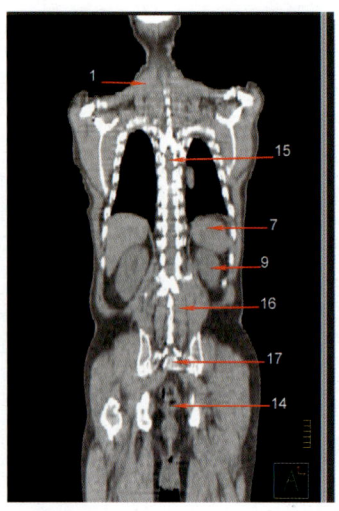
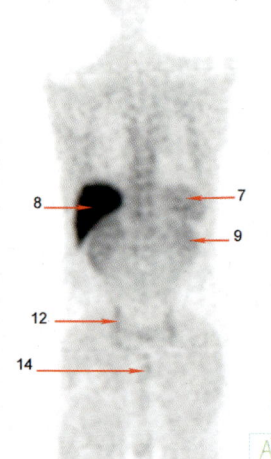
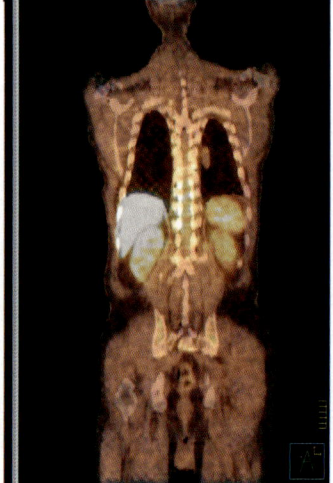

FIGURE 11-110: Coronal 22

1. Semispinalis cervicis muscle
2. Acromion
3. Deltoid muscle
4. Subscapularis muscle
5. Infraspinatus muscle
6. Descending aorta
7. Spleen
8. Right liver
9. Kidney
10. Psoas muscle
11. Sacral foramina
12. Iliac tuberosity
13. Ischium
14. Rectum
15. Spinal canal
16. Erector spinae muscle
17. Sacrum

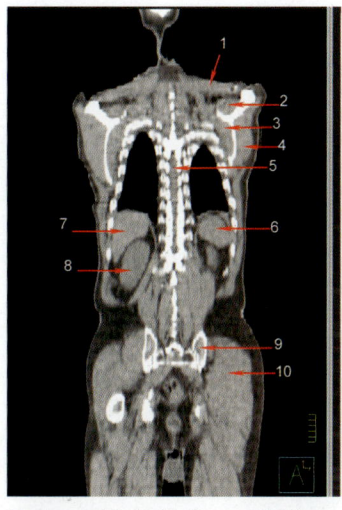
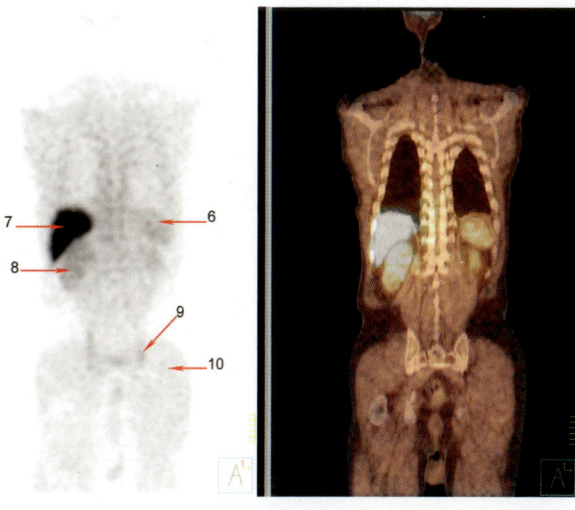

FIGURE 11-111: Coronal 23

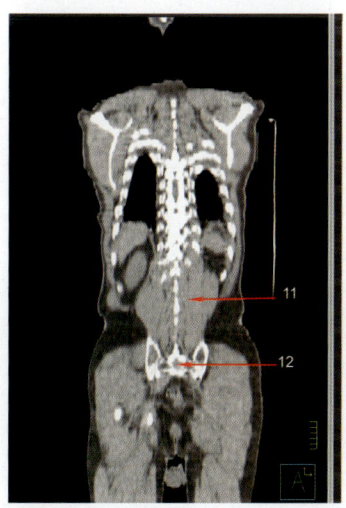
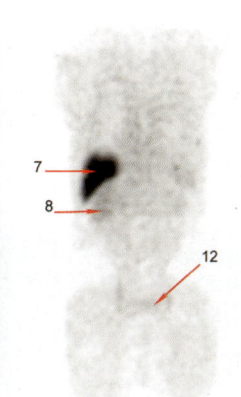

FIGURE 11-112: Coronal 24

1. Trapezius muscle
2. Supraspinatus muscle
3. Subscapularis muscle
4. Infraspinatus muscle
5. Spinal cord
6. Spleen
7. Right hepatic lobe
8. Kidney
9. Iliac tuberosity
10. Gluteus muscle
11. Erector spinae muscle
12. Sacrum

Sagittals

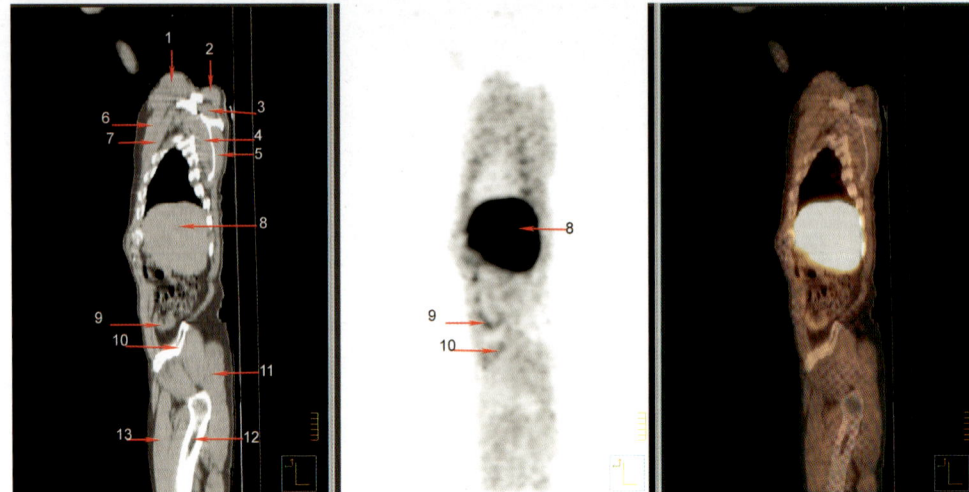

FIGURE 11-113: Sagittal 1

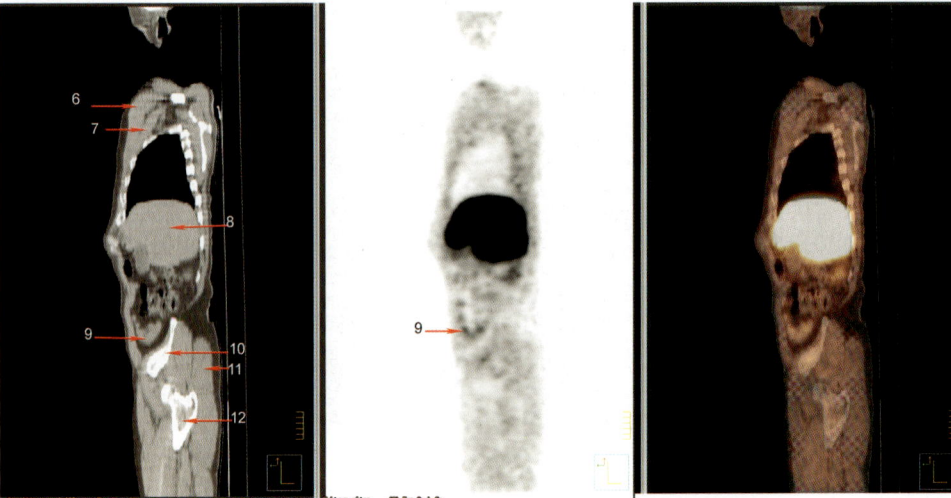

FIGURE 11-114: Sagittal 2

1. Deltoid muscle
2. Trapezius muscle
3. Supraspinatus muscle
4. Subscapularis
5. Infraspinatus
6. Pectoralis major muscle
7. Pectoralis minor muscle
8. Right hepatic lobe
9. Ascending colon
10. Ilium
11. Gluteus maximus muscle
12. Femur
13. Rectus femoris muscle

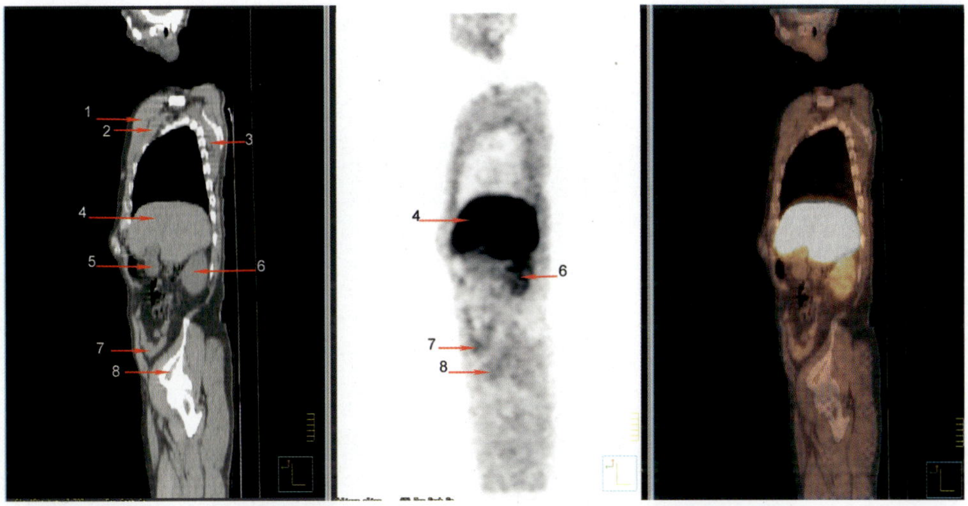

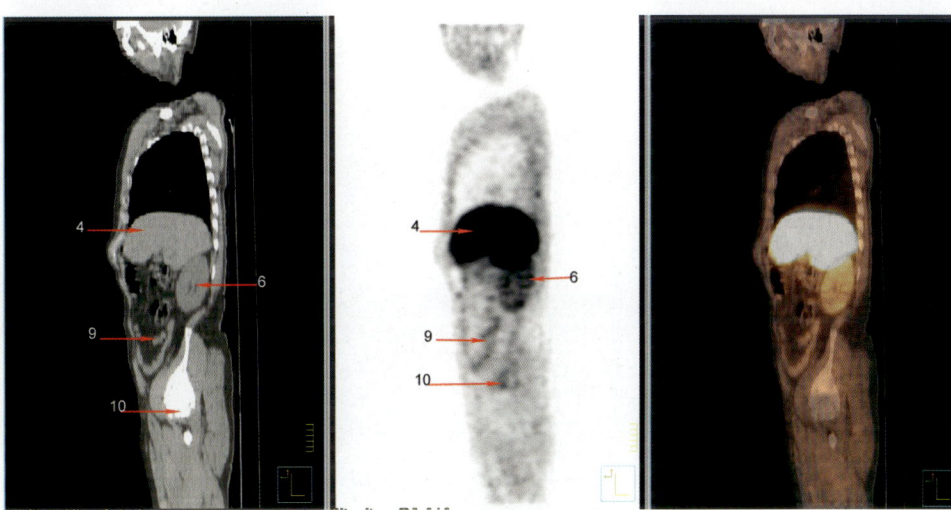

FIGURE 11-115: Sagittal 3

FIGURE 11-116: Sagittal 4

1. Pectoralis major muscle
2. Pectoralis minor muscle
3. Subscapularis muscle
4. Right hepatic lobe
   (segment VIII)
5. Gallbladder
6. Kidney
7. Ascending colon
8. Ilium
9. Ileum
10. Femoral head

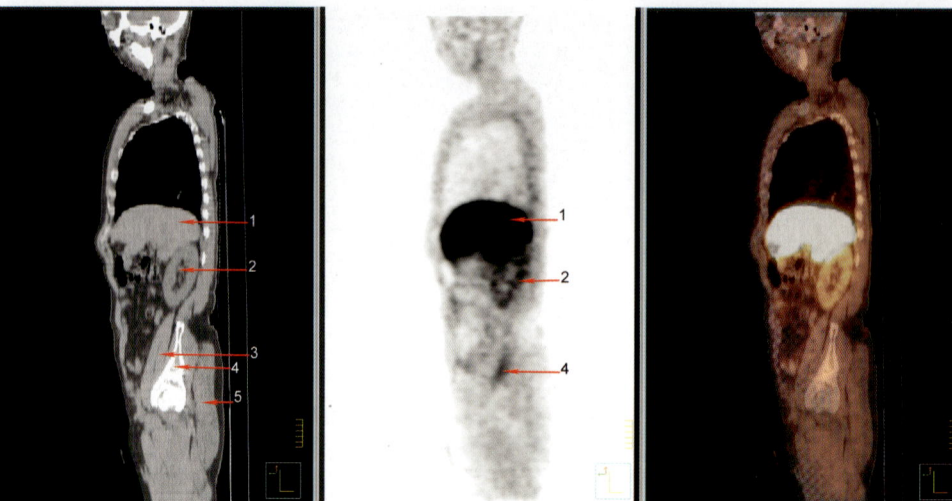

FIGURE 11-117: Sagittal 5

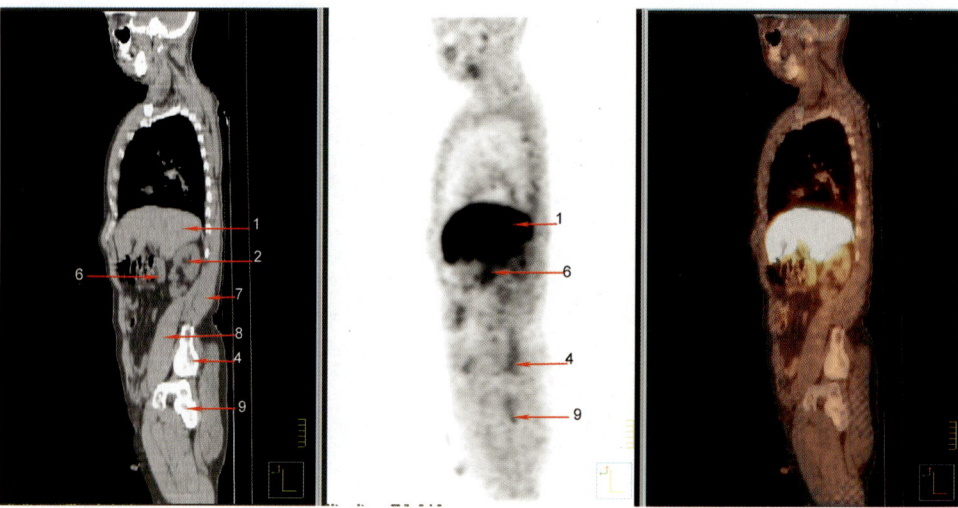

FIGURE 11-118: Sagittal 6

1. Right hepatic lobe
   (segment VII)
2. Kidney
3. Iliacus muscle

4. Ilium
5. Gluteus maximus muscle
6. Pancreas (head)
7. Quadratus lumborum

8. Iliopsoas muscle
9. Trochanter of femur

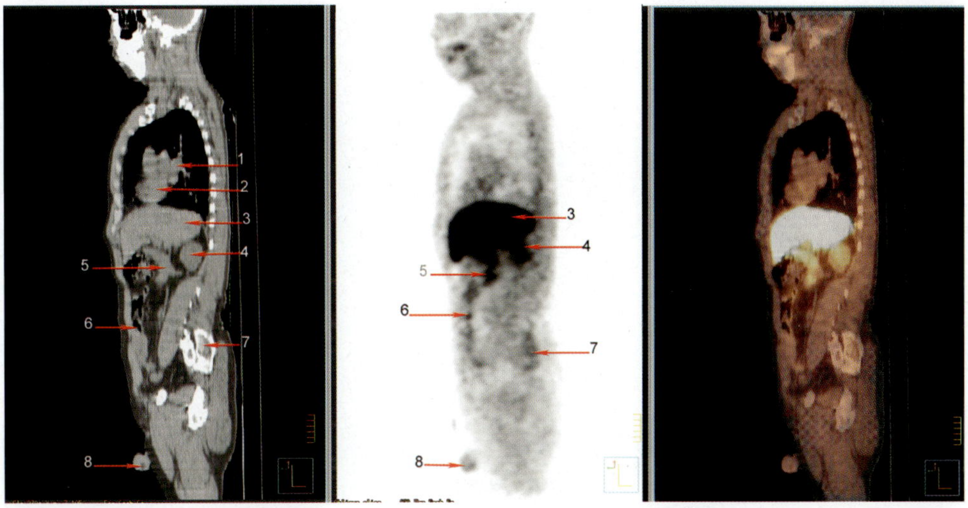

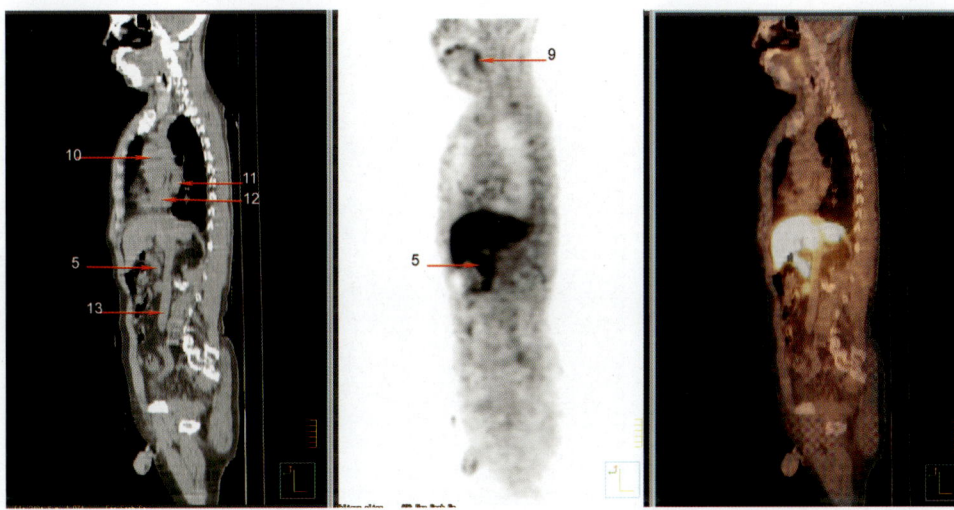

FIGURE 11-119: Sagittal 7

FIGURE 11-120: Sagittal 8

1. Right pulmonary artery
2. Right ventricle
3. Right hepatic lobe
   (segment VII)
4. Kidney
5. Pancreas (head)
6. Ascending colon
7. Iliac tuberosity
8. Testis
9. Tongue base
10. Main pulmonary artery
11. Left atrium
12. Right ventricle
13. Inferior vena cava

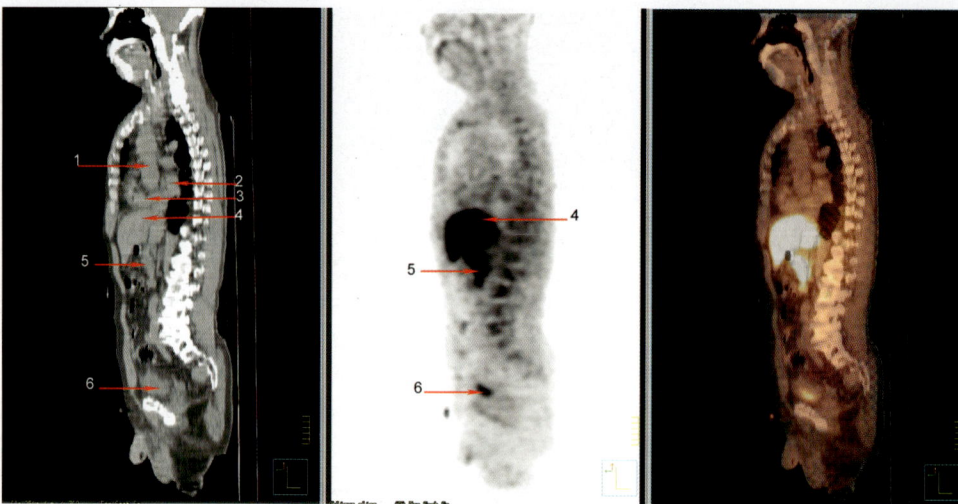

FIGURE 11-121: Sagittal 9

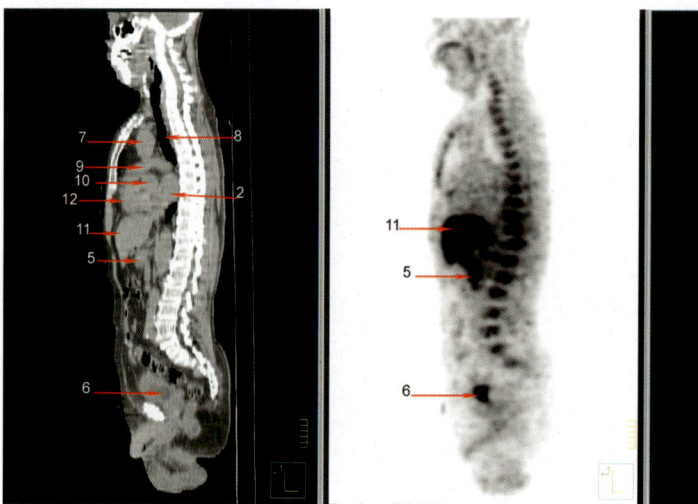

FIGURE 11-122: Sagittal 10

1. Ascending aorta
2. Left atrium
3. Left ventricle
4. Left hepatic lobe (segment II)
5. Pancreas (head)
6. Bladder
7. Right brachiocephalic vein
8. Trachea
9. Aortic arch
10. Right pulmonary artery
11. Left hepatic lobe (segment III)
12. Right ventricle

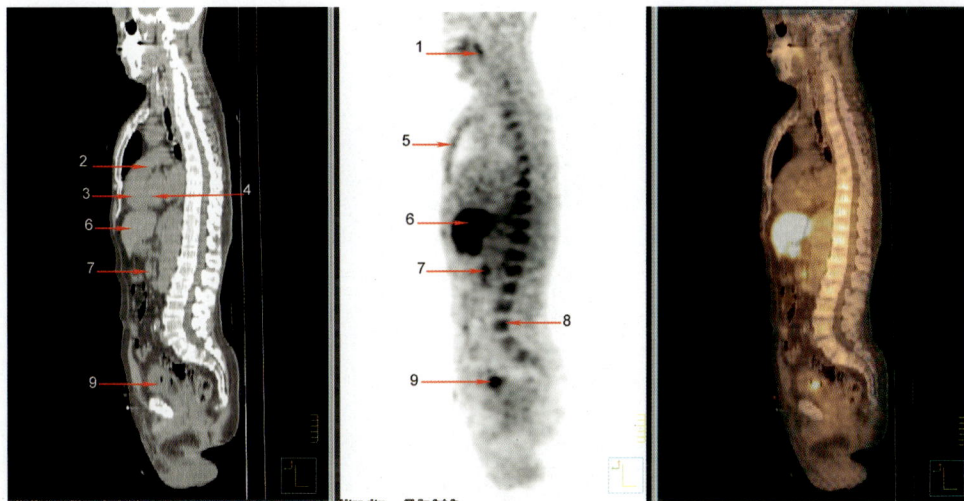

FIGURE 11-123: Sagittal 11

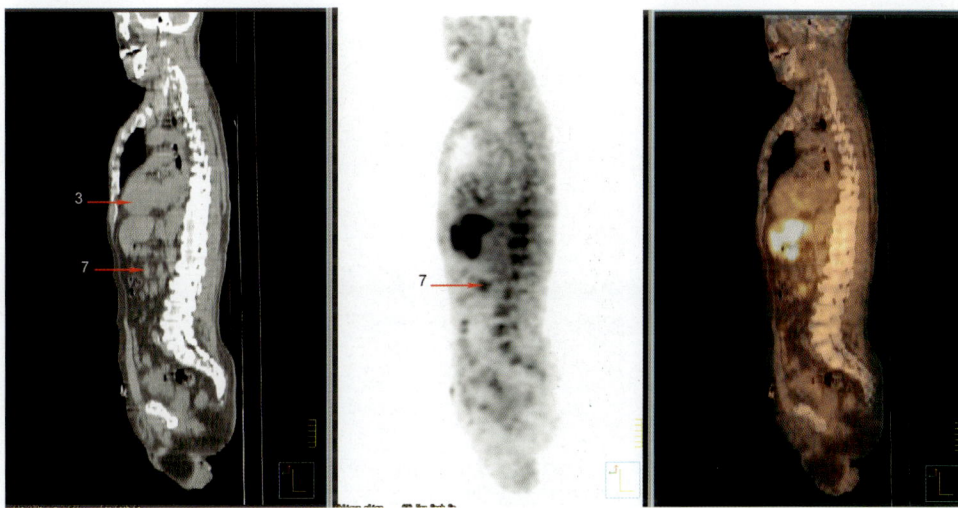

FIGURE 11-124: Sagittal 12

1. Tongue base
2. Aortic arch
3. Left ventricle
4. Left atrium
5. Sternum
6. Left hepatic lobe
7. Pancreas (body)
8. L5 marrow
9. Bladder

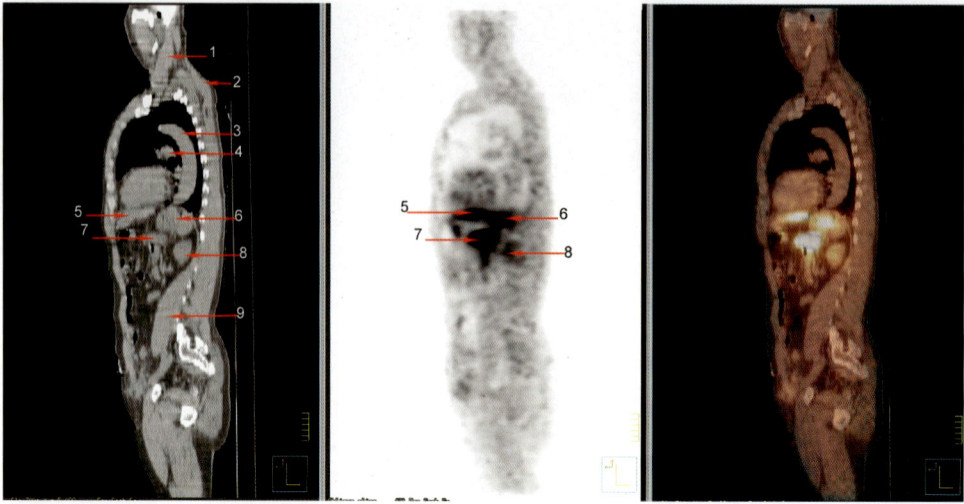

FIGURE 11-125: Sagittal 13

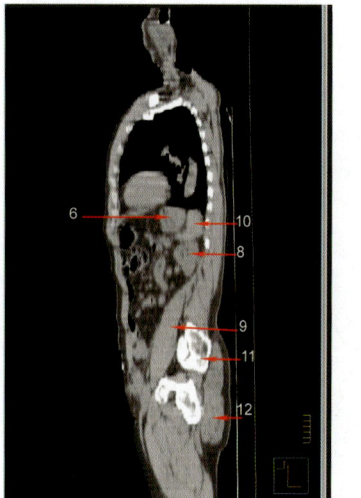 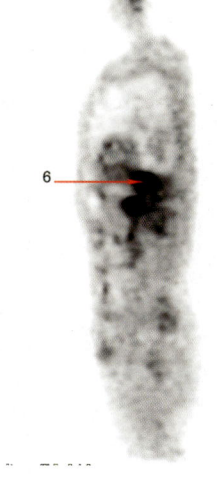 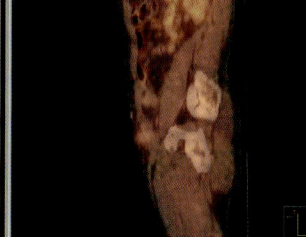

FIGURE 11-126: Sagittal 14

1. Sternocleidomastoid
   muscle
2. Trapezius muscle
3. Aortic arch
4. Left pulmonary artery

5. Left hepatic lobe
6. Stomach
7. Pancreas (body)
8. Kidney
9. Psoas muscle

10. Spleen
11. Iliac tuberosity
12. Gluteus muscle

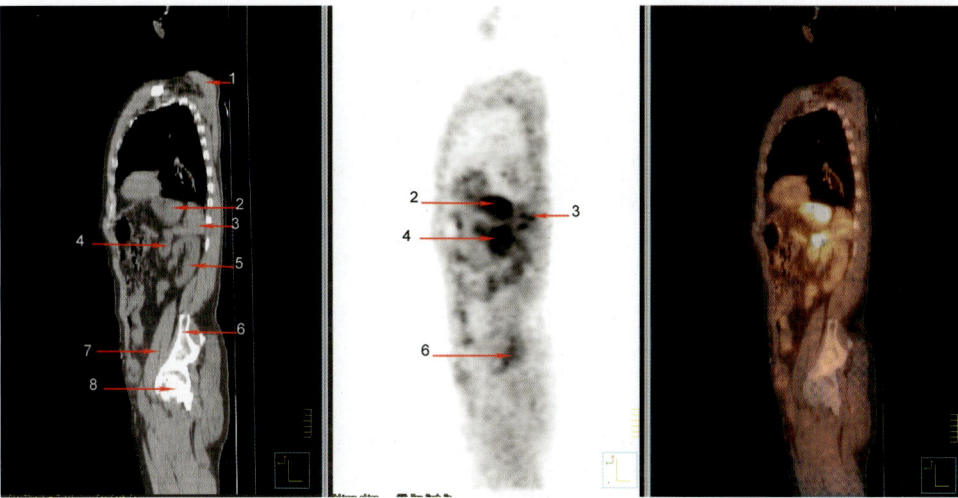

FIGURE 11-127: Sagittal 15

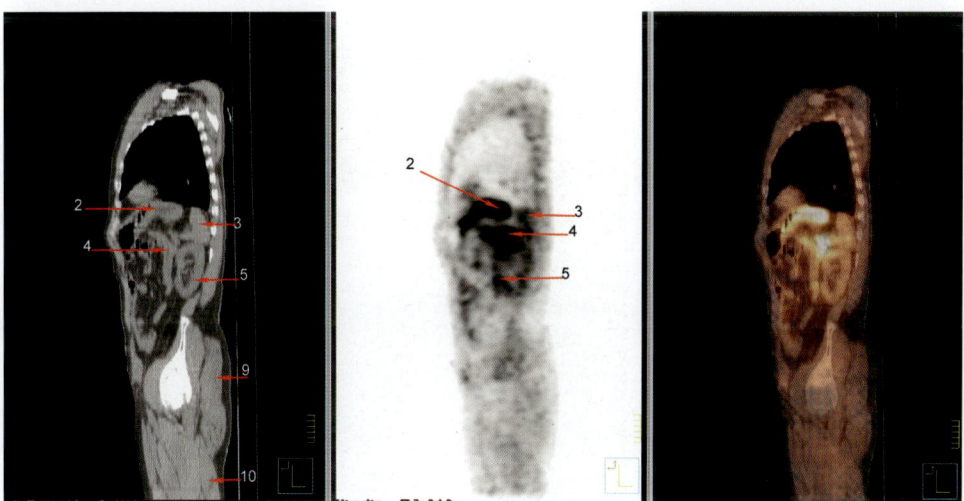

FIGURE 11-128: Sagittal 16

1. Trapezius muscle
2. Stomach
3. Spleen
4. Pancreas (body)

5. Kidney
6. Iliac ala
7. Iliacus muscle
8. Femoral head

9. Gluteus maximus muscle
10. Biceps femoris muscle

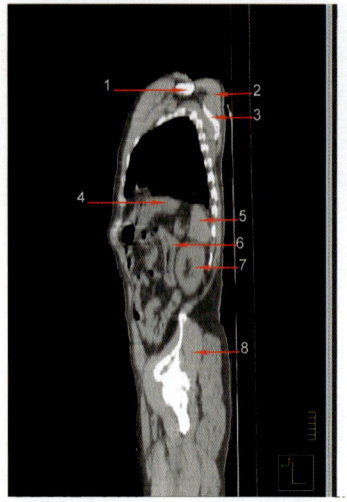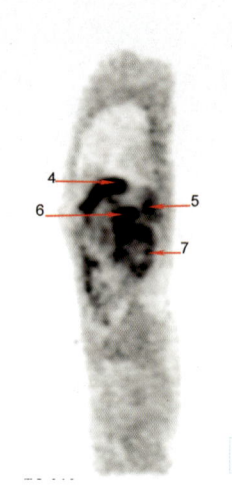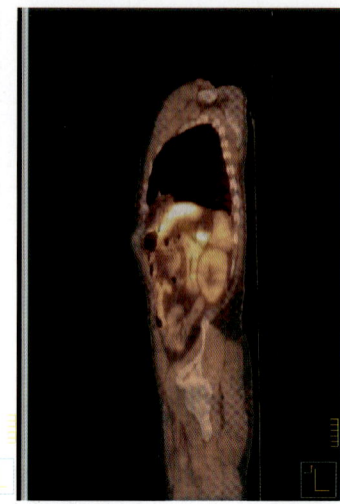

FIGURE 11-129: Sagittal 17

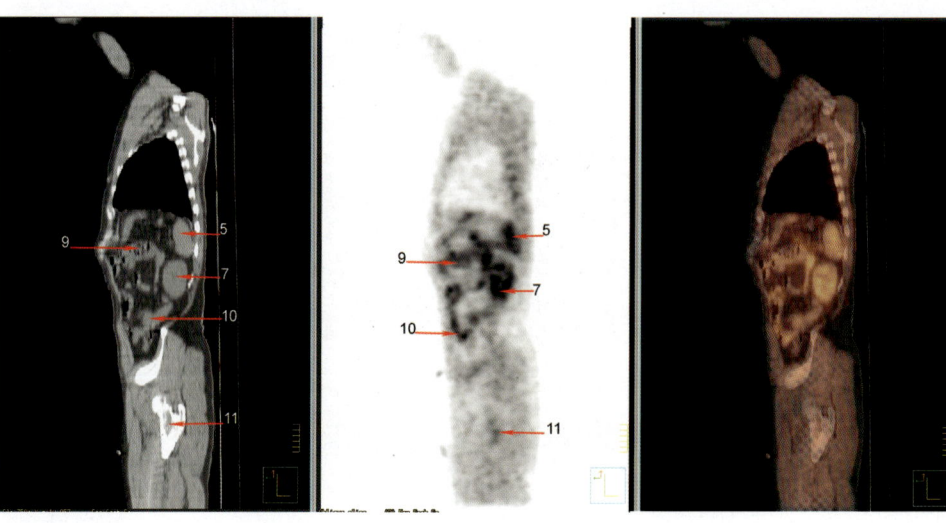

FIGURE 11-130: Sagittal 18

1. Clavicle	5. Spleen	9. Jejunum
2. Trapezius muscle	6. Pancreas (tail)	10. Ilium
3. Scapula	7. Kidney	11. Trochanter of femur
4. Left hepatic lobe	8. Gluteus medius muscle	

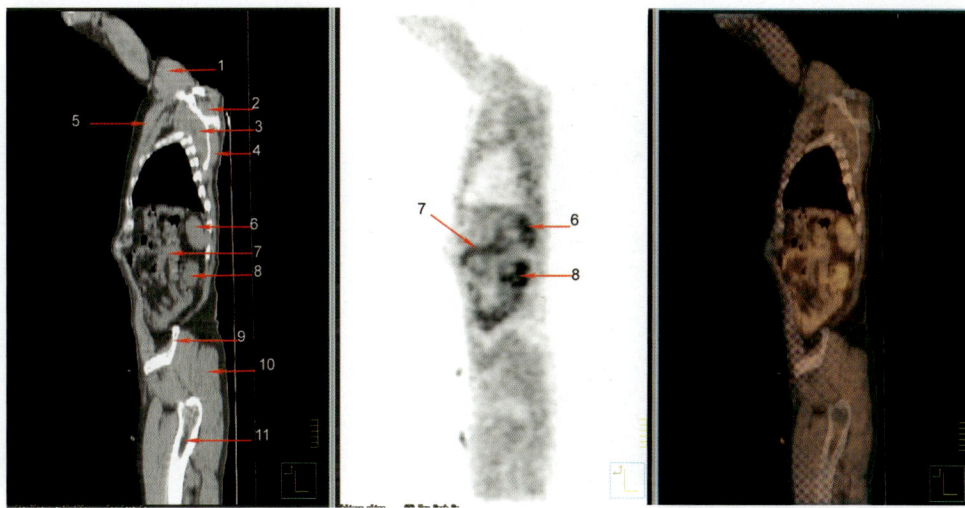

FIGURE 11-131: Sagittal 19

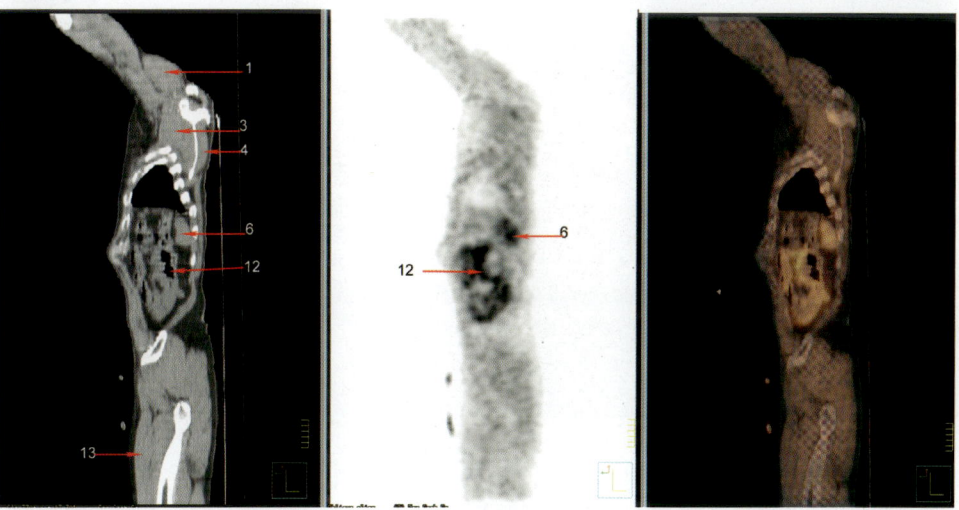

FIGURE 11-132: Sagittal 20

1. Deltoid muscle
2. Supraspinatus muscle
3. Subscapularis muscle
4. Infraspinatus muscle
5. Pectoralis major muscle
6. Spleen
7. Jejunum
8. Kidney
9. Iliac ala
10. Gluteus maximus muscle
11. Femur
12. Descending colon
13. Rectus femoris muscle

Transaxials

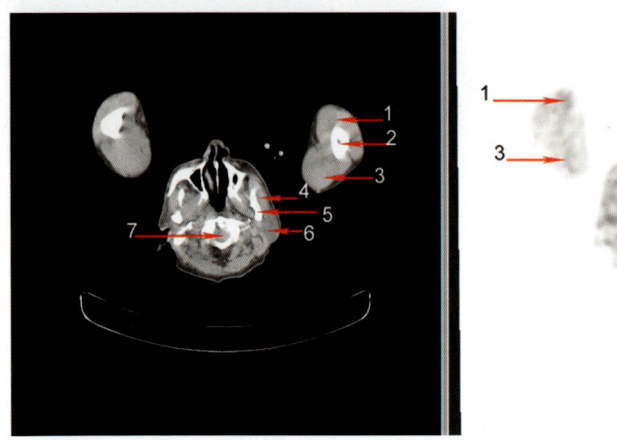 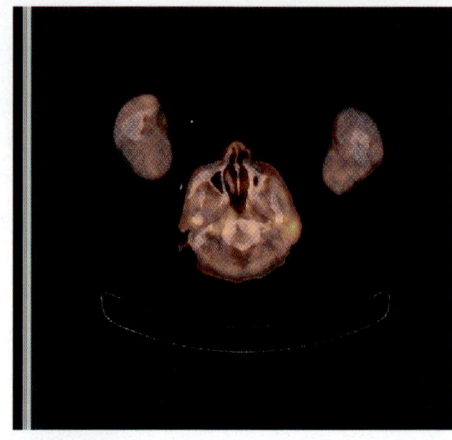

FIGURE 11-133: Transaxial 1

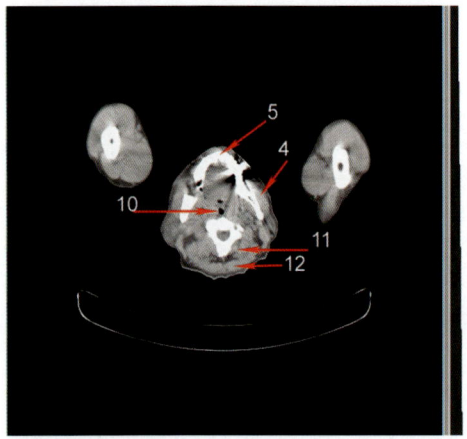 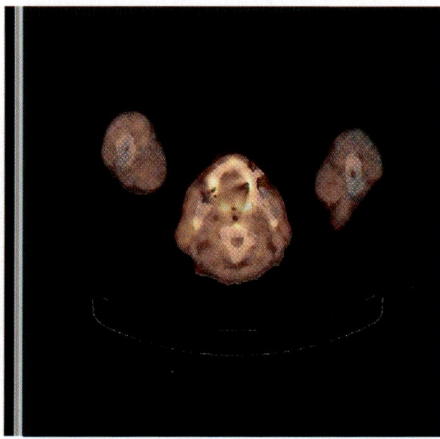

FIGURE 11-134: Transaxial 2

1. Biceps brachii muscle
2. Humerus
3. Triceps brachii muscle
4. Masseter muscle
5. Mandible
6. Parotid gland
7. Spinal cord
8. Periodontal muscle (omohyoid muscle)
9. Lingual tonsil
10. Oropharynx
11. Semispinalis cervicis muscle
12. Splenius capitis muscle

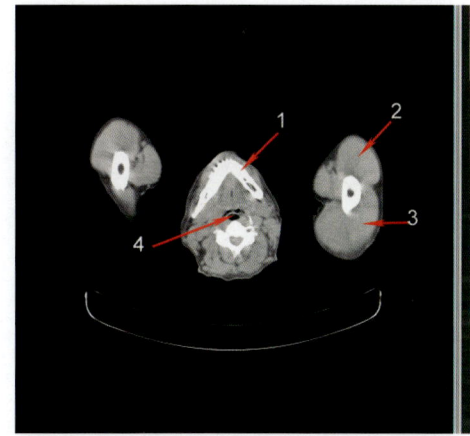
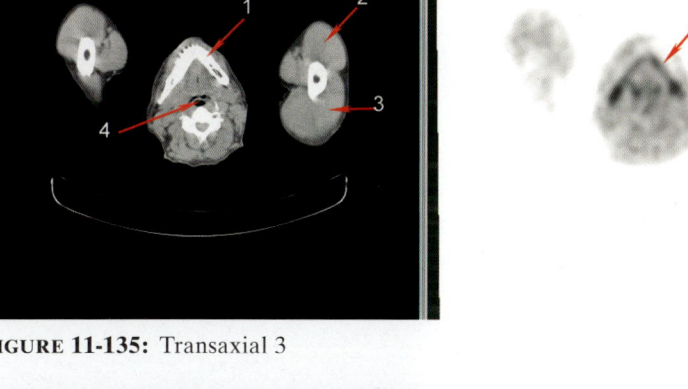
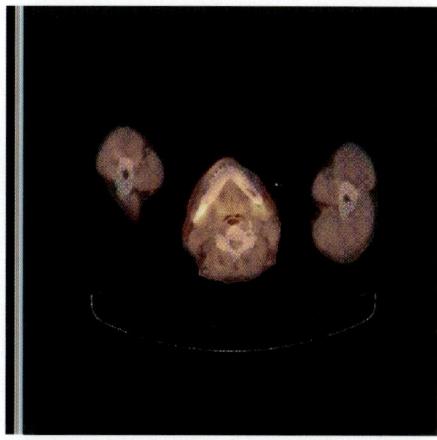

FIGURE 11-135: Transaxial 3

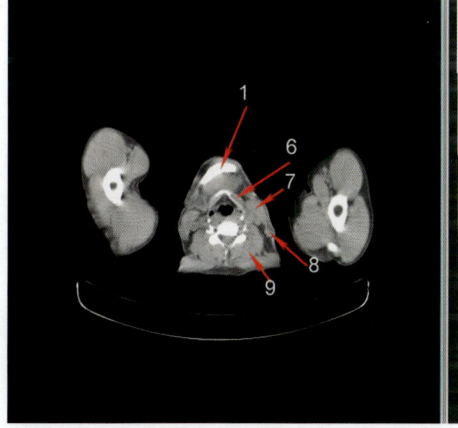
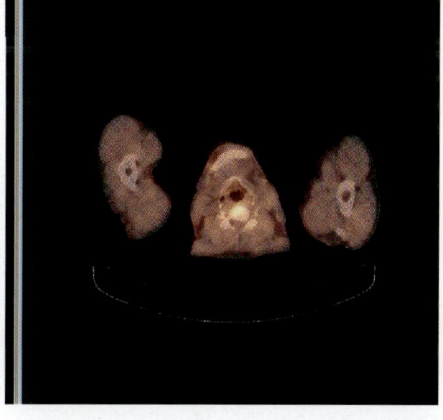

FIGURE 11-136: Transaxial 4

1. Mandible
2. Biceps brachii muscle
3. Triceps brachii muscle
4. Oropharynx

5. Periodontal muscle (omohyoid muscle)
6. Hyoid bone
7. Parotid gland

8. Sternocleidomastoid muscle
9. Semispinalis capitis
10. Marrow of cervical vertebral body

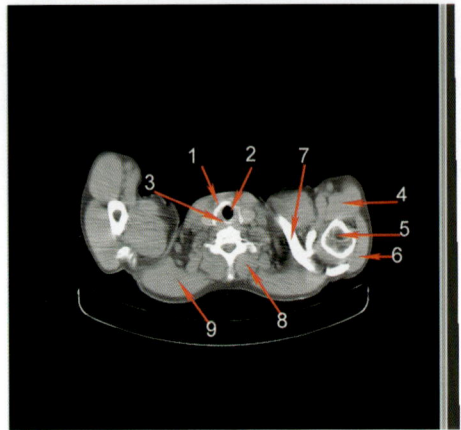

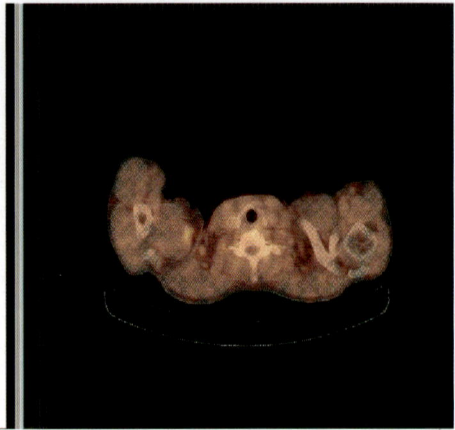

FIGURE 11-137: Transaxial 5

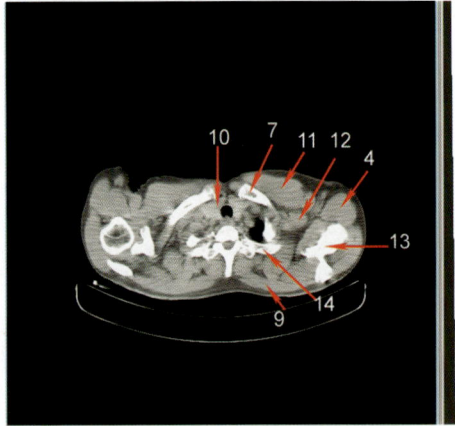
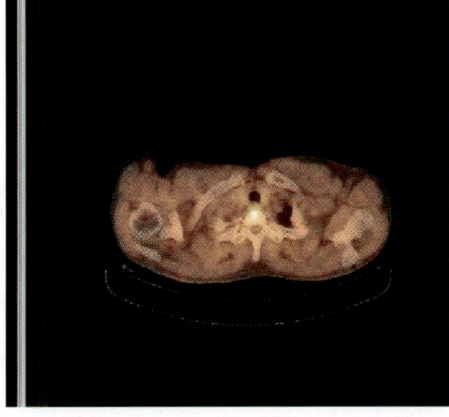

FIGURE 11-138: Transaxial 6

1. Thyroid cartilage
2. Trachea
3. Cricoid cartilage
4. Deltoid muscle
5. Humeral head
6. Supraspinatus muscle
7. Clavicle
8. Semispinalis cervicis muscle
9. Trapezius muscle
10. Thyroid lobe
11. Pectoralis major muscle
12. Scalene muscle
13. Scapula
14. Rib

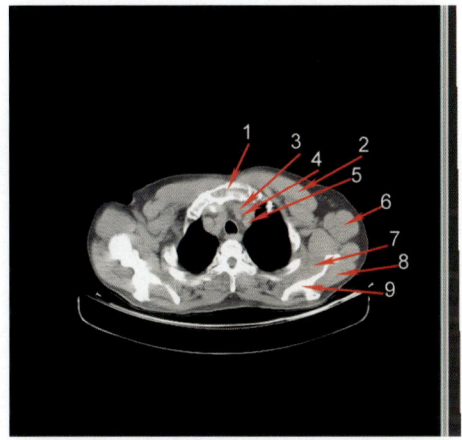
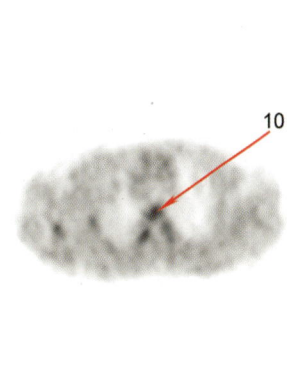
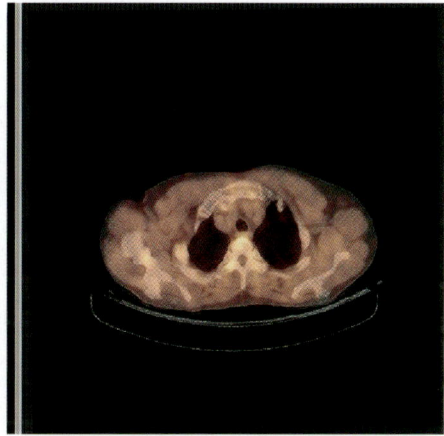

FIGURE 11-139: Transaxial 7

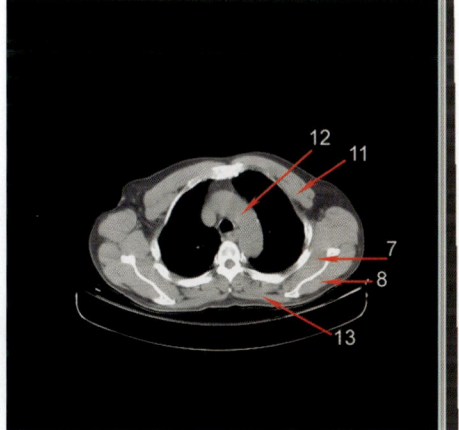
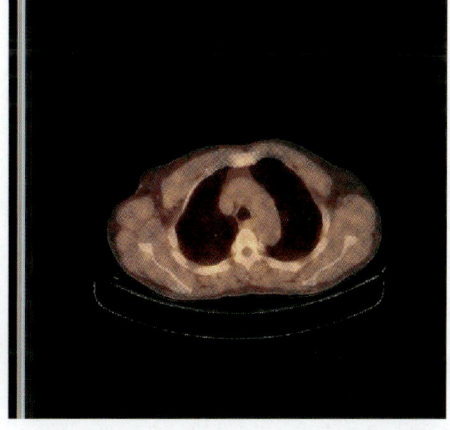

FIGURE 11-140: Transaxial 8

1. Manubrium
2. Pectoralis major muscle
3. Left innominate vein
4. Left common carotid artery
5. Left subclavian artery
6. Deltoid muscle
7. Subscapularis muscle
8. Infraspinatus muscle
9. Scapula
10. Marrow of thoracic vertebral body
11. Pectoralis minor muscle
12. Aortic arch
13. Trapezius muscle

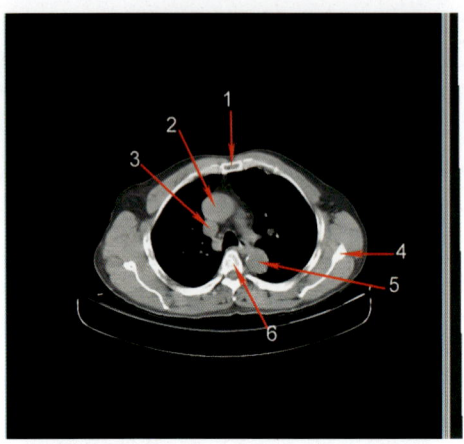 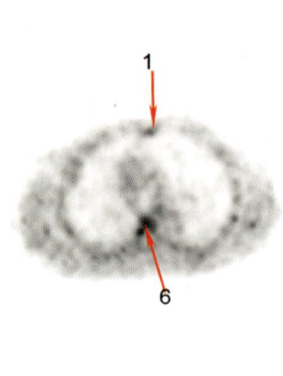 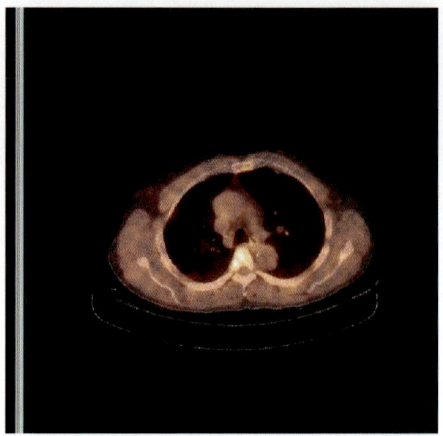

FIGURE 11-141: Transaxial 9

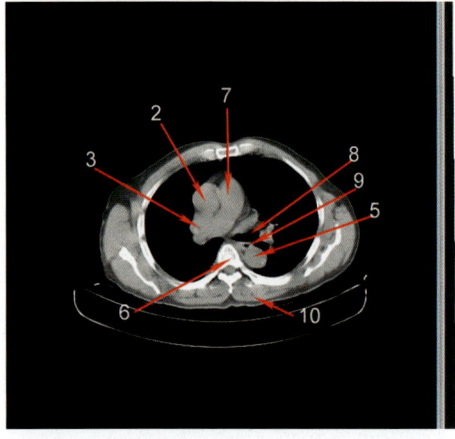 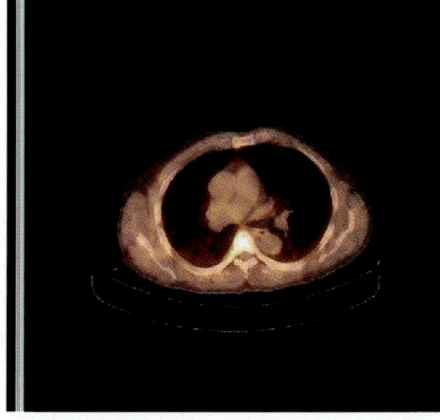

FIGURE 11-142: Transaxial 10

1. Sternum
2. Ascending aorta
3. Superior vena cava
4. Scapula
5. Descending aorta
6. Vertebral body
7. Main pulmonary artery
8. Left main bronchus
9. Esophagus
10. Trapezius muscle

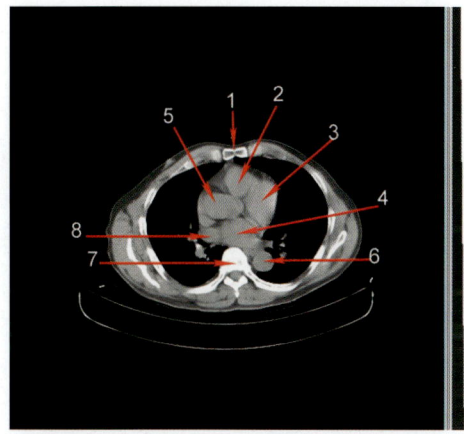
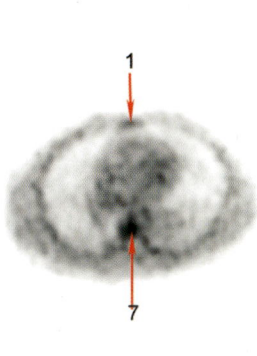
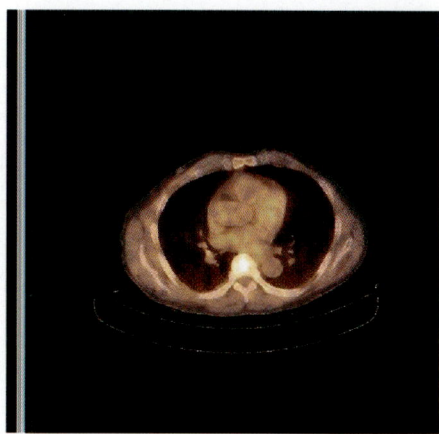

FIGURE 11-143: Transaxial 11

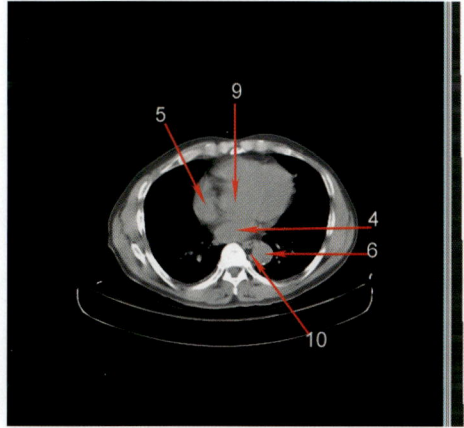
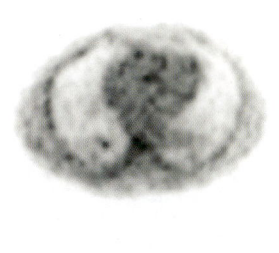
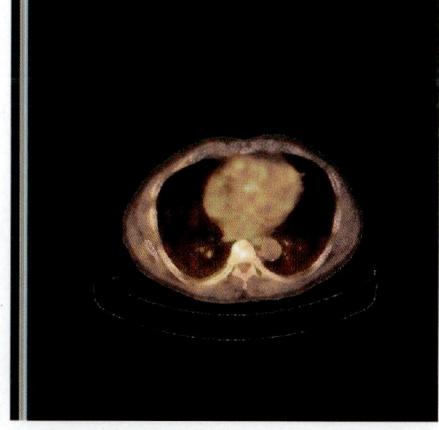

FIGURE 11-144: Transaxial 12

1. Sternum
2. Right ventricle
3. Left ventricle
4. Left atrium

5. Right atrium
6. Descending aorta
7. Vertebral body
8. Pulmonary vein

9. Ascending aorta
10. Esophagus

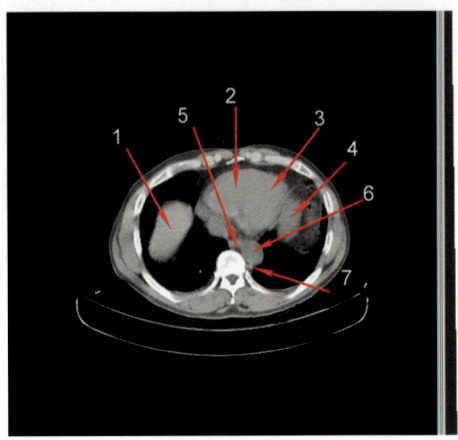
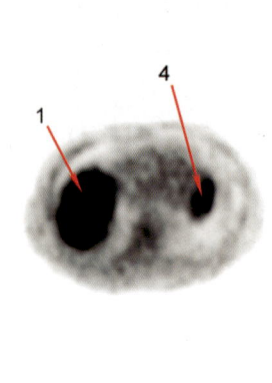
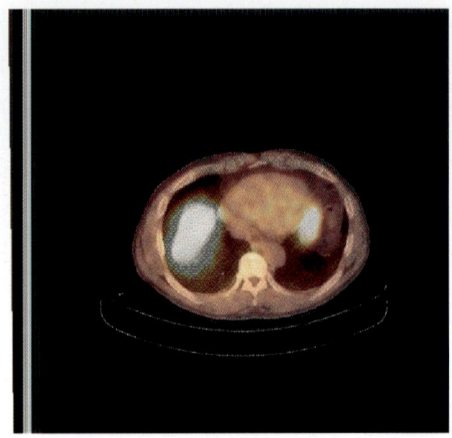

FIGURE 11-145: Transaxial 13

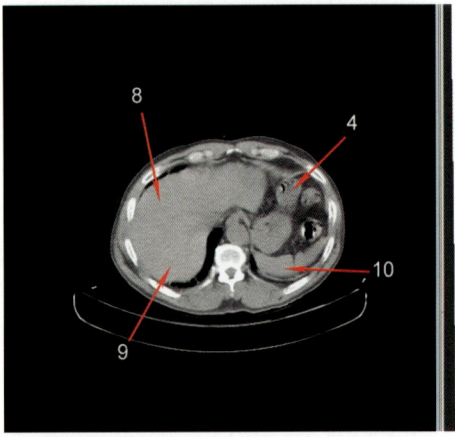
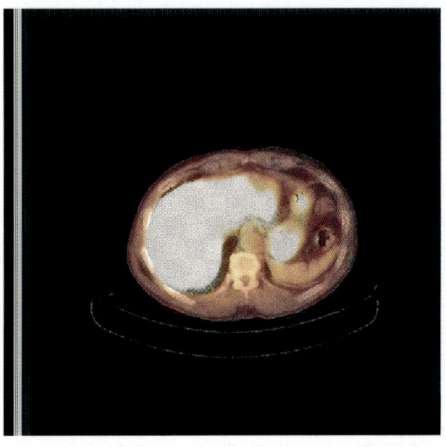

FIGURE 11-146: Transaxial 14

1. Hepatic dome
2. Right ventricle
3. Left ventricle
4. Stomach

5. Esophagus
6. Descending aorta
7. Hemiazygos vein
8. Right hepatic lobe (segment VIII)

9. Right hepatic lobe (segment VII)
10. Spleen

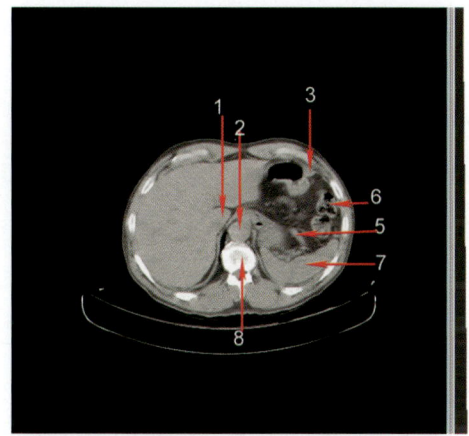
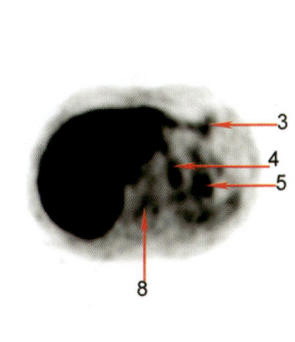
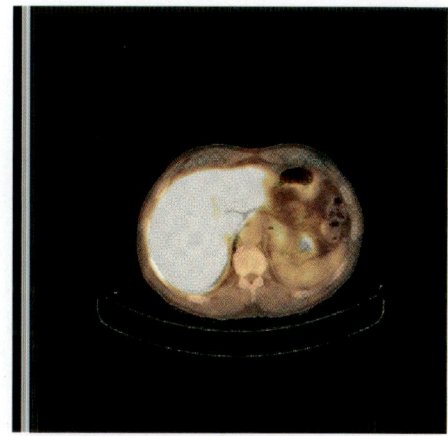

FIGURE 11-147: Transaxial 15

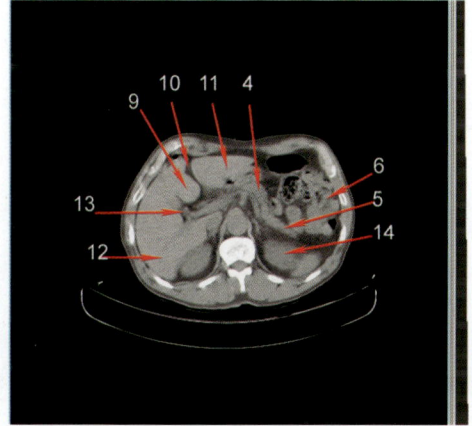
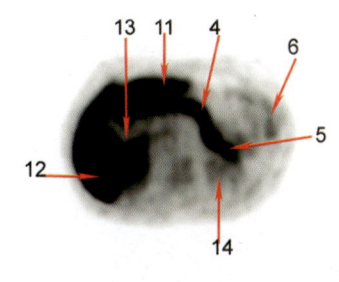

FIGURE 11-148: Transaxial 16

1. Inferior vena cava
2. Aorta
3. Stomach
4. Pancreas (body)
5. Pancreas (tail)
6. Jejunum
7. Spleen
8. Vertebral body
9. Left hepatic lobe (segment IV)
10. Falciform ligament
11. Left hepatic lobe (segment III)
12. Right hepatic lobe (segment VI)
13. Porta hepatis
14. Kidney

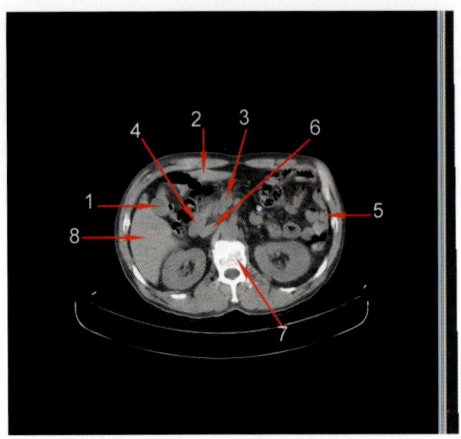 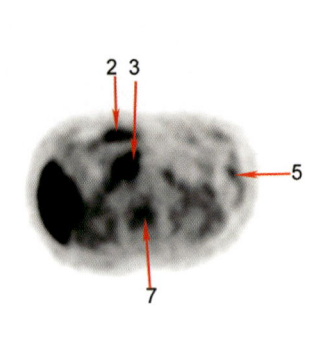 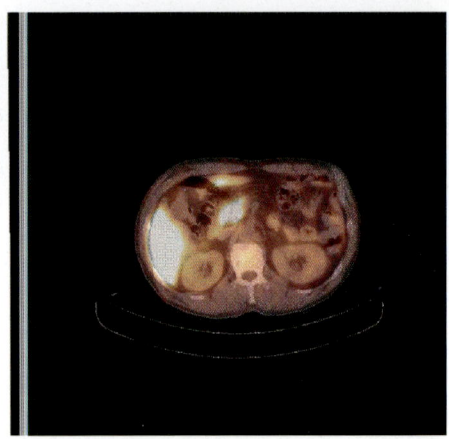

FIGURE 11-149: Transaxial 17

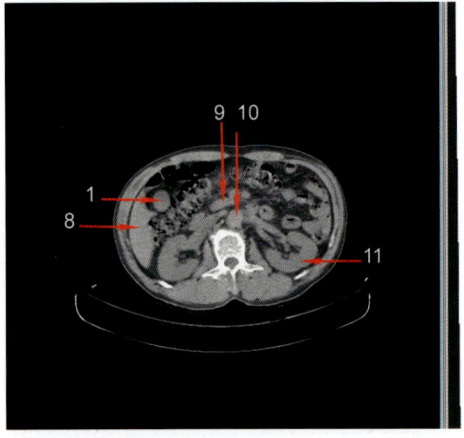 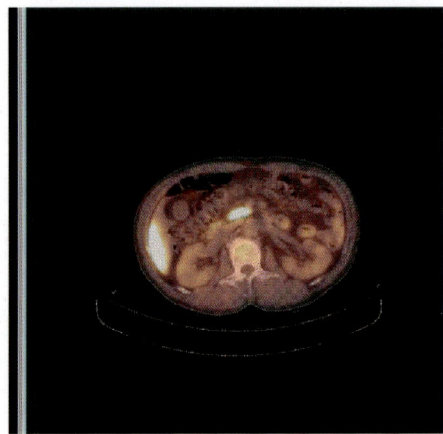

FIGURE 11-150: Transaxial 18

1. Gallbladder
2. Gastric antrum
3. Pancreas (head)
4. Duodenum (second portion)
5. Jejunum
6. Inferior vena cava
7. Vertebral body
8. Right hepatic lobe (segment V)
9. Pancreas (body)
10. Aorta
11. Kidney

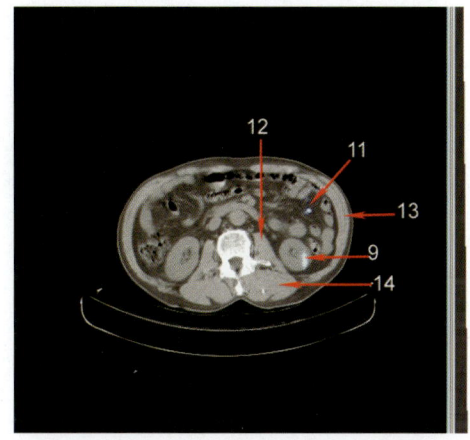

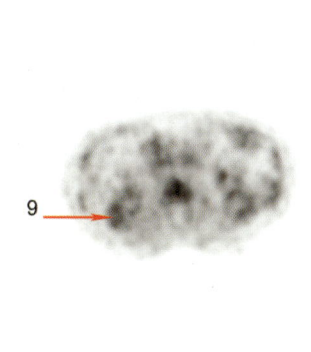

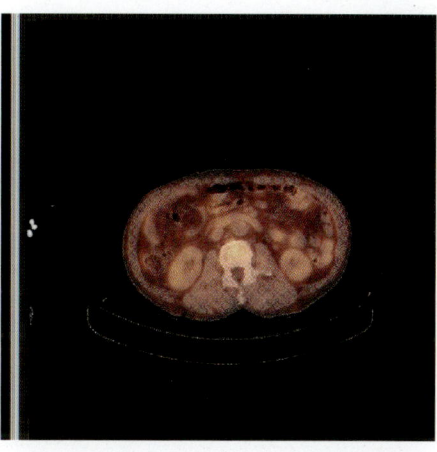

FIGURE 11-151: Transaxial 19

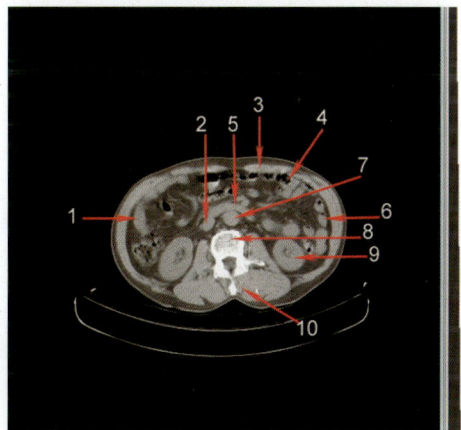

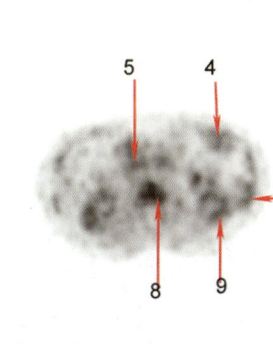

FIGURE 11-152: Transaxial 20

1. Tip of right hepatic lobe
2. Inferior vena cava
3. Rectus abdominis muscle
4. Transverse colon
5. Duodenum (third portion)

6. Jejunum
7. Aorta
8. Vertebral body
9. Kidney
10. Erector spinae muscle

11. Mesentery
12. Psoas muscle
13. Oblique muscles
14. Quadratum lumborum muscle

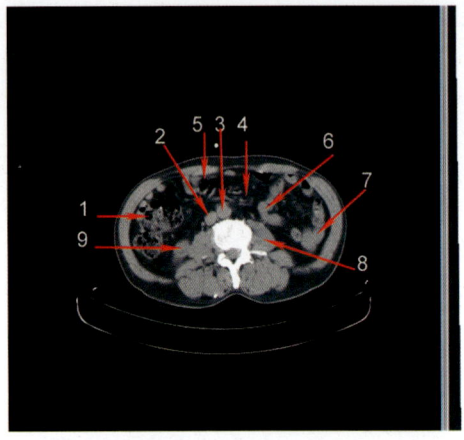 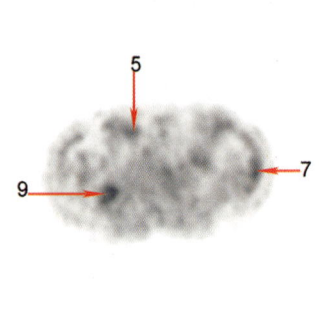 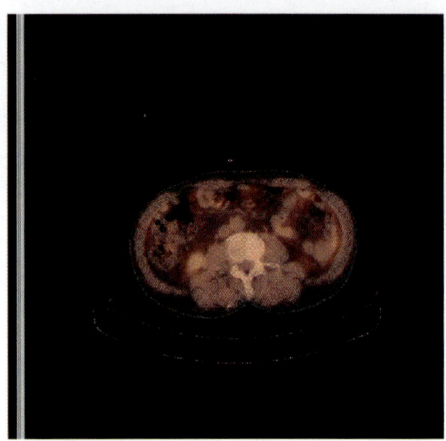

FIGURE 11-153: Transaxial 21

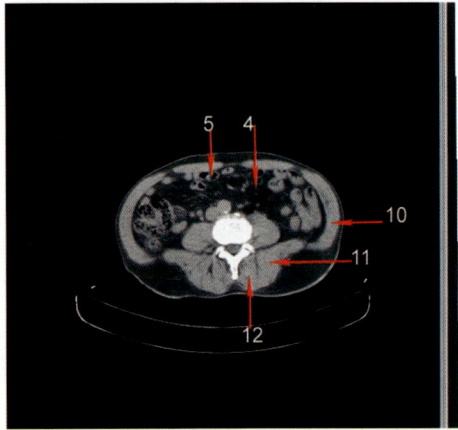 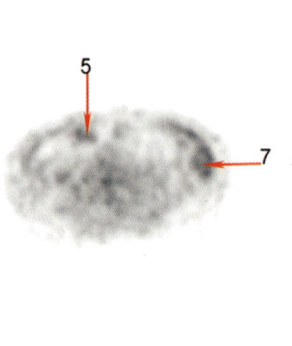 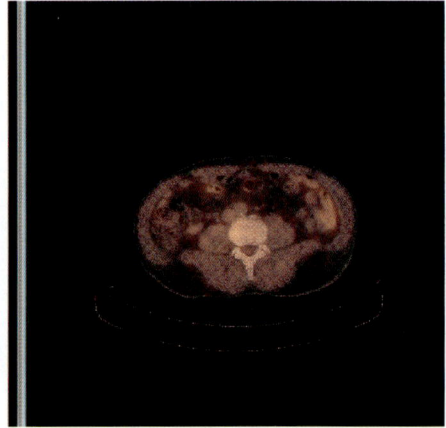

FIGURE 11-154: Transaxial 22

1. Ascending colon
2. Inferior vena cava
3. Aorta
4. Mesentery
5. Transverse colon
6. Jejunum
7. Descending colon
8. Psoas muscle
9. Tip of right kidney
10. Oblique muscles
11. Quadratus lumborum muscle
12. Erector spinae muscle

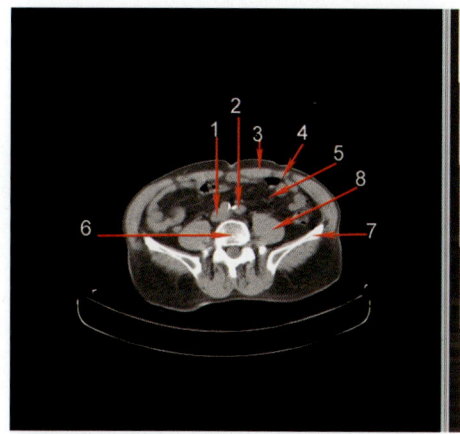
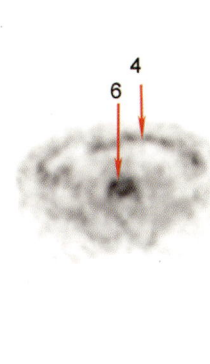
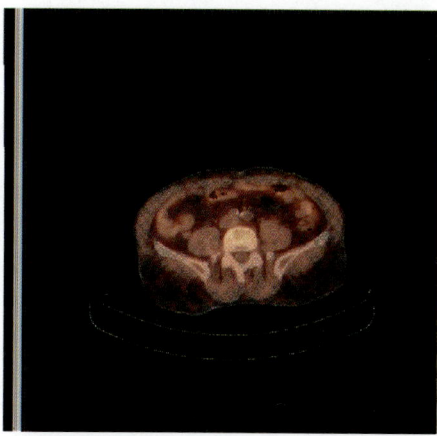

FIGURE 11-155: Transaxial 23

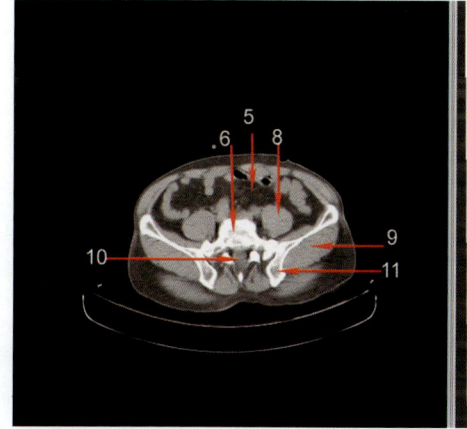
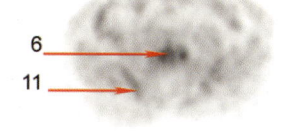
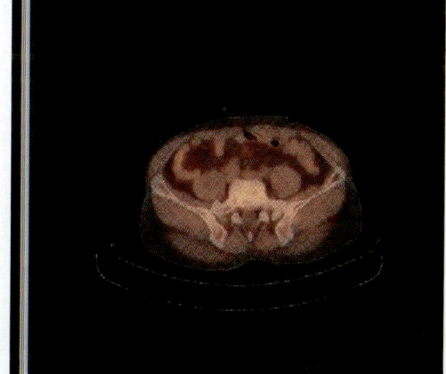

FIGURE 11-156: Transaxial 24

1. Inferior vena cava
2. Aorta
3. Rectus abdominis muscle
4. Transverse colon
5. Mesentery
6. Vertebral body
7. Iliac ala (crest)
8. Psoas muscle
9. Gluteus medius muscle
10. Sacral canal
11. Iliac tuberosity

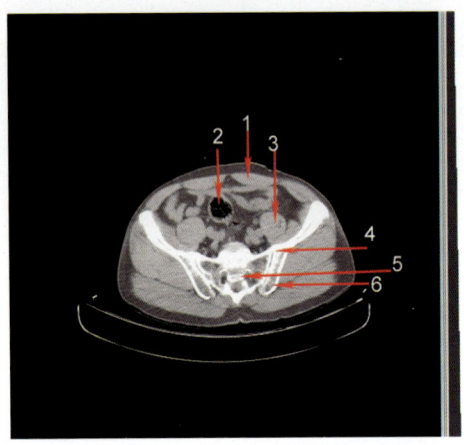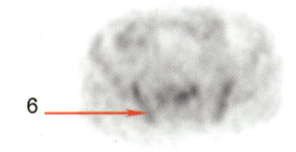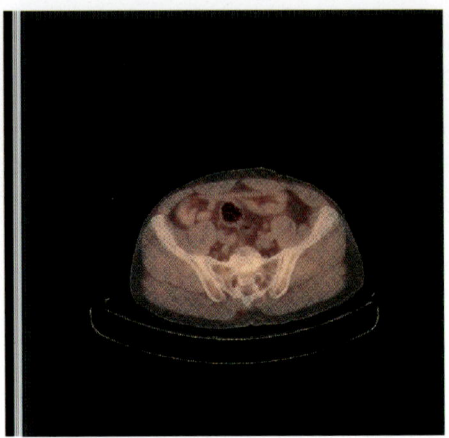

FIGURE 11-157: Transaxial 25

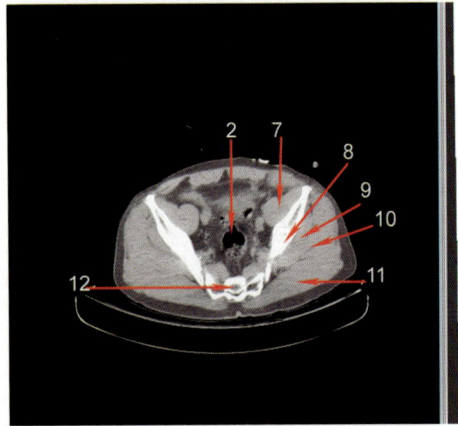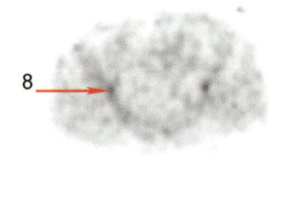

FIGURE 11-158: Transaxial 26

1. Rectus abdominis muscle
2. Sigmoid colon
3. Psoas muscle
4. Sacroiliac joint

5. Neural foramen of sacrum
6. Iliac tuberosity
7. Iliopsoas muscle
8. Iliac ala

9. Gluteus minimus muscle
10. Gluteus medius muscle
11. Gluteus maximus muscle
12. Sacral promontory

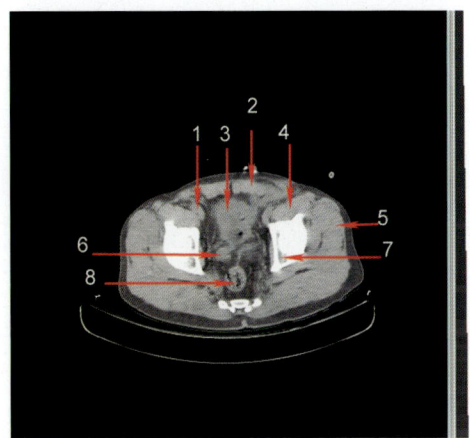
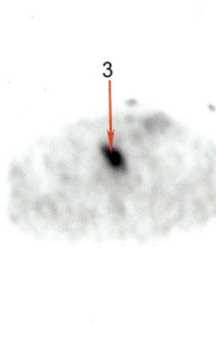
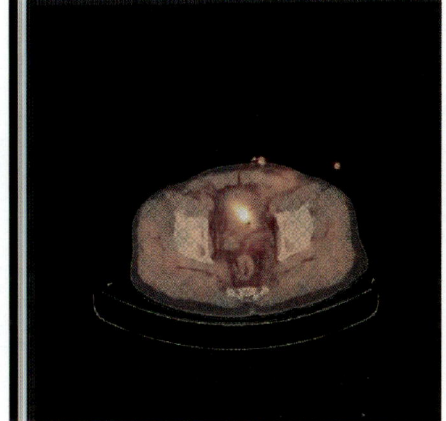

FIGURE 11-159: Transaxial 27

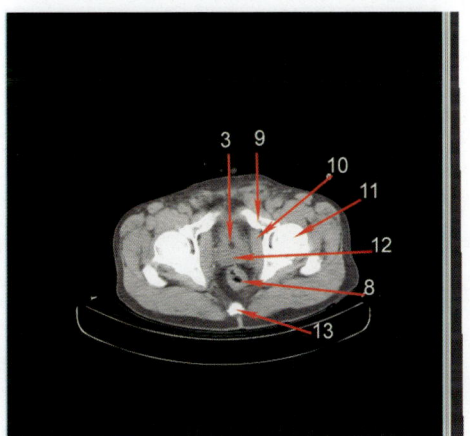
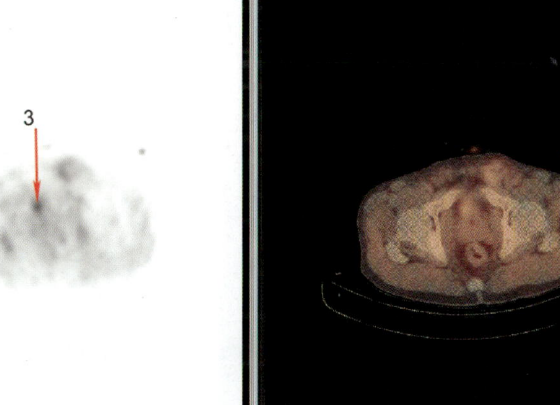

FIGURE 11-160: Transaxial 28

1. External iliac vessels
2. Rectus abdominis muscle
3. Bladder
4. Iliacus muscle
5. Gluteus medius muscle
6. Seminal vesicle
7. Acetabulum (posterior lip)
8. Rectum
9. Superior ramus of pubic bone
10. Obturator interna muscle
11. Femoral head
12. Prostate gland
13. Coccyx

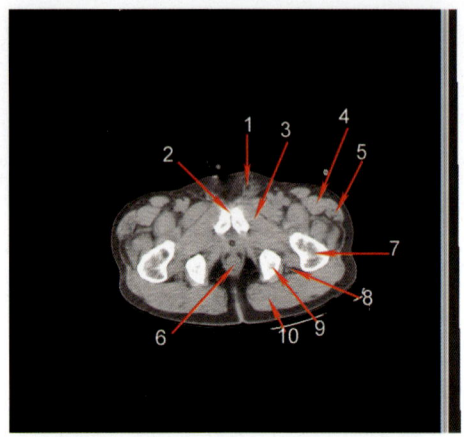
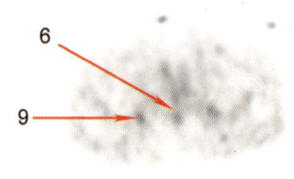
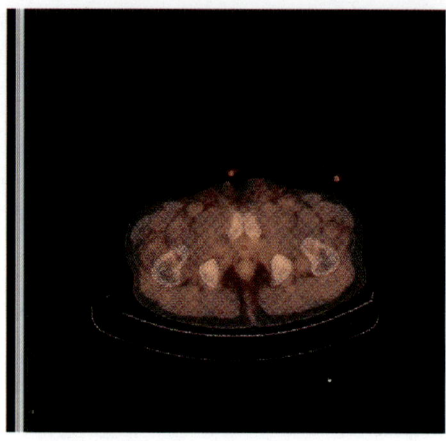

FIGURE 11-161: Transaxial 29

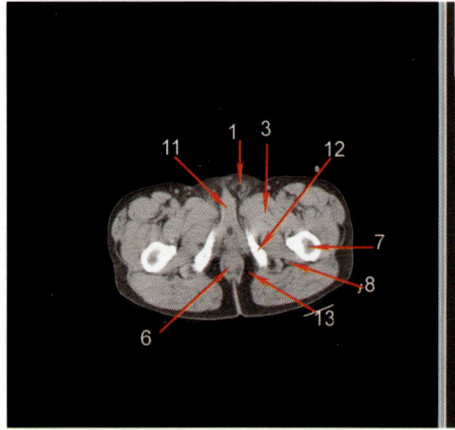
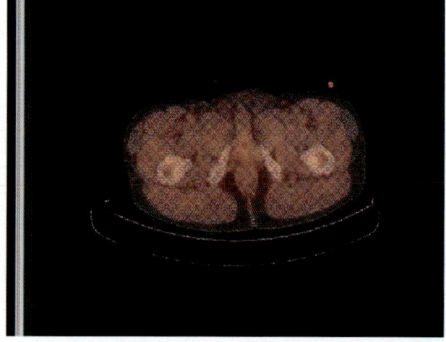

FIGURE 11-162: Transaxial 30

1. Spermatic cord
2. Symphysis pubis
3. Adductor muscles
4. Biceps femoris muscle
5. Fascia lata muscle
6. Anus
7. Trochanter of femur
8. Sciatic nerve
9. Ischial tuberosity
10. Gluteus maximus muscle
11. Corpora muscles of penis
12. Ischium
13. Perianal fat

12 SPECT/CT Anatomy: Variations and Artifacts

Successful interpretation of scintigraphic images depends on, to a great extent, the ability of the individual to appreciate deviations from the norm. In some cases, alterations in the biodistribution of a radiopharmaceutical or a gamma camera or computer-associated artifact may simulate a disease process and create considerable confusion for the less-experienced physician interpreting the study. Problems caused by the frequently used radiopharmaceuticals are related to preparation of radiopharmaceutical, factors associated with the radionuclide such as carrier technetium-99, components such as stanerous ion and particle, preparation procedures, and miscellaneous factors such as aluminum and pH. Artifacts and pitfalls can be caused by improper radiopharmaceutical administration.[1] Most artifacts related to the integrity of the detector head, computer system, and hard copy device can be detected on the uniformity image, because the most sensitive indicator of gamma camera performance is uniformity.[2]

Section 1: SPECT/CT Anatomy Variations

Bone Scan

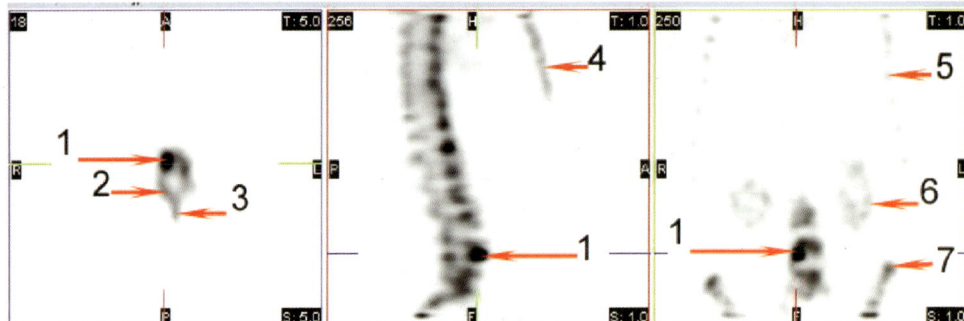

FIGURE 12-1: Bone SPECT 1

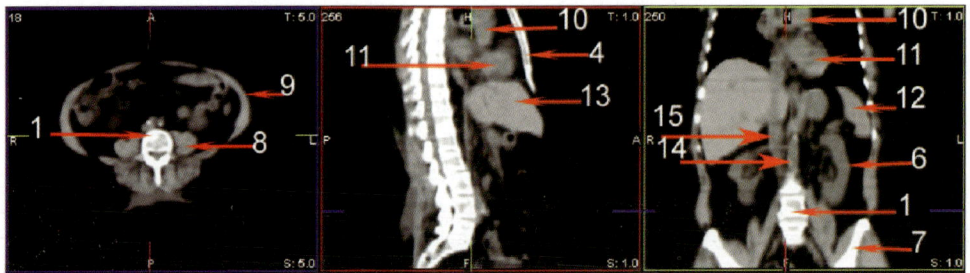

FIGURE 12-2: Bone CT 1

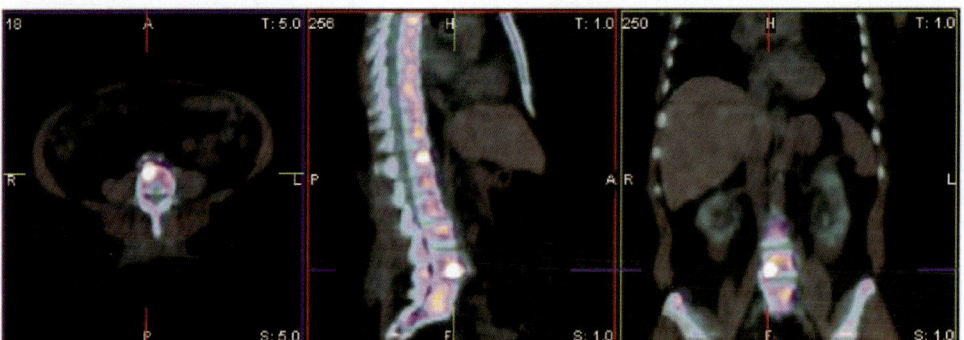

FIGURE 12-3: Bone Fused 1

1. Metastasis in L5 vertebral body
2. Vertebral pedicle
3. Vertebral spinous process
4. Sternum
5. Rib
6. Kidney
7. Ilium
8. Psoas muscle
9. Oblique muscles
10. Perihilar mass
11. Heart
12. Spleen
13. Left hepatic lobe
14. Abdominal aorta
15. Inferior vena cava

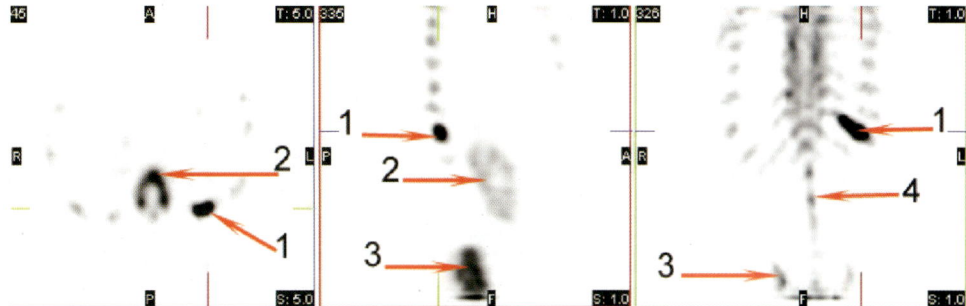

FIGURE 12-4: Bone SPECT 2

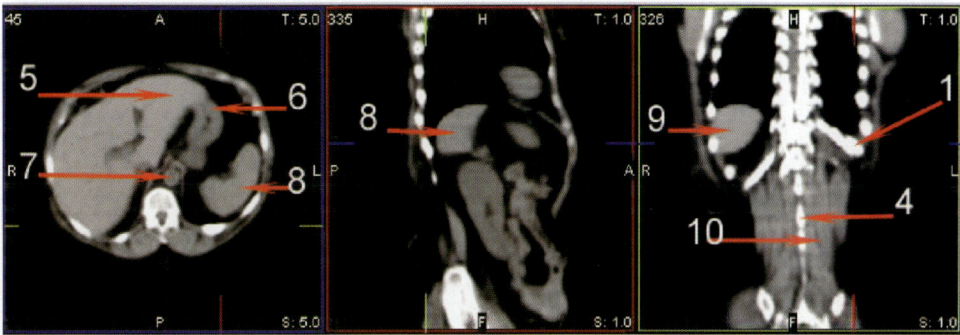

FIGURE 12-5: Bone CT 2

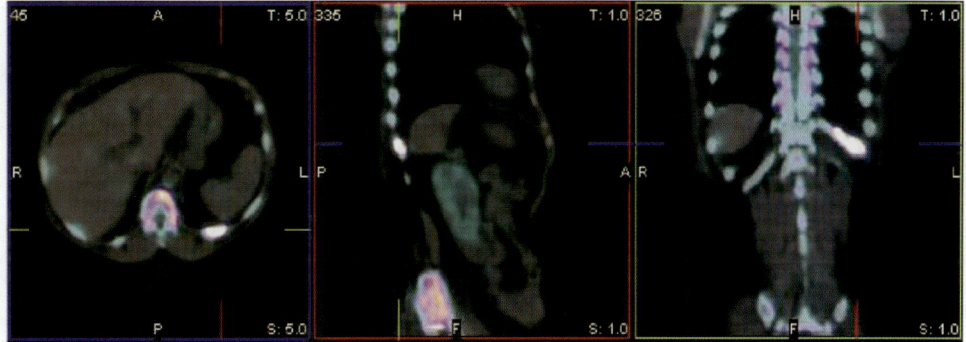

FIGURE 12-6: Bone Fused 2

1. Metastasis in left posterior 11th rib	4. Spinous process	8. Spleen
	5. Left hepatic lobe	9. Right hepatic lobe
2. T12 vertebral body	6. Stomach	10. Quadratus lumborum muscle
3. Right iliac tuberosity	7. Abdominal aorta	

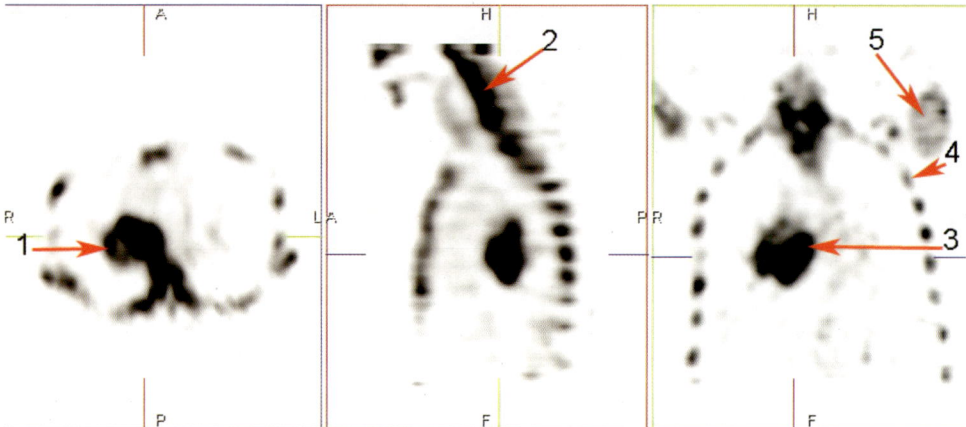

FIGURE 12-7: Bone Calcification ECT 1 (emission computed tomography)

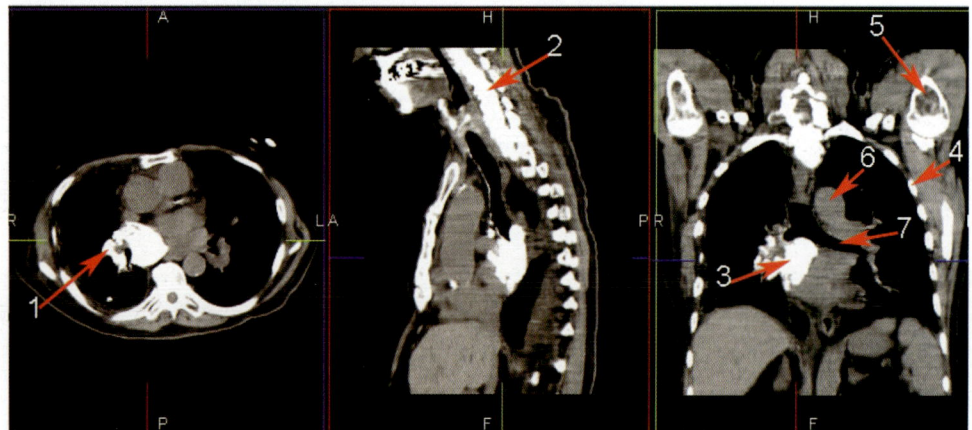

FIGURE 12-8: Bone Calcification CT 1

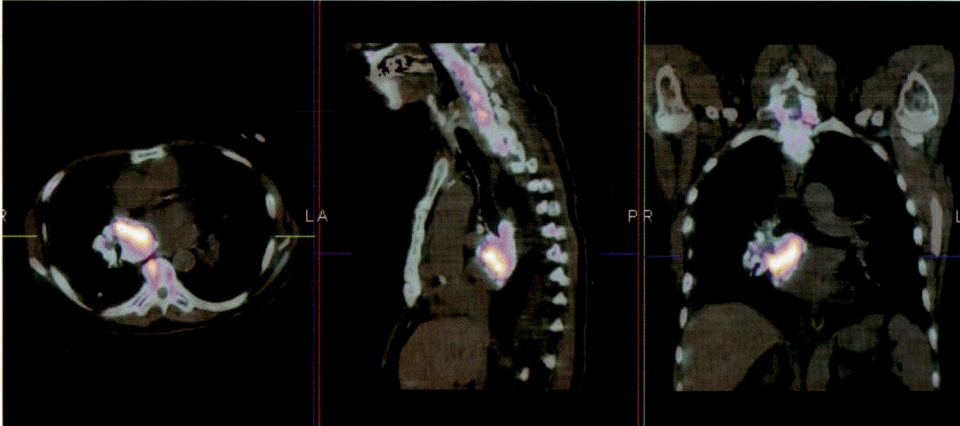

FIGURE 12-9: Bone Calcification Fused 1

1. Calcified granuloma in right hilum
2. Cervical vertebral body
3. Calcified granuloma in right subcarinal area
4. Rib
5. Left humerus
6. Main pulmonary artery
7. Left main bronchus

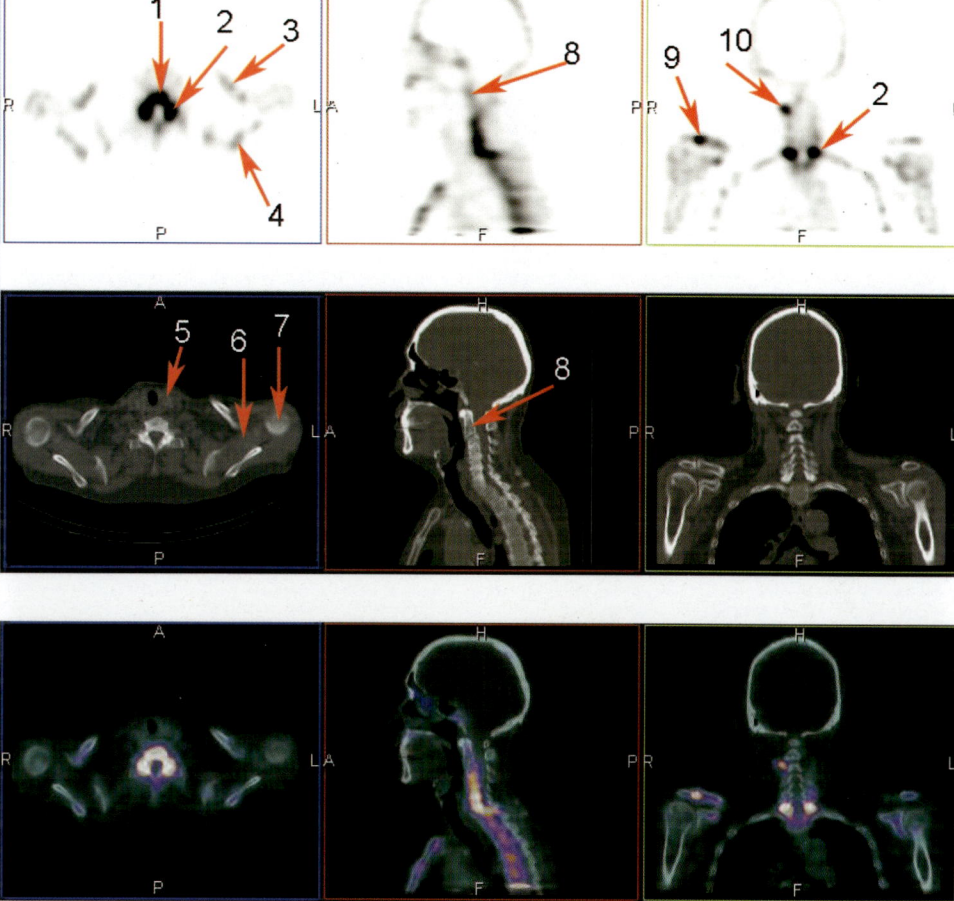

FIGURE 12-10: Bone Head/Neck 1

1. T1 vertebral body
2. Facet joint with arthritic changes
3. Left clavicle
4. Left scapula
5. Thyroid
6. Subscapularis muscle
7. Humeral head
8. C2 body
9. Right acro-clavicular joint with degenerative changes
10. Degenerative changes on the right C2–3

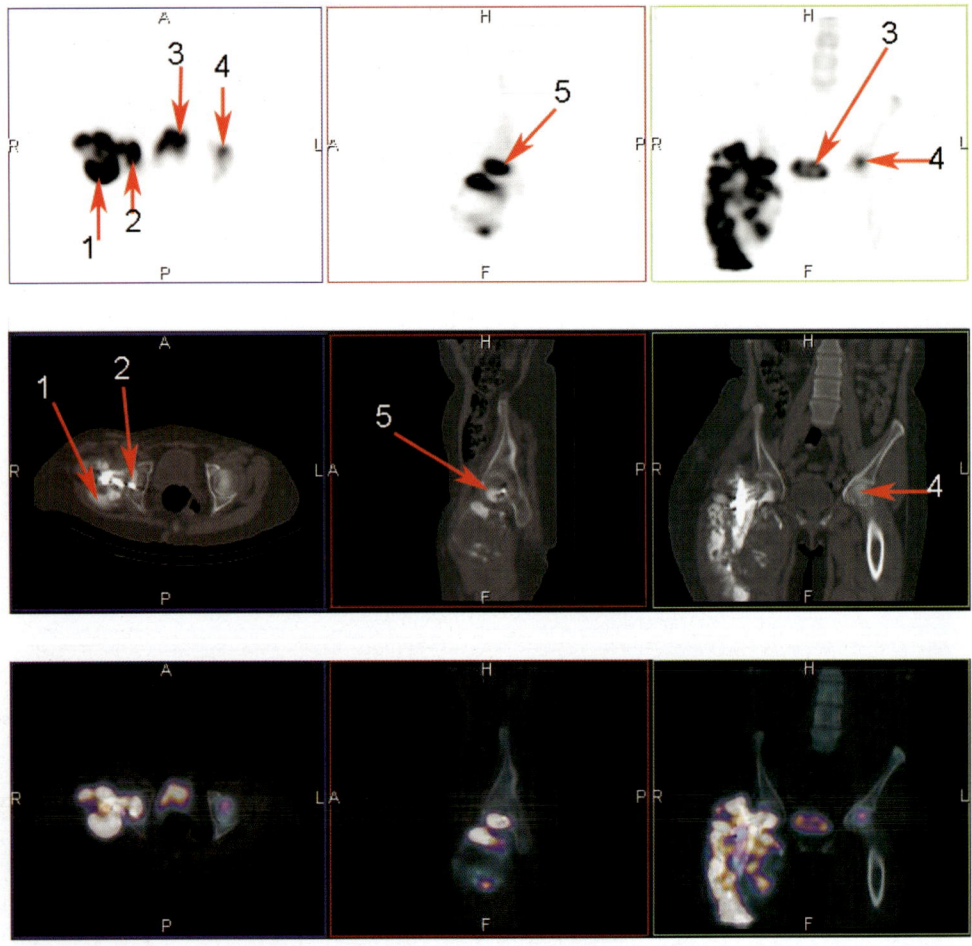

FIGURE 12-11: Bone Pelvis 1

1. Calcified sarcoma in right gluteus muscle
2. Lesion in right acetabulum
3. Bladder
4. Lesion in left femoral head
5. Lesion in right femoral head

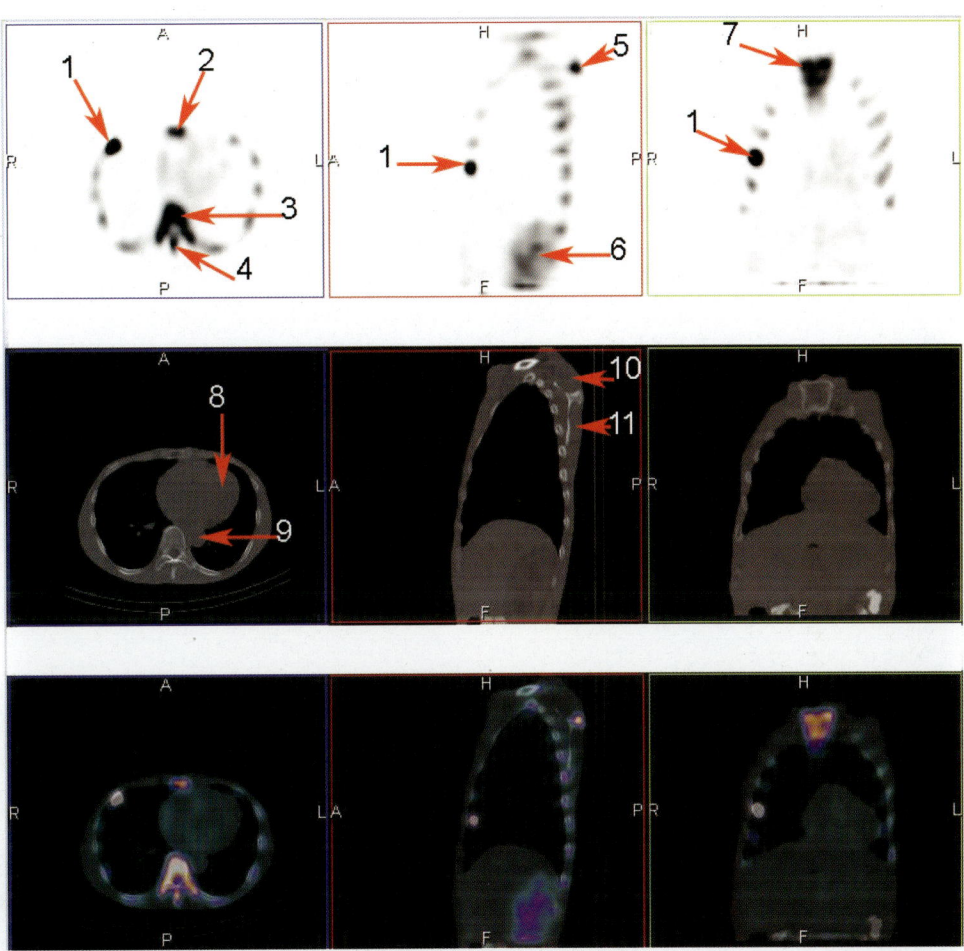

FIGURE 12-12: Bone Rib 1

1. Metastasis in rib end
2. Sternum
3. Vertebral body
4. Spinous process
5. Scapula
6. Kidney
7. Manubrium
8. Left heart
9. Descending aorta
10. Supraspinatus muscle
11. Infraspinatus muscle

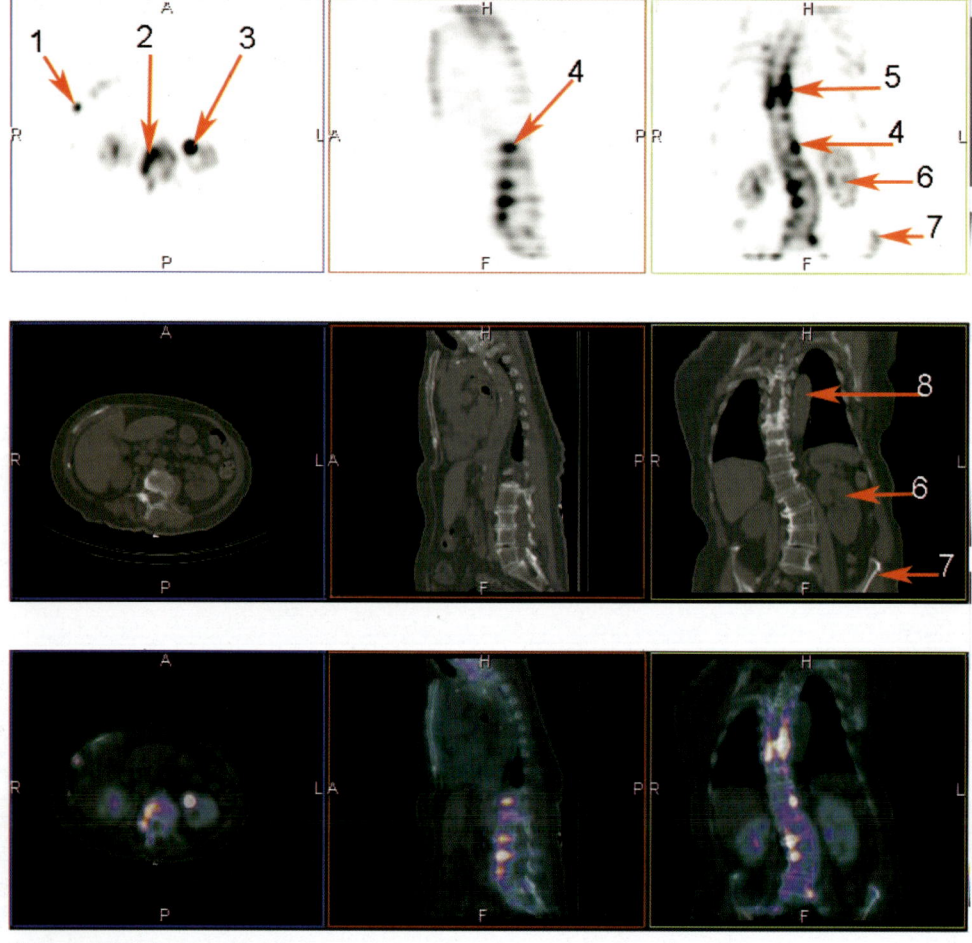

FIGURE 12-13: Bone Spine 1

1. Lesion in rib
2. Scoliotic vertebral body
3. Left renal pelvis
4. Degenerative changes in the left side of T12
5. Degenerative changes in the left side of T8
6. Left kidney
7. Left ilium
8. Descending aorta

Gallium Scan

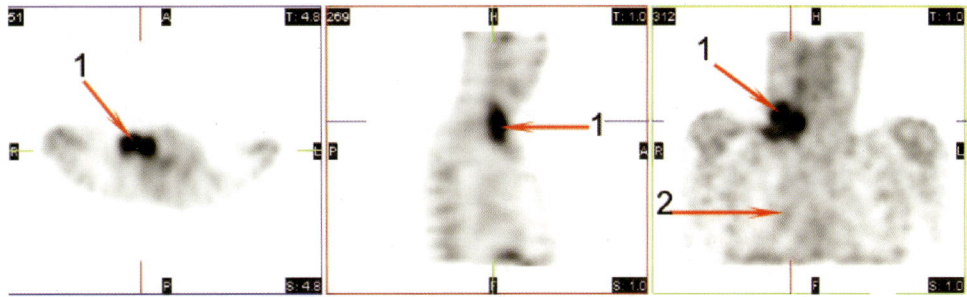

FIGURE 12-14: Gallium SPECT 1

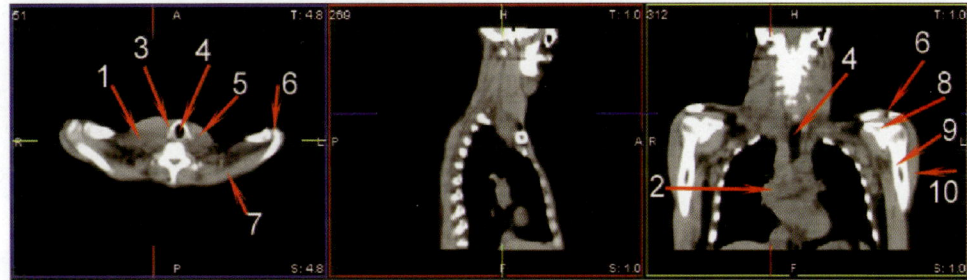

FIGURE 12-15: Gallium CT 1

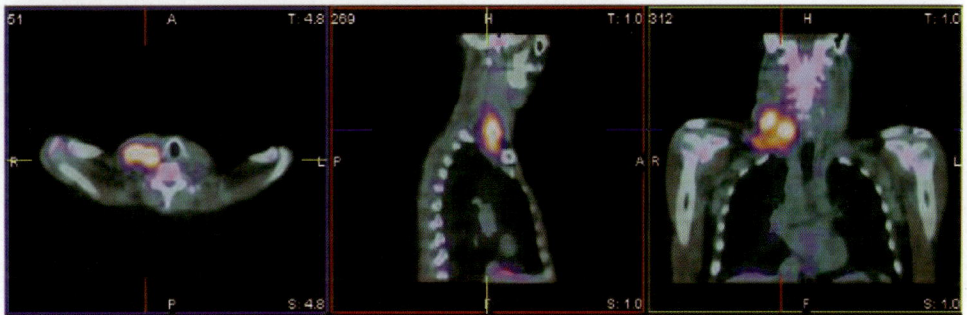

FIGURE 12-16: Gallium Fused

1. Lymphoma in supraclavicular nodes
2. Right hilum
3. Thyroid cartilage
4. Trachea
5. Sternocleidomastoid muscle
6. Acromioclavicular joint
7. Trapezius muscle
8. Coracoid process of scapula
9. Humerus
10. Deltoid muscle

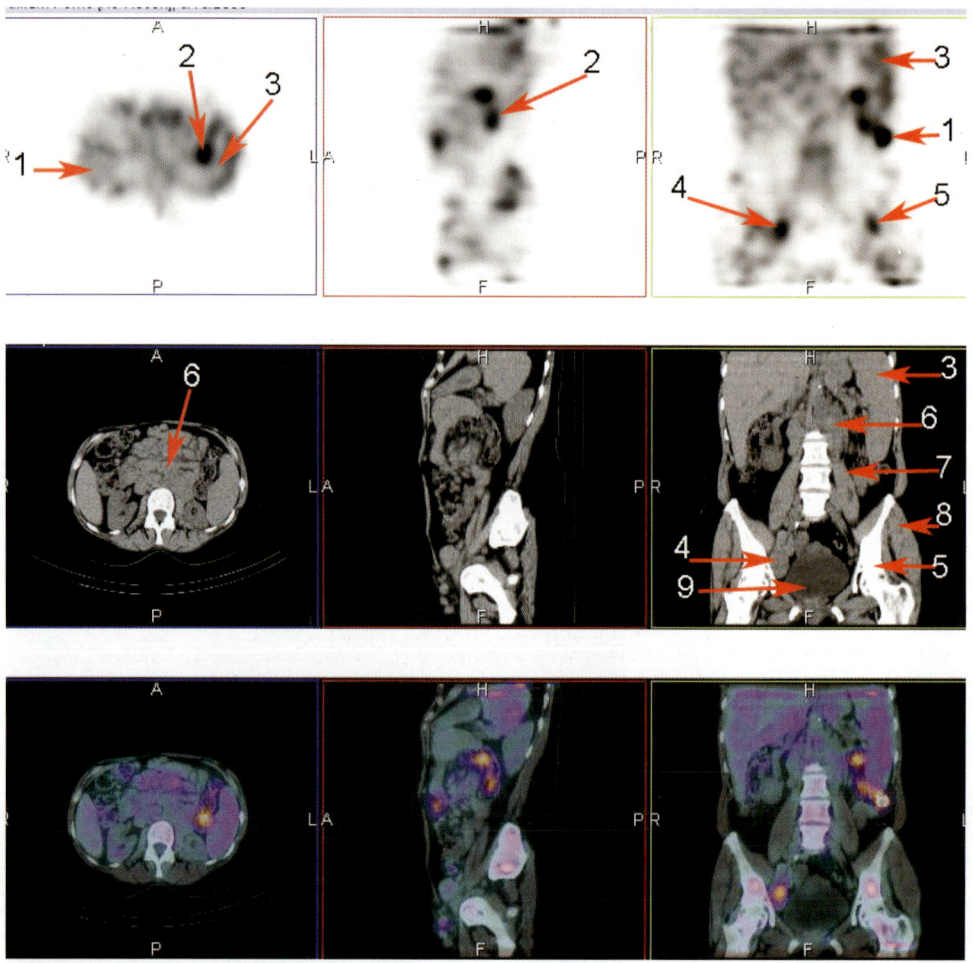

FIGURE 12-17: Gallium Abdomen/Pelvis 1

1. Right liver	4. Right external iliac node	7. Left psoas muscle
2. Splenic flexure of colon	5. Left acetabulum	8. Gluteus muscle
3. Spleen	6. Periaortic nodes	9. Bladder

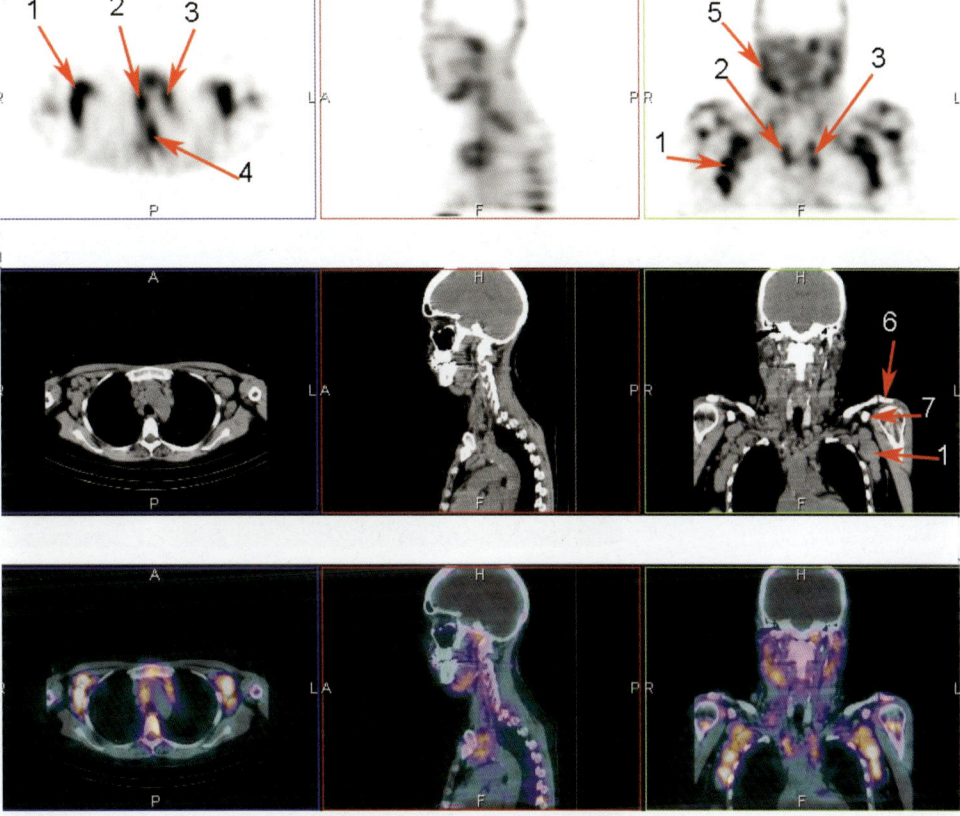

FIGURE 12-18: Gallium Head/Chest 1

1. Axillary nodes
2. Right paratracheal nodes
3. Prevascular nodes
4. Vertebral body
5. Right jugular nodes
6. Acromion of left scapula
7. Coracoid process of left scapula

Octreotide Scan

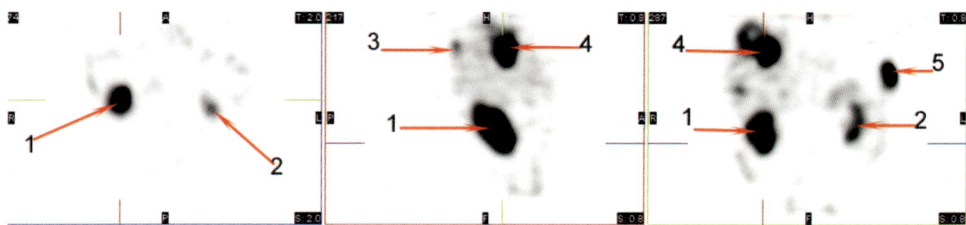

FIGURE 12-19: Octreotide SPECT 1

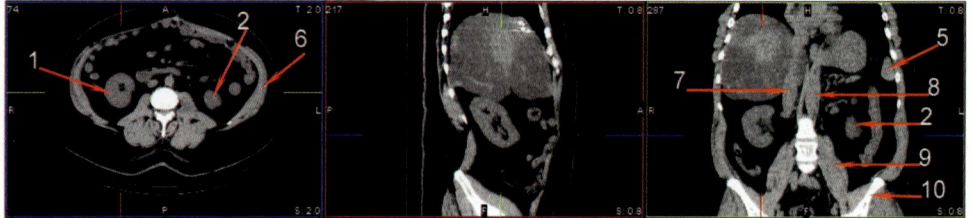

FIGURE 12-20: Octreotide CT 1

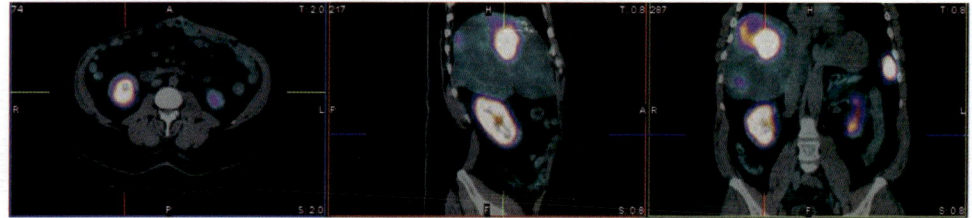

FIGURE 12-21: Octreotide Fused 1

1. Right kidney
2. Left kidney
3. Metastatic carcinoid in right liver (segment 7)
4. Metastatic carcinoid in right liver(segment 8)
5. Spleen
6. Oblique muscles
7. Inferior vena cava
8. Abdominal aorta
9. Psoas muscle
10. Ilium

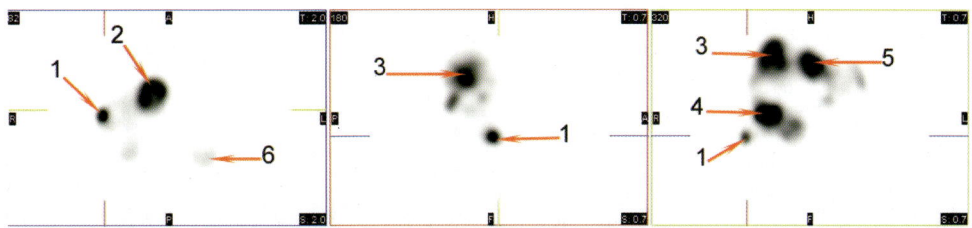

FIGURE 12-22: Octreotide SPECT 2

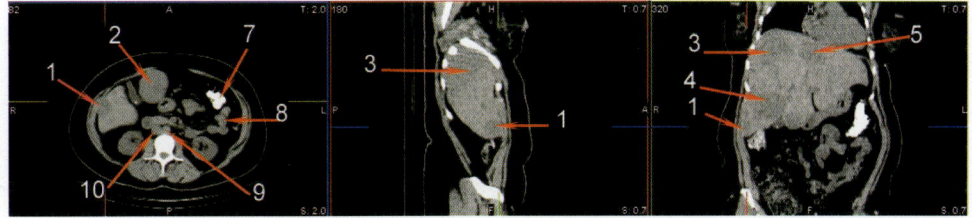

FIGURE 12-23: Octreotide CT 2

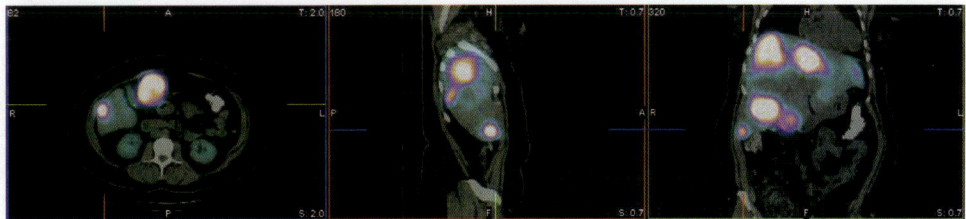

FIGURE 12-24: Octreotide Fused 2

1. Metastatic carcinoid in right hepatic lobe (segment V)
2. Metastatic carcinoid in right upper mesentery
3. Metastatic carcinoid in right hepatic lobe (segment VIII)
4. Metastatic carcinoid in right hepatic lobe (segment V)
5. Metastatic carcinoid in right hepatic lobe (segment VIII)
6. Kidney
7. Splenic flexure of colon
8. Jejunum
9. Abdominal aorta
10. Inferior vena cava

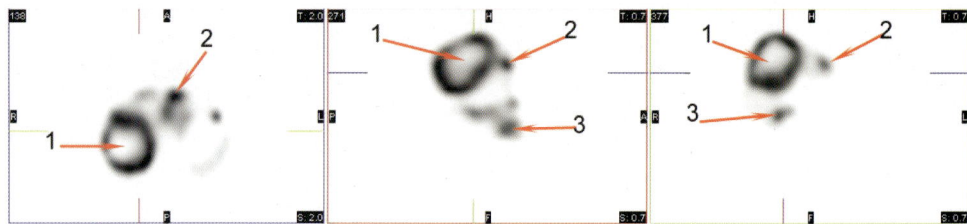

FIGURE 12-25: Octreotide SPECT 3

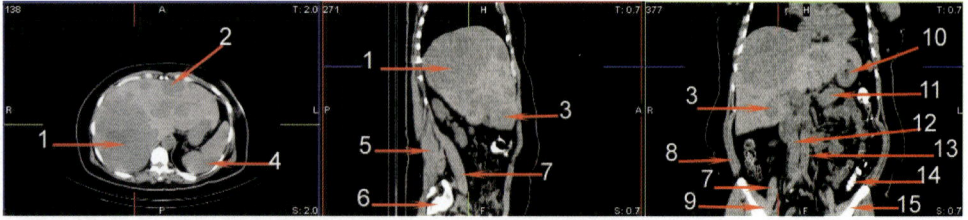

FIGURE 12-26: Octreotide CT 3

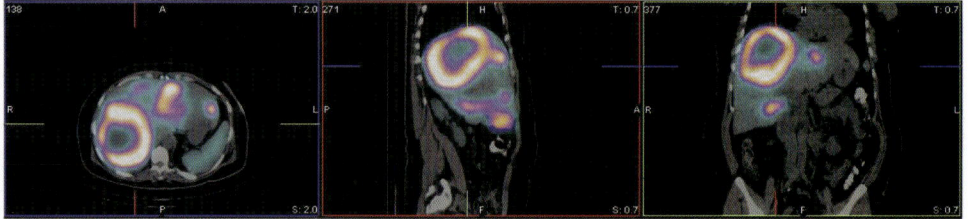

FIGURE 12-27: Octreotide Fused 3

1. Central necrosis of metastatic carcinoid in right hepatic lobe (segment VIII)
2. Metastatic carcinoid in left hepatic lobe (segment III)
3. Metastatic carcinoid in right hepatic lobe (segment V)
4. Spleen
5. Quadratus lumborum muscle
6. Iliac tuberosity
7. Psoas muscle
8. Oblique muscle
9. Iliacus muscle
10. Stomach
11. Pancreas
12. Inferior vena cava
13. Abdominal aorta
14. Sigmoid colon with barium
15. Ilium

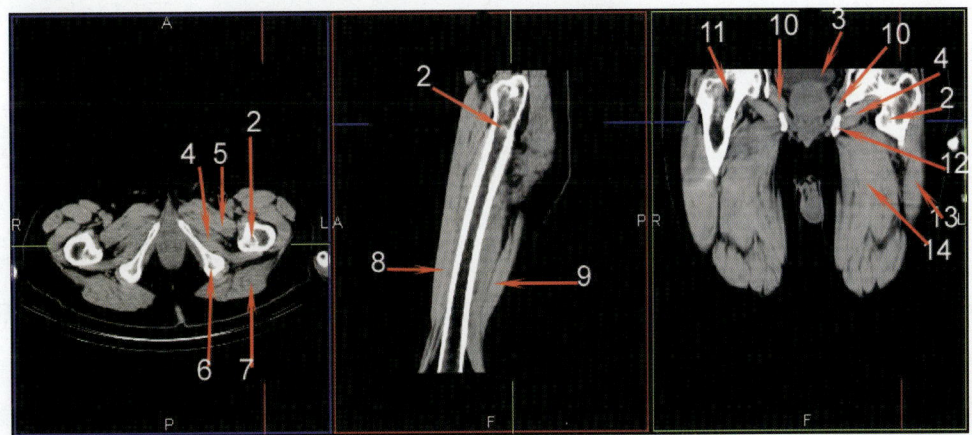

FIGURE 12-28: Octreotide Femur SPECT 1

FIGURE 12-29: Octreotide Femur CT 1

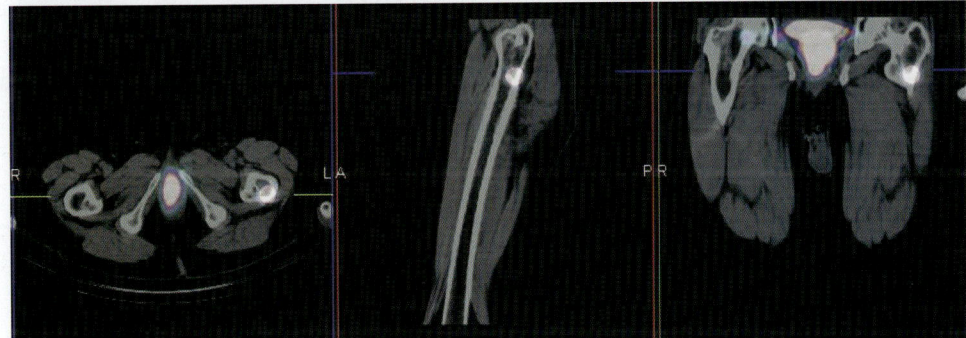

FIGURE 12-30: Octreotide Femur Fused 1

1. Corpus muscle and urethra
2. Metastatic carcinoid in left femur
3. Bladder
4. External obturator muscle
5. Adductor brevis muscle
6. Ischial tuberosity
7. Gluteus maximus muscle
8. Rectus femoris muscle
9. Biceps femoris muscle
10. Internal obturator muscle
11. Femoral neck
12. Pubic ramus
13. Vastus lateralis muscle
14. Adductor magna muscle

FIGURE 12-31: Octreotide Scoliosis SPECT 1

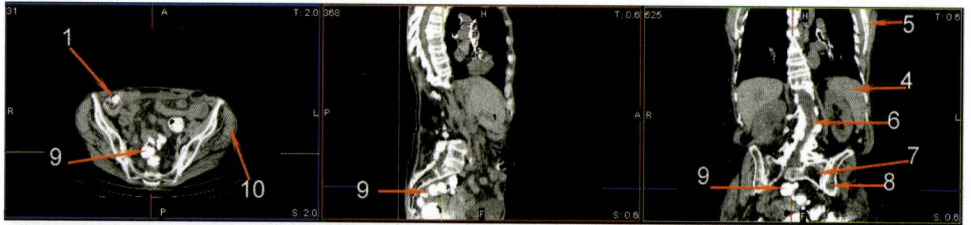

FIGURE 12-32: Octreotide Scoliosis CT 1

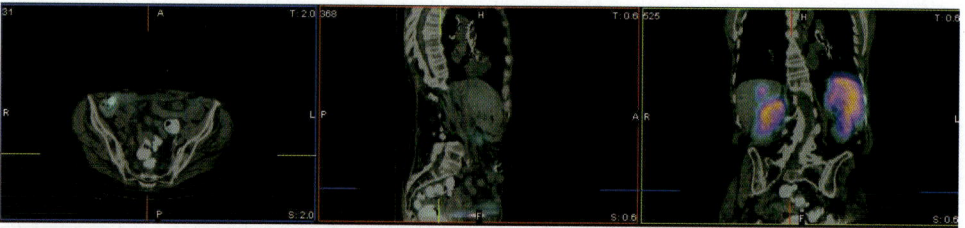

FIGURE 12-33: Octreotide Scoliosis Fused 1

1. Cecum
2. Metastatic carcinoid in right hepatic lobe (segment VII)
3. Right kidney
4. Spleen
5. Scapula
6. Scoliosis of lumbar spine
7. Sacral ala
8. Iliac tuberosity
9. Sigmoid colon with barium
10. Gluteus medius muscle

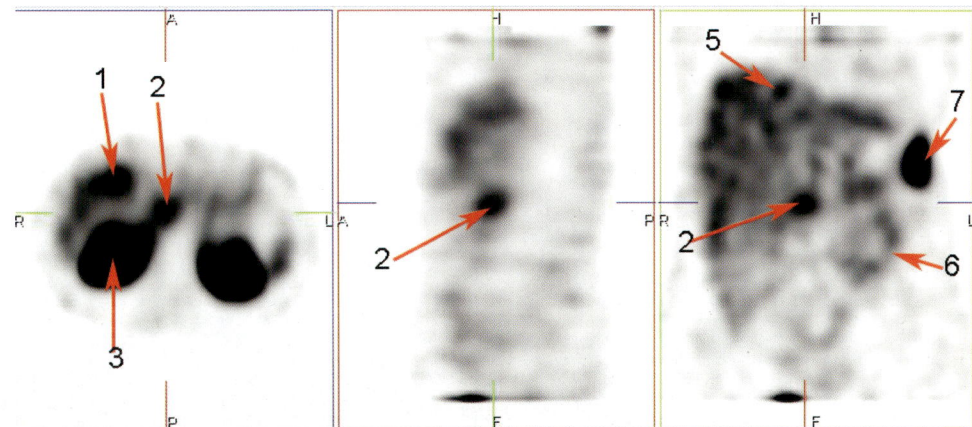

FIGURE 12-34: Octreotide Abdominal ECT 1

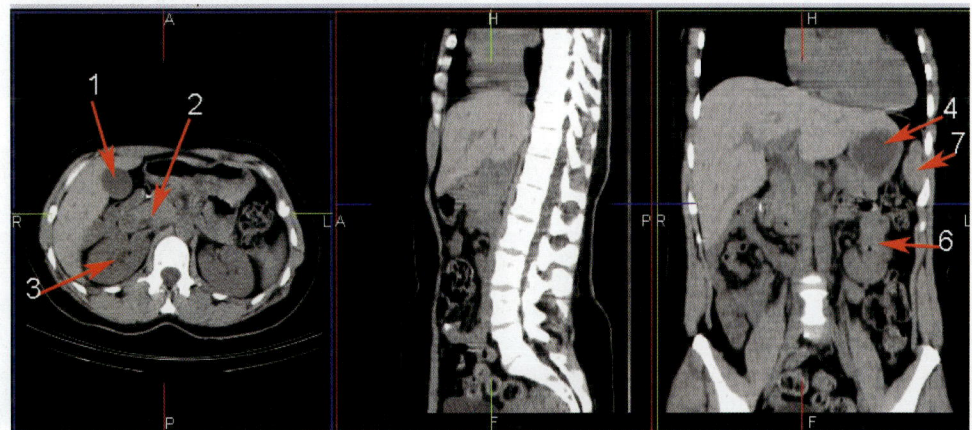

FIGURE 12-35: Octreotide Abdominal CT 1

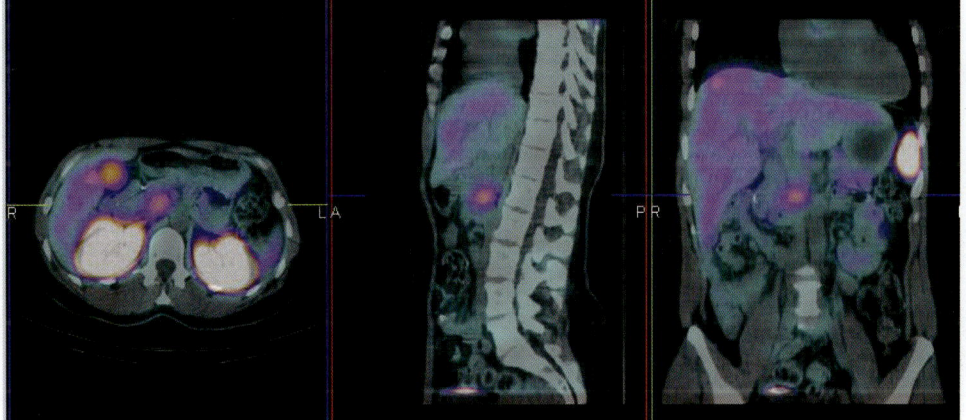

FIGURE 12-36: Octreotide Abdominal Fused 1

1. Gallbladder	4. Stomach	6. Jejunum
2. Pancreas head tumor	5. Heterogeneous liver	7. Spleen
3. Kidney	activity	

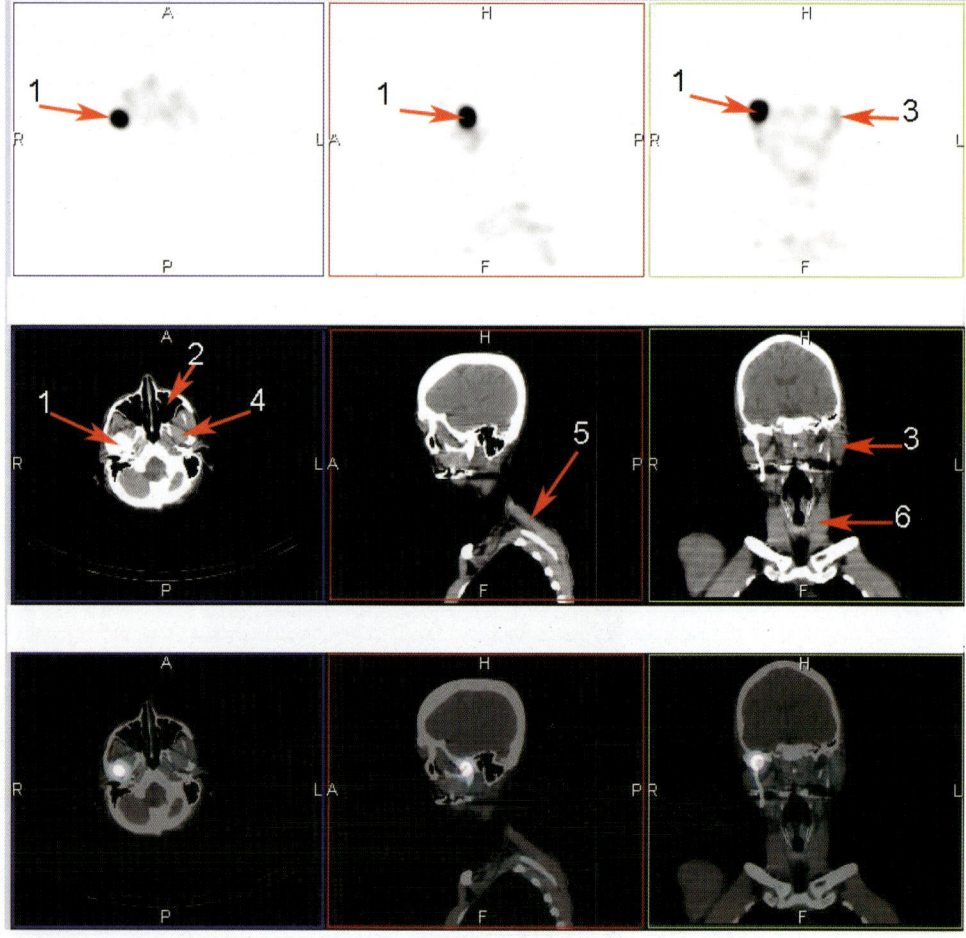

FIGURE 12-37: Octreotide Head/Neck 1

1. Meningioma in sphenoid wing
2. Maxillary sinus
3. Parotid gland
4. Pterygoid muscle
5. Trapezius muscle
6. Sternocleidomastoid muscle

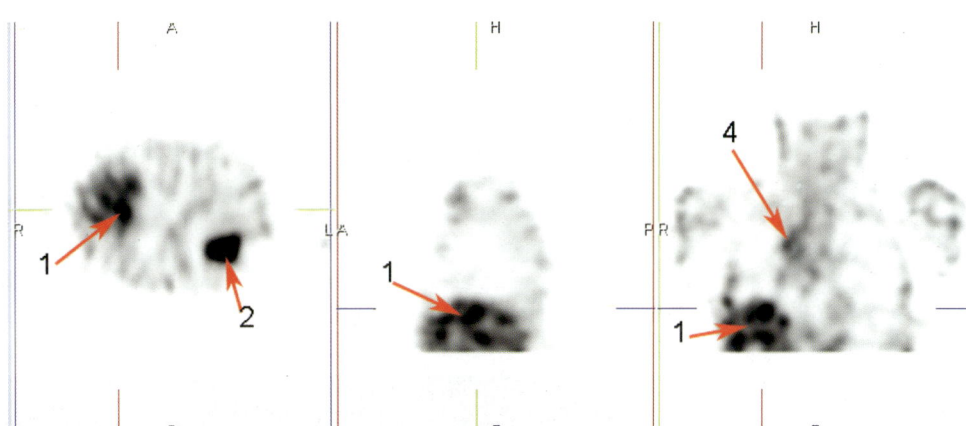

FIGURE 12-38: Octreotide Thorax ECT 1

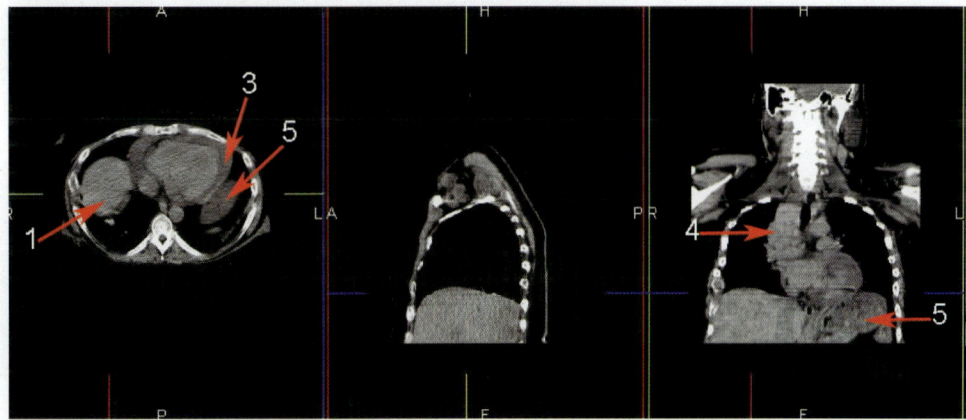

FIGURE 12-39: Octreotide Thorax CT 1

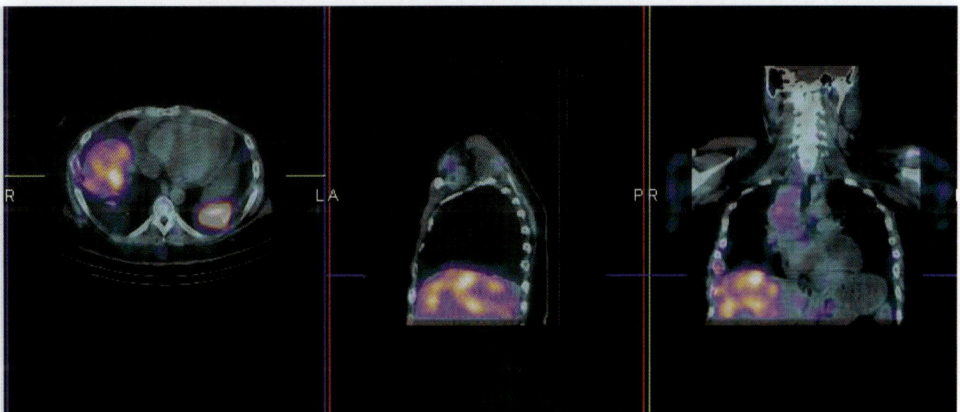

FIGURE 12-40: Octreotide Thorax Fused 1

1. Heterogeneous liver activity
2. Spleen
3. Pericardial effusion
4. Right paratracheal paragangliomas
5. Stomach

I-131 Scan

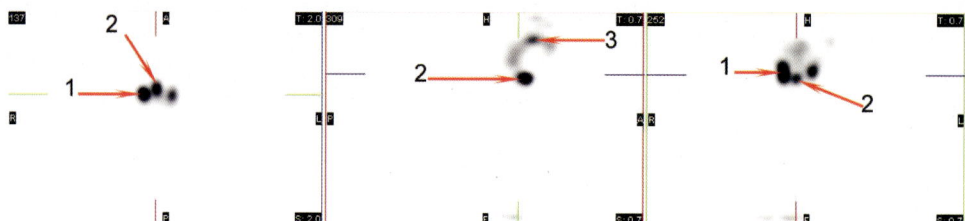

FIGURE 12-41: I-131 Thyroid SPECT 1

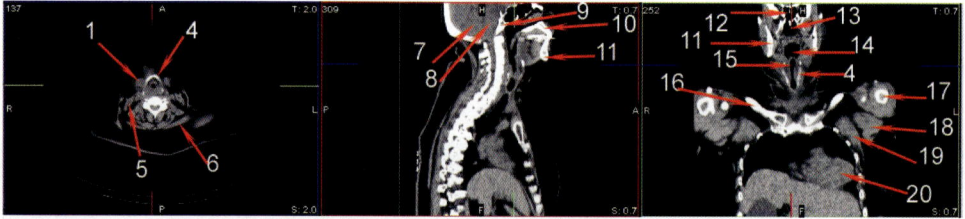

FIGURE 12-42: I-131 Thyroid CT 1

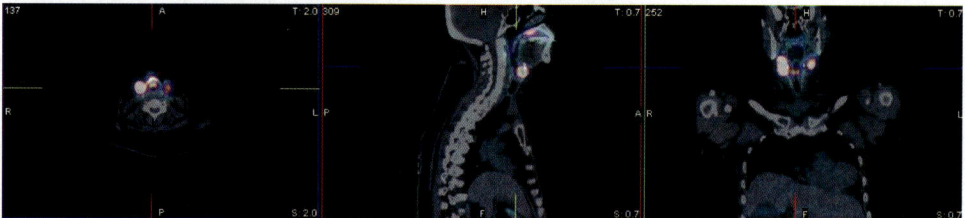

FIGURE 12-43: I-131 Thyroid Fused 1

1. Submandibular gland
2. Thyroglossal duct
3. Soft palate
4. Thyroid cartilage
5. Sternocleidomastoid muscle
6. Trapezius muscle
7. Cerebellum
8. Pons
9. Clivus
10. Maxilla
11. Mandible
12. Sphenoid sinus
13. Nasopharynx
14. Oropharynx
15. Hypopharynx
16. Clavicle
17. Humeral head
18. Infraspinatus muscle
19. Subscapularis muscle
20. Right ventricle of heart

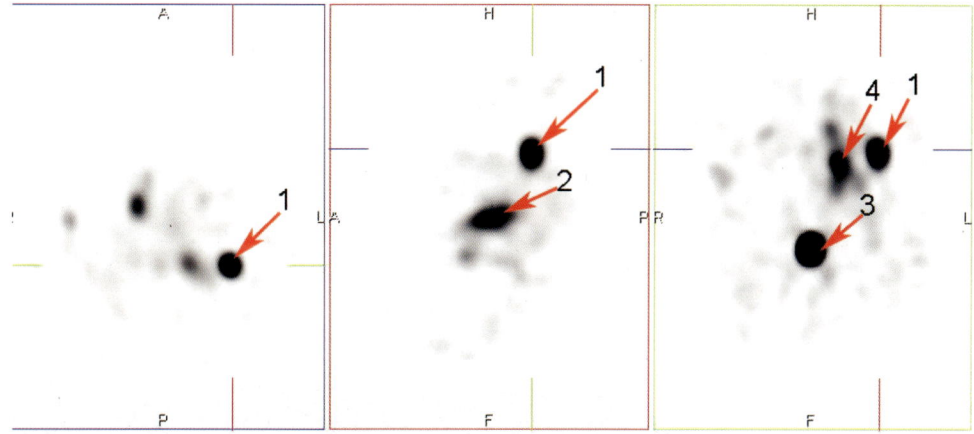

FIGURE 12-44: I-131 Abdomen SPECT 1

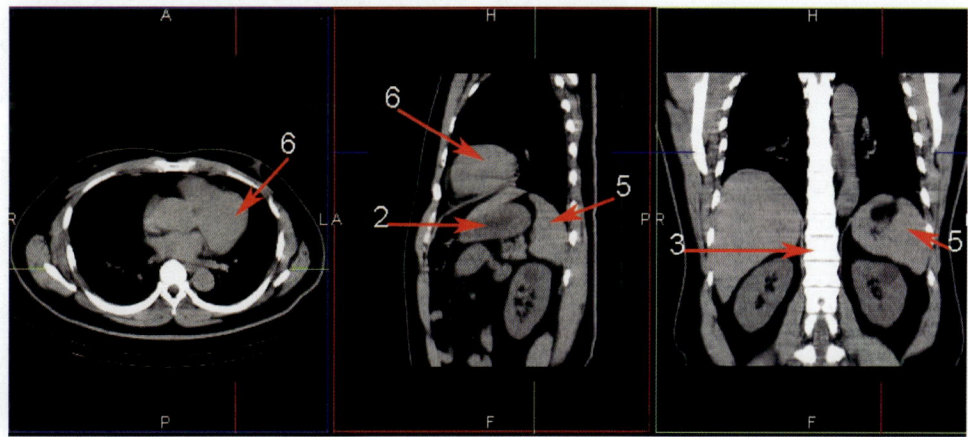

FIGURE 12-45: I-131 Abdomen CT 1

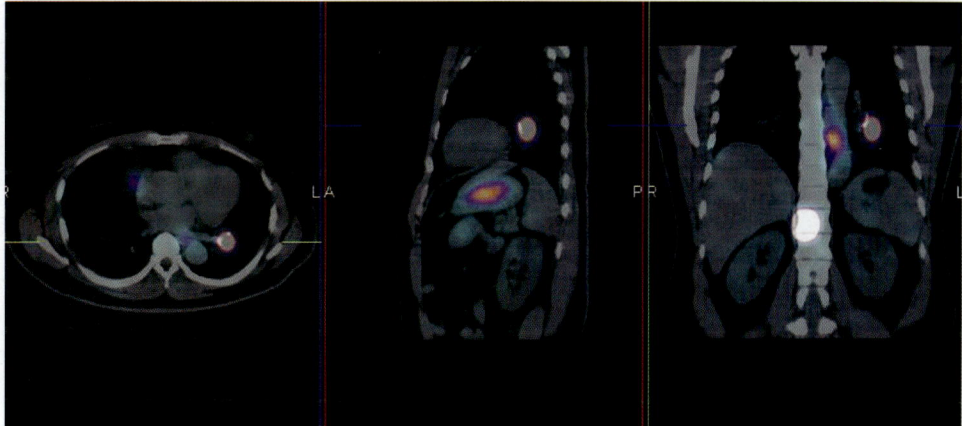

FIGURE 12-46: I-131 Abdomen Fused 1

1. Metastasis in left lower lung
2. Stomach
3. Metastasis in T12
4. Left paraspinal metastasis
5. Spleen
6. Left heart

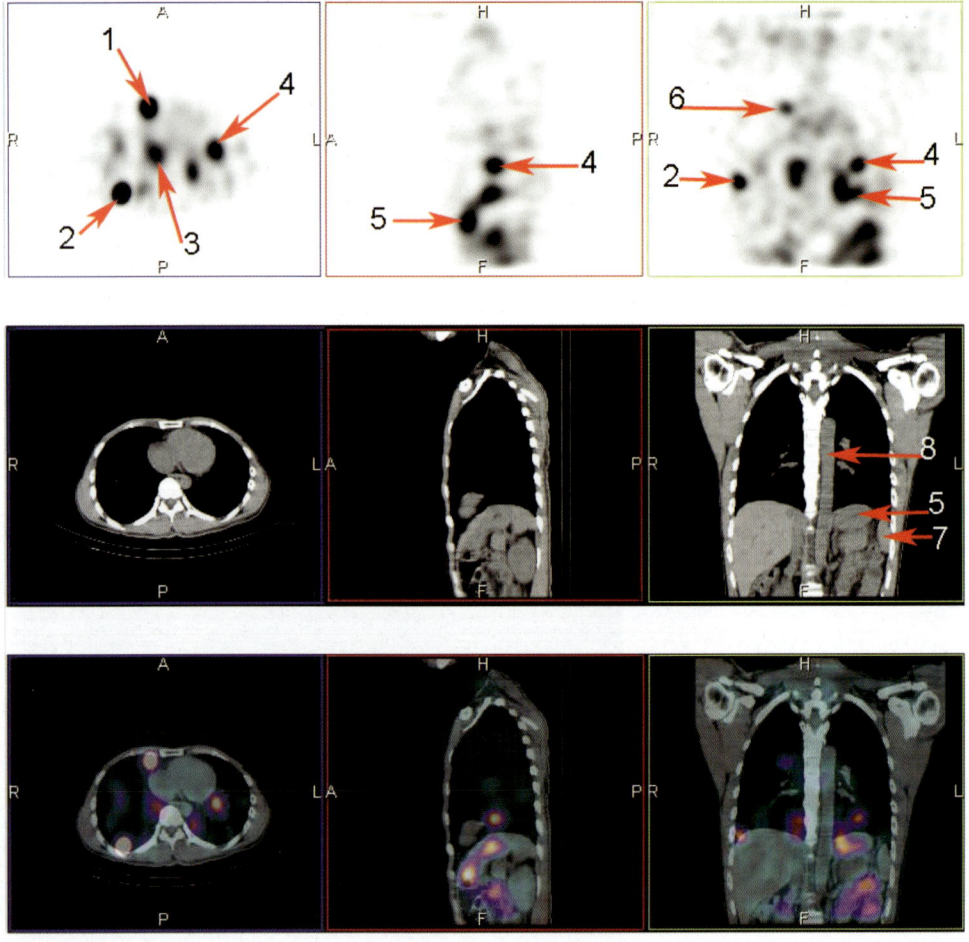

FIGURE 12-47: I-131 Thorax 1

1. Metastasis in right parasternal rib
2. Metastasis in right lower posterior rib
3. Metastasis in T10 body
4. Metastasis in left lower lung
5. Stomach
6. Metastasis in right hilum
7. Spleen
8. Descending aorta

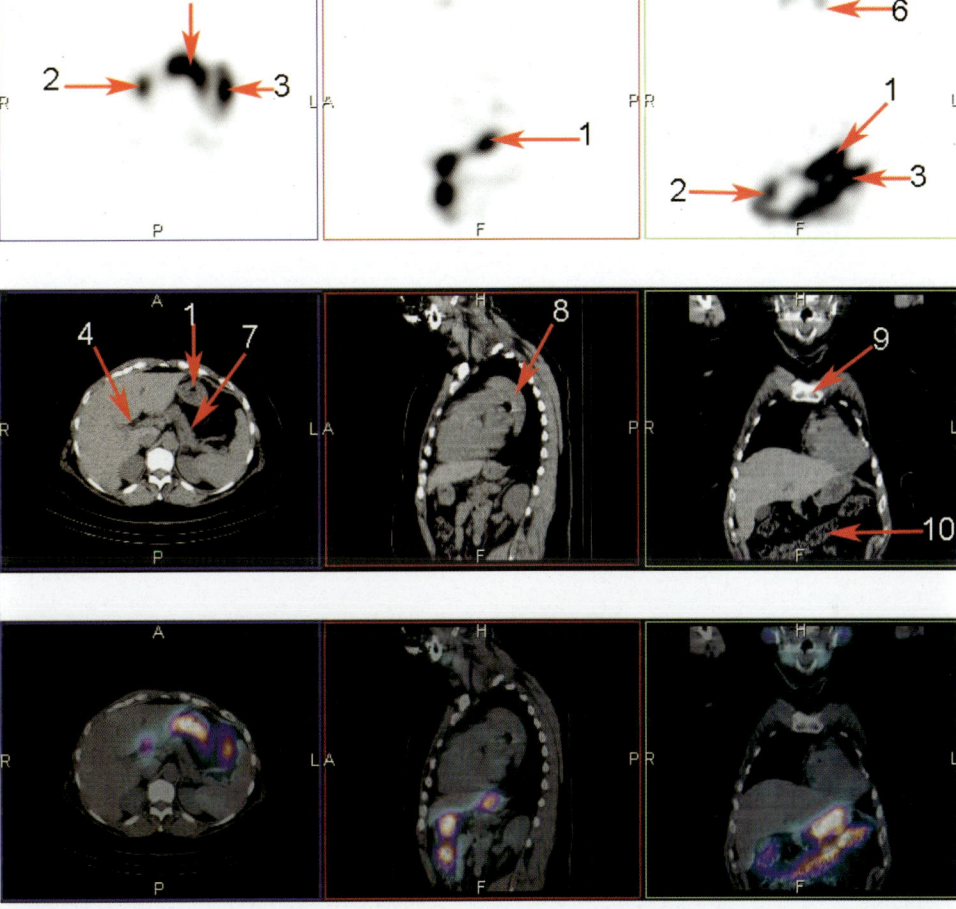

FIGURE 12-48: I-131 Thorax 2

1. Stomach
2. Duodenum
3. Jejunum
4. Porta hepatis
5. Parotid gland
6. Submandibular gland
7. Pancreas
8. Descending aorta
9. Manubrium
10. Transverse colon

Lymphoscintigraphy Scan

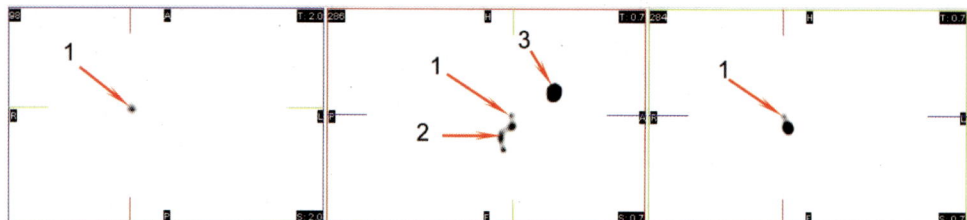

FIGURE 12-49: Lymphoscintigraphy SPECT 1

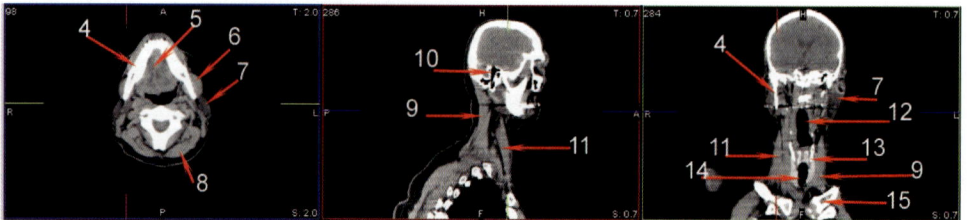

FIGURE 12-50: Lymphoscintigraphy CT 1

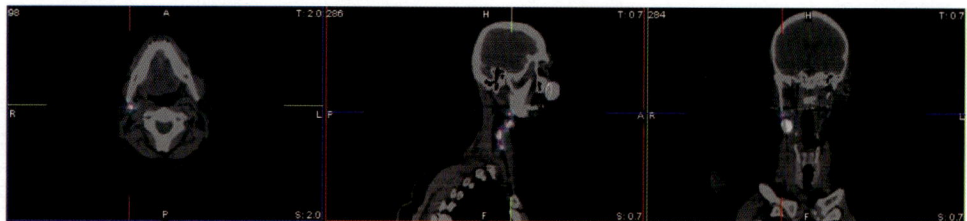

FIGURE 12-51: Lymphoscintigraphy Fused 1

1. Sentinel node in upper jugular lymphatic chain (level #II)
2. Nodes in middle jugular lymphatic chain (level #III)
3. Injection site at nose
4. Mandible
5. Tongue
6. Masseter muscle
7. Parotid gland
8. Semispinalis capitis muscle
9. Longus colli muscle
10. Mastoid air cells
11. Sternocleidomastoid muscle
12. Oropharynx
13. Thyroid cartilage
14. Hypopharynx
15. Clavicle

Metaiodobenzylguanidine (MIBG) Scan

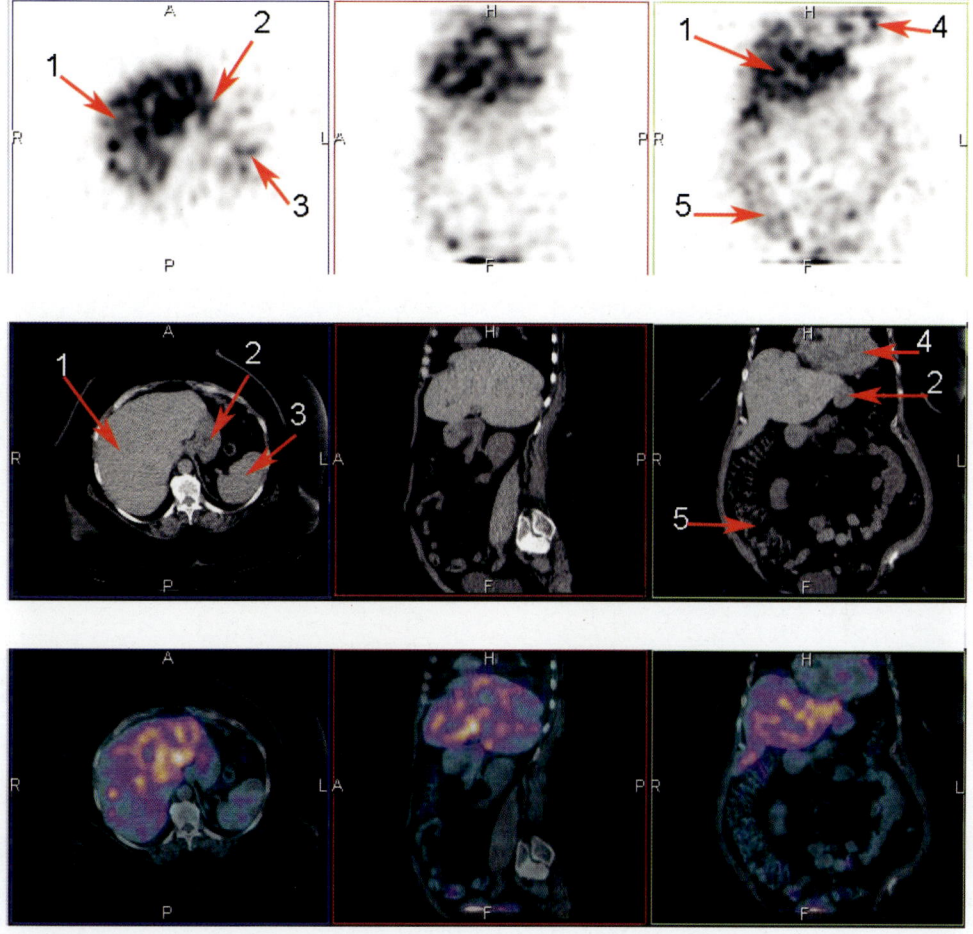

FIGURE 12-52: MIBG

1. Right hepatic lobe with
   heterogeneous activity
2. Stomach

3. Left kidney
4. Left ventricular
   myocardium

5. Ascending colon

Lung Scan

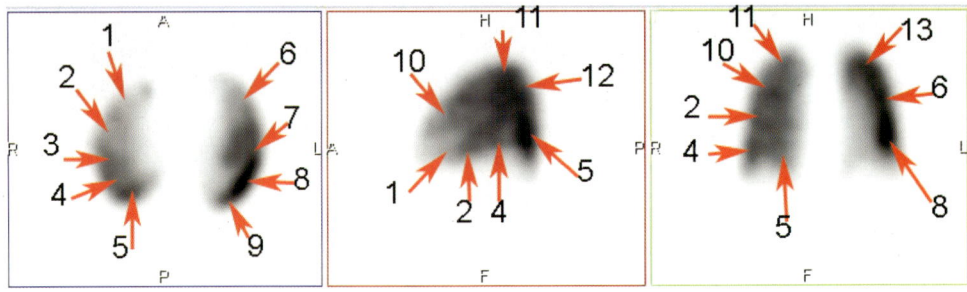

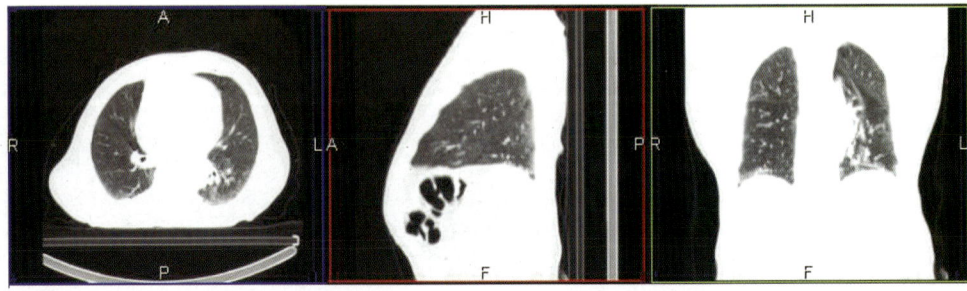

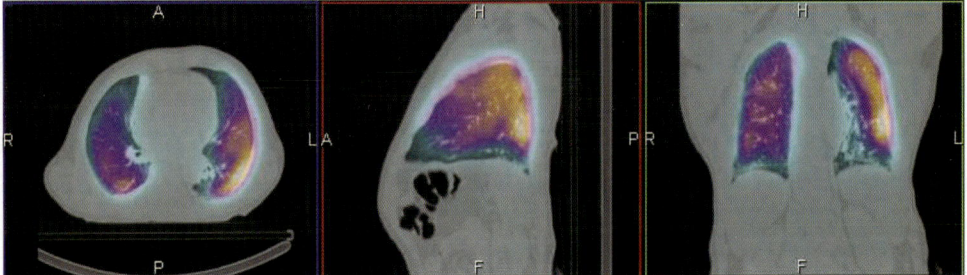

FIGURE 12-53: Lung 1

1. Medial segment of right middle lobe
2. Lateral segment of right middle lobe
3. Anterior basal segment of right lower lobe
4. Lateral basal segment of right lower lobe
5. Posterior basal segment of right lower lobe
6. Inferior lingual segment of left upper lobe
7. Anteromedial basal segment of left lower lobe
8. Lateral basal segment of left lower lobe
9. Posterior basal segment of left lower lobe
10. Anterior segment of right upper lobe
11. Apical segment of right upper lobe
12. Posterior segment of right upper lobe
13. Apicoposterior segment of left upper lobe

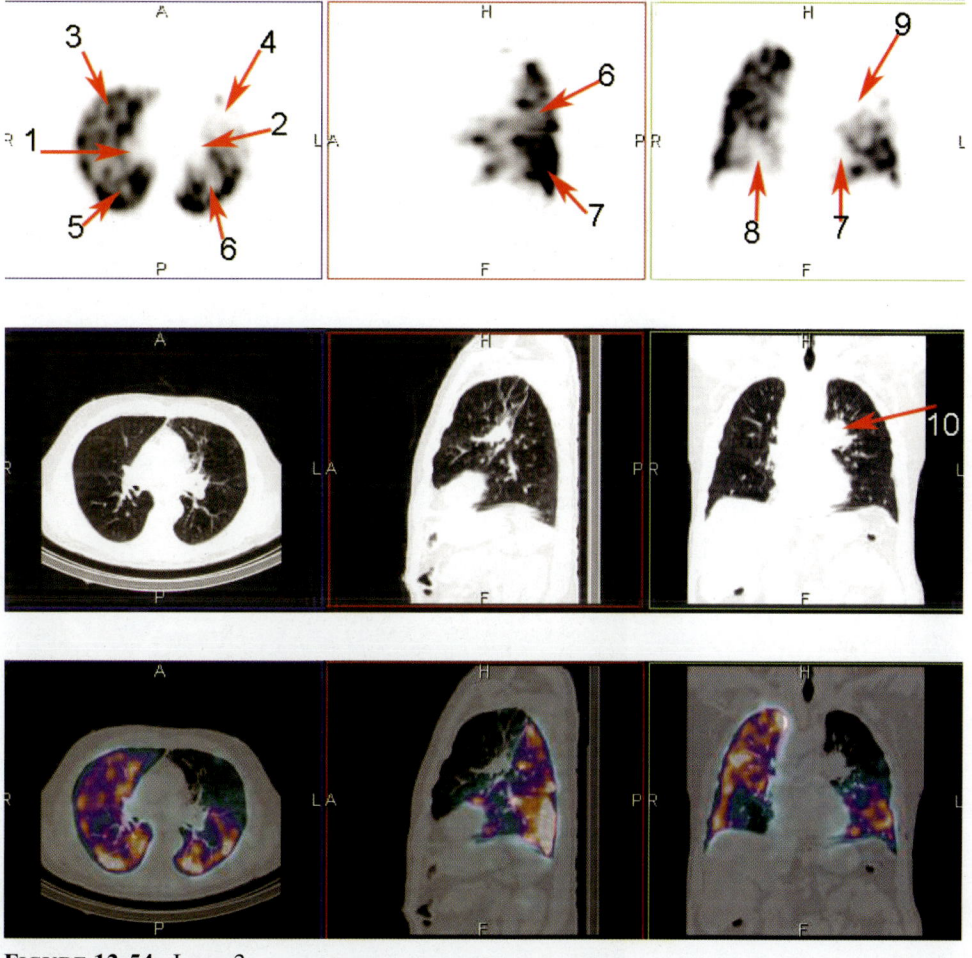

FIGURE 12-54: Lung 2

1. Right hilum
2. Left hilum
3. Anterior basal segment of right lower lobe
4. Anterior basal segment of left lower lobe
5. Superior segment of right lower lobe
6. Superior segment of left lower lobe
7. Posterior basal segment of left lower lobe
8. Posterior basal segment of right lower lobe
9. Left upper lobe
10. Main pulmonary artery

Parathyroid Scan

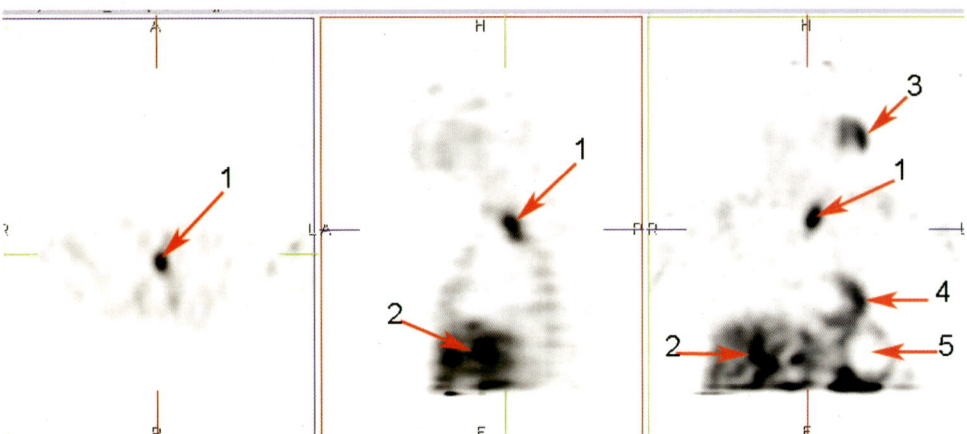

FIGURE 12-55: Parathyroid ECT 1

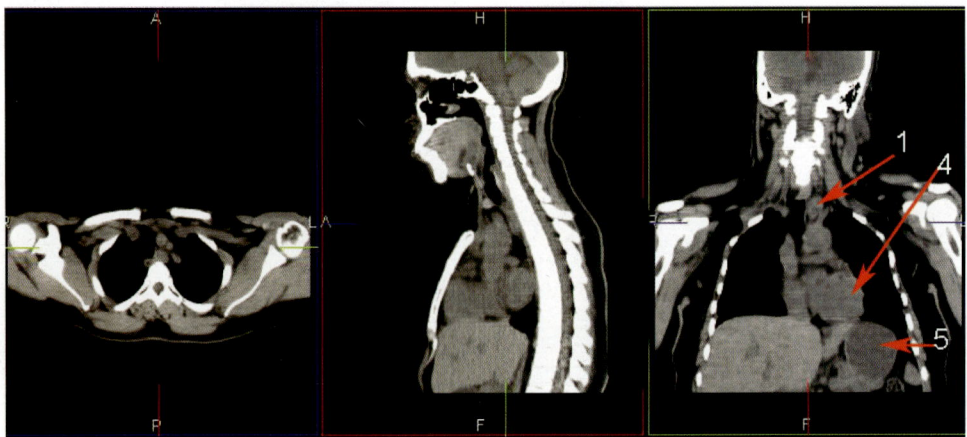

FIGURE 12-56: Parathyroid CT 1

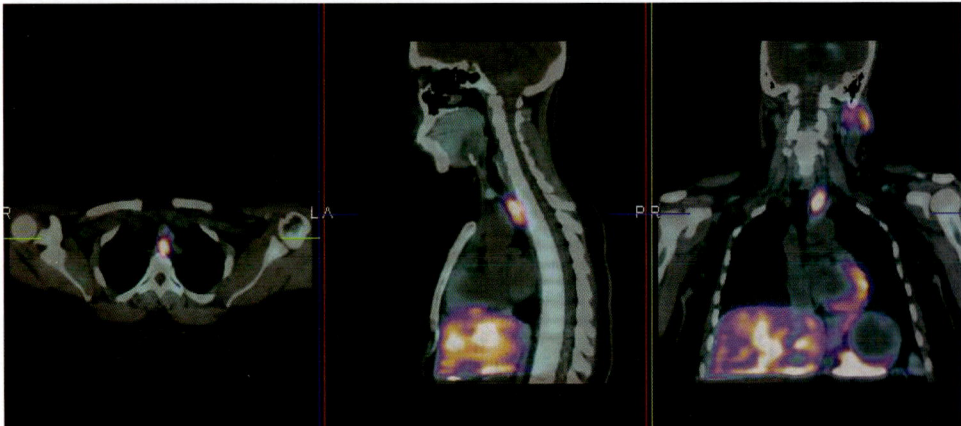

FIGURE 12-57: Parathyroid Fused 1

1. Parathyroid adenoma in left superior paraesophageal lymphatic chain
2. Heterogeneous hepatic activity
3. Left submandibular gland
4. Left ventricular myocardium
5. Stomach

Prostascintigraphy Scan

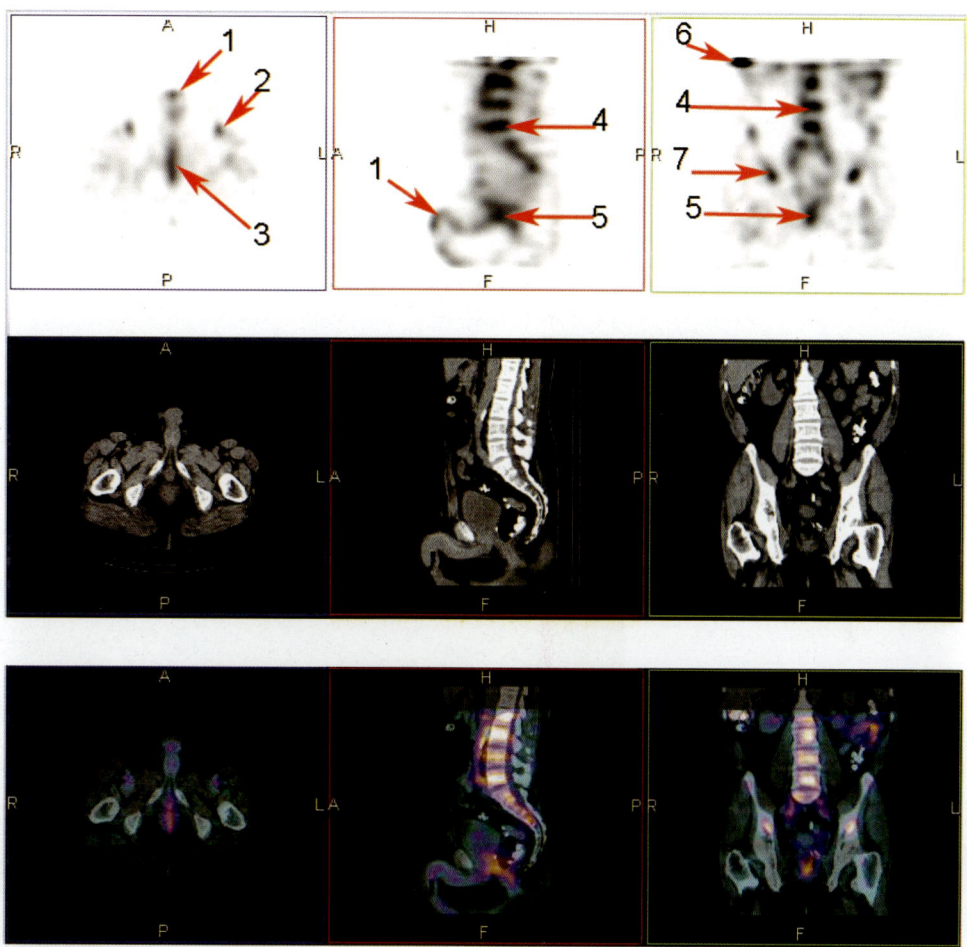

FIGURE 12-58: Prostascint 1

1. Penis
2. Femoral vein
3. Perineum (anus)
4. L5 body
5. Prostate with recurrent cancer
6. Inferior hepatic tip
7. Acetabulum

Section 2: SPECT/CT Anatomy Artifacts

Octreotide Scan Misregistration (Shifted Leg Inward)

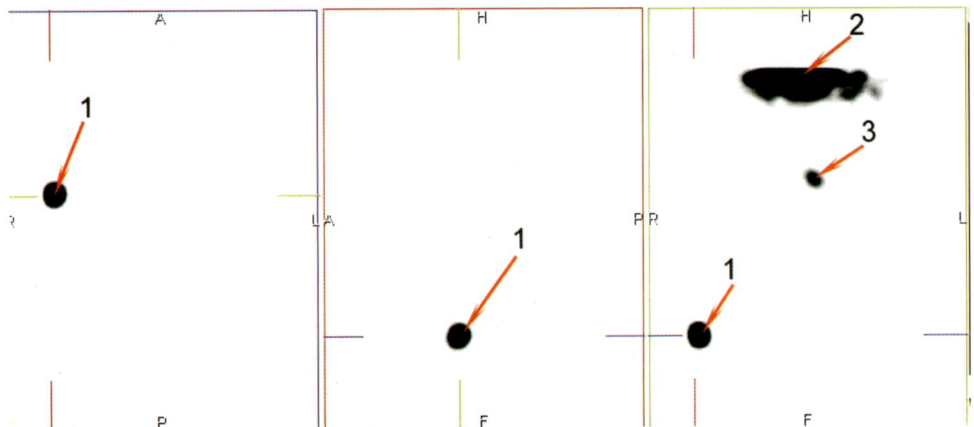

FIGURE 12-59: Octreotide Leg SPECT 1

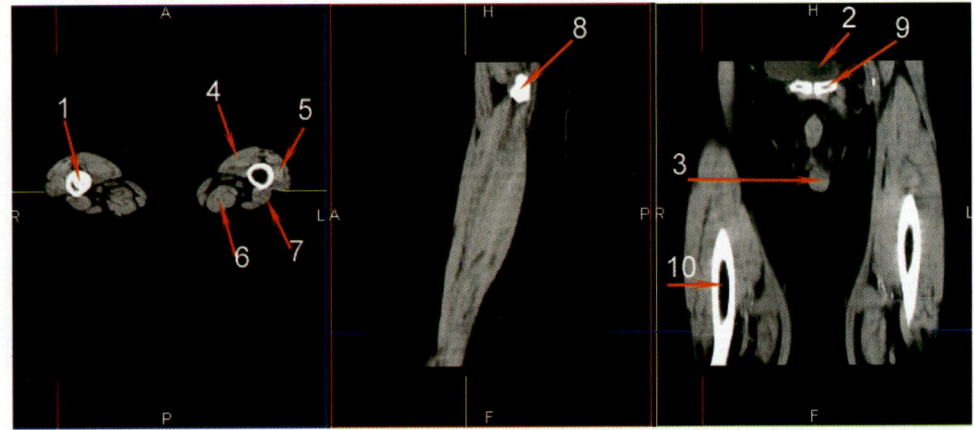

FIGURE 12-60: Octreotide Leg CT 1

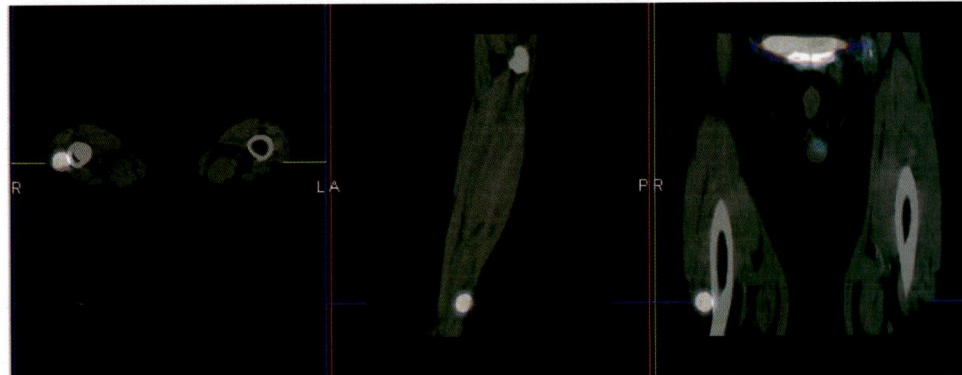

FIGURE 12-61: Octreotide Leg Fused 1

1. Metastatic carcinoid in right distal femur
2. Bladder
3. Scrotum
4. Vastus medialis muscle
5. Vastus lateralis muscle
6. Semimembranous muscle
7. Biceps femoris muscle
8. Trochanter of femur
9. Superior ramus of pubis
10. Midshaft of femur

Parathyroid Misregistration (Moved Head During SPECT)

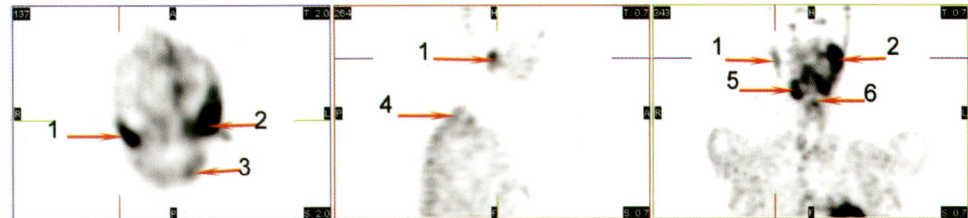

FIGURE 12-62: Parathyroid Motion SPECT 1

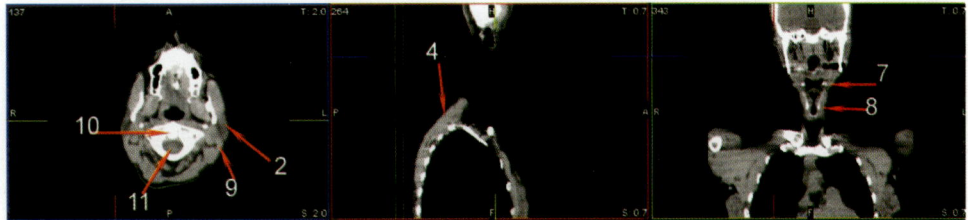

FIGURE 12-63: Parathyroid Motion CT 1

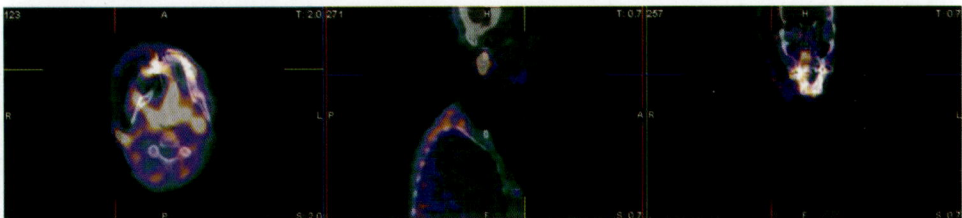

FIGURE 12-64: Parathyroid Motion Fused 1

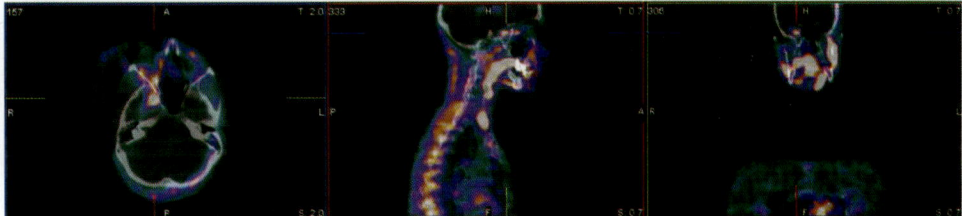

FIGURE 12-65: Parathyroid Motion Fused 2

1. Right parotid gland
2. Left parotid gland
3. Splenius capitis muscle
4. Trapezius muscle
5. Right submandibular gland
6. Vocal muscle
7. Hyoid bone
8. Thyroid cartilage
9. Sternocleidomastoid muscle
10. Dens
11. Spinal cord

Octreotide Scan with Heart Artifact

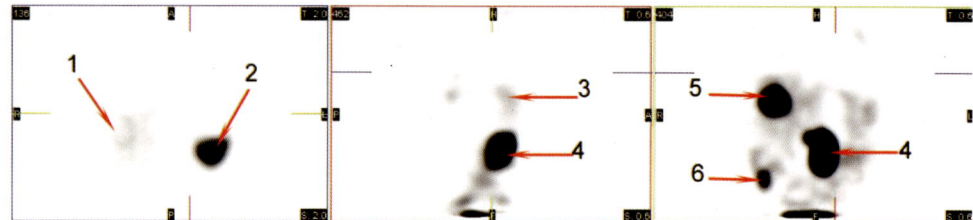

FIGURE 12-66: Heart Artifact SPECT 1

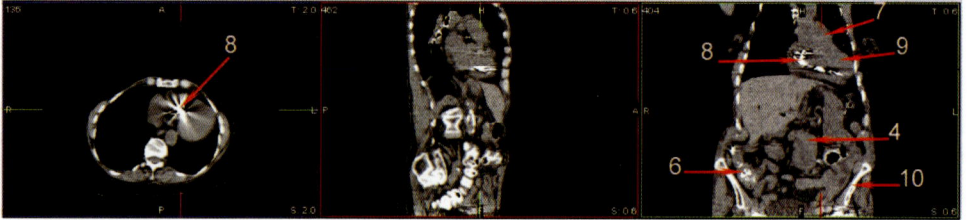

FIGURE 12-67: Heart Artifact CT 1

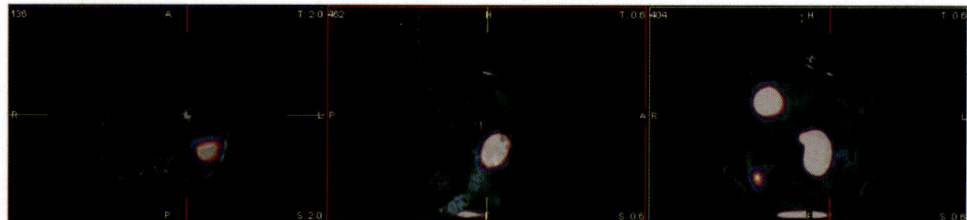

FIGURE 12-68: Heart Artifact and Misregistration of Liver and Spleen

1. Right liver
2. Spleen
3. Left liver
4. Metastatic carcinoid in jejunal mesentery
5. Metastatic carcinoid in right hepatic lobe
6. Cecum
7. Aortic arch
8. Metallic artifact from coronary surgery
9. Left ventricle
10. Iliac crest

Lymphoscintigraphy Denture Artifact

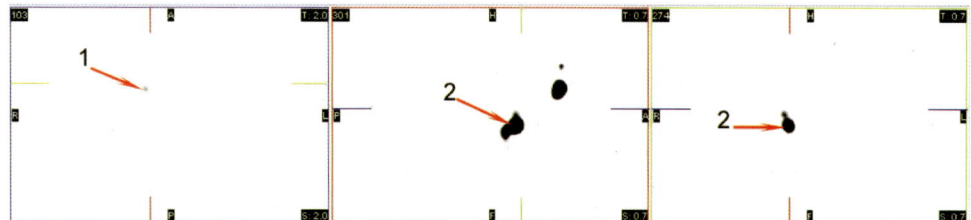

FIGURE 12-69: Lymphoscintigraphy Denture Artifact SPECT 1

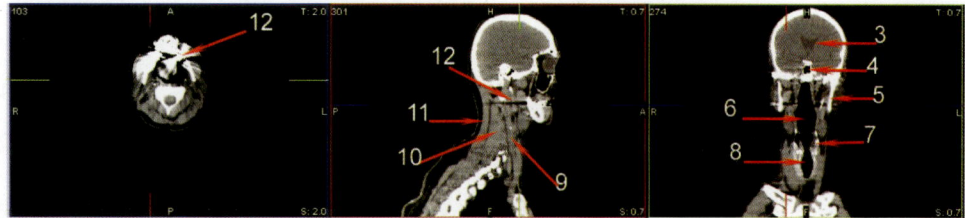

FIGURE 12-70: Lymphoscintigraphy Denture Artifact CT 1

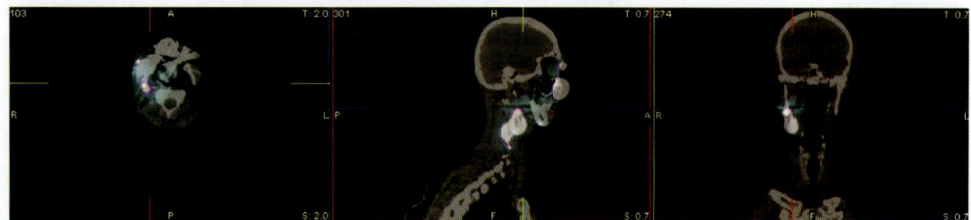

FIGURE 12-71: Lymphoscintigraphy Denture Artifact Fused 1

1. Sentinel node in right upper jugular chain (level #II)
2. Nodes in right middle jugular chain (level #III)
3. Frontal horn of lateral ventricle in brain
4. Sphenoid sinus
5. Mandible
6. Oropharynx
7. Thyroid cartilage
8. Hypopharynx
9. Sternocleidomastoid muscle
10. Splenius capitis muscle
11. Trapezius muscle
12. Metallic artifacts from dentures

Parathyroid Scan with Hair Clips

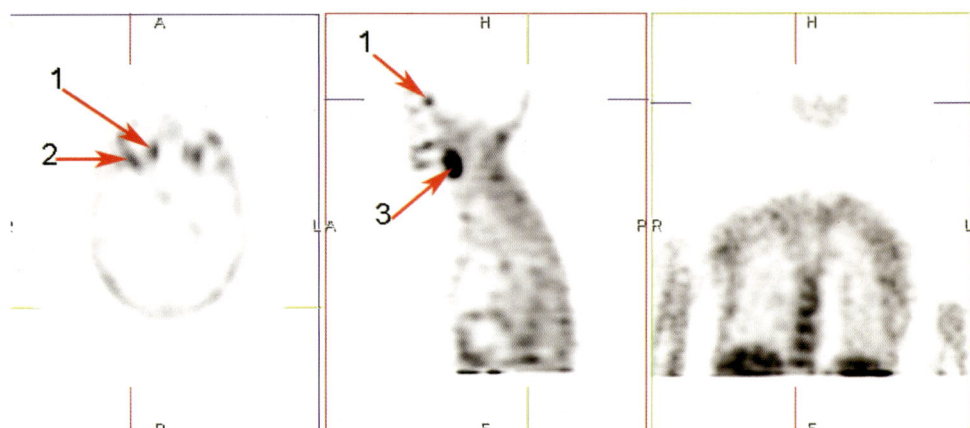

FIGURE 12-72: Parathyroid Hair Clip ECT 1

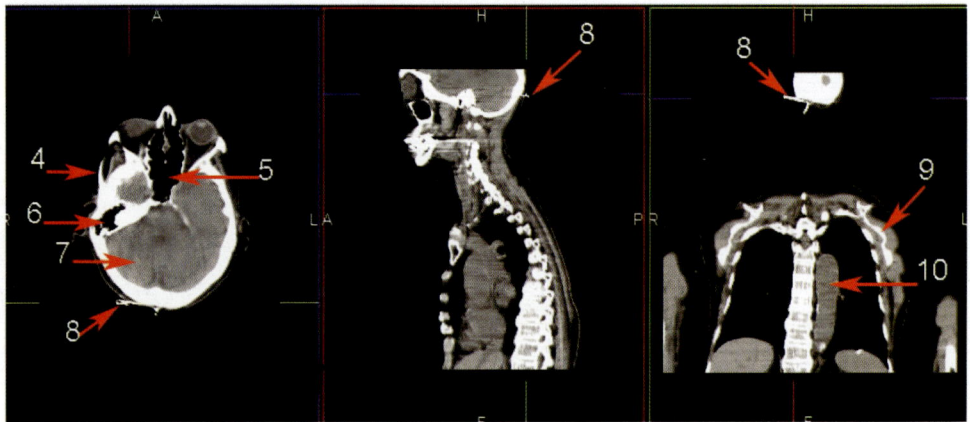

FIGURE 12-73: Parathyroid Hair Clip CT 1

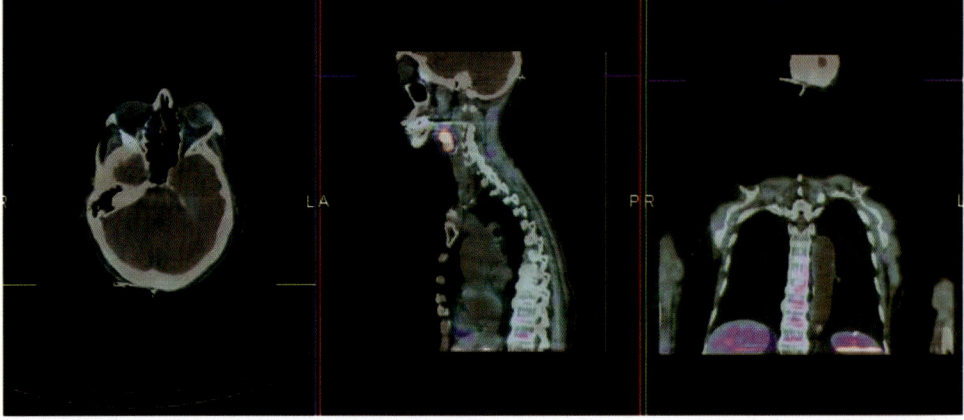

FIGURE 12-74: Parathyroid Hair Clip Fused 1

1. Medial rectus muscle of right eye
2. Lateral rectus muscle of right eye
3. Right submandibular gland
4. Right zygomatic arch
5. Nasopharynx
6. Right mastoid air cells
7. Right occipital brain
8. Hair clip
9. Left scapula
10. Descending aorta

Parathyroid with Pacemaker

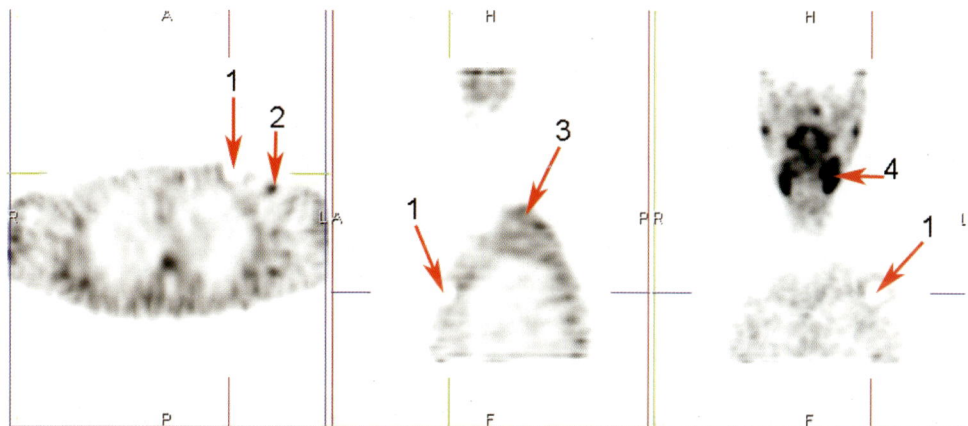

FIGURE 12-75: Parathyroid with Pacemaker ECT 1

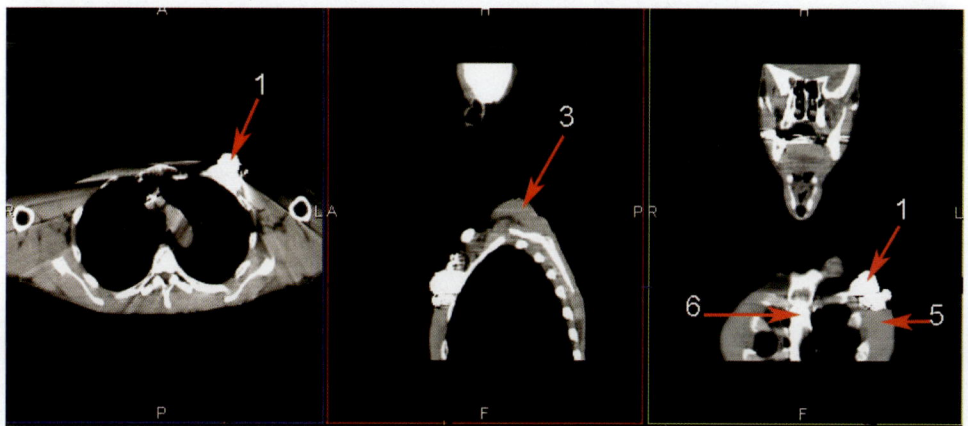

FIGURE 12-76: Parathyroid with Pacemaker CT 1

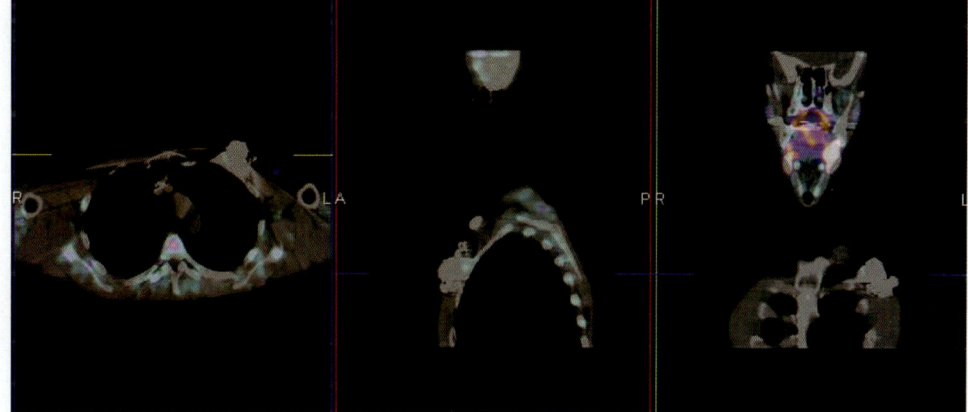

FIGURE 12-77: Parathyroid with Pacemaker Fused 1

1. Pacemaker artifact
2. Left axillary node
3. Trapezius muscle
4. Submandibular gland
5. Serratus anterior muscle
6. Sternum

Notes

Chapter 1: FDG PET/CT

1. Yap C, Quon A, Schiepers C, et al. Additional value of PET/CT over PET for cancer staging and lesion localization. Radiology 2003;229(suppl):487–492.
2. Cohade C, Osman M, Leal J, Wahl RL. Direct comparison of F-18 FDG PET and PET/CT in patients with colorectal carcinoma. J Nucl Med 2003;44(11):1997–1803.

Chapter 2: Non-FDG PET/CT

1. O'Tuama LA, Phillips PC, Smith IR, et al. L-Methionine uptake by human cerebral cortex. Maturation from infancy to old age. J Nucl Med 1991;32:16–20.
2. Mankoff DA, Shields AF, Link JM, et al. Kinetic analysis of 2-C-11 thymidine PET imaging studies: validation studies. J Nucl Med 1999;40:614–624.
3. Armbrecht JJ, Buxton DB, Bremken RC, et al. Regional myocardial oxygen consumption determined noninvasively in humans with C-11 acetate and dynamic positron tomography. Circulation 1989;80:863–872.
4. Cook GJR, Gogelman I. The role of PET in skeletal disease. Semin Nucl Med 2001;31: 50–61.

Chapter 3: Lymphoscintigraphy SPECT/CT

1. Gemnari R, Bartolomei M, Testori A, et al. Sentinel node localization in primary melanoma: preoperative dynamic lymphoscintigraphy, intraoperative gamma probe, and vital dye guidance. Surgery 1000;127:19–25.
2. Summer WE, Ross MI, Mansfield PF, et al. Implications of lymphatic drainage to unusual sentinel node sites in patients with primary cutaneous melanoma. Cancer 2002;95: 354–360.
3. Even-Sapir E, Lerman H, Lievshitz G, et al. Lymphoscintigraphy for sentinel node mapping using a hybrid SPECT/CT system. J Nucl Med 2003;44(9):1413–1420.
4. Klien M, Bocher M, Chisin R. Contribution of combined SPECT-CT in localization by lymphoscintigraphy [abstract]. J Nucl Med 2002;43:157.

Chapter 4: 3D Lung

1. Ling CC, Humm J, Larson S, et al. Towards multidimensional radiotherapy (MD-CRT): biological imaging and biological conformality. Int J Radiat Oncol Biol Phys 2000;47: 551–560.
2. Seppenwoolde Y, Engelsman M, De Jager K, et al. Optimizing radiation treatment plans for lung cancer using lung perfusion information. Radiother Oncol 2002;63:165–177.

3. Munkey MT, Marks LB, Scarfone C, et al. Multimodality nuclear medicine imaging in three-dimensional radiation treatment planning for lung cancer: challenges and prospects. Lung Cancer 1999;23:105–114.

Chapter 5: Parathyroid SPECT/CT

1. Rubello D, Casara D, Fiore D, et al. An ectopic mediastinal parathyroid adenoma accurately located by a single-day imaging protocol of Tc-99m pertechnetate-MIBI subtraction scintigraphy and MIBI-SPECT-computed tomographic fusion. Clin Nucl Med 2002;27:186–190.
2. Even-Sapir E, Keidar Z, Sachs J, et al. The new technology of combined transmission and emission tomography in evaluation of endocrine neoplasms. J Nucl Med 2001;42:998–1004.
3. Kienast O, Dobrozemsky G, Kaczirek K, et al. Combined XCT/SPECT technology in patients with hyperthyroidism [abstract]. Eur J Nucl Med Mol Imaging 2002;29:580.

Chapter 6: Bone SPECT/CT

1. Eschmann SM, Horger M, Pfannemberg AC, et al. Improved specificity on bone scintigraphy by combined transmission and emission tomography [abstract]. J Nucl Med 2002;43:340.

Chapter 7: 131-I SPECT/CT

1. Chung JK. Sodium iodide symporter: its role is nuclear medicine. J Nucl Med 2002;43:1188–1200.
2. Yamamoto Y, Nishiyama Y, Monden T, et al. Clinical usefulness of fusion image of I-131 SPECT and CT in patients with differentiated thyroid adenoma [abstract]. J Nucl Med 2001;43:321.
3. Bernier MD, Leenhardt L, Hoang C, et al. Survival and therapeutic modalities in patients with bone metastases of differentiated thyroid carcinomas. J Clin Endocrinol Metab 2001;86:1568–1573.
4. Shapiro B, Rufini V, Jarwan A, et al. Artifacts, anatomical and physiological variants and unrelated diseases that might cause false-positive whole-body I-131 scans in patients with thyroid cancer. Semin Nucl Med 2000;30:115–132.

Chapter 8: MIBG SPECT/CT

1. Beierwaltes WH. Endocrine imaging: parathyroid, adrenal cortex and medulla, and other endocrine tumors. Part II. J Nucl Med 1991;32:1627–1639.
2. Freitas JC. Adrenal cortical and medullary imaging. Semin Nucl Med 1995;25:235–250.
3. Ozer S, Kienast O, Dobrozemsky G, et al. Combined XCT/SPECT in a single device in patients with suspected or known pheochromocytoma [abstract]. Eur J Nucl Med Mol Imag 2002;29L586.
4. Fujita A, Hyodoh H, Kawamura Y, et al. Use of fusion images of I-131 MIBG, SPECT, and MR studies to identify a malignant pheochromocytoma. Clin Nucl Med 2000;25:440–442.

Chapter 9: Gallium SPECT/CT

1. Front D, Israel O, Epelbaunm R, et al. Ga-67 SPECT before and after treatment of lymphoma. Radiology 1990;175:515–519.

2. Front D, Bon-Shalom R, Israel O. The continuing role of Ga-67 scintigraphy in the age of receptor imaging. Semin Nucl Med 1997;27:68–74.
3. Kaplan WD, Jochelson MS, Herman TS, et al. Gallium-67 imaging: a predictor of residual tumor viability and clinical outcome in patients with diffuse large-cell lymphoma. J Clin Oncol 1990;8:1966–1970.
4. Frohlich DE, Chen JL, Neuberg D, et al. When is hilar uptake of Ga-67 citrate indicative of residual disease after CHOP chemotherapy? J Nucl Med 2000;41:269–274.
5. Chajari M, Lacroix J, Peng AM, et al. Ga-67 scintigraphy in lymphoma: is there a benefit of image fusion with CT? Eur J Nucl Med Mol Imaging 2002;29:380–382.
6. Koral KF, Dewaraja Y, Li J, et al. Initial results for hybrid SPECT-conjugate-view tumor dosimetry in I-131 anti-B1 antibody therapy of previously untreated patients with lymphoma. J Nucl Med 2000;41:1579–1586.

Chapter 10: Octreotide SPECT/CT

1. Norton JA, Fraker DL, Alexander HR, et al. Surgery to cure the Zollinger-Ellison syndrome. N Engl J Med 1999;341:635–644.
2. Kwekkeboom DJ, Krenning EP. Somatostatin receptor imaging. Semin Nucl Med 2002;32:84–91.
3. Even-Sapir E, Keidar Z, Sachs J, et al. The new technology of combined transmission and emission tomography in evaluation of endocrine neoplasms. J Nucl Med 2001;42:998–1004.
4. Alexander HR, Fraker DL, Norton JA, et al. Prospective study of somatostatin receptor scintigraphy and its effect on operative outcome in patients with Zollinger-Ellison syndrome. Ann Surg 1998;228:228–238.

Part II: Anatomic Variations and Artifacts of PET/CT and SPECT/CT

1. Shreve PD, Anzai Y, Wahl R. Pitfalls in oncologic diagnosis with FDG PET imaging: physiologic and benign variants. Radiographics 1999;19(1):61–77.
2. Hany TF, Gharehpapagh E, Kamel EM, et al. Brown adipose tissue: a factor to consider in symmetrical tracer uptake in the neck and upper chest region. Eur J Nucl Med 2002;29:1393–1398.
3. Yasuda S, Fuji H, Takahashi W, et al. Elevated F-18 FDG uptake in the psoas muscle. Clin Nucl Med 1998;23:716–717.
4. Cook GJR, Fogelman I, Maisay MN. Normal physiological and benign pathological variants of F-18 deoxyglucose PET scanning: potential for error in interpretation. Semin Nucl Med 1996;26:308–314.
5. Sugawara Y, Fisher SJ, Zasadny KR, et al. Preclinical and clinical studies of bone marrow uptake of F-18 FDG with or without granulocyte colony stimulating factor during chemotherapy. J Clin Oncol 1998;16:173–180.
6. Goerres GW, Ziegler SI, Burger C, et al. Artifacts at PET and PET/CT caused by metallic hip prosthetic material. Radiology 2003;226:577–584.
7. Alibazoglu H, Megremis D, Ali A, et al. Injection artifact on FDG-PET imaging. Clin Nucl Med 1998;23:264–265.
8. Kubota R, Kubota K, Yamada S, et al. Methionine uptake by tumor tissue: a micro-autoradiographic comparison with FDG. J Nucl Med 1995;36:484–492.
9. Loubeyre P, Angelic E, Grozek F, et al. Spiral CT artifact that simulates aortic dissection: image reconstruction with use of 180 degrees and 360 degrees linear interpolation algorithms. Radiology 1997;205:153–157.
10. Ende JF, Huda W, Ros PR, Litwiller AL. Image mottle in abdominal CT. Invest Radiol 1999;34:282–286.
11. Wang G, Frei T, Vannier MW. Fast interactive algorithm for metal artifact reduction to X-ray CT. Acad Radiol 2000;7:607–614.

12. Brown SJ, Hayball MP, Coulden RA. Impact of motion artifact on the measurement of coronary calcium score. Br J Radiol 2000;73:956–965.
13. Köhler T, Proksa R, Bontus C, et al. Artifact analysis of approximate helical cone-beam CT reconstruction algorithm. Med Phys 2002;29:51–64.

Chapter 11: PET/CT Anatomy: Variations and Artifacts

1. Antoch G, Freudenbery LS, Egelhof T, et al. Focal tracer uptake: a potential artifact in contrast-enhanced dual-modality PET/CT scans. J Nucl Med 2002;43(10):1339–1342.
2. Nehmeh SA, Erdi YE, Kalaigian H, et al. Correction for oral contrast artifacts in CT attenuation-corrected PET images obtained by combined PET/CT. J Nucl Med 2003; 44(12):1940–1944.
3. Goerres GW, Burger C, Kamel E, et al. Respiration-induced attenuation artifact at PET/CT: technical considerations. Radiology 2003;226(3):906–910.

Chapter 12: SPECT/CT Anatomy: Variations and Artifacts

1. Hung JC, Ponto JA, Hammes RJ. Radiopharmaceutical-related pitfalls and artifacts. Semin Nucl Med 1996;16:208–255.
2. O'Connor MK. Instrument- and computer-related problems and artifacts in nuclear medicine. Semin Nucl Med 1996;16:256–277.

Index

Printed in China